This book provides a thorough introduction to the theory of classical integrable systems, discussing the various approaches to the subject and explaining their interrelations. The book begins by introducing the central ideas of the theory of integrable systems, based on Lax representations, loop groups and Riemann surfaces. These ideas are then illustrated with detailed studies of model systems. The connection between isomonodromic deformation and integrability is discussed, and integrable field theories are covered in detail. The KP, KdV and Toda hierarchies are explained using the notion of Grassmannian, vertex operators and pseudo-differential operators. A chapter is devoted to the inverse scattering method and three complementary chapters cover the necessary mathematical tools from symplectic geometry, Riemann surfaces and Lie algebras.

The book contains many worked examples and is suitable for use as a textbook on graduate courses. It also provides a comprehensive reference for researchers already working in the field.

OLIVIER BABELON has been a member of the Centre National de la Recherche Scientifique (CNRS) since 1978. He works at the Laboratoire de Physique Théorique et Hautes Energies (LPTHE) at the University of Paris VI-Paris VII. His main fields of interest are particle physics, gauge theories and integrables systems.

MICHEL TALON has been a member of the CNRS since 1977. He works at the LPTHE at the University of Paris VI-Paris VII. He is involved in the computation of radiative corrections and anomalies in gauge theories and integrable systems.

DENIS BERNARD has been a member of the CNRS since 1988. He currently works at the Service de Physique Théorique de Saclay. His main fields of interest are conformal field theories and integrable systems, and other aspects of statistical field theories, including statistical turbulence.

CAMBRIDGE MONOGRAPHS ON
MATHEMATICAL PHYSICS

General editors: P. V. Landshoff, D. R. Nelson, S. Weinberg

[†] Issued as a paperback

Introduction to Classical Integrable Systems

OLIVIER BABELON
Laboratoire de Physique Théorique et Hautes Energies, Universités Paris VI–VII

DENIS BERNARD
Service de Physique Théorique de Saclay, Gif-sur-Yvette

MICHEL TALON
Laboratoire de Physique Théorique et Hautes Energies, Universités Paris VI–VII

CAMBRIDGE UNIVERSITY PRESS
Cambridge, New York, Melbourne, Madrid, Cape Town, Singapore, São Paulo

Cambridge University Press
The Edinburgh Building, Cambridge CB2 2RU, UK

Published in the United States of America by Cambridge University Press, New York

www.cambridge.org
Information on this title: www.cambridge.org/9780521822671

First published 2003
This digitally printed first paperback version 2006

A catalogue record for this publication is available from the British Library

Library of Congress Cataloguing in Publication data
Babelon, Olivier, 1951–
Introduction to classical integrable systems / Olivier Babelon, Denis Bernard, Michel Talon.
p.   cm. – (Cambridge monographs on mathematical physics)
Includes bibliographical references and index.
ISBN 0 521 82267 X
1. Dynamics.   2. Hamiltonian systems.   I. Bernard, Denis, 1961–
II. Talon, Michel, 1952–   III. Title.   IV. Series.
QA845 .B32 2003
531′.163–dc21   2002034955

ISBN-13 978-0-521-82267-1 hardback
ISBN-10 0-521-82267-X hardback

ISBN-13 978-0-521-03670-2 paperback
ISBN-10 0-521-03670-4 paperback

Contents

1
Introduction

The aim of this book is to introduce the reader to classical integrable systems. Because the subject has been developed by several schools having different perspectives, it may appear fragmented at first sight. We develop here the thesis that it has a profound unity and that the various approaches are simply changes of point of view on the same underlying reality. The more one understands each approach, the more one sees their unity. At the end one gets a very small set of interconnected methods.

This fundamental fact sets the tone of the book. We hope in this way to convey to the reader the extraordinary beauty of the structures emerging in this field, which have illuminated many other branches of theoretical physics.

The field of integrable systems is born together with Classical Mechanics, with a quest for exact solutions to Newton's equations of motion. It turned out that apart from the Kepler problem which was solved by Newton himself, after two centuries of hard investigations, only a handful of other cases were found. In the nineteenth century, Liouville finally provided a general framework characterizing the cases where the equations of motion are "solvable by quadratures". All examples previously found indeed pertained to this setting. The subject stayed dormant until the second half of the twentieth century when Gardner, Greene, Kruskal and Miura invented the Classical Inverse Scattering Method for the Korteweg–de Vries equation, which had been introduced in fluid mechanics. Soon afterwards, the Lax formulation was discovered, and the connection with integrability was unveiled by Faddeev, Zakharov and Gardner. This was the signal for a revival of the domain leading to an enormous amount of results, and truly general structures emerged which organized the subject. More recently, the extension of these results to Quantum Mechanics already led to remarkable results and is still a very active field of research.

Let us give a general overview of the ideas we present in this book. They all find their roots in the notion of Lax pairs. It consists of presenting the equations of motion of the system in the form $\dot{L}(\lambda) = [M(\lambda), L(\lambda)]$, where the matrices $L(\lambda)$ and $M(\lambda)$ depend on the dynamical variables and on a parameter λ called the spectral parameter, and $[\ ,\ ]$ denotes the commutator of matrices. The importance of Lax pairs stems from the following simple remark: the Lax equation is an isospectral evolution equation for the Lax matrix $L(\lambda)$. It follows that the curve defined by the equation $\det(L(\lambda) - \mu I) = 0$ is time-independent. This curve, called the spectral curve, can be seen as a Riemann surface. Its moduli contain the conserved quantities. This immediately introduces the two main structures into the theory: groups enter through the Lie algebra involved in the commutator $[M, L]$, while complex analysis enters through the spectral curve.

As integrable systems are rather rare, one naturally expects strong constraints on the matrices $L(\lambda)$ and $M(\lambda)$. Constructing consistent Lax matrices may be achieved by appealing to factorization problems in appropriate groups. Taking into account the spectral parameter promotes this group to a loop group. The factorization problem may then be viewed as a Riemann–Hilbert problem, a central tool of this subject.

In the group theoretical setting, solving the equations of motion amounts to solving the factorization problem. In the analytical setting, solutions are obtained by considering the eigenvectors of the Lax matrix. At any point of the spectral curve there exists an eigenvector of $L(\lambda)$ with eigenvalue μ. This defines an analytic line bundle $\mathcal{L}$ on the spectral curve with prescribed Chern class. The time evolution is described as follows: if $\mathcal{L}(t)$ is the line bundle at time t then $\mathcal{L}(t)\mathcal{L}^{-1}(0)$ is of Chern class 0, i.e. is a point on the Jacobian of the spectral curve. It is a beautiful result that this point evolves linearly on the Jacobian. As a consequence, one can express the dynamical variables in terms of theta-functions defined on the Jacobian of the spectral curve. The two methods are related as follows: the factorization problem in the loop group defines transition functions for the line bundle $\mathcal{L}$.

The framework can be generalized by replacing the Lax matrix by the first order differential equation $(\partial_\lambda - M_\lambda(\lambda))\Psi = 0$, where $M_\lambda(\lambda)$ depends rationally on λ. The solution Ψ acquires non-trivial monodromy when λ describes a loop around a pole of M_λ. The isomonodromy problem consists of finding all M_λ with prescribed monodromy data. The solutions depend, in general, on a number of continuous parameters. The deformation equations with respect to these parameters form an integrable system. The theta-functions of the isospectral approach are then promoted to more general objects called the tau-functions.

One can study the behaviour around each singularity of the differential operator quite independently. In the group theoretical version, the above extension of the framework corresponds to centrally extending the loop groups. Around a singularity the most general extended group is the group $GL(\infty)$ which corresponds to the KP hierarchy. It can be represented in a fermionic Fock space. Fermionic monomials acting on the vacuum yield decomposed vectors, which describe an infinite Grassmannian introduced by Sato. In this setting, the time flows are induced by the action of commuting one-parameter subgroups, and the tau-function is defined on the Grassmannian, i.e. the orbit of the vacuum, and characterizes it. Finally the Plücker equations of the Grassmannian are identified with the equations of motion, written in the bilinear Hirota form.

We have tried, as much as possible, to make the book self-contained, and to achieve that each chapter can be studied quite independently. Generally, we first explain methods and then show how they can be applied to particular examples, even though this does not correspond to the historical development of the subject.

In Chapter 2 we introduce the classical definition of integrable systems through the Liouville theorem. We present the Lax pair formulation, and describe the symplectic structure which is encoded into the so-called r-matrix form. In Chapter 3 we explain how to construct Lax pairs with spectral parameter, for finite and infinite-dimensional systems. The Lax matrix may be viewed as an element of a coadjoint orbit of a loop group. This introduces immediately a natural symplectic structure and a factorization problem in the loop group. We also introduce, at this early stage, the notion of tau-functions. In Chapter 4 we discuss the abstract group theoretical formulation of the theory. We then describe the analytical aspects of the theory in Chapter 5. In this setting, the action variables are g moduli of the spectral curve, a Riemann surface of genus g, and the angle variables are g points on it. We illustrate the general constructions by the examples of the closed Toda chain in Chapter 6, and the Calogero model in Chapter 7.

The following two Chapters, 8 and 9, describe respectively the isomonodromic deformation problem and the infinite Grassmannian. Soliton solutions are obtained using vertex operators. Chapters 10 and 11 are devoted to the classical study of the KP and KdV hierarchies. We develop and use the formalism of pseudo-differential operators which allows us to give simple proofs of the main formal properties. Finite-zone solutions of KdV allow us to make contact with integrable systems of finite dimensionality and soliton solutions.

In the next Chapter, 12, we study the class of Toda and sine-Gordon field theories. We use this opportunity to exhibit the relations between their conformal and integrable properties. The sine-Gordon model is presented in the framework of the Classical Inverse Scattering Method in Chapter 13. This very ingenious method is exploited to solve the sine-Gordon equation.

The last three chapters may be viewed as mathematical appendices, provided to help the reader. First we present the basic facts of symplectic geometry, which is the natural language to speak about Classical Mechanics and integrable systems. Since mathematical tools from Riemann surfaces and Lie groups are used almost everywhere, we have written two chapters presenting them in a concise way. We hope that they will be useful at least as an introduction and to fix notations.

Let us say briefly how we have limited our discussion. First we choose to remain consistently at a relatively elementary mathematical level, and have been obliged to exclude some important developments which require more advanced mathematics. We put the emphasis on methods and we have not tried to make an exhaustive list of integrable systems. Another aspect of the theory we have touched only very briefly, through the Whitham equations, is the study of perturbations of integrable systems. All these subjects are very interesting by themselves, but the present book is big enough!

A most active field of recent research is concerned with quantum integrable systems or the closely related field of exactly soluble models in statistical mechanics. When writing this book we always had the quantum theory present in mind, and have introduced all classical objects which have a well-known quantum counterpart, or are semi-classical limits of quantum objects. This explains our emphasis on Hamiltonians methods, Poisson brackets, classical r-matrices, Lie–Poisson properties of dressing transformations and the method of separation of variables. Although there is nothing quantum in this book, a large part of the apparatus necessary to understand the literature on quantum integrable systems is in fact present.

The bibliography for integrable systems would fill a book by itself. We have made no attempt to provide one. Instead, we give, at the end of each chapter, a short list of references, which complements and enhances the material presented in the chapter, and we highly encourage the reader to consult them. Of course these references are far from complete, and we apologize to the numerous authors having contributed to the domain, and whose due credit is not acknowledged. Finally we want to thank our many colleagues from whom we learned so much and with whom we have discussed many parts of this book.

2

Integrable dynamical systems

We introduce the definition of integrable systems through the Liouville theorem, i.e. systems for which n conserved quantities in involution are known on a phase space of dimension $2n$. The Liouville theorem asserts that the equations of motion can then be solved by quadrature. The notion of Lax matrix is introduced. This is a matrix whose elements are dynamical and whose time evolution is isospectral, a central object in the theory. It is also shown that the Poisson brackets of the elements of the Lax matrix are expressed in the so-called r-matrix form. Finally, we present some historical examples of integrable systems which are solved by the method of separation of variables. This leads to linearization of the time evolution on the Jacobian of Riemann surfaces, another recurring theme in the book.

2.1 Introduction

In Classical Mechanics the state of the system is specified by a point in phase space. This is generally a space of even dimension with coordinates of position q_i and momentum p_i. The Hamiltonian is a function on phase space, denoted $H(p_i, q_i)$. The equations of motion are a first order differential system taking the Hamiltonian form:

$$\dot{q}_i = \frac{\partial H}{\partial p_i}, \quad \dot{p}_i = -\frac{\partial H}{\partial q_i} \tag{2.1}$$

Here and in the following, a dot will refer to a time derivative. For any function $F(p, q)$ on phase space, this implies that $F(p(t), q(t))$ obeys:

$$\dot{F} \equiv \frac{dF}{dt} = \{H, F\}$$

5

where for any functions F and G the Poisson bracket $\{F, G\}$ is defined as:

$$\{F, G\} \equiv \sum_i \frac{\partial F}{\partial p_i} \frac{\partial G}{\partial q_i} - \frac{\partial G}{\partial p_i} \frac{\partial F}{\partial q_i}$$

For the coordinates p_i, q_i themselves we have

$$\{q_i, q_j\} = 0, \quad \{p_i, p_j\} = 0, \quad \{p_i, q_j\} = \delta_{ij} \tag{2.2}$$

The quantity $H(p, q)$ is automatically conserved under time evolution, $\frac{d}{dt} H(p, q) = 0$, so that the motion takes place on the subvariety of phase space defined by $H = E$ constant.

Historically, it proved very difficult to find dynamical systems such that eqs. (2.1) could be solved exactly. However, there is a general framework where the explicit solutions can be obtained by solving a finite number of algebraic equations and computing finite number of integrals, i.e. the solution is obtained by quadratures. These dynamical systems are the Liouville integrable systems that we will consider in this book. A dynamical system on a phase space of dimension $2n$ is Liouville integrable if one knows n independent functions F_i on the phase space which Poisson *commute*, that is $\{F_i, F_j\} = 0$. The Hamiltonian is assumed to be a function of the F_i.

In order to understand the geometry of the situation, let us discuss a very simple example: the harmonic oscillator. The phase space is of dimension 2 and the Hamiltonian is $H = \frac{1}{2}(p^2 + w^2 q^2)$ with Poisson bracket $\{p, q\} = 1$. The phase space is fibred into ellipses $H = E$ except for the point $(0, 0)$ which is a stationary point. An adapted coordinate system ρ, θ is given by:

$$p = \rho \cos(\theta), \quad q = \frac{\rho}{\omega} \sin(\theta) \tag{2.3}$$

and the non-vanishing Poisson bracket is $\{\rho, \theta\} = \omega/\rho$. In these coordinates the flow reads:

$$\rho = \sqrt{2E}, \quad \theta = \omega t + \theta_0$$

i.e. the flow takes place on the above ellipsis.

This can be straightforwardly generalized to a direct sum of n harmonic oscillators with

$$H = \sum_{i=1}^{n} \frac{1}{2}(p_i^2 + \omega_i^2 q_i^2)$$

and Poisson bracket eq. (2.2). We do have n conserved quantities in involution, $F_i = \frac{1}{2}(p_i^2 + \omega_i^2 q_i^2)$, and the level manifold M_f, i.e. the set of points of phase space such that $F_i = f_i$, is an n-dimensional real torus, which is explicitly a cartesian product of n topological circles. The motion takes place on these tori which foliate the phase space. We can intoduce n angles θ_i as above which evolve linearly in time with frequency ω_i. An orbit of the dynamical flow is dense on the torus when the ω_i are rationally independent.

For Liouville integrable systems, we shall assume that the conserved quantities are well-behaved so that the n dimensional surfaces M_f defined by $F_i = f_i$ are generically regular, and foliate the phase space. This does not preclude the existence of singular points such as $p_i = q_i = 0$ in the above example of the harmonic oscillator. In this setting we are now going to prove the Liouville theorem and show that the geometry of the situation is analogous to that of the harmonic oscillator example.

2.2 The Liouville theorem

We consider a dynamical Hamiltonian system with phase space M of dimension $2n$. Introduce canonical coordinates p_i, q_i such that the non-degenerate Poisson bracket reads as in eq. (2.2). As usual a non-degenerate Poisson bracket on M is equivalent to the data of a non-degenerate closed 2-form ω, $d\omega = 0$, defined on M, called the symplectic form, see Chapter 14. In the canonical coordinates the symplectic form reads

$$\omega = \sum_j dp_j \wedge dq_j$$

Let H be the Hamiltonian of the system.

Definition. *The system is Liouville integrable if it possesses n independent conserved quantities F_i, $i = 1, \ldots, n$, $\{H, F_j\} = 0$, in involution*

$$\{F_i, F_j\} = 0$$

The independence means that at generic points (i.e. anywhere except on a set of measure zero), the dF_i are linearly independent, or that the tangent space of the surface $F_i = f_i$ exists everywhere and is of dimension n. There cannot be more than n independent quantities in involution otherwise the Poisson bracket would be degenerate. In particular, the Hamiltonian H is a function of the F_i.

The Liouville theorem. *The solution of the equations of motion of a Liouville integrable system is obtained by "quadrature".*

<u>Proof.</u> Let $\alpha = \sum_i p_i dq_i$ be the canonical 1-form and $\omega = d\alpha = \sum_i dp_i \wedge dq_i$ be the symplectic 2-form on the phase space M. We will construct

a canonical transformation $(p_i, q_i) \rightarrow (F_i, \Psi_i)$ such that the conserved quantities F_i are among the new coordinates:

$$\omega = \sum_i dp_i \wedge dq_i = \sum_i dF_i \wedge d\Psi_i$$

If we succeed in doing that, the equations of motion become trivial:

$$\dot{F}_j = \{H, F_j\} = 0$$
$$\dot{\psi}_j = \{H, \psi_j\} = \frac{\partial H}{\partial F_j} = \Omega_j \qquad (2.4)$$

The Ω_j depend only on F and so are constant in time. In these coordinates, the solution of the equations of motion read:

$$F_j(t) = F_j(0), \quad \psi_j(t) = \psi_j(0) + t\Omega_j$$

To construct this canonical transformation, we exhibit its so-called generating function S. Let M_f be the level manifold $F_i(p, q) = f_i$. Suppose that on M_f we can solve for p_i, $p_i = p_i(f, q)$, and consider the function

$$S(F, q) \equiv \int_{m_0}^{m} \alpha = \int_{q_0}^{q} \sum_i p_i(f, q) dq_i$$

where the integration path is drawn on M_f and goes from the point of coordinate $(p(f, q_0), q_0)$ to the point $(p(f, q), q)$, where q_0 is some reference value.

Suppose that this function exists, i.e. if it does not depend on the path from m_0 to m, then $p_j = \frac{\partial S}{\partial q_j}$. Defining ψ_j by

$$\psi_j = \frac{\partial S}{\partial F_j}$$

we have

$$dS = \sum_j \psi_j dF_j + p_j dq_j$$

Since $d^2 S = 0$ we deduce that $\omega = \sum_j dp_j \wedge dq_j = \sum_j dF_j \wedge d\psi_j$. This shows that if S is a well-defined function, then the transformation is canonical.

To show that S exists, we must prove that it is independent of the integration path. By Stokes theorem, we have to prove that:

$$d\alpha|_{M_f} = \omega|_{M_f} = 0$$

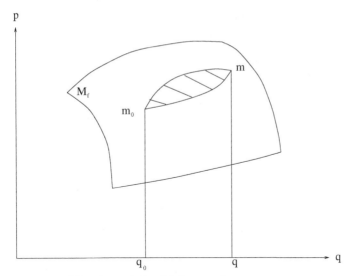

Fig. 2.1. A leaf M_f on phase space

Let X_i be the Hamiltonian vector field associated with F_i, defined by $dF_i = \omega(X_i, \cdot)$,

$$X_i = \sum_k \frac{\partial F_i}{\partial q_k} \frac{\partial}{\partial p_k} - \frac{\partial F_i}{\partial p_k} \frac{\partial}{\partial q_k}$$

These vector fields are tangent to the manifold M_f because the F_j are in involution,

$$X_i(F_j) = \{F_i, F_j\} = 0$$

Since the F_j are assumed to be independent functions, the tangent space to the submanifold M_f is generated at each point $m \in M$ by the vectors $X_i|_m$ $(i = 1, \ldots, n)$. But then $\omega(X_i, X_j) = dF_i(X_j) = 0$ and we have proved that $\omega|_{M_f} = 0$, and therefore S exists. ∎

We have effectively obtained the solution of the equations of motion through one quadrature (to calculate the function S) and some "algebraic manipulation" (to express the p as functions of q and F)

Remark 1. From the closedness of α on M_f, the function S is unchanged by *continuous* deformations of the path (m_0, m). However, if M_f has non-trivial cycles, which is generically the case, S is a multivalued function defined in a neighbourhood of M_f. The variation over a cycle

$$\Delta_{\text{cycle}} S = \int_{\text{cycle}} \alpha$$

is a function of F only. This induces a multivaluedness of the variables ψ_j: $\Delta_{\text{cycle}}\psi_j = \frac{\partial}{\partial F_j} \Delta_{\text{cycle}} S$. For instance, in the case of harmonic oscillators, we see that above each point $(q_1, \ldots, q_n)$ we have 2^n points on the M_f level surface, due to the independent choices of sign in $p_i = \pm\sqrt{2f_i - \omega_i^2 q_i^2}$. So we have many choices for the path of integration, reflecting the topology of the torus.

Remark 2. The definition we have given of a Liouville integrable system requires some care. Given any Hamiltonian H, the Darboux theorem, see Chapter 14, implies that we can always find *locally* a system of canonical coordinates on phase space $(P_1, \ldots, P_n; Q_1, \ldots, Q_n)$, with $H = P_1$, hence fulfilling the hypothesis of the Liouville theorem. For integrable systems we require that the conserved quantities are globally defined on a sufficiently large open set, and that the surfaces $F_i = f_i$ are well-behaved and foliate the phase space. This is not generally the case for the P_i constructed by the Darboux theorem. Moreover, in all known examples, the conserved quantities are even algebraic functions of canonical coordinates on some open domain and the solutions of the equations of motion are analytic.

Remark 3. Using the Poisson commuting functions F_i, one can solve *simultaneously* the n time evolution equations $dF/dt_i = \{F_i, F\}$, since:

$$\frac{\partial}{\partial t_i}\frac{\partial}{\partial t_j}F - \frac{\partial}{\partial t_j}\frac{\partial}{\partial t_i}F = \{F_i, \{F_j, F\}\} - \{F_j, \{F_i, F\}\} = \{\{F_i, F_j\}, F\} = 0$$

Since the Hamiltonian vector fields are well-defined and linearly independent everywhere, the flows define a locally free (no fixed points) and transitive (goes everywhere) action of a small open set in $\mathbb{R}^n$ on the surface M_f. Assuming that M_f is connected and compact, the flows extend to all values of the times t_i and fill the whole surface M_f, hence we have a surjective action of $\mathbb{R}^n$ on M_f. The stabilizer of a point is an Abelian discrete subgroup of $\mathbb{R}^n$ since the action is locally free, so it is of the form $\mathbb{Z}^n$. Thus M_f appears as the quotient of $\mathbb{R}^n$ by $\mathbb{Z}^n$, i.e. a torus. This refinement, due to Arnold, of the Liouville theorem shows that, under suitable global hypothesis, the phase space is indeed foliated by n dimensional tori, called the Liouville tori. It is remarkable that for small perturbations of integrable systems, there still exist Liouville tori "almost everywhere". This is the content of the famous Kolmogorov–Arnold–Moser (KAM) theorem.

2.3 Action–angle variables

As already noticed in the proof of the Liouville theorem, the level manifold M_f has non-trivial cycles. Under suitable compactness and connectivity conditions, the M_f are n-dimensional tori T_n. This points to the introduction of angle variables to describe the motion along the cycles. The torus T_n is isomorphic to a product of n circles C_i. We may choose special angular coordinates on M_f dual to the n fundamental cycles C_i (see eq. (2.5)).

The action variables I_j are defined as the integrals of the canonical 1-form over the cycles C_j,

$$I_j = \frac{1}{2\pi} \int_{C_j} \alpha$$

The I_j are functions of the constants of motion F_j and we suppose they are independent, so that if the values of I_j $(j = 1, \ldots, n)$ are known, then M_f is determined.

Let us consider the canonical transformation generated by the same function as above:

$$S(I, q) = \int_{m_0}^{m} \alpha$$

but expressed in terms of the variables I_i instead of F_i. Denoting by θ_j the variable conjugate to I_j, the canonical transformation generated by S is defined by

$$p_k = \frac{\partial S}{\partial q_k}, \quad \theta_k = \frac{\partial S}{\partial I_k}$$

The variables θ_k are canonically conjugated to the action variables I_j.

We show that they can be regarded as normalized angular variables on the cycles C_j. That is,

$$\frac{1}{2\pi} \int_{C_j} d\theta_k = \delta_{jk} \tag{2.5}$$

By definition of θ_k,

$$\int_{C_j} d\theta_k = \frac{\partial}{\partial I_k} \int_{C_j} dS, \quad dS = \sum_i \frac{\partial S}{\partial q_i} dq_i + \frac{\partial S}{\partial I_i} dI_i$$

Since on the manifold M_f, $dI_i = 0$, we have

$$\int_{C_j} d\theta_k = \frac{\partial}{\partial I_k} \int_{C_j} \frac{\partial S}{\partial q_i} dq_i = \frac{\partial}{\partial I_k} \int_{C_j} \alpha = 2\pi \delta_{jk}$$

This proves that θ_k are angle variables.

2.4 Lax pairs

The new concept which emerged from the modern studies of integrable systems is the notion of Lax pairs. A Lax pair L, M consists of two matrices, functions on the phase space of the system, such that the Hamiltonian evolution equations, eq. (2.1), may be written as

$$\frac{dL}{dt} \equiv \dot{L} = [M, L] \tag{2.6}$$

Here, $[M, L] = ML - LM$ denotes the commutator of the matrices M and L.

The immediate interest in the existence of such a pair lies in the fact that it allows for an easy construction of conserved quantities. Indeed, the solution of eq. (2.6) is of the form

$$L(t) = g(t)L(0)g(t)^{-1}$$

where the invertible matrix $g(t)$ is determined by the equation

$$M = \frac{dg}{dt}g^{-1}$$

It follows that if $I(L)$ is a function of L invariant by conjugation $L \rightarrow gLg^{-1}$, then $I(L(t))$ is a constant of the motion. Such functions are functions of the eigenvalues of L. We say that the evolution equation (2.6) is isospectral, which means that the spectrum of L is preserved by the time evolution.

Remark 1. Recall that integrability of the system in the sense of Liouville demands that (i) the number of independent conserved quantities equals the number of degree of freedom, and that (ii) these conserved quantities are in involution.

Remark 2. A Lax pair is by no means unique. Even the size of the matrices may be changed. There is also a natural gauge group acting on the Lax pair:

$$L \longrightarrow gLg^{-1}, \quad M \longrightarrow gMg^{-1} + \frac{dg}{dt}g^{-1}$$

where g is an invertible matrix, a function on phase space.

Let us present some simple examples showing that the equations of motion can indeed be recast into Lax form.

Example 1. For any integrable system in the sense of Liouville, one may construct a Lax pair in a tautological way. Consider a finite-dimensional Hamiltonian system, with n degrees of freedom, Poisson bracket $\{\ ,\ \}$ and Hamiltonian H. Suppose it is integrable in the sense of Liouville, which means that it possesses n independent integrals of the motion $F_i, i = 1, \ldots, n$, in involution. The Liouville theorem states that there exists, at least locally, a system of conjugate coordinates $I_i, \ \theta_i, i = 1, \ldots, n$, where the I_j are functions of the F_i only. In these coordinates, the equations of motion take the very simple form

$$\dot{I}_j = 0, \quad \dot{\theta}_j = \frac{\partial H}{\partial I_j} \tag{2.7}$$

Introduce the Lie algebra generated by $\{H_i, E_i,\ i = 1, \ldots, n\}$ with relations

$$[H_i, H_j] = 0, \quad [H_i, E_j] = 2\,\delta_{ij}E_j, \quad [E_i, E_j] = 0$$

This Lie algebra has a natural representation by $2n \times 2n$ matrices. Set:

$$L = \sum_{j=1}^{n} I_j\,H_j + 2I_j\theta_j E_j, \quad M = -\sum_{j=1}^{n} \frac{\partial H}{\partial I_j} E_j$$

The equation $\dot{L} = [M, L]$ is then equivalent to eq. (2.7). Thus L, M form a Lax pair. However, this construction is useless since it requires the knowledge of the action angle variables to build the Lax pair, but if these are known, there is no need for a Lax pair any more.

Example 2. As a second example we exhibit a Lax pair for the harmonic oscillator. Let:

$$L = \begin{pmatrix} p & \omega q \\ \omega q & -p \end{pmatrix}, \quad M = \begin{pmatrix} 0 & -\omega/2 \\ \omega/2 & 0 \end{pmatrix} \qquad (2.8)$$

We check immediately that the Lax equation, eq. (2.6), is equivalent to the equations of motion $\dot{q} = p, \dot{p} = -\omega^2 p$. Let us observe that the Hamiltonian H can be written as $\frac{1}{4}\mathrm{Tr}L^2$. This example can be generalized to n independent harmonic oscillators by writing the Lax matrices L, M in a block diagonal form where each block is a two by two matrix as above. Now the conserved quantities are $\mathrm{Tr}L^{2p} = 2\sum(2F_i)^p$, with $2F_i = p_i^2 + \omega^2 q_i^2$, and $\mathrm{Tr}L^{2p+1} = 0$, so that they are equivalent to the collection of the F_i.

2.5 Existence of an r-matrix

A Lax pair provides us with conserved quantities without referring to a Poisson structure. The notion of Liouville integrability requires the knowledge of a Poisson structure together with the involution property of the conserved quantities. We shall now present the general form of Poisson brackets between the matrix elements of the Lax matrix which ensures the involution property for the conserved quantities.

Suppose we are given a Lax pair L, M, which are $N \times N$ matrices, and suppose that the matrix L can be diagonalized:

$$L = U\Lambda U^{-1} \qquad (2.9)$$

The matrix elements λ_k of the diagonal matrix Λ are the conserved quantities. We will not consider here the question of the independence of these quantities.

Let us first introduce some notations. Let E_{ij} be the canonical basis of the $N \times N$ matrices, $(E_{ij})_{kl} = \delta_{ik}\delta_{jl}$. We can write

$$L = \sum_{ij} L_{ij} E_{ij}$$

The components L_{ij} of the Lax matrix are functions on the phase space. We can evaluate the Poisson brackets $\{L_{ij}, L_{kl}\}$ and gather the results as follows. Let

$$L_1 \equiv L \otimes 1 = \sum_{ij} L_{ij}(E_{ij} \otimes 1), \quad L_2 \equiv 1 \otimes L = \sum_{ij} L_{ij}(1 \otimes E_{ij})$$

The index 1 or 2 means that the matrix L sits in the first or second factor in the tensor product. Similarly, for T living in the tensor product of two copies of $N \times N$ matrices, we set

$$T = T_{12} = \sum_{ij,kl} T_{ij,kl} \, E_{ij} \otimes E_{kl}, \quad T_{21} = \sum_{ij,kl} T_{ij,kl} \, E_{kl} \otimes E_{ij}$$

More generally, when we have tensor products with more copies of $N \times N$ matrices, we denote by L_α the embedding of L in the α position, e.g. $L_3 = 1 \otimes 1 \otimes L \otimes 1 \otimes \cdots$, and $T_{\alpha\beta}$ the embedding of T in the α and β position.

We shall also denote by Tr_α the partial trace on the space in α position in a tensor product. For example

$$\mathrm{Tr}_1 T_{12} = \sum_{ij,kl} T_{ij,kl} \, \mathrm{Tr}(E_{ij}) \, E_{kl}$$

Define $\{L_1, L_2\}$ as the matrix of Poisson brackets between the elements of L:

$$\{L_1, L_2\} = \sum_{ij,kl} \{L_{ij}, L_{kl}\} E_{ij} \otimes E_{kl}$$

For an integrable system, the Poisson brackets between the elements of the Lax matrix L can be written in the following very special form:

Proposition. *The involution property of the eigenvalues of L is equivalent to the existence of a function, r_{12}, on the phase space such that:*

$$\{L_1, L_2\} = [r_{12}, L_1] - [r_{21}, L_2] \tag{2.10}$$

<u>Proof.</u> Assume first that the eigenvalues of L Poisson commute, $\{\lambda_i, \lambda_j\} = 0$. Recall that L is diagonalized by U, eq. (2.9). Since U is a function

on phase space, we compute directly the Poisson brackets $\{L_1, L_2\} = \{U_1\Lambda_1 U_1^{-1}, U_2\Lambda_2 U_2^{-1}\}$ using the Leibnitz rule. We get nine terms. Out of these, four terms involve the Poisson brackets $\{U_1, U_2\}$. Introducing the quantity $k_{12} = \{U_1, U_2\}U_1^{-1}U_2^{-1}$, these terms can be written as

$$[[k_{12}, L_2], L_1] = \tfrac{1}{2}[[k_{12}, L_2], L_1] - \tfrac{1}{2}[[k_{21}, L_1], L_2]$$

Four other terms involve $\{\Lambda_1, U_2\}$ and $\{U_1, \Lambda_2\}$. Introducing

$$q_{12} = U_2\{U_1, \Lambda_2\}U_1^{-1}U_2^{-1}$$

we can write them as $[q_{12}, L_1] - [q_{21}, L_2]$. Putting all this together, we get

$$\{L_1, L_2\} = U_1 U_2\{\Lambda_1, \Lambda_2\}U_1^{-1}U_2^{-1} + [r_{12}, L_1] - [r_{21}, L_2]$$

where $r_{12} = q_{12} + \tfrac{1}{2}[k_{12}, L_2]$. This proves one part of the equivalence when $\{\Lambda_1, \Lambda_2\} = 0$. Conversely, suppose we have eq. (2.10). Then, in any matrix representation

$$\{L_1^n, L_2^m\} = [a_{12}^{n,m}, L_1] + [b_{12}^{n,m}, L_2] \tag{2.11}$$

with

$$a_{12}^{n,m} = \sum_{p=0}^{n-1}\sum_{q=0}^{m-1} L_1^{n-p-1}L_2^{m-q-1}r_{12}L_1^p L_2^q$$

$$b_{12}^{n,m} = -\sum_{p=0}^{n-1}\sum_{q=0}^{m-1} L_1^{n-p-1}L_2^{m-q-1}r_{21}L_1^p L_2^q$$

Taking the trace of eq. (2.11), and using that the trace of a commutator is zero, we get that the quantities of $\mathrm{Tr}\,(L^n)$ are in involution. This is equivalent to the involution of the eigenvalues λ_k of L. ∎

Although simple to prove, this proposition is important for developing formal aspects of integrable systems since it allows us to control the Poisson brackets of the Lax matrix.

The Jacobi identity on the Poisson bracket, see Chapter 14, yields the following constraint on r:

$$[L_1, [r_{12}, r_{13}] + [r_{12}, r_{23}] + [r_{32}, r_{13}] + \{L_2, r_{13}\} - \{L_3, r_{12}\}] + \text{cyc. perm.} = 0 \tag{2.12}$$

where cyc. perm. means cyclic permutations of tensor indices $1, 2, 3$. In a sense, solving this equation amounts to classifying integrable Hamiltonian systems.

If r happens to be constant, the only remaining terms in eq. (2.12) are the first ones. In particular, the Jacobi identity is satisfied if a constant r-matrix satisfies

$$[r_{12}, r_{13}] + [r_{12}, r_{23}] + [r_{32}, r_{13}] = 0$$

When r is antisymmetric, $r_{12} = -r_{21}$, this is called the classical Yang–Baxter equation. This case will be extensively studied in Chapter 4.

Remark 1. The form of the bracket is preserved by gauge transformations. If

$$\{L_1, L_2\} = [r_{12}, L_1] - [r_{21}, L_2]$$

and $L' = gLg^{-1}$, then there exists a matrix function r'_{12} such that

$$\{L'_1, L'_2\} = [r'_{12}, L'_1] - [r'_{21}, L'_2]$$

The function r'_{12} can be expressed in terms of the functions r_{12} and of the Poisson brackets between g and the Lax matrix L.

$$r'_{12} = g_1 g_2 \left(r_{12} + g_1^{-1}\{g_1, L_2\} + \frac{1}{2}[u_{12}, L_2] \right) g_1^{-1} g_2^{-1} \tag{2.13}$$

where $u_{12} = g_1^{-1} g_2^{-1}\{g_1, g_2\}$.

Remark 2. In the form (2.10) the antisymmetry property of the bracket is explicit, although r has no special symmetry property. Furthermore, we have the freedom to redefine r by

$$r_{12} \longrightarrow r_{12} + [\sigma_{12}, L_2] \tag{2.14}$$

where σ is symmetric, without changing the Poisson bracket.

Example. Let us give an example of this construction in the simple example of the harmonic oscillator. The Lax matrix L is given in eq. (2.8) and we introduce the action–angle coordinates ρ, θ as in eq. (2.3). In these coordinates the matrix L is diagonalized by:

$$U = U^{-1} = \begin{pmatrix} \cos\frac{\theta}{2} & \sin\frac{\theta}{2} \\ \sin\frac{\theta}{2} & -\cos\frac{\theta}{2} \end{pmatrix}$$

Since $\{U_1, U_2\} = 0$, $r_{12} = q_{12}$, which is easily computed to be:

$$r_{12} = \frac{\omega}{2\rho^2} \begin{pmatrix} 0 & 1 \\ -1 & 0 \end{pmatrix} \otimes L$$

It is easy to verify that this r-matrix indeed satisfies eq. (2.10). Let us notice that it is a dynamical r-matrix, which means that it depends explicitly on the dynamical variables.

2.6 Commuting flows

The Poisson brackets, eq. (2.10), are equivalent to the involution of the eigenvalues of L. An equivalent set of commuting Hamiltonians, H_n, is given by the traces of the powers of the Lax matrix:

$$H_n = \text{Tr}\,(L^n) \qquad (2.15)$$

The Hamiltonians H_n are in involution, $\{H_n, H_m\} = 0$, since they are symmetric polynomials in the eigenvalues. Furthermore, we show that the time evolution of the Lax matrix L with Hamiltonian H_n naturally takes the Lax form.

Proposition. *Suppose that* $\{L_1, L_2\} = [r_{12}, L_1] - [r_{21}, L_2]$. *If we take* $H_n = \text{Tr}\,(L^n)$ *as Hamiltonians, then the equations of motion admit a Lax representation:*

$$\frac{dL}{dt_n} \equiv \{H_n, L\} = [M_n, L]\,, \quad \text{with } M_n = -n\,\text{Tr}_1\,(L_1^{n-1} r_{21}) \qquad (2.16)$$

<u>Proof</u>. Set $m = 1$ in eq. (2.11), and take the trace over the first space, to get $dL/dt_n = [M_n, L]$ with $M_n = -n\,\text{Tr}_1(L_1^{n-1} r_{21})$. ∎

Note that the matrices M_n are unchanged under the transformation eq. (2.14). However, the matrices M_n are not unique since adding any matrix commuting with L does not change the equations of motion.

We close this chapter by presenting some of the few historical examples of integrable systems which were known at the end of the nineteenth century. Of course all systems with one degree of freedom (phase space of dimension 2) are integrable since the Hamiltonian H is conserved. We discuss below more sophisticated examples with higher dimensional phase spaces.

2.7 The Kepler problem

The first historical integrable system is the Kepler two-body problem. In the centre of mass frame, the equations of motion take the form:

$$\frac{d^2 x_i}{dt^2} = -\frac{\partial V(r)}{\partial x_i}, \quad r = \sqrt{x_1^2 + x_2^2 + x_3^2}$$

In the traditional Kepler problem, $V(r) = C/r$, but we will consider any centrally symmetric potential $V(r)$. This is a Hamiltonian system with

$$H = \frac{1}{2}\sum_{i=1}^{3} p_i^2 + V(r)$$

and Poisson brackets $\{p_i, x_j\} = \delta_{ij}$. The phase space is of dimension 6, and we have to exhibit three commuting conserved quantities. Due to central symmetry, the angular momentum

$$\vec{J} = (J_1, J_2, J_3), \quad J_{ij} = x_i p_j - x_j p_i = \epsilon_{ijk} J_k$$

is conserved. Here ϵ_{ijk} is the totally antisymmetric Levi–Civita tensor. The three components J_i are conserved but do not Poisson commute. However, a set of three independent Poisson commuting quantities is provided by H, $J_3 \equiv J_{12}$, $J^2 \equiv J_{12}^2 + J_{23}^2 + J_{13}^2$. At this point, one may follow the standard solution, which takes advantage of the conservation of $\vec{J}$, and restrict ourselves to the plane perpendicular to $\vec{J}$ where the motion takes place. Here we prefer to show the Liouville theorem at work, and use only the three commuting conserved quantities. Due to the spherical symmetry of the problem, it is convenient to use spherical coordinates:

$$x_1 = r \sin\theta \cos\phi, \quad x_2 = r\sin\theta \sin\phi, \quad x_3 = r\cos\theta$$

We introduce the conjugate momenta p_r, p_θ, p_ϕ by writing the canonical 1-form $\alpha = \sum p_i dx_i = p_r dr + p_\theta d\theta + p_\phi d\phi$. In these coordinates the conserved quantities read:

$$H = \frac{1}{2}\left(p_r^2 + \frac{1}{r^2}p_\theta^2 + \frac{1}{r^2\sin^2\theta}p_\phi^2\right) + V(r)$$

$$J^2 = p_\theta^2 + \frac{1}{\sin^2\theta}p_\phi^2$$

$$J_3 = p_\phi \tag{2.17}$$

On the surface M_f corresponding to fixed values of the conserved quantities, we solve for the p in terms of the the position variables, yielding:

$$p_r = \sqrt{2\left(H - V(r)\right) - \frac{J^2}{r^2}}, \quad p_\theta = \sqrt{J^2 - \frac{J_3^2}{\sin^2\theta}}, \quad p_\phi = J_3$$

Note that on M_f, p_r depends only on r, p_θ only on θ and p_ϕ only on ϕ (it is constant). The variables r, θ, ϕ are then called *separated* variables. The 1-form α restricted on M_f is then obviously closed. The action S appearing in the Liouville theorem reads:

$$S = \int^r \sqrt{2\left(H - V(r)\right) - \frac{J^2}{r^2}}\,dr + \int^\theta \sqrt{J^2 - \frac{J_3^2}{\sin^2\theta}}\,d\theta + \int^\phi J_3 d\phi$$

The angle variables corresponding to our action variables are given by

$$\psi_H = \frac{\partial S}{\partial H}, \quad \psi_{J^2} = \frac{\partial S}{\partial J^2}, \quad \psi_{J_3} = \frac{\partial S}{\partial J_3}$$

and have simple time evolution with respective frequencies $(1, 0, 0)$ by eq. (2.4). Hence ψ_{J^2} and ψ_{J_3} remain constant, while $\psi_H = t - t_0$. This gives the standard formula for the Kepler motion:

$$t - t_0 = \int^r \frac{dr}{\sqrt{2\left(H - V(r)\right) - \frac{J^2}{r^2}}}$$

Note that the constancy of ψ_{J_3} implies:

$$\dot{\phi} = \frac{J_3}{\sin^2 \theta \sqrt{J^2 - \frac{J_3^2}{\sin^2 \theta}}} \dot{\theta}$$

This, in turn, implies the conservation of J_1, J_2:

$$J_1 = -J_3 \cot \theta \cos \phi - \sin \phi \sqrt{J^2 - \frac{J_3^2}{\sin^2 \theta}}$$

$$J_2 = -J_3 \cot \theta \sin \phi + \cos \phi \sqrt{J^2 - \frac{J_3^2}{\sin^2 \theta}}$$

so that the motion takes place in the plane perpendicular to $\vec{J}$, as expected. It is worth noticing that the present approach is the one which prevails in Quantum Mechanics, where the three components of $\vec{J}$ cannot be measured simultaneously.

2.8 The Euler top

We consider a rotating solid body attached to a fixed point. The Euler top corresponds to the case where there is no external force. It is very convenient to consider the equations of motion in a frame rotating with the body, as discovered by Euler. We choose the moving frame with origin at the fixed point of the top (that is the point where the top is attached), and the axis being the principal inertia axis which diagonalizes, the inertia tensor computed with respect to the fixed point $I_{ij} = \int (\vec{x}^2 \delta_{ij} - x_i x_j) \rho(x) dx$ with $\rho(x)$ the mass density. Let $\vec{J}$ be the angular momentum of the top seen in the moving frame. We have $\vec{J} = I.\vec{\omega}$ where $I = \mathrm{Diag}(I_1, I_2, I_3)$ and $\vec{\omega}$ is the rotation vector of the moving frame. We shall assume that the principal moments of inertia I_i are all different. The equation of motion reads:

$$\frac{d\vec{J}}{dt} = -\vec{\omega} \wedge \vec{J}$$

It expresses the conservation of $\vec{J}$ in the absolute frame. This can be recast into the Hamiltonian framework by defining the Poisson brackets:

$$\{J_i, J_j\} = \epsilon_{ijk} J_k$$

where ϵ_{ijk} is the usual antisymmetric tensor. The Hamiltonian reads:

$$H = \frac{1}{2} \sum_{i=1}^{3} \frac{J_i^2}{I_i}$$

This Poisson bracket is degenerate because $\vec{J}^2$ Poisson commutes with everything. One must choose a symplectic leaf to get a well-defined Hamiltonian system. This is achieved by fixing the value of $\vec{J}^2$ to a numerical value. Then the phase space is of dimension 2, and the system is integrable with conserved quantity H. Note that the trajectories are immediately obtained as the intersection of the sphere $J_1^2 + J_2^2 + J_3^2 = \vec{J}^2$ and the ellipsoid $J_1^2/I_1 + J_2^2/I_2 + J_3^2/I_3 = 2H$. Using these relations to compute J_2 and J_3 in terms of J_1 and substituting into the equation of motion of J_1, i.e. $\dot{J}_1 = (I_3^{-1} - I_2^{-1}) J_2 J_3$, yields an equation of the form $\dot{J}_1 = \sqrt{\alpha + \beta J_1^2 + \gamma J_1^4}$, so that J_1 is an elliptic function of t.

2.9 The Lagrange top

When the top is in a gravitational field its weight has to be taken into account and the problem is more complicated. Let us assume that the rotating frame has its origin at the fixed point. The problem is integrable only in special cases. One case, found by Lagrange, is when two inertia moments are equal, for example $I_1 = I_2$, and the centre of mass is located at a position $(x_1 = 0, x_2 = 0, x_3 = h)$ with respect to the rotating frame. This situation is achieved when the top has an axis of symmetry (around the third axis) and is attached to a point on this axis.

For any top in a gravitational field, the equations of motion in the rotating frame take the form:

$$\frac{d\vec{J}}{dt} = -\vec{\omega} \wedge \vec{J} + \vec{h} \wedge \vec{P}, \quad \frac{d\vec{P}}{dt} = -\vec{\omega} \wedge \vec{P} \qquad (2.18)$$

where $\vec{P}$ is the weight of the top, which is constant in the absolute frame, and $\vec{h}$ is the vector from the fixed point to the centre of mass, which is constant in the rotating frame. From these equations one can check the conservation of three quantities: $\vec{P}^2$, $\vec{J} \cdot \vec{P}$ and the energy

$$H = \frac{1}{2}(\vec{J} \cdot I^{-1} \vec{J}) - \vec{P} \cdot \vec{h}$$

In order to formulate the equations of motion in a Hamiltonian framework let us introduce the following Poisson brackets between the six dynamical quantities J_i and P_i:

$$\{J_i, J_j\} = \epsilon_{ijk} J_k, \quad \{J_i, P_j\} = \epsilon_{ijk} P_k, \quad \{P_i, P_j\} = 0$$

The Hamilton equations of motion are precisely eqs. (2.18). The Poisson structure is degenerate, i.e. the two conserved quantities $\vec{P}^2$, $\vec{J} \cdot \vec{P}$ are in the centre. Hence the symplectic leaves are of dimension 4. The Hamiltonian H provides one conserved quantity defined on the leaves and that is all in general.

Using the particular hypothesis of the Lagrange top, namely the rotational symmetry around the third axis, it is easy to see that $\vec{J} \cdot \vec{h}$ is a second independent conserved quantity. Indeed, multiplying the first eq. (2.18) by $\vec{h}$ we find

$$\frac{d\,\vec{h} \cdot \vec{J}}{dt} = \vec{h} \cdot (\vec{\omega} \wedge \vec{J}) = h_3(\omega_1 J_2 - \omega_2 J_1) = h_3 \omega_1 \omega_2 (I_2 - I_1) = 0$$

where we have used that $\vec{h}$ is along the third axis. Since $\vec{h}$ is conserved, this quantity Poisson commutes with the Hamiltonian, hence the system is integrable.

To solve this integrable system we describe the top by Euler angles (θ, ϕ, ψ). Recall that the rotation vector and the weight vector can be expressed in the moving frame as:

$$\begin{pmatrix} \omega_1 \\ \omega_2 \\ \omega_3 \end{pmatrix} = \begin{pmatrix} \dot{\phi} \sin\theta \sin\psi + \dot{\theta} \cos\psi \\ \dot{\phi} \sin\theta \cos\psi - \dot{\theta} \sin\psi \\ \dot{\phi} \cos\theta + \dot{\psi} \end{pmatrix}, \quad \begin{pmatrix} P_1 \\ P_2 \\ P_3 \end{pmatrix} = \begin{pmatrix} P \sin\theta \sin\psi \\ P \sin\theta \cos\psi \\ P \cos\theta \end{pmatrix}$$

The two quantities in the centre of the Poisson bracket are $\vec{P}^2 = P^2$ and $\vec{P} \cdot \vec{J} \equiv PJ_z = P(I_1 \sin^2\theta + I_3 \cos^2\theta)\dot{\phi} + PI_3 \cos\theta\dot{\psi}$. Moreover, the Lagrange conserved quantity reads $\vec{h} \cdot \vec{J} \equiv hJ_3 = hI_3(\dot{\phi}\cos\theta + \dot{\psi})$. This allows us to eliminate $\dot{\phi}$ and $\dot{\psi}$ in the Hamiltonian:

$$H = \frac{1}{2}I_1(\sin^2\theta\dot{\phi}^2 + \dot{\theta}^2) + \frac{1}{2}I_3(\dot{\phi}\cos\theta + \dot{\psi})^2 - P\cos\theta$$

We get a one-dimensional system in the variable θ with Hamiltonian:

$$H = \frac{1}{2}I_1\dot{\theta}^2 + \frac{1}{2I_1}\frac{(J_z - J_3\cos\theta)^2}{\sin^2\theta} - P\cos\theta + \frac{1}{2}\frac{J_3^2}{I_3}$$

It follows that θ is an elliptic function of the time t. Note that the choice of the Euler angle coordinates has disentangled the dynamics, since ϕ and ψ no longer appear. This is related to the symmetry of the problem with respect to both the vertical axis and the axis of the top.

2.10 The Kowalevski top

There is another much hidden case of integrability of the top, which was discovered by S. Kowalevski. As before we consider the motion of the top in a moving frame with origin at the fixed point of the top. Assume that the moments of inertia obey $I_1 = I_2 = 2I_3$. Assume further that the centre of mass is in the plane $x_3 = 0$, but away from the origin, so that the top has no rotational symmetry. However, we are free to choose the inertia axis up to a rotation around the third one, hence to assume that the fixed point is on the first axis. We introduce the traditional notations

$$\vec{\omega} = \begin{pmatrix} p \\ q \\ r \end{pmatrix}, \quad \vec{P} = \begin{pmatrix} \gamma_1 \\ \gamma_2 \\ \gamma_3 \end{pmatrix}, \quad \vec{h} = \begin{pmatrix} h \\ 0 \\ 0 \end{pmatrix}$$

and we write eqs. (2.18) in components, with $c_0 = h/I_3$:

$$\begin{array}{ll} 2\dot{p} = qr & \dot{\gamma}_1 = r\gamma_2 - q\gamma_3 \\ 2\dot{q} = -pr - c_0\gamma_3 & \dot{\gamma}_2 = p\gamma_3 - r\gamma_1 \\ \dot{r} = c_0\gamma_2 & \dot{\gamma}_3 = q\gamma_1 - p\gamma_2 \end{array}$$

The Hamiltonian and Poisson brackets are the same as in the Lagrange case. Again the Hamiltonian

$$H = \frac{I_3}{2}(2p^2 + 2q^2 + r^2) - h\gamma_1$$

and the following quantities

$$\vec{P}^2 = \gamma_1^2 + \gamma_2^2 + \gamma_3^2, \quad \vec{P} \cdot \vec{J} = I_3(2p\gamma_1 + 2q\gamma_2 + r\gamma_3)$$

are conserved. The last two are in the centre of the Poisson bracket, so the symplectic leaves are of dimension 4, and we need one further conserved quantity to prove integrability. To introduce it naturally, consider $z = p + iq$ and $\xi = \gamma_1 + i\gamma_2$. The equations of motion give:

$$2\dot{z} = -irz - ic_0\gamma_3, \quad \dot{\xi} = -ir\xi + i\gamma_3 z$$

We can eliminate γ_3 by considering the combination $z^2 + c_0\xi$ which obeys:

$$\frac{d}{dt}(z^2 + c_0\xi) = -ir(z^2 + c_0\xi), \quad \frac{d}{dt}(\bar{z}^2 + c_0\bar{\xi}) = ir(\bar{z}^2 + c_0\bar{\xi})$$

where the second equation is obtained from the first one by complex conjugation. It is then clear that $|z^2 + c_0\xi|^2$ is conserved. In terms of the original variables, we have obtained the Kowalevski conserved quantity:

$$K = (p^2 - q^2 + c_0\gamma_1)^2 + (2pq + c_0\gamma_2)^2 \tag{2.19}$$

Note that the conditions $I_1 = I_2 = 2I_3$ are essential in this calculation. The solution of this model has been obtained by S. Kowalevski and is considerably more involved than in the previous cases. The main steps of the solution will be presented in Chapters 4 and 5.

2.11 The Neumann model

This model deals with the motion of a particle on a sphere S^{N-1} submitted to harmonic forces with generically different frequencies in each direction. It was first introduced by Neumann. An easy formulation is achieved by introducing a Lagrange multiplier, Λ, and writing the Lagrangian:

$$\mathcal{L} = \sum_{k=1}^{N} \frac{1}{2}(\dot{x}_k^2 - a_k x_k^2) + \frac{1}{2}\Lambda \left(\sum_{l=1}^{N} x_l^2 - 1 \right)$$

The equations of motion are:

$$\ddot{x}_k = -a_k x_k + \Lambda x_k, \quad \sum_l x_l^2 = 1$$

To compute Λ, we multiply by x_k and sum over k, which yields $\Lambda = -\sum_k (\dot{x}_k^2 - a_k x_k^2)$ where we used that the constraint implies $\sum_k (x_k \ddot{x}_k + \dot{x}_k^2) = 0$. This leads to the non-linear Newton equations of motion for the particle:

$$\ddot{x}_k = -a_k x_k - x_k \sum_l (\dot{x}_l^{\,2} - a_l x_l^2) \tag{2.20}$$

Conversely, if we start with initial conditions satisfying $\sum_k x_k^2 = 1$ and $\sum_k x_k \dot{x}_k = 0$, these conditions are preserved by the time evolution.

It is important for us to cast this model in the Hamiltonian formulation. This is achieved by introducing a larger phase space and then reducing by a symmetry. Consider a $2N$-dimensional phase space with coordinates $x_n, y_n, n = 1, \ldots, N$ and canonical Poisson brackets

$$\{x_n, y_m\} = \delta_{nm} \tag{2.21}$$

and introduce the "angular momentum" antisymmetric matrix: $J_{kl} = x_k y_l - x_l y_k$ and the Hamiltonian:

$$H = \frac{1}{4}\sum_{k \neq l} J_{kl}^2 + \frac{1}{2}\sum_k a_k x_k^2 \tag{2.22}$$

We shall assume in the following that $a_1 < a_2 < \cdots < a_N$. The Hamiltonian equations are, with $X = (x_k)$, $Y = (y_k)$, and the diagonal constant matrix $L_0 = (a_k \delta_{kl})$:

$$\dot{X} = -JX, \qquad \dot{Y} = -JY - L_0 X \qquad (2.23)$$

The Hamiltonian and the symplectic form have a symmetry

$$Y \to Y + \lambda X, \quad X \to X$$

and we can perform a Hamiltonian reduction under this symmetry group (see Chapter 14). The moment map is given by $M = \frac{1}{2}\sum_k x_k^2$ since $\{M, Y_k\} = X_k$, $\{M, X_k\} = 0$. We fix the moment to $M = \frac{1}{2}$. The reduced phase space is obtained by then taking the quotient by the group of stability of the moment which is here the whole group of symmetry. This amounts to imposing some gauge condition, e.g. $(X, Y) = 0$. The reduced phase space has the correct dimension $2n - 2$ for a point on a sphere.

Remark. The reduced equations of motion are equivalent to the equations of motion for the Neumann model. Indeed, the reduced system is characterized by the conditions ${}^t X X = 1$ and ${}^t X Y = 0$, but the equations of motion eq. (2.23) do not preserve the second condition. We need to perform simultaneously a time-dependent gauge transformation $Y \to Y + \lambda(t)X$ to keep the motion on the gauge surface. Writing:

$$0 = \frac{d}{dt}(X, Y + \lambda X) = (-JX, Y + \lambda X) + (X, -JY - L_0 X + \dot{\lambda}X - \lambda JX)$$

$$= -(X, L_0 X) + \dot{\lambda}$$

since J is antisymmetric, gives $\dot{\lambda} = (X, L_0 X)$. The equation of motion for $Y' = (Y + \lambda X)$ on the gauge surface is thus:

$$\dot{Y}' = (-JY - L_0 X) - \lambda JX + \dot{\lambda}X = -JY' - L_0 X + (X, L_0 X)X$$

Since $J = J'$ we have $\dot{X} = -J'X = Y'$ and $\dot{Y}' = -(Y', Y')X - L_0 X + (X, L_0 X)X$ so that eliminating Y' we finally get:

$$\ddot{X} = -L_0 X - \left((\dot{X}, \dot{X}) - (X, L_0 X) \right) X$$

which is identical to eq. (2.20).

The Liouville integrability of this system is a consequence of the existence of $(N - 1)$ independent quantities in involution, first found by K. Uhlenbeck:

$$F_k = x_k^2 + \sum_{l \neq k} \frac{J_{kl}^2}{a_k - a_l}, \quad \sum_k F_k = 1 \qquad (2.24)$$

Notice that the Hamiltonian of the Neumann model can be expressed in terms of the F_k as

$$H = \frac{1}{2} \sum_k a_k F_k$$

Alternatively we can implement the Hamiltonian reduction by considering functions of X and Y invariant under $Y \to Y + \lambda X$. Such invariant functions are functions of X and J.

Given X and J, an antisymmetric rank 2 matrix whose image contains X, we can always find a vector Y up to the above symmetry such that $J_{kl} = x_k y_l - x_l y_k$.

The equations of motion can be written in terms of the two gauge invariant matrices $J = X\,^t Y - Y\,^t X$ and $K = X\,^t X$

$$\dot{K} = -[J, K], \quad \dot{J} = [L_0, K]$$

That eq. (2.23) implies the above is a simple computation. Conversely, knowing K it is easy to compute X since K is a projector on X whose length is 1 and then one can compute Y knowing J, up to a gauge.

2.12 Geodesics on an ellipsoid

The Neumann problem was shown by Moser to contain, in particular, the geodesic motion on an ellipsoid which was found to be integrable by Jacobi. Consider the Hamiltonian belonging to the Neumann conserved quantities, but different from the usual Neumann Hamiltonian:

$$H = \sum_k \frac{1}{a_k} F_k = Q(X, X) - Q(X, X)Q(Y, Y) + Q^2(X, Y)$$

where we have defined the quadratic form:

$$Q(X, Y) = \sum_k \frac{1}{a_k} x_k y_k$$

The Hamiltonian H is of course conserved, so we can restrict ourselves on the surface $H = 0$. Note that if one defines the quantity ξ, invariant under the gauge transformation $Y \to Y + \lambda X$ by

$$\xi = Y - \frac{Q(X, Y)}{Q(X, X)} X$$

we have $H = Q(X, X)(1 - Q(\xi, \xi))$ so that the condition $H = 0$ is equivalent to the condition $Q(\xi, \xi) = 1$. The flow of H, for $H = 0$, leaves ξ on the ellipsoid $Q(\xi, \xi) = 1$.

We want to show that the trajectories of ξ are geodesics on the ellipsoid. We compute the time evolution of ξ, and for this we remark that since ξ is gauge invariant, one can use the unreduced equations of motion:

$$\dot{x}_k = \frac{\partial H}{\partial y_k} = -2Q(X,X)\frac{y_k}{a_k} + 2Q(X,Y)\frac{x_k}{a_k} = -2Q(X,X)\frac{\xi_k}{a_k}$$

$$\dot{y}_k = -\frac{\partial H}{\partial x_k} = -2(1-Q(Y,Y))\frac{x_k}{a_k} - 2Q(X,Y)\frac{y_k}{a_k}$$

Note that, since X is gauge independent, so is its time derivative. Denoting $s = Q(X,Y)/Q(X,X)$ we then have:

$$\dot{\xi}_k = -2\left(1 - Q(Y,Y) + \frac{Q^2(X,Y)}{Q(X,X)}\right)\frac{x_k}{a_k} - \dot{s}x_k$$

On the surface $H = 0$ this becomes simply:

$$\frac{d\xi_k}{dt} = -\dot{s}x_k \implies \frac{d\xi_k}{ds} = -x_k$$

Since $\vec{x}^2 = 1$ the length of the vector $d\vec{\xi}/ds$ is 1, which means that s is the length parameter on the trajectory of ξ. Next we compute:

$$\frac{d^2\xi_k}{ds^2} = \frac{1}{\dot{s}}\frac{dx_k}{dt} = -2\frac{Q(X,X)}{\dot{s}}\frac{\xi_k}{a_k} \qquad (2.25)$$

The vector with components ξ_k/a_k is the gradient of $Q(\xi,\xi)$ hence is normal to the ellipsoid $Q(\xi,\xi) = 1$. Equation (2.25) shows that the second derivative of $\vec{\xi}$ with respect to the length parameter s is normal to the surface. This characterizes geodesics, as we now show. Indeed, to find geodesics on the surface $f(\xi) = 0$, we have to minimize the arc length:

$$\int \left(\sqrt{\dot{\vec{\xi}}\cdot\dot{\vec{\xi}}} + \Lambda f(\xi)\right) dt$$

where Λ is a Lagrange parameter. The Euler–Lagrange equation reads:

$$\frac{d}{dt}\frac{\ddot{\vec{\xi}}}{\sqrt{\dot{\vec{\xi}}\cdot\dot{\vec{\xi}}}} = \Lambda\nabla f(\xi)$$

So the derivative of the normalized velocity vector is perpendicular to the surface.

The geodesic motion on an ellipsoid was solved originally by Jacobi by introducing ellipsoidal coordinates which separate the variables for this problem. As they separate the variables of the Neumann model as well, we now explain this method for the Neumann model.

2.13 Separation of variables in the Neumann model

Following Jacobi and Neumann, we introduce $(N-1)$ parameters on the sphere $\zeta_1, \ldots, \zeta_{N-1}$. They are the roots of the equation:

$$u(\zeta) \equiv \sum_k \frac{x_k^2}{\zeta - a_k} = 0$$

This equation is invariant by $x_k \longrightarrow \lambda x_k$ so that $\zeta_1 < \zeta_2 < \cdots < \zeta_{N-1}$ are indeed defined on the sphere. Conversely, by definition of the ζ_j we have for $x \in S^{(N-1)}$:

$$u(\zeta) = \frac{\prod_j (\zeta - \zeta_j)}{\prod_k (\zeta - a_k)} \implies x_k^2 = \frac{\prod_j (a_k - \zeta_j)}{\prod_{l \neq k} (a_k - a_l)} \tag{2.26}$$

Considering the graph of $u(\zeta)$ it is easy to see that:

$$a_1 < \zeta_1 < a_2 < \zeta_2 < a_3 < \cdots < \zeta_{N-1} < a_N$$

and we have a bijection of this domain $\mathcal{D}$ of the ζ_j on the "quadrant" $x_k > 0 \; \forall k$ of the sphere.

The (ζ_j) define an orthogonal system of coordinates on the sphere, of ellipsoidal type.

Let us consider, for each root ζ_j, the vector:

$$\frac{\partial \vec{x}}{\partial \zeta_j} = \frac{1}{2} \vec{v}_j, \quad \vec{v}_j = \left(\frac{x_1}{\zeta_j - a_1}, \ldots, \frac{x_N}{\zeta_j - a_N} \right) \tag{2.27}$$

Since ζ_j solves $u(\zeta) = 0$, we have $\vec{x} \cdot \vec{v}_j = 0$. Moreover,

$$\vec{v}_j \cdot \vec{v}_{j'} = \sum_k \frac{x_k^2}{(\zeta_j - a_k)(\zeta_{j'} - a_k)} = -\frac{u(\zeta_j) - u(\zeta_{j'})}{\zeta_j - \zeta_{j'}} = 0, \quad \text{if } j \neq j'$$

Therefore the vectors $\vec{v}_j$ are $(N-1)$ orthogonal vectors in the tangent plane to the sphere $S^{(N-1)}$ at the point $\vec{x}$. As a byproduct, since $\vec{v}_j^2 = -u'(\zeta_j)$ we also get the metric tensor:

$$g_{jj'} = \frac{\partial \vec{x}}{\partial \zeta_j} \cdot \frac{\partial \vec{x}}{\partial \zeta_{j'}} = -\frac{1}{4} \delta_{jj'} u'(\zeta_j)$$

To compute the momenta conjugated to the variables ζ_j, we consider the canonical 1-form $\alpha = \sum_k y_k dx_k$ associated with the Poisson bracket eq. (2.21). We write it as $\alpha = \sum_j p_j d\zeta_j$. One gets

$$p_j = \sum_k y_k \frac{\partial x_k}{\partial \zeta_j} = \frac{1}{2} \vec{y} \cdot \vec{v}_j$$

These $(N-1)$ equations determine $\vec{y}$ up to a vector proportional to $\vec{x}$ which does not affect the value of J_{kl}. A solution is $\vec{y} = 1/2 \sum_j g^{jj} p_j \vec{v}_j$ which easily gives:

$$J_{kl} = -\frac{1}{2} \sum_j (a_k - a_l) v_j^k v_j^l g^{jj} p_j$$

With this, we can compute the conserved quantities F_k, eq. (2.24), in terms of the new canonical coordinates ζ_i, p_j:

$$F_k = x_k^2 + \frac{1}{4} \sum_{jj'} \left(a_k \vec{v}_j \cdot \vec{v}_{j'} - \sum_l a_l v_j^l v_{j'}^l \right) v_j^k v_{j'}^k g^{jj} g^{j'j'} p_j p_{j'}$$

Noting that $\vec{v}_j \cdot \vec{v}_{j'} = 4g_{jj}\delta_{jj'}$ and $\sum_l a_l v_j^l v_{j'}^l = 4\zeta_j g_{jj}\delta_{jj'}$, we obtain:

$$F_k = x_k^2 \left(1 - \sum_j \frac{g^{jj} p_j^2}{\zeta_j - a_k} \right)$$

It is convenient to introduce the generating function for the F_k:

$$\mathcal{H}(\lambda) \equiv \sum_k \frac{F_k}{\lambda - a_k} = \frac{\prod_n (\lambda - b_n)}{\prod_k (\lambda - a_k)} \tag{2.28}$$

for appropriate b_n, $n = 1, \ldots, N-1$, and we have used $\sum F_k = 1$ to normalize the leading coefficient in the numerator. By a simple calculation we find:

$$\mathcal{H}(\lambda) = u(\lambda) \left(1 - \sum_j \frac{g^{jj} p_j^2}{\zeta_j - \lambda} \right) \tag{2.29}$$

Following the general strategy of the Liouville theorem, we express the momenta p_j in terms of the conserved quantities F_k and the ζ_j. We have:

$$g^{jj} p_j^2 = \lim_{\lambda \to \zeta_j} \frac{\lambda - \zeta_j}{u(\lambda)} \mathcal{H}(\lambda) \implies p_j^2 = -\frac{1}{4}\mathcal{H}(\zeta_j)$$

where we have taken into account the value of the metric tensor and eq. (2.26). Notice that, on M_f, the momentum p_j is a function of ζ_j only, so that the coordinates ζ_j form a set of separated variables for the Neumann model.

The function $S = \int \sum_j p_j d\zeta_j$ reads

$$S = \frac{1}{2} \sum_j \int^{\zeta_j} \sqrt{-\mathcal{H}(\zeta)}\, d\zeta = \frac{1}{2} \sum_j \int^{\zeta_j} \sqrt{-\frac{\Pi_n(\zeta - b_n)}{\Pi_k(\zeta - a_k)}}\, d\zeta \qquad (2.30)$$

and is a sum of terms, one for each separated variable.

Choosing as independent action variables the $(N-1)$ independent quantities b_n instead of the N dependent quantities F_k, the conjugate variables ψ_n are:

$$\psi_n = \frac{\partial S}{\partial b_n} = -\frac{1}{4} \sum_j \int^{\zeta_j} \sqrt{-\mathcal{H}(\zeta)}\, \frac{d\zeta}{\zeta - b_n} \qquad (2.31)$$

By the Liouville theorem, the time evolution of the ψ_n under the Hamiltonian $\mathcal{H}(\lambda)$ is linear:

$$\dot{\psi}_n = \frac{\partial \mathcal{H}(\lambda)}{\partial b_n} = -\frac{\mathcal{H}(\lambda)}{\lambda - b_n} = -\frac{1}{4} \sum_j \frac{\dot{\zeta}_j \sqrt{-\mathcal{H}(\zeta_j)}}{\zeta_j - b_n} \qquad (2.32)$$

where the last equality is obtained by differentiating eq. (2.31) with respect to time. Equation (2.32) gives the evolution of the variables ζ_j.

This can be formulated in a more geometrical way by introducing the polynomial of degree $(2N-1)$:

$$P(\zeta) = \prod_{i=1}^{N}(\zeta - a_i) \prod_{n=1}^{N-1}(\zeta - b_n)$$

We can rewrite eq. (2.32) as the set of $N-1$ equations:

$$\sum_j \frac{Q_n(\zeta_j) d\zeta_j}{\sqrt{-P(\zeta_j)}} = 4 \frac{dt}{\Pi_i(\lambda - a_i)} Q_n(\lambda), \quad n = 1, \ldots, N-1$$

with

$$Q_n(\lambda) = \prod_{m \neq n}(\lambda - b_m)$$

Since the $Q_n(\lambda)$ are $N-1$ linearly independent polynomials of degree $N-2$ this system is equivalent to the following one:

$$\sum_j \frac{\zeta_j^k d\zeta_j}{\sqrt{-P(\zeta_j)}} = 4 \frac{\lambda^k}{\Pi_i(\lambda - a_i)} dt, \quad k = 0, \ldots, N-2 \qquad (2.33)$$

The quantities $\zeta^k d\zeta / \sqrt{-P(\zeta)}$, $k = 0, \ldots, N - 2$, are the Abelian differentials of the first kind on the hyperelliptic Riemann surface of genus $g = N - 1$ given by the equation $s^2 + P(\zeta) = 0$, see Chapter 15. The sums appearing in the left-hand side are thus Abel sums, and describe a point in the Jacobian of the curve. Equation (2.33) shows that this point moves linearly in time.

This relationship between integrability, separation of variables, Riemann surfaces and linear flows on their Jacobian will reappear in a much broader context in the following chapters.

References

[1] J.L. Lagrange, *Mécanique analytique.* (1788) In *Oeuvres de Lagrange,* **XII**, Gauthier–Villars (1889), Paris.

[2] C. Jacobi, *Vorlesungen über Dynamik,* Gesammelte Werke, Supplement band (1884) Berlin.

[3] J. Liouville, Note sur l'intégration des équations différentielles de la dynamique. *Journal de Mathématiques (Journal de Liouville)* **XX** (1855) 137.

[4] C. Neumann, De problemate quodam mechanico, quod ad primam integralium ultraellipticorum classem revocatur. *Crelle Journal* **56** (1859) 46–63.

[5] Sophie Kowalevski, Sur le problème de la rotation d'un corps solide autour d'un point fixe. *Acta Mathematica* **12** (1889) 177–232.

[6] C.S. Gardner, J.M. Greene, M.D. Kruskal and R.M. Miura, Method for solving the Korteweg–de Vries equation. *Phys. Rev. Lett.* **19** (1967) 1095.

[7] P.D. Lax, Integrals of non-linear equations of evolution and solitary waves. *Comm. Pure Appl. Math.* **21** (1968) 467.

[8] L. Landau and E. Lifchitz, *Mécanique.* MIR (1969) Moscow.

[9] K. Uhlenbeck, *Minimal 2-spheres and tori in S^k.* Preprint (1975).

[10] V. Arnold, *Méthodes mathématiques de la mécanique classique.* MIR (1976) Moscow.

[11] J. Moser, *Various aspects of integrable Hamiltonian systems.* Proc. *CIME Bressanone,* Progress in Mathematics, 8 Birkhauser (1978) 233.

[12] E.K. Sklyanin, *On the complete integrability of the Landau–Lifchitz equation*. Preprint LOMI E-3-79. Leningrad, 1979.

[13] M. Semenov-Tian-Shansky, What is a classical r-matrix? *Funct. Anal. and Appl.* **17** 4 (1983) 17.

[14] D. Mumford, *Tata lectures on Theta II*. Progress in Mathematics, **43** (1984) Birkhäuser.

3

Synopsis of integrable systems

In this chapter, we introduce Lax pairs with spectral parameters. These are Lax matrices $L(\lambda)$ and $M(\lambda)$ depending analytically on a parameter λ. The study of the analytical properties of the Lax equation $\dot{L}(\lambda) = [M(\lambda), L(\lambda)]$ yields considerable insight into its structure, and in fact, quickly introduces many of the major objects and concepts, which will be developed in depth in the subsequent chapters.

The first important result is that the possible forms of $M(\lambda)$ are completely determined by eq. (3.15). This form of $M(\lambda)$ is such that the commutator $[M(\lambda), L(\lambda)]$ has the same polar structure as $L(\lambda)$. The Lax equation has then a natural interpretation as a flow on a coadjoint orbit of a loop group. This has in turn the important consequence of introducing a symplectic structure into the theory allowing us to connect with Liouville integrability. Moreover, this geometric interpretation of the Lax equation lends itself to its solution by factorization in a loop group, which is a Riemann–Hilbert problem. Studying the analytic structure of $M(\lambda)$, we are led to consider an infinite family of elementary flows, depending on the order of the poles. This introduces a connection between time flows and the spectral parameter dependence, which finds a striking expression in Sato's formula expressing the wave function in terms of tau-functions, eq. (3.61).

The same ideas are exploited to analyse field theories. Here the Lax equation is replaced by a zero curvature equation for a Lax connection depending on a spectral parameter. We show that the role of the Lax matrix is played by the so-called monodromy matrix. Starting from a linear Poisson bracket in the r-matrix form for the Lax connection, we get a quadratic Poisson bracket for the monodromy matrix. The factorization problem allows us to define a group of transformations, the dressing group, acting on the solutions of the equations of motion. It is shown that this action is a Lie–Poisson action whose generator is the monodromy matrix. Finally, we use simple dressing elements to produce the so-called soliton solutions.

3.1 Examples of Lax pairs with spectral parameter

We begin our study of Lax pairs depending on a complex parameter λ, called the spectral parameter, by giving a few examples.

Example 1. Our first example will be provided by the Euler top, see section (2.8) in Chapter 2. In this case, a Lax pair appears naturally. Let us introduce the 3×3 matrices $J_{ij} = \epsilon_{ijk} J_k$ and $\Omega_{ij} = \epsilon_{ijk} \omega_k$. Then the equation of motion $\frac{d\vec{J}}{dt} = -\vec{\omega} \wedge \vec{J}$ can be recast in matrix form:

$$\frac{dJ}{dt} = [\Omega, J]$$

This is a Lax pair with $L = J$, and $M = \Omega$, but unfortunately the conserved quantities, $\mathrm{Tr}\, L^n$, either vanish or are functions of $\vec{J}^2$, and therefore the Hamiltonian is not included in this set of conserved quantities (recall that $\vec{J}^2$ is in the centre of the Poisson bracket).

To cure this problem some modifications are needed. Let us introduce a diagonal matrix: $\mathcal{I} = \mathrm{Diag}(\mathcal{I}_1, \mathcal{I}_2, \mathcal{I}_3)$ with $\mathcal{I}_k = \frac{1}{2}(I_i + I_j - I_k)$, where (i, j, k) is a cyclic permutation of $(1, 2, 3)$. With these notations we have

$$J = \mathcal{I}\Omega + \Omega\mathcal{I}$$

We assume that all $\mathcal{I}_j$ are different and we set:

$$L(\lambda) = \mathcal{I}^2 + \frac{1}{\lambda} J, \quad M(\lambda) = \lambda\mathcal{I} + \Omega \tag{3.1}$$

where λ is a free arbitrary parameter, the so-called spectral parameter. To check that the Lax equation gives back the equations of motion, we compute:

$$\dot{L}(\lambda) - [M(\lambda), L(\lambda)] = [J, \mathcal{I}] + [\mathcal{I}^2, \Omega] + \frac{1}{\lambda}(\dot{J} + [J, \Omega])$$

The first two terms cancel, while the vanishing of the $1/\lambda$-term gives the equations of motion. This Lax pair is much better than the previous one because:

$$\mathrm{Tr}\, L^2(\lambda) = \mathrm{Tr}\, \mathcal{I}^4 - \frac{2}{\lambda^2} \vec{J}^2$$

$$\mathrm{Tr}\, L^3(\lambda) = \mathrm{Tr}\, \mathcal{I}^6 + \frac{3}{\lambda^2} \mathrm{Tr}\, \mathcal{I}^2 J^2 = \mathrm{Tr}\, \mathcal{I}^6 - \frac{3}{\lambda^2}\left(\frac{1}{4}(\mathrm{Tr}\, I)^2 \vec{J}^2 - I_1 I_2 I_3\, H\right)$$

hence we now do have the Hamiltonian among the conserved quantities of the form $\operatorname{Tr} L^n(\lambda)$.

The new important point is that the Lax matrix depends on a spectral parameter λ and this was necessary to generate the proper conserved quantities. Furthermore, the Lax equation holds true identically in λ.

Example 2. As a second example we consider the Lagrange top. The matrices $L(\lambda)$ and $M(\lambda)$ are written as 4×4 matrices in block form,

$$L(\lambda) = \begin{pmatrix} 0 & I\,{}^t h + \lambda^{-2}\,{}^t P \\ Ih + \lambda^{-2}P & \lambda^{-1}J \end{pmatrix}, \quad M(\lambda) = \begin{pmatrix} 0 & \lambda\,{}^t h \\ \lambda h & \Omega \end{pmatrix} \quad (3.2)$$

where the 3×3 matrices J and Ω are as in the previous example, and h and P are 3×1 matrices corresponding to the vectors $\vec{h}$ and $\vec{P}$ of the Lagrange top, see section (2.9) in Chapter 2. Moreover, I stands for the two equal moments of inertia of the top $I = I_1 = I_2$. Let us write the Lax equation $0 = \dot{L} - [M, L]$ or:

$$0 = \begin{pmatrix} 0 & I\,{}^t h \Omega - {}^t h J + \lambda^{-2}\,({}^t \dot{P} + {}^t P \Omega) \\ -I\Omega h + Jh + \lambda^{-2}\,(\dot{P} - \Omega P) & \lambda^{-1}(\dot{J} + [J, \Omega] + P\,{}^t h - h\,{}^t P) \end{pmatrix}$$

Due to the Lagrange condition $I = I_1 = I_2$ we have $I\Omega h = Jh$ and the vanishing of the other elements reduces to the equations of motion $\dot{J} + [J, \Omega] + P\,{}^t h - h\,{}^t P = 0$ and $\dot{P} = \Omega P$.

Example 3. Finally consider the Neumann model. As we have seen in section (2.11) in Chapter 2 the equations of motion on gauge invariant quantities are:

$$\dot{K} = -[J, K], \quad \dot{J} = [L_0, K]$$

To recast these two relations into the Lax form we introduce the matrices

$$L(\lambda) = L_0 + \frac{1}{\lambda}J - \frac{1}{\lambda^2}K, \quad M(\lambda) = -\frac{1}{\lambda}K \quad (3.3)$$

and compute:

$$\dot{L}(\lambda) - [M(\lambda), L(\lambda)] = \frac{1}{\lambda}(\dot{J} - [L_0, K]) - \frac{1}{\lambda^2}(\dot{K} + [J, K])$$

The Lax equation with spectral parameter is equivalent to the vanishing of the two coefficients $\dot{J} - [L_0, K]$ and $\dot{K} + [J, K]$. Hence the Lax equation is equivalent to the equations of motion of the Neumann model.

There also exists a Lax pair with spectral parameter for the Kowalevski top, however, its construction is more involved and will be discussed in Chapter 4.

3.2 The Zakharov–Shabat construction

Given an integrable system, there does not yet exist a useful algorithm to construct a Lax pair. There does exist, however, a general procedure, due to Zakharov and Shabat, to construct consistent Lax pairs giving rise to integrable systems. This is a general method to construct matrices $L(\lambda)$ and $M(\lambda)$, depending on a spectral parameter λ, such that the Lax equation

$$\partial_t L(\lambda) = [M(\lambda), L(\lambda)] \tag{3.4}$$

is equivalent to the equations of motion of an integrable system. The method consists of specifying the analytical properties of the matrices $L(\lambda)$ and $M(\lambda)$, $\lambda \in \mathbb{C}$. We consider here systems with a finite number of degrees of freedom. The main result is eq. (3.15) expressing the possible forms of the matrix M in the Lax pair. We will end the section by showing that the previous examples do fit into this framework.

We first introduce a notation. For any matrix valued rational function $f(\lambda)$ with poles of order n_k at points λ_k at finite distance, we can decompose $f(\lambda)$ as

$$f(\lambda) = f_0 + \sum_k f_k(\lambda), \quad \text{with} \quad f_k(\lambda) = \sum_{r=-n_k}^{-1} f_{k,r}(\lambda - \lambda_k)^r$$

with f_0 a constant. The quantity $f_k(\lambda)$ is called the polar part at λ_k. When there is no ambiguity about the pole we are considering, we will often use the alternative notation $f_-(\lambda) \equiv f_k(\lambda)$. Around one of the points λ_k, $f(\lambda)$ may be decomposed as follows:

$$f(\lambda) = f(\lambda)_+ + f(\lambda)_- \tag{3.5}$$

with $f(\lambda)_+$ regular at the point λ_k and $f(\lambda)_- = f_k(\lambda)$ being the polar part.

Let us now consider matrices $L(\lambda)$ and $M(\lambda)$ of dimension $N \times N$. We will assume that the matrices $L(\lambda)$ and $M(\lambda)$ are rational functions of the parameter λ. Let $\{\lambda_k\}$ be the set of their poles, namely the poles of $L(\lambda)$ and those of $M(\lambda)$. With the above notations, assuming no pole at infinity, we can write quite generally:

$$L(\lambda) = L_0 + \sum_k L_k(\lambda), \quad \text{with} \quad L_k(\lambda) \equiv \sum_{r=-n_k}^{-1} L_{k,r}(\lambda - \lambda_k)^r \tag{3.6}$$

and

$$M(\lambda) = M_0 + \sum_k M_k(\lambda) \quad \text{with} \quad M_k(\lambda) \equiv \sum_{r=-m_k}^{-1} M_{k,r}(\lambda - \lambda_k)^r \quad (3.7)$$

Here n_k and m_k refer to the order of the poles at the corresponding point λ_k. The coefficients $L_{k,r}$ and $M_{k,r}$ are matrices. We will assume that the positions of the poles λ_k are constants independent of time.

The Lax equation (3.4), with $L(\lambda)$ and $M(\lambda)$ given by eqs. (3.6, 3.7), must hold identically in λ. Looking at eqs. (3.4) we see that the pole λ_k in the left-hand side is a priori of order n_k while in the right-hand side it is potentially of order $n_k + m_k$. Hence we have two types of equation. The first type does not contain time derivatives and comes from setting to zero the coefficients of the poles of order greater than n_k in the right-hand side of the equation. This will be interpreted as m_k constraint equations on M_k. The equations of the second type are obtained by matching the coefficients of the poles of order less or equal to n_k on both sides of the equation. These equations contain time derivatives and are thus the true dynamical equations.

Proposition. *Assuming that $L(\lambda)$ has distinct eigenvalues in a neighbourhood of λ_k, one can perform a regular similarity transformation $g^{(k)}(\lambda)$ diagonalizing $L(\lambda)$ in a vicinity of λ_k:*

$$L(\lambda) = g^{(k)}(\lambda) \ A^{(k)}(\lambda) \ g^{(k)-1}(\lambda) \qquad (3.8)$$

where $A^{(k)}(\lambda)$ is diagonal and has a pole of order n_k at λ_k. As a result, the decomposition of $L(\lambda)$ and $M(\lambda)$ in polar parts reads:

$$L = L_0 + \sum_k L_k, \quad \text{with} \quad L_k = \left(g^{(k)} A^{(k)} g^{(k)-1}\right)_- \qquad (3.9)$$

$$M = M_0 + \sum_k M_k, \quad \text{with} \quad M_k = \left(g^{(k)} B^{(k)} g^{(k)-1}\right)_- \qquad (3.10)$$

where $B^{(k)}(\lambda)$ has a pole of order m_k at λ_k. Moreover, the Lax equation implies that $B^{(k)}(\lambda)$ is diagonal.

Proof. If λ_k is a pole of $L(\lambda)$, demanding that $L(\lambda)$ has distinct eigenvalues in a neighbourhood of λ_k means that $L_{k,-n_k}$ has distinct eigenvalues. Then the matrix $Q(\lambda) = (\lambda - \lambda_k)^{n_k} L(\lambda)$, which is regular at λ_k, can be diagonalized in vicinity of λ_k with a regular matrix $g^{(k)}(\lambda)$. This proves eq. (3.8). Then defining $B^{(k)}(\lambda)$ by

$$M(\lambda) = g^{(k)}(\lambda) \ B^{(k)}(\lambda) \ g^{(k)-1}(\lambda) + \partial_t g^{(k)}(\lambda) \ g^{(k)-1}(\lambda) \qquad (3.11)$$

the Lax equation becomes:

$$\dot{A}^{(k)}(\lambda) = [B^{(k)}(\lambda), A^{(k)}(\lambda)]$$

This implies $\dot{A}^{(k)} = 0$ as expected (because the commutator with a diagonal matrix has no element on the diagonal), and, moreover, if we assume that the diagonal elements of $A^{(k)}$ are all distinct this equation implies that $B^{(k)}$ is also *diagonal*. Finally, the term $\partial_t g^{(k)} \, g^{(k)-1}$ is regular and does not contribute to the singular part M_k of M at λ_k. Hence $M_k = (g^{(k)} \, B^{(k)} \, g^{(k)-1})_-$ which only depends on $B_-^{(k)}$. It is worth noting that the first n coefficients of the expansion of $g^{(k)}(\lambda)$ only depend on the first n coefficients of the expansion of $Q(\lambda)$. The matrix $g^{(k)}(\lambda)$ is defined up to a right multiplication by an arbitrary analytic diagonal matrix. Note that this simultaneous diagonalization of $L(\lambda)$ and $M(\lambda)$ works around any point where $L(\lambda)$ has distinct eigenvalues. ∎

This proposition clarifies the structure of the Lax pair. Only the singular parts of $A^{(k)}$ and $B^{(k)}$ contribute to L_k and M_k. The independent parameters in $L(\lambda)$ are thus L_0, the singular diagonal matrices $A_-^{(k)}$ of the form

$$A_-^{(k)} = \sum_{r=-n_k}^{-1} A_{k,r}(\lambda - \lambda_k)^r \qquad (3.12)$$

and jets of regular matrices $\widehat{g}^{(k)}$ of order $n_k - 1$, defined up to right multiplication by a regular diagonal matrix $d^{(k)}(\lambda)$:

$$\widehat{g}^{(k)} = \sum_{r=0}^{n_k-1} g_{k,r}(\lambda - \lambda_k)^r \qquad (3.13)$$

From these data, we can reconstruct the Lax matrix $L(\lambda)$ by defining $L = L_0 + \sum_k L_k$ with

$$L_k \equiv \left(\widehat{g}^{(k)} \, A_-^{(k)} \, \widehat{g}^{(k)-1}\right)_- \qquad (3.14)$$

Then around each λ_k one can diagonalize $L(\lambda) = g^{(k)} A^{(k)} g^{(k)-1}$. This yields an extension of the matrices $A_-^{(k)}$ and $\widehat{g}^{(k)}$ to complete series $A^{(k)}$ and $g^{(k)}$ in $(\lambda - \lambda_k)$. Finally, to define $M(\lambda) = M_0 + \sum_k M_k$, we choose a set of polar matrices $(B^{(k)}(\lambda))_-$ and use the series $g^{(k)}$ to define M_k by eq. (3.10).

In the vicinity of a singularity, $L(\lambda)$ and $M(\lambda)$ can be simultaneously diagonalized if the Lax equation holds true. In this diagonal gauge, the

Lax equation simply states that the matrix $A^{(k)}(\lambda)$ is conserved and that $B^{(k)}(\lambda)$ is diagonal. When we transform these results into the original gauge, we get the general solution of the non-dynamical constraints on $M(\lambda)$:

Proposition. *Let $L(\lambda)$ be a Lax matrix of the form eq. (3.6). The general form of the matrix $M(\lambda)$ such that the orders of the poles match on both sides of the Lax equation is $M = M_0 + \sum_k M_k$ with*

$$M_k = \left(P^{(k)}(L, \lambda) \right)_- \qquad (3.15)$$

where $P^{(k)}(L, \lambda)$ is a polynomial in $L(\lambda)$ with coefficients rational in λ and ( )$_-$ denotes the singular part at $\lambda = \lambda_k$.

<u>Proof.</u> It is easy to show that this is indeed a solution. We have to check that the order of the poles is correct. Let us look at what happens around $\lambda = \lambda_k$. Using a beautiful argument first introduced by Gelfand and Dickey we write:

$$[M_k, L]_- = \left[\left(P^{(k)}(L, \lambda) \right)_-, L \right]_-$$

$$= \left[P^{(k)}(L, \lambda) - \left(P^{(k)}(L, \lambda) \right)_+, L \right]_- = - \left[\left(P^{(k)}(L, \lambda) \right)_+, L \right]_-$$

where we used that a polynomial in L commutes with L. From this we see that the order of the pole at λ_k is less than n_k. To show that this is a general solution, recall eqs. (3.8, 3.10). Since $A^{(k)}(\lambda)$ is a diagonal $N \times N$ matrix with all its elements distinct in a vicinity of λ_k, its powers 0 up to $N - 1$ span the space of diagonal matrices and one can write

$$B^{(k)} = P^{(k)}(A^{(k)}, \lambda) \qquad (3.16)$$

where $P^{(k)}(A^{(k)}, \lambda)$ is a polynomial of degree $N - 1$ in $A^{(k)}$. The coefficients of $P^{(k)}$ are rational combinations of the matrix elements of $A^{(k)}$ and $B^{(k)}$, hence admit Laurent expansions in $\lambda - \lambda_k$ in a vicinity of λ_k. Inserting eq. (3.16) into eq. (3.10) one gets $M_k = \left(P^{(k)}(L, \lambda) \right)_-$. Moreover, in this formula the Laurent expansions of the coefficients of $P^{(k)}$ can be truncated at some positive power of $\lambda - \lambda_k$ since a high enough power cannot contribute to the singular part, yielding a polynomial with coefficients Laurent *polynomials* in $\lambda - \lambda_k$. ∎

It is important to realize that the dynamical variables are the matrix elements of the Lax matrix, or the matrix elements of the $L_{k,r}$. Choosing the number and the order of the poles of the Lax matrix amounts to specifying a particular model. Choosing the polynomials $P^{(k)}(L, \lambda)$ amounts to specifying the dynamical flows.

The above propositions give the general form of $M(\lambda)$ as far as the matrix structure and the λ-dependence is concerned. One should keep in mind however that the coefficients of the polynomials $P^{(k)}(L, \lambda)$ are a priori functions of the matrix elements of L and require further characterizations in order to get an integrable system. In the setting of the next section these coefficients will be *constants*.

Remark 1. If λ_k is a pole of $M(\lambda)$ and not a pole of $L(\lambda)$, one can redefine $M(\lambda)$ without changing the Lax equations of motion so as to eliminate the singularities of $M(\lambda)$ at λ_k. Indeed, redefining:

$$M(\lambda) \rightarrow M(\lambda) - P^{(k)}(L, \lambda)$$

does not change the Lax equation. The new $M(\lambda)$ is regular at λ_k. Of course we cannot eliminate the poles common to $L(\lambda)$ and $M(\lambda)$ by this procedure.

Remark 2. The Lax equation is invariant under similarity transformations,

$$L \rightarrow L' = gLg^{-1}, \quad M \rightarrow M' = gMg^{-1} + \partial_t gg^{-1} \tag{3.17}$$

If this similarity transformation is independent of λ, it will not spoil the analytic properties of $L(\lambda)$ and $M(\lambda)$. We can use the gauge freedom eq. (3.17) to diagonalize L_0,

$$L_0 = \mathrm{Diag}(a_1, \ldots, a_N)$$

Consistency of eq. (3.4) then requires M_0 to be diagonal also and thus $\dot{L}_0 = [M_0, L_0] = 0$. Hence M_0 is a polynomial P of L_0, so that replacing $M(\lambda) \rightarrow M(\lambda) - P(L(\lambda))$ gets rid of M_0.

Remark 3. For Lax matrices $L(\lambda)$ and $M(\lambda)$ rational functions of λ, we can easily compare the number of variables to the number of equations contained in eq. (3.4). The variables are the matrices L_0, $L_{k,r}$ and M_0, $M_{k,r}$. A naive counting, assuming that $L(\lambda)$ and $M(\lambda)$ are generic and independent matrices, gives in units of N^2:

$$\text{number of variables} = 2 + \sum_k n_k + \sum_k m_k = 2 + l + m$$

$$\text{number of equations} = 1 + \sum_k (n_k + m_k) = 1 + l + m$$

where l and m are the total order (degree of the divisor) of the poles of $L(\lambda)$ and $M(\lambda)$ respectively. Therefore there is one more variable than the number of equations, which reflects the gauge invariance eq. (3.17) of the Lax equation.

If we assume however that λ belongs to a higher genus Riemann surface with genus g, the situation is very different. Indeed, suppose that $L(\lambda)$ and $M(\lambda)$ have poles of total multiplicity l and m respectively. Let us count the number of meromorphic functions on which $L(\lambda)$ and $M(\lambda)$ can be expanded with constant matrix coefficients. By the

Riemann–Roch theorem, $L(\lambda)$ can be expanded on a basis of $(l - g + 1)$ independent meromorphic functions (in the generic case), and $M(\lambda)$ on $(m - g + 1)$ functions. So we have

$$\text{number of variables} = 2 + l + m - 2g$$

Similarly, the commutator $[M(\lambda), L(\lambda)]$ has poles of total multiplicity $l + m$ and can be expanded on a basis of $(l + m - g + 1)$ independent functions. Therefore

$$\text{number of equations} = 1 + l + m - g$$

So (number of equations – number of variables) $= g - 1$. Taking into account the gauge symmetry of the Lax equation, we see that the number of equations is always greater than the number of unknowns when $g > 0$. This shows that if λ belongs to a Riemann surface of genus $g \geq 1$, (and such systems exist, at least for $g = 1$, see Chapter 7), one has to consider a non-generic situation.

Remark 4. As we already mentioned, the dynamical variables are the matrix elements of the $L_{k,r}$. In most cases this is too general and we may try to impose more restrictions on the matrix elements of the $L_{k,r}$. For instance, assuming that the λ_k are all real, then clearly one can impose that all the matrices L_k and M_k are anti-Hermitian. Another simple example is provided by the Neumann model whose Lax matrices (3.3) are such that ${}^t L(-\lambda) = L(\lambda)$, ${}^t M(-\lambda) = -M(\lambda)$. More generally, we may define the action of a reduction group R and impose that $L(\lambda)$ and $M(\lambda)$ are invariant under R. This may be done, for example, by demanding that the Lax pair $L(\lambda)$ and $M(\lambda)$ satisfies:

$$\widehat{R}^{-1} L(r(\lambda))\widehat{R} = L(\lambda), \quad \widehat{R}^{-1} M(r(\lambda))\widehat{R} = M(\lambda) \tag{3.18}$$

for some representations $\widehat{R}$ and $r(\lambda)$ of R. This type of restriction is always compatible with the Lax equation. It provides a way to lower the number of degrees of freedom. An example of this procedure can be found later in this chapter, see eq. (3.84).

We end this section by illustrating theses constructions on the examples of the Euler and Lagrange tops and of the Neumann model. We verify that the matrices $L(\lambda)$ and $M(\lambda)$ are indeed related as in eq. (3.15).

Example 1. Let us consider the Euler top. We see that $L(\lambda)$, eq. (3.1), has a pole at 0 and $M(\lambda)$ has a pole at ∞. Let us apply the above procedure to remove this pole. There exists a polynomial $P(x) = \alpha x^2 + \beta x + \gamma$ such that $P(\mathcal{I}^2) = \mathcal{I}$. We will need the coefficient $\alpha = -1/I_1 I_2 I_3$. Redefining $M(\lambda)$ to $M(\lambda) - \lambda P(L(\lambda))$ one gets $M = M_0 - (\alpha/\lambda)J^2$ with $M_0 = \Omega - \alpha(\mathcal{I}^2 J + J\mathcal{I}^2) - \beta J$. One can check that $M_0 = 0$. (Hint: for $i \neq j$ compute $(\mathcal{I}_i - \mathcal{I}_j)(M_0)_{ij}$ using $P(\mathcal{I}_i^2) = \mathcal{I}_i$). Hence for the Euler top we can choose

$$M(\lambda) = -\frac{\alpha}{\lambda}J^2 \tag{3.19}$$

We see that this new $M(\lambda)$ is such that $M(\lambda) = -\alpha(\lambda L^2)_-$. The Lax matrix of the Euler top $L(\lambda) = \mathcal{I}^2 + \lambda^{-1}J$ is of the form $L_0 + L_-$ with

L_0 diagonal non-dynamical. The eigenvalues of J are $(0, i\sqrt{\vec{J}^2}, -i\sqrt{\vec{J}^2})$, which are non-dynamical since $\vec{J}^2$ belongs to the centre of the Poisson bracket and has been fixed to a numerical value.

Example 2. For the Lagrange top $L(\lambda)$, eq. (3.2), has a pole at 0 and $M(\lambda)$ has a pole at ∞. One can remove this pole by redefining $M(\lambda) \rightarrow M(\lambda) - I^{-1}\lambda L(\lambda)$. Notice however that since the eigenvalues of L_0 are degenerate one cannot express M_0 as a polynomial in L_0. The new $M(\lambda)$ can be expressed as $M = M_0 + M_-$ with $M_-(\lambda) = -I^{-1}(\lambda L(\lambda))_-$. For the Lagrange top the Lax matrix is again of the form $L_0 + L_-$ where obviously L_0 is non-dynamical. For the singular part one gets, since J is antisymmetric, $A_- = \lambda^{-2}\mathrm{Diag}(\sqrt{\vec{P}^2}, -\sqrt{\vec{P}^2}, 0, 0)$ which again belongs to the centre of the Poisson bracket and is non-dynamical.

Example 3. Finally, let us consider the Neumann model. We see from eq. (3.3) that we have $M(\lambda) = (\lambda L(\lambda))_-$ where we project on the singular part at $\lambda = 0$. The Lax matrix of the Neumann model is of the form $L = L_0 + L_-$ with L_0 a numerical diagonal matrix, and the singular part L_- at the only pole $\lambda = 0$ is given by: $L_- = \lambda^{-1}J - \lambda^{-2}K$. This is a rank 2 matrix whose image is spanned by the vectors x and y. It is easy to diagonalize in this subspace and one gets the singular diagonal part $A_- = \lambda^{-2}\mathrm{Diag}(1, 0, \dots, 0)$. It is again a numerical matrix.

We have seen in these three examples that the singular parts, $A_-^{(k)}$, of the matrix $A^{(k)}(\lambda)$ are independent of the dynamical variables. We will show in the next section that, in this case, eq. (3.14) admits an important interpretation as a coadjoint orbit.

3.3 Coadjoint orbits and Hamiltonian formalism

In this section we show that the Zakharov–Shabat construction, when the matrices $A_-^{(k)}$ are non-dynamical, can be interpreted as coadjoint orbits. This introduces a natural symplectic structure in the problem and gives a Hamiltonian interpretation to the Lax equation. This also allows us to compute the Poisson brackets of the matrix elements of the Lax matrix in terms of an r-matrix.

We first recall some notions about adjoint and coadjoint actions of Lie algebras and Lie groups, see Chapter 14. Let G be a connected Lie group with Lie algebra $\mathcal{G}$. The group G acts on $\mathcal{G}$ by the adjoint action denoted Ad:

$$X \longrightarrow (\mathrm{Ad}\, g)(X) = gXg^{-1}, \quad g \in G, \ X \in \mathcal{G}$$

Similarly the coadjoint action of G on the dual $\mathcal{G}^*$ of the Lie algebra $\mathcal{G}$ (i.e. the vector space of linear forms on the Lie algebra) is defined by:

$$(\mathrm{Ad}^* g.\Xi)(X) = \Xi(\mathrm{Ad}\, g^{-1}(X)), \quad g \in G, \ \Xi \in \mathcal{G}^*, \ X \in \mathcal{G}$$

The infinitesimal version of these actions provides actions of the Lie algebra $\mathcal{G}$ on $\mathcal{G}$ and $\mathcal{G}^*$, denoted ad and ad* respectively and given by:

$$\mathrm{ad}\, X(Y) = [X, Y], \quad X, Y \in \mathcal{G},$$
$$\mathrm{ad}^* X.\Xi(Y) = -\Xi([X, Y]), \quad X, Y \in \mathcal{G}, \Xi \in \mathcal{G}^*$$

To see how these notions relate to our problem, let us consider first a Lax matrix with only one polar singularity at $\lambda = 0$:

$$L(\lambda) = \left(g(\lambda)\, A_-(\lambda)\, g^{-1}(\lambda)\right)_- \tag{3.20}$$

with $A_-(\lambda) = \sum_{r=-n}^{-1} A_r \lambda^r$, and $g(\lambda)$ has a regular expansion around $\lambda = 0$.

Let G be the loop group of invertible matrix valued power series expansion around $\lambda = 0$. The elements of G are regular series $g(\lambda) = \sum_{r=0}^{\infty} g_r \lambda^r$. The product law is the pointwise product: $(gh)(\lambda) = g(\lambda)h(\lambda)$. Formally, the Lie algebra $\mathcal{G}$ of G consists of elements of the form $X = \sum_{r=0}^{\infty} X_r \lambda^r$. Its Lie bracket is given by the pointwise commutator.

The dual $\mathcal{G}^*$ of $\mathcal{G}$ can be identified with the set of polar matrices $\Xi(\lambda) = \sum_{r \geq 1} \Xi_r \lambda^{-r}$, where the sum contains a finite but arbitrary large number of terms, by the pairing:

$$\langle \Xi, X \rangle \equiv \mathrm{Tr}\ \mathrm{Res}_{\lambda=0}\left(\Xi(\lambda)X(\lambda)\right) = \sum_r \mathrm{Tr}\left(\Xi_{r+1}X_r\right)$$

where $\mathrm{Res}_{\lambda=0}$ is defined to be the coefficient of λ^{-1}.

The coadjoint action of G on $\mathcal{G}^*$ is defined by $\left((\mathrm{Ad}^*g) \cdot \Xi\right)(X) = \Xi(g^{-1}Xg)$ for $\Xi \in \mathcal{G}^*$ and any $X \in \mathcal{G}$. Using the above model for $\mathcal{G}^*$, and since $\langle \Xi, g^{-1}Xg \rangle = \langle g\Xi g^{-1}, X \rangle = \langle (g\Xi g^{-1})_-, X \rangle$, we get

$$(\mathrm{Ad}^*g) \cdot \Xi(\lambda) = \left(g \cdot \Xi \cdot g^{-1}\right)_-$$

This is precisely eq. (3.20). The Lax matrix can thus be interpreted as belonging to the coadjoint orbit of the element $A_-(\lambda)$ of $\mathcal{G}^*$ under the loop group G.

With this interpretation, the Lax equation reads:

$$\dot{L} = \mathrm{ad}^* M \cdot L = [M, L] \tag{3.21}$$

This shows that the equation of motion is a flow on the coadjoint orbit.

Coadjoint orbits in $\mathcal{G}^*$ are equipped with the canonical Kostant–Kirillov symplectic structure. Choosing two linear functions $h_1(\Xi) = \Xi(X)$ and $h_2(\Xi) = \Xi(Y)$ with X, Y $\in \mathcal{G}$, so that $dh_1 = X$ and $dh_2 = Y$, the Kostant–Kirillov Poisson bracket reads:

$$\{\Xi(X), \Xi(Y)\} = \Xi([X, Y])$$

where the right-hand side is the linear function $\Xi \rightarrow \Xi([X, Y])$. This Poisson bracket is very natural but one has to be aware that it is degenerate. The kernel is the set of Ad*-invariant functions.

Let us specialize this construction to our case. We identify $\mathcal{G}^*$ with series expansions singular at $\lambda = 0$ using the linear form induced by Tr Res$_{\lambda=0}$. We parametrize the orbit of the element $A_-(\lambda)$ by the group element $g(\lambda)$. Consider the 1-form α on the group given by

$$\alpha = -\text{Tr Res}_{\lambda=0} \left(A_- g^{-1} \delta g \right)$$

The pullback on the group of the Kostant–Kirillov symplectic form reads (see Chapter 14):

$$\omega = \delta\alpha = \text{Tr Res}_{\lambda=0} \left(A_- g^{-1} \delta g \wedge g^{-1} \delta g \right) \tag{3.22}$$

This interpretation of $L(\lambda)$ as a coadjoint orbit assumes that $A_-(\lambda)$ is not a dynamical variable.

This construction can be extended to the multi-pole case. We consider the direct sum of loop algebras $\mathcal{G}_k$, around $\lambda = \lambda_k$:

$$\mathcal{G} \equiv \bigoplus_k \mathcal{G}_k$$

An element of this Lie algebra has the form of a multiplet

$$X(\lambda) = (X_1(\lambda), X_2(\lambda), \ldots)$$

where $X_k(\lambda)$, defined around λ_k, is of the form $X_k(\lambda) = \sum_{n \geq 0} X_{k,n} (\lambda - \lambda_k)^n$. The Lie bracket is such that $[X_k(\lambda), X_l(\lambda)] = 0$ if $k \neq l$. The group G is the direct product of the groups G_k of regular invertible matrices at λ_k:

$$G \equiv (G_1, G_2, \ldots) \tag{3.23}$$

The dual $\mathcal{G}^*$ of this Lie algebra consists of multiplets

$$\Xi = (\Xi_1(\lambda), \Xi_2(\lambda), \ldots)$$

where $\Xi_k(\lambda)$ around λ_k is of the form $\Xi_k(\lambda) = \sum_{r \geq 1} \Xi_{k,r} (\lambda - \lambda_k)^{-r}$. In this sum the number of terms is finite but arbitrary. The pairing is simply

$$\langle \Xi, X \rangle \equiv \sum_k \langle \Xi_k, X_k \rangle = \sum_k \text{Tr Res}_{\lambda_k} (\Xi_k(\lambda) X_k(\lambda))$$

The coadjoint action of G on $\mathcal{G}^*$ is given by the usual formula: if $g = (g_1, g_2, \ldots) \in G$ and $\Xi = (\Xi_1, \Xi_2, \ldots) \in \mathcal{G}^*$

$$(\mathrm{Ad}^* g).\Xi(\lambda) = ((g_1 \Xi_1 g_1^{-1})_-, (g_2 \Xi_2 g_2^{-1})_-, \ldots)$$

A coadjoint orbit consists of elements Ξ_k with a fixed maximal order of the pole. Then, we can interpret eq. (3.9) as the coadjoint orbit of the element $((A_1)_-, (A_2)_-, \ldots)$.

Alternatively, we can consider the function on $\mathcal{G}^*$

$$L(\lambda) = L_0 + \sum_k \Xi_k \tag{3.24}$$

with poles at the points λ_k. Given this function we can recover the Ξ_k by extracting the polar parts. The constant matrix L_0 is added to match the formula for the Lax matrix, eq. (3.9). By choice it is assumed to be invariant by coadjoint action. The pairing can be rewritten as

$$\langle L, X \rangle = \sum_k \mathrm{Tr}\, \mathrm{Res}_{\lambda_k} L(\lambda) X_k(\lambda)$$

Note that only Ξ_k contributes to the residue at λ_k and the formula is compatible with the matrix L_0 being invariant by coadjoint action.

It is interesting to compute the dimension of this coadjoint orbit. In the formula $L_k = (g^{(k)} A_-^{(k)} g^{(k)-1})_-$, the matrices $A_-^{(k)}$ characterize the orbit and are non-dynamical. The dynamical variables are the jets of order $(n_k - 1)$ of the $g^{(k)}$, which gives $N^2 n_k$ parameters. But L_k is invariant under $g^{(k)} \to g^{(k)} d^{(k)}$ with $d^{(k)}$ a jet of diagonal matrices of the same order. Hence the dimension of the L_k orbit is $(N^2 - N) n_k$, and the dimension of the orbit is the even number:

$$\dim \mathcal{M} = (N^2 - N) \sum_k n_k$$

In the multi-pole case, the pullback of the symplectic form reads:

$$\omega = \sum_k \mathrm{Tr}\, \mathrm{Res}_{\lambda_k} \left(A_-^{(k)} g^{(k)-1} \delta g^{(k)} \wedge g^{(k)-1} \delta g^{(k)} \right) \tag{3.25}$$

We can now use this symplectic form to evaluate the Poisson brackets of the elements of the Lax matrix. To write them we use the tensor notation of section (2.5) in Chapter 2 and show that they take the r-matrix form. We assume that each $L_k(\lambda)$ is a generic element of an orbit of the loop group $GL(N)[\lambda]$, that L_0 and the $A_-^{(k)}$ are non-dynamical, and the symplectic form is given by eq. (3.25).

Proposition. *With the symplectic structure eq. (3.25), the Poisson brackets of the matrix elements of $L(\lambda)$ can be written as:*

$$\{L_1(\lambda), L_2(\mu)\} = -\left[\frac{C_{12}}{\lambda - \mu}, L_1(\lambda) + L_2(\mu)\right] \qquad (3.26)$$

with $C_{12} = \sum_{i,j} E_{ij} \otimes E_{ji}$, where the E_{ij} are the canonical basis matrices. The commutator in the right-hand side of eq. (3.26) is the usual matrix commutator.

<u>Proof.</u>   Let us first assume that we have only one-pole and $L = (gA_-g^{-1})_-$. Because we are dealing with a Kostant–Kirillov bracket for the loop algebra of $gl(N)$, we can immediately write the Poisson bracket of the Lax matrix using the defining relation $\{L(X), L(Y)\} = L([X, Y])$. Using $L(X) = \text{Tr Res}_{\lambda=0}(L(\lambda)X(\lambda))$, this gives:

$$\{L(X), L(Y)\} = \text{Tr Res}_{\lambda=0} \left(L(\lambda)[X(\lambda), Y(\lambda)]\right) \qquad (3.27)$$

By definition of the notation $\{L_1, L_2\}$, we have:

$$\{L(X), L(Y)\} = \langle \{L_1(\lambda), L_2(\mu)\}, X(\lambda) \otimes Y(\mu)\rangle$$

where $\langle,\rangle = \text{Tr}_{12}\text{Res}_{\lambda}\text{Res}_{\mu}$. We need to factorize $X(\lambda) \otimes Y(\mu)$ in eq. (3.27). To this end, we introduce a Casimir operator

$$\mathcal{C}_{12} = \sum_{\alpha} E_{\alpha} \otimes E_{\alpha}^* \in \mathcal{G} \otimes \mathcal{G}^*$$

where E_{α} and E_{α}^* are two dual bases of $\mathcal{G}$ and $\mathcal{G}^*$ respectively. We choose

$$E_{ij}^n = \lambda^n E_{ij}, \quad E_{ij}^{*n} = \lambda^{-n-1} E_{ji}, \quad n \geq 0$$

so that under the pairing Tr Res we have $\langle E_{ij}^{*n}, E_{kl}^m \rangle = \delta_{ik}\delta_{jl}\delta_{nm}$. The Casimir operator is such that for $Y \in \mathcal{G}$, we have $Y_1 = \mathcal{C}_{12}(Y_2)$. Then we want to write $\langle L, [X, Y]\rangle = \langle [\mathcal{C}_{12}, L_1], X \otimes Y\rangle$, however, this formula does not make sense as it stands because $[\mathcal{C}_{12}, L_1] = \sum_{\alpha}[E_{ij}^n, L] \otimes E_{ij}^{*n}$ and $E_{ij}^n \in \mathcal{G}$ while $L \in \mathcal{G}^*$ and the commutator is not defined. To overcome this problem we embed $\mathcal{G}$ and its dual $\mathcal{G}^*$ into the full loop algebra $\tilde{\mathcal{G}}$ generated by E_{ij}^n, $n \in \mathbb{Z}$.

$$\tilde{\mathcal{G}} = \mathcal{G} + \mathcal{G}^* \qquad (3.28)$$

Note that in this sum, $\mathcal{G}$ and $\mathcal{G}^*$ do not commute. Let us compute $\mathcal{C}_{12}$, assuming $|\lambda| < |\mu|$:

$$\mathcal{C}_{12}(\lambda, \mu) = C_{12} \sum_{n=0}^{\infty} \frac{\lambda^n}{\mu^{n+1}} = -\frac{C_{12}}{\lambda - \mu}, \quad C_{12} = \sum_{i,j} E_{ij} \otimes E_{ji}$$

We can now write $\langle L(\lambda)[X(\lambda), Y(\lambda)]\rangle = \langle [\mathcal{C}_{12}(\lambda, \mu), L(\lambda) \otimes 1], X(\lambda) \otimes Y(\mu)\rangle$. Consider the rational function of λ: $\varphi(\lambda) = \{L_1(\lambda), L_2(\mu)\} - [\mathcal{C}_{12}(\lambda, \mu), L(\lambda) \otimes 1]$. By inspection φ contains only negative powers of μ, and we have $\langle \varphi, X(\lambda) \otimes Y(\mu)\rangle = 0$. Hence φ contains only *positive* powers of λ and is regular at $\lambda = 0$. It has a pole at $\lambda = \mu$, due to the form of $\mathcal{C}(\lambda, \mu)$. We remove this pole by subtracting to φ the quantity $[\mathcal{C}_{12}(\lambda, \mu), 1 \otimes L(\mu)]$ which contains only positive powers of λ and is therefore in the kernel of $\langle \cdot, X(\lambda) \otimes Y(\mu)\rangle$. The pole at $\lambda = \mu$ disappears since $[\mathcal{C}_{12}, L(\mu) \otimes 1 + 1 \otimes L(\mu)] = 0$. The redefined φ is regular everywhere and vanishes for $\lambda \to \infty$, hence vanishes identically. This proves eq. (3.26) in the one-pole case.

We can now study the multi-pole situation occuring in eq. (3.9). Consider $L = L_0 + \sum_{k=1} L_k$. Each L_k lives in a coadjoint orbit as above equipped with its own symplectic structure. From eq. (3.25) they have vanishing mutual Poisson brackets $\{L_{j1}, L_{k2}\} = 0$ for $j, k = 0, \ldots, N$ and $j \neq k$. We assume further that L_0 does not contain dynamical variables

$$\{L_{01}, L_{02}\} = 0, \quad \{L_{01}, L_{k2}\} = 0$$

(here the indices 1 and 2 refer to the tensorial notation). Then since $\mathcal{C}_{12}/(\lambda - \mu)$ is *independent* of the pole λ_k, it is obvious that the r-matrix relations for each orbit combine by addition to give eq. (3.26) for the complete Lax matrix $L(\lambda)$. ∎

Remark. The quantity

$$C_{12} = \sum_{i,j} E_{ij} \otimes E_{ji} \tag{3.29}$$

often occurs when calculating r-matrices. It is called the tensor Casimir of $gl(N)$. Its main properties are

$$[C_{12}, g \otimes g] = 0, \quad \mathrm{Tr}_2\, C_{12} g_2 = g_1, \quad \forall g \in GL(N) \tag{3.30}$$

This proposition shows that the generic Zakharov–Shabat system, equipped with this symplectic structure, is an *integrable Hamiltonian system* (the precise counting of independent conserved quantities will be done in Chapter 5). It also gives us a very simple formula for the r-matrix specifying the Poisson bracket of $L(\lambda)$:

$$r_{12}(\lambda, \mu) = -r_{21}(\mu, \lambda) = -\frac{C_{12}}{(\lambda - \mu)} \tag{3.31}$$

The Jacobi identity is satisfied because this r-matrix verifies the classical Yang–Baxter equation (see eq. (2.12) in Chapter 2):

$$[r_{12}, r_{13}] + [r_{12}, r_{23}] + [r_{13}, r_{23}] = 0$$

where r_{ij} stands for $r_{ij}(\lambda_i, \lambda_j)$. Note that r_{12} is antisymmetric: $r_{12}(\lambda_1, \lambda_2) = -r_{21}(\lambda_2, \lambda_1)$.

As in Chapter 2, these Poisson brackets for the Lax matrix ensure that one can define commuting quantities. The associated equations of motion take the Lax form.

Proposition. *The functions on phase space:*

$$H^{(n)}(\lambda) \equiv \mathrm{Tr}\left(L^n(\lambda)\right)$$

are in involution. The equations of motion associated with $H^{(n)}(\mu)$ can be written in the Lax form with $M = \sum_k M_k$:

$$M_k(\lambda) = -n\left(\frac{L^{n-1}(\lambda)}{\lambda - \mu}\right)_k \tag{3.32}$$

<u>Proof.</u> The quantities $H^{(n)}(\lambda)$ are in involution because

$$\{\mathrm{Tr}\, L^n(\lambda), \mathrm{Tr}\, L^m(\mu)\} = nm\,\mathrm{Tr}_{12}\{L_1(\lambda), L_2(\mu)\}L_1^{n-1}(\lambda)L_2^{m-1}(\mu)$$
$$= -\frac{nm}{\lambda - \mu}\mathrm{Tr}_{12}\left([C_{12}, L_1^n(\lambda)]L_2^{m-1}(\mu) + [C_{12}, L_2^m(\mu)]L_1^{n-1}(\lambda)\right) = 0$$

where we have used that the trace of a commutator vanishes. Similarly, we have:

$$\dot{L}(\lambda) = \{H^{(n)}(\mu), L(\lambda)\} = n\,\mathrm{Tr}_2\left[\frac{C_{12}}{\lambda - \mu}L_2^{n-1}(\mu), L_1(\lambda)\right]$$

Performing the trace and remembering that $\mathrm{Tr}_2\left(C_{12}M_2\right) = M_1$, we get

$$\dot{L}(\lambda) = [M^{(n)}(\lambda, \mu), L(\lambda)], \quad M^{(n)}(\lambda, \mu) = n\frac{L^{n-1}(\mu)}{\lambda - \mu} \tag{3.33}$$

This $M^{(n)}(\lambda, \mu)$ has a pole at $\lambda = \mu$ and is otherwise regular. According to the general procedure we can remove this pole by subtracting some polynomial in $L(\lambda)$ without changing the equations of motion. Obviously one can redefine:

$$M^{(n)}(\lambda, \mu) \to M^{(n)}(\lambda, \mu) - n\frac{L^{n-1}(\lambda)}{\lambda - \mu} = -n\frac{L^{n-1}(\lambda) - L^{n-1}(\mu)}{\lambda - \mu}$$

This new M has poles at all λ_k and is regular at $\lambda = \mu$. Decomposing it into its polar parts, we write $M = \sum_k M_k$ with

$$M_k(\lambda) = -n \left(\frac{L^{n-1}(\lambda)}{\lambda - \mu} \right)_k$$

This is of the form eq. (3.15) with

$$P^{(k)}(L, \lambda) = -\frac{n}{\lambda - \mu} L^{n-1}(\lambda) \tag{3.34}$$

Notice that the coefficients of the polynomial $P^{(k)}(L, \lambda)$ are pure numerical constants. ∎

Example 1. In the case of the Euler top $L(\lambda) = \mathcal{I}^2 + \frac{1}{\lambda}J$. The singular part satisfies ${}^t L_-(-\lambda) = L_-(\lambda)$. This is *not* preserved by a general coadjoint action $L_-(\lambda) = (g(\lambda)L_-(\lambda)g^{-1}(\lambda))_-$. To overcome this problem we consider the subgroup of matrices satisfying ${}^t g^{-1}(-\lambda) = g(\lambda)$ which may be called graded orthogonal. Its Lie algebra consists of matrices $X(\lambda)$ such that ${}^t X(-\lambda) = -X(\lambda)$. Its dual under the pairing Tr Res consists of matrices $L(\lambda)$ such that ${}^t L(-\lambda) = L(\lambda)$, and having an expansion in a finite sum of strictly negative powers of λ. The matrix $L_-(\lambda)$ is an orbit under this coadjoint action.

The symplectic structure on this orbit is obtained by applying eq. (3.27):

$$\{J_{ij}, J_{kl}\} = -\frac{1}{2} \left(\delta_{jk}J_{il} - \delta_{jl}J_{ik} + \delta_{il}J_{jk} - \delta_{ik}J_{jl} \right) \tag{3.35}$$

The computation of the r-matrix of the Euler top is similar to the proof of eq. (3.26), except that the loop algebra being different, the Casimir operator has to be recomputed. We keep the factor $-1/2$ in eq. (3.35) in order to match the general considerations. Of course if this factor is omitted we must multiply the final r-matrix by -2.

A basis E_α of the graded loop algebra is given for $n = 0, 1, \ldots$ by:

$$(E_{ij} - E_{ji})\lambda^{2n}, \ i < j, \quad (E_{ij} + E_{ji})\lambda^{2n+1}, \ i < j, \quad E_{ii}\lambda^{2n+1}$$

The dual basis E_α^* under Tr Res is given respectively for $n = 0, 1, \ldots$ by:

$$-\frac{1}{2}(E_{ij} - E_{ji})\lambda^{-2n-1}, \ i < j, \quad \frac{1}{2}(E_{ij} + E_{ji})\lambda^{-2n-2}, \ i < j, \quad E_{ii}\lambda^{-2n-2}$$

Then one gets for $\mathcal{C}_{12} = \sum_\alpha E_\alpha \otimes E_\alpha^*$:

$$\mathcal{C}_{12}(\lambda, \mu) = -\frac{1}{2}\frac{1}{\lambda - \mu}\sum_{ij} E_{ij} \otimes E_{ji} - \frac{1}{2}\frac{1}{\lambda + \mu}\sum_{ij} E_{ij} \otimes E_{ij}$$

This implies that poles at $\lambda = \pm\mu$ appear in $\varphi(\lambda) = \{L_1(\lambda), L_2(\mu)\} - [\mathcal{C}_{12}(\lambda, \mu), L(\lambda) \otimes 1]$. To cancel these poles we now subtract $[\mathcal{C}_{21}(\mu, \lambda), 1 \otimes L(\mu)]$ which only contains positive powers of λ. Indeed, the residue at $\lambda = \mu$ is $[\mathcal{C}_{12}, L_1(\mu) + L_2(\mu)]$, with $\mathcal{C}_{12} = \sum_{ij} E_{ij} \otimes E_{ji}$, and therefore vanishes as previously. The residue at $\lambda = -\mu$ reads $[D_{12}, L_1(-\mu) - L_2(\mu)]$ with $D_{12} = \sum_{ij} E_{ij} \otimes E_{ij}$. Now $L(-\mu) = {}^{t}L(\mu)$ and one checks that for any matrix A the commutator $[D_{12}, A_1 - {}^{t}A_2]$ vanishes. Hence $\varphi(\lambda) - [\mathcal{C}_{21}(\mu, \lambda), 1 \otimes L(\mu)] = 0$. Strictly speaking, one should have done this calculation on the polar part $L_-(\lambda)$ and added the L_0 part afterwards. However, since the full Lax matrix satisfies ${}^{t}L(-\lambda) = L(\lambda)$, the reasoning is actually valid for the full Lax matrix. Finally one gets:

$$\{L_1(\lambda), L_2(\mu)\} = [r_{12}(\lambda, \mu), L_1(\lambda)] - [r_{21}(\mu, \lambda), L_2(\mu)]$$

with $r_{12}(\lambda, \mu) = \mathcal{C}_{12}(\lambda, \mu)$. Note that this is a two-poles r-matrix with poles at $\lambda = \pm\mu$.

Example 2. We consider next the Neumann model. Recall that the Lax matrix reads $L(\lambda) = L_0 + \frac{1}{\lambda}J - \frac{1}{\lambda^2}K$. As in the Euler top it satisfies ${}^{t}L(-\lambda) = L(\lambda)$. Hence we are dealing with the graded orthogonal group ${}^{t}g^{-1}(-\lambda) = g(\lambda)$. Let us check that the matrix $L_-(\lambda)$ is an orbit under the coadjoint action. As a matter of fact:

$$(g(\lambda)L_-(\lambda)g^{-1}(\lambda))_- = -\frac{1}{\lambda^2}\, g_0 K \,{}^{t}g_0 + \frac{1}{\lambda}(g_0 J \,{}^{t}g_0 - g_1 K \,{}^{t}g_0 + g_0 K \,{}^{t}g_1)$$

with $g(\lambda) = g_0 + \lambda g_1 + \dots$. Recalling that $K = X\,{}^{t}X$ and $J = X\,{}^{t}Y - Y\,{}^{t}X$, we see that this is exactly of the same form as $L_-(\lambda)$ with

$$X \to g_0 X, \quad Y \to g_0 Y + g_1 X$$

One can check that the Kostant–Kirillov bracket on this orbit reproduces the canonical Poisson bracket on the variables X and Y. It follows that the Neumann model has the same r-matrix as the Euler top.

3.4 Elementary flows and wave function

We have found that the time evolution of a Zakharov–Shabat system is given by matrices $M(\lambda)$ of the form eq. (3.10). This leaves an infinite number of choices for $M(\lambda)$. We introduce an infinite number of elementary times corresponding to these choices. We will show that these flows are pairwise commuting. This defines a so-called integrable hierarchy.

The elementary flows correspond to diagonal matrices $B_-^{(k)}$ having a single pole of order n at λ_k in matrix diagonal entry α. Here, and in the following, we use the multi-index $i = (k, n, \alpha)$. We thus define matrices M_i by:

$$M_i \equiv \left(g^{(k)} \xi_i g^{(k)-1} \right)_k, \qquad \xi_i \equiv \xi_{(k,n,\alpha)} = \frac{1}{(\lambda - \lambda_k)^n} E_{\alpha\alpha} \qquad (3.36)$$

We call $t_i = t_{(k,n,\alpha)}$ the time variable associated with M_i through the Lax equation:

$$\partial_{t_i} L = [M_i, L] \qquad (3.37)$$

A general flow is a linear combination of these elementary ones. Note that $M_i(\lambda)$, which is a priori defined around λ_k, is a rational fonction of λ, with only a polar part at λ_k, hence is defined in the whole λ-plane. The Lax equation, eq. (3.37), has a meaning in the whole λ-plane and defines the time evolution of the quantities locally defined at $\lambda_{k'}$, such as $g^{(k')}$, with respect to the times associated with λ_k.

We will need some notations. Let:

$$\xi^{(k)}(\lambda, t) = \sum_{n,\alpha} \xi_{(k,n,\alpha)} t_{(k,n,\alpha)} \qquad (3.38)$$

$$\xi(\lambda, t) = \sum_k \xi^{(k)}(\lambda, t) = \sum_{i=(k,n,\alpha)} \xi_i t_i \qquad (3.39)$$

These are generating functions with coefficients rational in λ. The function $\xi^{(k)}(\lambda, t)$ involves all the times above the singularity λ_k, while $\xi(\lambda, t)$ involves all the times of the hierarchy.

It is easy to find the Hamiltonians generating these elementary flows.

Proposition. *The Hamiltonian generating the flow t_i is*

$$H_i = \oint_{\Gamma^{(k)}} \frac{d\lambda}{2i\pi} \frac{\mathrm{Tr}(A^{(k)}(\lambda) E_{\alpha\alpha})}{(\lambda - \lambda_k)^n} = \mathrm{Tr}\ \mathrm{Res}_{\lambda_k} \left(\frac{A^{(k)}(\lambda) E_{\alpha\alpha}}{(\lambda - \lambda_k)^n} \right)$$

where $A^{(k)}(\lambda)$ is the diagonal form of $L(\lambda)$, with singular part $A_-^{(k)}(\lambda)$, and $\Gamma^{(k)}$ a small contour around λ_k.

Proof. Let us introduce the differential $dH_i(L)$ defined by $\delta H_i = \langle \delta L, dH_i \rangle$ for any variation δL of the Lax matrix. To compute it we start from eq. (3.8), written as $A^{(k)} = g^{(k)-1} L g^{(k)}$, so that $\delta A^{(k)} = g^{(k)-1} \delta L g^{(k)} + [A^{(k)}, g^{(k)-1} \delta g^{(k)}]$. Hence $\delta \mathrm{Tr}(A^{(k)}(\lambda) E_{\alpha\alpha}) = \mathrm{Tr}(\delta L\, g^{(k)} E_{\alpha\alpha} g^{(k)-1})$, where we used that $[E_{\alpha\alpha}, A^{(k)}] = 0$. We get the formula:

$$dH_i(L) = (\lambda - \lambda_k)^{-n} g^{(k)} E_{\alpha\alpha} g^{(k)-1} \qquad (3.40)$$

Next, $\partial_{t_i} L(\mu) = \{H_i, L(\mu)\}$, so we have:

$$\partial_{t_i} L(\mu) = \mathrm{Tr}_1 \mathrm{Res}_{\lambda = \lambda_k} \Big(dH_i(\lambda) \otimes 1[r_{12}(\lambda, \mu), L_1(\lambda) + L_2(\mu)] \Big)$$
$$= \mathrm{Tr}_1 \mathrm{Res}_{\lambda = \lambda_k} (r_{12}(\lambda, \mu)[L(\lambda), dH_i(\lambda)]_1)$$
$$+ \Big[\mathrm{Tr}_1 \mathrm{Res}_{\lambda = \lambda_k} (r_{12}(\lambda, \mu) dH_i(\lambda) \otimes 1), L_2(\mu) \Big]$$

with $r_{12}(\lambda, \mu)$ given by eq. (3.31). In the first term, $[L(\lambda), dH_i(\lambda)]$ is proportional to $g^{(k)}[A^{(k)}, E_{\alpha\alpha}]g^{(k)-1}$ and vanishes since $A^{(k)}$ and $E_{\alpha\alpha}$ are both diagonal. In the second term we expand $r_{12}(\lambda, \mu)$ in positive powers of $(\lambda - \lambda_k)/(\mu - \lambda_k)$ to get something polar in $(\mu - \lambda_k)$:

$$\mathrm{Tr}_1 \mathrm{Res}_{\lambda = \lambda_k} (r_{12}(\lambda, \mu) dH_i(\lambda) \otimes 1) = \sum_{m=0}^{\infty} \mathrm{Res}_{\lambda = \lambda_k} \left(\frac{(\lambda - \lambda_k)^m}{(\mu - \lambda_k)^{m+1}} dH_i(\lambda) \right)$$
$$= (dH_i(\mu))_k = \left(g^{(k)} \frac{E_{\alpha\alpha}}{(\mu - \lambda_k)^n} g^{(k)-1} \right)_k = M_i(\mu)$$

In the last step we used that for any function $f(\lambda) = \sum_{m=-\infty}^{+\infty} \lambda^m f_m$, one has the identity

$$\sum_{m=0}^{\infty} \mathrm{Res}(\lambda^m f(\lambda)) \mu^{-m-1} = \sum_{m=0}^{\infty} f_{-m-1} \mu^{-m-1} = (f(\mu))_-$$

■

Comparing eq. (3.36) and eq. (3.40) we find the useful relation:

$$dH_i(L) = g^{(k)} \xi_i g^{(k)-1}, \quad M_i = (dH_i)_- \tag{3.41}$$

We now verify directly that the flows ∂_{t_i} defined by eqs. (3.37) and (3.36) all commute. This amounts to showing that $[L, \partial_{t_i} M_j - \partial_{t_j} M_i - [M_i, M_j]] = 0$. We get even the stronger result:

Proposition. *The matrices M_i defining the time evolution t_i satisfy the zero curvature condition*

$$\partial_{t_i} M_j - \partial_{t_j} M_i - [M_i, M_j] = 0 \tag{3.42}$$

As a consequence, the flows defined by eqs. (3.36, 3.37) are all commuting.

Proof. Let $i = (k, n, \alpha)$ and $j = (k', n', \alpha')$. Diagonalizing $L = g^{(k')} A^{(k')} g^{(k')-1}$ around $\lambda_{k'}$, the Lax equation gives

$$\partial_{t_i} g^{(k')} = M_i g^{(k')} + g^{(k')} d_i^{(k')} \tag{3.43}$$

where $d_i^{(k')}$ is an unknown diagonal matrix. This equation holds true in a vicinity of $\lambda_{k'}$. If $i = (k, n, \alpha)$ and $k' = k$, this implies that $d_i^{(k)} = -\xi_i + \text{regular}$, because $\partial_{t_i} g^{(k)}$ is regular and $M_i = (g^{(k)} \xi_i g^{(k)-1})_-$, while if $k' \neq k$, we only conclude that $d_i^{(k')}$ is regular around k'. Note that $g^{(k)}$ is known only up to a right multiplication by a regular diagonal matrix $g^{(k)} \to g^{(k)} d^{(k)}$ and this changes $d_i^{(k')} \to d_i^{(k')} - d^{(k')-1} \partial_{t_i} d^{(k')}$.
From eqs. (3.36) and (3.43), we get

$$\partial_{t_j} M_i = [M_j, g^{(k)} \xi_i g^{(k)-1}]_k + (g^{(k)} [d_j^{(k)}, \xi_i] g^{(k)-1})_k$$

Since the commutator in the second term involves diagonal matrices, it vanishes. Hence we have $\partial_{t_j} M_i = [M_j, g^{(k)} \xi_i g^{(k)-1}]_k$. Let us assume first that $\lambda_k \neq \lambda_{k'}$. Then M_j is regular at λ_k and only $(g^{(k)} \xi_i g^{(k)-1})_k = M_i$ contributes to the polar part of the above commutator, yielding

$$\partial_{t_j} M_i = [M_j, M_i]_k$$

Similarly we have

$$\partial_{t_i} M_j = [M_i, M_j]_{k'}$$

The zero curvature condition follows because $[M_i, M_j]$ is a rational function with poles only at λ_k and $\lambda_{k'}$ and vanishes at infinity, so that

$$[M_i, M_j] = [M_i, M_j]_k + [M_i, M_j]_{k'}$$

Assume next that $\lambda_k = \lambda_{k'}$. We still have

$$\partial_{t_j} M_i = [M_j, g^{(k)} \xi_i g^{(k)-1}]_k, \quad \partial_{t_i} M_j = [M_i, g^{(k)} \xi_j g^{(k)-1}]_k$$

where now all projections are at λ_k. But $M_i - g^{(k)} \xi_i g^{(k)-1} = O(1)$ and $M_j - g^{(k)} \xi_j g^{(k)-1} = O(1)$ so that $[M_i - g^{(k)} \xi_i g^{(k)-1}, M_j - g^{(k)} \xi_j g^{(k)-1}]_k = 0$, or

$$[M_i, M_j] - [g^{(k)} \xi_i g^{(k)-1}, M_j]_k - [M_i, g^{(k)} \xi_j g^{(k)-1}]_k$$
$$+ [g^{(k)} \xi_i g^{(k)-1}, g^{(k)} \xi_j g^{(k)-1}]_k = 0$$

The last term vanishes because $[\xi_i, \xi_j] = 0$. From this the zero curvature condition readily follows.   ∎

When parametrizing $L(\lambda)$ as in eq. (3.9), the dynamical variables are the $g^{(k)}(\lambda)$, modulo gauge transformations consisting of right multiplication by regular diagonal matrices. Let us write the equations of motion on the variables $g^{(k)}(\lambda)$.

Proposition. *There exists a gauge choice such that the equations of motion read, for each $i = (k, n, \alpha)$:*

$$\partial_{t_i} g^{(k')} = M_i g^{(k')} - g^{(k')} \partial_{t_i} \xi^{(k')}(\lambda, t)\delta_{kk'} \qquad (3.44)$$

Proof.   Equation (3.43) means that $d_i^{(k')} = g^{(k')-1}(\partial_{t_i} - M_i)g^{(k')}$ is the gauge transform of M_i by $g^{(k')}$. The zero curvature equation being invariant under gauge transformation implies that $\partial_{t_i} d_j^{(k')} - \partial_{t_j} d_i^{(k')} - [d_i^{(k')}, d_j^{(k')}] = 0$ for any indices (i, j, k'). Since the matrices $d_i^{(k)}$ are diagonal, the commutator vanishes, and the condition implies that $d_i^{(k')} = \partial_{t_i} h^{(k')}$ for some diagonal matrix $h^{(k')}$. We can now use the freedom $g^{(k')} \rightarrow g^{(k')}d^{(k')}$ to suppress the regular part of $h^{(k')}$ around $\lambda_{k'}$ by choosing $d^{(k')} = \exp(h_+^{(k')})$. This is a choice of gauge. We already noticed that the singular part of $d_i^{(k')}$ around $\lambda_{k'}$ is fixed, $d_i^{(k')} = 0$ if $k \neq k'$ and $d_i^{(k')} = -\xi_i$ if $k = k'$, this determines $d_i^{(k')}$ completely: $d_i^{(k)} = -\partial_{t_i}\xi^{(k')}(\lambda, t)\delta_{kk'}$. ∎

The set of Lax equations, eq. (3.37), for all the times t_i is what is called an integrable hierarchy. Written on the variables $g^{(k)}$, eq. (3.44) reads in detail, when $i = (k', n, \alpha)$ and $k' = k$:

$$\partial_{t_i} g^{(k)} = \left(g^{(k)}\xi_i g^{(k)-1}\right)_- g^{(k)} - g^{(k)}\partial_{t_i}\xi^{(k)} = -\left(g^{(k)}\xi_i g^{(k)-1}\right)_+ g^{(k)} \quad (3.45)$$

and when $i = (k', n, \alpha)$ and $k' \neq k$:

$$\partial_{t_i} g^{(k)} = M_i g^{(k)} \qquad (3.46)$$

In this equation, M_i, which is a rational function of λ with only one-pole at $\lambda_{k'}$, is regular around λ_k.

The zero curvature condition also allows to introduce the "wave function":

Definition. *The wave function $\Psi(\lambda; t_1, t_2, \ldots)$, is a matrix function depending on all the times simultaneously, and satisfying*

$$\partial_{t_i}\Psi = M_i\Psi, \quad \Psi(\lambda, t)|_{t=0} = 1 \qquad (3.47)$$

Locally around each λ_k we have

$$\Psi(\lambda, t) = g^{(k)}(\lambda, t)e^{\xi^{(k)}(\lambda, t)}g^{(k)-1}(\lambda, 0) \qquad (3.48)$$

The compatibility conditions of eqs. (3.47) are precisely the zero curvature equations, eqs. (3.42). Equation (3.48) follows because $g^{(k)}(\lambda, t)e^{\xi^{(k)}(\lambda, t)}$ is easily seen to satisfy eq. (3.47). Multiplying on the right by $g^{(k)-1}(\lambda, 0)$ enforces the initial condition $\Psi(\lambda, t)|_{t=0} = 1$.

3.5 Factorization problem

We now show that solving the hierarchy amounts to solving a factorization problem in a loop group. This is in fact solving a Riemann–Hilbert
factorization problem. These two aspects, group theory and analytic properties, will be fundamental in Chapters 4 and 5. At the end of the section
we make contact with the wave function.

In the construction of coadjoint orbits, we introduced a loop algebra
$\mathcal{G}$ of elements $X = \sum_{n\geq0} X_n\lambda^n$ regular at $\lambda = 0$. Its dual space $\mathcal{G}^*$ was
identified with the set of elements $\Xi = \sum_{n<0} \Xi_n\lambda^n$, regular at $\lambda = \infty$. It
so happens that $\mathcal{G}^*$ is itself a loop algebra, and in the computation of the
r-matrix we had to embed $\mathcal{G}$ and $\mathcal{G}^*$ into a single larger loop algebra $\tilde{\mathcal{G}}$,
see eq. (3.28):

$$\tilde{\mathcal{G}} = \mathcal{G} + \mathcal{G}^*$$

To adapt the notations to a more algebraic setting, in this section we
denote by $\mathcal{G}_+$ the subalgebra $\mathcal{G}$ of $\tilde{\mathcal{G}}$, and by $\mathcal{G}_-$ the subalgebra $\mathcal{G}^*$. Any
element $\tilde{X} \in \tilde{\mathcal{G}}$ can be decomposed uniquely as

$$\tilde{X} = X_+ - X_-, \quad X_\pm \in \mathcal{G}_\pm$$

At the group level this corresponds, formally, to decomposing an element
$\tilde{g} \in \tilde{G} = \exp(\tilde{\mathcal{G}})$ as

$$\tilde{g} = g_-^{-1}g_+, \quad g_\pm \in \exp(\mathcal{G}_\pm)$$

To give a meaning to this formal factorization we interpret it as a
Riemann–Hilbert problem. To do this, we introduce a small contour Γ
around $\lambda = 0$. An element $\tilde{X} \in \tilde{\mathcal{G}}$ is then viewed as a matrix valued
function $\tilde{X}(\lambda)$ defined on Γ. An element $X_+ \in \mathcal{G}_+$ is an invertible matrix
valued function $X_+(\lambda)$ on Γ which can be analytically extended inside Γ.
An element $X_- \in \mathcal{G}_-$ is a matrix valued function $X_-(\lambda)$ on Γ which can
be analytically extended outside Γ. An element $\tilde{g}(\lambda) \in \exp(\tilde{\mathcal{G}})$ is an invertible matrix valued function on Γ. The Riemann–Hilbert problem consists
of factorizing $\tilde{g}(\lambda)$ as a product of two matrices $g_\pm(\lambda)$ analytic inside and
outside the contour respectively. The existence of such a decomposition
is ensured by the following:

Theorem. *Let Γ be the closed contour $|\lambda| = 1$ in the λ-plane and $\tilde{g}(\lambda)$
a matrix defined on Γ. There exist two matrices $g_\pm(\lambda)$, with $g_+(\lambda)$ analytic inside Γ and $g_-(\lambda)$ analytic outside Γ such that $\det g_\pm \neq 0$ in their
respective domain of definition, and:*

$$\tilde{g}(\lambda) = g_-^{-1}(\lambda)\Lambda(\lambda)g_+(\lambda) \tag{3.49}$$

Here $\Lambda(\lambda)$ is a diagonal matrix with entries of the form λ^{k_i}. The k_i are integers, uniquely determined up to order by $\tilde{g}(\lambda)$, and called the indices. This solution to the factorization problem is unique if we require $g_-(\lambda)|_{\lambda=\infty} = 1$.

This theorem, which amounts to a classification of holomorphic vector bundles on the Riemann sphere, has a long history and has been proved by D. Hilbert, G. Birkhoff, A. Grothendieck and many others. For a sketch of the proof, see Chapter 15.

Remark 1. To understand the occurence of indices, consider the scalar case, and assume that $\tilde{g}$ is analytic in a ring surrounding Γ. The factorization problem is easily solved taking logarithms, but one has to be careful about the multivaluedness of $\log \tilde{g}$. Suppose that it jumps by $2ik\pi$ when λ describes Γ. Then the function $\log \tilde{g}(\lambda) - k \log(\lambda)$ is analytic and monovalued in the ring, hence can be expanded in a Laurent series:

$$\log \tilde{g}(\lambda) = k \log(\lambda) + X_+(\lambda) - X_-(\lambda)$$

where $X_+(\lambda)$ is the series of positive powers of λ in this Laurent expansion, and converges in the disc $|\lambda| \leq 1$, while $X_-(\lambda)$ is the series of strictly negative powers of λ, which converges for $|\lambda| \geq 1$. Then $g_\pm = \exp X_\pm$ are uniquely determined and non-vanishing. The integer k is the index. In the case where $\tilde{g}$ is a priori given as an exponential of an element of $\tilde{\mathcal{G}}$ the index naturally vanishes.

We treat first the case of one-pole which we assume to be located at $\lambda_k = 0$. We call g_+ the element $g^{(k)}$ around $\lambda_k = 0$ appearing in equations of the hierarchy, eq. (3.45). Since we have only one-pole, the full set of equations reads, with $i = (n, \alpha)$,

$$\partial_{t_i} g_+ = -\left(g_+ \xi_i g_+^{-1}\right)_+ g_+ \tag{3.50}$$

Proposition. *For small enough time, the solution of the system eq. (3.50) is obtained by solving the factorization problem*

$$g_-^{-1}(\lambda, t) g_+(\lambda, t) = e^{\xi(\lambda, t)} g_+(\lambda, 0) e^{-\xi(\lambda, t)} \tag{3.51}$$

Note that in the right-hand side, the time dependence is explicit.

Proof. We want to show that $g_+(\lambda, t)$, defined through the factorization problem eq. (3.51), satisfies eq. (3.50). Taking the time derivative of eq. (3.51) with respect to t_j, and multiplying on the left by $g_- \equiv g_-(\lambda, t)$ and on the right by $g_+^{-1} \equiv g_+^{-1}(\lambda, t)$, we get with, $\xi_j = \partial_{t_j} \xi$:

$$-\partial_{t_j} g_- g_-^{-1} + \partial_{t_j} g_+ g_+^{-1} = g_- \xi_j g_-^{-1} - g_+ \xi_j g_+^{-1}$$

Identifying the $+$ and $-$ parts, we have:

$$\partial_{t_j} g_+ g_+^{-1} = -\left(g_+ \xi_j g_+^{-1}\right)_+, \quad \partial_{t_j} g_- g_-^{-1} = -g_- \xi_j g_-^{-1} + \left(g_+ \xi_j g_+^{-1}\right)_-$$

The first equation is just eq. (3.50).   ■

This is a remarkable result, as it shows that the solution of the integrable hierarchy is reduced to solving a factorization problem in group theory. This will be promoted to an abstract algebraic setting for integrable systems in Chapter 4.

Remark 2. In eq. (3.51) we assumed that there are no indices. This is certainly true for small enough times because in the equivalent formulation eq. (3.52), the right-hand side is close to the identity for t small. As times get larger, indices may jump and may cause singularities in the solution. In Chapter 5 we will show that, indeed, poles appears at finite complex time.

The factorization problem, eq. (3.51), can be stated in a slightly different but equivalent way. Since $\xi(\lambda, t)$ contains only a polar part, $\xi(\lambda, t) \in \mathcal{G}_-$ and it can be reabsorbed in $g_-(\lambda, t)$. Let us define $\theta_-^{-1}(\lambda, t)$ and $\theta_+(\lambda, t)$ by

$$\theta_-^{-1}(\lambda, t) = e^{-\xi(\lambda,t)} g_-^{-1}(\lambda, t), \quad \theta_+(\lambda, t) = g_+(\lambda, t) g_+^{-1}(\lambda, 0)$$

Equation (3.51) now takes the form

$$\theta_-^{-1}(\lambda, t)\theta_+(\lambda, t) = g_+(\lambda, 0)e^{-\xi(\lambda,t)} g_+^{-1}(\lambda, 0) \tag{3.52}$$

Since $g_+(\lambda, 0)$ is the matrix which diagonalizes the Lax matrix $L(\lambda, 0) = g_+(\lambda, 0)A(\lambda)g_+^{-1}(\lambda, 0)$, we can formulate the factorization problem in the form

$$\theta_-^{-1}(\lambda, t)\,\theta_+(\lambda, t) = e^{-\sum_i dH_i(L(\lambda,0))\, t_i}$$

where dH_i is the differential of the Hamiltonian generating the flow t_i given in eq. (3.40).

It is this last formulation which lends itself to an easy generalization to the multi-pole case. We introduce small contours $\Gamma^{(k)}$ around each pole λ_k. As for the one-pole case, we define the group G_+ as the set of invertible matrices analytic inside *all* the contours $\Gamma^{(k)}$, and the group G_- as the set of invertible matrices analytic outside these contours, and normalized to the identity at $\lambda = \infty$. It is important to understand that the group G_+ and the group G, the direct product of the groups G_k appearing in eq. (3.23), may be identified. Indeed, an element of G_+ is known if we give its restrictions, $g^{(k)}$, to the interiors of the $\Gamma^{(k)}$. Hence an element of G_+ can be viewed as a multiplet of elements $g^{(k)}$ analytic inside $\Gamma^{(k)}$. This is a group homomorphism. The Lie algebra of G_+ is then identified with $\mathcal{G}$. Similarly, the Lie algebra of G_- can be identified with $\mathcal{G}^*$, as a vector space, using the non-degenerate pairing:

$$\langle X_-, X_+ \rangle = \oint_\Gamma \frac{d\lambda}{2i\pi} \text{Tr}(X_- X_+) = \sum_k \oint_{\Gamma_k} \frac{d\lambda}{2i\pi} \text{Tr}(X_- X_+^{(k)})$$

In the multi-pole case, the Riemann–Hilbert problem can be formulated as follows. As in the one-pole case, we introduce the loop group $\tilde{G}$ of matrices defined on Γ. An element $\tilde{g}(\lambda) \in \tilde{G}$ is a collection of elements $\tilde{g}^{(k)}(\lambda)$ given on the contours $\Gamma^{(k)}$. The Riemann–Hilbert problem now consists of factorizing elements $\tilde{g}(\lambda)$ as follows:

$$g_-^{-1}(\lambda)g_+(\lambda) = \tilde{g}^{(k)}(\lambda), \quad \lambda \in \Gamma^{(k)} \tag{3.53}$$

where $g_+(\lambda)$ is analytic inside all the $\Gamma^{(k)}$ and $g_-(\lambda)$ is analytic outside all the $\Gamma^{(k)}$. As in eq. (3.49), non-trivial indices may occur, but this does not happen for $\tilde{g}^{(k)}(\lambda)$ close enough to the identity.

This problem is solved by reducing it recursively to Riemann–Hilbert problems with only one contour. Indeed, let $h_-^{-1}h_+ = g^{(1)}$ be the solution for the contour $\Gamma^{(1)}$ with $h_-(\infty) = 1$. We seek the complete solution in the form $g = fh_-$ outside $\Gamma^{(1)}$ and $g = fh_+$ inside $\Gamma^{(1)}$. On $\Gamma^{(1)}$ we get $f_-^{-1}f_+ = 1$ while on $\Gamma^{(k)}$ for $k \geq 2$ we get $f_-^{-1}f_+ = h_-\tilde{g}^{(k)}h_-^{-1}$, so that f is obtained by solving a modified Riemann–Hilbert problem on the contours $\Gamma^{(2)}$ and so on.

Proposition. *The solution of the hierarchy equations eq. (3.37) is obtained by solving the multi-pole Riemann–Hilbert problem*

$$\theta_-^{-1}(\lambda, t)\theta_+(\lambda, t) = e^{-\sum_i dH_i(L(\lambda,0))\, t_i} \tag{3.54}$$

In the right-hand side the sum extends over all times of the hierarchy. The Lax matrix at time t is reconstructed from the initial condition $L(\lambda, 0)$ by

$$L(\lambda, t) = \theta_+(\lambda, t)L(\lambda, 0)\theta_+^{-1}(\lambda, t) = \theta_-(\lambda, t)L(\lambda, 0)\theta_-^{-1}(\lambda, t) \tag{3.55}$$

Proof. We first have to check that eq. (3.55) is consistent. The formula with θ_+ gives a definition of $L(\lambda, t)$ with λ inside the $\Gamma^{(k)}$, while the formula with θ_- gives a definition of $L(\lambda, t)$ with λ outside the $\Gamma^{(k)}$. They coincide for $\lambda \in \Gamma^{(k)}$ since by the factorization problem $\theta_-^{-1}(\lambda, t)\theta_+(\lambda, t)$ commutes with $L(\lambda, 0)$. Since $\theta_\pm$ are regular inside their respective domain of definition, the singularities of $L(\lambda, t)$ are the same as the singularities of $L(\lambda, 0)$. Taking the derivative of eq. (3.54) with respect to t_i, we get for $\lambda \in \Gamma^{(k)}$

$$\partial_{t_i}\theta_+(t)\theta_+^{-1}(t) - \partial_{t_i}\theta_-(t)\theta_-^{-1}(t)$$
$$= -\theta_-(t)\, dH_i(L(0))e^{-\sum_i dH_i(L(0))\, t_i}\, \theta_+^{-1}(t)$$
$$= -\theta_-(t)dH_i(L(0))\, \theta_-^{-1}(t) = -dH_i(L(t))$$

Hence we need to decompose $dH_i(L(t)) \in \tilde{\mathcal{G}}$ into its $+$ and $-$ parts. The solution of this factorization problem is (see eq. (3.41))

$$\partial_{t_i}\theta_-(t)\theta_-^{-1}(t) = M_i, \quad \partial_{t_i}\theta_+(t)\theta_+^{-1}(t) = -dH_i(L) + M_i$$

Indeed M_i is analytic everywhere outside λ_k, while $-dH_i(L) - M_i$ is analytic inside all the $\Gamma^{(l)}$. Since the decomposition is unique, this is the solution. Using the expression of L in terms of $\theta_-(t)$ we get

$$\partial_{t_i}L = [\partial_{t_i}\theta_-\theta_-^{-1}, L] = [M_i, L]$$

■

With the solution $\theta_\pm$ of the factorization problem, we can reconstruct the wave function Ψ defined in eq. (3.47). Recall that around λ_k the wave function admits the expansion eq. (3.48). The matrix $g^{(k)}(\lambda, t)$ in this formula diagonalizes $L(\lambda, t)$ and, in view of eq. (3.55), it is related to $\theta_+(\lambda, t)$ by $g^{(k)}(\lambda, t) = \theta_+(\lambda, t)g^{(k)}(\lambda, 0)d^{(k)}(\lambda, t)$, where $d^{(k)}(\lambda, t)$ is a diagonal matrix regular inside $\Gamma^{(k)}$. Note that $d^{(k)}(\lambda, t)|_{t=0}$ is the identity matrix. To find it for other values of t, we write its time evolution:

$$\partial_{t_j}\xi^{(k)}(\lambda, t) = g^{(k)-1}(\lambda, t)\, dH_j(L(\lambda, t))\, g^{(k)}(\lambda, t) - d^{(k)-1}\partial_{t_j}d^{(k)}$$

By eq. (3.40), we have $g^{(k)-1}(\lambda, t)\, dH_j(L(\lambda, t))\, g^{(k)}(\lambda, t) = \xi_j(\lambda)$. The time evolution of $d^{(k)}$ is therefore dictated by the simple equation $d^{(k)-1}\partial_{t_j}d^{(k)} = -\partial_{t_j}\xi^{(k)}(\lambda, t) + \xi_j(\lambda)$ whose solution is

$$d^{(k)}(\lambda, t) = e^{\xi(\lambda,t)-\xi^{(k)}(\lambda,t)}$$

Inserting into eq. (3.48), $\Psi(\lambda, t) = \theta_+(\lambda, t)g^{(k)}(\lambda, 0)e^{\xi(\lambda,t)}g^{(k)-1}(\lambda, 0)$, or:

$$\Psi(\lambda, t) = \theta_+(\lambda, t)e^{\sum_i dH_i(L(\lambda,0))t_i} \tag{3.56}$$

This provides a formula for $\Psi(\lambda, t)$ inside the contours $\Gamma^{(k)}$. Outside these contours we simply have

$$\Psi(\lambda, t) = \theta_-(\lambda, t) \tag{3.57}$$

because, by the factorization problem, the two expressions coincide on the $\Gamma^{(k)}$. Hence, we have found a global expression for the wave function. For completeness, we write the relation between the wave function and the Lax matrix

$$L(\lambda, t)\Psi(\lambda, t) = \Psi(\lambda, t)L(\lambda, 0)$$

The wave function is analytic in λ except at the λ_k where it has essential singularities. This could be expected from Poincaré's theorem on differential equations, applied to the system eq. (3.47). The time dependence of the essential singularities is also extremely simple: they are exponentials with linear time dependence, see eq. (3.56). This fact will be important in Chapter 5 where the function $\Psi(\lambda, t)$ will be reconstructed from its analytical properties.

3.6 Tau-functions

We now introduce the important concept of tau-functions. The elementary flows induce a coupling between the λ dependence and the corresponding time dependence, which is best expressed in terms of tau-functions.

The existence of tau-functions relies on the following main observation.

Proposition. *Let $d = \sum_i dt_i \partial_{t_i}$. In the gauge of eq. (3.44), the 1-form*

$$\Upsilon \equiv -\sum_k \mathrm{Tr}\ \mathrm{Res}_{\lambda_k} \left(g^{(k)-1} \partial_\lambda g^{(k)} d\xi^{(k)} \right) \tag{3.58}$$

is closed, $d\Upsilon = 0$. As a consequence we can define a function $\tau(t_1, t_2, \ldots)$ of the infinite series of times $t_{(k,n,\alpha)}$ for all k, n, α such that $\Upsilon = d\log\tau$ or:

$$\partial_{t_i} \log\tau = -\sum_k \mathrm{Tr}\ \mathrm{Res}_{\lambda_k} \left(g^{(k)-1} \partial_\lambda g^{(k)} \xi_i \right) \tag{3.59}$$

Proof. The evolution equations eq. (3.44) can be written as $dg^{(k)} = \mathcal{M}g^{(k)} - g^{(k)}d\xi^{(k)}$, where we have set $\mathcal{M} = \sum_i M_i dt_i$. Hence by differentiating eq. (3.58), we have

$$d\Upsilon = \sum_k \mathrm{Tr}\ \mathrm{Res}_{\lambda=\lambda_k} \left(d\partial_\lambda \xi^{(k)} \wedge d\xi^{(k)} - \partial_\lambda \mathcal{M} \wedge g^{(k)} d\xi^{(k)} g^{(k)-1} \right)$$

The first term vanishes, because the order of the pole at λ_k of $\mathrm{Tr}\ (d\partial_\lambda \xi^{(k)} \wedge d\xi^{(k)})$ is at least 3. To transform the second term, we notice that $\mathcal{M}$ has the same polar part as $g^{(k)}d\xi^{(k)}g^{(k)-1}$ at λ_k so that we can write $g^{(k)}d\xi^{(k)}g^{(k)-1} = \mathcal{M} + \mathcal{N}^{(k)}$ where $\mathcal{N}^{(k)}$ is regular at λ_k. We get:

$$d\Upsilon = -\sum_k \mathrm{Tr}\ \mathrm{Res}_{\lambda=\lambda_k} \left(\partial_\lambda \mathcal{M} \wedge (\mathcal{M} + \mathcal{N}^{(k)}) \right)$$

We now show that:

$$\mathrm{Tr}\ \mathrm{Res}_{\lambda=\lambda_k} (\partial_\lambda \mathcal{M} \wedge \mathcal{N}^{(k)}) = -\frac{1}{2} \mathrm{Tr}\ \mathrm{Res}_{\lambda=\lambda_k} (\partial_\lambda \mathcal{M} \wedge \mathcal{M}) \tag{3.60}$$

This implies that $d\Upsilon = -\frac{1}{2}\sum_k \text{Tr Res}_{\lambda=\lambda_k}(\partial_\lambda \mathcal{M} \wedge \mathcal{M})$ vanishes because $\text{Tr}(\partial_\lambda \mathcal{M} \wedge \mathcal{M})$ is a rational 1-form on the λ Riemann sphere, hence the sum of its residues vanishes.

Equation (3.60) results from two local computations around λ_k. First we have:

$$\text{Tr Res}_{\lambda=\lambda_k}(\partial_\lambda(\mathcal{M} + \mathcal{N}^{(k)}) \wedge (\mathcal{M} + \mathcal{N}^{(k)}))$$
$$= \text{Tr Res}_{\lambda=\lambda_k}(\partial_\lambda(g^{(k)}d\xi^{(k)}g^{(k)-1}) \wedge (g^{(k)}d\xi^{(k)}g^{(k)-1}))$$
$$= \text{Tr Res}_{\lambda=\lambda_k}(d\partial_\lambda\xi^{(k)} \wedge d\xi^{(k)}) = 0$$

Next, using that the residue of a derivative of a function of λ vanishes, $0 = \text{Tr Res}_{\lambda=\lambda_k}\partial_\lambda(\mathcal{N}^{(k)} \wedge \mathcal{M})$, we obtain $\text{Tr Res}_{\lambda=\lambda_k}(\partial_\lambda\mathcal{N}^{(k)} \wedge \mathcal{M}) = \text{Tr Res}_{\lambda=\lambda_k}(\partial_\lambda\mathcal{M} \wedge \mathcal{N}^{(k)})$. Inserting this relation into the left-hand side of the above formula yields eq. (3.60). ■

The tau-function has remarkable and beautiful implications in the theory of integrable systems. Here we shall only present its fundamental relation to the wave-function Ψ. Around a singularity λ_k, we can expand Ψ as

$$\Psi = g^{(k)}(\lambda, t)e^{\xi^{(k)}(\lambda,t)}g^{(k)-1}(\lambda, 0), \quad g^{(k)}(\lambda) = g_0^{(k)}h^{(k)}(\lambda)$$

with $h^{(k)}(\lambda) = 1 + O(\lambda - \lambda_k)$. We will relate the matrix elements $h_{\beta\alpha}^{(k)}(\lambda)$ to tau-functions.

Theorem. *The matrix $h^{(k)}(t, \lambda)$ can be expressed in the form:*

$$h_{\alpha\alpha}^{(k)}(t, \lambda) = \frac{\tau\left(t - [\lambda]_\alpha^{(k)}\right)}{\tau(t)} \tag{3.61}$$

$$h_{\beta\alpha}^{(k)}(t, \lambda) = (\lambda - \lambda_k)\frac{\tau_{\beta\alpha}^{(k)}\left(t - [\lambda]_\alpha^{(k)}\right)}{\tau(t)}, \quad \alpha \neq \beta$$

where the notation $t - [\lambda]_\alpha^{(k)}$ means the shift of the time $t_{(k,n,\alpha)}$

$$t - [\lambda]_\alpha^{(k)} \equiv t_{(k',n,\gamma)} - \delta_{kk'}\delta_{\gamma\alpha}\frac{(\lambda - \lambda_k)^n}{n}$$

Equation (3.61) is a famous formula discovered by Sato. Its main feature is that the λ dependence in the numerator is expressed by translations of the appropriate time variables. More information on tau-functions, including expressions of $\tau_{\beta\alpha}^{(k)}(t)$ in terms of $\tau(t)$, will be given in Chapter 8.

The proof is long and will be done in several steps.

Lemma 1. *Let $f(\lambda) = \sum_{j=0}^{\infty} f_j \lambda^j$. We have*

$$\sum_{n=1}^{\infty} \lambda^n \left(\lambda^{-n} f(\lambda)\right)_+ = \lambda \partial_\lambda f(\lambda), \quad \sum_{n=1}^{\infty} \lambda^n \mathrm{Res}(\lambda^{-n} f(\lambda)) = \lambda f(\lambda) \quad (3.62)$$

<u>Proof.</u> Indeed,

$$\sum_{n=1}^{\infty} \lambda^n \left(\lambda^{-n} f(\lambda)\right)_+ = \sum_{n=1}^{\infty} \sum_{j=n}^{\infty} f_j \lambda^j = \sum_{j=1}^{\infty} \sum_{n=0}^{j} f_j \lambda^j = \sum_{j=1}^{\infty} j f_j \lambda^j = \lambda \partial_\lambda f(\lambda)$$

This shows the first equality. The second one is simpler:

$$\sum_{n=1}^{\infty} \lambda^n \mathrm{Res}(\lambda^{-n} f(\lambda)) = \sum_{n=1}^{\infty} \lambda^n f_{n-1} = \lambda f(\lambda)$$

∎

Let us consider the vicinity of a given pole which we assume to be at the origin $\lambda_k = 0$ (the case $\lambda_k \neq 0$ is simply recovered by translating λ in the formulae below).

Introduce the generating differential operator of time derivatives $\nabla_\alpha^{(k)}$:

$$\nabla_\alpha^{(k)} = \sum_{n>0} (\lambda - \lambda_k)^n \frac{\partial}{\partial t_{(k,n,\alpha)}} = \sum_{n>0} \lambda^n \frac{\partial}{\partial t_{(k,n,\alpha)}} \quad (3.63)$$

Lemma 2. *The action of $\nabla_\alpha^{(k)}$ on the tau-function is given by:*

$$\nabla_\alpha^{(k)} \log \tau = -\lambda \,\mathrm{Tr}\left(E_{\alpha\alpha} h^{(k)-1} \partial_\lambda h^{(k)}\right) \quad (3.64)$$

Its action on $h^{(k)}$ is given by:

$$h^{(k)-1} \nabla_\alpha^{(k)} h^{(k)} = -\lambda [h^{(k)-1} \partial_\lambda h^{(k)}, E_{\alpha\alpha}] - h^{(k)-1} E_{\alpha\alpha} h^{(k)} + E_{\alpha\alpha} \quad (3.65)$$

<u>Proof.</u> In the definition of $\log \tau$, eq. (3.59), we can replace $g^{(k)}$ by $h^{(k)}$ because $g_0^{(k)}$ is independent of λ. Hence, using eq. (3.62),

$$-\nabla_\alpha^{(k)} \log \tau = \sum_{n=1}^{\infty} \lambda^n \mathrm{Tr}\,\mathrm{Res}(\lambda^{-n} E_{\alpha\alpha} h^{(k)-1} \partial_\lambda h^{(k)}) = \lambda \mathrm{Tr}(E_{\alpha\alpha} h^{(k)-1} \partial_\lambda h^{(k)})$$

To prove the second formula, we start from the equations of the hierarchy, eq. (3.45), with $\xi_i = E_{\alpha\alpha} \lambda^{-n}$. They read:

$$\partial_{t_i} g_0^{(k)} = -g_0^{(k)} \left(h^{(k)} E_{\alpha\alpha} \lambda^{-n} h^{(k)-1}\right)_0$$

$$\partial_{t_i} h^{(k)} = -\left(h^{(k)} E_{\alpha\alpha} \lambda^{-n} h^{(k)-1}\right)_{++} h^{(k)}$$

where for any series $f(\lambda) = \sum_{i \geq 0} f_i \lambda^i$ we have defined $(f(\lambda))_0 = f_0$ and $f(\lambda)_{++} = \sum_{i \geq 1} f_i \lambda^i$.

Using eq. (3.62) again and $\sum_{n=1}^{\infty} \lambda^n (\lambda^{-n} f(\lambda))_0 = f(\lambda) - f_0$, we get

$$\nabla_\alpha^{(k)} h^{(k)} = -\sum_{n=1}^{\infty} \lambda^n \left(h^{(k)} E_{\alpha\alpha} \lambda^{-n} h^{(k)-1} \right)_{++} h^{(k)}$$

$$= \left(-\lambda \partial_\lambda (h^{(k)} E_{\alpha\alpha} h^{(k)-1}) + h^{(k)} E_{\alpha\alpha} h^{(k)-1} - E_{\alpha\alpha} \right) h^{(k)}$$

This is equivalent to eq. (3.65).   ∎

We can now give the proof of the theorem:

<u>Proof.</u> Let us take the matrix element β, α of eq. (3.65). Separating the cases $\beta = \alpha$ and $\beta \neq \alpha$, we can write it as:

$$(h^{(k)-1} \nabla_\alpha^{(k)} h^{(k)})_{\beta\alpha} = -\lambda (h^{(k)-1} \partial_\lambda h^{(k)})_{\beta\alpha} - (h^{(k)-1} E_{\alpha\alpha} h^{(k)} - E_{\alpha\alpha})_{\beta\alpha}$$
$$+ \delta_{\beta\alpha} \lambda (h^{(k)-1} \partial_\lambda h^{(k)})_{\alpha\alpha}$$

Multiplying on the left by $h^{(k)}$ and remembering that, by eq. (3.64), we have $\lambda (h^{(k)-1} \partial_\lambda h^{(k)})_{\alpha\alpha} = -\nabla_\alpha^{(k)} \log \tau$, we obtain

$$\nabla_\alpha^{(k)} h_{\beta\alpha}^{(k)} = -\lambda \partial_\lambda h_{\beta\alpha}^{(k)} - (E_{\alpha\alpha} h^{(k)})_{\beta\alpha} + (h^{(k)} E_{\alpha\alpha})_{\beta\alpha} - (\nabla_\alpha^{(k)} \log \tau) h_{\beta\alpha}^{(k)}$$

If $\beta = \alpha$ we set $X_{\alpha\alpha}^{(k)} = \tau h_{\alpha\alpha}^{(k)}$ and if $\beta \neq \alpha$, we set $X_{\beta\alpha}^{(k)} = \lambda^{-1} \tau h_{\beta\alpha}^{(k)}$. The equation becomes $(\nabla_\alpha^{(k)} + \lambda \partial_\lambda) X_{\beta\alpha}^{(k)} = 0$. This means that $X_{\beta\alpha}^{(k)}(t, \lambda)$ is of the form $X_{\beta\alpha}^{(k)}(t, \lambda) = \tau_{\beta\alpha}^{(k)}(\ldots, t_{(k,n,\alpha)} - \lambda^n/n, \ldots)$. Since $h_{\alpha\alpha}^{(k)}(\lambda)|_{\lambda=0} = 1$, the function $\tau_{\alpha\alpha}^{(k)}(t)$ is in fact equal to $\tau(t)$.   ∎

3.7 Integrable field theories and monodromy matrix

For a system with a finite number of degrees of freedom, we have seen that a Lax matrix could be interpreted as a coadjoint orbit. It is possible to adapt this interpretation to field theory by properly choosing the Lie algebra involved. We shall consider two-dimensional field theory on a cylinder with space variable $x \in [0, 2\pi]$ and time variable $t \in [-\infty, +\infty]$. To introduce the space variable x, we consider the loop algebra $\widetilde{\mathcal{G}}$ of (periodic) maps from the circle S^1 to the some Lie algebra $\mathcal{G}$, i.e. maps $S^1 \to \mathcal{G}$. The simplest case corresponds to choosing $\mathcal{G}$ to be the algebra of $N \times N$ matrices, but more frequently, it will be an element of a loop algebra with spectral parameter λ as in the finite-dimensional case. So we are dealing with double loop algebras. In order to introduce some structure in the x direction, we consider the central extension of the x–loop algebra (see Chapter 16):

$$\widehat{\mathcal{G}} = \widetilde{\mathcal{G}} + \mathbb{C}K$$

The commutator of two elements $X_i = \widetilde{X}_i(x) + c_i K$ reads by definition:

$$[X_1, X_2] = [\widetilde{X}_1(x), \widetilde{X}_2(x)] + \int_0^{2\pi} (\widetilde{X}_1(x)\partial_x \widetilde{X}_2(x))dx \ K$$

where $(\ , \ )$ is an invariant non-degenerate bilinear form on $\mathcal{G}$ (such as Tr or Tr Res). The dual space $\widehat{\mathcal{G}}^*$ of $\widehat{\mathcal{G}}$ can be identified with the space of pairs of elements of the form $\Xi = (\widetilde{\Xi}(x), \zeta)$ with pairing

$$\Xi(X) = \int_0^{2\pi} (\widetilde{\Xi}(x), \widetilde{X}(x))dx + \zeta c$$

The coadjoint action is defined as usual $(\mathrm{ad}^* X \cdot \Xi)(Y) = -\Xi([X, Y])$ and takes the form

$$(\mathrm{ad}^* X \cdot \Xi)(Y) = -\int_0^{2\pi} (\widetilde{\Xi}(x), [\widetilde{X}(x), \widetilde{Y}(x)])dx - \zeta \int_0^{2\pi} (\widetilde{X}(x)\partial_x \widetilde{Y}(x))dx$$

$$= \int_0^{2\pi} (-[\widetilde{\Xi}(x), \widetilde{X}(x)] + \zeta \partial_x \widetilde{X}(x), \widetilde{Y}(x))dx$$

so that

$$\mathrm{ad}^* X \cdot \Xi = (-[\widetilde{\Xi}(x), \widetilde{X}(x)] + \zeta \partial_x \widetilde{X}(x), 0) \qquad (3.66)$$

We see that the element ζ is invariant by coadjoint action, and we will choose orbits with $\zeta = 1$.

In this setting, the Lax equation eq. (3.21) reads, for $L = (U, 1)$ and $M = V$:

$$\partial_t U - \partial_x V - [V, U] = 0 \qquad (3.67)$$

This is a zero curvature condition.

Alternatively, one can say that the variable x behaves like one of the times of finite-dimensional systems, see eq. (3.42). It is important, however, to realise that in the field theory case, the construction of commuting quantities is more complicated because we have to construct functions invariant under the coadjoint action eq. (3.66). For this, the right object to consider is the so-called monodromy matrix which we now introduce.

The zero curvature condition (3.67) expresses the compatibility condition of the associated linear system

$$(\partial_x - U)\Psi = 0, \quad (\partial_t - V)\Psi = 0 \qquad (3.68)$$

The matrices U and V can be thought of as the x and t components of a connection. This connection will be called the Lax connection. Given U and V, the linear system (3.68) determines the matrix Ψ up to multiplication on the right by a constant matrix, which we can fix by requiring $\Psi(\lambda, 0, 0) = 1$. This Ψ will be called the wave function.

Choosing a path γ from the origin to the point (x, t), the wave function can be written symbolically as

$$\Psi(x, t) = \overleftarrow{\exp}\left[\int_\gamma (U\, dx + V\, dt) \right] \tag{3.69}$$

where $\overleftarrow{\exp}$ denotes the path-ordered exponential. This is just the parallel transport along the curve γ with the connection (U, V). Since the Lax connection satisfies the zero curvature relation (3.67) the value of the path-ordered exponential is independent of the choice of this path. In particular, if γ is the path $x \in [0, 2\pi]$ with fixed time t, we call $\Psi(2\pi, t)$ the monodromy matrix $T(\lambda, t)$:

$$T(\lambda, t) \equiv \overleftarrow{\exp}\left[\int_0^{2\pi} U(\lambda, x, t)dx \right] \tag{3.70}$$

where we assume that $U(\lambda, x, t)$ and $V(\lambda, x, t)$ depend on a spectral parameter λ.

Proposition. *Assume that all fields are periodic in x with period 2π. Let $T(\lambda, t)$ be the monodromy matrix and let*

$$H^{(n)}(\lambda) = \mathrm{Tr}\left(T^n(\lambda, t) \right) \tag{3.71}$$

Then, $H^{(n)}(\lambda)$ is independent of time. Hence traces of powers of the monodromy matrix generate conserved quantities.

<u>Proof.</u> Thinking of the path-ordered exponential on $[a, b]$ as

$$\overleftarrow{\exp}\left[\int_a^b U(x)dx \right] \sim (1 + \delta x U(x_n)) \cdots (1 + \delta x U(x_1))$$

with a subdivision $x_1 = a < x_2 < \cdots < x_n = b$ such that $x_{i+1} - x_i = \delta x \to 0$, we get (all exponentials are path-ordered exponentials):

$$\partial_t T(\lambda, t) = \int_0^{2\pi} dx\, e^{\int_x^{2\pi} U\,dx} \dot U(\lambda, x) e^{\int_0^x U\,dx}$$

$$= \int_0^{2\pi} dx\, e^{\int_x^{2\pi} U\,dx} (\partial_x V + [V, U]) e^{\int_0^x U\,dx}$$

$$= \int_0^{2\pi} dx\, \partial_x \left(e^{\int_x^{2\pi} U\,dx} V e^{\int_0^x U\,dx} \right)$$

Performing the integral,

$$\partial_t T(\lambda, t) = V(\lambda, 2\pi, t) T(\lambda, t) - T(\lambda, t) V(\lambda, 0, t) \qquad (3.72)$$

So, if the fields are periodic, we have $V(\lambda, 2\pi, t) = V(\lambda, 0, t)$ and the relation becomes

$$\partial_t T(\lambda, t) = [V(\lambda, 0, t), T(\lambda, t)]$$

This is a Lax equation. It implies that $H^{(n)}(\lambda)$ is time-independent. Expanding in λ we obtain an infinite set of conserved quantities. ∎

It is the monodromy matrix which plays the role of the Lax matrix in the field theoretical context.

3.8 Abelianization

We now discuss the analogue of the Zakharov–Shabat construction for field theory. We consider the linear system eq. (3.68) where $U(\lambda, x, t)$ and $V(\lambda, x, t)$ are matrices depending in a rational way on a parameter λ having poles at *constant* values λ_k.

$$U = U_0 + \sum_k U_k \quad \text{with} \quad U_k = \sum_{r=-n_k}^{-1} U_{k,r} (\lambda - \lambda_k)^r \qquad (3.73)$$

$$V = V_0 + \sum_k V_k \quad \text{with} \quad V_k = \sum_{r=-m_k}^{-1} V_{k,r} (\lambda - \lambda_k)^r \qquad (3.74)$$

The compatibility condition of the linear system (3.68) is the zero curvature condition (3.67). We demand that it holds identically in λ. These conditions are always compatible, since by the same naive counting argument as for finite-dimensional systems there is one more variable than the number of equations. The origin of this indeterminacy is the same: the zero curvature condition is invariant by gauge transformations. If the gauge transformation is independent of λ, it will not spoil the analytic properties of U and V. Notice that eq. (3.67) implies that U_0 and V_0 are pure gauge, i.e. there exists a group valued function h such that

$$U_0 = \partial_x h h^{-1} \quad \text{and} \quad V_0 = \partial_t h h^{-1} \qquad (3.75)$$

Remark 1. Using a λ independent gauge transformation *periodic* in x, we can always choose a gauge in which U_0 is constant diagonal and $V_0 = 0$. To show this we start from eq. (3.75). Writing that $U_0(x)$ is periodic implies that $\partial_x (h^{-1}(x) h(x+2\pi)) = 0$. We change basis so that $h^{-1}(x) h(x+2\pi)$ is diagonal and denote it by $\exp(2\pi P)$ with

P diagonal. Hence we can write $h(x,t) = \tilde{h}(x,t)\exp(Px)$ with $\tilde{h}(x+2\pi,t) = \tilde{h}(x,t)$ and we gauge transform under $\tilde{h}$. Then we have $\tilde{U}_0 = P$ and $\tilde{V}_0 = x\partial_t P$. But $\tilde{V}_0$ is periodic in x, hence $\partial_t P = 0$.

As for finite-dimensional systems, we first make a local analysis around each pole λ_k in order to understand solutions of eq. (3.67). We show that around each singularity λ_k, one can perform a gauge transformation bringing simultaneously $U(\lambda)$ and $V(\lambda)$ to a diagonal form. The important new feature we want to emphasize, as compared to the finite-dimensional case, is that this construction is *local* in x.

Let us assume that the pole is located at $\lambda = 0$. The rational functions $U(\lambda, x, t)$, $V(\lambda, x, t)$ can be expanded in a Taylor series in a neighbourhood of this pole:

$$U(\lambda, x, t) = \sum_{r=-n}^{\infty} U_r(x,t)\lambda^r, \quad V(\lambda, x, t) = \sum_{r=-m}^{\infty} V_r(x,t)\lambda^r \qquad (3.76)$$

We have:

Proposition. *There exists a local, periodic, gauge transformation*

$$\partial_x - U = g(\partial_x - A)g^{-1}, \quad \partial_t - V = g(\partial_t - B)g^{-1} \qquad (3.77)$$

where $g(\lambda)$, $A(\lambda)$ *and* $B(\lambda)$ *are formal series in* λ:

$$g = \sum_{r=0}^{\infty} g_r \lambda^r, \quad A = \sum_{r=-n}^{\infty} A_r \lambda^r, \quad B = \sum_{r=-m}^{\infty} B_r \lambda^r$$

such that the matrices $A(\lambda)$ *and* $B(\lambda)$ *are diagonal. Moreover* $\partial_t A(\lambda) - \partial_x B(\lambda) = 0$.

<u>Proof</u>. Let g_0 be the matrix diagonalizing the leading term in eq. (3.76),

$$U_{-n} = g_0 \, A_{-n} \, g_0^{-1}$$

Let $\partial_x - U = g_0(\partial_x - \tilde{U})g_0^{-1}$, $\tilde{U} = \sum_{r=-n}^{\infty} \tilde{U}_r \lambda^r$. Since $\lambda = 0$ is a pole of U and g_0 is regular we have $\tilde{U}_{-n} = A_{-n}$. We set $g = g_0 h$. Equation (3.77) becomes $\partial_x - \tilde{U} = h(\partial_x - A)h^{-1}$, or $\partial_x h - \tilde{U}h + hA = 0$. Expanding in powers of λ, we get

$$\partial_x h_l - \sum_{r=-n}^{l} (\tilde{U}_r \, h_{l-r} - h_{l-r} \, A_r) = 0, \quad l = -n, \ldots, \infty \qquad (3.78)$$

Of course $h_l = 0$ if $l < 0$; $h_0 = 1$. In the sum, we separate the first and the last term

$$\partial_x h_l - [A_{-n}, h_{l+n}] - \sum_{r=-n+1}^{l-1} (\tilde{U}_r h_{l-r} - h_{l-r} A_r) - \tilde{U}_l + A_l = 0, \quad l = -n, \dots, \infty$$

(3.79)

Projecting this equation on the diagonal matrices, we determine A_l in terms of A_k, $k < l$, and h_k, $k < l + n$. (The term $[A_{-n}, h_{l+n}]$ does not contribute to the diagonal since A_{-n} is itself diagonal). Similarly, the off-diagonal part of this equation determines the off-diagonal part of h_{l+n} in terms of the same variables. We can make the solution unique by requiring, for instance, diag $(h_{l+n}) = 0$. Therefore h and A are determined recursively. Note that this is a purely algebraic computation, so g and A are periodic functions of x and are algebraic functions of the matrix elements of U and their derivatives.

Under the gauge transformation g we obviously have $\partial_t A - \partial_x B = [B, A]$, or $\partial_t A_l - \partial_x B_l = \sum_{r=-n}^{\infty} [B_{l-r}, A_r]$. If B is regular at $\lambda = 0$ the expansion of B starts at B_0 and $\partial_t A_{-n} = [B_0, A_{-n}]$. Hence the off-diagonal part of the commutator is zero and therefore B_0 is diagonal. If, however, B is singular, the expansion of B starts at B_{-m} and the most singular term in the commutator is $\lambda^{-n-m}[B_{-m}, A_{-n}]$, and this has to vanish because $n + m > \max(n, m)$. Hence B_{-m} is diagonal. We finish the proof by induction on l. Assume B_r is diagonal until B_{l+n-1}. Then $\partial_t A_l - \partial_x B_l = \sum_{r=-n}^{l+m} [B_{l-r}, A_r] = [B_{l+n}, A_{-n}]$, hence B_{l+n} is diagonal. It is important to notice that this procedure only requires local computations. There is no differential equation to be integrated when recursively diagonalizing the Lax connection around its poles. ∎

As for finite-dimensional systems, we can reconstruct all the matrices U_k and V_k, and therefore the Lax connection, from simple data.

$$U = U_0 + \sum_k U_k, \quad \text{with} \quad U_k \equiv \left(g^{(k)} A_-^{(k)} g^{(k)-1} \right)_- \qquad (3.80)$$

$$V = V_0 + \sum_k V_k, \quad \text{with} \quad V_k \equiv \left(g^{(k)} B_-^{(k)} g^{(k)-1} \right)_- \qquad (3.81)$$

Remark 2. Let us check that the reconstruction formulae (3.80, 3.81) for U and V are such that the order of the poles in the commutator appearing in the zero curvature condition matches the order of the poles in the derivative terms. Indeed the polar part at λ_k of the commutator is $[U_k, V]_- + [V_k, U]_-$ and can be higher than the order of the poles of the derivatives $\partial_t U_k$ and $\partial_x V_k$. So we have to show that the formula ensures that the poles of these commutators, which are naively of order $n_k + m_k$, are actually of order less than $\max(n_k, m_k)$. Indeed, consider for example the commutator $[U_k, V]_-$. Similarly as for finite-dimensional systems, we may write it as:

$$
\begin{aligned}
[U_k, V]_- &= \left[\left(g^{(k)} A_-^{(k)} g^{(k)-1} \right)_- , V \right]_- \tag{3.82}\\
&= \left[g^{(k)} A_-^{(k)} g^{(k)-1} - \left(g^{(k)} A_-^{(k)} g^{(k)-1} \right)_+ , V \right]_-\\
&= \left[g^{(k)} A_-^{(k)} g^{(k)-1}, g^{(k)} \partial_t g^{(k)-1} \right]_- - \left[\left(g^{(k)} A_-^{(k)} g^{(k)-1} \right)_+ , V \right]_-
\end{aligned}
$$

In the last line we use the fact that $\partial_t - V$ is diagonalized by $g^{(k)}$, i.e. $V = -g^{(k)} \partial_t g^{(k)-1} + g^{(k)} B^{(k)} g_k^{-1}$ with $B^{(k)}$ diagonal. All the terms of the last line have a pole of order at most n_k. Similarly one shows that the order of the pole of $[V_k, U]_-$ is at most m_k. This shows that the constraints hidden in eq. (3.67) are solved by the formulas (3.80, 3.81).

We now use this diagonal gauge to compute the conserved quantities.

Proposition. *The quantities $Q^{(k)}(\lambda) = \int_0^{2\pi} A^{(k)}(\lambda, x, t)\, dx$ are local conserved quantities of the field theory. They are related to eq. (3.71) by*

$$
H^{(n)}(\lambda) = \operatorname{Tr} \exp\left[n \int_0^{2\pi} A^{(k)}(\lambda, x, t) dx \right] = \operatorname{Tr} \exp\left[n Q^{(k)}(\lambda) \right]
$$

Proof. Around each pole, in the diagonal gauge, the zero curvature condition reduces to

$$
\partial_t A^{(k)}(\lambda, x, t) - \partial_x B^{(k)}(\lambda, x, t) = 0
$$

It is the equation of conservation of a current. The charge $Q^{(k)}(\lambda) = \int_0^{2\pi} A^{(k)}(\lambda, x, t)\, dx$ is conserved because

$$
\partial_t Q_k(\lambda) = \int_0^{2\pi} \partial_x B^{(k)}(\lambda, x, t) = B^{(k)}(\lambda, 2\pi, t) - B^{(k)}(\lambda, 0, t) = 0
$$

where we have used the fact that $A^{(k)}(\lambda, x, t)$ and $B^{(k)}(\lambda, x, t)$ are local in terms of the coefficients of the Lax connection and are therefore periodic in x. Expanding $Q^{(k)}(\lambda)$ in powers of $\lambda - \lambda_k$ produces an infinite number of conserved quantities. Under a gauge transformation, $U \to {}^g U = g^{-1} U g - g^{-1} \partial_x g$, the monodromy matrix is transformed into ${}^g T(\lambda, t)$ with

$$
{}^g T(\lambda, t) = g^{-1}(2\pi, t) T(\lambda, t) g(0, t)
$$

Thus, if $g(x,t)$ is periodic, $g(2\pi, t) = g(0, t)$, one has

$$\text{Tr } ({}^g T(\lambda, t)) = \text{Tr } (T(\lambda, t))$$

In the diagonal gauge around $\lambda = \lambda_k$, the monodromy matrix is easily computed and one gets:

$$H^{(n)}(\lambda) = \text{Tr exp} \left[n \int_0^{2\pi} A^{(k)}(\lambda, x, t) dx \right]$$

There is no problem of ordering in the exponential since the matrices $A^{(k)}$ are diagonal. ∎

Remark 3. The abelianization procedure can be used to give a local expression of the wave function,

$$\Psi(\lambda, x, t) = g^{(k)}(\lambda, x, t) e^{\int_0^{(x,t)} A^{(k)}(\lambda) dx + B^{(k)}(\lambda) dt} g^{(k)-1}(\lambda, 0, 0)$$

As for finite-dimensional systems, this shows that $\Psi(\lambda, x, t)$ possesses essential singularities at the points λ_k. By the Poincaré theorem on differential equations, these are the only singularities of $\Psi(\lambda, x, t)$.

We now give some examples of two-dimensional field theories having a zero curvature representation.

Example 1. The first example is the non-linear σ model. For simplicity, we look for a Lax connection in which U and V have only one simple pole at two different points and $U_0 = V_0 = 0$. Choosing these points to be at $\lambda = \pm 1$, we can thus parametrize U and V as:

$$U = \frac{1}{\lambda - 1} J_x, \quad V = -\frac{1}{\lambda + 1} J_t \tag{3.83}$$

with J_x and J_t taking values in some Lie algebra. Decomposing the zero curvature condition $[\partial_x - U, \partial_t - V] = 0$ over its simple poles gives two equations:

$$\partial_t J_x - \tfrac{1}{2}[J_x, J_t] = 0,$$
$$\partial_x J_t + \tfrac{1}{2}[J_x, J_t] = 0.$$

Taking the difference implies that $[\partial_t + J_t, \partial_x + J_x] = 0$. Thus J is a pure gauge and there exists g such $J_t = g^{-1}\partial_t g$ and $J_x = g^{-1}\partial_x g$. Taking now the sum of the two equations implies $\partial_t J_x + \partial_x J_t = 0$, or equivalently,

$$\partial_t (g^{-1}\partial_x g) + \partial_x (g^{-1}\partial_t g) = 0$$

This is the field equation of the so-called non-linear sigma model, with x, t as light-cone coordinates.

Example 2. Another important example is the sinh-Gordon model. It also has a two-poles Lax connection, one-pole at $\lambda = 0$, the other at $\lambda = \infty$. Moreover, we require that in the light-cone coordinates, $x_\pm = x \pm t$, $U(\lambda, x_\pm)$ has a simple pole at $\lambda = 0$ and $V(\lambda, x_\pm)$ a simple pole at $\lambda = \infty$. The most general 2×2 system of this form is:

$$(\partial_{x_+} - U)\Psi = 0, \quad U = U_0 + \lambda^{-1} U_1$$
$$(\partial_{x_-} - V)\Psi = 0, \quad V = V_0 + \lambda V_1$$

The matrices U_i, V_i are taken to be traceless matrices, so contain 12 parameters. One can reduce this number by imposing a symmetry condition under a discrete group, as in eq. (3.18). Namely, we consider the group $\mathbb{Z}_2$ acting by:

$$\Psi(\lambda) \longrightarrow \sigma_z \Psi(-\lambda)\sigma_z^{-1}, \quad \sigma_z = \begin{pmatrix} 1 & 0 \\ 0 & -1 \end{pmatrix} \qquad (3.84)$$

and we demand that Ψ be invariant by this action. This restriction means that the wave function belongs to the twisted loop group. It follows that: $\sigma_z U(-\lambda)\sigma_z = U(\lambda)$ and $\sigma_z V(-\lambda)\sigma_z = V(\lambda)$. We still have the possibility of performing a gauge transformation by an element g, independent of λ, in order to preserve the pole structure of the connection, and commuting with the action of $\mathbb{Z}_2$, i.e. g diagonal. This gauge freedom can be used to set $(V_0)_{ii} = 0$. The symmetry condition then gives:

$$U = \begin{pmatrix} u_0 & \lambda^{-1} u_1 \\ \lambda^{-1} u_2 & -u_0 \end{pmatrix}, \quad V = \begin{pmatrix} 0 & \lambda v_1 \\ \lambda v_2 & 0 \end{pmatrix}$$

In this gauge, the zero curvature equation reduces to:

$$\partial_{x_-} u_0 - u_1 v_2 + v_1 u_2 = 0 \qquad (3.85)$$
$$\partial_{x_-} u_1 = 0, \quad \partial_{x_-} u_2 = 0 \qquad (3.86)$$
$$\partial_{x_+} v_1 - 2v_1 u_0 = 0, \quad \partial_{x_+} v_2 + 2v_2 u_0 = 0 \qquad (3.87)$$

From eq. (3.86) we have $u_1 = \alpha(x_+)$, $u_2 = \beta(x_+)$. We set $u_0 = \partial_{x_+}\varphi$. Then, from eq (3.87) we have $v_1 = \gamma(x_-) \exp 2\varphi$ and $v_2 = \delta(x_-) \exp -2\varphi$. Finally, eq (3.85) becomes :

$$\partial_{x_+} \partial_{x_-} \varphi + \beta(x_+)\gamma(x_-)e^{2\varphi} - \alpha(x_+)\delta(x_-)e^{-2\varphi} = 0$$

This is the sinh-Gordon equation. The arbitrary functions $\alpha(x_+), \beta(x_+)$ and $\gamma(x_-), \delta(x_-)$ are irrelevant: they can be absorbed into a redefinition of the field φ and a change of the coordinates x_+, x_-. Taking them as constants, equal to m, we finally get

$$U = \begin{pmatrix} \partial_{x_+}\varphi & m\lambda^{-1} \\ m\lambda^{-1} & -\partial_{x_+}\varphi \end{pmatrix}; \quad V = \begin{pmatrix} 0 & m\lambda e^{2\varphi} \\ m\lambda e^{-2\varphi} & 0 \end{pmatrix} \quad (3.88)$$

Hence the Lax connection of the sinh-Gordon model is naturally recovered from two-poles systems with $\mathbb{Z}_2$ symmetry. This construction generalizes to other Lie algebras, the reduction group being generated by the Coxeter automorphism, and yields the Toda field theories.

Remark 4. There is a relation between the linear system eq. (3.88) and what is called Bäcklund transformations. These transformations produce new solutions of a non-linear partial differential equation from old ones. Assume that φ satisfies a second order non-linear partial differential equation. The Bäcklund transformation requires that the new function $\widehat{\varphi}$ is obtained by solving a first order system:

$$\partial_{x_+}\widehat{\varphi} = P(\widehat{\varphi}, \varphi, \partial_{x_+}\varphi, \partial_{x_-}\varphi), \quad \partial_{x_-}\widehat{\varphi} = Q(\widehat{\varphi}, \varphi, \partial_{x_+}\varphi, \partial_{x_-}\varphi)$$

Of course the compatibility condition of this system puts strong constraints on P and Q. We say the transformation is auto-Bäcklund if $\widehat{\varphi}$ satisfies the same equation as φ. In the sinh-Gordon case, the transformation is defined by:

$$\partial_{x_+}(\varphi + \widehat{\varphi}) = 2m\lambda^{-1}\sinh(\varphi - \widehat{\varphi}), \quad \partial_{x_-}(\varphi - \widehat{\varphi}) = 2m\lambda\sinh(\varphi + \widehat{\varphi}) \quad (3.89)$$

where λ is an arbitrary parameter. The compatibility condition reads:

$$\begin{aligned}
\partial_{x_+}\partial_{x_-}\widehat{\varphi} &= -\partial_{x_+}\partial_{x_-}\varphi + 4m^2\cosh(\widehat{\varphi} - \varphi)\sinh(\widehat{\varphi} + \varphi) \\
&= \partial_{x_+}\partial_{x_-}\varphi + 4m^2\cosh(\widehat{\varphi} + \varphi)\sinh(\widehat{\varphi} - \varphi)
\end{aligned}$$

and reduces to the sinh-Gordon equation $\partial_{x_+}\partial_{x_-}\varphi = 2m^2\sinh(2\varphi)$. Moreover, we then find $\partial_{x_+}\partial_{x_-}\widehat{\varphi} = 2m^2\sinh(2\widehat{\varphi})$, so if φ solves the sinh-Gordon equation, so does the transformed field $\widehat{\varphi}$.

The relation between eq. (3.89) and the linear system eq. (3.88) is obtained by setting $e^{\widehat{\varphi}} = e^{\varphi}\frac{u}{v}$. The Bäcklund transformation then reads:

$$u(-\partial_{x_+}v + \partial_{x_+}\varphi\, v + m\lambda^{-1}u) + v(\partial_{x_+}u + \partial_{x_+}\varphi\, u - m\lambda^{-1}v) = 0$$
$$u(\partial_{x_-}v - -m\lambda e^{2\varphi}u) + v(-\partial_{x_-}u + m\lambda e^{-2\varphi}v) = 0$$

Requiring the vanishing of the four terms in the parenthesis yields exactly the linear system:

$$(\partial_{x_+} - U)\begin{pmatrix} v \\ u \end{pmatrix} = 0, \quad (\partial_{x_-} - V)\begin{pmatrix} v \\ u \end{pmatrix} = 0$$

where the connection is given in eq. (3.88). Conversely, if we have a solution (u, v) of the linear sytem associated with φ, then $e^{\widehat{\varphi}} = e^{\varphi}\frac{u}{v}$ is a solution of the Bäcklund transformation.

In general the relation between φ and $\widehat{\varphi}$ is non-local. However, when expanding $\widehat{\varphi}$ in formal power of either λ or $1/\lambda$ each term of the infinite series is a local function of φ. This remark can be used to deduce an infinite set of local conserved currents in the sinh-Gordon model. Indeed, from the defining relations (3.89) of the Bäcklund transformation, we see that the current J_{x_+}, J_{x_-} with components:

$$J_{x_+} = \lambda^{-1} \cosh(\varphi - \widehat{\varphi}), \quad J_{x_-} = -\lambda \cosh(\varphi + \widehat{\varphi})$$

is conserved: $\partial_{x_-} J_{x_+} + \partial_{x_+} J_{x_-} = 0$. Expanding it in power series of either λ or $1/\lambda$ gives two infinite series of local conserved currents.

3.9 Poisson brackets of the monodromy matrix

As we just saw, the zero curvature equation leads to the construction of an infinite set of conserved currents. We want to compute the Poisson brackets of the conserved charges associated with these conserved currents. For this we will compute the Poisson brackets of the matrix elements of the monodromy matrix.

In order to do it we assume the existence of an r-matrix relation such that:

$$\{U_1(\lambda, x), U_2(\mu, y)\} = [r_{12}(\lambda - \mu), U_1(\lambda, x) + U_2(\mu, y)]\delta(x - y) \quad (3.90)$$

We assume that r is a non-dynamical r-matrix such as eq. (3.31). We say that the Poisson bracket eq. (3.90) is ultralocal due to the presence of $\delta(x - y)$ only. This hypothesis actually covers a large class of integrable field theories.

Since we are computing Poisson brackets, let us fix the time t, and consider the transport matrix from x to y

$$T(\lambda; y, x) = \overleftarrow{\exp} \left(\int_x^y U(\lambda, z)dz \right)$$

In particular the monodromy matrix is $T(\lambda) = T(\lambda; 2\pi, 0)$. The matrix elements $[T]_{ij}$ of $T(\lambda; y, x)$ are functions on phase space. As in Chapter 2, section (2.5), we use the tensor notation to arrange the table of their Poisson brackets.

Proposition. *If eq. (3.90) holds, we have the fundamental Sklyanin relation for the transport matrix:*

$$\{T_1(\lambda; y, x), T_2(\mu; y, x)\} = [r_{12}(\lambda, \mu), T_1(\lambda; y, x)T_2(\mu; y, x)] \quad (3.91)$$

As a consequence, the traces of powers of the monodromy matrix $H^{(n)}(\lambda) = \mathrm{Tr}\,(T^n(\lambda))$, *generate Poisson commuting quantities:*

$$\{H^{(n)}(\lambda), H^{(m)}(\mu)\} = 0 \quad (3.92)$$

Proof. Let us first prove the relation (3.91) for the Poisson brackets of the transport matrices. Notice that λ is attached to T_1 and μ to T_2, so that there is no ambiguity if we do not write explicitly the λ and μ dependence. The transport matrix $T(y, x)$ verifies the differential equations

$$\partial_x T(y, x) + T(y, x) U(x) = 0 \tag{3.93}$$
$$\partial_y T(y, x) - U(y) T(y, x) = 0$$

Since Poisson brackets satisfy the Leibnitz rules, we have

$$\{T_1(y, x), T_2(y, x)\} = \tag{3.94}$$
$$\int_x^y \int_x^y du\, dv\, T_1(y, u) T_2(y, v) \{U_1(u), U_2(v)\} T_1(u, x) T_2(v, x)$$

Replacing $\{U_1(u), U_2(v)\}$ by eq. (3.90), and using the differential equation satisfied by $T(y, x)$, this yields:

$$\{T_1(y, x), T_2(y, x)\}$$
$$= \int_x^y \int_x^y du\, dv\, \delta(u - v) . \Big(T_1(y, u) T_2(y, v)\, r_{12}\, (\partial_u + \partial_v) T_1(u, x) T_2(v, x)$$
$$+ (\partial_u + \partial_v)(T_1(y, u) T_2(y, v))\, r_{12}\, T_1(u, x) T_2(v, x) \Big)$$
$$= \int_x^y dz\, \partial_z \Big(T_1(y, z) T_2(y, z) . r_{12} . T_1(z, x) T_2(z, x) \Big)$$

Integrating this exact derivative gives the relation (3.91). Let us now show that the trace of the monodromy matrix $H^{(n)}(\lambda)$ generates Poisson commuting quantities. Equation (3.91) implies

$$\{T_1^n(\lambda), T_2^m(\mu)\} = [r_{12}(\lambda, \mu), T_1^n(\lambda) T_2^m(\mu)]$$

We take the trace of this relation. In the left-hand side we use the fact that $\mathrm{Tr}_{12}(A \otimes B) = \mathrm{Tr}(A)\mathrm{Tr}(B)$ and get $\{H^{(n)}(\lambda), H^{(m)}(\mu)\}$. The right-hand side gives zero because it is the trace of a commutator. ∎

Let us emphasize that it is the integration process involved in the transport matrix which leads from the linear Poisson bracket eq. (3.90) to the quadratic Sklyanin Poisson bracket eq. (3.91).

The proposition shows that we may take as Hamiltonian any element of the family generated by $H^{(n)}(\mu)$. We show that the corresponding equations of motion take the form of a zero curvature condition.

Proposition. *Taking $H^{(n)}(\mu)$ as Hamiltonian, we have*

$$\dot{U}(\lambda, x) \equiv \{H^{(n)}(\mu), U(\lambda, x)\} = \partial_x V^{(n)}(\lambda, \mu, x) + [V^{(n)}(\lambda, \mu, x), U(\lambda, x)] \tag{3.95}$$

where

$$V^{(n)}(\lambda, \mu; x) = n \mathrm{Tr}_1 \Big(T_1(\mu; 2\pi, x) r_{12}(\mu, \lambda) T_1(\mu; x, 0) T_1^{n-1}(\mu, 2\pi, 0) \Big)$$

This provides the equations of motion for a hierarchy of times, when we expand in μ.

<u>Proof.</u> To simplify the notation, we do not explicitly write the λ, μ dependence as above, noting that μ is attached to the tensorial index 1 and λ to the tensorial index 2. We have:

$$\{T_1(2\pi, 0), U_2(x)\} = \int_0^{2\pi} dy \, T_1(2\pi, y) \, \{U_1(y), U_2(x)\} \, T_1(y, 0)$$
$$= T_1(2\pi, x) \, [r_{12}, U_1(x) + U_2(x)] \, T_1(x, 0)$$

Expanding the commutator we get four terms

$$\{T_1(2\pi, 0), U_2(x)\} =$$
$$T_1(2\pi, x) \cdot r_{12} \cdot \underbrace{U_1(x) T_1(x, 0)}_{\text{use diff. eq.}} + T_1(2\pi, x) \cdot r_{12} \cdot \underbrace{U_2(x) \, T_1(x, 0)}_{\text{commute}}$$
$$- \underbrace{T_1(2\pi, x) \, U_1(x)}_{\text{use diff. eq.}} \cdot r_{12} \cdot T_1(x, 0) - \underbrace{T_1(2\pi, x) \, U_2(x)}_{\text{commute}} \cdot r_{12} \cdot T_1(x, 0)$$

Using the differential equations (3.93) and commuting factors as indicated gives

$$\{T_1(2\pi, 0), U_2(x)\} = \partial_x V_{12}(x) + [V_{12}(x), U_2(x)]$$

where we have introduced $V_{12}(x) = T_1(2\pi, x) \cdot r_{12} \cdot T_1(x, 0)$. From this we get $\{T_1^n(2\pi, 0), U_2(x)\} = \partial_x V_{12}^{(n)}(x) + [V_{12}^{(n)}(x), U_2(x)]$ with $V_{12}^{(n)}(x) = \sum_i T_1^{n-i-1} V_{12}(x) T_1^i$. Taking the trace over the first space, remembering that $H^{(n)}(\mu) = \mathrm{Tr}\, T^n(\mu)$, and setting $V^{(n)}(\lambda, \mu, x) = \mathrm{Tr}_1 V_{12}^{(n)}(x)$, we find eq. (3.95). ∎

3.10 The group of dressing transformations

We now introduce a very important notion, the group of dressing transformations, which is related to the Zakharov–Shabat construction. These transformations provide a way to construct new solutions of the field equations of motion from old ones. It defines a group action on the space of classical solutions of the model, and therefore on the phase space of the model.

Dressing transformations are special non-local gauge transformations preserving the analytical structure of the Lax connection. These transformations are intimately related to the Riemann–Hilbert problem which we have discussed in the section on factorization.

We choose a contour Γ in the λ-plane such that none of the poles λ_k of the Lax connection are on Γ. We will take for Γ the sum of contours $\Gamma^{(k)}$, each one surrounding a pole λ_k as in the factorization problem.

To define the dressing transformation, we pick a group valued function $g(\lambda) \in \tilde{G}$ on Γ. From the Riemann–Hilbert problem, eqs. (3.49, 3.53), $g(\lambda)$ can be factorized as:

$$g(\lambda) = g_-^{-1}(\lambda)g_+(\lambda)$$

where $g_+(\lambda)$ and $g_-(\lambda)$ are analytic inside and ouside the contour Γ respectively. In the following discussion we assume that $g(\lambda)$ is close enough to the identity so that there are no indices.

Let U, V be a solution of the zero curvature equation eq. (3.67) with the prescribed singularities specified in eqs. (3.73, 3.74). Let $\Psi \equiv \Psi(\lambda; x, t)$ be the solution of the linear system (3.68) normalized by $\Psi(\lambda; 0, 0) = 1$. We set:

$$\theta(\lambda; x, t) = \Psi(\lambda; x, t) \cdot g(\lambda) \cdot \Psi(\lambda; x, t)^{-1} \qquad (3.96)$$

At each space–time point (x, t), we perform a λ decomposition of $\theta(\lambda, x, t)$ according to the Riemann–Hilbert problem as:

$$\theta(\lambda; x, t) = \theta_-^{-1}(\lambda; x, t) \cdot \theta_+(\lambda; x, t) \qquad (3.97)$$

with θ_+ and θ_- analytic inside and outside the contour Γ respectively. Then,

Proposition. *The following function, defined for λ on the contour Γ,*

$$\Psi^g(\lambda; x, t) = \theta_\pm(\lambda; x, t) \cdot \Psi(\lambda; x, t) \cdot g_\pm^{-1}(\lambda) \qquad (3.98)$$

extends to a function Ψ_+^g, defined inside Γ except at the points λ_k where it has essential singularities, and a function Ψ_-^g defined outside Γ. On Γ we have $\Psi_-^{g\,-1}\Psi_+^g|_\Gamma = 1$. So $\Psi_\pm^g$ define a unique function Ψ^g which is normalized by $\Psi^g(\lambda, 0) = 1$ and is a solution of the linear system (3.68) with Lax connection U^g and V^g given by

$$U^g(\lambda; x, t) = \theta_\pm \cdot U \cdot \theta_\pm^{-1} + \partial_x \theta_\pm \cdot \theta_\pm^{-1} \qquad (3.99)$$

$$V^g(\lambda; x, t) = \theta_\pm \cdot V \cdot \theta_\pm^{-1} + \partial_t \theta_\pm \cdot \theta_\pm^{-1} \qquad (3.100)$$

The matrices U^g and V^g, which satisfy the zero curvature equation (3.67), are meromorphic functions on the whole complex λ plane with the same analytic structure as the components $U(\lambda)$ and $V(\lambda)$ of the original Lax connection.

Proof. First it follows directly from the definitions of $g_\pm$ and $\theta_\pm$ that for λ on Γ,

$$\theta_+(\lambda; x, t) \cdot \Psi(\lambda; x, t) \cdot g_+^{-1}(\lambda) = \theta_-(\lambda; x, t) \cdot \Psi(\lambda; x, t) \cdot g_-^{-1}(\lambda)$$

so that the two expressions of the right-hand side of eq. (3.98) with the $+$ and $-$ signs are equal, and effectively define a unique function Ψ^g on Γ. It is clear that this function can be extended into two functions $\Psi_\pm^g$ respectively defined inside and outside this contour by:

$$\Psi_\pm^g = \theta_\pm \cdot \Psi \cdot g_\pm^{-1}$$

These functions have the same essential singularities as Ψ at the points λ_k. By construction, they are such that $\Psi^{g\,-1}_-\Psi^g_+|_\Gamma = 1$.

We may use $\Psi_\pm^g$ to define the Lax connection $U_\pm^g$, $V_\pm^g$ inside and outside the contour Γ. Explicitly:

$$U_\pm^g = \partial_x \Psi_\pm^g \cdot \Psi_\pm^{g-1} = \partial_x \theta_\pm \theta_\pm^{-1} + \theta_\pm U \theta_\pm^{-1}$$
$$V_\pm^g = \partial_t \Psi_\pm^g \cdot \Psi_\pm^{g-1} = \partial_t \theta_\pm \theta_\pm^{-1} + \theta_\pm V \theta_\pm^{-1}$$

Since $\Psi^{g\,-1}_-\Psi^g_+|_\Gamma = 1$ we see that U_+ coincides with U_- on the contour Γ and similarly $V_+ = V_-$ for $\lambda \in \Gamma$ and hence the pairs $U_\pm^g$, $V_\pm^g$ define a conection U^g, V^g on the whole λ-plane. Since $\theta_\pm$ are regular in their respective domains of definition, we see that U^g, V^g have the same singularities as U, V. ∎

This proposition effectively states that the dressing transformations (3.98) map solutions of the equations of motion into new solutions. Given a solution U, V of the zero curvature equation with the prescribed pole structure and an element of the loop group $\tilde{G}$, we produce a new solution of the zero curvature equation with same analytical structure. But, since this analytic structure is the main information which specifies the model, we have produced a new solution of the equations of motion.

Thus, the dressing transformations $\Psi \to \Psi^g$ act on the solution space. They form a group, called the dressing group and denoted by G_R. This group is modeled on $\tilde{G}$ but is not isomorphic to it since its composition law is different. Indeed,

Proposition. *Let $g = g_-^{-1}g_+$ and $h = h_-^{-1}h_+$ be two elements of the dressing group G_R. The composition law of dressing transformations is given by*

$$h \bullet g = (h_-g_-)^{-1}(h_+g_+) \tag{3.101}$$

Representing the elements of the dressing group by the pairs (g_-, g_+) and (h_-, h_+) we may write the composition law as: $(h_-, h_+) \bullet (g_-, g_+) = (h_- g_-, h_+ g_+)$. In particular the plus and minus components commute.

<u>Proof.</u> Consider two elements $g = g_-^{-1} g_+$ and $h = h_-^{-1} h_+$ and transform successively $\Psi \to \Psi^g \to (\Psi^g)^h$; we have:

$$\begin{aligned}
\Psi^g &= \theta_\pm^g \, \Psi \, g_\pm^{-1} &&\text{with} &&\theta_\pm^g = (\Psi g \Psi^{-1})_\pm \\
(\Psi^g)^h &= \theta_\pm^{hg} \, \Psi^g \, h_\pm^{-1} &&\text{with} &&\theta_\pm^{hg} = (\Psi^g h \Psi^{g-1})_\pm \quad (3.102)
\end{aligned}$$

The factorization of $(\Psi^g h \Psi^{g-1})$ can be written as follows:

$$(\theta_-^{hg})^{-1} \theta_+^{hg} \equiv \Psi^g h \Psi^{g-1} = \theta_-^g \, \Psi \, (h_- g_-)^{-1} (h_+ g_+) \, \Psi^{-1} \, \theta_+^{g \,-1}$$

or, equivalently,

$$\theta_\pm^{hg} \theta_\pm^g = \left(\Psi \, (h_- g_-)^{-1} (h_+ g_+) \, \Psi^{-1} \right)_\pm$$

Inserting this formula into eq. (3.102) proves the multiplication law for the dressing transformations. ∎

The Lie algebra $\mathcal{G}_R$ of the dressing group is composed of the two commuting subalgebras $\mathcal{G}_\pm$ that we have introduced in the section on factorization. Recall that $\mathcal{G}_-$ consists of maps $X_-(\lambda)$ extendable ouside Γ, while $\mathcal{G}_+ = \oplus_k \mathcal{G}_k$, $[\mathcal{G}_k, \mathcal{G}_{k'}] = 0$, $k \neq k'$, consists of a collection of maps $X_k(\lambda)$ regular inside $\Gamma^{(k)}$. As a vector space, $\mathcal{G}_R$ is isomorphic to $\tilde{\mathcal{G}}$, but in the dressing Lie algebra, $[\mathcal{G}_-, \mathcal{G}_+] = 0$.

The infinitesimal form of the dressing transformation, eq. (3.98), for any $\tilde{X} = X_+ - X_-$, with $X_\pm \in \mathcal{G}_\pm$ is:

$$\delta_{\tilde{X}} \, \Psi = (\Psi \tilde{X} \Psi^{-1})_\pm \Psi - \Psi \, X_\pm \quad (3.103)$$

We end these general considerations on dressing transformations by clarifying their relation to the Poisson structure of the theory. We shall assume the Poisson bracket eq. (3.90) for the Lax connection with

$$r_{12}(\lambda, \mu) = -\frac{C_{12}}{\lambda - \mu}, \quad C_{12} = \sum_{ij} E_{ij} \otimes E_{ji}$$

The Poisson bracket of the wave function is thus, from eqs. (3.69, 3.91):

$$\{\Psi_1(\lambda; x), \Psi_2(\mu; x)\} = [r_{12}(\lambda, \mu), \Psi_1(\lambda; x) \Psi_2(\mu; x)] \quad (3.104)$$

The r-matrix is related to the factorization problem in the loop algebra $\tilde{\mathcal{G}}$ whose elements are maps $\tilde{X}(\lambda)$ defined on Γ. Recall that $\tilde{X}(\lambda)$ can be decomposed as $\tilde{X} = X_+ - X_-$ with $X_\pm \in \mathcal{G}_\pm$. Its component $X_-(\lambda)$ can be computed by:

$$X_-(\lambda) = \oint_\Gamma \frac{d\mu}{2i\pi} \, \mathrm{Tr}_2 r_{12}(\lambda, \mu) \tilde{X}_2(\mu), \quad \lambda \text{ outside } \Gamma$$

Its component $X_+(\lambda) = (X_1(\lambda), X_2(\lambda), \ldots)$ reads:

$$X_k(\lambda) = \oint_\Gamma \frac{d\mu}{2i\pi} \, \mathrm{Tr}_2 r_{12}(\lambda, \mu) \tilde{X}_2(\mu), \quad \lambda \text{ inside } \Gamma^{(k)}$$

We have to verify that $(X_k - X_-)|_{\Gamma^{(k)}} = \tilde{X}_k$, where $\tilde{X}_k$ denotes the component of $\tilde{X}(\lambda)$ on $\Gamma^{(k)}$. Recalling the formula

$$\frac{1}{x + i0} - \frac{1}{x - i0} = -2i\pi\delta(x)$$

and taking $\lambda_\pm$ be two values of λ pinching the contour $\Gamma^{(k)}$ from inside and outside, we can write:

$$r_{12}(\lambda_+, \mu) - r_{12}(\lambda_-, \mu) = 2i\pi C_{12}\delta(\lambda - \mu) \qquad (3.105)$$

with $\delta(\lambda - \mu)$ the Dirac measure. Then we have:

$$(X_k - X_-)(\lambda)|_{\Gamma^{(k)}} = \sum_l \oint_{\Gamma^{(l)}} \frac{d\mu}{2i\pi} \, \mathrm{Tr}_2 \left((r_{12}(\lambda_+, \mu) - r_{12}(\lambda_-, \mu)) \tilde{X}_{2,l}(\mu) \right)$$

$$= \tilde{X}_k(\lambda)$$

Introducing two maps $R^\pm$ acting on the loop algebra $\tilde{\mathcal{G}}$ by

$$R^\pm(\tilde{X})(\lambda) = X_\pm(\lambda) = \oint_\Gamma \frac{d\mu}{2i\pi} \mathrm{Tr}_2 r_{12}(\lambda_\pm, \mu) \tilde{X}_2(\mu) \qquad (3.106)$$

we have shown that $\tilde{X} = X_+ - X_-$. This is equivalent to $R^+ - R^- = \mathrm{Id}$, the identity operator.

With this result at hand we can spell out the Poisson property of dressing transformations. The action eq. (3.103) does not naively preserve the Poisson brackets eq. (3.104). However, a good Poisson action is recovered if the dressing group itself is equipped with a non-trivial Poisson structure. It is then called a Poisson–Lie group and the action a Lie–Poisson action. The theory of Poisson–Lie groups and Lie–Poisson actions is sketched in Chapter 14. Here we only need to know that infinitesimal actions are generated by so-called non-Abelian Hamiltonians T. This means that there exists a function on phase space, T, taking value in the dual group, such that for any function f on phase space

$$\delta_X f = \langle X, T^{-1}\{T, f\}\rangle \qquad (3.107)$$

Here X is an element of the Lie algebra of the Poisson–Lie group acting on the manifold, and $T^{-1}\{T, f\}$ belongs to the dual of this Lie algebra; $\langle\,\rangle$ is the pairing, see Chapter 14. In the Abelian case, writing $T = \exp\mathcal{P}$, we get $\delta_X f = \langle X, \{\mathcal{P}, f\}\rangle = \{H(X), f\}$, where $H(X) = \mathcal{P}(X)$. This is the standard formula showing that the action is symplectic in this case. In our situation X is an element of the Lie algebra of the dressing group $\mathcal{G}_R = (\mathcal{G}_+, \mathcal{G}_-)$ and T an element of the loop group $\exp\tilde{\mathcal{G}}$.

Proposition. *The action of dressing transformations is a Lie–Poisson action. The non–Abelian generator is the monodromy matrix.*

<u>Proof.</u> Introduce the monodromy matrix $T(\mu) = \Psi(\mu, 2\pi)$. Using the ultralocality property, its Poisson bracket with the wave function is, for $0 \le x \le 2\pi$:

$$\{\Psi_1(\lambda, x), T_2(\mu)\} = T_2(\mu)\Psi_2^{-1}(\mu, x)[r_{12}(\lambda, \mu), \Psi_1(\lambda, x)\Psi_2(\mu, x)]$$

In this formula, we can freely replace λ on a contour $\Gamma^{(k)}$ by either of the two values $\lambda_\pm$ pinching it from inside and outside $\Gamma^{(k)}$. This is because for μ outside $\Gamma^{(k)}$, this replacement has no effect since $r_{12}(\lambda, \mu)$ is regular, while for μ on $\Gamma^{(k)}$, by eq. (3.105), the difference is the product of the Dirac measure $\delta(\lambda - \mu)$ and the commutator $[C_{12}, \Psi_1(\lambda)\Psi_2(\lambda)]$ which vanishes. Therefore, for any $\tilde{X} \in \mathcal{G}_R$ we have:

$$\oint_\Gamma \frac{d\mu}{2i\pi} \, \mathrm{Tr}_2\left(\tilde{X}_2(\mu)T_2^{-1}(\mu)\{\Psi_1(\lambda_\pm), T_2(\mu)\}\right)$$

$$= R^\pm\left(\Psi\tilde{X}\Psi^{-1}\right)(\lambda)\,\Psi(\lambda) - \Psi(\lambda, x)\,R^\pm(\tilde{X})(\lambda)$$

where the two signs $\pm$ give the same answer. Since the maps $R^\pm$ project on the subalgebras $\mathcal{G}_\pm$, see eq. (3.106), this reads:

$$\oint_\Gamma \frac{d\mu}{2i\pi} \, \mathrm{Tr}_2\left(\tilde{X}_2(\mu)T_2^{-1}(\mu)\{\Psi_1(\lambda_\pm, x), T_2(\mu)\}\right) = \delta_{\tilde{X}}\,\Psi(\lambda, x)$$

where $\delta_{\tilde{X}}\,\Psi(\lambda, x)$ is the infinitesimal form of the dressing transformation, eq. (3.103). Comparing with eq. (3.107), this proves that $T(\mu)$ is the non–Abelian Hamiltonian and shows that the action is Lie–Poisson. ∎

3.11 Soliton solutions

In general, a matrix Riemann–Hilbert problem like eq. (3.49) cannot be solved explicitly by analytical methods. This statement applies to the fundamental solution of the Riemann–Hilbert problem, i.e. the one satisfying the conditions $\det\theta_\pm \ne 0$. However, once the fundamental solution is known, new solutions "with zeroes" can easily be constructed from it. This can be used to produce new solutions to the equations of motion. Starting from a trivial vacuum solution, we obtain in this way the so-called soliton solutions.

To define the Riemann–Hilbert with zeroes, we first introduce a definition. We say that a matrix function $\theta(\lambda)$ has a zero at the point λ_0 if $\det \theta(\lambda_0) = 0$ and if in the vicinity of this point

$$\theta(\lambda) = F_0 + (\lambda - \lambda_0)F_1 + O(\lambda - \lambda_0)^2, \quad \theta^{-1}(\lambda) = \frac{1}{\lambda - \lambda_0}C_0 + C_1 + O(\lambda - \lambda_0)$$

Since $\theta(\lambda)\theta^{-1}(\lambda) = \theta^{-1}(\lambda)\theta(\lambda) = \mathrm{Id}$, we have $F_0 C_0 = C_0 F_0 = 0$, $C_0 F_1 + C_1 F_0 = \mathrm{Id}$. In particular

$$\mathrm{Ker}\, F_0 = \mathrm{Im}\, C_0, \quad \mathrm{Ker}\, C_0 = \mathrm{Im}\, F_0$$

Let now Γ be a closed contour in the λ-plane. As in the previous section Γ could be a sum of small contours $\Gamma^{(k)}$ around each point λ_k, but here we can also consider more general contours provided no point λ_k sits on them. Let $g(\lambda)$ be a matrix defined on Γ, and consider the Riemann–Hilbert problem

$$g(\lambda) = \theta_-^{-1}(\lambda)\theta_+(\lambda) \qquad (3.108)$$

where $\theta_+(\lambda)$ is analytic inside Γ and has N zeroes located at the points $\mu_1, \ldots, \mu_N$, and $\theta_-^{-1}(\lambda)$ is analytic outside Γ and has N zeroes at the points $\lambda_1, \ldots, \lambda_N$. We emphasize that it is $\theta_-^{-1}(\lambda)$ which has zeroes, and not $\theta_-(\lambda)$. Let us fix the two set of subspaces:

$$\mathcal{V}_n = \mathrm{Im}\, \theta_-^{-1}(\lambda)|_{\lambda = \lambda_n}, \quad \mathcal{W}_n = \mathrm{Ker}\, \theta_+(\lambda)|_{\lambda = \mu_n}$$

Then we have:

Proposition. *The choice of the subspaces $\mathcal{V}_n, \mathcal{W}_n$, specifies uniquely the solution of the factorization problem eq. (3.108), up to a left multiplication by a* constant *matrix. This factorization problem is called a Riemann–Hilbert problem with zeroes.*

<u>Proof.</u> Suppose $\theta_\pm$ and $\tilde{\theta}_\pm$ are two solutions of the Riemann–Hilbert problem. Then the function $\chi = \tilde{\theta}_-\theta_-^{-1} = \tilde{\theta}_+\theta_+^{-1}$ is a meromorphic function in the whole complex plane. Its possible poles are located at λ_n, μ_n. But around such a pole, let us say μ_n, we have

$$\theta_+(\lambda) = F_n + O(\lambda - \mu_n), \quad \theta_+^{-1}(\lambda) = \frac{1}{\lambda - \mu_n}C_n + O(1),$$

$$\tilde{\theta}_+(\lambda) = \tilde{F}_n + O(\lambda - \mu_n), \quad \tilde{\theta}_+^{-1}(\lambda) = \frac{1}{\lambda - \mu_n}\tilde{C}_n + O(1)$$

Since $\operatorname{Ker} F_n = \mathcal{W}_n = \operatorname{Im} C_n$, $\operatorname{Ker} \tilde{F}_n = \mathcal{W}_n = \operatorname{Im} \tilde{C}_n$, we see that around $\lambda = \mu_n$, $\chi = \frac{1}{\lambda - \mu_n} \tilde{F}_n C_n + O(1)$ is regular because $\tilde{F}_n C_n = 0$. The same analysis holds true at λ_n, and so $\chi(\lambda)$ is regular everywhere, hence a constant. ∎

There is a simple way to construct the solution of the Riemann–Hilbert problem with zeroes, from its fundamental solution. We begin by adding a pair of zeroes at (μ_N, λ_N).

Proposition. *Let θ_+ and θ_-^{-1} be the solution of the Riemann–Hilbert problem eq. (3.108) with zeroes at μ_N and λ_N respectively such that $\mathcal{W}_N = \operatorname{Ker} \theta_+(\mu_N)$ and $\mathcal{V}_N = \operatorname{Im} \theta_-^{-1}(\lambda_N)$ are fixed. Let $\tilde{\theta}_\pm$ be a solution of the Riemann–Hilbert problem, without zeroes at these points. Then*

$$\theta_+(\lambda) = \chi_0^{-1}\left(1 - \frac{\mu_N - \lambda_N}{\lambda - \lambda_N} P_N\right) \tilde{\theta}_+(\lambda)$$

$$\theta_-^{-1}(\lambda) = \tilde{\theta}_-^{-1}(\lambda)\left(1 - \frac{\lambda_N - \mu_N}{\lambda - \mu_N} P_N\right) \chi_0 \qquad (3.109)$$

where χ_0 is a constant matrix and P_N is a projector such that

$$\operatorname{Im} P_N = \tilde{\theta}_+(\mu_N)\mathcal{W}_N, \quad \operatorname{Ker} P_N = \tilde{\theta}_-(\lambda_N)\mathcal{V}_N$$

Proof. Introduce the matrix $\chi(\lambda) = \tilde{\theta}_- \theta_-^{-1} = \tilde{\theta}_+ \theta_+^{-1}$ as above. It is a meromorphic function in the λ-plane with possible simple poles at λ_N and μ_N, so we can parametrize χ as

$$\chi = \chi_0 + \frac{1}{\lambda - \mu_N}\chi_1, \quad \chi^{-1} = \chi_2 + \frac{1}{\lambda - \lambda_N}\chi_3$$

From the condition $\chi(\lambda)\chi^{-1}(\lambda) = 1$, we find $\chi_2 = \chi_0^{-1}$ and the two relations

$$\chi_1\left(\chi_0^{-1} + \frac{1}{\mu_N - \lambda_N}\chi_3\right) = 0, \quad \left(\chi_0 + \frac{1}{\lambda_N - \mu_N}\chi_1\right)\chi_3 = 0$$

Adding these two equations gives $\chi_0 \chi_3 = -\chi_1 \chi_0^{-1}$. Let us define

$$P_N = \frac{\chi_0 \chi_3}{\lambda_N - \mu_N} = -\frac{\chi_1 \chi_0^{-1}}{\lambda_N - \mu_N}$$

This matrix is a projector since $P_N^2 = P_N$, because $\chi_1 \chi_3 = (\lambda_N - \mu_N)\chi_1 \chi_0^{-1}$. We can rewrite $\chi(\lambda)$ in terms of P_N:

$$\chi(\lambda) = \left(1 - \frac{\lambda_N - \mu_N}{\lambda - \mu_N}P_N\right)\chi_0; \quad \chi^{-1}(\lambda) = \chi_0^{-1}\left(1 - \frac{\mu_N - \lambda_N}{\lambda - \mu_N}P_N\right)$$

from which eqs. (3.109) follow. Next we demand $\mathcal{W}_N = \text{Ker}\,\theta_+(\mu_N) = \text{Ker}\left((1-P_N)\tilde{\theta}_+(\mu_N)\right)$ so that $(1-P_N)\tilde{\theta}_+(\mu_N)\mathcal{W}_N = 0$ or $\text{Ker}\,(1-P_N) = \text{Im}\,P_N = \tilde{\theta}_+(\mu_N)\mathcal{W}_N$. Similarly, $\mathcal{V}_N = \text{Im}\,\theta_-^{-1}(\lambda_N) = \text{Im}\,\tilde{\theta}_-^{-1}(\lambda_N)(1-P_N)\chi_0$, or what is the same, $\mathcal{V}_N = \tilde{\theta}_-^{-1}(\lambda_N)\text{Ker}\,P_N$, so that $\text{Ker}\,P_N = \tilde{\theta}_-(\lambda_N)\mathcal{V}_N$. $\blacksquare$

Repeated applications of this result allows one to build a solution of the Riemann–Hilbert problem with zeroes at $\mu_1,\ldots,\mu_N,\lambda_1,\ldots,\lambda_N$.

$$\theta_+(\lambda) = \chi_N^{-1}\left(1 - \frac{\mu_N - \lambda_N}{\lambda - \lambda_N}P_N\right)\cdots\chi_1^{-1}\left(1 - \frac{\mu_1 - \lambda_1}{\lambda - \lambda_1}P_1\right)\tilde{\theta}_+(\lambda)$$

$$\theta_-^{-1}(\lambda) = \tilde{\theta}_-^{-1}(\lambda)\left(1 - \frac{\lambda_1 - \mu_1}{\lambda - \mu_1}P_1\right)\chi_1\cdots\left(1 - \frac{\lambda_N - \mu_N}{\lambda - \mu_N}P_N\right)\chi_N$$

Here $\tilde{\theta}_\pm(\lambda)$ refers to the fundamental solution of the Riemann–Hilbert problem.

We now extend the method of dressing transformations to the case of a Riemann–Hilbert problem with zeroes.

Proposition. *Given a Lax connection satisfying the zero curvature condition and the associated wave function $\Psi(\lambda, x, t)$, and given vector spaces $\mathcal{V}_n(0), \mathcal{W}_n(0)$, we define*

$$\mathcal{V}_n(x,t) = \Psi(\lambda_n, x, t)\mathcal{V}_n(0), \quad \mathcal{W}_n(x,t) = \Psi(\mu_n, x, t)\mathcal{W}_n(0) \qquad (3.110)$$

We use $\mathcal{V}_n(x,t), \mathcal{W}_n(x,t)$ to define a Riemann–Hilbert problem with zeroes at $\lambda_1, \lambda_2, \ldots$ and $\mu_1, \mu_2, \ldots$. Then for any $g(\lambda) = g_-^{-1}(\lambda)g_+(\lambda)$ on Γ, the transformation $\Psi \to \Psi^g$,

$$\Psi^g = \theta_\pm \Psi g_\pm^{-1}, \quad \theta_-^{-1}\theta_+ = \Psi^{-1}g\Psi$$

is a dressing transformation, i.e. preserves the analytic structure of the Lax connection.

<u>Proof.</u> We start with the linear system

$$(\partial_x - U(\lambda, x, t))\Psi = 0, \quad (\partial_t - V(\lambda, x, t))\Psi = 0$$

and dress it with a solution with zeroes of the Riemann–Hilbert problem, according to eqs. (3.99, 3.100):

$$U^g = \theta_\pm \cdot U \cdot \theta_\pm^{-1} + \partial_x\theta_\pm \cdot \theta_\pm^{-1}, \quad V^g = \theta_\pm \cdot V \cdot \theta_\pm^{-1} + \partial_t\theta_\pm \cdot \theta_\pm^{-1}$$

In general, the components of the dressed connection will have simple poles at the points μ_n, λ_n. We must require that the residues of these poles vanish. At $\lambda = \mu_n$ we have

$$\theta_+(\lambda) = F_n + O(\lambda - \mu_n), \; \theta_+^{-1}(\lambda) = \frac{1}{\lambda - \mu_n} C_n + O(1)$$

Requiring that the residue at $\lambda = \mu_n$ vanishes yields

$$F_n(\partial_x - U|_{\lambda=\mu_n})C_n = 0, \quad F_n(\partial_t - V|_{\lambda=\mu_n})C_n = 0$$

This means that the space $\mathcal{W}_n = \operatorname{Ker} F_n = \operatorname{Im} C_n$ should be invariant under the action of the operators $\partial_x - U|_{\lambda=\mu_n}$ and $\partial_t - V|_{\lambda=\mu_n}$. Similarly, at $\lambda = \lambda_n$ we have

$$\theta_-^{-1}(\lambda) = F_n + O(\lambda - \lambda_n), \; \theta_-(\lambda) = \frac{1}{\lambda - \lambda_n} C_n + O(1)$$

and setting the residues to zero gives

$$C_n(\partial_x - U|_{\lambda=\lambda_n})F_n = 0, \quad C_n(\partial_t - V|_{\lambda=\lambda_n})F_n = 0$$

This means that the space $\mathcal{V}_n = \operatorname{Im} F_n = \operatorname{Ker} C_n$ should be invariant under the action of the operators $\partial_x - U|_{\lambda=\lambda_n}$ and $\partial_t - V|_{\lambda=\lambda_n}$. The simplest solution is to choose the spaces $\mathcal{V}_n(x,t), \mathcal{W}_n(x,t)$ as in eq. (3.110). ∎

The interest of this procedure is that it yields non-trivial results even if the Riemann–Hilbert problem is trivial, i.e. $g(\lambda) = \operatorname{Id}$. Then its fundamental solution is also trivial, $\tilde{\theta}_\pm(\lambda) = \operatorname{Id}$, and the solutions with zeroes are constructed by purely algebraic means. The resulting $\theta_\pm(\lambda)$ are rational functions of λ.

To make this method effective, we need a simple solution of the zero curvature condition $\partial_t U - \partial_x V - [V, U] = 0$ to start with. Simple solutions can be found in the form

$$U = A(\lambda, x), \quad V = B(\lambda, t), \quad [A, B] = 0$$

Then $\Psi = \exp\left(\int_0^x A dx + \int_0^t B dt\right)$. The solutions obtained by dressing this simple type of solutions by the trivial Riemann–Hilbert problem with zeroes are called soliton solutions.

Example. Let us illustrate this construction on the non-linear sigma model. There, one has (see eq. (3.83))

$$U = \frac{1}{\lambda - 1} J_x, \quad V = -\frac{1}{\lambda + 1} J_t$$

We consider the case of 2×2 matrices. The non-linear sigma model field is related to J_x, J_t, by $J_x = g^{-1}\partial_x g, J_t = g^{-1}\partial_t g$. The matrix g can be easily related to the solution Ψ of the linear system:

$$g = \Psi^{-1}(\lambda = 0, x, t)$$

A simple solution of the equations of motion is

$$J_x = a\sigma_3, \quad J_t = b\sigma_3, \quad \Psi_0 = e^{\left[\left(\frac{ax}{\lambda-1} - \frac{bt}{\lambda+1}\right)\sigma_3\right]}, \quad g_0 = e^{-(ax+bt)\sigma_3}$$

We want to dress this solution by solving a Riemann–Hilbert problem with zeroes at λ_1 and μ_1. According to the general formulae, we have

$$\theta_+(\lambda) = \chi_0^{-1}\left(1 - \frac{\mu_1 - \lambda_1}{\lambda - \lambda_1}P\right), \quad \theta_-^{-1}(\lambda) = \left(1 - \frac{\lambda_1 - \mu_1}{\lambda - \mu_1}P\right)\chi_0$$

with χ_0 an constant matrix and P a projector. It can be parametrized by two vectors w and n:

$$P_{ij} = \frac{w_i n_j}{w \cdot n}$$

The spaces $\mathcal{V}_1$ and $\mathcal{W}_1$ are defined by

$$\mathcal{V}_1 = \mathrm{Im}\,\theta_-^{-1}(\lambda)|_{\lambda=\lambda_1} = \mathrm{Im}\,(1 - P)\chi_0 = \mathrm{Ker}\,P$$
$$\mathcal{W}_1 = \mathrm{Ker}\,\theta_+(\lambda)|_{\lambda=\mu_1} = \mathrm{Ker}\,(1 - P)\chi_0 = \mathrm{Im}\,P$$

Hence $\mathcal{W}_1$ is spanned by the vector w, and $\mathcal{V}_1$ is spanned by the vector $n^{\perp}$ perpendicular to n. The invariance properties eq. (3.110) are ensured if we set

$$w(x, t) = \Psi(x, t, \mu_1)w(0), \quad n^{\perp}(x, t) = \Psi(x, t, \lambda_1)n^{\perp}(0),$$

In components, this reads

$$w_1(x, t) = e^{\left(\frac{ax}{\mu_1-1} - \frac{bt}{\mu_1+1}\right)}w_1(0), \quad w_2(x, t) = e^{-\left(\frac{ax}{\mu_1-1} - \frac{bt}{\mu_1+1}\right)}w_2(0)$$

and

$$n_1(x, t) = e^{-\left(\frac{ax}{\lambda_1-1} - \frac{bt}{\lambda_1+1}\right)}n_1(0), \quad n_2(x, t) = e^{\left(\frac{ax}{\lambda_1-1} - \frac{bt}{\lambda_1+1}\right)}n_2(0)$$

These formulae completely determine the projector $P(x, t)$. The dressed field is then reconstructed by

$$g(x, t) = \Psi^{-1}|_{\lambda=0} = \Psi_0^{-1}|_{\lambda=0}\,\theta_-^{-1}|_{\lambda=0} = g_0(x, t)\left(1 + \frac{\lambda_1 - \mu_1}{\mu_1}P(x, t)\right)\chi_0$$

The generalization to the N-soliton case is clear.

References

[1] I.M. Gelfand and L.A. Dickey, Fractional powers of operators and Hamiltonian systems. *Funct. Anal. Appl.* **10** (1976) 259.

[2] V.E. Zakharov and A.B. Shabat, Integration of Non Linear Equations of Mathematical Physics by the Method of Inverse Scattering. II *Funct. Anal. Appl.* **13** (1979) 166.

[3] M. Adler, On a trace functional for formal pseudodifferential operators and symplectic structure of the Korteweg–de Vries type equations. *Inv. Math.* **50** (1979) 219.

[4] M. Jimbo and T. Miwa, Monodromy preserving deformation of linear ordinary differential equations with rational coefficients. III *Physica D* **4** (1981) 26–46.

[5] M. Semenov-Tian-Shansky. What is a classical r-matrix? *Funct. Anal. Appl.* **17** 4 (1983) 17.

[6] L.D. Faddeev, *Integrable Models in 1 + 1 Dimensional Quantum Field Theory.* Les Houches Lectures 1982. Elsevier Science Publishers (1984).

[7] M. Semenov-Tian-Shansky, Dressing transformations and Poisson group actions. *Publ. RIMS* **21** (1985) 1237.

[8] L.D. Faddeev and L.A. Takhtajan, *Hamiltonian Methods in the Theory of Solitons.* Springer (1986).

[9] L.A. Dickey, *Soliton Equations and Hamiltonian Systems.* World Scientific (1991).

[10] L.A. Dickey, On the τ–function of matrix hierarchies of integrable equations. *J. Math. Phys.* **32** (1991) 2996–3002.

4

Algebraic methods

We abstract the group theoretical settings of integrable systems. In this framework, the Lax matrix is viewed as a coadjoint orbit of a Lie algebra $\mathcal{G}$. When the r-matrix is non-dynamical the Jacobi identity simplifies and one can use it to define a second Lie algebra structure on $\mathcal{G}$. Hence $\mathcal{G}$ has a structure of Lie bi-algebra, and conversely, such a structure defines an r-matrix. One can then build dynamical systems admitting Lax representations and conserved quantities in involution. Furthermore, the solution of the equations of motion is reduced to a factorization problem in Lie group theory. We illustrate these constructions in the case of finite-dimensional Lie groups, with the open Toda chain model which we solve completely by algebraic methods. Finally, we demonstrate the versatility of the algebraic setting by constructing the Lax pair with spectral parameter for the Kowalevski top.

4.1 The classical and modified Yang–Baxter equations

In Chapter 2, we have computed the Poisson brackets of the entries L_{ij} of the Lax matrix $L = \sum_{ij} L_{ij} E_{ij}$, and have shown that they can be expressed with an r-matrix (see section (2.5) in Chapter 2):

$$\{L_1, L_2\} = [r_{12}, L_1] - [r_{21}, L_2] \qquad (4.1)$$

with $r_{12} = \sum_{ij,kl} r_{ij,kl} E_{ij} \otimes E_{kl}$, and E_{ij} is the canonical basis of $gl(N)$. So r_{12} can be viewed as an element of $gl(N) \otimes gl(N)$.

We can generalize this setup immediately by considering a Lie algebra $\mathcal{G}$ equipped with a non-degenerate invariant scalar product (,), also denoted by Tr ( ). We will use a basis $\{T_a\}$ of $\mathcal{G}$, and denote the matrix of scalar products by $g_{ab} = (T_a, T_b) = \text{Tr}\,(T_a T_b)$ and its inverse by g^{ab}.

The proper interpretation of the Lax matrix L is as an element of $\mathcal{G}^*$, i.e. a linear form on $\mathcal{G}$. It can also be viewed as an element of $\mathcal{G}$, since the invariant scalar product allows us to identify $\mathcal{G}$ and its dual $\mathcal{G}^*$:

$$X \in \mathcal{G} \longrightarrow L(X) \equiv (L, X)$$

With L viewed as an element of $\mathcal{G}$, eq. (4.1) shows that r_{12} is an element of $\mathcal{G} \otimes \mathcal{G}$.

The nice structural aspects, however, appear when one interprets the r-matrix as a linear map $R : \mathcal{G} \longrightarrow \mathcal{G}$. If $r_{12} = \sum_{ab} r^{ab} T_a \otimes T_b$, then

$$R(X) = \sum_{ab} r^{ab} T_a (T_b, X) = \mathrm{Tr}_2 \left(r_{12} X_2 \right) \qquad (4.2)$$

The r-matrix relation, eq. (4.1), can be presented in a dual form:

$$\{L(X), L(Y)\} = L([X, Y]_R) \qquad (4.3)$$

with the R-bracket, $[\,,\,]_R$, defined as

$$[X, Y]_R = [R(X), Y] + [X, R(Y)] \qquad (4.4)$$

To prove these formulae, take the invariant scalar product of both sides of eq. (4.1) by $X \otimes Y$. On the left-hand side we get $\{L(X), L(Y)\}$, while on the right-hand side we get $([R(Y), L], X) - ([R(X), L], Y)$, which is equal to $(L, [X, Y]_R)$ by invariance of the scalar product.

In this dual formalism, the Jacobi identity, eq. (2.12) in Chapter 2, becomes an equation on R:

$$L([X, J(Y, Z)] + \text{cyc.perm.}) = 0 \qquad (4.5)$$

where cyc.perm. means cyclic permutation of (X, Y, Z) and the quantity $J(Y, Z)$ is defined as

$$J(Y, Z) = \{L(Y), R(Z)\} - \{L(Z), R(Y)\} + [R(Y), R(Z)] - R([Y, Z]_R)$$

If the R matrix is a constant on phase space (independent of the dynamical variables), the Jacobi identity becomes:

$$L([X, [R(Y), R(Z)] - R([Y, Z]_R)] + \text{cyc.perm.}) = 0$$

A particular way to fulfil this equation is to set

$$[R(X), R(Y)] - R([X, R(Y)] + [R(X), Y]) = -\frac{1}{4}[X, Y] \qquad (4.6)$$

so that it reduces to the Jacobi identity in $\mathcal{G}$. Equation (4.6) is called the *modified* Yang–Baxter equation and will be extensively studied below. The factor $1/4$ is conventional and can be changed by a rescaling of R.

Example. The simplest example of an equation of the type eq. (4.3) is provided by the case in which the map R is proportional to the identity map. It corresponds to the Kostant–Kirillov bracket on $\mathcal{G}^*$, which is defined by:

$$\{L(X), L(Y)\}_K = L([X,Y])$$

for any X, Y in $\mathcal{G}$. Comparing with eq. (4.3) we see that this bracket corresponds to $R_K = 1/2\,\mathrm{Id}$. The modified Yang–Baxter equation (4.6) is satisfied with this value of R_K. Under dualization, the Poisson bracket can be written in the form (4.1) with $r_{12} = \frac{1}{2}C_{12}$, where C_{12} is the tensor Casimir of $\mathcal{G}$,

$$C_{12} = \sum_{ab} g^{ab} T_a \otimes T_b \tag{4.7}$$

Recall that the tensor Casimir has the two main properties:

$$[C_{12}, X_1 + X_2] = 0, \quad X_1 = \mathrm{Tr}_2(C_{12}X_2), \quad X \in \mathcal{G}$$

where we used the tensorial notations of section (2.5) in Chapter 2.

Before studying the modified Yang–Baxter equation, we would like to explain its relation with another important equation appearing in this domain: the *classical* Yang–Baxter equation. For any r in $\mathcal{G} \otimes \mathcal{G}$ it reads:

$$[r_{12}, r_{13}] + [r_{12}, r_{23}] + [r_{13}, r_{23}] = 0 \tag{4.8}$$

This equation is important because it is the classical limit of the quantum Yang–Baxter equation, which is one of the main tools in the study of many quantum integrable models. In dualized form eq. (4.8) reads

$$[R(X), R(Y)] - R([X, R(Y)] - [{}^tR(X), Y]) = 0 \tag{4.9}$$

where we defined ${}^tR(X) = \mathrm{Tr}_2(r_{21}X_2)$. Please, notice the subtle difference between the left-hand sides of eq. (4.6) and eq. (4.9). The two expressions agree if ${}^tR = -R$, i.e. if the r-matrix is antisymmetric: $r_{12} = -r_{21}$.

The relation between the solutions of the modified Yang–Baxter equation and of the classical Yang–Baxter equation is as follows:

Proposition. *Let R be an antisymmetric solution of the modified Yang–Baxter equation, then $R^{\pm} = R \pm \frac{1}{2}\,\mathrm{Id}$ both satisfy the classical Yang–Baxter equation.*

<u>Proof.</u> We compute

$$R^{\pm}([X,Y]_R) = R([X,Y]_R) \pm \tfrac{1}{2}[X,Y]_R = [R^{\pm}(X), R^{\pm}(Y)] \qquad (4.10)$$

The statement follows from the observation that

$$[X,Y]_R = [R^{\mp}(X), Y] + [X, R^{\pm}(Y)]$$

and the fact that $R^{\mp} = -{}^t R^{\pm}$ when R is antisymmetric. ∎

Viewed as an element of $\mathcal{G} \otimes \mathcal{G}$, $R^{\pm}$ corresponds to $r_{12}^{\pm}$ with

$$r_{12}^{\pm} = r_{12} \pm \tfrac{1}{2} C_{12} \qquad (4.11)$$

where r_{12} is the dualized form of the antisymmetric solution of the modified Yang–Baxter equation, and C_{12} is the tensor Casimir element in $\mathcal{G} \otimes \mathcal{G}$.

4.2 Algebraic meaning of the classical Yang–Baxter equations

We now undertake an abstract study of the bracket eq. (4.3), for a generic linear mapping $R : \mathcal{G} \longrightarrow \mathcal{G}$ solution of the modified Yang–Baxter equation. This will naturally lead us to introduce a factorization problem in the Lie group associated with $\mathcal{G}$.

Proposition. *Let R be a solution of the modified Yang–Baxter equation (4.6), then the antisymmetric bracket $[\, , \,]_R$ satisfies the Jacobi identity. It thus defines a second Lie algebra structure on $\mathcal{G}$.*

<u>Proof.</u> We have to prove that the bracket $[\, , \,]_R$ satisfies the Jacobi identity:

$$[X, [Y,Z]_R]_R + [Z, [X,Y]_R]_R + [Y, [Z,X]_R]_R = 0$$

Expanding the external R-brackets, we get:

$$[R(X), [Y,Z]_R] + [R(Z), [X,Y]_R] + [R(Y), [Z,X]_R]$$
$$+ [X, R([Y,Z]_R)] + [Z, R([X,Y]_R)] + [Y, R([Z,X]_R)] = 0$$

Developing the terms in the first line and using the Jacobi identity on the Lie bracket $[\, , \,]$, we can rewrite them as

$$-[X, [R(Y), R(Z)]] - [Y, [R(Z), R(X)]] - [Z, [R(X), R(Y)]]$$

They combine with the terms in the second line to yield:

$$[X, [R(Y), R(Z)] - R([Y,Z]_R)] + \text{cyclic permutations} = 0$$

This is satisfied if R is a solution of the modified Yang–Baxter equation because the original Lie bracket $[\ ,\ ]$ obeys the Jacobi identity. ∎

We are then in a very special situation where the vector space $\mathcal{G}$ is equipped with two Lie algebra structures defined by the two brackets $[\ ,\ ]$ and $[\ ,\ ]_R$. This is called a Lie bi-algebra. We will denote by $\mathcal{G}_R$ the Lie algebra with underlying vector space $\mathcal{G}$ but with Lie bracket $[\ ,\ ]_R$.

The relation between these two Lie algebra structures is described by the following:

Proposition. *Let us define $R^\pm = R \pm \frac{1}{2}\mathrm{Id}$, $\mathcal{K}_\pm = \mathrm{Ker}\,(R^\mp)$ and $\mathcal{G}_\pm = \mathrm{Im}\,(R^\pm)$, then*
(i) $(R^\pm) : \mathcal{G}_R \longrightarrow \mathcal{G}$ are Lie algebra homomorphisms,
(ii) $\mathcal{G}_\pm \subset \mathcal{G}$ are Lie subalgebras of $\mathcal{G}$,
(iii) $\mathcal{K}_\pm \subset \mathcal{G}_R$ are ideals of $\mathcal{G}_R$ and $\mathcal{G}_\pm \simeq \mathcal{G}_R/\mathcal{K}_\mp$.

Proof. This proposition is a straightforward consequence of the modified Yang–Baxter equation written as in eq. (4.10). ∎

Given the maps R^+ and R^-, we thus construct two subalgebras $\mathcal{G}_\pm \in \mathcal{G}$. Since $R^+ - R^- = \mathrm{Id}$, any element $X \in \mathcal{G}$ can be written, perhaps not uniquely, as

$$X = X_+ - X_- \quad \text{with} \quad X_\pm = R^\pm(X) \in \mathcal{G}_\pm = \mathrm{Im}\,(R^\pm) \qquad (4.12)$$

Note that the bracket $[\ ,\ ]_R$ takes the simple form:

$$[X,Y]_R = [X_+,Y_+] - [X_-,Y_-] \qquad (4.13)$$

We define the Lie algebra $\mathcal{G}_+ \oplus \mathcal{G}_-$ as the Cartesian product of $\mathcal{G}_+$ and $\mathcal{G}_-$ in which $[\mathcal{G}_+,\mathcal{G}_-] = 0$. We can embed $\mathcal{G}_R$ into $\mathcal{G}_+ \oplus \mathcal{G}_-$ by the map $X \to (R^+(X), R^-(X))$. From eq. (4.13) we see that $\mathcal{G}_R$ is a subalgebra of $\mathcal{G}_+ \oplus \mathcal{G}_-$ through this embedding. The question then arises to determine the image $\tilde{\mathcal{G}}_R$ of $\mathcal{G}_R$ in $\mathcal{G}_+ \oplus \mathcal{G}_-$. This is the object of the next two propositions.

Proposition.
(i) $\mathcal{K}_\pm \subset \mathcal{G}_\pm$ are ideals in $\mathcal{G}_\pm$, hence $\mathcal{G}_\pm/\mathcal{K}_\pm$ are Lie algebras.
(ii)   The mapping $\nu : \mathcal{G}_+/\mathcal{K}_+ \longrightarrow \mathcal{G}_-/\mathcal{K}_-$ defined by $\nu : R^+X \longrightarrow R^-X$ is a Lie algebra isomorphism.

Proof. To prove the first part of the proposition, we remark that $\mathcal{K}_\pm \subset \mathcal{G}_\pm$ since, for $X \in \mathcal{K}_\pm$, we have by definition $RX = \pm\frac{1}{2}X$, so that $X = \pm R^\pm X$. For the same reason, on $\mathcal{K}_\pm$ we have $[\ ,\ ]_R = \pm[\ ,\ ]$ and $\mathcal{K}_\pm$ are indeed subalgebras in $\mathcal{G}_\pm$. To prove that they are ideals, we consider $X \in \mathcal{K}_\pm$, $Y \in \mathcal{G}_\pm$. We can write $Y = R^\pm Z$, then

$$[X,Y] = [X, R^\pm Z] = [X, RZ] \pm \tfrac{1}{2}[X,Z] = [X,RZ] + [RX,Z] = [X,Z]_R$$

but $[X, Z]_R \in \mathcal{K}_\pm$ since $\mathcal{K}_\pm$ are ideals in $\mathcal{G}_R$.
To prove the second part let us denote by $\overline{X}_\pm$ the equivalence class

$$\overline{X}_\pm = R^\pm X \ [\mathrm{mod}\ \mathcal{K}_\pm]$$

First $\nu : \mathcal{G}_+/\mathcal{K}_+ \longrightarrow \mathcal{G}_-/\mathcal{K}_-$ is well-defined, since an element of $\mathcal{K}_+$ is mapped to 0. The mapping ν is surjective because it is induced by the surjective mapping $\mathcal{G}_+ \to \mathcal{G}_-$ given by $R^+X \to R^-X$. It is injective because if $R^-X \in \mathcal{K}_-$ one has $R^+(X) = 0$ by definition of $\mathcal{K}_-$. Finally we prove that ν is a Lie algebra isomorphism, i.e: $[\nu(x), \nu(y)] = \nu[x, y]$ for any x, y of the form $x = \overline{R^+X}$, $y = \overline{R^+Y}$. Recalling that $[R^\pm X, R^\pm Y] = R^\pm[X, Y]_R$ and the definition of the Lie algebra bracket on $\mathcal{G}_\pm/\mathcal{K}_\pm$, we have:

$$\nu([\overline{R^+X}, \overline{R^+Y}]) = \nu(\overline{R^+[X, Y]_R}) = \overline{R^-[X, Y]_R}$$
$$= [\overline{R^-X}, \overline{R^-Y}] = [\nu(\overline{R^+X}), \nu(\overline{R^+Y})]$$

∎

Finally, we have the following important result:

Proposition. *Consider the two maps:*

$$\mathcal{G}_R \xrightarrow{R^+, R^-} \mathcal{G}_+ \oplus \mathcal{G}_- \xrightarrow{1, -1} \mathcal{G}$$

that is $X \to (R^+(X), R^-(X))$ and $(X_+, X_-) \to X_+ - X_-$ respectively. The first map is a Lie algebra injective homomorphism. Let $\tilde{\mathcal{G}}_R$ be its image. The second map, when restricted to $\tilde{\mathcal{G}}_R$, is bijective. Finally, $\tilde{\mathcal{G}}_R$ is characterized by the set of couples (R^+X, R^-Y) such that $\nu(\overline{R^+X}) = \overline{R^-Y}$.

Proof. The first map is injective and the second one is surjective because $R^+ - R^- = \mathrm{Id}$. The elements of $\tilde{\mathcal{G}}_R$ are of the form (R^+X, R^-X) hence satisfy the condition $\nu(\overline{R^+X}) = \overline{R^-X}$. Conversely, let us start from a pair (R^+X, R^-Y) such that $\nu(\overline{R^+X}) = \overline{R^-Y}$. This means $\overline{R^-Y} = \overline{R^-X}$, hence there exists $K_- \in \mathcal{K}_-$ such that $R^-(X - Y) = K_- = -R^-K_-$. It follows that $(X - Y + K_-) = K_+$ belongs to $\mathrm{Ker}\,R^- = \mathcal{K}_+$. Then we have $R^+X = R^+\tilde{X}$, $R^-Y = R^-\tilde{X}$ with $\tilde{X} = X + K_- = Y + K_+$ so that the point (R^+X, R^-Y) belongs to $\tilde{\mathcal{G}}_R$. ∎

This proposition shows that we can decompose uniquely any element $X \in \mathcal{G}$ as $X = X_+ - X_-$ with $X_\pm \in \mathcal{G}_\pm$ and $\nu(\overline{X_+}) = \overline{X_-}$. This will be the basis for the factorization problem below.

Another important consequence of these results is that the algebraic structure we have exhibited is equivalent to the existence of an r-matrix.

This is the starting point of the classification theorems of the solutions of the Yang–Baxter equation by Belavin–Drinfeld and Semenov-Tian-Shansky.

In the next sections we content ourselves with giving simple examples of this algebraic setting. In the Adler–Kostant–Symes scheme ν is trivial, and in the open Toda chain ν is non-trivial.

4.3 Adler–Kostant–Symes scheme

A class of solutions of the modified Yang–Baxter equation is produced by the Adler–Kostant–Symes scheme. Let $\mathcal{G}$ be a Lie algebra and assume that we have two Lie subalgebras $\mathcal{A}$ and $\mathcal{B}$ such that, *as a vector space*, $\mathcal{G}$ is the direct sum

$$\mathcal{G} = \mathcal{A} + \mathcal{B}$$

Note that $\mathcal{A}$ and $\mathcal{B}$ are Lie subalgebras but they are not assumed to commute ($[\mathcal{A}, \mathcal{B}] \neq 0$). We denote by $P_{\mathcal{A}}$ the projection on $\mathcal{A}$ along $\mathcal{B}$ and $P_{\mathcal{B}} = 1 - P_{\mathcal{A}}$.

Proposition. *The linear map* $R = \frac{1}{2}(P_{\mathcal{A}} - P_{\mathcal{B}})$ *satisfies the modified Yang–Baxter equation.*

Proof. By definition we have $[X, Y]_R = [P_{\mathcal{A}}X, P_{\mathcal{A}}Y] - [P_{\mathcal{B}}X, P_{\mathcal{B}}Y]$, where we used $P_{\mathcal{A}} + P_{\mathcal{B}} = 1$, thus:

$$R([X, Y]_R) = \tfrac{1}{2}\left([P_{\mathcal{A}}X, P_{\mathcal{A}}Y] + [P_{\mathcal{B}}X, P_{\mathcal{B}}Y]\right)$$

Since

$$[RX, RY] = \frac{1}{4}\left([P_{\mathcal{A}}X, P_{\mathcal{A}}Y] - [P_{\mathcal{A}}X, P_{\mathcal{B}}Y] - [P_{\mathcal{B}}X, P_{\mathcal{A}}Y] + [P_{\mathcal{B}}X, P_{\mathcal{B}}Y]\right)$$

we obtain

$$[RX, RY] - R([X, Y]_R) = -\frac{1}{4}[X, Y]$$

This is the modified Yang–Baxter equation.   ∎

This construction is a particular example of the general discussion. With the notations of the previous section, $R_+ = R + \frac{1}{2} = P_{\mathcal{A}}$ and $R_- = R - \frac{1}{2} = -P_{\mathcal{B}}$, and $\mathcal{G}_+ = \mathcal{K}_+ = \mathcal{A}$, $\mathcal{G}_- = \mathcal{K}_- = \mathcal{B}$. Thus $\mathcal{G}_\pm/\mathcal{K}_\pm = \{0\}$ so we have the decomposition of $\mathcal{G}_R$ as a direct sum *of Lie algebras*:

$$\mathcal{G}_R = \mathcal{A} \oplus \mathcal{B}$$

In the Lie algebra $\mathcal{G}_R$, $\mathcal{A}$ and $\mathcal{B}$ commute. It is then clear that any element $X \in \mathcal{G}$ can be decomposed as $X = X_+ - X_-$ with $X_+ = P_{\mathcal{A}}X$ and $X_- = -P_{\mathcal{B}}X$.

Example. Let $\tilde{\mathcal{G}}$ be the loop algebra $\tilde{\mathcal{G}} = \mathcal{G} \otimes \mathbb{C}[[\lambda, \lambda^{-1}]]$ with $\mathcal{G}$ a finite-dimensional simple Lie algebra. Elements of $\tilde{\mathcal{G}}$ are linear combinations of elements $X \otimes \lambda^n$ with $X \in \mathcal{G}$ and $n \in \mathbb{Z}$. The commutation relations in $\tilde{\mathcal{G}}$ are:

$$[X \otimes \lambda^n, Y \otimes \lambda^m] = [X, Y] \otimes \lambda^{n+m} \tag{4.14}$$

The two subalgebras $\mathcal{A} = \mathcal{G}_+$, $\mathcal{B} = \mathcal{G}_-$ are spanned by elements of the form $X \otimes \lambda^n$ with $n \geq 0$ and $n < 0$ respectively. Clearly $[\mathcal{A}, \mathcal{B}] \neq 0$ for the Lie algebra structure of $\tilde{\mathcal{G}}$. For any $X \in \mathcal{G}$ and any formal power series $f(\lambda) = \sum_{n \in \mathbb{Z}} f_n \lambda^n$, the maps $R^\pm$ are defined by:

$$
\begin{aligned}
R^+(X \otimes f(\lambda)) &= X \otimes f_+(\lambda) \\
R^-(X \otimes f(\lambda)) &= -X \otimes f_-(\lambda)
\end{aligned}
\tag{4.15}
$$

where $f_+(\lambda) = \sum_{n \geq 0} f_n \lambda^n$ and $f_-(\lambda) = \sum_{n < 0} f_n \lambda^n$ denote the regular and singular part of $f(\lambda)$ around $\lambda = 0$ respectively. Since by definition $f(\lambda) = f_+(\lambda) + f_-(\lambda)$, we have $R^+ - R^- = \mathbf{1}$. Elements of the subalgebras $\mathcal{G}_\pm$ are of the form $X \otimes f_\pm(\lambda)$ with $f_\pm(\lambda)$ regular (resp. singular) at the origin. We have $\mathcal{K}_\pm = \mathrm{Ker}\, R^\mp = \mathcal{G}_\pm$. Thus $\mathcal{G}_\pm/\mathcal{K}_\pm = \{0\}$ and the map ν is trivial. The decomposition eq. (4.12) for $X \otimes f(\lambda)$ consists of writing $X \otimes f(\lambda)$ as the sum of two functions, one analytic around the origin and the other one analytic around infinity. Introducing the scalar product in $\tilde{\mathcal{G}}$:

$$(X(\lambda), Y(\lambda)) = \mathrm{Res}\, \mathrm{Tr}\, (X(\lambda)Y(\lambda)) = \oint \frac{d\lambda}{2i\pi} \mathrm{Tr}\, (X(\lambda)Y(\lambda)) \tag{4.16}$$

where $\mathrm{Tr}\, ()$ is the invariant bilinear form on $\mathcal{G}$, we can write the maps $R^\pm$ in a more operatorial way: for any $X(\lambda)$ in $\tilde{\mathcal{G}}$

$$(R^\pm X)(\lambda) = \oint_{\Gamma_\pm} \frac{d\mu}{2i\pi} \frac{\mathrm{Tr}_2(C_{12} X_2(\mu))}{\mu - \lambda}$$

where C_{12} is the tensor Casimir for $\mathcal{G}$ given in eq. (4.7). The integration contours $\Gamma_\pm$ are for R^+ a path enclosing λ and $\mu = 0$, and for R^- a path enclosing $\mu = 0$ but not λ. These formulae are of the form eq. (4.2) with $r_{12}^+ \in \tilde{\mathcal{G}} \otimes \tilde{\mathcal{G}}$ given by $r_{12}^+ = -C_{12}/(\lambda - \mu) = C_{12} \sum_{n=0}^{\infty} \lambda^n/\mu^{n+1}$, where we expand in powers of λ/μ because $|\lambda| < |\mu|$. Similarly $r_{12}^- = -C_{12}/(\lambda - \mu) = -C_{12} \sum_{n=0}^{\infty} \mu^n/\lambda^{n+1}$. Finally we get $r_{12} = (r_{12}^+ + r_{12}^-)/2$ in the form:

$$r_{12}(\lambda, \mu) = -\frac{C_{12}}{\lambda - \mu}$$

This is the r-matrix that we met in Chapter 3.

4.4 Construction of integrable systems

We now show that the setting of section (4.2) can be used to construct integrable systems. Starting from a Lie algebra structure on $\mathcal{G}$ one can construct a Poisson bracket on its dual $\mathcal{G}^*$. Better, starting from a Lie bialgebra, we have two Lie algebra structures on $\mathcal{G}$, namely $[\,,\,]$ and $[\,,\,]_R$, and one obtains two Poisson brackets $\{\,,\,\}$ and $\{\,,\,\}_R$ on $\mathcal{G}^*$. As we know, the functions on $\mathcal{G}^*$, invariant under the action $[\,,\,]$, are in the centre of the Poisson structure $\{\,,\,\}$ and cannot provide useful Hamiltonians. However, they will not be in the centre of the second Poisson structure $\{\,,\,\}_R$ and we will show that they then provide commuting Hamiltonians. Moreover, the equations of motion take the Lax form.

We start from an r-matrix solution of the modified Yang–Baxter equation. On $\mathcal{G}$, we have two Lie algebra structures, one associated with the original Lie bracket $[\,,\,]$ and one corresponding to the bracket $[\,,\,]_R$. We denote by $\mathcal{G}_R$ the Lie algebra equipped with this second bracket. Similarly, we denote by G and G_R the corresponding simply connected Lie groups.

On $\mathcal{G}$, we have two adjoint actions

$$\mathrm{ad}X(Y) = [X, Y], \quad \mathrm{ad}_R X(Y) = [X, Y]_R = [RX, Y] + [X, RY]$$

and also two coadjoint actions of $\mathcal{G}$ on $\mathcal{G}^*$:

$$\mathrm{ad}^* X \cdot L(Y) = -L([X, Y]), \quad \mathrm{ad}_R^* X \cdot L(Y) = -L([X, Y]_R)$$

One can express ad_R^* in terms of ad^*

$$\mathrm{ad}_R^* X \cdot L = \mathrm{ad}^*(RX) \cdot L + R^* \mathrm{ad}^* X \cdot L \qquad (4.17)$$

We then have two Poisson brackets on $\mathcal{F}(\mathcal{G}^*)$, $\{\,,\,\}$ and $\{\,,\,\}_R$, which are the Kostant–Kirillov brackets for the two Lie algebra structures, i.e.

$$\{f_1, f_2\}\,(L) = L([df_1, df_2]), \quad \{f_1, f_2\}_R(L) = L([df_1, df_2]_R)$$

For linear functions $f_1(L) = L(X)$ and $f_2(L) = L(Y)$ the second Poisson bracket becomes exactly eq. (4.3). Thus the Poisson bracket structure of the Lax matrix is given by the *Kostant–Kirillov bracket for the algebra* $\mathcal{G}_R$. In accordance, we view L as an element of a coadjoint orbit of G_R.

Theorem.
(i)   The Ad^*-invariant functions on $\mathcal{G}^*$ are in involution with respect to both Poisson brackets.

(ii) *Choosing an* Ad*-*invariant Hamiltonian* H, *the equation of motion on* $\mathcal{G}^*$ *with respect to the Poisson bracket* $\{\ ,\ \}_R$ *may be written in the two equivalent forms*

$$\frac{dL}{dt} = -\mathrm{ad}_R^* dH \cdot L \tag{4.18}$$

$$\frac{dL}{dt} = \mathrm{ad}^* M \cdot L \quad \text{with} \quad M = -R(dH) \tag{4.19}$$

Proof. Let us choose two ad*-invariant functions on $\mathcal{G}^*$, f_1 and f_2, i.e. $\mathrm{ad}^* df_i \cdot L = 0$ (for $i = 1, 2$). As shown in Chapter 14, f_1 and f_2 are in the kernel of $\{\ ,\ \}$, and a fortiori $\{f_1, f_2\} = 0$. On the other hand,

$$\{f_1, f_2\}_R(L) = L([df_1, df_2]_R) = L([Rdf_1, df_2] + [df_1, Rdf_2])$$
$$= \mathrm{ad}^* df_2.L(Rdf_1) - \mathrm{ad}^* df_1.L(Rdf_2) = 0$$

This proves the first part of the proposition. For a general function f on phase space, the equation of motion reads $\dot{f} = \{H, f\}_R$, where H is the Hamiltonian which is chosen to be an ad*-invariant function on $\mathcal{G}^*$. By definition of df we have $\dot{f} = \dot{L}(df)$. Thus:

$$\dot{L}(df)(L) = \{H, f\}_R(L) = L([dH, df]_R) = -\mathrm{ad}_R^* dH \cdot L(df)$$

Since this is true for any f we get eq. (4.18). We may also use eq. (4.17) to express ad_R^* in terms of ad*. We get $\mathrm{ad}_R^* dH \cdot L = -\mathrm{ad}^* M \cdot L$ with $M = -RdH$, where we have used the invariance of H which implies $\mathrm{ad}^* dH \cdot L = 0$. This proves eq. (4.19). ∎

This theorem allows one to build an integrable system, provided we have enough Ad*-invariant functions, from the algebraic data $\mathcal{G}$, $\mathcal{G}_R$. In the case where $\mathcal{G}$ is equipped with an invariant non-degenerate quadratic form $(\ ,\ )$, $\mathcal{G}$ and its dual $\mathcal{G}^*$ may be identified, and ad* becomes ad. The condition for a function on $\mathcal{G}^*$ to be ad*-invariant may then be written as $[dH, L] = 0$. Moreover, we see that the equation of motion of L, eq. (4.19), takes the form of a Lax equation:

$$\frac{dL}{dt} = [M, L], \quad M = -R(dH) \tag{4.20}$$

Remark 1. Notice that M is not uniquely defined, as we can add to it anything which commutes with L, in particular any polynomial in L.

Remark 2. By construction the Lax matrix L lies on a coadjoint orbit of G_R. However, the Lax form of the equation of motion shows that the evolution also takes place on an orbit of G. Hence the flow occurs on the intersection of orbits of G and G_R.

4.5 Solving by factorization

We consider a Hamiltonian system with a Lax matrix $L \in \mathcal{G}^*$, with $\mathcal{G}^*$ the dual of $\mathcal{G}$. We assume that $\mathcal{G}$ is equipped with an invariant bilinear form, so that $\mathcal{G}^*$ and $\mathcal{G}$ are identified. We also assume that the Poisson bracket of L is given as above by $\{L(X), L(Y)\}_R = L([X, Y]_R)$, where R is a solution of the modified Yang–Baxter equation.

As explained in the previous section, the Hamiltonians H_n in involution $\{H_n, H_m\}_R = 0$ are taken among the Ad^*-invariant functions on $\mathcal{G}^* \sim \mathcal{G}$, for example, $H_n = \mathrm{Tr}\,(L^n)$. Let us choose one of these functions as the Hamiltonian H. The equations of motion have a Lax representation, eq(4.20). Since H is an invariant function, one has $[L, dH] = 0$ and we can write as well, using $R^\pm = R \pm \frac{1}{2}\mathrm{Id}$:

$$\frac{\partial L}{\partial t} = [M_+, L] = [M_-, L], \quad M_\pm = -R^\pm(dH) \qquad (4.21)$$

The fact that the equations of motion admit two Lax representations, with $M_\pm \in \mathcal{G}_\pm$ and $(M_+, M_-) \in \tilde{\mathcal{G}}_R$, is the key point of the following discussion.

Proposition. *The solution of the equations of motion with Hamiltonian H is given, for t small enough, by:*

$$L(t) = \theta_+(t)\, L_0\, \theta_+^{-1}(t) = \theta_-(t)\, L_0\, \theta_-^{-1}(t)$$

where $\theta_+(t)$ and $\theta_-(t)$ are the solutions of the factorization problem

$$\exp(-tdH(L_0)) = \theta_-^{-1}(t)\, \theta_+(t) \qquad (4.22)$$

with $\theta_\pm \in G_\pm$, $\nu(\bar{\theta}_+) = \bar{\theta}_-$.

<u>Proof</u>. We want to solve

$$\frac{dL}{dt} = [M_+, L], \quad \frac{dL}{dt} = [M_-, L]$$

where $M_\pm$ are defined in eq. (4.21). Since $M_\pm \in \mathcal{G}_\pm$ we know that there are $\theta_+(t) \in G_+$ and $\theta_-(t) \in G_-$ such that $\dot{\theta}_+ \theta_+^{-1} = M_+$ and $\dot{\theta}_- \theta_-^{-1} = M_-$, with initial value $\theta_\pm(0) = 1$. Then we have:

$$L(t) = \theta_+(t)\, L_0\, \theta_+^{-1}(t) = \theta_-(t)\, L_0\, \theta_-^{-1}(t)$$

Using $M_+ - M_- = -dH$ we get:

$$dH(L) = -\dot{\theta}_+ \theta_+^{-1} + \dot{\theta}_- \theta_-^{-1} = -\theta_- \left(\frac{d}{dt}(\theta_-^{-1}\theta_+) \right) \theta_+^{-1}$$

From the invariance of H by coadjoint action on L, we see that $dH(L) = dH(\theta_- \cdot L_0 \cdot \theta_-^{-1}) = \theta_- \cdot dH(L_0) \cdot \theta_-^{-1}$, which implies:

$$\frac{d}{dt}(\theta_-^{-1}\theta_+) \cdot (\theta_-^{-1}\theta_+)^{-1} = -dH(L_0)$$

yielding $\theta_-^{-1}(t)\theta_+(t) = \exp\left(-tdH(L_0)\right)$. We are led to the algebraic problem of decomposing the group element $\exp(-tdH(L_0))$, whose time dependence is explicit, as a product of elements in $G_\pm$, the Lie groups corresponding to $\mathcal{G}_\pm$. Since there is a unique decomposition at the Lie algebra level, see section (4.2), the solution of this factorization problem exists and is unique, at least in the vicinity of the unit element. The solution of this algebraic problem yields the solution of the evolution equation, at least for small t. ∎

Remark 1. The factorization problem in the Proposition is of the same form as in eq. (3.54) of Chapter 3. The reconstruction formula for the L matrix is also identical to eq. (3.55) and the proofs are parallel. Hence the present discussion is a more algebraic presentation of the material in Chapter 3.

Remark 2. Viewing L as a matrix, a set of Ad^*-invariant functions is provided by the traces of the powers of L: $H_n = \mathrm{Tr}\,(L^n)$. In this case, we have, as in eq. (3.33) in Chapter 3:

$$dH_n(L) = n\,\mathrm{Tr}_2\left(L_2^{n-1}C_{12}\right) \quad \text{and} \quad M_\pm = -n\,\mathrm{Tr}_2\left(L_2^{n-1}r_{12}^\pm\right)$$

with C_{12} the tensor Casimir.

4.6 The open Toda chain

In this section we introduce the Toda chains. They are integrable systems associated with Lie algebras. The open chains that we consider here are associated with finite-dimensional Lie algebras, and provide simple examples to illustrate the algebraic constructions presented above. In particular, the solution of the model will be achieved by solving the factorization problem. In contrast, the study of the closed Toda chain requires the consideration of loop algebras, and will be presented in Chapter 6.

We will first review some notations for Lie algebras introduced in Chapter 16. Let $\mathcal{G}$ be a finite-dimensional simple Lie algebra of rank r, equipped with an invariant scalar product denoted by $(\ ,\ )$ or $\mathrm{Tr}\,(\ )$. We choose a Cartan subalgebra with an orthonormal basis $\{H_i\}$ (not to be confused with the Hamiltonians!). We have the Cartan decomposition of $\mathcal{G}$:

$$\mathcal{G} = \mathcal{N}_- \oplus \mathcal{H} \oplus \mathcal{N}_+, \quad \text{with} \quad \mathcal{N}_\pm = \bigoplus_{\beta\ \text{positive}} \mathcal{G}_{\pm\beta}$$

where $\mathcal{G}_{\pm\beta}$ are one-dimensional, generated by the root vectors $E_{\pm\beta}$ for any positive root β. We will need the commutation relations:

$$[H_i, H_j] = 0$$
$$[H, E_{\pm\alpha}] = \pm\alpha(H)\, E_{\pm\alpha}$$
$$[E_\alpha, E_{-\alpha}] = (E_\alpha, E_{-\alpha})\, H_\alpha,$$

where $H_\alpha = \sum_i \alpha(H_i) H_i$. We will often use the constants n_α such that

$$n_\alpha^2\, (E_\alpha, E_{-\alpha}) = 1$$

Toda chains are associated with any simple Lie algebra $\mathcal{G}$ as follows. Consider two elements in the Cartan subalgebra:

$$q = \sum_{i=1}^{r} q_i H_i, \quad p = \sum_{i=1}^{r} p_i H_i, \quad r = \operatorname{rank}\mathcal{G}$$

The coefficients (q_i, p_i), $i = 1, \ldots, r$ are the coordinates of a $(2r)$-dimensional phase space with Poisson brackets

$$\{p_i, q_j\} = \frac{1}{2}\delta_{ij} \tag{4.23}$$

where the factor $1/2$ is introduced to simplify later formulae. The Hamiltonian of the open Toda chain is by definition:

$$H = (p, p) + 2 \sum_{\alpha \text{ simple}} \exp\left(2\alpha(q)\right)$$

The equations of motion are then:

$$\frac{dq}{dt} = \{H, q\} = p$$
$$\frac{dp}{dt} = \{H, p\} = -2 \sum_{\alpha \text{ simple}} H_\alpha \exp\left(2\alpha(q)\right)$$

or, combining the two,

$$\frac{d^2 q}{dt^2} = -2 \sum_{\alpha \text{ simple}} H_\alpha \exp\left(2\alpha(q)\right) \tag{4.24}$$

In the usual $sl(N+1)$ case, the Lie algebra of traceless $(N+1) \times (N+1)$ matrices, the Cartan algebra can be taken as traceless diagonal matrices. The simple root vectors E_{α_i} are the canonical matrices $E_{i,i+1}$, and

$E_{-\alpha_i} = E_{i+1,i}$ for $i = 1, \ldots, N$. In this case it is not very convenient to use an orthonormal basis of the Cartan algebra under the scalar product $(X, Y) = \mathrm{Tr}\,(XY)$. Instead we write an element in the Cartan subalgebra as $q = \sum_{i=1}^{N+1} q_i E_{ii}$ with the constraint $\sum q_i = 0$. In this description the simple roots are $\alpha_i(q) = q_i - q_{i+1}$. The dynamical variables of the open Toda chain are the two elements $q = \sum_{i=1}^{N+1} q_i E_{ii}$ and $p = \sum_{i=1}^{N+1} p_i E_{ii}$ with the constraints $\sum q_i = 0$ and $\sum p_i = 0$. These constraints are not compatible with the canonical Poisson bracket, eq. (4.23), so we use instead the Dirac bracket:

$$\{p_i, q_j\} = \frac{1}{2}\left(\delta_{ij} - \frac{1}{N+1}\right), \quad \{p_i, p_j\} = \{q_i, q_j\} = 0$$

The Hamiltonian

$$H = \sum_{i=1}^{N+1} p_i^2 + 2\sum_{i=1}^{N} e^{2(q_i - q_{i+1})}$$

generates the equations of motion:

$$\dot{q}_i = p_i, \qquad\qquad\qquad i = 1, \ldots, N+1$$
$$\dot{p}_1 = -2e^{2(q_1 - q_2)}$$
$$\dot{p}_i = -2e^{2(q_i - q_{i+1})} + 2e^{2(q_{i-1} - q_i)}, \quad i = 2, \ldots, N$$
$$\dot{p}_{N+1} = 2e^{2(q_N - q_{N+1})}$$

The particular form of the equations for $\dot{p}_1$ and $\dot{p}_{N+1}$ explains the terminology "open" Toda chain.

The equations of motion of the open Toda chain can be written in Lax form and the system is integrable.

Proposition. *The equations of motion of the open Toda chain admit a Lax pair representation* $\dot{L} = [M, L]$ *with:*

$$L = p + \sum_{\alpha \text{ simple}} n_\alpha\, e^{\alpha(q)}\, (E_\alpha + E_{-\alpha}), \quad M = -\sum_{\alpha \text{ simple}} n_\alpha\, e^{\alpha(q)}\, (E_\alpha - E_{-\alpha})$$

$$(4.25)$$

<u>Proof.</u> We compute successively the time derivative of L and the commutator $[M, L]$. We have

$$\frac{dL}{dt} = \frac{dp}{dt} + \sum_\alpha n_\alpha\, \alpha\left(\frac{dq}{dt}\right) e^{\alpha(q)}(E_\alpha + E_{-\alpha})$$

$$[M, L] = \sum_\alpha n_\alpha\, \alpha(p)\, e^{\alpha(q)}\, (E_\alpha + E_{-\alpha})$$
$$- \sum_\alpha \sum_\beta n_\alpha n_\beta\, e^{(\alpha+\beta)(q)}[E_\alpha - E_{-\alpha}, E_\beta + E_{-\beta}]$$

Since $[E_{\pm\alpha}, E_{\pm\beta}]$ is antisymmetric in the exchange of α and β, the second term reduces to

$$-\sum_\alpha \sum_\beta n_\alpha n_\beta \, e^{(\alpha+\beta)(q)} \left([E_\alpha, E_{-\beta}] - [E_{-\alpha}, E_\beta]\right)$$

Since α and β are simple roots, we have $[E_\alpha, E_{-\beta}] = \delta_{\alpha\beta}(E_\alpha, E_{-\alpha})H_\alpha$ and therefore

$$[M, L] = -2\sum_\alpha e^{2\alpha(q)} H_\alpha + \sum_\alpha n_\alpha \, \alpha(p) \, e^{\alpha(q)} \, (E_\alpha + E_{-\alpha})$$

from which the result follows.   ■

It is important to notice that this calculation is performed using only the Lie algebra structure of $\mathcal{G}$ and never refers to a representation. It thus extends naturally to the case of a Kac–Moody algebra.

This Lax representation provides conserved quantities, namely any invariant polynomial in L. To count the independent conserved quantities, we recall the Chevalley theorem which states that, for a simple Lie algebra, the ring of invariant polynomials is freely generated by $r = \operatorname{rank} \mathcal{G}$ primitive polynomials which are of degree n_i, $i = 1, \ldots, r$, the so-called exponents of the Lie algebra. We have thus the r independent conserved quantities at our disposal for a phase space of dimension $2r$. It remains to show that they are in involution. This is the aim of the next section.

4.7 The r-matrix of the Toda models

Since the r-matrix of the Toda chain is the canonical example of solution of the classical Yang–Baxter equation associated with simple Lie algebras, and plays an important role in the following as well as in many integrable models, we present its computation in detail.

Proposition. *Let L be the Lax matrix (4.25), with p and q satisfying the Poisson bracket eq. (4.23). Then, there exists a unique antisymmetric r-matrix $r_{12} = -r_{21} \in \mathcal{G} \otimes \mathcal{G}$, independent of p and q, such that $\{L_1, L_2\} = [r_{12}, L_1 + L_2]$.*

$$r_{12} = \frac{1}{2} \sum_{\alpha \text{ positive}} \frac{E_\alpha \otimes E_{-\alpha} - E_{-\alpha} \otimes E_\alpha}{(E_\alpha, E_{-\alpha})} \tag{4.26}$$

This implies that the conserved Hamiltonians are in involution.

<u>Proof.</u> We have

$$\{L_1, L_2\} = \frac{1}{2} \sum_{\alpha \text{ simple}} n_\alpha e^{\alpha(q)} [H_\alpha \otimes (E_\alpha + E_{-\alpha}) - (E_\alpha + E_{-\alpha}) \otimes H_\alpha]$$

and

$$[r_{12}, L_1 + L_2] = [r_{12}, p \otimes 1 + 1 \otimes p]$$
$$+ \sum_{\alpha \text{ simple}} n_\alpha e^{\alpha(q)} [r_{12}, (E_\alpha + E_{-\alpha}) \otimes 1 + 1 \otimes (E_\alpha + E_{-\alpha})]$$

Looking for r_{12} independent of p and q, one can identify the coefficients of p and q in these equations, and we find for the r-matrix the conditions:

$$[r_{12}, H_i \otimes 1 + 1 \otimes H_i] = 0$$
$$[r_{12}, (E_\alpha + E_{-\alpha}) \otimes 1 + 1 \otimes (E_\alpha + E_{-\alpha})]$$
$$= \frac{1}{2}(H_\alpha \otimes (E_\alpha + E_{-\alpha}) - (E_\alpha + E_{-\alpha}) \otimes H_\alpha)$$

when α is a simple root. Taking the commutator of the second relation with $H \otimes 1 + 1 \otimes H$, and using the first one, we see that it splits into two independent equations involving E_α and $E_{-\alpha}$ separately, i.e:

$$[r_{12}, H_i \otimes 1 + 1 \otimes H_i] = 0 \tag{4.27}$$
$$[r_{12}, E_{\pm\alpha} \otimes 1 + 1 \otimes E_{\pm\alpha}] = \frac{1}{2}(H_\alpha \otimes E_{\pm\alpha} - E_{\pm\alpha} \otimes H_\alpha) \quad (\alpha \text{ simple})$$

To solve these equations, we notice that the Casimir element C_{12} in $\mathcal{G} \otimes \mathcal{G}$ obeys $[C_{12}, 1 \otimes X + X \otimes 1] = 0$ for all $X \in \mathcal{G}$. Therefore, eq. (4.27) tells us that r_{12} is determined only up to the addition of a multiple of C_{12}. Next we observe that if r_{12} is a solution, then $-r_{21}$ is another solution, hence $(r_{12} - r_{21})/2$ is an antisymmetric solution. So we can assume without loss of generality that r_{12} is antisymmetric. Note that the Casimir element is symmetric, and therefore does not affect such a solution. To find r_{12}, we dualize the equations using $R(X) = \text{Tr}_2 (r_{12} \cdot 1 \otimes X)$. Denoting $(X, Y) = \text{Tr}(XY)$, we obtain for α simple:

$$[R(X), H_i] - R([X, H_i]) = 0 \tag{4.28}$$
$$[R(X), E_{\pm\alpha}] - R([X, E_{\pm\alpha}]) = \tfrac{1}{2}((E_{\pm\alpha}, X)H_\alpha - (H_\alpha, X)E_{\pm\alpha}) \tag{4.29}$$

The first equation implies $R(\mathcal{H}) \subset \mathcal{H}$ and $R(\mathcal{G}_{\pm\beta}) \subset \mathcal{G}_{\pm\beta}$. The most general form of r_{12} compatible with these requirements and the antisymmetry property is:

$$r = \sum_{ij} r^{ij} H_i \otimes H_j + \sum_{\beta \text{ positive}} \frac{r_\beta}{(E_\beta, E_{-\beta})} (E_\beta \otimes E_{-\beta} - E_{-\beta} \otimes E_\beta)$$

If $X \in \mathcal{H}$, eq. (4.29) reduces to

$$\pm\alpha(X) \left[r_\alpha - \frac{1}{2} \right] E_{\pm\alpha} = \alpha(R(X))E_{\pm\alpha}, \quad (\alpha \text{ simple})$$

which implies $R(X) = 0$ for all $X \in \mathcal{H}$, i.e. $r^{ij} = 0$, and $r_\alpha = \frac{1}{2}$ for α a simple root. Finally, for β any positive root, we have $R(E_\beta) = r_\beta E_\beta$, and eq. (4.29) gives

$$r_\beta [E_\beta, E_\alpha] - r_{\beta+\alpha} [E_\beta, E_\alpha] = 0$$

Therefore, we deduce that $r_\beta = \frac{1}{2}$, for all positive roots β. Gathering all this information, we obtain:

$$R(H) = 0 \tag{4.30}$$

$$R(E_{\pm\beta}) = \pm \frac{1}{2} E_{\pm\beta} \quad (\beta \text{ positive}) \tag{4.31}$$

which is equivalent to eq. (4.26).   ■

This result proves the integrability of the Toda chain models. We have found a constant antisymmetric r-matrix. Let us see how it fits into the general framework explained above.

Proposition. *Consider the r-matrix eq. (4.26). Define the map $R(X) = \mathrm{Tr}_2 (r_{12} \cdot 1 \otimes X)$. It satisfies the modified Yang–Baxter equation:*

$$[R(X), R(Y)] - R([X,Y]_R) = -\frac{1}{4}[X,Y] \tag{4.32}$$

Proof. This is checked by a case by case analysis, using eqs. (4.30, 4.31).   ■

The antisymmetric r-matrix r_{12} is a solution of the modified Yang–Baxter equation (4.6). Hence $r_{12}^\pm = r_{12} \pm \frac{1}{2} C_{12}$ (C_{12} is the Casimir) are solutions of the classical Yang–Baxter equations:

$$[r_{12}^\pm, r_{13}^\pm] + [r_{12}^\pm, r_{23}^\pm] + [r_{13}^\pm, r_{23}^\pm] = 0$$

For simple Lie algebras, C_{12} reads:

$$C_{12} = \sum_i H_i \otimes H_i + \sum_{\alpha \text{ positive}} \frac{E_\alpha \otimes E_{-\alpha} + E_{-\alpha} \otimes E_\alpha}{(E_\alpha, E_{-\alpha})}$$

Thus, the matrices $r_{12}^\pm$ are given explicitly by:

$$r_{12}^+ = \frac{1}{2} \sum_i H_i \otimes H_i + \sum_{\alpha \text{ positive}} \frac{E_\alpha \otimes E_{-\alpha}}{(E_\alpha, E_{-\alpha})} \tag{4.33}$$

$$r_{12}^- = -\frac{1}{2} \sum_i H_i \otimes H_i - \sum_{\alpha \text{ positive}} \frac{E_{-\alpha} \otimes E_\alpha}{(E_\alpha, E_{-\alpha})} \tag{4.34}$$

The general structural results of the algebraic approach allow one to understand the Lax pair of the open Toda chain, starting from this Toda r-matrix. The action of $R^{\pm}$ on the generators of $\mathcal{G}$ is given by

$$
\begin{cases}
R^+(H) = \frac{1}{2}H \\
R^+(E_\alpha) = E_\alpha \\
R^+(E_{-\alpha}) = 0
\end{cases}
\qquad
\begin{cases}
R^-(H) = -\frac{1}{2}H \\
R^-(E_\alpha) = 0 \\
R^-(E_{-\alpha}) = -E_{-\alpha}
\end{cases}
\tag{4.35}
$$

for α any positive root. The subalgebras $\mathcal{G}_{\pm} = \text{Im } R^{\pm}$, are just the two standard Borel subalgebras of $\mathcal{G}$, $\mathcal{G}_{\pm} = \mathcal{B}_{\pm} = \mathcal{H} \oplus \mathcal{N}_{\pm}$. The subalgebras $\mathcal{K}_{\pm} = \text{Ker } R^{\mp}$ are $\mathcal{K}_{\pm} = \mathcal{N}_{\pm}$. Therefore $\mathcal{G}_{\pm}/\mathcal{K}_{\pm} = \mathcal{H}$, and the isomorphism $\nu : \mathcal{H} \to \mathcal{H}$ is just $\nu(H) = -H$; $H \in \mathcal{H}$. In this example the isomorphism ν is non-trivial. The image $\tilde{\mathcal{G}}_R$ of $\mathcal{G}_R$ in $\mathcal{B}_+ \oplus \mathcal{B}_-$ consists of the couples (X_+, X_-) such that the components of X_+ and X_- on the Cartan subalgebra $\mathcal{H}$ are opposite. Any element $X \in \mathcal{G}$ can be decomposed uniquely as:

$$
X = X_+ - X_- \quad \text{with} \quad X_{\pm} \in \mathcal{B}_{\pm}, \quad X_+|_{\mathcal{H}} = -X_-|_{\mathcal{H}}
$$

In the simple Lie group G, the corresponding decomposition is a factorization problem, $g = g_-^{-1} g_+$, where $g_{\pm} \in \exp(\mathcal{B}_{\pm})$ and the factors of $g_{\pm}$ on the Cartan torus are inverse to each other, i.e. $g_{\pm} = h^{\pm 1} n_{\pm}$ with $h \in \exp \mathcal{H}$ and $n_{\pm} \in \exp \mathcal{N}_{\pm}$.

Let us write explicitly the R-bracket for the antisymmetric r-matrix eq. (4.26):

$$
[H_i, H_j]_R = 0, \quad [H_i, E_{\pm \alpha}]_R = \frac{1}{2}\alpha(H_i)E_{\pm \alpha}, \quad [E_\alpha, E_{-\beta}]_R = 0
$$

We identify $\mathcal{G}$ and $\mathcal{G}^*$ via the scalar product $(\ ,\ )$. We can write a generic element of $\mathcal{G}^*$ as

$$
L = \sum_i x_i H_i + \sum_{\alpha \text{ positive}} \left[\frac{x_{-\alpha}}{(E_\alpha, E_{-\alpha})} E_\alpha + \frac{x_\alpha}{(E_\alpha, E_{-\alpha})} E_{-\alpha} \right]
$$

Notice that $x_i = x_i(L) = (L, H_i)$ and $x_{\pm \alpha} = x_{\pm \alpha}(L) = (L, E_{\pm \alpha})$. Next, in order to construct an integrable system, we must choose an orbit of G_R in $\mathcal{G}^*$. This is the object of the following:

Proposition. *The elements of $\mathcal{G}^*$ of the form*

$$
L = \sum_i x_i H_i + \sum_{\alpha \text{ simple}} \frac{x_\alpha}{(E_\alpha, E_{-\alpha})}(E_\alpha + E_{-\alpha})
\tag{4.36}
$$

describe an orbit under the coadjoint action of G_R.

<u>Proof</u>. We must show that this form of L is stable under an infinitesimal transformation $\operatorname{ad}_R^* X$, $\forall X \in \mathcal{G}$. Recall that

$$\operatorname{ad}_R^* X \cdot L = \operatorname{ad}^*(RX) \cdot L + R^* \operatorname{ad}^* X \cdot L$$

Using the identification of $\mathcal{G}$ and $\mathcal{G}^*$, and the fact that R is antisymmetric, we have

$$\operatorname{ad}_R^* X \cdot L = -[RX, L] + R[X, L]$$

If $X = H \in \mathcal{H}$, we have

$$\operatorname{ad}_R^* H \cdot L = R[H, L] = \frac{1}{2} \sum_{\alpha \text{ simple}} \frac{\alpha(H) x_\alpha}{(E_\alpha, E_{-\alpha})} (E_\alpha + E_{-\alpha})$$

This shows that $\operatorname{ad}_R^* H$ only modifies the x_α coordinates of L. If $X = E_\beta \in \mathcal{G}_\beta$, with β positive, we have

$$\operatorname{ad}_R^* E_\beta \cdot L = (R - \frac{1}{2})[E_\beta, L]$$

Since L contains only roots of height ≥ -1, we have

$$[E_\beta, L] \in \mathcal{H} \oplus \sum_{\alpha \text{ positive}} \mathcal{G}_\alpha$$

and therefore $(R - \frac{1}{2})[E_\beta, L] \in \mathcal{H}$ because $\mathcal{G}_\alpha \subset \operatorname{Ker}(R - \frac{1}{2})$ $\forall \alpha$ positive. Similarly $\operatorname{ad}_R^* E_{-\beta} \cdot L = (R + \frac{1}{2})[E_{-\beta}, L] \in \mathcal{H}$. This shows that $\operatorname{ad}_R^* E_{\pm\beta}$ only modify the x_i coordinates of L. ∎

This result gives a description of the orbit in terms of the coordinates x_i, x_α. Computing the Kostant–Kirillov bracket with respect to $\mathcal{G}_R$ we get on the orbit (4.36):

$$\{x_i, x_j\}_R = 0, \quad \{x_i, x_\alpha\}_R = \frac{1}{2} \alpha(H_i) x_\alpha, \quad \{x_\alpha, x_\beta\}_R = 0$$

Indeed, since $dx_i = H_i$, $dx_{\pm\alpha} = E_{\pm\alpha}$, the Poisson bracket $\{\ ,\ \}_R$ is given by:

$$\{x_i, x_{\pm\alpha}\}_R = (L, [dx_i, dx_{\pm\alpha}]_R) = \frac{1}{2} \alpha(H_i)(L, E_{\pm\alpha}) = \frac{1}{2} \alpha(H_i) x_{\pm\alpha}$$

and similarly: $\{x_\alpha, x_{-\beta}\}_R = 0$, $\{x_\alpha, x_\beta\}_R = C_{\alpha,\beta}\, x_{\alpha+\beta}$, where $C_{\alpha,\beta}$ are the structure constants of $\mathcal{G}$. Moreover, on the orbit (4.36) we have $x_{\alpha+\beta} = 0$ since only simple roots appear.

The orbit being a symplectic manifold, we know by the theorem of Darboux, see Chapter 14, that there exists (locally) a set of canonical coordinates p_i, q_i such that $\{p_i, q_j\}_R = \frac{1}{2}\delta_{ij}$. Here these coordinates are given by:

$$x_i = p_i, \quad x_\alpha = n_\alpha(E_\alpha, E_{-\alpha})\exp(\alpha(q))$$

Inserting this parametrization into eq. (4.36) one recovers the Lax matrix of the open Toda chain. The model is completely specified once we choose a Hamiltonian $H \in \mathcal{I}(\mathcal{G}^*)$. Taking $H = (L, L)$ yields:

$$M = -R(dH) = -2R(L) = -\sum_{\alpha \text{ simple}} n_\alpha e^{\alpha(q)}(E_\alpha - E_{-\alpha})$$

which is the second element of the Lax pair of the open Toda chain.

4.8 Solution of the open Toda chain

To solve the equations of motion of the open Toda chain, we have to solve the factorization problem eq. (4.22). Specifically,

$$e^{-tdH(L_0)} = \theta_-(t)^{-1}\theta_+(t); \quad L(t) = \theta_+(t)L_0\theta_+^{-1}(t) = \theta_-(t)L_0\theta_-^{-1}(t) \tag{4.37}$$

with L_0 the Lax matrix at time $t = 0$. The factorization problem is specified by the r-matrix of the Toda chain. It is a Gauss decomposition with the diagonal part equally shared between the two factors. Even though this is a purely algebraic problem, it may not be an easy task to perform explicitly. Fortunately, if $\mathcal{G}$ is a finite-dimensional simple Lie algebra, there exists a suitable parametrization of the Lax operator, found by Kostant, in which the factorization can be done. We will explain this method in the case of $\mathcal{G} = sl(N+1)$.

We write the Lax matrix of the Toda chain as:

$$L_{\text{Toda}} = e^{-\text{ad } q}\mathcal{E}_- + p + e^{\text{ad } q}\mathcal{E}_+, \quad \text{with} \quad \mathcal{E}_\pm = \sum_{\alpha \text{ simple}} n_\alpha E_{\pm\alpha}$$

Note that if we assign degree 0 to the generators of the Cartan subalgebra $\mathcal{H}$ and degree 1 or -1 to the generators E_α or $E_{-\alpha}$ with α simple, then $\mathcal{E}_-$ is of degree -1, p of degree 0 and $\mathcal{E}_+$ of degree $+1$.

The Lax matrix is a traceless symmetric Jacobi matrix, i.e. $L_{ij} = 0$ except for $j = i\pm 1, i$. We will parametrize the space of traceless symmetric $(N+1) \times (N+1)$ Jacobi matrices, which is of dimension $2N$, with two diagonal matrices: $\Omega \in sl(N+1)$, and $\Delta \in SL(N+1)$. To this end, we introduce the matrix w which is the unique matrix such that $\det w = 1$, $w\mathcal{H}w^{-1} \subset \mathcal{H}$ and

$$wE_{i,i+1}w^{-1} = E_{N+2-i,N+1-i} \tag{4.38}$$

Explicitly, its matrix elements are $[w]_{ij} = (-1)^{\frac{N}{2}} \delta_{N+2-i,j}$. The action eq. (4.38) represents the action of the longest element in the Weyl group. Its main property in what follows is the relation

$$w^{-1} \mathcal{E}_- \, w = \mathcal{E}_+ \tag{4.39}$$

Lemma. *Let Ω be a traceless diagonal matrix, and Δ be a diagonal matrix with determinant 1. Considering the unique element $u \in N_- = \exp(\mathcal{N}_-)$ which diagonalizes $\mathcal{E}_- + \Omega$:*

$$\mathcal{E}_- + \Omega = u\Omega u^{-1} \tag{4.40}$$

and introducing the unique solution $b_\pm \in B_\pm = \exp(\mathcal{B}_\pm)$ of the factorization problem

$$b_-^{-1} b_+ = w^{-1} \, u\Delta u^{-1}$$

where we require that the diagonal is equally shared between b_+ and b_-^{-1} as in the previous section, we have

$$b_+ \left(\mathcal{E}_- + \Omega \right) b_+^{-1} = b_- \left(\mathcal{E}_+ + w^{-1}\Omega \, w \right) b_-^{-1}$$

<u>Proof.</u> From the factorization condition, we have $b_+ = b_- w^{-1} u\Delta u^{-1}$. Then

$$b_+ \left(\mathcal{E}_- + \Omega \right) b_+^{-1} = b_- w^{-1} u\Delta u^{-1} \left(\mathcal{E}_- + \Omega \right) u\Delta^{-1} u^{-1} w b_-^{-1}$$
$$= b_- w^{-1} \left(\mathcal{E}_- + \Omega \right) w b_-^{-1} = b_- \left(\mathcal{E}_+ + w^{-1}\Omega \, w \right) b_-^{-1}$$

because $u\Delta u^{-1}$ commutes with $\mathcal{E}_- + \Omega = u\Omega u^{-1}$.   ∎

Note that the factorization problem is non-trivial due to the presence of w which does not belong to N_-. As a straightforward consequence of this Lemma, we have the:

Proposition. *Choose Ω and Δ as above, then the matrix L defined by*

$$L = b_+ \left(\mathcal{E}_- + \Omega \right) b_+^{-1} = b_- \left(\mathcal{E}_+ + w^{-1}\Omega \, w \right) b_-^{-1} \tag{4.41}$$

is a symmetric traceless Jacobi matrix.

<u>Proof.</u> The first expression of L shows that it expands on degrees ≥ -1, while the second form shows that it expands on degrees ≤ 1, hence L is a Jacobi matrix. To show that it is symmetric, write $b_\pm = h^{\pm 1} n_\pm$ with $n_\pm \in N_\pm$ and h in the Cartan torus. We implemented the condition that $b_\pm$ have inverse components on the Cartan torus. We write $h = \exp(-q)$ for

$q \in \mathcal{H}$. The $E_{-\alpha}$ term in L is given by $\exp\left(-\operatorname{ad} q\right)E_{-\alpha} = \exp\left(\alpha(q)\right)E_{-\alpha}$. Similarly, the E_α term is given by $\exp\left(\operatorname{ad} q\right)E_\alpha = \exp\left(\alpha(q)\right)E_\alpha$. The two coefficients are equal. Finally, since Ω is traceless, so is L. ∎

The fact that we have at our disposal $2N$ parameters in Ω and Δ indicates that we can parametrize all symmetric traceless Jacobi matrices in this way. The diagonal matrices Ω, Δ form a system of coordinates on the space of symmetric Jacobi matrices.

In the $sl(N+1)$ case we have, since $h = \exp(-q)$,

$$\frac{h_{i+1}^2}{h_i^2} = e^{2(q_i - q_{i+1})} \tag{4.42}$$

where h_i are the entries of the diagonal matrix h. Moreover, denoting $\Omega = \operatorname{diag}(\omega_i)$, we have explicitly

$$u_{ij} = \prod_{l=j+1}^{i} \frac{1}{\omega_j - \omega_l}, \quad u_{ij}^{-1} = \prod_{l=j}^{i-1} \frac{1}{\omega_i - \omega_l} \tag{4.43}$$

This is obtained by writing $u\,\Omega = (\mathcal{E}_- + \Omega)\,u$ or $(\omega_j - \omega_i)u_{ij} = u_{i-1,j}$. The solution of this recursion relation with the boundary condition $u_{ii} = 1$ yields eq. (4.43). To compute u^{-1} we proceed in the same way with the equation $\Omega u^{-1} = u^{-1}(\mathcal{E}_- + \Omega)$.

Coming back to the open Toda chain, we parametrize the Lax matrix $L(t)$ by two matrices $\Omega(t)$ and $\Delta(t)$. We call $L_0 = L(t=0)$ the Lax matrix at time $t = 0$, and let Ω_0 and Δ_0 denote its coordinates. Note that the matrix elements of $\Omega(t)$ are the eigenvalues of $L(t)$, due to eqs. (4.40, 4.41), and so are *constant* in time: $\Omega(t) = \Omega_0$. Similarly, the matrix $u(t) = u$ is independent of time because it depends only on the matrix elements of $\Omega(t)$ by eq. (4.43).

Lemma. *In the above coordinates we have*

$$L(t) = \theta_+(t)L_0\theta_+^{-1}(t) = \theta_-(t)L_0\theta_-^{-1}(t)$$

where $\theta_\pm$ have inverse components on the Cartan torus and

$$\theta_+(t) = b_+(t)b_+^{-1}(0), \quad \theta_-(t) = b_-(t)b_-^{-1}(0)$$

Proof. The result follows by writing in eq. (4.41) the time-independent matrix $\Omega(t)$ in terms of L_0. We have $(\mathcal{E}_- + \Omega_0) = b_+^{-1}(0)L_0 b_+(0)$ and also $(\mathcal{E}_+ + w^{-1}\Omega_0 w) = b_-^{-1}(0)L_0 b_-(0)$. The components of $\theta_\pm$ on the Cartan torus are $(h(t)h^{-1}(0))^{\pm 1}$ and are clearly inverse to each other. ∎

We now find the time evolution of the coordinate $\Delta(t)$ by requiring that eq. (4.37) holds with the above $\theta_\pm(t)$. This yields the:

Proposition. *The time evolution of the Toda chain, in the coordinates* $\Omega(t), \Delta(t)$, *reads*

$$\Omega(t) = \Omega_0, \quad \Delta(t) = \Delta_0 e^{-tdH(\Omega_0)} \tag{4.44}$$

Proof. The first assertion has already been proved. To prove the second one, we start from $\exp(-tdH(L_0)) = \theta_-(t)^{-1}\theta_+(t)$. The left-hand side is easy to evaluate using eq. (4.41):

$$\exp(-tdH(L_0)) = b_+(0)u\exp(-tdH(\Omega_0))u^{-1}b_+^{-1}(0) \tag{4.45}$$

On the other hand, using the explicit form of $\theta_\pm(t)$, the right-hand side is equal to:

$$\theta_-^{-1}(t)\theta_+(t) = b_-(0)b_-^{-1}(t)b_+(t)b_+^{-1}(0) = b_-(0)w^{-1}u\Delta(t)u^{-1}b_+^{-1}(0)$$

where in the second expression we used the factorization problem at time t for $b_\pm(t)$. However, at $t = 0$ the same problem yields $b_-(0) = b_+(0)u\Delta_0^{-1}u^{-1}w$. Plugging into the previous expression we get:

$$\theta_-^{-1}(t)\theta_+(t) = b_+(0)u\Delta_0^{-1}\Delta(t)u^{-1}b_+^{-1}(0)$$

Comparing with eq. (4.45), we get $\Delta_0^{-1}\Delta(t) = \exp(-tdH(\Omega_0))$. ∎

This proposition solves the equations of motion of the open Toda chain since it gives the time evolution of the coordinates used to write the Lax matrix. We can give a more explicit form of the solution of the equations of motion in terms of determinants. To do that, consider the matrix

$$B = b_-^{-1}b_+ = n_-^{-1}h^2n_+ = w^{-1}u\Delta u^{-1}$$

We want to compute h^2 directly. If we have a representation V of $sl(N+1)$ with a highest weight vector $|v\rangle$ and its dual V^* with the dual highest weight vector $\langle v|$, we have $\langle v|B|v\rangle = \langle v|h^2|v\rangle$, because by definition $n_+|v\rangle = |v\rangle$ and $\langle v|n_- = \langle v|$. Since h lives in a space of dimension $r = \operatorname{rank} G$, using r such representations with independent highest weights, one completely reconstructs h. In the $sl(N+1)$ case we can take the N representations $\Lambda^k\mathbb{C}^{N+1}$, $k = 1, \ldots, N$, where Λ denotes the wedge product. Let $|\epsilon_1\rangle, \ldots, |\epsilon_{N+1}\rangle$ be the canonical basis of $\mathbb{C}^{N+1}$. Then the highest weight vector $|v_k\rangle$ of $\Lambda^k\mathbb{C}^{N+1}$ and its dual $\langle v_k|$ are:

$$|v_k\rangle = |\epsilon_1\rangle \wedge \cdots \wedge |\epsilon_k\rangle, \quad \langle v_k| = \langle \epsilon_1| \wedge \cdots \wedge \langle \epsilon_k|$$

Since by definition, in the representation $\Lambda^k \mathbb{C}^{N+1}$, $B|v_k\rangle = B|\epsilon_1\rangle \wedge \cdots \wedge$ $B|\epsilon_k\rangle$ it is easy to check that $n_+|v_k\rangle = |v_k\rangle$. Moreover, we have for any $(N+1) \times (N+1)$ matrix B:

$$\langle v_k|B|v_k\rangle = \det \begin{pmatrix} B_{11} & \cdots & B_{1k} \\ \vdots & & \vdots \\ B_{k1} & \cdots & B_{kk} \end{pmatrix}, \quad \text{with } B = (B_{ij})$$

Proposition. *Let $|v_k\rangle$ be the highest weight vectors of $sl(N+1)$ defined above. Let $B = w^{-1}u\Delta(t)u^{-1}$ and define the N functions of t,*

$$\tau_k(t) \equiv \langle v_k|B|v_k\rangle$$

then the solution of the open Toda chain is given by

$$e^{2(q_i - q_{i+1})} = \frac{\tau_{i+1}(t)\tau_{i-1}(t)}{\tau_i^2(t)} \tag{4.46}$$

<u>Proof.</u> Let h_i be the matrix elements of the diagonal $(N+1) \times (N+1)$ matrix h in $b_-^{-1}b_+ = n_-^{-1}h^2 n_+$. The Toda position coordinates are given by eq. (4.42). According to the previous analysis we have $\tau_k = \langle v_k|B|v_k\rangle = \langle v_k|h^2|v_k\rangle$, because $B = n_-^{-1}h^2 n_+$, so that $\tau_k = h_1^2 \cdots h_k^2$. Setting $\tau_0 = \tau_{N+1} = 1$ by convention, we deduce

$$h_i^2 = \frac{\tau_i}{\tau_{i-1}}$$

∎

Let us summarize the way to find the solution: given the initial data L_0 one computes the coordinates (Ω_0, Δ_0) and we let these coordinates evolve according to eq. (4.44). Then, one computes u and evaluates the τ_i by the determinant formula. Finally, the Toda coordinates q_i are given by (4.46). All these steps are simple algebraic computations.

4.9 Toda system and Hamiltonian reduction

Complementary to the algebraic approach to integrable systems presented in the previous sections, there exists a more geometrical approach, advocated by Olshanetsky and Perelomov, which we want to present on the example of the open Toda chain. The idea is to start from the geodesic motion on a Lie group G and to perform a suitable Hamiltonian reduction, which leads to the open Toda chain.

Geodesics on the group G correspond to left translations of one-parameter groups (the tangent vector is transported parallel to itself), so $\frac{d}{dt}(g^{-1}\dot{g}) = 0$. The solution is of the form $g(t) = g(0)\exp(Xt)$, for X in the Lie algebra of G. Of course this leaves open the problem of effectively computing the exponential, which was the main concern of the previous section.

We first recall some notions about Lie groups and symmetric spaces, see Chapter 16. Let G be a complex simple Lie group with Lie algebra $\mathcal{G}$. Let $\{H_i\}$ be the generators of a Cartan subalgebra $\mathcal{H}$, and let $\{E_{\pm\alpha}\}$ be the corresponding root vectors, chosen to form a Weyl basis, i.e. all the structure constants are real, see Chapter 16. The real normal form of $\mathcal{G}$ is the real Lie algebra $\mathcal{G}_0$ generated over $\mathbb{R}$ by the H_i and $E_{\pm\alpha}$. Let σ be the Cartan involution: $\sigma(H_i) = -H_i$, $\sigma(E_{\pm\alpha}) = -E_{\mp\alpha}$. The fixed points of σ form a Lie subalgebra $\mathcal{K}$ of $\mathcal{G}_0$ generated by $\{E_\alpha - E_{-\alpha}\}$. We have the decomposition:

$$\mathcal{G}_0 = \mathcal{K} \oplus \mathcal{M}$$

where $\mathcal{M}$ is the real vector space generated by the $\{E_\alpha + E_{-\alpha}\}$ and the $\{H_i\}$. Notice that due to the choice of the real normal form, $\mathcal{A} = \mathcal{H} \cap \mathcal{G}_0$ is a maximal Abelian subalgebra of $\mathcal{G}_0$ and it is entirely contained in $\mathcal{M}$. Finally, we need the real nilpotent subalgebras $\mathcal{N}_\pm$ generated respectively by the $\{E_{\pm\alpha}\}$.

Let G_0 be the connected Lie group corresponding to the Lie algebra $\mathcal{G}_0$, and similarly K corresponding to $\mathcal{K}$ and $N_\pm$ corresponding to $\mathcal{N}_\pm$. The Cartan algebra $\mathcal{A}$ exponentiates to A. The coset space G_0/K is a symmetric space of the non-compact type.

The connected Lie group G_0 admits the following Iwasawa decomposition:

$$G_0 \simeq N_+ \times A \times K \text{ as a manifold}$$

that is, any element g in G_0 can be written *uniquely* as $g = nQk$.

Remark. In the case $G_0 = SL(N+1, \mathbb{R})$, let $\sigma : g \to {}^t g^{-1}$ be the canonical automorphism of G. At the Lie algebra level, σ reads $\sigma(X) = -{}^t X$. The subgroup K is the group of orthogonal matrices with determinant 1, and the symmetric space G_0/K can be viewed as the space of symmetric matrices, with determinant 1. Finally, A is the set of diagonal matrices, and N_+ is the group of upper triangular matrices with 1 on the diagonal. To show that any g can be written as nQk, recall the Gram–Schmidt orthogonalization procedure. Starting from a basis $(f_1, \ldots, f_{N+1})$ one constructs inductively an orthogonal basis by setting $u_1 = f_1$, $u_2 = f_2 - f_1 (f_2, f_1)/(f_1, f_1)$, etc., that is $(u_1, \ldots, u_{N+1}) = (f_1, \ldots, f_{N+1})n$ with $n \in N_+$. One then normalizes this basis by setting $(u_1', \ldots, u_{N+1}') = (u_1, \ldots, u_{N+1})Q$, with Q the diagonal matrix with entries

$Q_j = 1/\sqrt{(u_j, u_j)}$. Applying this procedure to the basis f_j of columns of g^{-1}, we have $(f_1, \ldots, f_{N+1})g = (e_1, \ldots, e_{N+1})$ where e_j is the canonical basis of $\mathbb{R}^{N+1}$. On the other hand:

$$(f_1, \ldots, f_{N+1})n = (u'_1, \ldots, u'_{N+1})$$
$$(u_1, \ldots, u_{N+1})Q = (u'_1, \ldots, u'_{N+1})$$
$$(u'_1, \ldots, u'_{N+1})k = (e_1, \ldots, e_{N+1})$$

for a unique orthogonal matrix k. Comparing, we get the unique decomposition $g = nQk$.

We next recall some generalities about the symplectic structure on the cotangent bundle of a Lie group, and Hamiltonian reduction, see Chapter 14. The tangent space of G at $g \in G$ consists of vectors $g \cdot X$, where $X \in \mathcal{G}$, the Lie algebra of G. The cotangent bundle $N = T^*G$ is identified with $G \times \mathcal{G}^*$ by $p \in T_g^*(G) \longrightarrow (g, \xi)$, where $p(g \cdot X) = \xi(X)$. Using the invariant bilinear form on $\mathcal{G}$, we can identify $\mathcal{G}$ and $\mathcal{G}^*$, so that $\xi(X) = (\xi, X)$.

If $(\delta g, \delta \xi)$ is a vector tangent to T^*G at the point (g, ξ), the canonical 1-form is given by:

$$\alpha\,(\delta g, \delta \xi) = 2\xi\,(g^{-1} \cdot \delta g) \tag{4.47}$$

thereby providing the symplectic form $\omega = d\alpha$. The factor 2 has been introduced to match the conventions of the previous sections. Notice that α and ω are invariant under both left and right translations.

One can then attempt to reduce this phase space using Lie subgroups H_L and H_R of G with Lie algebras $\mathcal{H}_L$ and $\mathcal{H}_R$, acting respectively on the left and on the right on T^*G by:

$$(\,(h_L, h_R), (g, \xi)\,) \to (h_L g h_R^{-1}, h_R \xi h_R^{-1})$$

We are interested in the computation of Poisson brackets of functions on phase space. It is important to remember that, if we have a group action on a manifold M, $(g, m) \to gm$, the action on functions on M reads $f \to g \cdot f$, where $g \cdot f(m) = f(g^{-1}m)$. In our case we can take as elementary functions the collection of the matrix elements of g in a faithful representation of G, and the components of ξ. Any other function can be expressed in these coordinates.

So it is enough to give the Poisson brackets of these elementary functions. The general theory, see Chapter 14, shows that these Poisson brackets are expressed as:

$$\{\xi(X), \xi(Y)\} = \frac{1}{2}\xi([X, Y]), \quad \{\xi(X), g\} = \frac{1}{2}g\,X, \quad \{g, g\} = 0 \tag{4.48}$$

In these equations g is an abuse of notation for the matrix $\rho(g)$ in some faithful representation. What eq. (4.48) really means is that $\{\xi(X), \rho(g)_{ij}\} = \frac{1}{2}\rho(gX)_{ij}$, and $\{\rho(g)_{ij}, \rho(g)_{kl}\} = 0$. On these elementary *functions*, g and $\xi(Y)$, the infinitesimal group action $\delta_{(X_L, X_R)}$ reads:

$$\delta_{(X_L, X_R)}g = -X_L g + g X_R, \quad \delta_{(X_L, X_R)}\xi(Y) = -([X_R, \xi], Y)$$

It is straightforward to check, using the Poisson brackets (4.48), that the generating Hamiltonians are:

$$H_{X_L} = -2\xi(g^{-1}X_L g), \quad H_{X_R} = 2\xi(X_R)$$

where we recall that for any function f, we have by definition of the Hamiltonian H_X generating the group action, $\delta_X f = \{H_X, f\}$. This means that the moments, defined by $H_X = (\mathcal{P}, X)$, are:

$$\mathcal{P}^L(g, \xi) = -2P_{\mathcal{H}_L^*}\, g\xi g^{-1}, \quad \mathcal{P}^R(g, \xi) = 2P_{\mathcal{H}_R^*}\, \xi \qquad (4.49)$$

where we have introduced the projectors on the spaces $\mathcal{H}_{L,R}^*$ in $\mathcal{G}^*$ by taking the restriction of $\xi \in \mathcal{G}^*$ to $\mathcal{H}_{L,R}$, the Lie algebras of $H_{L,R}$.

The geodesic motion on G is a Hamiltonian system on T^*G whose Hamiltonian is $H = (\xi, \xi)$. Indeed, one finds using eq. (4.48), $\dot{\xi} = 0$ and $g^{-1}\dot{g} = \xi$ as equations of motion. Notice that H is bi-invariant, so one can reduce by the group actions on the left and on the right.

To get the Toda chain we shall perform the reduction of the geodesic motion on T^*G_0 by the action of the group N_+ on the left and K on the right according to the Iwasawa decomposition. More precisely, this is achieved by a suitable choice of the momentum $(\mu = \mu^R, \mu^L)$. We take:

$$P_{\mathcal{K}^*}(\xi) = \mu^R = 0 \qquad (4.50)$$

$$-2P_{\mathcal{N}_+^*}(g\xi g^{-1}) = \mu^L \equiv -2\sum_{\alpha \text{ simple}} E_{-\alpha} \qquad (4.51)$$

In the following we shall identify $\mathcal{N}_+^*$ with $\mathcal{N}_-$, and $\mathcal{K}^*$ with $\mathcal{K}$ through the invariant scalar product.

The isotropy group G_μ of the momentum μ, is $N_+ \times K$. This is obvious for the right component since $\mu^R = 0$. The isotropy group of μ^L is by definition the set of elements $g \in N_+$ such that:

$$(\mu^L, g^{-1}Xg) = (\mu^L, X), \quad \forall X \in \mathcal{N}_+$$

Since μ^L only contains roots of degree -1 (see section (4.8)), the only contribution to (μ^L, X) comes from $X^{(1)}$, the degree one component of X. But

$(g^{-1}Xg)^{(1)} = X^{(1)} \; \forall g \in N_+$, because, for $Y \in \mathcal{N}_+$, $\exp{(Y)}X \exp(-Y) = X + [Y, X] + \cdots$, and Y is of degree ≥ 1. Hence the isotropy group of μ^L is N_+ itself.

The reduced phase space is $\mathcal{F}_\mu = N_\mu/G_\mu$, where N_μ is the level manifold defined by the equations (4.50, 4.51). Let us compute its dimension. Let $d = \dim \mathcal{G}_0$ and $r = \dim \mathcal{A}$. Then we have:

$$\dim \mathcal{K} = \dim \mathcal{N}_+ = \frac{d-r}{2}, \quad \dim T^*G_0 = 2d$$

The dimension of the submanifold N_μ is $\dim N_\mu = 2d - \dim \mathcal{K} - \dim \mathcal{N}_+ = d + r$ and the dimension of the reduced phase space is $\dim \mathcal{F}_\mu = \dim N_\mu - \dim \mathcal{K} - \dim \mathcal{N}_+ = 2r$, which is the correct dimension of the phase space of the Toda chain.

Since the isotropy group is the whole of $N_+ \times K$, any point (g, ξ) of N_μ can be brought to the form

$$(g, \xi) \rightarrow (Q, L)$$

with $Q \in A$ (due to the Iwasawa decomposition) by the action of the isotropy group. We shall identify $\mathcal{G}_0$ and $\mathcal{G}_0^*$ under the Killing form for convenience. Equation (4.50) implies that $L \in \mathcal{M}$, which is orthogonal to $\mathcal{K}$. Thus we can write:

$$L = p + \sum_{\alpha \text{ positive}} l_\alpha(E_\alpha + E_{-\alpha}), \quad p \in \mathcal{A}$$

Inserting this form into eq. (4.51) and setting $Q = \exp(q)$, we get:

$$P_{\mathcal{N}_-} \, QLQ^{-1} = \sum_{\alpha > 0} l_\alpha e^{-\alpha(q)} E_{-\alpha} = \sum_{\alpha \text{ simple}} E_{-\alpha}$$

hence $l_\alpha = \exp(\alpha(q))$ for α simple and $l_\alpha = 0$ otherwise. We have obtained the Lax matrix of the Toda chain:

$$L = p + \sum_{\alpha \text{ simple}} e^{\alpha(q)} \, (E_\alpha + E_{-\alpha}) \tag{4.52}$$

The set of the (Q, L) is obviously a $(2r)$-dimensional subvariety $\mathcal{S}$ of N_μ. For any point (g, ξ) in N_μ one can write uniquely $g = nQk$ and $\xi = k^{-1}Lk$.

To compute the reduced symplectic form in the coordinates (Q, L), it is enough to evaluate the canonical 1-form:

$$\alpha = 2\xi(g^{-1}\delta g) = 2L(Q^{-1}\delta Q) = 2L(\delta q) = 2\sum_i p_i \delta q_i$$

This completes the identification of the reduced geodesic flow with the open Toda chain.

An interesting feature of this approach is that it allows us to compute naturally the r-matrix of the Toda model. We refer to Chapter 14 for the exposition of the general method. The function $L(X)$ (for any $X \in \mathcal{M}$) on the reduced phase space has a uniquely defined extension on T^*G_0, invariant under the action of the group $N_+ \times K$:

$$F_X(g, \xi) = (\xi, k^{-1}Xk) \tag{4.53}$$

where $k = k(g)$ is uniquely determined by the Iwasawa decomposition $g = nQk$. In this situation one has simply the reduced Poisson bracket:

$$\{L(X), L(Y)\} = \{F_X, F_Y\} = \{(k\xi k^{-1}, X), (k\xi k^{-1}, Y)\}$$

We evaluate this bracket with the help of eqs. (4.48), and then restrict to $n = k = e$. For a function $f : G \to G$ we have:

$$\{\xi(X), f(g)\} = \frac{1}{2}\frac{d}{dt}f(ge^{tX})|_{t=0}$$

Applying this to the function $k = k(g)$, we get:

$$\{\xi(X), k(g)\} = \frac{1}{2}\nabla k(X), \quad \text{where } \nabla k(X) = \frac{\partial}{\partial t}k(Q\exp(tX))|_{t=0}$$

We then immediately obtain:

$$\{L(X), L(Y)\} = \frac{1}{2}\left(L, [X, Y] - [X, \nabla k(Y)] - [\nabla k(X), Y]\right)$$

Notice that $[X, Y] \in \mathcal{K}$ because $X, Y \in \mathcal{M}$, hence $L([X, Y]) = 0$, because $L \in \mathcal{M}$. From this equation we get an r-matrix for the Toda system given by:

$$R'X = -\frac{1}{2}\nabla k(X) \in \mathcal{K}$$

We can compute this r-matrix as follows: Given an element $X \in \mathcal{M}$, we write the Iwasawa decomposition of Qe^{tX} as:

$$Qe^{tX} = e^{tX_+} \cdot Qe^{tX_a} \cdot e^{tX_K}$$

This uniquely defines the quantities, (a priori depending on t), $X_+ \in \mathcal{N}_+$, $X_a \in \mathcal{A}$, and $X_K \in \mathcal{K}$. Letting t tend to zero, we see that $\nabla k(X) = X_K$ and moreover

$$QX = X_+Q + X_aQ + QX_K \Rightarrow X = Q^{-1}X_+Q + X_a + X_K \tag{4.54}$$

Any $X \in \mathcal{M}$ can be written in the form $X = \sum_{\alpha > 0} x_\alpha (E_\alpha + E_{-\alpha}) + \sum_i x_i H_i$ which we rewrite as

$$X = 2 \sum_{\alpha > 0} x_\alpha E_\alpha + \sum_i x_i H_i - \sum_{\alpha > 0} x_\alpha (E_\alpha - E_{-\alpha}) \qquad (4.55)$$

Noticing that $Q^{-1} X_+ Q \in \mathcal{N}_+$, we get by comparison of eqs. (4.54) and (4.55), $Q^{-1} X_+ Q = 2 \sum_{\alpha > 0} x_\alpha E_\alpha$, $X_a = \sum_i x_i H_i$, and finally

$$R'X = -\frac{1}{2} \nabla k(X) = -\frac{1}{2} X_K = \frac{1}{2} \sum_{\alpha > 0} x_\alpha (E_\alpha - E_{-\alpha})$$

By dualization, $R'X = \mathrm{Tr}_2(r'_{12} . \mathbf{1} \otimes X)$, we obtain the r-matrix:

$$r'_{12} = \frac{1}{4} \sum_{\alpha > 0} \frac{(E_\alpha - E_{-\alpha}) \otimes (E_\alpha + E_{-\alpha})}{(E_\alpha, E_{-\alpha})} \qquad (4.56)$$

We want to relate r'_{12} to the r-matrix, r_{12}, we computed in eq. (4.26). For this, we recall that L and r_{12} satisfy the equation $\{L_1, L_2\} = [r_{12}, L_1] - [r_{21}, L_2]$. Applying the automorphism $\sigma_1 = \sigma \otimes 1$ to this equation and noting that $\sigma(L) = -L$, we get $\{L_1, L_2\} = [\sigma_1 r_{12}, L_1] + [\sigma_1 r_{21}, L_2]$. Adding the two equations, we get a similar r-matrix relation:

$$\{L_1, L_2\} = [\frac{1}{2}(r_{12} + \sigma_1 r_{12}), L_1] - [\frac{1}{2}(r_{21} - \sigma_1 r_{21}), L_2]$$

It is straightforward to check that:

$$\frac{1}{2}(r_{12} + \sigma_1 r_{12}) = r'_{12}, \quad \frac{1}{2}(r_{21} - \sigma_1 r_{21}) = r'_{21}$$

hence we recover eq. (4.56).

4.10 The Lax pair of the Kowalevski top

Applying the algebraic setting of this chapter to classical Lie groups and low-dimensional coadjoint orbits, one can find a rich variety of integrable systems. This pragmatic approach was rewarded by the discovery of a Lax pair with spectral parameter for the Kowalevski top by Reyman and Semenov-Tian-Shansky. We now explain this construction.

Consider a Lie group G with an involutive automorphism σ. This involution defines a linear involution on the Lie algebra $\mathcal{G}$ that we also call σ. Let K be the subgroup of fixed points of σ, and $\mathcal{K}$ its Lie algebra. We have

$$\mathcal{G} = \mathcal{K} \oplus \mathcal{R}$$

where $\sigma = +1$ on $\mathcal{K}$ and $\sigma = -1$ on $\mathcal{R}$. Since σ is a Lie algebra automorphism of order 2, we have

$$[\mathcal{K}, \mathcal{K}] \subset \mathcal{K}, \quad [\mathcal{K}, \mathcal{R}] \subset \mathcal{R}, \quad [\mathcal{R}, \mathcal{R}] \subset \mathcal{K}$$

By exponentiation, the vector space $\mathcal{R}$ is a representation space for the Lie group K acting by conjugation. From these data we first construct the loop algebra $\tilde{\mathcal{G}} = \mathcal{G} \otimes \mathbb{C}[[\lambda, \lambda^{-1}]]$ with the commutation relations eq. (4.14), and then the twisted loop algebra $\tilde{\mathcal{G}}_\sigma$ as the set of fixed points of the induced involution $\hat{\sigma}(X(\lambda)) = (\sigma \cdot X)(-\lambda)$. Elements of the twisted loop algebra are of the form $X(\lambda) = \sum X_n \lambda^n$ with $X_n \in \mathcal{K}$ for n even, and $X_n \in \mathcal{R}$ for n odd. As in section (4.3), this algebra decomposes as a sum of two subalgebras:

$$\tilde{\mathcal{G}}_\sigma = \tilde{\mathcal{G}}_{\sigma+} + \tilde{\mathcal{G}}_{\sigma-}$$

where elements of $\tilde{\mathcal{G}}_{\sigma+}$ have vanishing or positive powers of λ while elements of $\tilde{\mathcal{G}}_{\sigma-}$ have strictly negative powers of λ. Applying the Adler–Kostant–Symes scheme, we define a pair of matrices $R^\pm$ whose actions are given by

$$R^+(X(\lambda)) = -X_+(\lambda)$$
$$R^-(X(\lambda)) = X_-(\lambda)$$

As compared with eq. (4.15), we have introduced a minus sign to match the conventions of Chapter 2.

Hence we have two Lie algebra brackets on $\tilde{\mathcal{G}}_\sigma$, the original one, and the one in which $\tilde{\mathcal{G}}_{\sigma\pm}$ commute, defining the Lie algebra $\tilde{\mathcal{G}}_{\sigma R}$, with which is associated the Lie group $\tilde{G}_R$.

The invariant bilinear form on $\tilde{\mathcal{G}}_\sigma$ is still given by eq. (4.16). It allows us to write elements of the dual $\tilde{\mathcal{G}}_\sigma^*$ as series $L(\lambda) = \sum_n l_n \lambda^{n-1}$ with $l_n \in \mathcal{K}$ for n even and $l_n \in \mathcal{R}$ for n odd. On this dual we consider the Poisson algebra defined by the Poisson brackets $\{\ ,\ \}_R$. To construct an integrable system we select a particularly simple orbit:

Proposition. *Let A and B be two fixed elements of $\mathcal{R}$.*
(i) The set of elements

$$L(\lambda; \xi, k) = A + \xi \lambda^{-1} + (k^{-1} B k) \lambda^{-2}, \quad \xi \in \mathcal{K}, \ k \in K \qquad (4.57)$$

is a coadjoint orbit of $\tilde{G}_R$.
(ii) Setting $L = l_{-1} \lambda^{-2} + l_0 \lambda^{-1} + l_1$, with $l_0 \in \mathcal{K}$ and $l_{\pm 1} \in \mathcal{R}$, the Poisson brackets of this Lax matrix are given for $x, y \in \mathcal{K}$, $z \in \mathcal{R}$ by:

$$\{(l_0, x), (l_0, y)\}_R = -(l_0, [x, y]), \quad \{(l_0, x), (l_{-1}, z)\}_R = -(l_{-1}, [x, z])$$
$$(4.58)$$

*All other Poisson brackets vanish. In particular, l_1 is in the centre of the
Poisson algebra.*
*(iii) Equation (4.57) shows that there is a map from $T^*(K)$ to the orbit.
This is a Poisson map.*

<u>Proof.</u> We first show that the elements of the form $L = l_{-1}\lambda^{-2} + l_0\lambda^{-1} + l_1$
form a coadjoint orbit of $\tilde{G}_R$. It is enough to show that such elements are
stable under the coadjoint action of $\tilde{G}_{\sigma R}$. For $X = \sum x_n\lambda^n$ we have by
eq. (4.13) $(\mathrm{ad}_R^* X \cdot L)(Y) = L([X,Y]_R) = -L([X_+,Y_+] + [X_-,Y_-])$, so
that:

$$(\mathrm{ad}_R^* X \cdot L, Y) = -([X_+,L],Y_+) + ([X_-,L],Y_-)$$
$$= -([x_0,l_{-1}]\lambda^{-2} - ([x_0,l_0] + [x_1,l_{-1}])\lambda^{-1}, Y_+)$$

We see that under this coadjoint action l_1 remains invariant, l_{-1} is transformed by the coadjoint action of $\mathcal{K}$, and l_0 is transformed to a generic
element of $\mathcal{K}$:

$$\delta l_1 = 0, \quad \delta l_{-1} = -[x_0,l_{-1}], \quad \delta l_0 = -[x_0,l_0] - [x_1,l_{-1}]$$

Next we compute the Poisson brackets of L induced by the Kostant–
Kirillov bracket $\{\,,\,\}_R$. In the formula $\{(L,X),(L,Y)\}_R = (L,[X,Y]_R)$,
we choose $X = x_p\lambda^p$, $p = -1, 0, 1$, and similarly for Y, to probe the three
components of L. For example $\{(l_1,x_{-1}),(l_0,y_0)\}_R = (L,[x_{-1},y_0]_R) = 0$
because x_{-1} and y_0 lie in the two subspaces of $\tilde{G}_\sigma$ which commute in
$\tilde{G}_{\sigma R}$. Similarly, $\{(l_0,x_0),(l_{-1},y_1)\}_R = -(L,[x_0,y_1]\lambda) = -(l_{-1},[x_0,y_1])$,
etc. To show that these Poisson brackets are the same as those of $T^*(K)$,
recall the parametrization (k,ξ) of the generic element of the cotangent
bundle $T^*(K)$ and the Poisson bracket formulae (with an extra minus
sign compared to Chapter 14):

$$\{\xi(x),\xi(y)\} = -\xi([x,y]), \quad \{\xi(x),k\} = -kx, \quad \{k,k\} = 0 \qquad (4.59)$$

The first of eq. (4.59) reproduces the Kostant–Kirillov bracket of $\mathcal{K}$ as in
the first of eq. (4.58) with the identification $\xi = l_0$. The second of eq. (4.59)
is equivalent to $\{\xi(x),k^{-1}Bk\} = -[k^{-1}Bk,x]$, which coincides with the
second of eq. (4.58) since $(l_{-1},[x,z]) = ([l_{-1},x],z)$ and $l_{-1} = k^{-1}Bk$. It
follows that the map from $T^*(K)$ to the orbit preserves the symplectic
structure. ∎

In the general Adler–Kostant–Symes construction, any function invariant under the coadjoint action of $\tilde{G}_\sigma$ yields an integrable flow on our
coadjoint orbit. Such functions are given by Res Tr $(\lambda^m L^n(\lambda))$ for any
m, n. In particular, taking $m = 1$ and $n = 2$ we get the Hamiltonian:

$$H = -\mathrm{Tr}\,(\xi^2) - 2\mathrm{Tr}\,(k^{-1}BkA) \qquad (4.60)$$

where the minus sign is appropriate to get a positive Hamiltonian when K is a compact group. Recall that K acts on the left and on the right on $T^*(K)$ by $((h_L, h_R), (k, \xi)) \to (h_L k h_R^{-1}, h_R \xi h_R^{-1})$, so this Hamiltonian is invariant under the subgroup of K_L of elements which stabilize B, $h_L^{-1} B h_L = B$, and under the subgroup of K_R of elements which stabilize A, $h_R^{-1} A h_R = A$. More generally, under these special subgroups, the Lax matrix L is invariant under h_L and gets conjugated under h_R, so that any Hamiltonian of the hierarchy is *invariant*. One can then perform a Hamiltonian reduction under the action of such subgroups. These reduced Hamiltonians are in involution using the reduced Poisson bracket, because reduced Poisson brackets coincide with unreduced ones for invariant functions.

This scheme is particularly interesting when G is a real Lie group and σ is a Cartan involution. Then K is a maximal compact subgroup of G, see Chapter 16, so that G/K is a symmetric space. The situation which leads to the Kowalevski top is obtained by considering $G = SO(p, q)$ and $K = SO(p) \times SO(q)$ with $p \geq q$. The Kowalevski case corresponds to $p = 3, q = 2$. The group $SO(p, q)$ is the real Lie group of pseudo-orthogonal transformations of $\mathbb{R}^{p+q}$ leaving invariant the metric $x_1^2 + \cdots + x_p^2 - x_{p+1}^2 - \cdots - x_{p+q}^2$. Elements of its Lie algebra may be represented as matrices of the form:

$$X = \begin{pmatrix} X_p & D \\ {}^t D & X_q \end{pmatrix} \quad (X_p, X_q) \in so(p) \oplus so(q) \qquad (4.61)$$

where D is an arbitrary (p, q) matrix. The subalgebra $\mathcal{K}$ consists of block diagonal matrices (X_p, X_q), while the subspace $\mathcal{R}$ consists of the off-diagonal terms. Let us write the L matrix eq. (4.57) in this context, with $k = (k_p, k_q) \in SO(p) \times SO(q)$ and $\xi = (\xi_p, \xi_q) \in so(p) \oplus so(q)$:

$$L = \begin{pmatrix} 0 & a \\ {}^t a & 0 \end{pmatrix} + \begin{pmatrix} \xi_p & 0 \\ 0 & \xi_q \end{pmatrix} \lambda^{-1} + \begin{pmatrix} 0 & k_p^{-1} b k_q \\ k_q^{-1} {}^t b k_p & 0 \end{pmatrix} \lambda^{-2} \qquad (4.62)$$

Here, we have written the matrices $A, B \in \mathcal{R}$ in eq. (4.57) in the form

$$A = \begin{pmatrix} 0 & a \\ {}^t a & 0 \end{pmatrix}, \quad B = \begin{pmatrix} 0 & b \\ {}^t b & 0 \end{pmatrix}$$

where a and b are rectangular $p \times q$ matrices. We have also computed $k^{-1} B k$ explicitly.

To obtain the Kowalevski top we choose a specific orbit, i.e. we specify B. This amounts to choosing the rectangular $p \times q$ matrix b, which we take as:

$$b = \begin{pmatrix} 1_q \\ 0_{p-q} \end{pmatrix}$$

where 1_q means the $q \times q$ identity matrix, and 0_{p-q} means the $(p-q) \times q$ zero matrix. The subgroup of K_L which stabilizes B consists of matrices of the form:

$$h_L = \begin{pmatrix} \widehat{h}_q & 0 \\ 0 & h_q \end{pmatrix}, \quad \widehat{h}_q \equiv \begin{pmatrix} h_q & 0 \\ 0 & 1_{p-q} \end{pmatrix} \in SO(p) \quad \text{for } h_q \in SO(q)$$

where the map $h_q \to \widehat{h}_q$ embeds $SO(q)$ into $SO(p)$ and the map $h_q \to h_L$ embeds the group $SO(q)$ into $SO(p) \times SO(q)$. Hence the reduction subgroup identifies with $SO(q)$. At the Lie algebra level, it is realized into the Lie algebra $so(p) \oplus so(q)$ by pairs of matrices

$$X_L = (\widehat{X}_q, X_q)$$

Using these embeddings we can write $k_p^{-1} b k_q = rb$, where $r = k_p^{-1} \widehat{k}_q \in SO(p)$. The quantity r is invariant under the action of the reduction group since $h_q \cdot (k_p, k_q) = (\widehat{h}_q k_p, h_q k_q)$. Moreover, the action of K_L leaves ξ invariant, so we have the manifestly invariant expression for L:

$$L = \begin{pmatrix} 0 & a \\ {}^t a & 0 \end{pmatrix} + \begin{pmatrix} \xi_p & 0 \\ 0 & \xi_q \end{pmatrix} \lambda^{-1} + \begin{pmatrix} 0 & rb \\ {}^t b r^{-1} & 0 \end{pmatrix} \lambda^{-2}, \quad r = k_p^{-1} \widehat{k}_q$$

The moment map is given by eq. (4.49). For any X_L in the Lie algebra of the reduction group it is given by:

$$\begin{aligned} (\mathcal{P}^L(k,\xi), X_L) &= (k_p \xi_p k_p^{-1}, \widehat{X}_q) + (k_q \xi_q k_q^{-1}, X_q) \\ &= (k_q \xi_q k_q^{-1} + \pi_q(k_p \xi_p k_p^{-1}), X_q) \end{aligned}$$

where $X_q \in so(q)$, $k = (k_p, k_q) \in K_L$, $\xi = (\xi_p, \xi_q)$ and $\pi_q(M)$ is the projection operator which restricts the $p \times p$ matrix M to its $q \times q$ upper left corner. This projector appears because $\widehat{X}_q$ is a matrix with X_q in the upper left $q \times q$ corner and 0 in the lower right $(p-q) \times (p-q)$ corner.

We choose to reduce on the surface of zero momentum which is given by $k_q \xi_q k_q^{-1} + \pi_q(k_p \xi_p k_p^{-1}) = 0$, that is:

$$\xi_q = -\frac{1}{2} \pi_q(J), \quad J = 2r^{-1} \xi_p r \in so(p) \tag{4.63}$$

The factor of 2 in the definition of J is introduced for later convenience. We still need to quotient by the stability group of the moment, i.e. the whole group $SO(q)$. However, since we deal only with invariant quantities like L, r, the Poisson brackets on the reduced phase space can be computed by simply using the Poisson brackets on the unreduced phase space, see Chapter 14. From eq. (4.59), we easily compute the Poisson brackets that we will need later:

$$\{\xi_p(r^{-1}Xr), \xi_p(r^{-1}Yr)\} = \xi_p(r^{-1}[X,Y]r)$$

$$\{\xi_p(X), r\} = Xr, \quad \{\xi_q(X), r\} = -r\widehat{X} \tag{4.64}$$

The Hamiltonian eq. (4.60) takes the form

$$H = -\operatorname{Tr}(\xi^2) - 4\operatorname{Tr}(^tbF), \quad F = r^{-1}a \tag{4.65}$$

Using the explicit form of b, we have $\operatorname{Tr}(^tbF) = \operatorname{Tr}\pi_q(F)$, while $\operatorname{Tr}(\xi^2) = \operatorname{Tr}(\xi_p^2) + \operatorname{Tr}(\xi_q^2)$. By eq. (4.63), we have $\operatorname{Tr}(\xi_p^2) = 1/4\operatorname{Tr}(J^2)$ and $\operatorname{Tr}(\xi_q^2) = 1/4\operatorname{Tr}\pi_q(J)^2$. In particular when $(p=3, q=2)$ we use that J is a 3×3 antisymmetric matrix which we write as $J_{ij} = \epsilon_{ijk}J_k$, and we express the 3×2 matrix F as $F = (\vec{P}, \vec{P}')$ for two vectors $\vec{P}$, $\vec{P}'$ of components γ_i, γ_i', $i = 1, 2, 3$. Then $\operatorname{Tr}\pi_q(F) = \gamma_1 + \gamma_2'$, and $\operatorname{Tr}\pi_q(J)^2 = -2J_3^2$, while $\operatorname{Tr}J^2 = -2\vec{J}^2$, so the Hamiltonian eq. (4.60) takes the form:

$$H = \frac{1}{2}(J_1^2 + J_2^2 + 2J_3^2) - 4(\gamma_1 + \gamma_2')$$

The Poisson brackets eq. (4.64) translate into the following non-zero Poisson brackets for the quantities J_i, γ_i, γ_i':

$$\{J_i, J_j\} = \epsilon_{ijk}J_k, \quad \{J_i, \gamma_j\} = \epsilon_{ijk}\gamma_k, \quad \{J_i, \gamma_j'\} = \epsilon_{ijk}\gamma_k'$$

We recover exactly the dynamical data of the Kowalevski top, except that it is generalized to a situation with two external forces $\vec{\gamma}$ and $\vec{\gamma}'$. The special Kowalevski case corresponds to $\vec{\gamma}' = 0$.

It is now clear that $\vec{J}$ is the angular momentum. The matrix r can be identified as the rotation relating the absolute and the moving frames, so that $\Omega = -r^{-1}\dot{r}$ is the rotation vector. We have $r^{-1}\dot{r} = r^{-1}\{H, r\} = -r^{-1}\{\operatorname{Tr}(\xi_p^2) + \operatorname{Tr}(\xi_q^2), r\}$, where in the last equality we dropped the F term in the Hamiltonian, eq. (4.65), because $\{r_1, r_2\} = 0$. Using eq. (4.64), we get $r^{-1}\dot{r} = -2r^{-1}\xi_p r + 2\widehat{\xi_q} = -J + 2\widehat{\xi_q}$. It follows that, on the zero momentum surface, we have:

$$\Omega = J + \pi_q(J)$$

For $(p = 3, q = 2)$ we denote $\Omega_{ij} = \epsilon_{ijk}\omega_k$ and we see that $J_k = I_k\omega_k$ with $I_1 = I_2 = 1$ and $I_3 = 1/2$. Finally, $F = r^{-1}a$ represents external forces in the rotating frame. These forces are constant in the absolute frame and given by the constant matrix a.

It is convenient to conjugate the Lax matrix on the zero momentum surface by the block diagonal $(p+q) \times (p+q)$ matrix $D = \mathrm{Diag}(r, 1_q)$ so that the expression of the rotated Lax matrix is:

$$\widehat{L} \equiv D^{-1}LD = \begin{pmatrix} 0 & F \\ {}^tF & 0 \end{pmatrix} + \frac{1}{2}\begin{pmatrix} J & 0 \\ 0 & -\pi_q(J) \end{pmatrix}\lambda^{-1} + \begin{pmatrix} 0 & b \\ {}^tb & 0 \end{pmatrix}\lambda^{-2} \quad (4.66)$$

In this way the Lax matrix is naturally expressed only in terms of quantities defined in the moving frame.

The Lax equation describing the motion of the matrix $\widehat{L}$ can be written in the form:

$$\frac{d}{dt}\widehat{L}(\lambda) = [\widehat{M}(\lambda), \widehat{L}(\lambda)], \quad \widehat{M}(\lambda) = 2\begin{pmatrix} 0 & b \\ {}^tb & 0 \end{pmatrix}\lambda^{-1} + \begin{pmatrix} \Omega & 0 \\ 0 & 0 \end{pmatrix}$$

In fact we have $\widehat{M}(\lambda) = 2(\lambda\widehat{L}(\lambda))_- - D^{-1}\dot{D}$, the first term follows from the general theory, and the second term is produced by the conjugation by D in $\widehat{L}$. Explicitly, this Lax equation reads:

$$\dot{F} = \Omega F, \quad \dot{J} = -[J,\Omega] - 4(F\,{}^tb - b\,{}^tF), \quad \pi_q(\dot{J}) = 4({}^tFb - {}^t bF)$$

which in components gives:

$$\begin{array}{lll} 2\dot{p} = qr + 8\gamma_3' & \dot{\gamma}_1 = r\gamma_2 - q\gamma_3 & \dot{\gamma}'_1 = r\gamma_2' - q\gamma_3' \\ 2\dot{q} = -pr - 8\gamma_3 & \dot{\gamma}_2 = p\gamma_3 - r\gamma_1 & \dot{\gamma}'_2 = p\gamma_3' - r\gamma_1' \\ \dot{r} = -8(\gamma_1' - \gamma_2') & \dot{\gamma}_3 = q\gamma_1 - p\gamma_2 & \dot{\gamma}'_3 = q\gamma_1' - p\gamma_2' \end{array}$$

These equations generalize the equations of the Kowalevski top given in Chapter 2, which are recovered if $\vec{\gamma}' = 0$ and $c_0 = 8$ or $h = 4$.

Finally, it is convenient to switch to a more compact representation of the $so(3,2)$ Lie algebra by 4×4 matrices instead of 5×5 matrices. We first consider the complex Lie algebra $so(5,\mathbb{C})$ and view $so(3,2)$ as one of its non-compact real forms, obtained using the conjugation $X \to \eta\bar{X}\eta$ where $\eta = \mathrm{Diag}\,(1,1,1,-1,-1)$ is the metric left-invariant by $so(3,2)$. Any element of the Lie algebra $so(3,2)$, in the vector representation, can be written as in eq. (4.61) with:

$$X_p = \begin{pmatrix} 0 & x_3 & -x_2 \\ -x_3 & 0 & x_1 \\ x_2 & -x_1 & 0 \end{pmatrix}, \quad X_q = \begin{pmatrix} 0 & y \\ -y & 0 \end{pmatrix}, \quad D = (\vec{\gamma}, \vec{\gamma}')$$

This matrix is mapped to an *antisymmetric* matrix of $so(5,\mathbb{C})$ by a change of basis, namely conjugation by $\mathrm{Diag}\,(1,1,1,i,i)$. In this basis the matrix $\widehat{X}$ reads:

$$\widehat{X} = \begin{pmatrix} X_p & iD \\ -i\,{}^tD & X_q \end{pmatrix} \quad (4.67)$$

Let $T_{\mu\nu} = E_{\mu\nu} - E_{\nu\mu}$ for $\mu < \nu$ be the standard basis of antisymmetric matrices, obeying the $so(5)$ algebra:

$$[T_{\mu\nu}, T_{\rho\sigma}] = T_{\mu\sigma}\delta_{\nu\rho} - T_{\rho\nu}\delta_{\mu\sigma} + T_{\sigma\nu}\delta_{\mu\rho} - T_{\mu\rho}\delta_{\nu\sigma}$$

It is well known, and easy to check, that this algebra can be realized with Γ matrices satisfying the anticommutation relations $\{\Gamma_\mu, \Gamma_\nu\} = 2\delta_{\mu\nu}$, $\mu, \nu = 1, \ldots, 5$. This is called the spinorial representation of $so(5, \mathbb{C})$. The $T_{\mu\nu}$ are given by $T_{\mu\nu} = 1/4[\Gamma_\mu, \Gamma_\nu]$. It remains to notice that the Γ_μ can be represented by 4×4 matrices as follows:

$$\Gamma_j = \sigma_1 \otimes \sigma_j, \quad \Gamma_4 = \sigma_2 \otimes \sigma_0, \quad \Gamma_5 = \sigma_3 \otimes \sigma_1$$

where $\sigma_0 = 1$ and the σ_j are the 2×2 Pauli matrices:

$$\sigma_1 = \begin{pmatrix} 0 & 1 \\ 1 & 0 \end{pmatrix}, \quad \sigma_2 = \begin{pmatrix} 0 & -i \\ i & 0 \end{pmatrix}, \quad \sigma_3 = \begin{pmatrix} 1 & 0 \\ 0 & -1 \end{pmatrix} \tag{4.68}$$

obeying the relations $\sigma_i^2 = 1$ and $\sigma_i\sigma_j = i\epsilon_{ijk}\sigma_k$ for $(i, j, k) = (1, 2, 3)$. The matrix $\widehat{X}$ in eq. (4.67) expands on the $T_{\mu\nu}$ as

$$\widehat{X} = x_3 T_{12} - x_2 T_{13} + x_1 T_{23} + y T_{45} + i \sum_{j=1}^{3} (\gamma_j T_{j4} + \gamma_j' T_{j5})$$

Plugging the representation of $T_{\mu\nu}$ in terms of the Γ_μ, we get:

$$\widehat{X} = \frac{1}{2} \begin{pmatrix} i\vec{x} \cdot \vec{\sigma} - \vec{\gamma} \cdot \vec{\sigma} & iy\sigma_0 - i\vec{\gamma}' \cdot \vec{\sigma} \\ iy\sigma_0 + i\vec{\gamma}' \cdot \vec{\sigma} & i\vec{x} \cdot \vec{\sigma} + \vec{\gamma} \cdot \vec{\sigma} \end{pmatrix}$$

We finally write the Lax pair in this 4×4 representation:

$$\widehat{L} = \frac{1}{2} \begin{pmatrix} -\vec{\gamma} \cdot \vec{\sigma} & -i\vec{\gamma}' \cdot \vec{\sigma} \\ i\vec{\gamma}' \cdot \vec{\sigma} & \vec{\gamma} \cdot \vec{\sigma} \end{pmatrix}$$
$$+ \frac{1}{4\lambda} \begin{pmatrix} i\vec{J} \cdot \vec{\sigma} & -iJ_3\sigma_0 \\ -iJ_3\sigma_0 & i\vec{J} \cdot \vec{\sigma} \end{pmatrix} + \frac{1}{2\lambda^2} \begin{pmatrix} -\sigma_1 & -i\sigma_2 \\ i\sigma_2 & \sigma_1 \end{pmatrix}$$
$$\widehat{M} = \frac{1}{2} \begin{pmatrix} i\vec{\omega} \cdot \vec{\sigma} & 0 \\ 0 & i\vec{\omega} \cdot \vec{\sigma} \end{pmatrix} + \frac{1}{\lambda} \begin{pmatrix} -\sigma_1 & -i\sigma_2 \\ i\sigma_2 & \sigma_1 \end{pmatrix} \tag{4.69}$$

It is a simple computation to check directly that the Lax equation with this Lax pair reproduces the equations of motion of the Kowalevski top.

References

[1] M. Toda, Vibration of a chain with non-linear interaction. *J. Phys. Soc. Japan* **22** (1967) 431.

[2] B. Kostant, The solution to the generalized Toda lattice and representation theory. *Adv. Math.* **34** (1979) 195–338.

[3] M. Adler, On a trace functional for formal pseudodifferential operators and symplectic structure of the Korteweg–de Vries type equations. *Inv. Math.* **50** (1979) 219.

[4] E.K. Sklyanin, *On the complete integrability of the Landau–Lifchitz equation.* Preprint LOMI E-3-79. Leningrad, 1979.

[5] W. Symes, Systems of Toda type, inverse spectral problems and representation theory. *Inv. Math.* **59** (1980) 13.

[6] M. Olshanetsky and A. Perelomov, Classical integrable finite-dimensional systems related to Lie algebras. *Phys. Rep.* **71** (1981) 313–400.

[7] A. Belavin and V. Drinfeld, On the solutions of the classical Yang–Baxter equation for simple Lie algebras. *Funct. Anal. Appl.* **16**, 3 (1982) 1–29.

[8] V. Drinfeld, Hamiltonian structures on Lie groups, Lie bialgebras, and the geometrical meaning of the Yang–Baxter equations. *Dokl. Akad. Nauk SSSR* **268** (1983) 285–287.

[9] D. Olive and N. Turok, Algebraic structure of Toda systems. *Nucl. Phys.* **B220** (1983) 491–507.

[10] M. Semenov-Tian-Shansky, What is a classical *r*-matrix? *Funct. Anal. Appl.* **17** 4 (1983) 17.

[11] L. Ferreira and D. Olive, Non-compact symmetric spaces and the Toda molecule equations. *Comm. Math. Phys.* **99** (1985) 365–384.

[12] A. Bobenko, A. Reyman and M. Semenov-Tian-Shansky, The Kowalevski top 99 years later: a Lax pair, generalizations and explicit solutions. *Comm. Math. Phys.* **122** (1989) 321–354.

[13] A. Reyman and M. Semenov-Tian-Shansky, *Group-theoretical methods in the theory of finite-dimensional integrable systems.* Encyclopaedia of Mathematical Sciences, 16. Springer-Verlag (1990).

5

Analytical methods

In this chapter, we present the general ideas for solving the Lax equations with spectral parameter. The spectral curve Γ is the characteristic equation for the eigenvalues of the Lax matrix: $\det(L(\lambda) - \mu) = 0$. Since the Lax equation $\dot{L}(\lambda) = [M(\lambda), L(\lambda)]$ is isospectral, the eigenvalues of $L(\lambda)$ are time-independent and so is the spectral curve. At any point of the spectral curve there exists, by definition, an eigenvector of $L(\lambda)$ with eigenvalue μ. We explain how we can reconstruct the eigenvector from its analyticity properties on Γ. In particular all the dynamical information is contained in the divisor of its poles which we call the dynamical divisor. The time evolution of this divisor is equivalent to a linear flow on the Jacobian of Γ. We give three proofs of this result. The first one proceeds by explicitly computing the time evolution of the image of the dynamical divisor by the Abel map in the Jacobian. The second one uses a special type of functions on the Riemann surface, the Baker–Akhiezer functions, to reconstruct the eigenvectors explicitly. Finally, the linearization property also follows very directly by properly interpreting the group theoretical factorization method in its Riemann–Hilbert incarnation. As a result, one can express the dynamical variables in terms of θ functions defined on the Jacobian of the spectral curve. We then show that the symplectic structure can be nicely written in terms of coordinates of the points of the dynamical divisor, hence exhibiting the interplay between analytical data and separation of variables. We finally present the application of theses ideas to sketch the solution of the Kowalevski top.

5.1 The spectral curve

Let us consider an $N \times N$ Lax matrix $L(\lambda)$, depending, as in Chapter 3, rationally on a spectral parameter $\lambda \in \mathbb{C}$ with poles at points λ_k:

$$L(\lambda) = L_0 + \sum_k L_k(\lambda) \tag{5.1}$$

L_0 is independent of λ and $L_k(\lambda)$ is the polar part of $L(\lambda)$ at λ_k, i.e. $L_k = \sum_{r=-n_k}^{-1} L_{k,r}(\lambda - \lambda_k)^r$.

The analytical method of solution of integrable systems is based on the study of the eigenvector equation:

$$(L(\lambda) - \mu \mathbf{1}) \, \Psi(\lambda, \mu) = 0 \tag{5.2}$$

where $\Psi(\lambda, \mu)$ is the eigenvector with eigenvalue μ. The characteristic equation for the eigenvalue problem (5.2) is:

$$\Gamma \quad : \quad \Gamma(\lambda, \mu) \equiv \det(L(\lambda) - \mu \, \mathbf{1}) = 0 \tag{5.3}$$

This defines an algebraic curve in $\mathbb{C}^2$ which is called the spectral curve. We are considering here the smooth compact curve obtained from this equation by the desingularization procedure explained in Chapter 15, even if we do not mention it explicitly. A point on Γ is a pair (λ, μ) satisfying eq. (5.3). If N is the dimension of the Lax matrix, the equation of the curve is of the form:

$$\Gamma \quad : \quad \Gamma(\lambda, \mu) \equiv (-\mu)^N + \sum_{q=0}^{N-1} r_q(\lambda) \mu^q = 0 \tag{5.4}$$

The coefficients $r_q(\lambda)$ are polynomials in the matrix elements of $L(\lambda)$ and therefore have poles at λ_k. Since the Lax equation $\dot{L} = [M, L]$ is isospectral, these coefficients are time-independent and are related to the action variables.

From eq. (5.4) we see that the spectral curve appears as an N-sheeted covering of the Riemann sphere. To a given point λ on the Riemann sphere there correspond N points on the curve whose coordinates are $(\lambda, \mu_1), \ldots, (\lambda, \mu_N)$, where the μ_i are the solutions of the algebraic equation $\Gamma(\lambda, \mu) = 0$. By definition μ_i are the eigenvalues of $L(\lambda)$.

Our goal is to determine the analytical properties of the eigenvector $\Psi(\lambda, \mu)$ and see how much of $L(\lambda)$ can be reconstructed from them. The result is that one can reconstruct $L(\lambda)$ up to global (independent of λ) similarity transformations. This is not too surprising since the analytical properties of $L(\lambda)$ and the spectral curve are invariant under *global* gauge transformations consisting of similarity transformations by constant invertible matrices. So from analyticity we can only hope to recover the system where global gauge transformations have been factored away.

In general, we may fix the gauge by diagonalizing $L(\lambda)$ for one value of λ. To be specific, we choose to diagonalize at $\lambda = \infty$, i.e. we diagonalize the coefficient L_0:

$$L_0 = \lim_{\lambda \to \infty} L(\lambda) = \mathrm{diag}(a_1, \dots, a_N) \tag{5.5}$$

We assume for simplicity that all the a_i are different. Then on the spectral curve we have N points above $\lambda = \infty$:

$$Q_i \equiv (\lambda = \infty, \mu_i = a_i)$$

In the gauge (5.5) there remains a residual action which consists of conjugating the Lax matrix by constant *diagonal* matrices. Generically, these transformations form a group of dimension $N - 1$ and we will have to factor it out.

Before doing complex analysis on Γ, one has to determine its genus. A general strategy is as follows. As we have seen, Γ is an N-sheeted covering of the Riemann sphere. There is a general formula expressing the genus g of an N-sheeted covering of a Riemann surface of genus g_0 (in our case $g_0 = 0$). It is the Riemann–Hurwitz formula:

$$2g - 2 = N(2g_0 - 2) + \nu \tag{5.6}$$

where ν is the branching index of the covering, see Chapter 15. Let us assume for simplicity that the branch points are all of order 2. To compute ν we observe that this is the number of values of λ where $\Gamma(\lambda, \mu)$ has a double root in μ. This is also the number of zeroes of $\partial_\mu \Gamma(\lambda, \mu)$ on the surface $\Gamma(\lambda, \mu) = 0$. But $\partial_\mu \Gamma(\lambda, \mu)$ is a meromorphic function on Γ, and therefore the number of its zeroes is equal to the number of its poles and it is enough to count the poles. These poles can only be located where the matrix $L(\lambda)$ itself has a pole. So we are down to a local analysis around the points of Γ such that $L(\lambda)$ has a pole.

Let us apply this idea to the matrix (5.1). Above a pole λ_k, we have N branches of the form $\mu_j = l_j/(\lambda - \lambda_k)^{n_k} + \cdots$, where l_j are the eigenvalues of $L_{k,-n_k}$ that are assumed all distinct. On such a branch we have $\partial_\mu \Gamma(\lambda, \mu)|_{(\lambda, \mu_j(\lambda))} = \prod_{i \neq j} (\mu_j(\lambda) - \mu_i(\lambda))$, which thus has a pole of order $(N-1)n_k$. Summing on all branches, the total order of the poles over λ_k is $N(N-1)n_k$. Summing on all poles λ_k of $L(\lambda)$, we see that the total branching index is $\nu = N(N-1)\sum_k n_k$. This gives:

$$g = \frac{N(N-1)}{2} \sum_k n_k - N + 1$$

For consistency of the method it is important to observe that the genus is related to the dimension of the phase space and to the number of action variables occuring as independent parameters in eq. (5.4), which should also be half the dimension of the phase space.

As we have seen in Chapter 3, $L_k = (g_k \cdot (A_k)_- \cdot g_k^{-1})_-$, where the $(A_k)_-$ characterize the orbit and are non-dynamical. The g_k are defined modulo right multiplication by diagonal matrices, and we have in addition to quotient by global gauge transformations.

Proposition. *The phase space $\mathcal{M}$ has dimension $2g$ and there are g proper action variables in eq. (5.4).*

<u>Proof.</u> The dynamical variables are the jets of order $(n_k - 1)$ of the g_k. This gives $N^2 n_k$ parameters. But L_k is invariant under $g_k \rightarrow g_k d_k$ with d_k a jet of diagonal matrices of the same order. Hence the dimension of the L_k orbit is $(N^2 - N)n_k$. We also have the residual global gauge invariance by diagonal matrices acting as $g_k \rightarrow dg_k$, or $L(\lambda) \rightarrow dL(\lambda)d^{-1}$. This preserves the diagonal form of L_0. The orbits of this action are of dimension $(N - 1)$, since the identity does not act. The phase space $\mathcal{M}$ is obtained by Hamiltonian reduction by this action (see Chapter 14). First one fixes the momentum, yielding $(N - 1)$ conditions, and then one takes the quotient by the stabilizer of the momentum which is here the whole group since it is Abelian. As a result, the dimension of the phase space is reduced by $2(N - 1)$, yielding:

$$\dim \mathcal{M} = (N^2 - N) \sum_k n_k - 2(N - 1) = 2g$$

Let us now count the number of independent coefficients in eq. (5.4). It is clear that $r_j(\lambda)$ is a rational function of λ. The value of r_j at ∞ is known since $\mu_j \rightarrow a_j$. Note that r_j is the symmetrical function $\sigma_j(\mu_1, \ldots, \mu_N)$, where μ_i are the eigenvalues of $L(\lambda)$. Above $\lambda = \lambda_k$, they can be written as

$$\mu_j = \sum_{n=1}^{n_k} \frac{c_n^{(j)}}{(\lambda - \lambda_k)^n} + \text{regular} \tag{5.7}$$

where all the coefficients $c_1^{(j)}, \ldots, c_{n_k}^{(j)}$ are fixed and non-dynamical because they are the matrix elements of the diagonal matrices $(A_k)_-$, while the regular part is dynamical. We see that $r_j(\lambda)$ has a pole of order jn_k at $\lambda = \lambda_k$, and so can be expressed using $j \sum_k n_k$ parameters, namely the coefficients of all these poles. Summing over j we have altogether $\frac{1}{2}N(N + 1) \sum_k n_k$ parameters. They are not all independent however, because in eq. (5.7) the coefficients $c_n^{(j)}$ are non-dynamical. This implies that the n_k highest order terms in r_j are fixed and yields Nn_k constraints on the coefficients of r_j. We are left with $\frac{1}{2}N(N-1)\sum_k n_k$ parameters, that is $g + N - 1$ parameters.

It remains to take the symplectic quotient by the action of constant diagonal matrices. We assume that the system is equipped with the Poisson bracket (3.26) of Chapter 3. Consider the Hamiltonians $H_n = (1/n)\operatorname{Res}_{\lambda=\infty}\operatorname{Tr}(L^n(\lambda))\,d\lambda$, i.e. the term in $1/\lambda$ in $\operatorname{Tr}(L^n(\lambda))$. These are functions of the $r_j(\lambda)$ in eq. (5.4). We show that they are the generators of the diagonal action. First we have:

$$\operatorname{Res}_{\lambda=\infty}\operatorname{Tr}(L^n(\lambda))d\lambda = n\operatorname{Res}_{\lambda=\infty}\operatorname{Tr}(L_0^{n-1}\sum_k L_k(\lambda))d\lambda$$

$$= n\operatorname{Res}_{\lambda=\infty}\operatorname{Tr}(L_0^{n-1}L(\lambda))d\lambda \qquad (5.8)$$

since all $L_k(\lambda)$ are of order $1/\lambda$ at ∞. Using the Poisson bracket we get

$$\{H_n, L(\mu)\} = -\operatorname{Res}_{\lambda=\infty}\operatorname{Tr}_1 L_0^{n-1}\otimes 1\left[\frac{C_{12}}{\lambda-\mu}, L(\lambda)\otimes 1 + 1\otimes L(\mu)\right]d\lambda$$

The term $L(\lambda)\otimes 1$ in the commutator does not contribute because the L_0 part produces a vanishing contribution by cyclicity of the trace and all other terms are of order at least $1/\lambda^2$. The term $1\otimes L(\mu)$ yields $-[L_0^{n-1}, L(\mu)]$, which is the coadjoint action of a diagonal matrix on $L(\mu)$. Since L_0 is generic, the L_0^n generate the space of all diagonal matrices, so we get exactly $N-1$ generators $H_1,\ldots,H_{N-1}$. In the Hamiltonian reduction procedure, the H_n are the moments of the group action and are to be set to fixed (non-dynamical) values. Hence when the system is properly reduced we are left with exactly g action variables. ∎

Example. Let us consider the example of the Neumann model. Recall the Lax matrix (see Chapter 3):

$$L(\lambda) = L_0 + \frac{1}{\lambda}J - \frac{1}{\lambda^2}K$$

The spectral curve can be computed as follows:

$$\det(L(\lambda)-\mu) = \det\left((L_0-\mu)+\frac{1}{\lambda}J-\frac{1}{\lambda^2}K\right)$$

$$= \det(L_0-\mu)\det\left(1+(L_0-\mu)^{-1}(\frac{1}{\lambda}J-\frac{1}{\lambda^2}K)\right)$$

$$= \prod_i(a_i-\mu)\left[1+\frac{1}{\lambda^2}\sum_k\frac{F_k}{\mu-a_k}\right] \qquad (5.9)$$

where the conserved quantities F_k are given by:

$$F_k = x_k^2 + \sum_{j \neq k} \frac{J_{kj}^2}{a_k - a_j} \tag{5.10}$$

This is because J is a projector of rank 2 and K is a projector of rank 1. The matrix $P = (L_0 - \mu)^{-1}(\frac{1}{\lambda}J - \frac{1}{\lambda^2}K)$ is of rank 2. Its image is spanned by the two vectors $v_1 = (L_0 - \mu)^{-1}X$ and $v_2 = (L_0 - \mu)^{-1}Y$ while its kernel is the $(N-2)$-dimensional space orthogonal to X and Y, which is generically supplementary to the image. We have $P = 0$ on the kernel, and:

$$Pv_1 = \left(\frac{1}{\lambda}V(\mu) - \frac{1}{\lambda^2}U(\mu)\right)v_1 - \frac{1}{\lambda}U(\mu)v_2$$

$$Pv_2 = \left(\frac{1}{\lambda}(W(\mu) + 1) - \frac{1}{\lambda^2}V(\mu)\right)v_1 - \frac{1}{\lambda}V(\mu)v_2$$

where the functions $U(\mu), V(\mu), W(\mu)$ are defined as:

$$U(\mu) = \sum_i \frac{x_i^2}{a_i - \mu}, \quad V(\mu) = \sum_i \frac{x_i y_i}{a_i - \mu}, \quad W(\mu) = -1 + \sum_i \frac{y_i^2}{a_i - \mu} \tag{5.11}$$

From this it follows that:

$$\det(1 + P) = 1 - \frac{1}{\lambda^2}\left(V^2(\mu) - U(\mu)W(\mu)\right) = 1 + \frac{1}{\lambda^2}\sum_k \frac{F_k}{\mu - a_k} \tag{5.12}$$

which yields the result. Incidentally this proves formula (2.24) in Chapter 2. Since we have already found an r-matrix for the Neumann system this proves its integrability.

The spectral curve can be written in the form:

$$\lambda^2 = -\frac{\sum_k \prod_{i \neq k}(\mu - a_i)F_k}{\prod_i(\mu - a_i)} = -\frac{\prod_i(\mu - b_i)}{\prod_i(\mu - a_i)} \tag{5.13}$$

Performing the birational transformation (see Chapter 15) $\lambda' = \lambda\prod_i(\mu - a_i)$, we get:

$$\lambda'^2 = -\prod_{i=1}^{N}(\mu - a_i)\prod_{i=1}^{N-1}(\mu - b_i) \tag{5.14}$$

which is a hyperelliptic curve of genus $g = N - 1$. Note that the phase space is of dimension $2(N-1)$ and that we have $(N-1)$ independent conserved quantities, namely the N quantities F_k modulo the relation

$\sum_k F_k = 1$. Let us remark that in this case we do not quotient by the diagonal action. This is because the diagonal action does not preserve the particular form of the Lax matrix in terms of the vectors X and Y. In this case the Hamiltonians $H_n = \text{Tr}\,(L_0^{n-1} J)$ identically vanish.

To illustrate the discussion of N-sheeted coverings, it is instructive to consider the covering projection $(\lambda, \mu) \to \lambda$ which allows us to see the spectral curve as an N-sheeted covering of the Riemann sphere of the variable λ. To compute the branching index of this covering we have to find the total number of poles of $\partial_\mu \Gamma(\lambda, \mu)$. Such poles can only occur when $\lambda = \infty$ or $\mu = \infty$. First, above $\lambda = \infty$ we have the N points $\mu = a_i$. These are not branch points and the local parameter is $1/\lambda$. Around such a point we have by eq. (5.9):

$$(\lambda, \mu) \to Q_i = (\infty, a_i), \quad \mu = a_i - F_i/\lambda^2 + O(1/\lambda^4) \qquad (5.15)$$

hence $\partial_\mu \Gamma = \lambda^2 \prod_{j \neq i}(a_i - a_j) + O(1)$. We thus have N double poles at these points. When $\mu \to \infty$ we have, by eq. (5.9), $\lambda^2 = -1/\mu \to 0$ and λ is again a local parameter,

$$(\lambda, \mu) \to (0, \infty), \quad \mu = -\frac{1}{\lambda^2} + O(1) \qquad (5.16)$$

At this point we have $\partial_\mu \Gamma = -(-1/\lambda^2)^{N-2} + \cdots$, hence we have a pole of order $2N - 4$. So the branching index is $\nu = 4N - 4$. This yields $g = N - 1$. Note that $\Gamma(\lambda, \mu)$ is a very non-generic polynomial of degree $2N - 1$ and the orbit to which $L(\lambda)$ belongs is a very low-dimensional one, nevertheless all the numbers fit nicely.

5.2 The eigenvector bundle

The aim of this chapter is to present the general procedure for solving integrable models using analytical properties on the spectral curve. In order to simplify the exposition we shall assume that all functions or matrices are generic.

We wish to examine how the Lax matrix can be reconstructed from the analytic data characterizing its eigenvectors. This analysis will exhibit the special role played by the divisor D of the poles at finite distances of the eigenvector $\Psi(P)$. *This divisor contains all the dynamical information.*

Let P be a point on the spectral curve. We assume that $P = (\lambda, \mu)$ is not a branch point so that all eigenvalues of $L(\lambda)$ are distinct and the eigenspace at P is one-dimensional. Let $\Psi(P)$ be an eigenvector, and

$\psi_j(P)$ its N components:

$$\Psi(P) = \begin{pmatrix} \psi_1(P) \\ \vdots \\ \psi_N(P) \end{pmatrix}$$

Since the normalization of the eigenvector $\Psi(\lambda, \mu)$ is arbitrary, one has to make a choice before making a statement about its analytical properties. We choose to normalize it such that its first component is equal to one, i.e.

$$\psi_1(P) = 1, \quad \text{at any point} \quad P \in \Gamma.$$

It is then clear that the $\psi_j(P)$ depend locally analytically on P. As a matter of fact:

Proposition. *With the above normalization, the components of the eigenvectors $\Psi(P)$ at the point $P = (\lambda, \mu)$ are meromorphic functions on the spectral curve Γ.*

Proof. For a generic point P on the curve Γ, i.e. for a pair $P = (\lambda, \mu)$ satisfying eq. (5.3), there exists a unique eigenvector $\Psi(P)$ of the matrix $L(\lambda)$ normalized by the condition $\psi_1(P) = 1$. The un-normalized components $\psi_i(P)$ can be taken as suitable minors $\Delta_i(P)$ of the matrix $L(\lambda) - \mu\mathbf{1}$, and are thus meromorphic functions on Γ. After dividing by $\Delta_1(\lambda, \mu)$ to normalize the first component, all the other components $\psi_j(P)$ are still meromorphic functions on Γ. ∎

With each point $P(\lambda, \mu)$ on Γ we associate a meromorphic eigenvector $\Psi(P)$. At a branch point however, special care must be taken since there could be several eigenvectors associated with that point. We show that, for a generic Lax matrix, the eigenspaces are one-dimensional even at a branch point P. Moreover, the eigenspaces around P admit a unique analytic continuation at P, irrespective of the branch chosen.

Proposition. *With each point P in Γ we associate the eigenspace at P. This allows us to define an analytic line bundle that we call the eigenvector bundle.*

Proof. The first point is to show that the eigenspace at P is of dimension 1 at each point of Γ, even at a branch point, in the generic case.

Consider the matrix $A \equiv L(\lambda) - \mu$. The fact that we are on a branch point of the curve Γ is expressed by two algebraic equations in the coefficients of A: $\Gamma(\lambda, \mu) = 0$ and $\partial_\mu \Gamma(\lambda, \mu) = 0$. To say that the kernel of

A at P is of dimension ≥ 2, means that the dimension of its image is $\leq (N-2)$. Let us show that this implies at least three algebraic equations on the coefficients of A. Let $v_1, \ldots, v_N$ be the columns of A. We assume for simplicity that the kernel of A at P is of dimension 2 and that $v_1, \ldots, v_{N-2}$ are independent. First we impose that $v_1, \ldots, v_N$ are linearly dependent, producing one condition $\det A = 0$, which is the equation of the spectral curve. Then we impose that $v_1, \ldots, v_{N-1}$ are dependent on $v_1, \ldots, v_{N-2}$. This is expressed by the vanishing of two $(N-1) \times (N-1)$ minors. One of these conditions is equivalent to $\partial_\mu \Gamma(\lambda, \mu) = 0$ but there remains another independent one, hence the variety of such matrices is of codimension 1. Another way of saying this is that when a matrix has coinciding eigenvalues it can be put in the Jordan form, but is not diagonalizable in general. We now construct an abstract line bundle starting from the dimension 1 eigenspace E_P at each point P, see Chapter 15. We call $e_1, \ldots, e_N$ the canonical basis of ambient space in which the eigenvectors live. Define N open sets U_i on the curve Γ by the conditions $U_i = \{P \in \Gamma |$ exists $V \in E_P$ s.t. $V_i = (V, e_i) \neq 0\}$, meaning that the eigenspace E_P is not perpendicular to e_i. Obviously the U_i form an open covering of Γ. On each intersection $U_i \cap U_j$ we define transition functions $t_{ij}(P)$ by $t_{ij}(P) = V_i/V_j$ where V is *any non-zero* eigenvector at P. The quotient is independent of the choice of V and the components V_i, V_j do not vanish on $U_i \cap U_j$. In view of the argument of the previous proposition it is clear that $t_{ij}(P)$ is analytic with respect to the point P on Γ and non-vanishing. Finally, the cocycle condition $t_{ij}t_{jk} = t_{ik}$ is trivially satisfied on $U_i \cap U_j \cap U_k$. Hence these transition functions define an analytic line bundle which we call the eigenvector bundle. Any meromorphic section of this bundle can be described as a collection $(V_1(P), \ldots, V_N(P))$, where $V_i(P)$ is defined and meromorphic on U_i and $V_i(P) = t_{ij}(P)V_j(P)$ on $U_i \cap U_j$. One can see this collection as a P-dependent vector lying in the eigenspace E_P with components $V_i(P)$ on e_i.   ∎

Remark 1. Alternatively one can define normalized eigenvectors $\psi^{(i)}(P)$ with $\psi_i^{(i)}(P) = 1$. Then U_i is the open set where $\psi^{(i)}(P)$ remains finite. The transition function $t_{ij}(P) = \psi_i^{(j)}(P) = 1/\psi_j^{(i)}(P)$ has no zero nor pole on $U_i \cap U_j$.

Remark 2. It may be useful to understand the situation at branch points on the simple example of 2×2 matrices. Let

$$L(\lambda) = \begin{pmatrix} a(\lambda) & b(\lambda) \\ c(\lambda) & d(\lambda) \end{pmatrix}$$

which has eigenvalues $\mu_\pm(\lambda) = \frac{1}{2}(a(\lambda) + d(\lambda)) \pm \frac{1}{2}\sqrt{\Delta(\lambda)}$ with $\Delta(\lambda) = (a(\lambda) - d(\lambda))^2 + 4b(\lambda)c(\lambda)$. The corresponding normalized eigenvectors are:

$$\Psi_\pm = \begin{pmatrix} 1 \\ \psi_\pm \end{pmatrix}, \quad \psi_\pm = \frac{d(\lambda) - a(\lambda)}{2b(\lambda)} \pm \frac{\sqrt{\Delta(\lambda)}}{2b(\lambda)}$$

Assume that λ_0 is a root of $\Delta(\lambda) = 0$. It is obvious that when $\lambda \to \lambda_0$, Ψ_+ and Ψ_- tend smoothly to the same limit except if one has also $b(\lambda_0) = 0$. If $b(\lambda_0) \neq 0$ one can express $L(\lambda_0)$ in the basis given by $\Psi(\lambda_0)$ and $(0, 1)$ getting

$$L(\lambda_0) \to \begin{pmatrix} \frac{1}{2}(a(\lambda_0) + d(\lambda_0)) & b(\lambda_0) \\ 0 & \frac{1}{2}(a(\lambda_0) + d(\lambda_0)) \end{pmatrix}$$

from which it is obvious that $L(\lambda_0)$ is of the Jordan form and has just one eigenvector. If, however, $b(\lambda_0) = 0$ then we also have $a(\lambda_0) = d(\lambda_0)$. Assuming that $d(\lambda) - a(\lambda)$, $b(\lambda)$ vanish to first order in $\lambda - \lambda_0$, then $(d(\lambda) - a(\lambda))/2b(\lambda)$ tends to some limit ψ_e. We see that $\psi_\pm \sim \psi_e \pm \sqrt{c(\lambda)/b(\lambda)}$. Hence if $c(\lambda_0) \neq 0$ we still have only one eigenvector of the form $(0, 1)$, while if $c(\lambda)$ also vanishes to first order in $\lambda - \lambda_0$ the matrix $L(\lambda_0)$ is diagonalizable, and the eigenvectors $\Psi_\pm$ tend generically to different limits at λ_0. However, in this case we have $\Delta \sim (\lambda - \lambda_0)^2$ so that the corresponding point (λ_0, μ_0) of the spectral curve is not a branch point, but a singular point. Upon desingularization it blows up to two points and the two values of Ψ are perfectly natural. Of course this analysis clearly covers what happens at a branch point of order 2 in the general case.

We now compute the Chern class of the eigenvector bundle. To do that we view $\Psi(P)$ as a meromorphic section of our bundle in the above way, i.e. as the collection defined respectively on $U_1, \ldots, U_N$: $(\psi_1(P) = 1, \psi_2(P), \ldots, \psi_N(P))$. Notice that this section does not vanish because $\psi_i(P)$ does not vanish on U_i by definition. We compute the number of poles of this section, which yields the Chern class.

The number of poles of the normalized eigenvectors cannot be deduced by simply counting the number of zeroes of minors. Indeed, let $\widehat{\Delta}(\lambda, \mu)$ be the matrix of cofactors of $(L(\lambda) - \mu\mathbf{1})$, which, by definition, is such that $(L(\lambda) - \mu\mathbf{1})\widehat{\Delta} = \Gamma(\lambda, \mu)\mathbf{1}$. Therefore at $P = (\lambda, \mu) \in \Gamma$, the matrix $\widehat{\Delta}(P)$ is a matrix of rank 1, since the kernel of $(L(\lambda) - \mu\mathbf{1})$ is of dimension 1. Hence, for $P \in \Gamma$ the matrix elements of $\widehat{\Delta}(P)$ are of the form $\alpha_i(P)\beta_j(P)$ and the components of the normalised eigenvector are $\psi_i(P) = \frac{\alpha_i(P)\beta_1(P)}{\alpha_1(P)\beta_1(P)} = \frac{\alpha_i(P)}{\alpha_1(P)}$. We thus expect cancellations to occur when we take the ratio of the minors and we cannot deduce the number of poles of the normalized eigenvector by simply counting the number of zeroes of the first minor.

Proposition. *We say that the vector $\Psi(P)$ possesses a pole if one of its components has a pole. The number of poles of the normalized vector $\Psi(P)$ is:*

$$m = g + N - 1 \tag{5.17}$$

<u>Proof.</u> Let us introduce the function $W(\lambda)$ of the complex variable λ defined by:

$$W(\lambda) = \left(\det \widehat{\Psi}(\lambda) \right)^2$$

where $\widehat{\Psi}(\lambda)$ is the matrix of eigenvectors of $L(\lambda)$ defined as follows:

$$\widehat{\Psi}(\lambda) = \begin{pmatrix} \psi_1(P_1) & \psi_1(P_2) & \cdots & \psi_1(P_N) \\ \vdots & \vdots & \vdots & \vdots \\ \psi_N(P_1) & \psi_N(P_2) & \cdots & \psi_N(P_N) \end{pmatrix} \tag{5.18}$$

where the points P_i are the N points above λ. In this formula $\psi_1(P_j) = 1$. Changing the normalization of the eigenvectors $\Psi(P_j)$ amounts to multiplying $\widehat{\Psi}(\lambda)$ on the right by a diagonal matrix. By definition $\widehat{\Psi}(\lambda)$ is the matrix diagonalizing $L(\lambda)$. The function $W(\lambda)$ is well-defined as a rational function of λ on the Riemann sphere since the square of the determinant does not depend on the order of the P_j. It has a double pole where $\Psi(P)$ has a simple pole. To count its poles, we count its zeroes.

We show that $W(\lambda)$ has a simple zero for values of λ corresponding to a branch-point of the covering, therefore $m = \nu/2$. Recall that from eq. (5.6) the number of branch points is $\nu = 2(N + g - 1)$.

First notice that $W(\lambda)$ only vanishes on branch points where there are at least two identical columns. Indeed, let $P_i = (\mu_i, \lambda)$ be the N points above λ. Then the $\Psi(P_i)$ are the eigenvectors of $L(\lambda)$ corresponding to the eigenvalues μ_i are thus linearly independent when all the μ_i are different. Therefore $W(\lambda)$ cannot vanish at such a point. The other possibility for the vanishing of $W(\lambda)$ would be that the vector $\Psi(P)$ itself vanish at some point (all components have a common zero at this point), but this is impossible because the first component is always 1. Let us assume now that λ_0 corresponds to a branch point, which is generically of order 2. At such a point $W(\lambda)$ has a simple zero. Indeed, let z be an analytical parameter on the curve around the branch point. The covering projection $P \to \lambda$ gets expressed as $\lambda = \lambda_0 + \lambda_1 z^2 + O(z^3)$. The determinant vanishes to order z, hence W vanishes to order z^2. This is precisely proportional to $\lambda - \lambda_0$. A similar analysis can be performed if the branch point is of higher order. ∎

We now need to examine the behaviour of the eigenvector around $\lambda = \infty$. At the N points Q_i above $\lambda = \infty$, the eigenvectors are proportional to the canonical vectors e_i, $(e_i)_k = \delta_{ik}$, since $L(\lambda = \infty)$ is diagonal, cf. eq. (5.5). While this is compatible with the normalization $\psi_1(P) = 1$ at the point Q_1, it is not compatible at the points Q_i, $i \geq 2$, if the proportionality factor remains finite. The situation is described more precisely by the following:

Proposition. *The kth component $\psi_k(P)$ of $\Psi(P)$ has a simple pole at Q_k and vanishes at Q_1 for $k = 2, 3, \ldots, N$.*

<u>Proof.</u> Around $Q_k(\lambda = \infty, \mu = a_k)$, $k = 1, \ldots, N$, the eigenspace of $L(\lambda)$ is spanned by a vector of the form $V_k(\lambda) = e_k + O(1/\lambda)$. The first component of V_k is $V_k^1 = \delta_{1k} + O(1/\lambda)$. To get the normalized Ψ one has to divide V_k by V_k^1. So we get:

$$\Psi(P)|_{P \sim Q_1} = \begin{pmatrix} 1 \\ O(1/\lambda) \\ \vdots \\ \vdots \\ \vdots \\ O(1/\lambda) \end{pmatrix}, \quad \Psi(P)|_{P \sim Q_k} = \begin{pmatrix} 1 \\ O(1) \\ \vdots \\ O(\lambda) \\ O(1) \\ \vdots \\ O(1) \end{pmatrix}, \quad k \geq 2 \quad (5.19)$$

where $O(\lambda)$ is the announced pole of the kth component of $\Psi(P)|_{P \sim Q_k}$. ∎

The previous proposition shows that fixing the gauge by imposing that $L(\lambda)$ is diagonal at $\lambda = \infty$ introduces $N - 1$ poles at the positions Q_i, $i = 2, \ldots, N$. The location of these poles is independent of time, and is really part of the choice of the gauge condition. These poles do not contain any dynamical information. Only the positions of the other g poles have a dynamical significance. Let D be the divisor of these dynamical poles. We call it the dynamical divisor. Recall that the vector $\Psi(P)$ possesses a pole if one of its components has a pole. Therefore the two previous propositions tell us that the divisor of the kth components of the eigenvector $\Psi(P)$ is bigger than $(-D + Q_1 - Q_k)$. This information is enough to reconstruct the eigenvectors and the Lax matrix.

Proposition. *Let D be a generic divisor on Γ of degree g. Up to normalization, there is a unique meromorphic function $\psi_k(P)$ with divisor $(\psi_k) \geq -D + Q_1 - Q_k$.*

<u>Proof.</u> This is a direct application of the Riemann–Roch theorem, since ψ_k is required to have $g + 1$ poles and one prescribed zero. Hence it is generically unique apart from multiplication by a constant $\psi_k \to d_k \psi_k$. ∎

Equipped with these functions $\psi_k(P)$ for $k = 2, \ldots, N$ we construct a vector function with values in $\mathbb{C}^N$:

$$\Psi(P) = \begin{pmatrix} 1 \\ \psi_2(P) \\ \vdots \\ \psi_N(P) \end{pmatrix}$$

The normalization ambiguity of the ψ_k translates into left multiplication of the vector $\Psi(P)$ by a *constant* diagonal matrix $d = \operatorname{diag}(1, d_2, \ldots, d_N)$. We have constructed a line bundle on the Riemann surface Γ, which is the line bundle associated to the divisor $-D - Q_2 - \cdots - Q_N$ of degree $-(g+N-1)$ (see Chapter 15). In fact we have constructed an embedding of this line bundle into $\Gamma \times \mathbb{C}^N$.

Theorem. *Given the spectral curve Γ, such that above the points λ_k the N branches satisfy eq. (5.7), there exists a unique matrix $L(\lambda)$, rational in λ, such that*

$$(L(\lambda) - \mu\mathbf{1})\Psi(P) = 0$$

This matrix has poles at the points λ_k and satisfies the boundary condition: $\lim_{\lambda \to \infty} L(\lambda) = \operatorname{diag}(a_1, \ldots, a_N)$.

Proof. Consider the matrix $\widehat{\Psi}(\lambda)$ whose columns are the vectors $\psi(P_i)$, where $P_i = (\lambda, \mu_i)$ are the N points above λ, cf. eq. (5.18). This matrix depends on the ordering of the columns, i.e. on the ordering of the points P_i. However, the matrix

$$L(\lambda) = \widehat{\Psi}(\lambda) \cdot \widehat{\mu} \cdot \widehat{\Psi}^{-1}(\lambda) \tag{5.20}$$

does not depend on this ordering and is a well-defined function on the base curve. Here $\widehat{\mu}$ is the diagonal matrix $\widehat{\mu} = \operatorname{diag}(\mu_1, \ldots, \mu_N)$. One has to examine the poles of the right-hand side of eq. (5.20). At a generic branch point two columns of the matrix $\widehat{\Psi}$ coalesce and its determinant has a simple zero with respect to the local parameter. These zeroes are the *only* zeroes of $\det \widehat{\Psi}(\lambda)$. This is because the meromorphic function $W(\lambda) = (\det \widehat{\Psi})^2(\lambda)$ is a function of λ and has $2(N+g-1)$ poles, since Ψ has $(g+N-1)$ poles. The function $W(\lambda)$ has the same number of zeroes and poles, hence has also $2(N+g-1)$ zeroes. At the branch points it behaves like $z^2 \sim (\lambda - \lambda_b)$ (where z is a local parameter), hence has a simple zero. Thus the branch points contribute to $\nu = 2(N+g-1)$ zeroes, which are all the zeroes of $W(\lambda)$.

We now show that, at a branch point, the matrix (5.20) is regular. Recall that if $\widehat{\Delta}$ is the matrix of cofactors of $\widehat{\Psi}$ we have:

$$\widehat{\Psi} \cdot \widehat{\Delta} = \det \widehat{\Psi} \; 1, \quad L(\lambda) = \frac{1}{\det \widehat{\Psi}} \; \widehat{\Psi} \cdot \widehat{\mu} \cdot \widehat{\Delta} \tag{5.21}$$

At the branch point, $\widehat{\Psi}\widehat{\Delta} = 0$, thus Im $\widehat{\Delta}$ = Ker $\widehat{\Psi}$ which is one-dimensional. We may assume without loss of generality that the two eigenvectors that coalesce are the first two columns of $\widehat{\Psi}$. So, at the branch point, the kernel of $\widehat{\Psi}$ is spanned by $e_1 - e_2$, where e_i are the canonical base vectors $(e_i)_j = \delta_{ij}$. Since the first two diagonal elements of $\widehat{\mu}$ also become equal we see that $\widehat{\mu}$ acts as a scalar on Im $\widehat{\Delta}$, so $\widehat{\Psi}(\lambda) \cdot \widehat{\mu} \cdot \widehat{\Delta} = 0$. Hence the numerator of eq. (5.21) has a simple zero at the branch point, cancelling the simple pole of the determinant. Therefore the matrix $L(\lambda)$ is a rational function of the parameter λ. It has poles only at the projections of the points where μ has poles, i.e. at the points λ_k, see eq. (5.7). At $\lambda = \infty$, the leading part of $\widehat{\Psi}(\lambda)$ is diagonal since it is dominated by the functions $\psi_k(P)$ with P approaching Q_k. Therefore at infinity $L(\lambda)$ goes to $\widehat{\mu}|_{\lambda=\infty} = \mathrm{diag}(a_1, \ldots, a_N)$. ∎

This theorem is a crucial step of this method of resolution. It says that, once the spectral curve has been given, which amounts to giving the values of the integrals of motion, *all remaining dynamical data are encoded into the divisor D*. In other words, this theorem teaches us that the dynamical variables are the action variables and the points of this divisor.

It should be emphasized, however, that $\Psi(P)$ is defined up to left multiplication by diagonal matrices $\psi_k \to d_k \psi_k$. On the Lax matrix $L(\lambda)$ this amounts to a conjugation by a constant diagonal matrix. Hence the object we reconstruct is actually the Hamiltonian reduction of the dynamical system by this group of diagonal matrices, as emphasized at the beginning of this chapter.

Remark. It is worth comparing the reconstruction formula (5.20) for the Lax matrix with the local analysis of section (3.2) in Chapter 3. Recall that in this chapter we explained that the pair of matrices $L(\lambda)$ and $M(\lambda)$ could be diagonalized simultaneously, locally around each pole λ_k. Explicitly, the diagonalization formula (3.8) for $L(\lambda)$ was $L(\lambda) = g_k A_k g_k^{-1}$ with g_k and A_k power series in $(\lambda - \lambda_k)$ and A_k diagonal. Of course g_k is determined up to right multiplication by a diagonal matrix. The expression (5.20) of $L(\lambda)$ in terms of the eigenvectors is simply a global version of the previous local statement.

Example. Let us illustrate the analytical properties of the eigenvector bundle on the example of the Neumann model. There is a simple description of the eigenvectors in this case. Indeed, from eq. (3.3) in Chapter 3,

$$L(\lambda)\Psi = \mu\Psi \implies \Psi = -(L_0 - \mu)^{-1}\left(\frac{1}{\lambda}J\Psi - \frac{1}{\lambda^2}K\Psi\right) \qquad (5.22)$$

Since $J\Psi = (Y \cdot \Psi)X - (X \cdot \Psi)Y$ and $K\Psi = (X \cdot \Psi)X$ we see that Ψ is known once we know its projections on X and Y. Projecting eq. (5.22) on X and Y one gets a 2×2 system:

$$\begin{pmatrix} 1 - \frac{1}{\lambda}V(\mu) - \frac{1}{\lambda^2}U(\mu) & \frac{1}{\lambda}U(\mu) \\ -\frac{1}{\lambda}(1 + W(\mu)) - \frac{1}{\lambda^2}V(\mu) & 1 + \frac{1}{\lambda}V(\mu) \end{pmatrix} \begin{pmatrix} X \cdot \Psi \\ Y \cdot \Psi \end{pmatrix} = 0$$

The vanishing of the determinant of this linear system is precisely the equation of the spectral curve:

$$\lambda^2 = V^2(\mu) - U(\mu)W(\mu)$$

Solving this system for $X \cdot \Psi$ and $Y \cdot \Psi$, and inserting back into eq. (5.22), one gets:

$$\frac{\psi_k}{\psi_1} = \frac{a_1 - \mu}{a_k - \mu} \cdot \frac{(\lambda - V(\mu))x_k + U(\mu)y_k}{(\lambda - V(\mu))x_1 + U(\mu)y_1}$$

Let us check the general results on these explicit formulae. Recalling the expansion (5.15), we see that $(a_1 - \mu)/(a_k - \mu)$ has a double zero at $Q_1(\lambda = \infty, \mu = a_1)$ and a double pole at $Q_k(\lambda = \infty, \mu = a_k)$. Consider the meromorphic function $\phi_k = (\lambda - V(\mu))x_k + U(\mu)y_k$. It has poles at the points Q_i. Using eq. (5.15) we see that it has double poles at Q_i, $i \neq k$ and a simple pole at Q_k. In fact $(-V(\mu)x_k + U(\mu)y_k) = \lambda^2 x_i J_{ki}/F_i + O(1)$. This show that ψ_k/ψ_1 has a *simple* zero at Q_1 and a *simple* pole at Q_k.

To find the other poles of ψ_k/ψ_1 we study the zeroes of ϕ_k which has $(2N-1)$ poles and therefore $(2N-1)$ zeroes. Among them, N are common to all functions ϕ_k and cancel in ψ_k/ψ_1. Indeed, a common zero is such that $\lambda - V(\mu) = U(\mu) = 0$. By eq. (5.12) the points satisfying these two equations are on the spectral curve. The equation $U(\mu) = 0$ has $(N-1)$ roots, μ_j, at finite μ, and $\lambda_j = V(\mu_j)$ selects one of the two points above μ_j. In addition we have the point $\lambda = 0$, $\mu = \infty$ which is a simple zero in view of eq. (5.16). Finally, ψ_k/ψ_1 has a zero at Q_1, a pole at Q_k, and $g = N - 1$ poles at finite distance depending on the dynamical data, in agreement with the general considerations of this section.

5.3 The adjoint linear system

In view of eq. (5.20) for the Lax matrix, it is important to compute $\widehat{\Psi}^{-1}(\lambda)$ in an efficient way. This is achieved by introducing the solution $\Psi^+(P)$ of the adjoint linear system:

$$\Psi^+(P)\,(L(\lambda) - \mu\mathbf{1}) = 0 \tag{5.23}$$

Here $\Psi^+(P)$ is a *row* vector. The precise relation between $\Psi^+(P)$ and the matrix $\widehat{\Psi}^{-1}(\lambda)$ is provided by the following:

Proposition. *Let $\Psi^+(P)$ be a solution of the adjoint system (5.23). The inverse of the matrix $\widehat{\Psi}(\lambda)$ defined in eq. (5.18) is the matrix whose rows are the N row vectors*

$$\Psi^{(-1)}(P_j) = \Psi^+(P_j)/\langle\Psi^+(P_j)\Psi(P_j)\rangle \tag{5.24}$$

with P_j the N points above λ and $\langle VW\rangle = \sum_i V_i W_i$.

<u>Proof.</u> One has to show that

$$\sum_{k=1}^{N} \frac{\psi_k^+(P_j)}{\langle\Psi^+(P_j)\Psi(P_j)\rangle}\psi_k(P_i) = \delta_{ij}$$

where P_i and P_j are two points of Γ above the same λ. This is obvious for $i = j$, and for $i \neq j$, $\Psi(P_i)$ and $\Psi^+(P_j)$ are orthogonal because computing $\langle\Psi^+(P_j)L(\lambda)\Psi(P_i)\rangle$ in two different ways we find $\mu(P_j)\langle\Psi^+(P_j)\Psi(P_i)\rangle = \langle\Psi^+(P_j)\Psi(P_i)\rangle\mu(P_i)$, hence the scalar product vanishes if $\mu(P_i)$ and $\mu(P_j)$ are different. So we get:

$$\widehat{\Psi}^{-1}(\lambda) = \begin{pmatrix} \frac{\Psi^+(P_1)}{\langle\Psi^+(P_1)\Psi(P_1)\rangle} \\ \vdots \\ \frac{\Psi^+(P_N)}{\langle\Psi^+(P_N)\Psi(P_N)\rangle} \end{pmatrix} \tag{5.25}$$

∎

We now use this relation between $\Psi^+(P)$ and $\widehat{\Psi}^{-1}(\lambda)$ to reconstruct them from their analyticity properties. One may perform on $\Psi^+(P)$ the same analysis as for the vector $\Psi(P)$. Normalizing the first component of $\Psi^+(P)$ to 1, one sees that the vector $\Psi^+(P)$ has g poles at a divisor D^+ and $(N-1)$ simple poles at $Q_2, \ldots, Q_N$ above $\lambda = \infty$. Moreover, $\psi_k^+(P)$, $k \geq 2$, has a zero at Q_1. Our first task is to relate the divisor D^+ to D.

Proposition. *Let $\Psi^+(P)$ be the solution of eq. (5.23) normalized by $\psi_1^+(P) = 1$. The differential form*

$$\Omega \equiv \frac{d\lambda}{\langle\Psi^+(P)\Psi(P)\rangle} \tag{5.26}$$

*is an Abelian differential of the second kind with a double pole at Q_1 and
zeroes at D and D^+. Conversely, there is a unique differential Ω of the
second kind with a double pole at the point Q_1 and having among its zeroes
the g points of D. Its g other zeroes are then completely fixed and define
D^+. Its image under the Abel map is given by:*

$$\mathcal{A}(D^+) = -\mathcal{A}(D) + \mathcal{A}(B) - 2\sum_{j=2}^{N} \mathcal{A}(Q_j) \tag{5.27}$$

where B is the divisor of branch points of the covering $(\lambda, \mu) \to \lambda$.

<u>Proof.</u> Consider the meromorphic function $f(p) = \langle \Psi^+(P)\Psi(P)\rangle$. It has
$2(g + N - 1)$ poles coming from the poles of Ψ and Ψ^+. Their divisor is
$D + D^+ + 2\sum_{j=2}^{N} Q_j$. Therefore it has also $2(g + N - 1)$ zeroes, which are
in fact the branch points of the covering, as we now see (recall that the
covering has $2(g+N-1)$ branch points by the Riemann–Hurwitz formula).
So let $P = (\lambda_0, \mu_0)$ be a branch point and consider two points P_1 and P_2
above the same λ close to λ_0 on the two sheets of the covering that coalesce
at P. Because $\Psi^+(P_1)$ and $\Psi(P_2)$ are dual eigenvectors corresponding to
different eigenvalues, they are orthogonal:

$$\langle \Psi^+(P_1)\Psi(P_2)\rangle = 0 \tag{5.28}$$

The assertion then follows by continuity, since $\Psi(P_2) \to \Psi(P)$ and
$\Psi^+(P_1) \to \Psi^+(P)$ (recall that the line bundles are analytic at P).

At a branch point P we have $(\mu - \mu_0)^2 \sim (\lambda - \lambda_0)$, so $\mu - \mu_0$ is a local
parameter and $d\lambda = (\mu - \mu_0)d\mu$ vanishes at P. Moreover, $d\lambda$ has double
poles at the points $Q_1, \ldots, Q_N$ above $\lambda = \infty$. We see that Ω is regular at
the branch points, has a double pole at Q_1, and has zeroes at $D + D^+$.

Recall that given a point $P \in \Gamma$ and a divisor $D = (\gamma_1, \ldots, \gamma_g)$ of degree
g on Γ, the Abel map with base point P_0 is defined by:

$$\mathcal{A}_j(P) = \int_{P_0}^{P} \omega_j \quad \text{and} \quad \mathcal{A}_j(D) = \sum_{j=1}^{g} \int_{P_0}^{\gamma_j} \omega_j \tag{5.29}$$

with ω_j is a normalized basis of Abelian differentials. Applying Abel's
theorem to the meromorphic function $\Omega/d\lambda$ one gets eq. (5.27).

Conversely, assume that we have two such forms Ω and Ω' with a double
pole at Q_1 and divisor of zeroes $D + D^+$ and $D + D'^+$ respectively. Their
quotient is a meromorphic function with a divisor of poles of degree g, i.e.
a constant generically. Hence the differential Ω is unique. $\blacksquare$

The outcome of this proposition is that Ω is uniquely characterized by
its behaviour at infinity and its zeroes at the points of D. Therefore, given

the dynamical divisor D, we know the form Ω and we can find the divisor D^+ as the complementary set of zeroes of Ω.

This information on the divisor D^+ can now be used to reconstruct the vector $\Psi^+(P)$. Its components $\psi_k^+(P)$ are uniquely determined up to normalization once we know the divisor D^+. Here, however, we have no freedom on these normalizations since we must preserve the orthogonality conditions (5.28). Let $\tilde{\psi}_k^+(P)$ be any choice of such meromorphic function and let $\psi_k^+(P) = (1/c_k)\,\tilde{\psi}_k^+(P)$. We want to determine the constants c_k. We require that $\psi_k^+(P)$ satisfies an orthogonality relation of the form $\langle\Psi^+(P_j)\Psi(P_i)\rangle = f(P_j)\delta_{ij}$. This means that the matrices of elements $\psi_i^+(P_j)/f(P_j)$ and $\psi_j(P_i)$ are inverse to each other. By uniqueness of the inverse matrix we also have:

$$\sum_k \psi_i^+(P_k)\psi_j(P_k)\frac{1}{f(P_k)} = \delta_{ij}, \quad \text{or} \quad \sum_k \tilde{\psi}_i^+(P_k)\psi_j(P_k)\frac{1}{f(P_k)} = c_i\delta_{ij}$$

We have an independent characterization of $f(P)$. By eq. (5.26) $f^{-1}(P)d\lambda = \Omega(P)$, and therefore $f(P)$ is known from its analyticity properties. This allows us to compute c_i as follows.

Consider on the Riemann surface the form $\tilde{\psi}_i^+(P)\psi_j(P)\Omega(P)$. For $i = j$, it has a double pole at Q_i and for $i \neq j$, it has simple poles at Q_i and Q_j and no other singularity. This is because the poles at D and D^+ in Ψ and $\tilde{\Psi}^+$ cancel against the zeroes of Ω, and the double pole of Ω at Q_1 combines with zeroes of Ψ and $\tilde{\Psi}^+$ at this point. Finally we define a form on the Riemann sphere λ by:

$$\omega_{ij} = \sum_k \tilde{\psi}_i^+(P_k)\psi_j(P_k)\Omega(P_k)$$

where the N points P_k are the points above λ. If there are no branch points among the P_k, λ is a local parameter around each P_k and $\omega_{ij} = g_{ij}(\lambda)d\lambda$. If there is a branch point a short computation shows that this still holds. Since ω_{ij} is regular for finite λ, the function $g_{ij}(\lambda)$ is in fact a polynomial in λ. Moreover, ω_{ij} has poles of order at most 1 for $i \neq j$ and 2 for $i = j$ at $\lambda = \infty$. Since $d\lambda$ has a double pole at ∞, this implies $\omega_{ij} = 0$ for $i \neq j$ and $\omega_{ii} = c_i d\lambda$ for some *constants* c_i. We have obtained orthogonality relations:

$$\omega_{ij} = c_i\delta_{ij}d\lambda, \quad c_i = \lim_{P\to Q_i} \tilde{\psi}_i^+(P)\psi_i(P)\frac{\Omega}{d\lambda}(P) \tag{5.30}$$

These orthogonality relations show that the inverse of the matrix $\widehat{\Psi}$ is given by:

$$\left(\widehat{\Psi}^{-1}\right)_{ij} = \frac{\Omega(P_i)}{d\lambda}\frac{1}{c_j}\tilde{\psi}_j^+(P_i) \tag{5.31}$$

It is worth noticing that this expression is invariant under a change of normalization of the components of $\tilde{\Psi}^+$, i.e. $\tilde{\psi}_j^+ \to d_j \tilde{\psi}_j^+$ yields $c_j \to d_j c_j$ and d_j cancels. On the other hand it transforms appropriately under a change of normalization of the components of Ψ. One then gets $\Omega = d\lambda/\langle\Psi^+\Psi\rangle$ if one sets the kth component of Ψ^+ to the *invariant* value $1/c_k\,\tilde{\psi}_k^+$. Let us summarize the situation in a proposition:

Proposition. *Given an effective divisor D of degree g, the functions $\psi_k(P)$ with divisor $\geq -(D + Q_k - Q_1)$ are unique up to normalization. There is a unique form Ω having a double pole at Q_1 and vanishing on D. It has g other zeroes at an effective divisor D^+ of degree g. There exists a* unique *set of functions $\psi_k^+(P)$ of divisor $\geq -(D^+ + Q_k - Q_1)$ such that $\sum_k \psi_i^+(P_k)\psi_j(P_k)\Omega(P_k) = \delta_{ij}d\lambda$. The inverse of the matrix $\widehat{\Psi}$ is $\widehat{\Psi}_{ij}^{-1} = \psi_j^+(P_i)\,\Omega(P_i)/d\lambda$.*

5.4 Time evolution

The aim of this section is to solve for the equations of motion by looking at the time evolution of the dynamical divisor D. The outcome is the beautiful fact that *the dynamical flow linearizes on the Jacobian of the spectral curve.*

Recall that the time evolution is governed by the Lax equation,

$$\frac{d}{dt}L(\lambda) = [M(\lambda), L(\lambda)] \qquad (5.32)$$

As we have seen in Chapter 3, the matrix $M(\lambda)$ is of the form

$$M = \sum_k M_k, \quad \text{with} \quad M_k(\lambda) = \left(P^{(k)}(L, \lambda)\right)_-$$

where $\left(P^{(k)}(L, \lambda)\right)$ is a polynomial in $L(\lambda)$ with *constant* rational coefficients in λ, and $\left(P^{(k)}(L, \lambda)\right)_-$ denotes its polar part at λ_k.

Suppose that, at time t, we made the analysis of the previous section and built the normalized eigenvector $\Psi(t, P)$. If $\Psi(t, P)$ is an eigenvector of $L(\lambda)$ with eigenvalue μ, the Lax equation eq. (5.32) implies $(L(\lambda) - \mu)(\frac{d\Psi}{dt} - M\Psi) = 0$. It follows that

$$\frac{d}{dt}\Psi(t, P) = (M(\lambda) - C(t, P)\mathbf{1})\,\Psi(t, P) \qquad (5.33)$$

where $C(t, P)$ is a scalar function. Normalizing the eigenvector $\Psi(t, P)$ such that its first component equals 1 gives:

$$C(t, P) = \sum_j M_{1j}(\lambda)\psi_j(t, P)$$

By the analysis of the previous section, the normalized eigenvector $\Psi(t, P)$ has poles at the dynamical divisor $D(t)$ and at the $N-1$ points Q_k, $k = 2, \ldots, N$. Consider the function $\mathcal{N}(t, dt, P) \equiv 1 + C(t, P)dt$, with dt infinitesimal:

$$\mathcal{N}(t, dt, P) = 1 + dt \sum_j M_{1j}(\lambda)\psi_j(t, P)$$

One can rewrite eq. (5.33) in the equivalent form:

$$\mathcal{N}(t, dt, P)\ \Psi(t+dt, P) = (\mathbf{1} + dtM(\lambda)) \cdot \Psi(t, P) + O(dt^2) \quad (5.34)$$

We see that the meromorphic function $\mathcal{N}(t, dt, P)$ of $P \in \Gamma$ normalizes the eigenvector whose time evolution is naturally induced by the Lax equation $\dot{\Psi} = M\Psi$. The divisor of this meromorphic function reads:

$$(\mathcal{N}) = D(t + dt) + \sum_{k,i} \sum_{\alpha=1}^{m_k} P_{k,i}^{\alpha} - D(t) - \sum_{k,i} m_k P_{k,i}$$

From eq. (5.34) we see that $\mathcal{N}$ cancels the poles of $\Psi(t+dt, P)$ at $D(t+dt)$ and produces the poles of $\Psi(t, P)$ at $D(t)$. The poles at $Q_2, \ldots, Q_N$ are the same on both sides and do not appear in $\mathcal{N}$. Moreover, since $M_k(\lambda)$ has a pole of order m_k at λ_k, $\mathcal{N}$ has poles of order m_k at the N points $P_{k,i}$ above λ_k. Finally, $\mathcal{N}$ has extra zeroes $P_{k,i}^{\alpha}$ to match the number of its poles. Since dt is small, and $\mathcal{N} = 1$ for $dt = 0$, the zeroes are close to the poles, $D(t + dt)$ is close to $D(t)$, and on each sheet i of the covering there are exactly m_k zeroes $P_{k,i}^{\alpha}$ close to the m_k^{th} order pole $P_{k,i}$.

Theorem. *Let $\gamma_j(t)$ with $j = 1, \ldots, g$ be the points of the dynamical divisor $D(t)$. Let ω be any holomorphic differential on Γ. The time evolution of the points $\gamma_j(t)$ induced by the Lax equation $\dot{L}(\lambda) = [M(\lambda), L(\lambda)]$ with $M(\lambda) = \sum_k \left(P^{(k)}(L, \lambda) \right)_-$ is such that:*

$$\frac{d}{dt} \sum_{j=1}^{g} \int^{\gamma_j(t)} \omega = \sum_k \sum_{i=1}^{N} \text{Res}_{P_{k,i}} \left(\omega P^{(k)}(\mu, \lambda) \right) \quad (5.35)$$

where the points $P_{k,i}$ are the N points above λ_k. Notice that the right-hand side is independent of time.

<u>Proof.</u> Since $\mathcal{N}(t+dt, P)$ is a meromorphic function, Abel's theorem tells us that:

$$\sum_{j=1}^{g} \int_{\gamma_j(t)}^{\gamma_j(t+dt)} \omega + \sum_{k,i,\alpha} \int_{P_{k,i}}^{P_{k,\alpha}(dt)} \omega = 0 \quad (5.36)$$

for any holomorphic differential ω. This equation will give us the time evolution of the divisor as in eq. (5.35) if we can evaluate the second sum. For this we need the following lemma:

Lemma. *Consider a point $P \in \mathbb{C}$, a holomorphic differential ω in the neigbourhood V of P and an analytic function u in V with a pole of order m at P. Consider $\epsilon \in \mathbb{C}$ small enough, so that the m points $P_\alpha(\epsilon)$, where $u(P_\alpha(\epsilon)) + \epsilon^{-1} = 0$, belong to V. Then*

$$\lim_{\epsilon \to 0} \frac{1}{\epsilon} \sum_{\alpha=1}^{m} \int_{P}^{P_\alpha(\epsilon)} \omega = -\mathrm{Res}_P(\omega u)$$

<u>Proof.</u> Let $\omega = d\sigma$ with $\sigma(P) = 0$. For any path π enclosing the zeroes $P_\alpha(\epsilon)$ of $u + \epsilon^{-1}$ and the point P, we have

$$\frac{1}{\epsilon} \sum_{\alpha=1}^{m} \int_{P}^{P_\alpha(\epsilon)} \omega = \frac{1}{\epsilon} \sum_{\alpha} \sigma(P_\alpha(\epsilon)) = \frac{1}{\epsilon} \sum_{\alpha} \mathrm{Res}_{P_\alpha(\epsilon)} \frac{u'}{u + \epsilon^{-1}} \sigma$$

$$= \frac{1}{2i\pi} \int_{\pi} \frac{u'}{1 + \epsilon u} \sigma dz$$

We used that the integrand is regular at P because $\sigma(P) = 0$. When ϵ tends to zero, the right-hand side tends to $\int_{\pi} u' \sigma \frac{dz}{2i\pi} = -\frac{1}{2i\pi} \int_{\pi} u\omega = -\mathrm{Res}_P(u\omega)$. ∎

Returning to the proof of the Theorem, we decompose the second sum in eq. (5.36) as a sum of terms associated with each pole of $M(\lambda)$, $P_{k,i}$ on Γ. The points $P_{k,i}^{\alpha}$, close to $P_{k,i}$, are by definition solutions of $1 + dt u_k(P_{k,i}^{\alpha}) = 0$ with $u_k = \sum_j (M_k)_{1j} \psi_j$. Thus, by the Lemma and eq. (5.36), one finds:

$$\sum_{j=1}^{g} \int_{\gamma_j(t)}^{\gamma_j(t+dt)} \omega = dt \sum_{k,i} \mathrm{Res}_{P_{k,i}} \left(\omega \sum_j (M_k)_{1j} \psi_j(t, P) \right) \qquad (5.37)$$

Recall that $M_k(\lambda)$ is the polar part of $P^{(k)}(L, \lambda)$, i.e.

$$M_k(\lambda) = \left(P^{(k)}(L, \lambda) \right)_{-} = P^{(k)}(L, \lambda) - \left(P^{(k)}(L, \lambda) \right)_{+}$$

with $\left(P^{(k)}(L, \lambda) \right)_{+}$ regular at λ_k, hence does not contribute to eq. (5.37). Therefore,

$$\mathrm{Res}_{P_{k,i}} \left(\omega \sum_j (M_k)_{1j} \psi_j(t, P) \right) = \mathrm{Res}_{P_{k,i}} \left(\omega \left(P^{(k)}(L, \lambda) \Psi(t, P) \right)_1 \right)$$

$$= \mathrm{Res}_{P_{k,i}} \left(\omega P^{(k)}(\mu, \lambda) \psi_1(t, P) \right) = \mathrm{Res}_{P_{k,i}} \left(\omega P^{(k)}(\mu, \lambda) \right)$$

where we have used the fact that $\Psi(t, P)$ is an eigenvector and the nor-malization $\psi_1(t, P) = 1$. ∎

Equation (5.35) can alternatively be written in terms of Abel's map.

Theorem. *The flow induced by the Lax equation (5.32) on the eigenvec-tor bundle is a linear flow on the Jacobian of the spectral curve:*

$$\mathcal{A}(D(t)) - \mathcal{A}(D(0)) = -tU^{(M)} \tag{5.38}$$

with

$$U_j^{(M)} = -\sum_{k,i} \operatorname{Res}_{P_{k,i}} \left(\omega_j P^{(k)}(\mu, \lambda) \right) \tag{5.39}$$

5.5 Theta-functions formulae

Since the motion linearizes on the Jacobian Jac (Γ), it is natural to express the solution in terms of Riemann's theta-functions. We use the notations and results of Chapter 15, in particular $\mathcal{K}$ is the vector of Riemann's constant.

We first recall the way to parametrize meromorphic functions on a Riemann surface Γ, in terms of theta-functions, on its Jacobian. For e in an open set of the divisor $\Theta = \{z \in \operatorname{Jac}(\Gamma) | \theta(z) = 0\}$, the function $\theta(e + \int_x^y \omega)$ does not vanish identically in y. By Riemann's theorem it has g zeroes $y_1, \ldots, y_g$ for given x. Since $e \in \Theta$, one of these zeroes is x and we choose $y_1 = x$. We now show that $y_2, \ldots, y_g$ are independent of x. Indeed, by Riemann's theorem we have: $\mathcal{A}(y_1) + \cdots + \mathcal{A}(y_g) = \mathcal{A}(x) - e - \mathcal{K}$, so that $y_2, \ldots, y_g$ are determined by $\mathcal{A}(y_2) + \cdots + \mathcal{A}(y_g) = -e - \mathcal{K}$. This equation is independent of x. As a side remark note that for such an y_j, $j \geq 2$, we have $\theta(-e + \int_{y_j}^x \omega) = 0$ for all x. This means that some translation of the curve Γ embedded into the Jacobian by the Abel map is entirely contained in Θ, hence one has to be careful in the choice of the vector e. To use this result to construct meromorphic functions on the Riemann surface, notice that the building block $\theta(e + \int_{x_1}^y \omega)/\theta(e + \int_{x_2}^y \omega)$ has a zero at $y = x_1$ and a pole at $y = x_2$. The extra $(g - 1)$ zeroes at the numerator and denominator cancel. We then assemble such blocks so that the product has no monodromy.

Explicit expressions for the matrix $\widehat{\Psi}(P)$ and its inverse are easily writ-ten. Let $D(t) = (\gamma_1(t), \ldots, \gamma_g(t))$ be the dynamical divisor. Let $U^{(M)}$ be the g-dimensional vector eq. (5.39) with components $U_j^{(M)}$ Then,

$$\psi_k(t, P) = d_k \frac{\theta(\mathcal{A}(P) - \mathcal{A}(Q_k) + \mathcal{A}(Q_1) - \zeta_{D(t)})}{\theta(\mathcal{A}(P) - \zeta_{D(t)})} \frac{\theta(e + \int_{Q_1}^P \omega)}{\theta(e + \int_{Q_k}^P \omega)} \tag{5.40}$$

In this equation $\zeta_D = \mathcal{A}(D) + \mathcal{K}$. Equation (5.38) implies $\zeta_{D(t)} = \zeta_{D(0)} - tU^{(M)}$. The d_k are constants ($d_1 = 1$) related to the residual diagonal action of the gauge group. Note that the sum of the arguments of the theta functions in the numerator is equal to the sum of the arguments in the denominator so that the whole expression has no monodromy when the point P loops around non-trivial cycles of Γ. Hence eq. (5.40) defines a meromorphic function on Γ with the correct zeroes and poles.

Similarly one has:

$$\psi_k^+(t, P) = \frac{1}{c_k} \frac{\theta(\mathcal{A}(P) - \mathcal{A}(Q_k) + \mathcal{A}(Q_1) - \zeta_{D^+(t)})}{\theta(\mathcal{A}(P) - \zeta_{D^+(t)})} \frac{\theta(e^+ + \int_{Q_1}^P \omega)}{\theta(e^+ + \int_{Q_k}^P \omega)}$$
(5.41)

Here D^+ is the divisor given by eq. (5.27) and the normalization constants c_k are defined in eq. (5.30). We now compute them. To do that we express the meromorphic function $\Omega/d\lambda$ in terms of theta-functions. Let B be the set of branch points of the covering $(\lambda, \mu) \to \lambda$. We decompose the set of $2(g + N - 1)$ points of B into four subsets B_0, B_0', B_1, B_1' such that card $B_0 = $ card $B_0' = N - 1$ and card $B_1 = $ card $B_1' = g$. This decomposition is arbitrary but does not affect the final formulae. Then we can write ($\prod_{B_i}$ means $\prod_{b_k \in B_i}$):

$$\frac{\Omega}{d\lambda}(P) = \qquad\qquad\qquad\qquad\qquad\qquad\qquad\qquad (5.42)$$

$$\frac{\prod_{j=2}^N \theta(e + \int_{Q_j}^P \omega)\theta(e^+ + \int_{Q_j}^P \omega)\ \theta(\mathcal{A}(P) - \zeta_{D(t)})\ \theta(\mathcal{A}(P) - \zeta_{D^+(t)})}{\prod_{B_0} \theta(e + \int_{b_k}^P \omega) \prod_{B_0'} \theta(e^+ + \int_{b_k}^P \omega)\ \theta(\mathcal{A}(P) - \zeta_{B_1})\ \theta(\mathcal{A}(P) - \zeta_{B_1'})}$$

This expression of $\Omega/d\lambda$ has the correct zeroes and poles, and has no monodromy in view of eq. (5.27).

To compute c_k we can now use eq. (5.30), yielding:

$$c_k = d_k \times \qquad\qquad\qquad\qquad\qquad\qquad\qquad\qquad (5.43)$$

$$\frac{\theta(\mathcal{A}(Q_1) - \zeta_{D(t)})\theta(\mathcal{A}(Q_1) - \zeta_{D^+(t)}) \prod_{j \neq k} \theta(e + \int_{Q_j}^{Q_k} \omega)\theta(e^+ + \int_{Q_j}^{Q_k} \omega)}{\prod_{B_0} \theta(e + \int_{b_k}^{Q_k} \omega) \prod_{B_0'} \theta(e^+ + \int_{b_k}^{Q_k} \omega)\theta(\mathcal{A}(Q_k) - \zeta_{B_1})\theta(\mathcal{A}(Q_k) - \zeta_{B_1'})}$$

This gives an expression of the elements of the matrix $\widehat{\Psi}^{-1}$ as products of theta-functions, through eq. (5.31).

Example. Let us apply the above formalism to find the solution of the Neumann model. In this case since ${}^t L(\lambda) = L(-\lambda)$ the normalized $\Psi^+(\lambda, \mu)$ is equal to ${}^t \Psi(-\lambda, \mu)$. The transformation $(\lambda, \mu) \to (-\lambda, \mu)$ is just the hyperelliptic involution σ on the spectral curve of the Neumann model. The fixed points of σ are the $(2g+2)$ points (λ, μ) lying above $\lambda = \infty$, namely the Q_j, and above $\lambda = 0$, namely the point $P_\infty(\lambda = 0, \mu = \infty)$ and the $N-1$ points of coordinates $(\lambda = 0, \mu = \beta_i)$, see eq. (5.14). We take the point P_∞ as base point of the Abel map, and note that the hyperelliptic involution changes the sign of the Abelian differentials which are of the form $p(\mu)d\mu/\lambda$, so that for any point P we get $\mathcal{A}(\sigma(P)) = -\mathcal{A}(P)$ modulo periods. The branch points B of the covering $(\lambda, \mu) \to \lambda$ are solutions of $\partial_\mu \Gamma(\lambda, \mu) = 0$, i.e. $\sum_k F_k/(\mu - a_k)^2 = 0$. This equation has $(2N-2)$ roots at finite distance, each one giving rise to two points related by the hyperelliptic involution. Hence the set B is globally invariant under σ and one can choose $B_0' = \sigma(B_0)$ and $B_1' = \sigma(B_1)$ in the above construction. Considering eq. (5.27), we see that $D^+ = \sigma(D)$. Indeed, since $B = \sum_i (b_i + \sigma(b_i))$ with $\sigma(b_i) \neq b_i$ we have $\mathcal{A}(B) = \sum_i(\mathcal{A}(b_i) + \mathcal{A}(\sigma(b_i))) = 0$ up to periods. Similarly, since $\sigma(Q_j) = Q_j$ we have $\mathcal{A}(Q_j) = -\mathcal{A}(Q_j)$ so that $2\mathcal{A}(Q_j) = 0$ modulo periods. This shows that $\mathcal{A}(Q_j)$ is a half-period on the Jacobian torus. Finally, we get $\mathcal{A}(D^+) = -\mathcal{A}(D)$ modulo periods so that $D^+ = \sigma(D)$ since D is generic.

From this we understand that the requirement $\psi_k^+(t, P) = \psi_k(t, \sigma(P))$ fixes the constants d_k and e^+ in the expressions (5.40) and (5.41). Let us compare for instance the theta-functions $\theta(\mathcal{A}(P) - \mathcal{A}(Q_k) + \mathcal{A}(Q_1) - \zeta_{D^+(t)})$ and $\theta(\mathcal{A}(\sigma(P)) + \mathcal{A}(Q_k) - \mathcal{A}(Q_1) - \zeta_{D(t)})$. The sum of the two arguments vanishes modulo periods, because in the hyperelliptic case the vector of Riemann's constants $\mathcal{K}$ is some half-period, hence these two theta-functions have the same zeroes. Applying a similar argument to the other theta-functions and choosing $e^+ = -e$, we see that:

$$\psi_k^+(t, P) = \frac{1}{c_k} \frac{\theta(-\mathcal{A}(P) + \mathcal{A}(Q_k) - \mathcal{A}(Q_1) - \zeta_{D(t)})}{\theta(-\mathcal{A}(P) - \zeta_{D(t)})} \frac{\theta(e - \int_{Q_1}^P \omega)}{\theta(e - \int_{Q_k}^P \omega)} \quad (5.44)$$

Starting from the expressions (5.40) for ψ_k, (5.44) for ψ_k^+ and (5.42) for $\Omega/d\lambda$, one computes c_k according to eq. (5.30). Then the relation $\psi_k^+(t, P) = \psi_k(t, \sigma(P))$ fixes the constant d_k.

To compute the solution of the Neumann model note that the diagonal element L_{kk} of the Lax matrix reads $L_{kk}(\lambda) = a_k - x_k^2/\lambda^2$. On the other hand, from the reconstruction formula eq. (5.20), we have $L_{kk}(\lambda) = \sum_i \psi_k(P_i)\mu(P_i)(\psi^{(-1)})_k(P_i)$, where the P_i are the N points above λ. When $\lambda \to 0$ only $\mu(P_\infty)$ diverges as $-1/\lambda^2$ while all the other terms remain finite. Hence:

$$x_k^2 = \psi_k(P_\infty)(\psi^{(-1)})_k(P_\infty)$$

Note that this formula implies $\sum_k x_k^2 = 1$ as it should be in the Neumann model, and moreover this expression is independent of the constant d_k. Inserting the above expressions we immediately find:

$$x_k^2(t) = \alpha_k \frac{\theta(\mathcal{A}(Q_k) - \mathcal{A}(Q_1) + \zeta_{D(t)})\, \theta(\mathcal{A}(Q_k) - \mathcal{A}(Q_1) - \zeta_{D(t)})}{\theta(\mathcal{A}(Q_1) - \zeta_{D(t)})\, \theta(\mathcal{A}(Q_1) + \zeta_{D(t)})}$$

where α_k is given by ratios of theta-functions *completely independent* of the dynamical divisor $D(t)$. It depends only on the geometry of the spectral curve.

It is convenient to express this result in terms of theta–functions with characteristics $\theta[\eta](z)$ which are essentially translates of the theta-function by half-periods:

$$\theta[\eta](z) = e^{i\pi\left({}^t\eta'\mathcal{B}\eta' + 2\,{}^t\eta'(z+\eta'')\right)}\, \theta(z + \eta) \tag{5.45}$$

where $\eta = \mathcal{B}\eta' + \eta''$ is a half-period, i.e. $\eta', \eta'' \in (\mathbb{Z}/2)^g$. Note that $\mathcal{A}(Q_i)$ are half-periods. By redefining $\zeta_D \to \zeta_D + \mathcal{A}(Q_1)$ one gets rid of $\mathcal{A}(Q_1)$ in the theta–functions. Indeed the first factors in the numerator and denominator get translated by the period $2\mathcal{A}(Q_1)$. This produces exponential factors whose dependence in ζ_D cancel. Similarly, the dependence in D in the exponential factor in eq. (5.45) cancels as well, this time between the two factors in the numerator, and separately between the two factors in the denominator of x_k^2. Finally $\mathcal{A}(Q_k) = \int_{P_\infty}^{Q_k} \omega = \mathcal{B}\eta'_{2k-1} + \eta''_{2k-1}$ is an even non-singular characteristic. Hence one gets

$$x_k^2(t) = \alpha'_k \frac{\theta^2[\eta_{2k-1}](\zeta_{D(t)})}{\theta^2[0](\zeta_{D(t)})}$$

One could in principle evaluate the coefficients α'_k directly by using the very special properties of theta-functions on hyperelliptic curves. However, we have a short cut by appealing to the Frobenius formula, only valid on hyperelliptic curves, which states that:

$$\sum_{k=1}^{g+1} \frac{\theta^2[\eta_{2k-1}](z)\, \theta^2[\eta_{2k-1}](0)}{\theta^2[0](z)\, \theta^2[0](0)} = 1$$

so we finally get:

$$x_k^2(t) = \frac{\theta^2[\eta_{2k-1}](z_0 - Ut)\, \theta^2[\eta_{2k-1}](0)}{\theta^2[0](z_0 - Ut)\, \theta^2[0](0)} \tag{5.46}$$

Here U is obtained by applying eq. (5.35) with $P^{(0)}(\mu, \lambda) = \lambda\mu$ (recall that for the Neumann model there is only a singularity at $\lambda = 0$ and

$M(\lambda) = (\lambda L(\lambda))_-$, see Chapter 3). This yields:

$$U_j = -\sum_{P_i} \mathrm{Res}_{P_i}(\lambda\mu)\omega_j(P) = -\mathrm{Res}_{P_\infty}(\lambda\mu)\omega_j(P) = \omega_j(P_\infty)$$

where P_i are the N points above $\lambda = 0$, and we used that for $\lambda = 0$, the product $(\lambda\mu)$ has only a simple pole at $P_\infty(\lambda = 0, \mu = \infty)$. The ω_j are the *normalized* Abelian differentials.

5.6 Baker–Akhiezer functions

Baker–Akhiezer functions are special functions with essential singularities on Riemann surfaces. With them, we have a very natural parametrization of eigenvectors of the linear system. We also get a very simple proof of the linearization theorem.

Definition. *Let $P_1, \ldots, P_l$ be points on a Riemann surface Γ of genus g. Let $w_i(P)$, with $w_i(P_i) = 0$, be local parameters around these points. Let $S_i(P) = \sum_{r=-m_i}^{-1} S_{i,r} w_i^r$ be some singular parts around P_i. Let D be a divisor on Γ. A Baker–Akhiezer function, $\Psi_{BA}(P)$, defined with these data, is a function such that:*
(1) it is meromorphic on Γ outside the points P_i with the divisor of its poles and zeroes satisfying $(\Psi_{BA}) + D \geq 0$,
(2) for $P \to P_i$ the product $\Psi_{BA}(P)e^{-S_i(w_i(P))}$ is analytic.

It is important to keep in mind the data involved in the definition of the Baker–Akhiezer functions. First we need a set of punctures P_i on the Riemann surface. Second a set of local parameters $w_i(P)$ allowing to define a set of singular parts S_i in the neighbourhood of each puncture. Notice that this definition is not invariant under change of local parameters w_i. Assuming for the moment that such functions exist it is worth doing a few remarks.

Remark 1. If Baker–Akhiezer functions associated with a given set of data exist, they form a vector space. However, the sum of two Baker–Akhiezer functions with different singular parts S_i is not a Baker–Akhiezer function.

Remark 2. The ratio of two Baker–Akhiezer functions associated with a given set of singular parts is a meromorphic function. This allows one to use standard analysis on Riemann surfaces to study Baker–Akhiezer functions.

Remark 3. Even though Baker–Akhiezer functions are not meromorphic functions, they have the same number of poles and zeroes. The differential form $d(\log f)$ is a meromorphic form. The sum of its residues is the number of zeroes minus the number of poles of f and has to vanish. Essential singularities do not contribute because around P_i we have $d(\log f) = dS_i + \text{regular}$ and dS_i has no residue.

We now give a fundamental formula expressing the Baker–Akhiezer functions in terms of Riemann theta-functions.

Recall that a differential of the second kind is a meromorphic differential with poles of order ≥ 2. See Chapter 15 for more details. Let $\Omega^{(S)}$ be the unique Abelian differential of the second kind, normalized with vanishing a-periods, and with singular part at the points P_i of the form $dS_i(w_i(P))$. Thus, near the points P_i,

$$\Omega^{(S)} = d\left(\sum_{r=-m_i}^{-1} S_{i,r} w_i^r + \text{regular} \right)$$

Denote by $2i\pi U^{(S)}$ the vector of b-periods of $\Omega^{(S)}$. Its g components $U_j^{(S)}$, $j = 1, \ldots, g$ are:

$$U_j^{(S)} = \frac{1}{2i\pi} \oint_{b_j} \Omega^{(S)} \tag{5.47}$$

Proposition. *If $D = \sum_{i=1}^{g} \gamma_i$ is a generic divisor of degree g, the following expression defines a Baker–Akhiezer function with D as divisor of poles:*

$$\Psi_{BA}(P) = \text{const. } \exp\left(\int_{P_0}^{P} \Omega^{(S)} \right) \frac{\theta(\mathcal{A}(P) + U^{(S)} - \zeta)}{\theta(\mathcal{A}(P) - \zeta)} \tag{5.48}$$

Here $\zeta = \mathcal{A}(D) + \mathcal{K}$, where $\mathcal{K}$ is the vector of Riemann's constants and $\mathcal{A}$ denotes the Abel map with based point P_0, cf. eq. (5.29).

Proof. It is enough to check that the function defined by the formula (5.48) is well-defined (i.e., it does not depend on the path of integration between P_0 and P) and has the desired analytical properties. Indeed, when P describes some a-cycle, nothing happens because the theta-functions are a-periodic and $\Omega^{(S)}$ is normalized. If P describes the b_j-cycle the quotient of theta-functions is multiplied by $\exp(-2i\pi U_j^{(S)})$ (see Chapter 15) while the exponential factor changes by $\exp(2i\pi U_j^{(S)})$, so that ψ_{BA} is well-defined. Clearly it has the right poles if $\deg D = g$. ∎

Remark 4. For a generic divisor D of degree $\geq g$, the dimension of the vector space of Baker–Akhiezer functions is equal to $\deg(D) - g + 1$. In particular for $\deg D = g$ the above formula gives the unique Baker–Akhiezer function having poles at D up to a constant. If we have two Baker–Akhiezer functions, their ratio is a meromorphic function with $d = \deg D$ poles. By the Riemann–Roch theorem the dimension of the space of such functions is $d - g + 1$.

It is worth noticing that generically we get a non-trivial Baker–Akhiezer function with only g poles, while to get a non-trivial meromorphic function we need generically $(g + 1)$ poles.

To understand why Baker–Akhiezer functions arise naturally in the construction of the eigenvectors, let us consider the unnormalized eigenvector $\Psi_{\text{un}}(t, P)$ whose time evolution is governed by the equation

$$\partial_t \Psi_{\text{un}}(t, P) = M(\lambda)\Psi_{\text{un}}(t, P) \tag{5.49}$$

The normalized eigenvector $\Psi(t, P)$ and the unnormalized eigenvector $\Psi_{\text{un}}(t, P)$ are related by multiplication by a scalar function: $\Psi_{\text{un}}(t, P) = f(t, P)\Psi(t, P)$ and $(\Psi_{\text{un}})_1(t, P) = f(t, P)$. Taking the first component of eq. (5.49), one gets:

$$\dot{f} = Cf \quad \text{with} \quad C = \sum_j M_{1j}\psi_j$$

where C is the same object appearing in eq. (5.33).

Let us describe the singularities of $f(t, P)$. Note that $C(P)$ has poles where Ψ has poles or at points above the poles λ_k of $M(\lambda)$ (recall that in general the poles of $M(\lambda)$ are a subset of the poles of $L(\lambda)$). Consider first the points $P_{k,i}$, $i = 1, \ldots, N$ above a point λ_k. In the vicinity of $P_{k,i}$ we have:

$$\dot{f} = (M\Psi)_1 f = \left(\left((P^{(k)}(L, \lambda))_- + \text{regular} \right) \Psi \right)_1 f$$

$$= \left(\left(P^{(k)}(L, \lambda) + \text{regular} \right) \Psi \right)_1 f = \left(\left(P^{(k)}(\mu, \lambda) \right)_- + \text{regular} \right) f$$

where we have used the fact that Ψ is an eigenvector of $L(\lambda)$ with eigenvalue μ and $\Psi_1 = 1$. The quantity $\dot{f}/f$ has poles at points (λ, μ) such that $\left(P^{(k)}(\mu, \lambda) \right)_- \neq 0$. The projection $()_-$ is computed using the local parameter λ. Notice that $\left(P^{(k)}(\mu(\lambda), \lambda) \right)_-$ is independent of time, therefore $f(t, P)$ has an essential singularity at $P_{k,i}$ of the form:

$$f(t, P) = e^{t\left(P^{(k)}(\mu,\lambda) \right)_-} \times \text{regular}$$

Let us now consider the poles of C coming from Ψ. First at $\lambda = \infty$, while Ψ has poles, M vanishes so that C is regular, and nothing special happens for f. At a point $\gamma(t)$ of the dynamical divisor $D(t)$ we have

$\psi_i \sim \alpha_i(t)/(\lambda - \gamma(t))$ and $C \sim r(t)/(\lambda - \gamma(t))$. Comparing the second order pole in both sides of eq. (5.33), we find $r(t) = -\dot{\gamma}(t)$. Thus $\partial_t \log f = \partial_t \log(\lambda - \gamma) + \text{regular}$, showing that $f(t, P)$ *vanishes* at the points of the dynamical divisor $D(t)$.

Finally, let us remark that the poles of $f(t, P)$ are independent of time. Indeed, assuming that $f(t, P)$ has a pole of order k at a point $\gamma(t)$, which by the previous argument is not a pole of C, the orders of the poles on both sides of the equation $\dot{f} = Cf$ are different if $\dot{\gamma} \neq 0$. Considering the solution such that $f(t = 0) = 1$ we see that the divisor of its zeroes is the dynamical divisor $D(t)$ and the divisor of its poles is $D(0)$. Moreover, f has essential singularities at the points $P_{k,i}$ with prescribed singular parts, hence f is the unique Baker–Akhiezer function with these essential singularities and the g poles corresponding to the divisor $D(0)$, so that:

$$f(t, P) = \exp\left(\int_{P_0}^{P} t\,\Omega^{(M)} \right) \frac{\theta(\mathcal{A}(P) + tU^{(M)} - \zeta_{D(0)})}{\theta(\mathcal{A}(P) - \zeta_{D(0)})} \qquad (5.50)$$

The linear time dependence in the theta-function in the numerator of this equation arises from the form of the singular exponential and the requirement that there is no monodromy. This provides another quick proof of the linearization of the flow on the Jacobian. If $D(t)$ is the divisor of the zeroes of $f(t, P)$, by the Riemann theorem it satisfies:

$$\mathcal{A}(D(t)) - \mathcal{A}(D_0) = -t\,U^{(M)} = -\frac{t}{2i\pi} \oint_b \Omega^{(M)} \qquad (5.51)$$

This shows that the flow is linear on the Jacobian! On the other hand we know that $\mathcal{A}(D(t))$ is given by eq. (5.38), which coincides with eq. (5.51) because by Riemann's bilinear indentity

$$\frac{1}{2i\pi} \oint_{b_j} \Omega^{(M)} = -\sum_{k}\sum_{i=1}^{N} \text{Res}_{P_{k,i}} \left(\omega_j P^{(k)}(\mu, \lambda) \right)$$

with ω_j the normalized Abelian differentials. See Chapter 15.

We finally give the expression of the components of the unnormalized eigenvector $\Psi_{\text{un}}(t, P) = f(t, P)\Psi(t, P)$:

$$(\Psi_{\text{un}})_k(t, P) = d_k \, \exp\left(t \int_{P_0}^{P} \Omega^{(M)} \right)$$

$$\times \frac{\theta(\mathcal{A}(P) - \mathcal{A}(Q_k) + \mathcal{A}(Q_1) - \zeta_{D(t)})}{\theta(\mathcal{A}(P) - \zeta_{D(0)})} \frac{\theta(e + \int_{Q_1}^{P} \omega)}{\theta(e + \int_{Q_k}^{P} \omega)}$$

In the product $f(t, P)\psi_k(t, P)$, the zeroes of f cancel the dynamical poles of ψ_k which are replaced by the *constant* poles of f.

Remark 5. As explained in Chapter 3, different functions $P^{(k)}(L, \lambda)$ correspond to different dynamical flows. Therefore, different Abelian differentials of the second kind with poles at the points above λ_k correspond to different time flows. In other words, all the different dynamics are encoded into the singular differentials $\Omega^{(M)}$.

5.7 Linearization and the factorization problem

We show that the solution of integrable systems by factorization can be interpreted as the time evolution of the eigenvector bundle. This gives a third very short proof that the flows linearize on the Jacobian of the spectral curve.

We consider small disks U_k around the poles λ_k of $L(\lambda)$ and define the open set U_+ as the union of U_k, while U_- is an open set slightly larger than the complement of U_+ in the complex plane. On these open sets we have defined in Chapter 3 a factorization problem

$$\theta_-^{-1}(\lambda, t)\theta_+(\lambda, t) = e^{-\sum_i t_i dH_i(L(\lambda, 0))}$$

where $\theta_\pm$ are analytic and invertible in $U_\pm$ respectively. Recall that we have shown that the matrix $\widehat{\Psi}(t)\widehat{\Psi}^{-1}(0)$ has different expressions on the patches U_+ and U_- given by eqs. (3.56, 3.57). Multiplying these equations on the right by $\widehat{\Psi}(0)$ we get on U_+ and U_- respectively:

$$\widehat{\Psi}(t) = \theta_+(\lambda, t)\widehat{\Psi}(0)e^{\sum_i t_i dH_i(\widehat{\mu})}, \quad \widehat{\Psi}(t) = \theta_-(\lambda, t)\widehat{\Psi}(0)$$

where $\widehat{\mu}$ is the diagonal matrix of eigenvalues of $L(\lambda, 0)$. We used that $dH(L) = \widehat{\Psi}dH(\widehat{\mu})\widehat{\Psi}^{-1}$ because $H(L)$ is an ad-invariant function.

We can interpret this matrix equation in λ as a vector equation on the Riemann surface Γ. First we lift each disc U_k around λ_k to N disks around the $P_{k,i}$, and still define U_+ as the union of these disks, and U_- as an open set in Γ containing the closure of U_+. Each column of these matrix equations can be viewed as vector equations at a point P above λ:

$$\Psi(t, P)e^{-\sum_i t_i dH_i(\mu(P))} = \theta_+(\lambda, t)\Psi(0, P), \quad \Psi(t, P) = \theta_-(\lambda, t)\Psi(0, P)$$
$$(5.52)$$

valid on the open sets $U_\pm$ respectively.

The vector $\Psi(0, P)$ is a section of the eigenvector bundle, E_0, at time $t = 0$. As explained in Chapter 15, $(\theta_+(\lambda, t)\Psi(0, P), \theta_-(\lambda, t)\Psi(0, P))$ defines a section of a line bundle *isomorphic* to E_0 due to the regularity properties of the matrices $\theta_\pm$. In the left-hand sides of eqs. (5.52), we write $\Psi(t, P) = f(t, P)\Psi_m(t, P)$, where $\Psi_m(t, P)$ is meromorphic with first component equal to 1, and $f(t, P)$ is the Baker–Akhiezer function (5.50). Note that, by definition, $\Psi_m(t, P)$ is a section of the eigenvector bundle E_t. We introduce the line bundle F_t with transition function $e^{-\sum_i t_i dH_i(\mu(P))}$ on $U_+ \cap U_-$ which possesses the section $(fe^{-\sum_i t_i dH_i(\mu(P))}, f)$ on (U_+, U_-) since the first term is regular in U_+ by eq. (5.52). Recall that the product of bundles admits the product of sections. It is now clear that $E_t \sim E_0 \otimes F_t$. The bundle F_t is of Chern class 0 because E_t and E_0 have the same Chern class at least for t small. Hence F_t defines a point in the Jacobian Jac (Γ). This point moves linearly in time since the addition law on the Jacobian corresponds to taking the product of transition functions.

5.8 Tau-functions

We now wish to relate the formula for the Baker–Akhiezer function to the so-called tau-function (see Chapter 3). Let us consider for simplicity the case of only one singular point P_∞, and let z be a local parameter for a point P in the vicinity of P_∞ such that $z(P_\infty) = 0$. The general case is similar but more cumbersome to present.

In order to be able to describe at once all possible singular parts, we introduce an infinite set of elementary time variables t_k. Denote by $\Omega^{(k)}$ the normalized differential of the second kind with singular part $d(z^{-k})$ at P_∞ and denote by $U^{(k)}$ its b-periods,

$$U_j^{(k)} = \frac{1}{2i\pi} \oint_{b_j} \Omega_\infty^{(k)}$$

Proposition. *Let $\psi_{BA}(P)$ be the Baker–Akhiezer function with divisor D of poles of degree g and singular part $\xi(t, z) = \sum_k t_k z^{-k}$ at P_∞ (this should be understood in the sense of formal series), normalized such that:*

$$\psi_{BA}(P) = e^{\xi(t,z)}(1 + O(z)), \quad z \sim 0$$

Define $t - [z] = \{t_k - \frac{1}{k}z^k\}$. Then we have in the vicinity of P_∞:

$$\psi_{BA}(P) = e^{\xi(t,z)} \frac{\tau(t - [z])}{\tau(t)} \tag{5.53}$$

The function $\tau(t)$ may be expressed in terms of the Riemann's theta-function,

$$\tau(t) = e^{\alpha(t)+\beta(t,t)} \theta\left(\mathcal{A}(P_\infty) + \sum_k t_k U^{(k)} - \zeta\right) \tag{5.54}$$

Here $\alpha(t)$ and $\beta(t, t)$ are a linear and a quadratic form in the times t respectively, and $\zeta = \mathcal{A}(D) + \mathcal{K}$. $\mathcal{A}(D)$ is the Abel map and $\mathcal{K}$ the vector of Riemann's constants.

<u>Proof</u>. From eq. (5.48), the normalized Baker–Akhiezer function can be written as:

$$\psi(P) = e^{\xi(t,z)} \, \exp\left[\sum_k t_k \left(\int_{P_\infty}^P (\Omega^{(k)} - d\, z^{-k})\right)\right] \tag{5.55}$$

$$\times \frac{\theta(\mathcal{A}(P) + \sum_k t_k U^{(k)} - \zeta)\theta(\mathcal{A}(P_\infty) - \zeta)}{\theta(\mathcal{A}(P) - \zeta)\theta(\mathcal{A}(P_\infty) + \sum_k t_k U^{(k)} - \zeta)}$$

Indeed, eq. (5.48) contains $\int_{P_0}^P \Omega$ which can be written $\xi(t, z) + \int_{P_\infty}^P (\Omega - d\xi) + C^{\text{st}}$ where $(\Omega - d\xi) = \sum_k t_k (\Omega^{(k)} - d\, z^{-k})$ is regular at P_∞. Moreover, eq. (5.55) is obviously correctly normalized.

On the other hand, if $\tau(t)$ is assumed to be of the form (5.54), we have

$$\frac{\tau(t - [z])}{\tau(t)} = e^{-\alpha([z]) - 2\beta(t,[z]) + \beta([z],[z])} \frac{\theta(\mathcal{A}(P_\infty) + \sum_k (t_k - \frac{z^{-k}}{k}) U^{(k)} - \zeta)}{\theta(\mathcal{A}(P_\infty) + \sum_k t_k U^{(k)} - \zeta)}$$

We need to compare eq. (5.53) and eq. (5.55). Recall that near P_∞, $\int_{P_\infty}^P (\Omega^{(k)} - d\, z^{-k}) = b_k(z)$ is regular. So we first choose $\alpha(t)$ and $\beta(t,t)$ such that:

$$\beta(t, [z]) = -\frac{1}{2}\sum_k t_k b_k(z), \quad \alpha([z]) = \beta([z], [z]) + \log\left(\frac{\theta(\mathcal{A}(P) - \zeta)}{\theta(\mathcal{A}(P_\infty) - \zeta)}\right)$$

This defines α and β uniquely and consistently. For this, we must check that the coefficients β_{kj} of β are symmetric. We have $\beta_{kj} = -1/2\, j b_{kj}$, where $b_k(z) = \sum_j b_{kj} z^j$. We apply the Riemann bilinear identity, eq. (15.8) in Chapter 15 to the two normalized second kind Abelian differentials Ω_k and Ω_l. The left-hand side in this identity vanishes because the integrals over a-cycles vanish, while the right-hand side yields $k b_{lk} = l b_{kl}$.

This choice takes care of the exponential prefactor and two of the theta functions in eq. (5.55). To deal with other two theta functions in (5.55) we Taylor expand the Abel map $\mathcal{A}(P) - \mathcal{A}(P_\infty)$ around P_∞. Writing $\omega_j = \sum_{i=0}^\infty c_i^{(j)} z^i \, dz$ in the vicinity of the point P_∞ and Taylor expanding using Riemann bilinear identities, one deduces:

$$\mathcal{A}_j(P) - \mathcal{A}_j(P_\infty) = \sum_{i=1}^\infty c_{i-1}^{(j)} \frac{z^i}{i} = -\frac{1}{2\pi i}\sum_{k=1}^\infty \frac{z^k}{k} \oint_{b_j} \Omega^{(k)} = -\sum_{k=1}^\infty \frac{z^k}{k} U_j^{(k)}$$

Using this relation we get:

$$\mathcal{A}(P) + \sum_k t_k U^{(k)} - \zeta = \mathcal{A}(P_\infty) + \sum_k (t_k - \frac{z^k}{k}) U^{(k)} - \zeta \tag{5.56}$$

Gathering all this we obtain eq. (5.53). ∎

The formula (5.53) relating Baker–Akhiezer functions to tau-functions is usually called the Sato formula, cf. eq. (3.61) in Chapter 3. It may easily be generalized for several punctures, see Chapter 8 for more details. It shows that the local parameter z can be generated from translations on the infinite set of times. The left-hand side of eqs. (5.56) gives a convergent expression for the formal series of the right-hand side. Moreover, the Baker–Akhiezer function provides a global meaning to Sato's formula in this case.

5.9 Symplectic form

Our aim is to express the symplectic form inherited from the coadjoint orbit structure in terms of the dynamical divisor.

We consider a rational Lax matrix of the form $L(\lambda) = L_0 + \sum_k L_k(\lambda)$, where each L_k may be written as:

$$L_k(\lambda) = \left(g_k \cdot (A_k)_- \cdot g_k^{-1}\right)_-$$

with $(A_k)_-$ diagonal matrices. Locally around λ_k, $L(\lambda)$ can be diagonalized as, see Chapter 3:

$$L(\lambda) = g_k \, A_k \, g_k^{-1}$$

Both matrices A_k and g_k depend on λ. A_k has poles at λ_k but g_k is *regular*. By definition L_0 is non-dynamical. The variables $(A_k)_-$ are also chosen to be not dynamical, and specify the coadjoint orbit. The dynamical variables are the matrix elements of the jets $\widehat{g}^{(k)}$, cf. eq. (3.13) in Chapter 3. The pullback on the loop group of the Kirillov symplectic form on the coadjoint orbit is:

$$\omega = \sum_k \text{Res}_{\lambda_k} \text{Tr}\left((A_k)_- g_k^{-1}\delta g_k \wedge g_k^{-1}\delta g_k\right) \, d\lambda$$

The dynamical variables $\widehat{g}^{(k)}$ and $\widehat{g}^{(k')}$ Poisson commute for $k \neq k'$.

We have seen that the Lax pair description of a dynamical system naturally provides coordinates on phase space, namely $g = $ genus (Γ) independent action variables F_i which parametrize the spectral curve Γ, and g points $\gamma_i = (\lambda_{\gamma_i}, \mu_{\gamma_i})$ on the spectral curve, which we called the dynamical divisor. It is important to express the symplectic form in these coordinates. The phase space appears as a fibred space whose base is the space of moduli of the spectral curve, explicitly described as coefficients of the equation $\Gamma(\lambda, \mu) = 0$ of the spectral curve, and the fibre at a given Γ is the Jacobian of the curve $\Gamma(\lambda, \mu) = 0$. On this space we introduce a differential δ which varies the dynamical variables F_i, λ_{γ_i}, μ_{γ_i} subjected to the constraint $\Gamma_{\{F_i\}}(\lambda_{\gamma_i}, \mu_{\gamma_i}) = 0$.

We will need an auxiliary fibre bundle above the same base whose fibre is $\Gamma \times \text{Jac}(\Gamma)$. We extend δ to this space by keeping the previous definition on the $\text{Jac}(\Gamma)$ part and on the Γ part, we differentiate any function of F_i, λ, μ with $\Gamma_{\{F_i\}}(\lambda, \mu) = 0$, by keeping λ *constant*. This definition makes sense because the bundle of curves is given by the family of equations $\Gamma(\lambda, \mu) = 0$, where the coefficients of Γ depend on the moduli which parametrize the base space. This provides a universal definition of the meromorphic function λ on the whole family of curves. So differentiating on the bundle of curves, keeping λ constant, provides a horizontal direction, i.e. a connection. Explicitly, for a function $f(P; F_i)$ we take λ as a local parameter, then $\delta f = \sum_i \partial_{F_i} f \delta F_i$. At a branch point, however, the local parameter is μ, and we have:

$$\delta f = \partial_\mu f \delta \mu + \sum_i \partial_{F_i} f \delta F_i \qquad (5.57)$$

To compute $\delta \mu$ we differentiate the equation $\Gamma_{\{F_i\}}(\lambda, \mu) = 0$ at λ constant, getting:

$$\delta \mu = -\frac{1}{\partial_\mu \Gamma_{\{F_i\}}(\lambda, \mu)} \sum_i \partial_{F_i} \Gamma_{\{F_i\}}(\lambda, \mu) \delta F_i \qquad (5.58)$$

At a branch point of the covering $(\lambda, \mu) \to \lambda$, we have $\partial_\mu \Gamma_{\{F_i\}}(\lambda, \mu) = 0$, hence the differential δf acquires a pole even though f is regular. Note, however, that if $f(P)$ depends on P only through $\lambda(P)$, δf is regular at the branch points.

Recall that at each point $P(\lambda, \mu)$ on Γ a column eigenvector $\Psi(P)$ of the Lax matrix is defined, up to normalization, and that we have defined a dual line eigenvector $\Psi^{(-1)}(P)$ such that $\langle \Psi^{(-1)}(P)\Psi(P) \rangle = 1$. This allows us to define a 3-form K on our extended fibre bundle. We regard it as a 1-form on Γ whose coefficients are 2-forms on phase space.

$$K = K_1 + K_2 \qquad (5.59)$$
$$K_1 = \langle \Psi^{(-1)}(P)\delta L(\lambda) \wedge \delta \Psi(P) \rangle \, d\lambda$$
$$K_2 = \langle \Psi^{(-1)}(P)\delta \mu \wedge \delta \Psi(P) \rangle \, d\lambda$$

Of course $\Psi(P)$ is defined, knowing the dynamical divisor, up to multiplication by a diagonal matrix independent of P. We normalize the eigenvectors at ∞ so that

$$\psi_i(Q_j) = \lambda \delta_{ij} + O(1), \quad \text{for } i, j = 2, \ldots, N \qquad (5.60)$$

Proposition. *Define the 2-form on phase space:* $\omega = \sum_{k,i} \mathrm{Res}_{P_{k,i}} K$, *where $P_{k,i}$ are the points above the poles λ_k of $L(\lambda)$. Then we have:*

$$\omega = 2 \sum_{i=1}^{g} \delta\lambda_{\gamma_i} \wedge \delta\mu_{\gamma_i} \qquad (5.61)$$

where $(\lambda_{\gamma_i}, \mu_{\gamma_i})$ are the coordinates of the points of the dynamical divisor D.

Proof. The sum of the residues of K, seen as a form on Γ, vanishes. The poles of K are located at four different places. First the dynamical poles of Ψ, then the poles at the $P_{k,i}$ coming from L and μ, next the poles above $\lambda = \infty$ coming from Ψ and $d\lambda$, and finally the poles at the branch points of the covering coming from the poles of $\Psi^{(-1)}$ and from eq. (5.58). Let us compute the residues at the dynamical poles $(\gamma_1, \dots, \gamma_g)$. We write the coordinates of these points as: $\gamma_i = (\lambda_{\gamma_i}, \mu_{\gamma_i})$ for $i = 1, \dots, g$. Near such a point we can choose λ as a universal local parameter and $\Psi = 1/(\lambda - \lambda_{\gamma_i}) \times \Psi_{\mathrm{reg}}$, hence:

$$\delta\Psi = \frac{\delta\lambda_{\gamma_i}}{\lambda - \lambda_{\gamma_i}} \left(\Psi + O(1)\right), \text{ so that } K_1 \sim \langle \Psi^{(-1)} \delta L \Psi \rangle \wedge \delta\lambda_{\gamma_i} \frac{d\lambda}{\lambda - \lambda_{\gamma_i}}$$

Since $(L-\mu)\Psi = 0$ and $\Psi^{(-1)}(L-\mu) = 0$, we have $(\delta L - \delta\mu)\Psi + (L-\mu)\delta\Psi = 0$. Multiplying by $\Psi^{(-1)}$ we get $\langle \Psi^{(-1)} \delta L \Psi \rangle = \delta\mu$, therefore:

$$\mathrm{Res}_{\gamma_i} K_1 = \delta\mu|_{\gamma_i} \wedge \delta\lambda_{\gamma_i} \qquad (5.62)$$

Here $\delta\mu$ is to be seen as a meromorphic function on Γ given by eq. (5.58), that is $\sum_j \partial_{F_j}\Gamma|_{\gamma_i}\delta F_j + \partial_\mu\Gamma|_{\gamma_i}\delta\mu|_{\gamma_i} = 0$. However, varying $\Gamma(\lambda_{\gamma_i}, \mu_{\gamma_i}) = 0$ we obtain $\sum_j \partial_{F_j}\Gamma\delta F_j + \partial_\lambda\Gamma\delta\lambda_{\gamma_i} + \partial_\mu\Gamma\delta\mu_{\gamma_i} = 0$. Comparing these equations we get:

$$\delta\mu|_{\gamma_i} = \delta\mu_{\gamma_i} + \frac{\partial_\lambda\Gamma}{\partial_\mu\Gamma}\bigg|_{\gamma_i} \delta\lambda_{\gamma_i}$$

and the second term does not contribute to the wedge product in eq. (5.62). The contribution of K_2 is exactly the same. So we finally get:

$$\mathrm{Res}_{\gamma_i} K = 2\delta\mu_{\gamma_i} \wedge \delta\lambda_{\gamma_i} \qquad (5.63)$$

We now show that there are no residues at the branch points due to the proper choice of K_2. Let us look at the term K_1. At a branch point b, $\Psi^{(-1)}$ has a simple pole, δL is regular, $\delta\Psi$ has a simple pole due to eq. (5.58) and the form $d\lambda$ has a simple zero, hence K_1 has a simple pole at b. To compute its residue it is enough to keep the polar part in $\delta\Psi$, i.e. to replace $\delta\Psi$ by $\partial_\mu\Psi\delta\mu$ (recall that μ is a good local parameter around b). We get:

$$\text{Res}_b K_1 = \text{Res}_b \langle \Psi^{(-1)}\delta L \partial_\mu\Psi\rangle \wedge \delta\mu\,d\lambda$$
$$= \text{Res}_b \langle \Psi^{(-1)}(\delta L - \delta\mu)\partial_\mu\Psi\rangle \wedge \delta\mu\,d\lambda$$

where in the last equation we have used the antisymmetry of the wedge product to replace δL by $\delta L - \delta\mu$. Using again the eigenvector equation $(L-\mu)\Psi = 0$, and varying the point (λ,μ) on the curve around b, one gets

$$(L-\mu)\partial_\mu\Psi = \Psi - \frac{d\lambda}{d\mu}\frac{dL}{d\lambda}\Psi \qquad (5.64)$$

where $d\lambda/d\mu$ vanishes at the branch point. We then differentiate with δ and multiply on the left by $\Psi^{(-1)}$ to get:

$$\text{Res}_b \langle \Psi^{(-1)}(\delta L - \delta\mu)\partial_\mu\Psi\rangle \wedge \delta\mu\,d\lambda = \text{Res}_b \langle \Psi^{(-1)}\delta\Psi\rangle \wedge \delta\mu\,d\lambda$$
$$- \text{Res}_b \langle \Psi^{(-1)}\delta \left(\frac{d\lambda}{d\mu}\frac{dL}{d\lambda}\Psi\right)\rangle \wedge \delta\mu\,d\lambda$$

It is easy to see that the first term is exactly cancelled by the term $\text{Res}_b K_2$. The second term gives a non-vanishing contribution

$$\text{Res}_b \frac{\delta\mu_b}{\mu - \mu_b} \wedge \delta\mu\,d\lambda \qquad (5.65)$$

To show it, note that the quantity $\zeta = (d\lambda/d\mu)(dL/d\lambda)\Psi$ vanishes at $b = (\lambda_b, \mu_b)$. Writing $\zeta = (\mu-\mu_b)\zeta_1$, we get $\delta\zeta = -\delta\mu_b/(\mu-\mu_b)\zeta + \delta\mu\,\zeta_1 + \zeta_2$ with ζ_2 regular. The ζ_1 term does not contribute due to the antisymmetry of the wedge product and the ζ_2 term has no residue. Using eq. (5.64) we have $\langle \Psi^{(-1)}\frac{d\lambda}{d\mu}\frac{dL}{d\lambda}\Psi\rangle = 1$ yielding eq. (5.65). This contribution is exactly cancelled by the contribution of a new form K_3:

$$K_3 = \delta\left(\log \partial_\mu\Gamma\right) \wedge \delta\mu\,d\lambda$$

We will see that K_3 has poles only at the branch points. At the branch point b, $\partial_\mu\Gamma$ has a zero, so we write $\partial_\mu\Gamma = (\mu - \mu_b)S$ with S regular. The contribution of the point b to K_3 is:

$$\text{Res}_b \frac{\delta\partial_\mu\Gamma}{\partial_\mu\Gamma} \wedge \delta\mu\,d\lambda$$

The variation of $\partial_\mu\Gamma$ reads $\delta\partial_\mu\Gamma = \delta(\mu-\mu_b)\,S + (\mu-\mu_b)\delta S$. The second term does not contribute to the residue because S is regular, while the

variation $\delta\mu$ cancels due to the antisymmetry of the wedge product, and
we are left with the contribution of $\delta\mu_b$ which exactly cancels eq. (5.65).

We now compute the residues above $\lambda = \infty$. Recall that we consider
a reduced Hamiltonian system under the action of diagonal matrices.
Recall the normalization of the eigenvectors at ∞, eq. (5.60). Notice
that $L = L_0 + O(1/\lambda)$, where L_0 is non-dynamical so $\delta L_0 = 0$, and that
$\mu = a_i + O(1/\lambda)$ around Q_i hence δL and $\delta\mu$ are $O(1/\lambda)$. Moreover,
$\Psi^{(-1)}$ vanishes at Q_i and $d\lambda$ has a double pole. Altogether K_1 and
K_2 are regular at Q_i since $(\delta\Psi)(Q_i) = O(1)$ due to the normalization
condition. Finally, K_3 is also regular since, on the sheet $\mu = \mu_i(\lambda)$,
one can write $\partial_\mu\Gamma = \prod_{j\neq i}(\mu_i - \mu_j)$ yielding $\delta\log\partial_\mu\Gamma = O(1/\lambda)$. Hence
$\delta\log\partial_\mu\Gamma \wedge \delta\mu = O(1/\lambda^2)$ has a double zero which compensates the double
pole of $d\lambda$ at infinity. All this shows that K has no residues above $\lambda = \infty$.

It remains to show that K_3 has no other poles. Obviously, K_3 is regular
at the points of the dynamical divisor and does not contribute to the
residues at these points. To compute the residue of K_3 at the points $P_{k,i}$
above λ_k, we note that if $\partial_\mu\Gamma$ has a pole of some order m, it can be
written $\partial_\mu\Gamma = c(\lambda)/(\lambda - \lambda_k)^m$, where $c(\lambda)$ is regular and non-vanishing.
Since $\delta\lambda = 0$ and $\delta\lambda_k = 0$ we get $\delta(\log\partial_\mu\Gamma) = \delta\log c(\lambda)$ which is
regular. At λ_k we remark that $\delta\mu$ is regular on all sheets above λ_k. This
is because, due to the form of $L(\lambda)$, we have $\widehat{\mu} = (A_k)_- + $ regular. Since
$(A_k)_-$ characterizes the coadjoint orbit and is not dynamical, one has to
take $\delta(A_k)_- = 0$. Hence K_3 has no residue.   ∎

Proposition. *The form eq. (5.61) is given by:*

$$\omega = 2\sum_{i=1}^{g}\delta\lambda_{\gamma_i} \wedge \delta\mu_{\gamma_i} = 2\sum_{k}\mathrm{Res}_{\lambda_k}\,\mathrm{Tr}\left((A_k)_- g_k^{-1}\delta g_k \wedge g_k^{-1}\delta g_k\right)\,d\lambda \,(5.66)$$

*where $(\lambda_{\gamma_i}, \mu_{\gamma_i})$, $i = 1,\dots,g$, are the coordinates of the points of the
dynamical divisor D. This shows that ω is the symplectic form on the
orbit.*

Proof. Let us compute the residues at the poles λ_k of K_1, where only
L_k contributes. Recall the local diagonalization theorem of Chapter 3,
eq. (3.8), which allows us to write the Lax matrix as $L = g_k A_k g_k^{-1}$ around
$\lambda = \lambda_k$. Thus locally around λ_k we may identify the matrix $\widehat{\Psi}(\lambda)$ with g_k.
More precisely, by eq. (5.20), we have $\widehat{\Psi}(\lambda) = g_k d_k$ and $\widehat{\Psi}^{-1}(\lambda) = d_k^{-1}g_k^{-1}$
with d_k a diagonal matrix. The residues are obtained by integrating over
small circles surrounding each of the N points $P_{k,i}$ above λ_k. We can
choose these small circles so that they project on the base λ on a single

small circle surrounding λ_k. Then we get

$$\sum_{i=1}^{N} \mathrm{Res}_{P_{k,i}} K_1 = \sum_{i=1}^{N} \frac{1}{2i\pi} \oint_{C_{k,i}} \langle \Psi^{(-1)}(P_i)\delta L(\lambda) \wedge \delta \Psi(P_i) \rangle \, d\lambda$$

$$= \frac{1}{2i\pi} \oint_{C_k} \mathrm{Tr}\left(\widehat{\Psi}^{-1}(\lambda)\delta L(\lambda) \wedge \delta\widehat{\Psi}(\lambda)d\lambda \right) \qquad (5.67)$$

where we used the fact that $\widehat{\Psi}^{-1}(\lambda)$ is equal to the matrix whose rows are the vectors $\Psi^{(-1)}(P_i)$. The trace has been reconstructed in eq. (5.67) because $\Psi(P_i)$, $i = 1, \ldots, N$, form a basis of eigenvectors. Using the identification of $\widehat{\Psi}(\lambda)$ in terms of g_k gives:

$$\mathrm{Res}_{\lambda_k} K_1 = \mathrm{Res}_{\lambda_k} \mathrm{Tr}\bigg($$

$$d_k^{-1}g_k^{-1}\left(\delta g_k (A_k)_- g_k^{-1} - g_k (A_k)_- g_k^{-1}\delta g_k g_k^{-1}\right) \wedge (\delta g_k d_k + g_k\delta d_k)\bigg)\, d\lambda$$

$$= -2\,\mathrm{Res}_{\lambda_k} \mathrm{Tr}\left((A_k)_- g_k^{-1}\delta g_k \wedge g_k^{-1}\delta g_k + g_k^{-1}\delta g_k[(A_k)_-, \delta d_k d_k^{-1}]\right)\, d\lambda$$

The last term vanishes because it involves the commutator of two diagonal matrices. Finally, K_2 is regular at λ_k because, as we already remarked, $\delta\mu$ is regular on all the sheets above λ_k. ∎

This proposition means that the coordinates $(\lambda_{\gamma_i}, \mu_{\gamma_i})$ of the point γ_i of the dynamical divisor are canonical coordinates.

Remark 1. This result shows the nice interplay between the analytical and the group-theoretical approaches to integrable systems. We are able to show that $(\lambda_{\gamma_i}, \mu_{\gamma_i})$ are canonical coordinates using *only* the fact that L parametrizes a coadjoint orbit, specified by *constant* matrices $(A_k)_-$ and L_0.

Remark 2. In practice, to perform this calculation, one has to compute at each pole of L the quantity $\omega_k = \mathrm{Res}_{\lambda_k} \mathrm{Tr}\left(\delta L \wedge \delta\widehat{\Psi}\,\widehat{\Psi}^{-1}\right)\, d\lambda$. In the residue at λ_k, only δL_k appears. From eq. (5.20) one has

$$\delta L_k = [(\delta\widehat{\Psi}\,\widehat{\Psi}^{-1})_{\mathrm{jet}}, L_k] \qquad (5.68)$$

where $()_{\mathrm{jet}}$ is the expansion to order $n_k - 1$. This equation determines $(\delta\widehat{\Psi}\,\widehat{\Psi}^{-1})_{\mathrm{jet}}$ up to a quantity commuting with L_k. It is easy to see that

$$\omega_k = \mathrm{Res}_{\lambda_k} \mathrm{Tr}\left([\delta\widehat{\Psi}\,\widehat{\Psi}^{-1}, L_k] \wedge \delta\widehat{\Psi}\,\widehat{\Psi}^{-1}\right)\, d\lambda$$

is not affected by this ambiguity, using the antisymmetry of the wedge product and the cyclicity of the trace.

Example. We consider the example of the Neumann model. The above analysis can be applied to this model, except for one feature. As we have already stressed, there is no residual action of diagonal matrices on L, hence one has to pay special attention to the residues above $\lambda = \infty$. The sum of residues at the poles above $\lambda = \infty$ is $\mathrm{Res}_\infty \mathrm{Tr}\left(\delta L \wedge \delta \widehat{\Psi}\, \widehat{\Psi}^{-1}\right) d\lambda$.

One can see that $\delta \widehat{\Psi}\, \widehat{\Psi}^{-1}$ is regular at ∞, while $d\lambda$ has a pole of order 2. Since δL has a zero of order 1, one generally gets a residue. However $\delta \widehat{\Psi}\, \widehat{\Psi}^{-1}$ is constrained by

$$\delta L = [\delta \widehat{\Psi}\, \widehat{\Psi}^{-1}, L] + \widehat{\Psi} \delta \widehat{\mu} \widehat{\Psi}^{-1}$$

At $\lambda = \infty$ the second term vanishes and L tends to D, hence the order 0 term in $\delta \widehat{\Psi}\, \widehat{\Psi}^{-1}$ is diagonal. The leading term in δL is $1/\lambda\, \delta J$ (see eq. (3.3) in Chapter 3) and has no diagonal element, consequently the considered trace vanishes. Similarly the term involving K_2 has no residue because $\delta \widehat{\mu}$ vanishes to order 2 at $\lambda = \infty$.

To compute the residue at $\lambda = 0$ (a second order pole of L) we remark that the jet:

$$\delta \widehat{\Psi}\, \widehat{\Psi}^{-1} = \delta X\,{}^t\!X - X \delta^t\!X + \lambda \left(\delta Y\,{}^t\!X + X \delta^t\!Y - \delta X\,{}^t\!Y - Y \delta^t\!X\right) + O(\lambda^2)$$

solves eq. (5.68) and has the correct symmetry properties. One gets the expression of the symplectic form of the Neumann model in terms of the dynamical divisor:

$$\omega = 4 \sum_{i=1}^{N} \delta y_i \wedge \delta x_i = 2 \sum_{j=1}^{N-1} \delta \lambda_{\gamma_j} \wedge \delta \mu_{\gamma_j} \tag{5.69}$$

5.10 Separation of variables and the spectral curve

Let us call F_i, $i = 1, \ldots, g$, the action variables which are also the moduli of the spectral curve. For fixed F_i, we have seen that the motion takes place on the Jacobian $\mathrm{Jac}\,(\Gamma_{\{F_i\}})$. When varying initial conditions, the $\{F_i\}$ will eventually vary and we get a foliation of the (complexified) phase space in terms of the Jacobian tori of $\Gamma_{\{F_i\}}$. So we are back to the situation described in Liouville's theorem, cf. Chapter 2. Let us check that the symplectic form does vanish when we restrict ourself to one of the tori of the foliation. We view $\mathrm{Jac}\,(\Gamma)$ as the gth symmetric product Γ^g. Solving the equation $\Gamma(\lambda_{\gamma_j}, \mu_{\gamma_j}; \{F_i\}) = 0$, one has $\mu_{\gamma_j} = \mu_{\gamma_j}(\{F_i\}, \lambda_{\gamma_j})$ which depends on λ_{γ_j} only and not on the other λ. The symplectic form can then be written as:

$$\omega = \delta \alpha = \sum_{j=1}^{g} \delta \mu_{\gamma_j} \wedge \delta \lambda_{\gamma_j} = \sum_{i,j} \left(\frac{\partial \mu_{\gamma_j}}{\partial F_i}\right) \delta F_i \wedge \delta \lambda_{\gamma_j}, \quad \alpha = \sum_{j=1}^{g} \mu_{\gamma_j} \delta \lambda_{\gamma_j}$$

If we restrict ourselves to a level manifold $F_i = f_i$, we have $\delta F_i = 0$ and $\omega|_f = 0$.

Let us explain why the conjugate variables $(\lambda_{\gamma_j}, \mu_{\gamma_j})$ form a set of separated variables. The construction is similar to the method used for proving the Liouville theorem. Consider the function

$$S(\{F_i\}, \{\lambda_{\gamma_j}\}) = \int_{m_0}^{m} \alpha = \sum_j \int_{\lambda_0}^{\lambda_{\gamma_j}} \mu(\lambda) d\lambda$$

The integration contour is drawn on the level manifold $F_i = f_i$. Just as in the Liouville case, this function does not depend on local variations of the integration path. It is explicitly separated since it can be written as a sum of functions each depending on only one variable λ_{γ_j}:

$$S(\{F_i\}, \{\lambda_{\gamma_j}\}) = \sum_j S_j(\{F_i\}, \lambda_{\gamma_j})$$

Since $\frac{\partial S_j}{\partial \lambda_{\gamma_j}} = \mu_{\gamma_j}$ and since the point $(\lambda_{\gamma_j}, \mu_{\gamma_j})$ belongs to the curve Γ with equation $\Gamma(\lambda, \mu) = 0$, each function S_j is a solution of the differential equation:

$$\Gamma\left(\lambda_{\gamma_j}, \frac{\partial S_j}{\partial \lambda_{\gamma_j}}; \{F_i\}\right) = 0, \quad k = 1, \ldots, g \tag{5.70}$$

Of course the coefficients of the function $\Gamma(\lambda, \mu)$ depend on the values of the integrals of motion F_i. This is an equation of the form

$$\left(-\frac{\partial S_j}{\partial \lambda_{\gamma_j}}\right)^N + \sum_{q=1}^{N-1} r_q(\lambda_{\gamma_j}) \left(\frac{\partial S_j}{\partial \lambda_{\gamma_j}}\right)^q = 0$$

where the coefficients $r_q(\lambda)$ are defined in eq. (5.4).

Remark. Equation (5.70) plays an important role in the quantum case. It is the separated Schroedinger equation also known as the Baxter equation in some cases.

The commuting Hamiltonians $\{F_i\}$ are functions of the $2g$ coordinates $\lambda_{\gamma_j}, \mu_{\gamma_j}$. To find them we write that the curve $\Gamma(\lambda, \mu; \{F_i\}) = 0$ passes through the g points $(\lambda_{\gamma_j}, \mu_{\gamma_j})$ of the dynamical divisor D. Hence the equations of the Liouville torus $F_i = f_i$ in these coordinates read

$$\Gamma(\lambda_{\gamma_j}, \mu_{\gamma_j}; \{f_i\}) = 0 \tag{5.71}$$

The standard Hamilton–Jacobi equation is obtained by setting $\mu_{\gamma_j} = \partial_{\lambda_{\gamma_j}} S$, where S is the action. Due to the form of eq. (5.71), it is clear that one can take $S(\{\lambda_{\gamma_j}\}) = \sum_j s(\lambda_{\gamma_j})$, where the unique function $s(\lambda)$ obeys the one-variable equation $\Gamma(\lambda, \partial_\lambda s; \{f_i\}) = 0$. This shows that the Hamilton–Jacobi equation separates into g identical one-variable equations, using the variables λ_{γ_j}. This is a particularly striking example of separation of variables.

Remark. It is sometime advantageous to consider λ_{γ_j} as a function of μ_{γ_j}. Defining $S' = \sum_j \int_{\mu_0}^{\mu_{\gamma_j}} \lambda_{\gamma_j} d\mu_{\gamma_j}$, we get $\lambda_{\gamma_j} = \frac{\partial S'}{\partial \mu_{\gamma_j}}$. The relation between S and S' is simply a Legendre transform: $S' = \sum_j \mu_{\gamma_j} \lambda_{\gamma_j} - S$.

Example. In the Neumann model, we see that eq. (2.30) in Chapter 2 is exactly of the form

$$ S = \frac{1}{2} \sum_j \int^{\mu_{\gamma_j}} \lambda_{\gamma_j} d\mu_{\gamma_j} $$

where the points $(\lambda_{\gamma_j}, \mu_{\gamma_j})$ belongs to the spectral curve eq. (5.13). So the results of Chapter 2 are particular cases of the general theory explained in this chapter. Moreover, we see in eq. (5.14) that the spectral curve depends on $g = N-1$ dynamical moduli b_i, while the a_i are non-dynamical. Asking that a curve of the form eq. (5.13) passes through the g points $(\lambda_{\gamma_j}, \mu_{\gamma_j})$ determines the symmetric functions of the coefficients b_i in terms of the $(\lambda_{\gamma_j}, \mu_{\gamma_j})$. In fact, setting $P(\mu) = \prod_i(\mu - b_i)$, we find the conditions $P(\mu_{\gamma_j}) = -\lambda_{\gamma_j}^2 \prod_i(\mu_{\gamma_j} - a_i)$. By the Lagrange interpolation formula we reconstruct $P(\mu)$:

$$ P(\mu) = \prod_j(\mu - \mu_{\gamma_j}) - \sum_j \lambda_{\gamma_j}^2 \frac{\prod_i(\mu_{\gamma_j} - a_i)}{\prod_{k\neq j}(\mu_{\gamma_j} - \mu_{\gamma_k})} \prod_{k\neq j}(\mu - \mu_{\gamma_k}) \qquad (5.72) $$

Using the canonical Poisson bracket eq. (5.69), it is a simple exercise to check that $\{P(\mu), P(\mu')\} = 0$, as it should be.

5.11 Action–angle variables

So far, we dealt with complexified dynamical sytems. We found that the phase space of this complexified system can be viewed as a fibration where the base is the moduli space of the spectral curve and the fibre is the Jacobian of the spectral curve corresponding to the specific values of the moduli parameters.

This is very similar to the situation in the Liouville theorem, but the Liouville tori are real tori of dimension g, while the Jacobian has real dimension $2g$. We need to choose a real slice of this complex phase space.

This can be done as follows. On Γ we choose a canonical basis of $2g$ cycles a_k, b_k. The cycles a_k are non-intersecting. Once they are chosen, we can adapt the basis of Abelian differentials of the first kind ω_k such that they are normalized by $\oint_{a_k} \omega_l = \delta_{kl}$.

The real slice can be defined by restricting the g points of the dynamical divisor D to move along these g non-intersecting cycles, each point on a different cycle. This obviously is a product of g real circles. One has to be aware that a real slice has in general several connected components, and the above description applies to one of them. Finally, explicit models correspond to specific cycles and not only to homology classes of cycles.

The g angle variables are given by

$$\theta_k = \sum_{i=1}^{g} \int_{m_0}^{\gamma_i} \omega_k = \sum_{i=1}^{g} \int_{\lambda_0}^{\lambda_{\gamma_i}} \sigma_k(\lambda) d\lambda \qquad (5.73)$$

where the integration paths are taken along the cycles a_k and the Abelian differentials ω_k are written in terms of the local parameter λ as $\omega_k = \sigma_k(\lambda) d\lambda$. With these assumptions the angles have real periods.

The angles being defined, one may find the conjugated action variables. To do this, we need a Lemma.

Lemma. *The conditions characterizing the dynamical moduli can be summarized into the single statement:*

$$\delta(\mu d\lambda) \text{ is a regular form} \qquad (5.74)$$

where δ is the differentiation with respect to the moduli, keeping λ constant.

<u>Proof</u>. Taking the variation at λ constant produces poles at the branch points of the covering which are cancelled by corresponding zeroes of $d\lambda$. The form $\mu d\lambda$ has poles at finite distance where $L(\lambda)$ has poles. Around a pole λ_k, we have

$$L_k(\lambda) = (g^{(k)}(\lambda) A_-^{(k)}(\lambda) g^{(k)-1}(\lambda))_-$$

and we assume that the diagonal polar part $A_-^{(k)}(\lambda)$ is non-dynamical. Hence the singular part of μ is kept fixed under δ, and $\delta(\mu d\lambda)$ is regular at λ_k. At $\lambda = \infty$, $d\lambda$ has a double pole. The dominant term of $\mu d\lambda$ is $a_i d\lambda$ when $\mu \to Q_i = (\infty, a_i)$, and is kept fixed because we assume that L_0 is non-diagonal. The subdominant term is also kept fixed because of the reduction by the group of conjugation by diagonal matrices.

The Hamiltonians H_n generating this group action are given in eq. (5.8). Setting $\mu_i = a_i + \frac{b_i}{\lambda} + \cdots$, we have

$$H_n = \sum_i a_i^{n-1} b_i$$

After Hamiltonian reduction, these quantities are to be kept fixed. So both a_i and b_i are non-dynamical and $\delta(\mu d\lambda)$ is regular at infinity. ∎

We emphasize that all the conditions specifying the non-dynamical variables in $L(\lambda)$ are accounted for by eq. (5.74). Under these conditions, we have seen at the beginning of this chapter that the counting of parameters leaves a phase space of dimension $2g$. The g action variables are now easily constructed:

Proposition. *Assume that $\delta(\mu d\lambda)$ is regular. Then we have:*

$$\omega = \sum_{i=1}^{g} \delta\mu_{\gamma_i} \wedge \delta\lambda_{\gamma_i} = \sum_{i=1}^{g} \delta I_i \wedge \delta\theta_i \qquad (5.75)$$

where the action variable I_k, canonically conjugated to θ_k, is

$$I_k = \oint_{a_k} \mu d\lambda \qquad (5.76)$$

Proof. By eq. (5.74), $\delta(\mu d\lambda)$ decomposes on the basis of holomorphic Abelian differentials: $\delta(\mu d\lambda) = \sum_i \alpha_i \omega_i$. To find the coefficients α_i, we integrate both sides on the cycles a_l. We get

$$\alpha_l = \oint_{a_l} \delta(\mu d\lambda) = \delta \oint_{a_l} \mu d\lambda = \delta I_l \qquad (5.77)$$

Hence we have, with $\omega_k = \sigma_k(\lambda) d\lambda$:

$$\delta(\mu d\lambda) = \sum_k \delta I_k \omega_k = \sum_k \delta I_k \sigma_k(\lambda) d\lambda$$

Since the variations are taken at λ constant so that $\delta(\mu d\lambda) = \delta(\mu) d\lambda$, and since $\delta\mu$ decomposes on the δI_k by eq. (5.58), we have

$$\delta\mu = \sum_k \frac{\partial\mu}{\partial I_k} \delta I_k, \qquad \frac{\partial\mu}{\partial I_k} = \sigma_k(\lambda)$$

By the definition of the angular variables in eq. (5.73) one has, using $\delta\sigma_i(\lambda) = \sum_k \frac{\partial\sigma_i(\lambda)}{\partial I_k} \delta I_k$:

$$\delta\theta_i = \sum_{j=1}^{g} \sigma_i(\lambda_{\gamma_j}) \delta\lambda_{\gamma_j} + \sum_{j=1}^{g} \int_{\lambda_0}^{\lambda_{\gamma_j}} \sum_k \frac{\partial^2\mu}{\partial I_i \partial I_k} d\lambda \, \delta I_k$$

Finally, we obtain:

$$\omega = \sum_i \delta\mu_{\gamma_i} \wedge \delta\lambda_{\gamma_i} = \sum_{i,j} \delta I_i \wedge \sigma_i(\lambda_{\gamma_j})\delta\lambda_{\gamma_j}$$

$$= \sum_i \delta I_i \wedge \left(\delta\theta_i - \sum_j \int_{\lambda_0}^{\lambda_{\gamma_j}} \sum_k \frac{\partial^2\mu}{\partial I_i \partial I_k} d\lambda\, \delta I_k \right) = \sum_i \delta I_i \wedge \delta\theta_i$$

where the second term vanishes because $\partial_{I_i}\partial_{I_k}\mu$ is symmetrical in the indices i, k and $\delta I_i \wedge \delta I_k$ is antisymmetric. ∎

This shows that the I_i are canonically conjugated to the θ_i. At the level of Poisson brackets, we have

$$\{I_i, I_j\} = 0, \quad \{I_i, \theta_j\} = \delta_{ij}, \quad \{\theta_i, \theta_j\} = 0$$

5.12 Riemann surfaces and integrability

We are now in a position to clarify the link between integrable systems and Riemann surfaces. Let Γ be a Riemann surface of genus g and let λ be a meromorphic function on it. We assume that λ takes each value N times. Any other meromorphic function μ on Γ is related to λ by an algebraic relation $\Gamma(\lambda, \mu) = 0$. One can choose μ such that this relation is irreducible. Then the field of meromorphic functions on Γ is the field of *rational* functions of λ and μ. The choice of these functions allows us to present Γ as an N sheeted covering of the Riemann sphere by $(\lambda, \mu) \to \lambda$.

We can interpret Γ as the spectral curve of a Lax matrix $L(\lambda)$ in the following way. Let $Q_1, Q_2, \ldots, Q_N$ be the N points above $\lambda = \infty, \mu(Q_i) = a_i$. Choose a divisor D of g points on Γ. From these data, we construct N linearly independent meromorphic functions, $\psi_1 = 1$ and ψ_k with a zero at Q_1 and poles at $D + Q_k$ for $k = 2, \ldots, N$. This determines ψ_k uniquely up to multiplication by a constant c_k. Let $P_i = (\lambda, \mu_i)$ be the N points above λ. Define the $N \times N$ matrices $\widehat{\Psi}_{ij} = \psi_i(P_j)$ and $\widehat{\mu} = \mathrm{diag}(\mu_i)$, and let

$$L = \widehat{\Psi}\, \widehat{\mu}\, \widehat{\Psi}^{-1}$$

This matrix is a rational function of λ because it is a rational function of $\lambda, \mu_1, \ldots, \mu_N$, invariant by permutations of the μ_j. It tends to the diagonal matrix $\mathrm{diag}(a_i)$ at ∞, and Γ is the spectral curve of $L(\lambda)$. Note that L is defined only up to conjugation by diagonal matrices due to the undeterminacy in the normalization of the functions ψ_k.

We now introduce time evolutions such that Γ is time-independent, but the divisor D depends on time. This is enough to assert the existence of a rational Lax equation

$$\dot{L}(\lambda) = [M(\lambda), L(\lambda)], \quad M(\lambda) = \dot{\widehat{\Psi}}\,\widehat{\Psi}^{-1}$$

We are thus exactly in the situation of the Zakharov–Shabat construction.

To relate to Liouville integrable systems, we have to introduce a symplectic structure on the dynamical variables. We have seen that imposing coadjoint orbit structure at the poles of $L(\lambda)$ automatically yields integrable systems once we have performed the Hamiltonian reduction by the diagonal group action of dimension $N - 1$. This produces a dynamical system of dimension $2g$. The g angle variables are given by the dynamical divisor, which evolves linearly on $\mathrm{Jac}(\Gamma)$, and the g action variables are contained in the moduli of the curve.

The conditions we impose on the moduli, coming from the coadjoint orbit structure and the Hamiltonian reduction, can be written in a very concise way:

$$\delta(\mu d\lambda) \text{ is a holomorphic differential}$$

where δ is the differential with respect to the dynamical moduli. This means that the polar parts of $\mu d\lambda$ are non-dynamical. Since $\delta \mu d\lambda = \sum_{k=1}^{g} \omega_k \delta I_k$, we see that we have exactly g dynamical modules.

In this setting, the standard symplectic form on the variables $\gamma_i = (\lambda_{\gamma_i}, \mu_{\gamma_i})$ is equal to the Kirillov symplectic form on $L(\lambda)$, as we have shown in eq. (5.61). Moreover, due to eq. (5.75), the angle variables $\theta_i = \sum_i \int^{\gamma_i} \omega_j$ are canonically conjugated to the action variables $I_i = \oint_{a_i} \mu d\lambda$, i.e.

$$\sum_{i=1}^{g} \delta\mu_{\gamma_i} \wedge \delta\lambda_{\gamma_i} = \omega_K = \sum_{i=1}^{g} \delta I_i \wedge \delta\theta_i \tag{5.78}$$

We want to emphasize the meaning of this result. Starting from a Riemann surface $\Gamma(\lambda, \mu) = 0$, we specify g dynamical moduli $F_1, \ldots, F_g$, by imposing that $\delta(\mu d\lambda)$ be regular. We take g arbitrary points $\gamma_i = (\lambda_{\gamma_i}, \mu_{\gamma_i})$ and impose the symplectic structure $\omega = \sum_i \delta\mu_{\gamma_i} \wedge \delta\lambda_{\gamma_i}$ on these data. We determine the g moduli F_i by solving the g equations meaning that the curve passes through the points γ_i:

$$\Gamma(\lambda_{\gamma_i}, \mu_{\gamma_i}; F_1, \ldots, F_g) = 0$$

This determines F_i as symmetric functions of the λ_j, μ_j. The beautiful result is that these functions Poisson commute, $\{F_i, F_j\} = 0$, because, by eq. (5.78), the action variables, I_i, Poisson commute and they are independent functions of the F_j. See eq. (5.72) for an example of this situation.

Remark 1. The above construction can be generalized by imposing conditions such as

$$\frac{\delta\mu}{\mu^n}\frac{d\lambda}{\lambda^m} = \sum_{k=1}^{g} \omega_k \delta I_k^{(n,m)}$$

for g modules $I_k^{(n,m)}$. This will modify the symplectic form as well. An example of this is given in Chapter 6.

Remark 2. If the Riemann surface Γ can be viewed as covering of the Riemann sphere in different ways, one can construct Lax matrices $L(\lambda)$ of different sizes in the above way. In particular if Γ is hyperelliptic, one can construct a 2×2 Lax matrix.

Remark 3. Lax matrices with elliptic dependence on the spectral parameter can be viewed as particular cases of this setup when the covering of the Riemann sphere $\lambda : \Gamma \to S$ factorizes as $\lambda : \Gamma \to T \to S$, where T is the torus. The rational Lax matrix has a size twice as big as the elliptic one

5.13 The Kowalevski top

We now briefly discuss the algebro-geometric solution of the Kowalevski top. It is more convenient to start from a slightly different Lax matrix from the one in eq. (4.69) in Chapter 4, obtained by conjugation $L(\lambda) \to \lambda P^{-1}L(\lambda)P$:

$$P = \begin{pmatrix} i\sigma_0 + \sigma_3 & \sigma_1 + \sigma_2 \\ \sigma_0 + i\sigma_3 & i\sigma_1 - i\sigma_2 \end{pmatrix}$$

The new Lax matrix reads:

$$L(\lambda) = \frac{i}{2}\begin{pmatrix} 0 & -\lambda\xi_2 & \frac{1}{2}z_2 & \lambda\gamma_3 \\ \lambda\xi_1 & 0 & -\lambda\gamma_3 & -\frac{1}{2}z_1 \\ \frac{1}{2}z_1 & \lambda\gamma_3 & -J_3 & 2\frac{1}{\lambda} + \lambda\xi_1 \\ -\lambda\gamma_3 & -\frac{1}{2}z_2 & -2\frac{1}{\lambda} - \lambda\xi_2 & J_3 \end{pmatrix} \qquad (5.79)$$

where we have used Kowalevski's variables $z_1 = J_1 + iJ_2$, $z_2 = J_1 - iJ_2$, $\xi_1 = \gamma_1 + i\gamma_2$, $\xi_2 = \gamma_1 - i\gamma_2$. We restrict ourselves to the pure Kowalevski case $\gamma' = 0$. In this basis the Lax matrix satisfies the symmetry properties:

$$L(-\lambda) = -\Sigma_1^{-1} {}^t L(\lambda)\, \Sigma_1, \ {}^t L(\lambda) = -\Sigma_2^{-1} L(\lambda)\, \Sigma_2, \ L(-\lambda) = \Sigma_3^{-1} L(\lambda)\, \Sigma_3 \qquad (5.80)$$

where the matrices Σ_1, Σ_2, Σ_3 are given by:

$$\Sigma_1 = \begin{pmatrix} \sigma_1 & 0 \\ 0 & \sigma_1 \end{pmatrix}, \quad \Sigma_2 = \begin{pmatrix} \sigma_2 & 0 \\ 0 & \sigma_2 \end{pmatrix}, \quad \Sigma_3 = \begin{pmatrix} \sigma_3 & 0 \\ 0 & \sigma_3 \end{pmatrix}$$

The matrix P has been chosen to simplify the expression of these symmetries and in particular to diagonalize Σ_3. The first of eqs. (5.80) expresses the fact that $L(\lambda)$ belongs to a twisted loop algebra. The second one says that $L(\lambda)$ belongs to $sp(4)$, which is well-known to be isomorphic to $so(3,2)$, while the third is a combination of the first two.

The equation of the spectral curve Γ : $\det(L(\lambda) - \mu) = 0$ reads:

$$\mu^4 - \left(\frac{\lambda^2}{2}\vec{\gamma}^2 - \frac{1}{4}H + \frac{1}{\lambda^2} \right)\mu^2 + \frac{\lambda^4}{16}(\vec{\gamma}^2)^2 + \frac{\lambda^2}{16}((\vec{J}\cdot\vec{\gamma})^2 - H\vec{\gamma}^2) + \frac{K}{256} = 0$$

$$(5.81)$$

The Hamiltonians H and K in this formula are given by:

$$H = \frac{1}{2}(J_1^2 + J_2^2 + 2J_3^2) - 4\gamma_1 = \frac{1}{2}z_1 z_2 + J_3^2 - 2(\xi_1 + \xi_2)$$

$$K = (J_1^2 - J_2^2 + 8\gamma_1)^2 + (2J_1 J_2 + 8\gamma_2)^2 = (z_1^2 + 8\xi_1)(z_2^2 + 8\xi_2)$$

while $\vec{\gamma}^2$ and $\vec{J}\cdot\vec{\gamma}$ are in the centre of the Poisson algebra. Note that the coordinates λ and μ on the spectral curve appear only through λ^2 and μ^2, which is a consequence of the symmetries eqs. (5.80).

It will be necessary in the following to have a clear picture of the solutions $\mu(\lambda)$ of eq. (5.81) around $\lambda = 0$ and $\lambda = \infty$. Around $\lambda = \infty$, we have four branches:

$$\mu = \epsilon \frac{\sqrt{\vec{\gamma}^2}}{2}\lambda + i\epsilon' \frac{\vec{J}\cdot\vec{\gamma}}{4\sqrt{\vec{\gamma}^2}} + O(\frac{1}{\lambda}) \qquad (5.82)$$

where ϵ, ϵ' are independent signs. Around $\lambda = 0$, we get two branches with $\mu \to 0$ and two branches with $\mu \to \infty$:

$$\mu = \epsilon \frac{\sqrt{K}}{16}\lambda + O(\lambda^3), \quad \mu = \epsilon\left(\frac{1}{\lambda} - \frac{H}{8}\lambda \right) + O(\lambda^3) \qquad (5.83)$$

Of course all these branches properly exchange under the symmetries $\lambda \to -\lambda$ and $\mu \to -\mu$.

At this point it is important to recall that Γ is defined as the desingularization of the curve defined by eq. (5.81). We are going to study Γ by considering it as successive coverings of simpler curves. Setting $\lambda^2 = z$ in eq. (5.81) yields a curve C of equation:

$$\mu^4 - \left(\frac{z}{2}\vec{\gamma}^2 - \frac{1}{4}H + \frac{1}{z} \right)\mu^2 + \frac{z^2}{16}(\vec{\gamma}^2)^2 + \frac{z}{16}((\vec{J}\cdot\vec{\gamma})^2 - H\vec{\gamma}^2) + \frac{K}{256} = 0$$

and Γ is a two-sheeted cover of C. Setting $\mu^2 = y$ we get the curve E of equation:

$$y^2 - \left(\frac{z}{2}\vec{\gamma}^2 - \frac{1}{4}H + \frac{1}{z}\right)y + \frac{z^2}{16}(\vec{\gamma}^2)^2 + \frac{z}{16}((\vec{J}\cdot\vec{\gamma})^2 - H\vec{\gamma}^2) + \frac{K}{256} = 0$$

and C is a two-sheeted branched cover of E. First, E is an elliptic curve of genus 1. Indeed, setting $t = 1/z$ and $Y = ty - \frac{1}{4}\vec{\gamma}^2 + \frac{H}{8}t - \frac{1}{2}t^2$, the equation of E takes the form $Y^2 = tP_3(t)$, where $P_3(t)$ is the polynomial of degree 3:

$$P_3(t) = \frac{1}{4}t^3 - \frac{H}{8}t^2 + \left(\frac{H^2}{64} + \frac{\vec{\gamma}^2}{4} - \frac{K}{256}\right)t - \frac{(\vec{J}\cdot\vec{\gamma})^2}{16}$$

The four branch points are obtained for $Y = 0$, so there is a branch point at $t = 0$ (or $z = \infty$) and three branch points at t (or z) finite.

We now study the covering $C \to E$, coming from $\mu \to y = \mu^2$. This two-sheeted covering can only be branched at $y = 0$ and $y = \infty$. The meromorphic function y on E takes each value three times because, given y, z is determined by a third degree equation. Hence it has three zeroes and three poles. Setting $y = 0$ in the equation of E, one gets

$$\frac{1}{16}(\vec{\gamma}^2)^2 + \frac{1}{16}t((\vec{J}\cdot\vec{\gamma})^2 - H\vec{\gamma}^2) + \frac{K}{256}t^2 = 0$$

yielding two points $(y = 0, t = t_1)$ and $(y = 0, t = t_2)$, where t_1, t_2 are the two roots of this second degree equation. The third point with $y = 0$ and the three points with $y = \infty$ occur when $t = 0$ and $t = \infty$. For $t \to \infty$ we have two points P_1, P_2 on the curve E corresponding to the branches:

$$P_1 : y = t - \frac{H}{4} + O\left(\frac{1}{t}\right), \quad P_2 : y = \frac{K}{256}\frac{1}{t} + O\left(\frac{1}{t^2}\right) \qquad (5.84)$$

Since t is a good local parameter at ∞, P_1 is a pole of y, while P_2 provides the third zero of y. For $t \to 0$, a good local parameter on E is $\sqrt{t}$ and we find two branches:

$$P_3 : y = \frac{\vec{\gamma}^2}{4}\frac{1}{t} \pm i\frac{\vec{J}\cdot\vec{\gamma}}{4}\frac{1}{\sqrt{t}} + O(1) \qquad (5.85)$$

showing that y has a double pole at this point $P_3(t = 0, y = \infty)$ on E. Of these six poles and zeroes of y only four are branch points of the covering $C \to E$. This is because at the point P_3 the equation of C is singular. Since C is the desingularized curve, P_3 blows up to two points $\tilde{P}_3$ and $\tilde{P}'_3$ of C, and the point P_3 has two pre-images as its neighbours. On the other hand, P_1 and P_2 are branch points and have just one pre-image each, $\tilde{P}_1$ and $\tilde{P}_2$ on C. Using the Riemann–Hurwitz formula $2g-2 = N(2g_0-2)+\nu$, where $g_0 = 1$, $N = 2$ and $\nu = 4$, we find that C has genus 3.

Finally, we study the covering $\Gamma \to C$, coming from $\lambda \to z = \lambda^2$ which can possibly be ramified only at $z = 0, \infty$. Generically, given z, there are four values of μ satisfying the equation of C, so that the meromorphic function z has four zeroes and four poles. We have already obtained the four branches of Γ above $\lambda = \infty$ in eq. (5.82) which correspond by definition to four points on the smooth curve Γ. They project on the two branches of C given by eq. (5.85), hence the covering at the two points $\tilde{P}_3$ and $\tilde{P}_3'$ is unbranched. Similarly, above $\lambda = 0$ we have the four branches of Γ given in eq. (5.83). The two points of Γ, Q_1 $(\mu \sim \lambda\sqrt{K}/16)$ and Q_2 $(\mu \sim -\lambda\sqrt{K}/16)$, project on $\tilde{P}_2$, and the two points Q_3 $(\mu \sim 1/\lambda)$ and Q_4 $(\mu \sim -1/\lambda)$ project on $\tilde{P}_1$, as seen from eq. (5.84). So the covering is unbranched at these points. This exhausts the zeroes and poles of z, hence the covering $\Gamma \to C$ is unbranched. Applying the Riemann–Hurwitz formula with $g_0 = 3$, $N = 2$, $\nu = 0$, we find that the genus of the spectral curve Γ is equal to 5.

We see that, in the case of the Kowalevski top, the Jacobian of the spectral curve is of dimension 5 while the Liouville tori are of dimension 2. This non-generic situation is related to the symmetry properties, eqs. (5.80), of the Lax matrix $L(\lambda)$ as we now show. Consider the eigenvector $\Psi(\lambda, \mu)$ satisfying $L(\lambda)\Psi = \mu\Psi$ at the point (λ, μ) of the spectral curve, normalized such that the first component is equal to 1. According to the general discussion Ψ has $g + N - 1 = 8$ poles on the spectral curve. It is easy to get the following expansions for the eigenvector at the four points Q_1, Q_2, Q_3, Q_4, above $\lambda = 0$. The matrix $\widehat{\Psi}(\lambda)$ reads

$$
\begin{pmatrix}
1 & 1 & 1 & 1 \\
i\zeta + O(\lambda) & -i\zeta + O(\lambda) & i\frac{z_1}{z_2} + O(\lambda) & -i\frac{z_1}{z_2} + O(\lambda) \\
\frac{-iz_2}{4}\zeta\lambda + O(\lambda^2) & \frac{iz_2}{4}\zeta\lambda + O(\lambda^2) & -\frac{4i}{z_2}\frac{1}{\lambda} + O(1) & \frac{4i}{z_2}\frac{1}{\lambda} + O(1) \\
-\frac{1}{4}z_1\lambda + O(\lambda^2) & -\frac{1}{4}z_1\lambda & -\frac{4}{z_2}\frac{1}{\lambda} + O(1) & -\frac{4}{z_2}\frac{1}{\lambda} + O(1)
\end{pmatrix}
$$
$$(5.86)$$

The column i corresponds to an expansion at the point Q_i. We have denoted $\zeta = \sqrt{(z_1^2 + 8\xi_1)/(z_2^2 + 8\xi_2)}$.

Note that the eigenvector $\Psi(\lambda, \mu)$ has simple poles at Q_3 and Q_4. This can be understood in the context of the general analysis of this chapter. Indeed, choosing a basis where the constant coefficient of $1/\lambda$ in $L(\lambda)$ (which plays the role of L_0 in the general discussion) is diagonal, we expect poles at the points above $\lambda = 0$. However, because we have two degenerate vanishing eigenvalues, we have only two-poles instead of the expected three. They are on the sheets corresponding to the non-vanishing eigenvalues, hence at the two points Q_3, Q_4. When returning to our basis, we have to make linear combinations of the last two components of Ψ and this explains our formulae.

Recall that the Lax matrix obeys eq. (5.80) so that if $L(\lambda)\Psi(\lambda,\mu) = \mu\Psi(\lambda,\mu)$ then $L(-\lambda)\Sigma_3\Psi(\lambda,\mu) = \mu\Sigma_3\Psi(\lambda,\mu)$. Since Σ_3 is diagonal and the first component of Ψ is equal to one, we get:

$$\Psi(-\lambda,\mu) = \Sigma_3\Psi(\lambda,\mu) \tag{5.87}$$

We have seen that Γ is a two-sheeted unbranched cover of C. The two sheets are exchanged under the involution $\tau : (\lambda,\mu) \rightarrow (-\lambda,\mu)$. Note that τ exchanges the points Q_1, Q_2 and also Q_3, Q_4. It exchanges the corresponding sheets in a vicinity of $\lambda = 0$. On the explicit solution, eq. (5.86), we have $\widehat{\Psi}(-\lambda) = \Sigma_3\widehat{\Psi}(\lambda)\Sigma_1$. The matrix Σ_1 on the right accounts for the exchange of the sheets under τ.

As a result of eq. (5.87) the meromorphic functions $\psi_i(P)$, $i = 2,3,4$ on the curve Γ obey the symmetry properties:

$$\psi_2(\tau \cdot P) = -\psi_2(P), \quad \psi_3(\tau \cdot P) = \psi_3(P), \quad \psi_4(\tau \cdot P) = -\psi_4(P)$$

In particular the divisor of the poles of Ψ is invariant under the involution τ. Two of these poles are the points Q_3, Q_4 which are exchanged by this involution. The remaining six poles thus come in three pairs $(\gamma_j, \tau \cdot \gamma_j)$, $j = 1,2,3$. These pairs can be seen as points on C, also denoted by γ_i. The dynamical divisor on C, $D(t) = \sum_{i=1}^{3} \gamma_i(t)$, is of degree 3 and moves linearly on the dimension 3 Jacobian of the curve C, as we now show. Considering the Lax pair L, M given in eq. (4.69) in Chapter 4 and remembering that L has been rescaled, $L \rightarrow \lambda L$, we see that the polar parts of L and M at $\lambda = 0$ are related by $M_- = 1/2\,L_-$. This relation is not affected by the further similarity $L \rightarrow P^{-1}LP$ that we have used in this section. Proceeding exactly as in eq. (5.35), we get:

$$\frac{d}{dt} \sum_{i=1}^{3} \left(\int^{\gamma_i(t)} \omega + \int^{\tau(\gamma_i(t))} \omega \right) = \sum_{i=3}^{4} \mathrm{Res}_{Q_i} \left(\frac{1}{2}\mu\omega \right) \tag{5.88}$$

where Q_3, Q_4 are the two points on Γ with $\lambda = 0$ and $\mu = \infty$, and ω is any Abelian differential of the first kind on Γ. In particular, choosing for ω the pullback on Γ of an Abelian differential ω on C, the right-hand side becomes twice the residue at the point $\tilde{P}_1$ on C of the form $(\frac{1}{2}\mu\omega)$. This is because in the vicinity of the corresponding points $\mu(-\lambda) = -\mu(\lambda)$, and the pullback of a form on C has a local expression $\sigma(\lambda)d\lambda$ with $\sigma(-\lambda) = -\sigma(\lambda)$. Note that non-vanishing higher order Hamiltonians in the Kowalevski hierarchy are traces of an even power of L ultimately yielding an odd power of μ in eq. (5.88), so the same conclusion applies to all these flows. Similarly, the left-hand side doubles between a pair of corresponding points γ_i, by definition of a pullback. Finally, we get:

$$\frac{d}{dt} \sum_{i=1}^{3} \int^{\gamma_i(t)} \omega = \operatorname{Res}_{\tilde{P}_1} \left(\frac{1}{2} \mu \omega \right) \qquad (5.89)$$

Since C is of genus 3, we have three independent forms ω and this proves that the flow is linear on the Jacobian of C.

One can view the functions $\psi_i(P)$ as functions on C in the following way. First $\psi_3(P)$ is a well-defined meromorphic function on C (since it is even under τ) with three poles at D, one-pole on $\tilde{P}_1$ and a zero at $\tilde{P}_2$, and is therefore uniquely determined. The functions $\psi_2(P)$ and $\psi_4(P)$ require special treatment since they are odd under τ, hence are multivalued on C. However, we can consider the function $\lambda = \sqrt{z}$ defined on Γ which is odd under τ and the functions $\lambda\psi_2(P)$ and $\lambda\psi_4(P)$ which, being even under τ, yield well-defined meromorphic functions on C. These functions are uniquely characterized by their analyticity properties: $\lambda\psi_2(P)$ has three simple poles at D, two simple zeroes at $\tilde{P}_1$ and $\tilde{P}_2$ and simple poles at $\tilde{P}_3$ and $\tilde{P}_3'$, while $\lambda\psi_4(P)$ has three simple poles at D, a double zero at $\tilde{P}_2$, and two simple poles at $\tilde{P}_3$ and $\tilde{P}_3'$. Hence we have shown that one can work only with the genus 3 curve C, together with the extra multivalued function $\lambda = \sqrt{z}$.

We still have three points in D, while we need only two degrees of freedom. A further restriction is provided by the other symmetry $(\lambda, \mu) \to (\lambda, -\mu)$ induced by the second eq. (5.80). Note that the right-hand side of eq. (5.89) contains only odd powers of μ for the general Kowalevski flow. At the point $\tilde{P}_1$ (which is a branch point of $C \to E$) μ is a good local parameter. Assume that ω is the pullback on a form on E, hence has a local expression $\sigma(\mu)d\mu$ with $\sigma(\mu)$ odd, then $\mu^{2k+1}\sigma(\mu)$ is even, and has no $1/\mu$ term, so that the right-hand side of eq. (5.89) vanishes. Since E is of genus 1, we get *one* condition which restricts the flow. We finally see that the flow occurs on a two-dimensional subvariety of Jac (C), the so-called Prym variety of the covering $C \to E$. It is defined as the subvariety such that any tangent vector is in the kernel of the pullback to C of any Abelian form of E. Here the action of ω on a tangent vector to Jac (C) is defined by the left-hand side of eq. (5.89). We have recovered a four-dimensional phase space for the Kowalevski top. At this point one can solve the equations of motion with theta-functions on Jac (C) and reduce them to two-dimensional theta-functions by using the Prym condition. Finally, Kowalevski has directly solved the system by using a curve of genus 2. For a study of these approaches and their relations we refer to the literature.

5.14 Infinite-dimensional systems

In the field theory case, we can use the previous constructions to find particular classes of solutions to the field equations, called finite-zone solutions. The equations we have to solve are the first order differential system:

$$(\partial_x - U(\lambda))\Psi = 0 \tag{5.90}$$
$$(\partial_t - V(\lambda))\Psi = 0 \tag{5.91}$$

whose compatibility conditions are equivalent to the field equations. The situation is very different as compared to the finite-dimensional case. As we saw in Chapter 3, the analogue of the spectral curve is

$$\det(T(\lambda) - \mu) = 0 \tag{5.92}$$

where $T(\lambda)$ is the monodromy matrix of the linear system (5.90, 5.91). This equation does not define an algebraic curve of finite genus. This had to be expected since, in field theory, we need an infinite number of action variables, which is incompatible with the finite genus of the spectral curve. Thus we cannot directly apply the previous construction.

However, if we restrict our goal to finding only particular solutions to eqs. (5.90, 5.91), then the knowledge acquired in this chapter becomes directly applicable.

In fact, the two equations (5.90, 5.91) are exactly of the type of eq. (5.33), whose solution was built in terms of Baker–Akhiezer functions. One can adapt this construction to solve them simultaneously. The idea consists of interpreting the two equations (5.90, 5.91) as evolution equations with respect to two different "times" for a system with a finite number of degrees of freedom associated with some Lax matrix $L(\lambda)$. This Lax matrix should satisfy:

$$[\partial_x - U(\lambda), L(\lambda)] = 0$$
$$[\partial_t - V(\lambda), L(\lambda)] = 0 \tag{5.93}$$

To exhibit such Lax matrices, we consider the higher order flows as described in eq. (3.95) of Chapter 3. They provide a family of compatible linear equations $(\partial_{t_i} - V_i)\Psi = 0$ for $i = 1, 2, 3, \ldots$, where we have identified $t_1 = x$, $V_1 = U$ and $t_2 = t$, $V_2 = V$. Since these equations are compatible they satisfy a zero-curvature condition:

$$F_{ij} \equiv \partial_{t_i} V_j - \partial_{t_j} V_i - [V_i, V_j] = 0, \quad \forall i, j = 1, \ldots, \infty$$

We now look for particular solutions which are *stationary* for some given time t_n, i.e. $\partial_{t_n} V_i = 0$ for all i. The zero-curvature conditions $F_{ni} = 0$

reduce to a system of Lax equations:

$$\frac{dL}{dt_i} = [M_i, L], \quad i = 1, \dots, \infty \quad \text{with } L = V_n, \ M_i = V_i$$

This is an integrable hierarchy for a finite-dimensional dynamical system described by the Lax matrix L. Taking n larger and larger, the genus of the corresponding spectral curve usually increases and we get families of solutions involving more and more parameters. We give an example of this procedure in Chapter 11.

The methods of this chapter deal in fact with systems with a finite number of degrees of freedom. In Chapter 13 we present the inverse scattering method which deals directly with systems with an infinite number of degrees of freedom.

References

[1] Sophie Kowalevski, Sur le problème de la rotation d'un corps solide autour d'un point fixe. *Acta Mathematica* **12** (1889) 177–232.

[2] B. Dubrovin, V. Matveev and S. Novikov, Non-linear equations of Korteweg–de Vries type, finite-zone linear operators, and Abelian varieties. *Russian Math. Surveys* **31** (1976) 59–146.

[3] P. van Moerbeke and D. Mumford, The spectrum of difference operators and algebraic curves. *Acta. Math.* **143** (1979) 93–154.

[4] A.G. Reyman and M.A. Semenov-Tian-Shansky, Reduction of Hamiltonian systems, affine Lie algebras and Lax equations. *Inventiones Mathematicae* **54** (1979) 81–100.

[5] M. Adler and P. van Moerbeke, Linearization of Hamiltonian systems, Jacobi varieties and representation theory. *Advances in Mathematics* **38** (1980) 318–379.

[6] A.G. Reyman and M.A. Semenov-Tian-Shansky, Reduction of Hamiltonian systems, affine Lie algebras and Lax equations II. *Inventiones Mathematicae* **63** (1981) 425–432.

[7] D. Mumford, *Tata Lectures on Theta* Vols. I and II, Birkhauser (1983–1984).

[8] A. Bobenko, A. Reyman and M. Semenov-Tian-Shansky, The Kowalevski top 99 years later: a Lax pair, generalizations and explicit solutions. *Comm. Math. Phys.* **122** (1989) 321–354.

[9] E. Horozov and P. van Moerbeke, The full geometry of Kowalevski's top and (1,2)-Abelian surfaces. *Comm. Pure Appl. Math.* **42** (1989) 357–407.

[10] B.A. Dubrovin, I.M. Krichever and S.P. Novikov, *Integrable Systems I.* Encyclopedia of Mathematical Sciences, Dynamical Systems IV, Springer (1990) 173–281.

[11] A. Beauville, Jacobienne des courbes spectrules et systèmes hamiltoniens complètement intégrables. *Act. Math.* **164** (1990) 211–235.

[12] I.M. Krichever and D.H. Phong, On the integrable geometry of soliton equations and N=2 supersymmetric gauge theories. *J. Diff. Geom.* **45** (1997) 349–389.

6

The closed Toda chain

In contrast to open Toda chains, the closed Toda chain is associated with loop algebras. This introduces a spectral parameter into the theory.

The aim of this chapter is to construct the general solution of the closed Toda chain by means of the analytical method. We shall do this in two ways.

The first method, based on an $(n{+}1)\times(n{+}1)$ Lax matrix, follows closely Chapter 5. This canonical example illustrates the general constructions of this chapter, for instance, linearization of the flows on the Jacobian of the spectral curve, separation of variables and the corresponding Hamilton–Jacobi equations.

The second method, which is based on 2×2 matrices, can be regarded as a lattice version of the field theoretical considerations of Chapter 3. A monodromy matrix is introduced which satisfies a quadratic Poisson bracket. This provides short cuts which in general are not available, and makes contact with Sklyanin's method of separation of variables.

We take advantage of this particular example to discuss the reality conditions, which are frequently quite subtle.

6.1 The model

We consider a chain of $(n + 1)$ points with positions q_i and momenta p_i with equations of motion given by:

$$\dot{q}_i = p_i, \quad \dot{p}_i = 2e^{2(q_{i-1}-q_i)} - 2e^{2(q_i-q_{i+1})}, \quad i = 1, \ldots, n+1 \qquad (6.1)$$

The fact that the chain is closed is implemented by setting $(p_0, q_0) = (p_{n+1}, q_{n+1})$ and $(p_{n+2}, q_{n+2}) = (p_1, q_1)$, which gives sense to the above equations for $i = 1$ and $i = n + 1$. Alternatively, one can view the points as sitting on a circle. This is a Hamiltonian system with canonical Poisson brackets $\{p_i, q_j\} = \frac{1}{2}\delta_{ij}$ and Hamiltonian:

$$H = \sum_i p_i^2 + 2\sum_i \exp 2(q_i - q_{i+1}) \tag{6.2}$$

The system has translational symmetry $q_i \to q_i + a$ and one can eliminate the centre of mass motion by imposing the two conditions $\sum_{i=1}^{n+1} p_i = 0$ and $\sum_{i=1}^{n+1} q_i = 0$, so we are left with a phase space of dimension $2n$.

In contrast to the open Toda chain which is associated with the finite-dimensional Lie algebra $sl(n+1)$, the closed Toda chain is associated with the infinite-dimensional Kac–Moody algebra. In fact we consider only its loop representation with vanishing central charge: $\widetilde{sl}_{n+1} \equiv sl_{n+1} \otimes \mathbb{C}(\lambda, \lambda^{-1})$. Its rank is $(n+1)$, see Chapter 16 for more details. The generators $E_{\pm\alpha_i}$ associated with the simple roots are represented by $E_{\alpha_i} = E_{i,i+1}$, $E_{-\alpha_i} = E_{i+1,i}$, for $i = 1,\ldots,n$, together with $E_{\alpha_{n+1}} = \lambda E_{n+1,1}$, and $E_{-\alpha_{n+1}} = \lambda^{-1} E_{1,n+1}$, where E_{jk} are the $(n+1) \times (n+1)$ canonical matrices $(E_{jk})_{mn} = \delta_{jm}\delta_{kn}$. The elements of the Cartan subalgebra are represented by $(n+1) \times (n+1)$ diagonal traceless matrices.

Applying the general construction of the Toda models, see Chapter 4, the Lax pair for the closed Toda chain is given by:

$$L(\lambda) = \sum_{i=1}^{n+1} p_i E_{ii} + \sum_{i=1}^{n+1} a_i(E_{\alpha_i} + E_{-\alpha_i}) \tag{6.3}$$

$$M(\lambda) = -\sum_{i=1}^{n+1} a_i(E_{\alpha_i} - E_{-\alpha_i})$$

where the coefficients a_i are given by $a_i = \exp(q_i - q_{i+1})$, $i = 1,\ldots,n$, and $a_{n+1} = \exp(q_{n+1} - q_1)$. Note that $(q_i - q_{i+1}) = \alpha_i(q)$, where q is the traceless diagonal matrix $\sum_i q_i E_{ii}$ and α_i is the simple root associated with the root vector $E_{i,i+1}$. The quantities a_i satisfy the condition $\prod_{i=1}^{n+1} a_i = 1$. Explicitly, the Lax pair reads:

$$L(\lambda) = \begin{pmatrix} p_1 & a_1 & 0 & \cdots & & \lambda^{-1}a_{n+1} \\ a_1 & p_2 & a_2 & \cdots & & 0 \\ \vdots & & \ddots & & & \vdots \\ 0 & & a_{i-1} & p_i & a_i & & 0 \\ \vdots & & & & \ddots & & \vdots \\ 0 & & \cdots & & a_{n-1} & p_n & a_n \\ \lambda a_{n+1} & & \cdots & & & 0 & a_n & p_{n+1} \end{pmatrix} \tag{6.4}$$

$$M(\lambda) = \begin{pmatrix} 0 & -a_1 & 0 & \cdots & & \lambda^{-1}a_{n+1} \\ a_1 & 0 & -a_2 & \cdots & & 0 \\ \vdots & & \ddots & & & \vdots \\ 0 & & a_{i-1} & 0 & -a_i & 0 \\ \vdots & & & \ddots & & \vdots \\ 0 & & \cdots & a_{n-1} & 0 & -a_n \\ -\lambda a_{n+1} & & \cdots & & a_n & 0 \end{pmatrix} \quad (6.5)$$

The matrices $L(\lambda)$ and $M(\lambda)$ have poles at $\lambda = 0$ and $\lambda = \infty$. The equations of motion eq. (6.1) are equivalent to the Lax equation:

$$\dot{L}(\lambda) = [M(\lambda), L(\lambda)]$$

as one can check easily by an explicit computation. Notice that the correct equations of motion for q_1 and q_{n+1} are obtained thanks to the λ-dependent terms in $L(\lambda)$ and $M(\lambda)$. As usual, the Lax equation ensures that $\mathrm{Tr}\,(L^p(\lambda))$ are conserved quantities, in particular the Hamiltonian eq. (6.2) reads $H = \mathrm{Tr}\,(L^2(\lambda))$ which is independent of λ.

As for all Toda models, the Poisson bracket of the Lax matrix can be written in terms of an r-matrix. This implies that the closed Toda chain is integrable. There are two natural r-matrices, $r^\pm$, which can be computed according to the general formula (4.34) in Chapter 4. In particular:

$$r_{12}^+(\lambda, \lambda') = \rho(\lambda) \otimes \rho(\lambda') \left(\frac{1}{2} \sum_i H_i \otimes H_i + \sum_{\alpha > 0} E_\alpha \otimes E_{-\alpha} \right)$$

where $\rho(\lambda)$ is the loop representation given above with spectral parameter λ. To compute r^+, we recall that the positive roots in $\tilde{sl}_{n+1}$ are $\lambda^n E_{ij}$ with $i < j;\, n \geq 0$ or $i \geq j;\, n > 0$, and the corresponding negative roots are $\lambda^{-n} E_{ji}$. We immediately get:

$$r_{12}^+(\lambda, \lambda') \qquad\qquad\qquad\qquad\qquad\qquad\qquad\qquad\qquad (6.6)$$

$$= \frac{1}{2} \frac{\lambda + \lambda'}{\lambda' - \lambda} \sum_i E_{ii} \otimes E_{ii} + \frac{1}{\lambda' - \lambda} \left(\lambda' \sum_{i<j} E_{ij} \otimes E_{ji} + \lambda \sum_{i>j} E_{ij} \otimes E_{ji} \right)$$

This expression is valid for $|\lambda| < |\lambda'|$. The formula for $r^-(\lambda, \lambda')$ is the same as for $r^+(\lambda, \lambda')$ but valid in the region $|\lambda| > |\lambda'|$. So we consider in general the rational function $r_{12}(\lambda, \lambda')$ which is the extension of the right-hand side of eq. (6.6). Notice that $r_{12}(\lambda, \lambda') = -r_{21}(\lambda', \lambda)$ so that

$$\{L_1(\lambda), L_2(\lambda')\} = [r_{12}(\lambda, \lambda'), L_1(\lambda) + L_2(\lambda')]$$

Using the dualization Tr Res, we define from $r_{12}^{\pm}$ two maps $R^{\pm}$. Let us define $M_{\pm} = -2R^{\pm}(L)$:

$$M_+ = -\sum_i p_i H_i - 2\sum_{i=1}^{n+1} a_i E_{\alpha_i}$$

$$M_- = \sum_i p_i H_i + 2\sum_{i=1}^{n+1} a_i E_{-\alpha_i}$$

we have from eqs. (6.4, 6.5):

$$M(\lambda) = \frac{1}{2}(M_+(\lambda) + M_-(\lambda)), \quad L(\lambda) = \frac{1}{2}(-M_+(\lambda) + M_-(\lambda))$$

so that the Lax equation can be written:

$$\dot{L}(\lambda) = [M(\lambda), L(\lambda)] = [M_+(\lambda), L(\lambda)] = [M_-(\lambda), L(\lambda)]$$

6.2 The spectral curve

The spectral curve Γ is the smooth algebraic curve defined by:

$$\Gamma \quad : \quad \det(L(\lambda) - \mu) = 0 \tag{6.7}$$

For a fixed λ, this is the equation for the eigenvalues of $L(\lambda)$. By expanding the determinant we see that it is of the form:

$$\Gamma(\lambda, \mu) \equiv (\lambda + \lambda^{-1}) - 2t(\mu) = 0 \tag{6.8}$$

where $2t(\mu) = \mu^{n+1} - \sum_{i=1}^{n+1} p_i \mu^n + \cdots$ is a polynomial of degree $(n+1)$. The spectral curve is a hyperelliptic curve since it can be written as

$$s^2 = t^2(\mu) - 1, \quad \text{with} \quad s = \lambda - t(\mu) \tag{6.9}$$

Let us compute the genus of the curve Γ. The polynomial $t^2(\mu)$ is of degree $2n+2$ and generically the equation $t^2(\mu) - 1 = 0$ has no double roots, so the genus of the curve Γ is $g = n$. This is equal to the number of degrees of freedom, i.e. half the dimension of phase-space. In the following, we shall always assume that we are in this generic situation. Notice that the hyperelliptic curve eq. (6.9) is very special because the polynomial of degree $2n+2$ in the right-hand side is expressed in terms of a polynomial of degree $n+1$. This also shows that the number of action variables is precisely $g = n$ when $\sum_{i=1}^{n+1} p_i = 0$.

Let us see how we can recover this result from the general analysis in Chapter 5, when looking at the curve as an $(n+1)$-sheeted cover of the λ plane. We recall the Riemann–Hurwitz formula for computing the genus:

$$2g - 2 = (2g_0 - 2)(n+1) + \nu$$

where $g_0 = 0$ and ν is the number of branch points. To find ν, we count the zeroes of $\frac{\partial}{\partial\mu}\Gamma(\lambda,\mu) = -2\frac{\partial}{\partial\mu}t(\mu)$. This is a polynomial of degree n in μ, independent of λ. Hence we have n zeroes at finite distance. To each value of μ correspond two values of λ, and the total contribution of the branch points at finite distance is $2n$. We now look at $\mu = \infty$ which corresponds to $\lambda = \infty$ or $\lambda = 0$. For $\lambda = \infty$, we have $\lambda \simeq \mu^{n+1}$, which means that we have a branch point of order $n+1$, contributing n to ν. Similarly, $\lambda = 0$ is a branch point of order $n+1$, contributing also n to ν. Adding everything, we find $\nu = 4n$ and $g = n$.

We will call P^+ and P^- the two points above $\lambda = \infty$ and $\lambda = 0$ respectively. In the neighbourhood of $P^\pm$ the local parameter is μ^{-1} and we have by direct expansion of eq. (6.8):

$$P^+ : \lambda = \mu^{n+1}\left(1 - \mu^{-1}\left(\sum_{j=1}^{n+1} p_j\right) + O(\mu^{-2})\right) \qquad (6.10)$$

$$P^- : \lambda = \mu^{-n-1}\left(1 + \mu^{-1}\left(\sum_{j=1}^{n+1} p_j\right) + O(\mu^{-2})\right) \qquad (6.11)$$

6.3 The eigenvectors

Equation (6.7) is the condition for μ to be an eigenvalue of $L(\lambda)$. Therefore, with each point P on Γ one can associate an $(n+1)$ dimensional eigenvector $\Psi(P)$:

$$(L(\lambda) - \mu)\Psi(P) = 0, \quad P = (\lambda, \mu) \in \Gamma$$

Writing this equation explicitly for

$$\Psi(P) = \begin{pmatrix} \psi_1 \\ \vdots \\ \psi_{n+1} \end{pmatrix}$$

we find the following system of linear equations:

$$\begin{aligned} p_1\psi_1 + a_1\psi_2 + \lambda^{-1}a_{n+1}\psi_{n+1} &= \mu\psi_1 \\ a_{i-1}\psi_{i-1} + p_i\psi_i + a_i\psi_{i+1} &= \mu\psi_i \\ \lambda a_{n+1}\psi_1 + a_n\psi_n + p_{n+1}\psi_{n+1} &= \mu\psi_{n+1} \end{aligned} \qquad (6.12)$$

We extend the definition of the coefficients a_i, p_i by periodicity, $a_{i+n+1} = a_i, p_{i+n+1} = p_i$, and introduce a second order difference operator $\mathcal{D}$:

$$(\mathcal{D}\Psi)_i \equiv a_{i-1}\psi_{i-1} + p_i\psi_i + a_i\psi_{i+1}$$

This operator is a discrete version of a Schroedinger operator with periodic potential. Equations (6.12) are then equivalent to:

$$(\mathcal{D}\Psi)_i = \mu\psi_i, \quad \text{with} \quad \psi_{i+n+1} = \lambda\psi_i \tag{6.13}$$

Thus, Ψ is a Bloch wave for the difference operator $\mathcal{D}$ with a Bloch momentum λ. We choose the normalization condition:

$$\psi_0(P) = 1 \quad \text{for all} \quad P \in \Gamma$$

or alternatively $\psi_{n+1}(P) = \lambda$. This is slightly different from the convention of Chapter 3 where we normalized $\psi_1(P) = 1$ but it will prove to be more convenient in the following.

We need to know the analyticity properties of Ψ at infinity.

Proposition. *Let us normalize $\psi_0(P) = 1$. Then, at the points P^+ and P^- above $\lambda = \infty$ and $\lambda = 0$, the eigenvector $\Psi(P)$ behaves as:*

$$\psi_i(P) = e^{q_i - q_0}\mu^i\left(1 - \mu^{-1}\left(\sum_{j=0}^{i-1}p_j\right) + O(\mu^{-2})\right), \quad P \sim P^+ \tag{6.14}$$

$$\psi_i(P) = e^{-q_i + q_0}\mu^{-i}\left(1 + \mu^{-1}\left(\sum_{j=1}^{i}p_j\right) + O(\mu^{-2})\right), \quad P \sim P^- \tag{6.15}$$

where the q_i are the Toda position parameters ($q_0 = q_{n+1}$, $p_0 = p_{n+1}$).

Proof. It is easy to check that this is consistent with eq. (6.12). The result then follows by the uniqueness of the eigenvector. ∎

From the general theory, see eq. (5.17) in Chapter 5, we expect $g + (n + 1) - 1 = 2n$ poles for the eigenvector. From eq. (6.14), we see that we have a fixed pole of order n at P^+. There remain $g = n$ poles at finite distance. These are the *dynamical poles*. We shall denote by $\gamma_1, \ldots, \gamma_n$ their positions. Note that their number equals the number of degrees of freedom. Recall from Chapter 5 that these dynamical poles contain all the relevant information to reconstruct the eigenvector.

We now consider the time evolution of the eigenvector. The Lax equation implies

$$\frac{d}{dt}\Psi(t, P) = (M(\lambda) - C(t, P)\, 1)\, \Psi(t, P)$$

with $C(t, P)$ some scalar function, determined by imposing the condition $\psi_0(t, P) = 1$ or equivalently $\psi_{n+1}(t, P) = \lambda$, which yields $C(t, P) = \sum_j M_{n+1,j}\psi_j$. Alternatively, one can use the natural time evolution

$$\frac{d}{dt}\Psi(t, P) = M(\lambda)\ \Psi(t, P) \tag{6.16}$$

but then, at $t \neq 0$ the eigenvector $\Psi(t, P)$ is a Baker–Akhiezer function. It has n poles at finite distance, independent of time, and essential singularities at the points $P^\pm$ given by:

Proposition. *Let $\Psi(t, P)$ be the eigenvector evolving according to natural eq. (6.16). Then*

$$\psi_i(t, \mu) = e^{q_i(t)}\ e^{-\mu t}\mu^i(1 + O(\mu^{-1})), \quad P \to P^+ \tag{6.17}$$
$$\psi_i(t, \mu) = e^{-q_i(t)}\ e^{\mu t}\mu^{-i}(1 + O(\mu^{-1})), \quad P \to P^- \tag{6.18}$$

<u>Proof.</u> The time evolution of the eigenvector $(L\psi)_i = \mu\psi_i$ is:

$$\frac{d}{dt}\psi_i = (M\Psi)_i = ((L + M_+)\Psi)_i = ((-L + M_-)\Psi)_i$$

Let us see what happens near P^+. Since $\Psi(t, P)$ is also a solution of $(L(\lambda) - \mu)\Psi(t, P) = 0$, the results of eqs. (6.14, 6.15) still apply with $q_i(t)$ now depending on t. Hence we can write

$$\psi_i(t, P) = f(t, P)e^{q_i(t)-q_0(t)}\mu^i(1 + O(\mu^{-1}))$$

There is an extra multiplicative factor $f(t, P)$, independent of n, because we relax the condition $\psi_0 = 1$ for $t \neq 0$. Writing

$$\dot{\psi}_{n+1} = \dot{f}\lambda = -((L - M_-)\Psi)_{n+1} = -\mu\lambda f + p_{n+1}\lambda f + 2a_n\psi_n$$

we get $\dot{f} = (-\mu + p_{n+1} + O(\mu^{-1}))f$, hence $f = \exp(-\mu t + q_0(t))(1 + O(\mu^{-1}))$, where we used $p_{n+1} = \dot{q}_0$. The analysis near P^- is similar, using $M = M_+ + L$. ∎

6.4 Reconstruction formula

Starting from a curve of the particular form eq. (6.8), we reconstruct the eigenvector $\Psi(P)$ from its analyticity properties. From this the whole closed Toda chain model and its solution is obtained.

Consider first what happens at time $t = 0$. Let the algebro-geometrical data be specified by the curve eq. (6.8), with a divisor of $g = n$ points on it $D = \gamma_1 + \ldots + \gamma_g$, and two punctures which are the points $P^\pm$ at infinity. From the previous section, the divisor of the component $\psi_i(P)$ of $\Psi(P)$ is:

$$(\psi_i) = -D + i\,P^- - i\,P^+ \qquad (6.19)$$

Applying the Riemann–Roch theorem with $\deg(\psi_i) = g$, we see that there exists a unique meromorphic function, up to a proportionality constant, having this divisor. This function is not constant, for $i > 0$, since it vanishes at P^-. We can fix the proportionality constant (up to a sign) by requiring that the coefficients of $\mu^{\mp i}$ at $P^\pm$ are inverse to each other. Denote these coefficients by $e^{\pm(q_i - q_0)}$. Note that this eliminates the residual gauge invariance by diagonal matrices which is present in the general theory.

The function $\psi_i(P)$ will have the form for all $i \geq 0$:

$$\psi_i(\mu) = e^{q_i - q_0}\,\mu^i(1 - \mu^{-1}\,\xi_i^+ + O(\mu^{-2})), \qquad P \to P^+ \qquad (6.20)$$

$$\psi_i(\mu) = e^{-q_i + q_0}\,\mu^{-i}(1 - \mu^{-1}\,\xi_i^- + O(\mu^{-2})), \qquad P \to P^- \qquad (6.21)$$

Here $\xi_i^\pm$ are just Taylor coefficients. Of course, since $\psi_i(P)$ is a meromorphic function, it also possesses g extra zeroes. We now show that these properties imply that the functions $\psi_i(P)$ constructed with these analyticity requirements are solutions of eq. (6.13):

Proposition. *Let $\psi_i(P)$ be the unique meromorphic function having simple poles at the g points γ_i, a pole of order i at P^+, and a zero of order i at P^-, and normalized as in eqs. (6.20, 6.21). Then,*
(i) The $\psi_i(P)$ satisfy the Schroedinger equations $(\mathcal{D} - \mu)\Psi = 0$, eq. (6.13), with coefficients a_i, p_i given by:

$$a_i = \exp(q_i - q_{i+1}), \qquad p_i = \xi_{i+1}^+ - \xi_i^+ \qquad (6.22)$$

(ii) The functions $\psi_i(P)$ are quasi-periodic:

$$\psi_{n+1+i}(P) = \lambda(P)\,\psi_i(P) \qquad (6.23)$$

Therefore, the $\psi_i(P)$ are the components of the eigenvector of a Lax matrix $L(\lambda)$ of the form eq. (6.4).

<u>Proof.</u> Let us consider the function $\widehat{\psi}_i(P) = ((\mathcal{D} - \mu)\Psi)_i(P)$. Since the coefficients of $(\mathcal{D} - \mu)$ are regular outside $P^\pm$, $\widehat{\psi}_i$ possesses g poles at the γ_i. Consider now the behaviour at $P^\pm$. We have:

$$((\mathcal{D} - \mu)\Psi)_i = \mu^{i+1}\left(a_i e^{q_{i+1} - q_0} - e^{q_i - q_0}\right)$$
$$- \mu^i\left(a_i e^{q_{i+1} - q_0}\xi_{i+1}^+ - e^{q_i - q_0}(p_i + \xi_i^+)\right) + O(\mu^{i-1}), \qquad P \to P^+$$

$$((\mathcal{D} - \mu)\Psi)_i = \mu^{-i+1}\left(a_{i-1} e^{q_0 - q_{i-1}} - e^{q_0 - q_i}\right) + O(\mu^{-i}), \qquad P \to P^-$$

If we choose the coefficients a_i and p_i as in eqs. (6.22), the function $(\mathcal{D} - \mu)\psi_i$ has a pole of order $(i-1)$ at P^+, and a zero of order i at P^-. Therefore the divisor of $\hat{\psi}_i(P)$ is greater than $-D' = -D + iP^- - (i-1))P^+$ and we have $\deg D' = g-1$. Thus, by the Riemann–Roch theorem, $(\mathcal{D} - \mu)\Psi = 0$. Notice that the coefficients a_i and p_i are chosen in order to decrease the degree of this divisor by one unit.

Let us now prove the periodicity relation $\psi_{n+1+i}(P) = \lambda\psi_i(P)$. We apply the Riemann–Roch theorem to the functions in each member of this equation. The divisor of the function λ is $(\lambda) = (n+1)P^- - (n+1)P^+$. The poles at finite distance of ψ_{n+1+i} and ψ_i are located at the same positions $\gamma_1, \ldots, \gamma_g$. So we have:

$$(\psi_{n+1+i}) \geq -D + (n+1+i)P^- - (n+1+i)P^+$$
$$(\lambda\psi_i) \geq -D + iP^- - iP^+ + (n+1)P^- - (n+1)P^+$$

Notice that these divisors are both greater than the degree g divisor $-D + (n+1+i)P^- - (n+1+i)P^+$. By the Riemann–Roch theorem, there is only one function satisfying this property and the two functions are therefore proportional. The proportionality constant is determined by comparing the behaviour at P^+ or P^-, and is found to be equal to one. ∎

Let us now consider what happens at $t \neq 0$. We define, for all integer $i \geq 0$, the function $\psi_i(t, P)$ as the Baker–Akhiezer function with g poles at the same positions $\gamma_1, \ldots, \gamma_n$ as above, and with essential singularities at $P^\pm$ given by:

$$\psi_i(t, \mu) = e^{q_i(t)} e^{-\mu t} \mu^i (1 - \mu^{-1} \xi_i^+(t) + \ldots), \quad P \to P^+ \quad (6.24)$$
$$\psi_i(t, \mu) = e^{-q_i(t)} e^{\mu t} \mu^{-i} (1 - \mu^{-1} \xi_i^-(t) + \ldots), \quad P \to P^- \quad (6.25)$$

with the normalization parameters $q_i(t)$ obtained by requiring that the expansions at P^+ and P^- start with inverse coefficients. Then the Taylor coefficients $\xi_i^\pm(t)$ are fixed. By the Riemann–Roch theorem this function exists and is unique. Moreover, we have $\psi_{n+1+i}(t, P) = \lambda\psi_i(t, P)$ by the same method as above. Taking into account the expansions eqs. (6.10, 6.11) of λ and the expansions eqs. (6.24, 6.25) of $\psi_i(t, P)$ near $P^\pm$, it follows that the Taylor coefficients $\xi_i^\pm(t)$ are periodic, i.e. $\xi_{i+n+1}^\pm(t) = \xi_i^\pm(t)$, when $\sum_i p_i = 0$ (centre of mass system).

Proposition. *The Baker–Akhiezer functions defined above satisfy the eigenvalue equation and the evolution equation:*

$$L(\lambda)\Psi(t, P) = \mu\,\Psi(t, P)$$
$$\frac{d}{dt}\Psi(t, P) = M(\lambda)\Psi(t, P)$$

with $M(\lambda)$ defined in eq. (6.3) with $a_i(t) = \exp\left(q_i(t) - q_{i+1}(t)\right)$.

Proof. The proof of the eigenvalue equation and the quasi-periodicity property (6.23) is the same as for the initial eigenvector at $t = 0$.
To prove the evolution equation let us consider for $i = 1, \dots, n$ the expression:

$$E_i \equiv \frac{d}{dt}\psi_i + (a_i\psi_{i+1} - a_{i-1}\psi_{i-1})$$

Since the poles of the ψ_i at finite distance are independent of time, and the a_i are constant on the Riemann surface, E_i has the same g poles at finite distance as ψ_i. Its behaviour at infinity is easily obtained as:

$$E_i = -e^{-\mu t}e^{q_i}\mu^i(\xi_{i+1}^+ - \xi_i^+ - \dot{q}_i) + O(\mu^{i-1}), \qquad P \to P^+ \quad (6.26)$$
$$E_i = e^{\mu t}e^{-q_i}\mu^{-i}(\xi_{i-1}^- - \xi_i^- - \dot{q}_i) + O(\mu^{-i+1}), \quad P \to P^- $$

By the Riemann–Roch theorem E_i is proportional to ψ_i, and we write $E_i = d_i(t)\psi_i$. Using the quasi-periodicity property eq. (6.23) we can re-state this result as:

$$\dot{\Psi} = M\Psi + d\Psi \qquad (6.27)$$

where M is the matrix given in eq. (6.5) and d is the time-dependent diagonal matrix $d = \mathrm{Diag}\,(d_1, \dots, d_{n+1})$, which is constant on the Riemann surface. Differentiating the relation $(L(\lambda) - \mu)\Psi = 0$ with respect to time we get, using eq. (6.27), $(\dot{L} - [M, L] - [d, L])\Psi = 0$. For any value of λ which is not a branch point of the covering $(\lambda, \mu) \to \lambda$, we have $n+1$ *independent* (see Chapter 5) vectors $\Psi(\lambda, \mu_k)$ for which this equation is true, hence we get the matrix equation $\dot{L}(\lambda) - [M(\lambda), L(\lambda)] = [d, L(\lambda)]$, which remains true for all values of λ by analytic continuation. In particular it is true when $\lambda \to 1/\lambda$. Note that ${}^tL(\lambda) = L(1/\lambda)$, ${}^tM(\lambda) = -M(1/\lambda)$, so taking the transpose of the above equation evaluated at $1/\lambda$ and comparing it with the original equation we get $[d, L(\lambda)] = 0$. This implies that d is proportional to the identity matrix, $d = \delta(t)I$. Then, comparing eq. (6.26) with $E_i = \delta(t)\psi_i$, we find $\dot{q}_i = \xi_{i+1}^+ - \xi_i^+ + \delta(t)$. In the centre of mass system, $\sum_{i=0}^n q_i = 0$, so that $\delta(t) = 0$ by the periodicity of ξ_i^+. ∎

At this point we have completely reconstructed the closed Toda chain, starting from an appropriate spectral curve. Moreover, the procedure also provides the solution of the equations of motion. To get explicit expressions for $q_i(t)$ we need explicit formulae for the Baker–Akhiezer functions. This is done using Riemann's theta-functions.

Let us fix a canonical set of cycles (a_i, b_j) on Γ, a base point Q_0 on Γ and a set of holomorphic Abelian differential ω_j, dual to the a-cycles (see Chapter 15). Let $\Omega^{(i)}$ be the meromorphic differential analytic on Γ outside the points $P^{\pm}$, and obeying the following normalization conditions:

$$\oint_{a_k} \Omega^{(i)} = 0,$$

$$\Omega^{(i)}(P) = \pm(\mu^{i-1} + O(\mu^{-2}))d\mu, \quad P \to P^{\pm} \tag{6.28}$$

The notation assumes that some multivalued primitives $\int_{Q_0}^{P} \Omega^{(i)}$ have been chosen for these differentials. Since the local parameter around $P^{\pm}$ is $1/\mu$, $\Omega^{(i)}$ has poles of order $(i+1)$ at the points $P^{\pm}$. Note that $\Omega^{(i)}$, $i \geq 1$, are Abelian differentials of the second kind, whereas $\Omega^{(0)}$ is a normalized Abelian differential of the third kind.

Let us also define, for each i, the g-dimensional vectors $U^{(i)}$ whose components are the b-periods of the forms $\Omega^{(i)}$:

$$U_k^{(i)} = \frac{1}{2\pi i} \oint_{b_k} \Omega^{(i)} \tag{6.29}$$

Given these data, the Baker–Akhiezer functions defined in eqs. (6.24, 6.25) are expressed as follows:

Proposition. *Let $\mathcal{A}(P)$ be the Abel map, $\mathcal{A}_k(P) = \int_{Q_0}^{P} \omega_k$. Then the Baker–Akhiezer function $\psi_i(t, P)$ has the following expression:*

$$\psi_i(P) = r_i(t)e^{\left(i \int_{Q_0}^{P} \Omega^{(0)} - t \int_{Q_0}^{P} \Omega^{(1)}\right)} \frac{\theta(\mathcal{A}(P) + i\, U^{(0)} - t U^{(1)} - \zeta_0)}{\theta(\mathcal{A}(P) - \zeta_0)} \tag{6.30}$$

where $r_i(t)$ is independent of P on Γ and

$$\zeta_0 = \sum_{i=1}^{g} \mathcal{A}(\gamma_i) + \mathcal{K} \tag{6.31}$$

with $\mathcal{K}$ the vector of Riemann constants.

Proof. First, one checks that this function is well-defined on Γ, i.e. it does not depend on the path of integration between Q_0 and P. This is done using the formulae of Chapter 15 on theta functions. Then one checks that it has the right poles at finite distance. They are given by the zeroes of the theta-function $\theta(\mathcal{A}(P) - \zeta_0)$. By Riemann's theorem, they are located at the points $\gamma_1, \ldots, \gamma_g$ because we chose the vector ζ_0 according to eq. (6.31). Finally, we check that it has the right behaviour in the neighbourhood of the points at infinity $P^{\pm}$. The theta functions are regular at infinity. Therefore, the behaviour at infinity is governed by the differentials $\Omega^{(0)}$ and $\Omega^{(1)}$. From eq. (6.28), we deduce that when $P \to P^\pm$:

$$\int_{Q_0}^{P} \Omega^{(0)} = \pm \log \mu + O(1)$$

$$\int_{Q_0}^{P} \Omega^{(i)} = \pm \frac{\mu^i}{i} + O(1), \quad i \geq 1$$

These expressions hold modulo periods, but as we have seen, this does not affect the global expression. Therefore, when $P \to P^\pm$:

$$\exp\left(i \int_{Q_0}^{P} \Omega^{(0)} - t \int_{Q_0}^{P} \Omega^{(1)} \right) = \text{const.} \; \mu^{\pm i} e^{\mp \mu t} \left(1 + O(\mu^{-1}) \right)$$

This proves the result. ∎

In eq. (6.30), the coefficient $r_i(t)$ is fixed up to a sign by the requirement that the leading term in the expansions of $\psi_i(P)$ at $P^\pm$ are inverse to each other. In the following we only need to consider ratios of the values of $\psi_i(P)$ at the vicinity of the points P^+ and P^-, in which $r_i(t)$ cancels out. So we do not give the value of $r_i(t)$.

Proposition. *Let $\tau_i(t)$ be the n tau–functions defined by:*

$$\tau_i(t) = e^{\frac{1}{2}\beta_0 i^2} \theta(iU^{(0)} - tU^{(1)} - \widehat{\zeta}_0) \tag{6.32}$$

where β_0 and $\widehat{\zeta}_0$ are constants. Then the solution of the equations of motion of the closed Toda chain is given by:

$$e^{2(q_i(t) - q_{i+1}(t))} = \frac{\tau_{i+1}(t)\tau_{i-1}(t)}{\tau_i^2(t)} \tag{6.33}$$

<u>Proof</u>. From eqs. (6.24, 6.25), we have:

$$\frac{\psi_i(P \to P^+)}{\psi_i(P \to P^-)} = e^{2q_i} \mu^{2i} e^{-2\mu t} (1 + O(\mu^{-1})), \quad \mu \to \infty$$

hence substituting the expression (6.30) of ψ_i we obtain an expression of the form:

$$e^{2q_i} = e^{\beta_0 i + \beta_1 t} \times$$
$$\left(\frac{\theta(A(P^+) + iU^{(0)} - tU^{(1)} - \zeta_0)}{\theta(A(P^+) - \zeta_0)} \right) \left(\frac{\theta(A(P^-) - \zeta_0)}{\theta(A(P^-) + iU^{(0)} - tU^{(1)} - \zeta)} \right)$$

The exponential prefactor comes from $\lim_{P \to P^\pm}(\int_{Q_0}^P \Omega^{(0)} \mp \log \mu) = \beta_0^\pm$ and $\lim_{P \to P^\pm}(\int_{Q_0}^P \Omega^{(1)} \mp \mu) = \beta_1^\pm$. Then $\beta_k = \beta_k^+ - \beta_k^-$ and the singular part cancels with $\mu^{2i}e^{-2\mu t}$. Taking the quotient of these expressions for i and $i+1$ gives:

$$e^{2(q_i - q_{i+1})} = e^{-\beta_0} \times$$
$$\frac{\theta(i\,U^{(0)} - tU^{(1)} - \zeta_0 + \mathcal{A}(P^+))\theta((i+1)U^{(0)} - tU^{(1)} - \zeta_0 + \mathcal{A}(P^-))}{\theta(i\,U^{(0)} - tU^{(1)} - \zeta_0 + \mathcal{A}(P^-))\theta((i+1)U^{(0)} - tU^{(1)} - \zeta_0 + \mathcal{A}(P^+))}$$

To end the proof of eq. (6.33), we show that:

$$\mathcal{A}(P^-) - \mathcal{A}(P^+) = U^{(0)}$$

This is a direct consequence of Riemann's bilinear relations which, for normalized Abelian differential of the third kind like $\Omega^{(0)}$, with residue -1 at P^+ and $+1$ at P^-, implies:

$$U^{(0)} \equiv \frac{1}{2i\pi} \oint_{b_k} \Omega^{(0)} = \int_{P^+}^{P^-} \omega_k = \mathcal{A}(P^-) - \mathcal{A}(P^+)$$

Then the result follows by defining $\widehat{\zeta}_0 = \zeta_0 - \mathcal{A}(P^+) - U^{(0)}$ and inserting the definition of the tau-function.   ∎

The explicit formula of the Baker–Akhiezer function in terms of theta-functions also shows that the Abel map linearizes the dynamics. Indeed, consider the component $\psi_0(t, P)$. It has poles at fixed position $\gamma_1, \ldots, \gamma_n$ with divisor ζ_0, and zeroes at n others points $\gamma_1(t), \ldots, \gamma_n(t)$. At $t = 0$, $\gamma_i(t = 0) = \gamma_i$ since $\psi_0(t = 0)$ is equal to one. The points $\gamma_i(t)$ are the zeroes of the theta-function which is in the numerator of eq. (6.30) taken for $i = 0$. Thus, by Riemann's theorem,

$$\sum_{i=1}^{n} \mathcal{A}(\gamma_i(t)) - \mathcal{A}(\gamma_i) = tU^{(1)} \tag{6.34}$$

This flow is linear.

Remark. The formula (6.30) can be generalized by considering all flows associated with the other conserved Hamiltonians. If we denote by t_p the time associated with these Hamiltonians, the generalized tau-functions are obtained by replacing $tU^{(1)}$ by $tU^{(1)} \to \sum_p t_p U^{(p)}$.

6.5 Symplectic structure

We now want to prove that the coordinates $(\lambda_{\gamma_i}, \mu_{\gamma_i})$ of the points of the dynamical divisor form a set of separated canonical coordinates. Recall that the Poisson bracket is the standard canonical one:

$$\{q_i, q_j\} = \{p_i, p_j\} = 0, \quad \{p_i, q_j\} = \frac{1}{2}\delta_{ij} \tag{6.35}$$

Proposition. *Let Γ be the spectral curve eq. (6.7). Let $(\lambda_{\gamma_i}, \mu_{\gamma_i}), i = 1, \ldots, g$, be the g points of the dynamical divisor D. Then*

$$\omega = 2\sum_i \delta q_i \wedge \delta p_i = \sum_i \frac{\delta\lambda_{\gamma_i}}{\lambda_{\gamma_i}} \wedge \delta\mu_{\gamma_i} \tag{6.36}$$

Proof. According to the general procedure explained in Chapter 5, we start from the 2-form K on phase space with values in the 1-forms on Γ defined by:

$$K = K_1 + K_2 + K_3 \tag{6.37}$$

$$K_1 = <\Psi^{(-1)}(P)\delta L(\lambda) \wedge \delta\Psi(P)> \frac{d\lambda}{\lambda}$$

$$K_2 = <\Psi^{(-1)}(P)\delta\mu \wedge \delta\Psi(P)> \frac{d\lambda}{\lambda}$$

$$K_3 = \delta\left(\log \partial_\mu \Gamma(\lambda, \mu)\right) \wedge \delta\mu \frac{d\lambda}{\lambda}$$

Here $\Gamma(\lambda, \mu) = 0$ is the equation of the spectral curve Γ. We have included the form K_3 in the definition of K, although its role is auxiliary. More importantly, notice the factor $1/\lambda$ as compared to the analysis of Chapter 5. This is necessary to get the right Poisson bracket on the variables q_i, p_i.

We write that the sum of the residues of the form K seen as a 1-form on Γ vanishes. The poles of K are located at three different places, first the dynamical poles of Ψ, then the poles at $P^\pm$ coming from Ψ, L and $d\lambda/\lambda$, and finally the poles at the branch points of the covering coming from the poles of $\Psi^{(-1)}$ and from the fact that the δ differential is taken at fixed λ. The evaluation of the residues at the dynamical poles and the branch points is exactly as in Chapter 5, yielding the sum $2\sum_k \delta\mu_{\gamma_k} \wedge \delta\lambda_{\gamma_k}/\lambda_{\gamma_k}$. So we are left with the computation of the residues at $P^\pm$.

The residue at the point P^+, for instance, is obtained by integrating over a small contour enclosing it. But the Riemann surface seen as a branched cover of the λ sphere has a branch point of order $n+1$ at P^+.

So, a closed contour around P^+ runs on the $n+1$ sheets of the covering before returning to its starting point. Therefore, we can write

$$\text{Res}_{P^+} K = \frac{1}{2i\pi} \oint K(P) = \frac{1}{2i\pi} \oint \sum_i K(P^i)$$

where in the last expression, the points P_i are the $n+1$ points of the contour over the point on the base with coordinate λ. This sum is independent of the order of the sheets and is therefore a 1-form on the base. The integral is also taken on the base.

The sum $\sum_i K_1(P_i)$ can be written as a trace since the vectors $\Psi(P_i)$, with P_i the $(n+1)$ points over λ, form a basis of eigenvectors of $L(\lambda)$. As in Chapter 5, it is convenient to introduce the $(n+1) \times (n+1)$ matrix $\widehat{\Psi}(P)$ whose columns are the $(n+1)$ vectors $\Psi(P_i)$. We thus have:

$$\text{Res}_{P^+}(K_1) = -\frac{1}{2i\pi} \oint \text{Tr}\left(\delta\widehat{\Psi}(\lambda)\widehat{\Psi}^{-1}(\lambda) \wedge \delta L(\lambda)\right)\frac{d\lambda}{\lambda}$$

The matrix $\delta\widehat{\Psi}(\lambda)\widehat{\Psi}^{-1}(\lambda)$ does not depend on the order of the sheets and is therefore a meromorphic function of λ which we now calculate.

Lemma.

$$\left(\delta\widehat{\Psi}\widehat{\Psi}^{-1}\right)_{ij}(\lambda) = -(\delta q_i - \delta q_0)\delta_{ij} \tag{6.38}$$

$$+ \delta\eta_i\left(\frac{1}{a_i}\delta_{i,j-1} + \frac{\lambda}{a_{n+1}}\delta_{i,n+1}\delta_{j,1}\right) + O(\lambda^2), \quad P \to P^-$$

$$\left(\delta\widehat{\Psi}\widehat{\Psi}^{-1}\right)_{ij}(\lambda) = (\delta q_i - \delta q_0)\delta_{ij} \tag{6.39}$$

$$- \delta\xi_i\left(\frac{1}{a_{i-1}}\delta_{i,j+1} + \frac{1}{\lambda a_{n+1}}\delta_{i,1}\delta_{j,n+1}\right) + O(\lambda^{-2}), \quad P \to P^+$$

where $\eta_i = \sum_{j=1}^{i} p_j$ *and* $\xi_i = \sum_{j=0}^{i-1} p_j$.

Proof. Let us first consider the behaviour near P^-. From eq. (6.15), we have

$$\widehat{\Psi}_{ij} = \widehat{\Psi}_i(P_j) = e^{-q_i+q_0}\mu_j^{-i}\left(1 + \eta_i\, \mu_j^{-1} + O(\mu_j^{-2})\right)$$

$$\widehat{\Psi}_{ij}^{-1} = \frac{1}{n+1}e^{q_j-q_0}\mu_i^{j}\left(1 + O(\mu_j^{-1})\right)$$

where μ_j^{-1} is the local coordinate of the point P_j above λ, i.e. $\mu_j = \lambda^{\frac{-1}{n+1}}\alpha_j + \cdots$ with α_j a $(n+1)$-th root of unity.

Taking variations, we limit ourselves to $O(\mu^{-2})$ so that we can set $\delta\mu = 0$ (using eq. (6.11) where $\sum_j p_j = 0$), and obtain: $\delta\widehat{\Psi}_{ij} = -(\delta q_i - \delta q_0)\widehat{\Psi}_{ij} + \delta\eta_i \widehat{\Psi}_{ij}\mu_j^{-1} + O(\mu_j^{-2})$. Hence

$$\left(\delta\widehat{\Psi}\widehat{\Psi}^{-1}\right)_{ij}(\lambda) = -(\delta q_i - \delta q_0)\delta_{ij} + \delta\eta_i \sum_k \widehat{\Psi}_{ik}\mu_k^{-1}\widehat{\Psi}_{kj}^{-1}$$

Note that this shows that $\widehat{\Psi}^{-1}$ needs to be computed to leading order only. The last term is equal to:

$$\sum_k \widehat{\Psi}_{ik}\mu_k^{-1}\widehat{\Psi}_{kj}^{-1} = \frac{e^{q_j-q_i}}{n+1}\sum_k \mu_k^{-i+j-1} = \frac{e^{q_j-q_i}}{n+1}\lambda^{\frac{i-j+1}{n+1}}\sum_k \alpha_k^{-i+j-1}$$

The last sum is over the roots of unity. It vanishes unless $i - j + 1 \equiv 0 \bmod [n+1]$, which is possible only if $i = j - 1$, $(i = 1, \ldots, n)$, and $i = n + 1, j = 1$. Evaluating these two types of terms yields eq. (6.38). The analysis at P^+ is similar. ∎

We now finish the proof of the proposition, by first analysing the residue of K_1 at $P^\pm$. It is clear that only the terms written in eqs. (6.38, 6.39) contribute to the residue. Indeed, if one keeps terms of order μ^{-2} the same computation as above yields contributions to $\delta\widehat{\Psi}\widehat{\Psi}^{-1}$ which are not vanishing only if $i - j \equiv \pm 2 \bmod [n+1]$, which cannot contribute to the trace with δL. The contributions to the residues at $P^\pm$ of the first terms in eqs. (6.38, 6.39) add to $2\sum_i \delta p_i \wedge \delta q_i$ (remember that $d\lambda/\lambda$ has residue ∓ 1 at $P^\pm$). Similarly, the contributions of the second terms at $P^\pm$ are also equal and add up to $2\sum_i \delta p_i \wedge \delta q_i$. Finally, since $\delta\mu$ has a simple zero at $P^\pm$ the forms K_2 and K_3 are regular at these points. ∎

The quantities $(\log \lambda_{\gamma_k}, \mu_{\gamma_k})$ are therefore canonical coordinates. Since the points $(\lambda_{\gamma_k}, \mu_{\gamma_k})$ of the dynamical divisor are on the spectral curve, we have the n equations $\Gamma(\lambda_{\gamma_k}, \mu_{\gamma_k}) = 0$, where $\Gamma(\lambda, \mu) = 0$ is the equation of the spectral curve:

$$\lambda_{\gamma_k} + \lambda_{\gamma_k}^{-1} = 2t(\mu_{\gamma_k}), \quad \text{for} \quad k = 1, \ldots, n$$

As explained in Chapter 5, this implies that the variables $(\lambda_{\gamma_k}, \mu_{\gamma_k})$ are separated and, furthermore, this allows to construct the action–angle variables.

6.6 The Sklyanin approach

We now introduce an equivalent description of the Toda chain. This approach can be viewed as a lattice version of the constructions of integrable field theories of Chapter 3. It is based on the use of a 2×2 transfer matrix whose Poisson brackets are quadratic, as in eq. (3.91) in Chapter 3. This provides a very simple way to find the separated canonical variables and the corresponding separated Hamilton–Jacobi equations. The linearization of the flow on the Jacobian of the spectral curve is also obtained in this approach.

We first introduce 2×2 matrices by replacing the linear second order system eq. (6.13) by the linear 2×2 first order system:

$$\begin{pmatrix} \psi_i \\ \psi_{i+1} \end{pmatrix} = \widehat{T}_i(\mu) \begin{pmatrix} \psi_{i-1} \\ \psi_i \end{pmatrix} \tag{6.40}$$

where the $\widehat{T}_i$ are given by:

$$\widehat{T}_i(\mu) = \begin{pmatrix} 0 & 1 \\ -a_{i-1}/a_i & -(p_i - \mu)/a_i \end{pmatrix} \tag{6.41}$$

The solution of eq. (6.40) with $\psi_0 = \lambda^{-1}\psi_{n+1}$ is:

$$\begin{pmatrix} \psi_i \\ \psi_{i+1} \end{pmatrix} = \widehat{T}_i(\mu)\widehat{T}_{i-1}(\mu) \cdots \widehat{T}_1(\mu) \begin{pmatrix} \psi_0 \\ \psi_1 \end{pmatrix}$$

Each matrix $\widehat{T}_i(\mu)$ may be viewed as an elementary transport matrix on a small segment at position i, and the product of such matrices is a transport matrix from site 1 to site i.

The periodicity condition, $\psi_{n+1+i} = \lambda\psi_i$, translates into the eigenvalue problem:

$$\widehat{T}(\mu) \begin{pmatrix} \psi_0 \\ \psi_1 \end{pmatrix} = \lambda \begin{pmatrix} \psi_0 \\ \psi_1 \end{pmatrix}$$

where $\widehat{T}(\mu)$ is the monodromy matrix defined by:

$$\widehat{T}(\mu) = \widehat{T}_{n+1}(\mu)\widehat{T}_n(\mu) \cdots \widehat{T}_1(\mu) \equiv \begin{pmatrix} \widehat{A}(\mu) & \widehat{B}(\mu) \\ \widehat{C}(\mu) & \widehat{D}(\mu) \end{pmatrix} \tag{6.42}$$

Here $\widehat{A}(\mu), \widehat{B}(\mu), \widehat{C}(\mu), \widehat{D}(\mu)$ are polynomials in μ of degrees deg $\widehat{A} = n-1$, deg $\widehat{B} =$ deg $\widehat{C} = n$, and deg $\widehat{D} = n+1$. Since det $\widehat{T}_i(\mu) = a_{i-1}/a_i$ one has det $\widehat{T}(\mu) = 1$, and the characteristic equation $\det(\widehat{T}(\mu) - \lambda) = 0$ reads

$$\lambda^2 - 2t(\mu)\lambda + 1 = 0, \tag{6.43}$$

with

$$2t(\mu) = \text{Tr}\,(\widehat{T}(\mu)) = \widehat{A}(\mu) + \widehat{D}(\mu)$$

Clearly $t(\mu)$ is a polynomial in μ of degree $(n+1)$: $t(\mu) = \frac{1}{2}\left(\mu^{n+1} + \cdots\right)$.
Moreover, eq. (6.43) also expresses the existence of an eigenvector of $L(\lambda)$,
hence is equivalent to eq. (6.7). By using $\widehat{T}(\mu)$ instead of $L(\lambda)$ we have
exchanged the role played by μ and λ. The spectral curve Γ is now presented as a two-sheeted covering of the μ-plane. Above each point μ there
are two points corresponding to the two roots

$$\lambda^\pm = t(\mu) \pm \sqrt{t^2(\mu) - 1}$$

of eq. (6.43) such that $\lambda^+\lambda^- = 1$. In particular, above $\mu = \infty$ we have
the two points $P^+(\lambda = \infty)$ and $P^-(\lambda = 0)$.

In the following, we shall choose the normalization condition $\psi_0(P) = 1$.
It is useful to introduce a standard basis of solutions of the linear system
eq. (6.40), which we denote by $\chi_i^{(0)}$, $\chi_i^{(1)}$, and specified by the boundary
conditions:

$$\begin{pmatrix} \chi_0^{(0)} \\ \chi_1^{(0)} \end{pmatrix} = \begin{pmatrix} 1 \\ 0 \end{pmatrix}, \quad \begin{pmatrix} \chi_0^{(1)} \\ \chi_1^{(1)} \end{pmatrix} = \begin{pmatrix} 0 \\ 1 \end{pmatrix} \tag{6.44}$$

These solutions are polynomials in μ with $\deg \chi_i^{(0)} = i-2$, $\deg \chi_i^{(1)} = i-1$.
We can expand any other solution on this basis. In particular:

$$\psi_i = \chi_i^{(0)} + \psi_1\chi_i^{(1)}$$

The coefficient ψ_1 is determined by the periodicity condition $\psi_{n+1+i} = \lambda\psi_i$. This is equivalent to the eigenvalue equation $\widehat{T}(\mu) \begin{pmatrix} 1 \\ \psi_1 \end{pmatrix} = \lambda \begin{pmatrix} 1 \\ \psi_1 \end{pmatrix}$

and gives:

$$\psi_i^\pm = \chi_i^{(0)} + \left(\frac{\lambda^\pm - \widehat{A}(\mu)}{\widehat{B}(\mu)}\right) \chi_i^{(1)} \tag{6.45}$$

We see that the two functions $\psi_i^\pm$ corresponding to the two Bloch waves
are in fact the values of a unique meromorphic function $\psi_i(P)$ evaluated
at the two points $(\mu, \lambda^\pm)$ above μ, on Γ.

The poles at finite distance of ψ_i are thus the sames as those of ψ_1
and are located above the n zeroes of $\widehat{B}(\mu)$. When $\widehat{B}(\mu_{\gamma_k}) = 0$, the two
eigenvalues of $\widehat{T}(\mu_{\gamma_k})$ are $\widehat{A}(\mu_{\gamma_k})$ and $\widehat{D}(\mu_{\gamma_k})$ so that the numerator $\lambda - \widehat{A}(\mu)$ vanishes on one of the two points above μ_{γ_k}. Therefore the function
$\psi_i(P)$ has only one-pole at $(\mu_{\gamma_k}, \lambda_{\gamma_k} = \widehat{D}(\mu_{\gamma_k}))$. Hence the dynamical
divisor is exactly the same as in the $(n+1) \times (n+1)$ matrix approach.
As we already know, the coordinates $(\mu_{\gamma_k}, \log \lambda_{\gamma_k})$ of these points form
a set of conjugated canonical variables. We present an alternative way of
deriving this result in the following section.

6.7 The Poisson brackets

We first establish an explicit formula for the Poisson brackets of the matrix elements of $\widehat{T}(\mu)$. It is actually more convenient to perform a gauge tranformation before computing these Poisson brackets. Let

$$\widehat{T}_i(\mu) \to T_i(\mu) = D_i \widehat{T}_i(\mu) D_{i-1}^{-1}$$

with D_i periodic, $D_{n+1+i} = D_i$. Since $\widehat{T}(\mu) = \widehat{T}_{n+1}(\mu) \cdots \widehat{T}_1(\mu)$ is the product of the matrices $\widehat{T}_i(\mu)$, it gets conjugated by D_0: $\widehat{T}(\mu) \to T(\mu) = D_0 \widehat{T}(\mu) D_0^{-1}$. In particular, the spectral curve $\det(\widehat{T}(\mu) - \lambda) = 0$ is preserved by such a gauge transformation.

We choose the matrices D_i so that the matrices $T_i(\mu)$ are local, i.e. $T_i(\mu)$ only depends on the canonical variables q_i and p_i. We take $D_i = \text{Diag}\left(d_i, d_{i+1}^{-1}\right)$ with $d_i = \exp(q_i)$. Notice that $a_i = d_i/d_{i+1}$. The explicit expressions for $T_i(\mu)$ are:

$$T_i(\mu) = \begin{pmatrix} 0 & e^{2q_i} \\ -e^{-2q_i} & (\mu - p_i) \end{pmatrix} \tag{6.46}$$

Note that $\det T_i(\mu) = 1$ for all i. The monodromy matrix $T(\mu)$ is equal to $T(\mu) = T_{n+1}(\mu) \cdots T_1(\mu)$.

Proposition. *The Poisson brackets of the matrix elements of $T(\mu)$ are given by:*

$$\{T_1(\mu), T_2(\mu')\} = [r_{12}(\mu - \mu'), T_1(\mu)T_2(\mu')] \tag{6.47}$$

where the r-matrix is given by

$$r_{12}(\mu - \mu') = \frac{C_{12}}{\mu - \mu'}, \quad C_{12} = \sum_{ij} E_{ij} \otimes E_{ji}$$

Proof. We first prove this relation for each individual $T_i(\mu)$. Specifically,

$$\{T_{1,i}(\mu), T_{2,i}(\mu')\} = [r_{12}(\mu - \mu'), T_{1,i}(\mu)T_{2,i}(\mu')] \tag{6.48}$$

This is shown by a direct computation using the explicit formula (6.46) for $T_i(\mu)$. We then prove that if two Poisson-commuting matrices $T_i(\mu)$ and $T_j(\mu')$ satisfy (6.48) then so does their product $T_i(\mu)T_j(\mu')$. Indeed, using the fact that the Poisson bracket and the Lie bracket are both derivations,

$$\{T_{1,j}(\mu)T_{1,i}(\mu), T_{2,j}(\mu')T_{2,i}(\mu')\} = T_{1,j}(\mu)T_{2,j}(\mu')\{T_{1,i}(\mu), T_{2,i}(\mu')\}$$
$$+ \{T_{1,j}(\mu), T_{2,j}(\mu')\}T_{1,i}(\mu)T_{2,i}(\mu')$$
$$= [r_{12}(\mu - \mu'), T_{1,j}(\mu)T_{1,i}(\mu)T_{2,j}(\mu')T_{2,i}(\mu')]$$

The claim then follows by induction because $\{T_{1,i}(\mu), T_{2,j}(\mu')\} = 0$ if $i \neq j$ by the locality of $T_i(\mu)$. Note that this also implies the integrability of the Toda chain. ∎

Let $A(\mu)$, $B(\mu)$, $C(\mu)$ and $D(\mu)$ be the matrix elements of $T(\mu)$:

$$T(\mu) = \begin{pmatrix} A(\mu) & B(\mu) \\ C(\mu) & D(\mu) \end{pmatrix}$$

In terms of the matrix elements of $\widehat{T}(\mu)$ introduced in the previous section, $A(\mu) = \widehat{A}(\mu)$, $D(\mu) = \widehat{D}(\mu)$ but $B(\mu) = (d_0 d_1)\widehat{B}(\mu)$ and $C(\mu) = \widehat{C}(\mu)/(d_0 d_1)$. In particular one finds (recalling that $\widehat{B}(\mu) = (d_0/d_1)\mu^n + \cdots$):

$$B(\mu) = d_0^2 \left(\mu^n - \mu^{n-1}(\sum_{j=1}^{n} p_j) + \cdots \right)$$

This is a polynomial of degree n with n zeroes, which we denote by μ_{γ_k}, $k = 1, \ldots, n$:

$$B(\mu) = d_0^2 \prod_{k=1}^{n} (\mu - \mu_{\gamma_k})$$

As explained in the previous section, these zeroes are the μ-coordinates of the poles of the Baker functions ψ_i.

Proposition. *Let μ_{γ_k} be the zeroes of $B(\mu)$ and λ_{γ_k} be the values of $D(\mu)$ at $\mu = \mu_{\gamma_k}$. The n points $(\lambda_{\gamma_k}, \mu_{\gamma_k})$ are the points of the dynamical divisor.*

$$B(\mu_{\gamma_k}) = 0 \quad \text{and} \quad \lambda_{\gamma_k} = D(\mu_{\gamma_k}) \tag{6.49}$$

These parameters obey the following Poisson brackets:

$$\{\mu_{\gamma_k}, \mu_{\gamma_{k'}}\} = \{\lambda_{\gamma_k}, \lambda_{\gamma_{k'}}\} = 0, \quad \{\lambda_{\gamma_k}, \mu_{\gamma_{k'}}\} = \lambda_{\gamma_k} \delta_{kk'} \tag{6.50}$$

($\log \lambda_{\gamma_k}, \mu_{\gamma_k}$) form a set of canonical coordinates.

Proof. Recall that the matrix $T(\mu)$ is lower triangular at $\mu = \mu_{\gamma_k}$, since $B(\mu_{\gamma_k}) = 0$. Therefore, at $\mu = \mu_{\gamma_k}$ we have: $1 = \det T(\mu_{\gamma_k}) = A(\mu_{\gamma_k})D(\mu_{\gamma_k})$ and $2t(\mu_{\gamma_k}) = A(\mu_{\gamma_k}) + D(\mu_{\gamma_k})$, hence the points $(\lambda_{\gamma_k}, \mu_{\gamma_k})$ are on the spectral curve:

$$\lambda_{\gamma_k} + \lambda_{\gamma_k}^{-1} = 2t(\mu_{\gamma_k}) \tag{6.51}$$

The Poisson brackets (6.47) for $T(\mu)$ imply the following:

$$\{A(\mu), A(\mu')\} = \{B(\mu), B(\mu')\} = \{C(\mu), C(\mu')\} = \{D(\mu), D(\mu')\} = 0$$

$$\{A(\mu), B(\mu')\} = \frac{A(\mu)B(\mu') - B(\mu)A(\mu')}{\mu - \mu'} \tag{6.52}$$

$$\{A(\mu), D(\mu')\} = \frac{C(\mu)B(\mu') - B(\mu)C(\mu')}{\mu - \mu'} \tag{6.53}$$

$$\{B(\mu), D(\mu')\} = \frac{D(\mu)B(\mu') - B(\mu)D(\mu')}{\mu - \mu'} \tag{6.54}$$

The first equation directly implies that the μ_{γ_k} Poisson commute.

Let $P(\mu)$ be a polynomial in μ with coefficients functions on phase space, and F be an arbitrary function on phase space. Then the Poisson bracket between F and the value of $P(\mu)$ at μ_{γ_k} is:

$$\{F, P(\mu_{\gamma_k})\} = \{F, P(\mu)\}\big|_{\mu = \mu_{\gamma_k}} + \{F, \mu_{\gamma_k}\}\, \partial_\mu P(\mu_{\gamma_k}) \tag{6.55}$$

We apply this to $F = D(\mu)$ and $P(\mu) = B(\mu)$. Since $B(\mu_{\gamma_k}) = 0$, we get:

$$0 = \{D(\mu), B(\mu_{\gamma_k})\} = \{D(\mu), B(\mu')\}\big|_{\mu' = \mu_{\gamma_k}} + \{D(\mu), \mu_{\gamma_k}\}\, B'(\mu_{\gamma_k})$$

where $B'(\mu) = \partial_\mu B(\mu)$. We evaluate the first term by using eq. (6.54), obtaining

$$\{D(\mu), \mu_{\gamma_k}\} = \frac{1}{\mu - \mu_{\gamma_k}} \frac{\lambda_{\gamma_k}}{B'(\mu_{\gamma_k})} B(\mu)$$

Next we apply the same formula to $F = \mu_{\gamma_k}$ and $P = D$, evaluated at $\mu = \mu_{\gamma_{k'}}$. Since $\{\mu_{\gamma_k}, \mu_{\gamma_{k'}}\} = 0$, one gets

$$\{\mu_{\gamma_k}, \lambda_{\gamma_{k'}}\} = \{\mu_{\gamma_k}, D(\mu_{\gamma_{k'}})\} = \{\mu_{\gamma_k}, D(\mu)\}|_{\mu = \mu_{\gamma_{k'}}}$$

So we have to evaluate $B(\mu)/(\mu - \mu_{\gamma_k})$ at $\mu = \mu_{\gamma_{k'}}$, which gives $\delta_{kk'} B'(\mu_{\gamma_k})$. The equation $\{\lambda_{\gamma_{k'}}, \mu_{\gamma_k}\} = \delta_{kk'} \lambda_{\gamma_k}$ follows.

Finally, we need to prove $\{\lambda_{\gamma_k}, \lambda_{\gamma_{k'}}\} = 0$. We compute $\{D(\mu_{\gamma_k}), D(\mu_{\gamma_{k'}})\}$ by expanding this expression completely using eq. (6.55), $\{D(\mu), D(\mu')\} = 0$ and $\{\mu_{\gamma_k}, \mu_{\gamma_{k'}}\} = 0$. We get:

$$\{\lambda_{\gamma_k}, \lambda_{\gamma_{k'}}\} = \partial_\mu D(\mu_{\gamma_{k'}})\{D(\mu), \mu_{\gamma_{k'}}\}|_{\mu = \mu_{\gamma_k}} - \partial_\mu D(\mu_{\gamma_k})\{D(\mu), \mu_{\gamma_k}\}|_{\mu = \mu_{\gamma_{k'}}}$$

We have already seen that each of these two terms vanishes when $k \neq k'$ and they obviously cancel each other when $k = k'$. ∎

In particular, the symplectic two-form ω can be written as:

$$\omega = \sum_{k=1}^{n} \frac{\delta \lambda_{\gamma k}}{\lambda_{\gamma k}} \wedge \delta \mu_{\gamma k}$$

where δ denote the differential on the phase space.

The fact that the equations of motion linearize on the Jacobian variety of the spectral curve may also be derived in this framework. The generating function for the conserved Hamiltonians is the trace of the monodromy matrix $\mathrm{Tr}\,(T(u)) = 2t(u) = A(u) + D(u)$. Thus the equations of motion of any function F on phase space for this generic Hamiltonian $t(u)$ are:

$$\dot{F} \equiv \{t(u), F\}$$

Proposition. *The equations of motion, relative to the Hamiltonian $t(u)$, of the zeroes $\mu_{\gamma k}$ of $B(\mu)$ are:*

$$\dot{\mu}_{\gamma k} = \{t(u), \mu_{\gamma k}\} = \frac{1}{2}(A(\mu_{\gamma k}) - D(\mu_{\gamma k})) \frac{B(u)}{(u - \mu_{\gamma k})B'(\mu_{\gamma k})} \qquad (6.56)$$

Their linearization, under the Abel map $\mathcal{A}$, is given by the following relations:

$$\dot{\mathcal{A}}_j = \sum_{k=1}^{n} \frac{\mu_{\gamma k}^j}{\sqrt{t^2(\mu_{\gamma k}) - 1}} \dot{\mu}_{\gamma k} = u^j, \qquad 0 \leq j \leq n-1 \qquad (6.57)$$

<u>Proof.</u> The Poisson bracket $\{t(u), \mu_{\gamma k}\}$ can be computed in the same way as in the previous proposition. One obtains:

$$\dot{\mu}_{\gamma k} = \{t(u), \mu_{\gamma k}\} = \frac{B(u)}{2(u - \mu_{\gamma k})B'(\mu_{\gamma k})}(A(\mu_{\gamma k}) - D(\mu_{\gamma k})) \qquad (6.58)$$

Hence $\dot{\mu}_k$ is a polynomial in u of degree $n-1$ which vanishes at the points $\mu_{\gamma k'}$ for $k' \neq k$ and takes the value $\frac{1}{2}(A(\mu_{\gamma k}) - D(\mu_{\gamma k})) = \sqrt{t^2(\mu_{\gamma k}) - 1}$ for $k' = k$. On a hyperelliptic curve the Abelian differentials are the

$$\omega_j = \frac{\mu^j}{\sqrt{t^2(\mu) - 1}} \, d\mu, \qquad 0 \leq j \leq n-1$$

so that the time derivative of the Abel map $\mathcal{A}_j \equiv \sum_k \int^{P_k} \omega_j$ is given by:

$$\dot{\mathcal{A}}_j = \sum_k \frac{\mu_{\gamma k}^j \dot{\mu}_{\gamma k}}{\sqrt{t^2(\mu_{\gamma k}) - 1}}$$

Replacing $\dot{\mu}_{\gamma k}$ by its value given in eq. (6.58), we see that $\dot{\mathcal{A}}_j$ is a polynomial in u of degree at most $n-1$. If one evaluates it at $u = \mu_{\gamma l}$, only the term $k = l$ contributes in the above sum, which gives $\mu_{\gamma l}^j$. Hence the polynomial itself is u^j. ∎

6.8 Reality conditions

Up to now the dynamical variables of the Toda chain were complex valued. We now come to the important and generally difficult question of choosing a real slice in the complexified phase space. This complex phase space is a fibred space with basis the action variables and fibre the Jacobian variety of the corresponding spectral curve. While it is easy to take real action variables, it is less trivial to choose a proper real slice on the Jacobian which will be identified with the Liouville torus.

Let us first remark that the equation of the spectral curve, eq. (6.8), shows that it is a hyperelliptic curve with hyperelliptic involution given by $(\lambda, \mu) \to (1/\lambda, \mu)$. Hence the branch points, which are invariant under this involution, are located at $\lambda = 1/\lambda$, i.e. $\lambda = \pm 1$. For such values of λ the Lax matrix $L(\lambda)$, eq. (6.4), is *symmetric*. If, moreover, the dynamical variables p_i, q_i, are real, the matrices $L(\lambda = \pm 1)$ are real symmetric, hence each have $(n + 1)$ real eigenvalues which are precisely the location of the branch points. We denote by $\beta_0 < \beta_1 \cdots < \beta_{2n+2}$ these $2(n + 1)$ *real* branch points which completely characterize the curve. This implies in particular that the action variables, i.e. the coefficients of eq. (6.8), are real.

Writing the equation of the spectral curve as:

$$\lambda + \frac{1}{\lambda} = 2t(\mu) \tag{6.59}$$

where $t(\mu)$ is a polynomial of degree $(n + 1)$ with real coefficients, the $2n + 2$ branch points are located at $t(\mu) = \pm 1$. It follows that the graph of $t(\mu)$ has the form shown in Fig. 6.1 (here assuming $n + 1$ even):

For $n+1$ odd, the left branch (where $\mu \to -\infty$) goes to $-\infty$. Note that for real μ the two values of λ such that $\lambda^2 - 2t(\mu)\lambda + 1 = 0$ are either real or complex conjugated with modulus one. Borrowing the terminology of solid state physics, the first case is called the forbidden zone, while the second case is called the allowed zone. We see that the forbidden zones correspond to the intervals $[\beta_i, \beta_{i+1}]$ for $i = 1, 3, \ldots, 2n - 1$ (for both n even and odd), to which is added one extra zone passing through $\mu = \infty$ given by $\mu < \beta_0$ or $\mu > \beta_{2n+2}$. So there are n compact forbidden zones.

We now have to locate the points of the dynamical divisor in agreement with the reality conditions. Recall that they are given by the points $B(\mu) = 0$ and $\lambda = D(\mu)$. We show that the zeroes of the polynomial $B(\mu)$ of degree n are real so that the corresponding λ are also real, since $D(\mu)$ has real coefficients. This immediately implies that the zeroes of $B(\mu)$ lie in the forbidden zones.

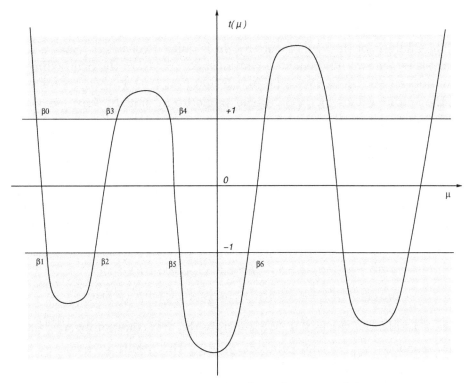

Fig. 6.1. The graph of the function $t(\mu)$

Proposition. *The points of the dynamical divisor have real coordinates, and there is exactly one such point in each of the n forbidden zones* $[\beta_i, \beta_{i+1}]$ *for* $i = 1, 3, \ldots, 2n - 1$.

<u>Proof.</u> We first show that the zeroes of $B(\mu)$ are real. This is because the zeroes of $B(\mu)$ are also the poles of the eigenvector of $L(\lambda)$ normalized by $\psi_0 = 1$. The eigenvector equation being:

$$\begin{pmatrix} p_0 - \mu & a_0 & \ldots & \lambda^{-1}a_n \\ a_0 & p_1 - \mu & \ldots & 0 \\ \vdots & \vdots & & \vdots \\ \lambda a_n & 0 & \ldots & p_n - \mu \end{pmatrix} \begin{pmatrix} \psi_0 \\ \psi_1 \\ \vdots \\ \psi_n \end{pmatrix} = 0$$

the poles of the normalized Ψ are included in the zero set of the minor of $(p_0 - \mu)$, see Chapter 5. This minor is the characteristic equation of a real symmetric $n \times n$ matrix, hence has n real eigenvalues μ_{γ_i}.

To show that there is just one zero of $B(\mu)$ in each $[\beta_i, \beta_{i+1}]$, we show that $B(\mu)$ has the same sign as $a_0 dt(\mu)/d\mu$ in the *allowed* zones, hence changes sign between β_i and β_{i+1}, i odd, because so does $dt(\mu)/d\mu$. Since

$B(\mu)$ has n zeroes the only possibility is that it has exactly one zero in each forbidden zone.

The analysis of the sign of $a_0 dt(\mu)/d\mu$ is quite involved. It rests on the analysis of the two fundamental solutions, eq. (6.44), of the Schroedinger equation, eq. (6.12). In particular, recall that $(\chi_{n+1}(\mu), \chi_{n+2}(\mu))$ is related to the initial values (χ_0, χ_1) by the monodromy matrix $\widehat{T}(\mu)$, so that we have $\widehat{A}(\mu) = \chi_{n+1}^{(0)}(\mu)$, $\widehat{B}(\mu) = \chi_{n+1}^{(1)}(\mu)$, $\widehat{C}(\mu) = \chi_{n+2}^{(0)}(\mu)$, $\widehat{D}(\mu) = \chi_{n+2}^{(1)}(\mu)$, and $2t(\mu) = \chi_{n+1}^{(0)}(\mu) + \chi_{n+2}^{(1)}(\mu)$. Moreover, since $\det \widehat{T}(\mu) = 1$, we get the Wronskian relation:

$$\chi_{n+1}^{(0)}(\mu)\chi_{n+2}^{(1)}(\mu) - \chi_{n+2}^{(0)}(\mu)\chi_{n+1}^{(1)}(\mu) = 1 \qquad (6.60)$$

We then consider two solutions $\chi_i^{(j)}(\mu)$ and $\chi_i^{(j')}(\mu')$ of the Schroedinger equation corresponding to different values μ and μ'.

$$a_{i-1}\chi_{i-1}^{(j)}(\mu) + p_i\chi_i^{(j)}(\mu) + a_i\chi_{i+1}^{(j)}(\mu) = \mu\chi_i^{(j)}(\mu)$$
$$a_{i-1}\chi_{i-1}^{(j')}(\mu') + p_i\chi_i^{(j')}(\mu') + a_i\chi_{i+1}^{(j')}(\mu') = \mu'\chi_i^{(j')}(\mu')$$

Here j, j' take the values $0, 1$. We multiply the first equation by $\chi_i^{(j')}(\mu')$, the second by $\chi_i^{(j)}(\mu)$ and subtract. Adding the resulting equations for $i = 1, 2, \ldots, n+1$ we get:

$$a_0\left(\chi_0^{(j)}(\mu)\chi_1^{(j')}(\mu') - \chi_1^{(j)}(\mu)\chi_0^{(j')}(\mu')\right)$$
$$-a_{n+1}\left(\chi_{n+1}^{(j)}(\mu)\chi_{n+2}^{(j')}(\mu') - \chi_{n+2}^{(j)}(\mu)\chi_{n+1}^{(j')}(\mu')\right)$$
$$= (\mu - \mu')\left(\chi^{(j)}(\mu), \chi^{(j')}(\mu')\right)$$

where we have denoted $(\chi^{(j)}(\mu), \chi^{(j')}(\mu')) = \sum_{i=1}^{n+1} \chi_i^{(j)}(\mu)\chi_i^{(j')}(\mu')$. Consider the four above equations for $(j, j') = (0,0), (1,1), (0,1), (1,0)$ and multiply them respectively by $\chi_{n+1}^{(1)}(\mu)$, $-\chi_{n+2}^{(0)}(\mu)$, $\chi_{n+2}^{(1)}(\mu)$ and $-\chi_{n+1}^{(0)}(\mu)$. Adding them and taking into account eq. (6.60), we get:

$$2a_0\frac{t(\mu) - t(\mu')}{\mu - \mu'} = \widehat{B}(\mu)(\chi^{(0)}(\mu), \chi^{(0)}(\mu')) - \widehat{C}(\mu)(\chi^{(1)}(\mu), \chi^{(1)}(\mu')) +$$
$$\widehat{D}(\mu)(\chi^{(0)}(\mu), \chi^{(1)}(\mu')) - \widehat{A}(\mu)(\chi^{(1)}(\mu), \chi^{(0)}(\mu'))$$

We can now set $\mu = \mu'$ and obtain our final relation:

$$2a_0\frac{dt(\mu)}{d\mu} = \widehat{B}(\mu)\left(\chi^{(0)}(\mu), \chi^{(0)}(\mu)\right) - \widehat{C}(\mu)\left(\chi^{(1)}(\mu), \chi^{(1)}(\mu)\right)$$
$$+(\widehat{D}(\mu) - \widehat{A}(\mu))\left(\chi^{(0)}(\mu), \chi^{(1)}(\mu)\right) \qquad (6.61)$$

The right-hand side is a sum over $i = 1, \ldots, n+1$ of quadratic forms in the variables $\chi_i^{(0)}(\mu)$ and $\chi_i^{(1)}(\mu)$, with discriminant $(\widehat{D}(\mu) - \widehat{A}(\mu))^2 + 4\widehat{B}(\mu)\widehat{C}(\mu) = 4(t^2(\mu) - 1)$. In the allowed zones we have $t^2(\mu) - 1 < 0$ and the quadratic form has a definite sign, which is the sign of $\widehat{B}(\mu)$. Equation (6.61) shows that this is also the sign of $a_0 dt(\mu)/d\mu$, as asserted above.   ∎

We can now describe the real part of the spectral curve. From eq. (6.59), we see that when $t(\mu)^2 > 1$, i.e. in the forbidden zones, we have two real solutions for λ, which are inverse to each other. For $\mu = \beta_j$, these two solutions coincide. In the allowed zones, $t(\mu)^2 < 1$, there is no real solution for λ. Hence the real slice of the spectral curve has n components $\mathcal{C}_k$ at finite distance, and one component which extends to ∞. See Fig. 6.2.

The dynamical divisor has n points γ_k with just one point in each of the components $\mathcal{C}_k$ at finite distance.

As time goes on, each of the points γ_k runs along the cycle $\mathcal{C}_k$, hence the whole motion lies on the real torus $\mathcal{C}_1 \times \mathcal{C}_2 \times \cdots \times \mathcal{C}_n$, which is the Liouville torus. For the Hamiltonian $t(u)$, the time evolution of μ_{γ_k} is

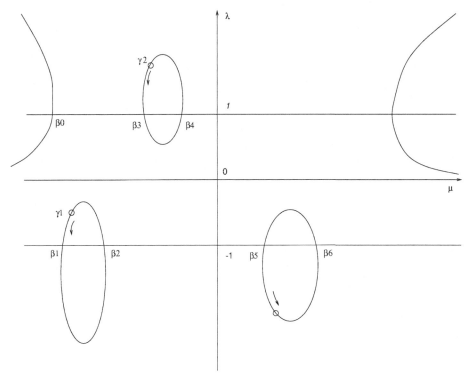

Fig. 6.2. The real slice of the spectral curve.

given by eq. (6.58) and the one of λ_{γ_k} is given similarly by:

$$\dot{\lambda}_k = -\frac{1}{u - \mu_{\gamma_k}} \frac{B(u)}{B'(\mu_{\gamma_k})} \lambda_{\gamma_k} t'(\mu_{\gamma_k})$$

Hence, when the point γ_k hits the line $\lambda = \pm 1$, we have $\dot{\mu}_k = 0$ and $\dot{\lambda}_k \neq 0$, so that the point γ_k continues in the same direction and loops around $\mathcal{C}_k$.

It is interesting to relate the Liouville torus to the real slice of the Jacobian variety of the spectral curve. One can define an antiholomorphic involution of the Jacobian by sending $\sum \gamma_k$ to $\sum \overline{\gamma}_k$, where the image of $\gamma = (\lambda, \mu)$ is $\overline{\gamma} = (\overline{\lambda}, \overline{\mu})$, which also lies on the spectral curve since $t(\mu)$ has real coefficients. A real slice of the Jacobian can be defined as the set of fixed points of this involution. Note that $\sum \gamma_k$ is invariant if the γ_k are real or occur in complex conjugate pairs. The various combinations of these two possibilities define several connected components of the real slice of the Jacobian. The Liouville torus is just one of these components. Note that we have discussed the reality condition for the dynamical variables a_i and p_i, but one should further ensure that the variables a_i are positive to get real q_i.

Alternatively, one can view the Liouville torus as some choice of a-cycles on the spectral curve. Consider it as a two-sheeted covering of the complex μ-plane, with cuts between branch points as shown in Fig. 6.3.

Here the a-cycles are drawn on the first sheet, so in the limit where the ellipses go to the segments $[\beta_i, \beta_{i+1}]$ we see that λ goes to the two real roots of the spectral equation, thereby producing the cycles in the previous drawing. The b-cycles are also drawn. Starting from ∞, they cut the corresponding a-cycle once, then pass on the second sheet through the cut, and return to ∞ without cutting any more of the a-cycles.

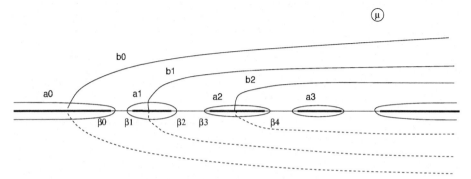

Fig. 6.3. Γ as the cut μ-plane.

References

[1] H. Flaschka, The Toda lattice I: Existence of integrals. *Phys. Rev.* **B9** (1974) 1924.

[2] H. Flaschka, The Toda lattice II: Inverse scattering solution. *Prog. Theor. Phys.* **51** (1974) 703–716.

[3] P. van Moerbeke, The spectrum of Jacobi Matrices. *Invent. Math.* (1976) 45–81.

[4] P. van Moerbeke and D. Mumford, The spectrum of difference operators and algebraic curves. *Acta. Math.* **143** (1979) 93–154.

[5] P. van Moerbeke, About Isospectral deformations of discrete Laplacians. *Lecture Notes in Math.* **755** (1979) 313–370.

[6] E. Sklyanin, The quantum Toda chain. *Lect. Notes Physics* **226** (1985) 196–233.

[7] B. Dubrovin, I. Krichever and S. Novikov, *Integrable Systems. I.* Encyclopedia of Mathematical Sciences, Dynamical Systems IV. Springer (1990).

7

The Calogero–Moser model

The elliptic Calogero–Moser model provides an example in which the Lax matrix is not a rational function of the spectral parameter but lives on an elliptic curve. As pointed out in Chapter 3, integrable systems whose spectral parameter belongs to Riemann surfaces of higher genus are highly non-generic. Furthermore, this model gives an example of an integrable system whose spectral curve is non-hyperelliptic. Nevertheless, most of the results obtained in the rational case extend to this case with slight but interesting adaptations. A special feature of this model is that its r-matrix explicitly contains dynamical variables. Another remarkable fact is the relation between doubly periodic solutions of the KP hierarchy and the Calogero–Moser model. Finally, we show that this model is a particular example of a general construction, due to Hitchin, which allows us to construct models with spectral parameter lying on higher genus Riemann surfaces.

7.1 The spin Calogero–Moser model

The Calogero–Moser model consists of N identical particles on a line, at positions q_i and momenta p_i, with pairwise interactions and Hamiltonian:

$$H = \frac{1}{2} \sum_i p_i^2 + \frac{\gamma^2}{2} \sum_{\substack{i,j=1 \\ i \neq j}}^{N} \frac{1}{(q_i - q_j)^2}$$

This dynamical system is integrable, moreover it remains integrable when the potential is replaced by an elliptic potential, i.e. $1/q^2 \rightarrow \wp(q)$ with $\wp(q)$ the Weierstrass elliptic function defined on a two dimensional torus with periods ω_1 and ω_2.

It is rewarding to consider a slightly generalized model, the so-called spin Calogero–Moser model, which contains extra dynamical spin variables. Let us introduce a set of dynamical variables (q_i, p_i) and (f_{ij}), $i, j = 1, \ldots, N$ together with the Poisson brackets:

$$\{p_i, q_j\} = \delta_{ij} \tag{7.1}$$

$$\{f_{ij}, f_{kl}\} = -\delta_{jk} \, f_{il} + \delta_{li} \, f_{kj} \tag{7.2}$$

The Poisson bracket eq. (7.2) is a Kostant–Kirillov bracket for the coadjoint action of the group $GL(N)$. The Hamiltonian reads

$$H = \frac{1}{2} \sum_{i=1}^{N} p_i^2 - \frac{1}{2} \sum_{\substack{i,j=1 \\ i \neq j}}^{N} f_{ij} f_{ji} V(q_i - q_j) \tag{7.3}$$

where the potential $V(q) \equiv \wp(q)$. The equations of motion are easily derived:

$$\dot{q}_i = p_i, \quad \dot{f}_{ii} = 0$$

$$\dot{p}_i = \sum_{\substack{j=1 \\ j \neq i}}^{N} f_{ij} \, f_{ji} \, V'(q_{ij}), \quad q_{ij} \equiv q_i - q_j$$

$$\dot{f}_{ij} = \sum_{\substack{k=1 \\ k \neq i,j}}^{N} f_{ik} \, f_{kj} \, [V(q_{ik}) - V(q_{jk})] + (f_{ii} - f_{jj}) f_{ij} V(q_{ij}), \quad i \neq j \tag{7.4}$$

From (7.4) we see that f_{ii} are integrals of motion, and we can restrict the system to the submanifold

$$f_{ii} = \alpha \tag{7.5}$$

where α is a constant *independent* of i. In this case the last term in eqs. (7.4) vanishes. These constraints are related to a Hamiltonian reduction, and we will see that it is this reduced system which admits a Lax pair and is integrable.

Let us count the number of degrees of freedom. This is not a completely trivial matter due to the degeneracy of the Poisson bracket eq. (7.2), and the necessary reduction to the manifold $f_{ii} = \alpha$.

The symplectic leaves of the Kostant–Kirillov bracket, eq. (7.2), are the coadjoint orbits of $GL(N)$ acting on the matrix $F = (f_{ij})$ by conjugation. These orbits are generically characterized by the eigenvalues of the matrix F which are in the centre of the Poisson bracket. Here we shall consider matrices F of rank l, with l different non-vanishing eigenvalues. Moreover, we shall assume that the matrix F is diagonalizable, so that the orbit is of the form

$$\{C\nu C^{-1}|C \in GL(N)\}, \quad \text{with } \nu = \text{Diag}(\nu_1, \ldots, \nu_l, 0, \ldots, 0) \quad (7.6)$$

The Hamiltonian eq. (7.3) is not invariant under the above $GL(N)$ but it is preserved by special subgroups. First we have the discrete subgroup of permutation matrices, i.e. the Weyl group of $GL(N)$, which simply operates by permutation of the N indices i. More importantly, we have the group of diagonal matrices, i.e. the Cartan torus, which operates by:

$$f_{ij} \to d_i^{-1} f_{ij} d_j \quad (7.7)$$

This action preserves the Hamiltonian which only depends on $f_{ij}f_{ji}$. We consider the dynamical system on an orbit of rank l, reduced under this diagonal action whose generators will be shown to be the f_{ii}.

Proposition. *The dimension of the reduced phase space $\mathcal{M}$ is*

$$\dim \mathcal{M} = 2\left[Nl - \frac{l(l+1)}{2} + 1\right] \quad (7.8)$$

<u>Proof.</u> The tangent space to the orbit (7.6) at $F = (f_{ij})$ is the set of matrices $U = [F, X]$ for any $X \in gl(N)$. In a basis where F is diagonal this equation reads $U_{ij} = (\nu_i - \nu_j)X_{ij}$, hence U_{ij} vanishes when $\nu_i = \nu_j$ but is otherwise arbitrary. So the dimension of the orbit is $2Nl - l^2 - l$. The action of the subgroup of diagonal matrices induces a fibring of the orbits with fibres of dimension $N-1$ because the identity does not act. The moment associated with this action is the collection of diagonal elements f_{ii}, that is $N - 1$ non-trivial moments since on the orbit, the eigenvalues of F being fixed, so is $\text{Tr}(F)$. Indeed, under infinitesimal action $d_i = 1 + \epsilon_i$, f_{ij} changes as $\delta f_{ij} = (\epsilon_j - \epsilon_i)f_{ij}$, and if $\mathcal{P} = \sum_i \epsilon_i f_{ii}$, we have $\{\mathcal{P}, f_{ij}\} = (\epsilon_j - \epsilon_i)f_{ij}$, so $\mathcal{P}$ is the corresponding Hamiltonian. We consider the reduced dynamical system obtained by first fixing the moments to a common value $f_{ii} = \alpha$, and then quotienting by the stabilizer of this moment, which is the whole diagonal group.

We can now count the number of degrees of freedom. We have $2N$ degrees of freedom for the q_i, p_i, plus $2Nl - l^2 - l$ for the orbit, minus $2(N - 1)$ due to the Hamiltonian reduction, which ends up with a phase space of dimension $2(Nl - l(l+1)/2 + 1)$. ∎

7.2 Lax pair

The first step in proving that the reduced spin Calogero–Moser model is integrable consists of finding a Lax pair formulation. Here we will just give the Lax pair, but in later sections we will explain methods to derive it.

We will need the Lamé function Φ:

$$\Phi(q, \lambda) = \frac{\sigma(\lambda - q)}{\sigma(\lambda)\sigma(q)} e^{\zeta(\lambda)q} \tag{7.9}$$

where σ and ζ are defined in Chapter 15. It is an elliptic function of the parameter λ and satisfies the equation

$$\left(\frac{d^2}{dx^2} - 2\wp(x) \right) \Phi(x, \lambda) = \wp(\lambda)\Phi(x, \lambda) \tag{7.10}$$

The Lamé function is used to construct the Lax pair of the spin Calogero–Moser model.

Proposition. *The equations of motion of the spin Calogero–Moser system are equivalent to the Lax equation*

$$\dot{L}(\lambda) = [M(\lambda), L(\lambda)] \tag{7.11}$$

where the Lax matrices, with spectral parameter λ, are given by:

$$L_{ij}(t, \lambda) = \dot{q}_i \delta_{ij} + (1 - \delta_{ij}) f_{ij} \Phi(q_i - q_j, \lambda) \tag{7.12}$$
$$M_{ij}(t, \lambda) = -(1 - \delta_{ij}) f_{ij} \Phi'(q_i - q_j, \lambda) \tag{7.13}$$

The prime in eq. (7.13) refers to the derivative with respect to q.

Proof. The Lax equation (7.11) reads:

$$\ddot{q}_i \delta_{ij} + (1 - \delta_{ij}) \left[\dot{f}_{ij} \Phi(q_{ij}, \lambda) + f_{ij} \dot{q}_{ij} \Phi'(q_{ij}, \lambda) \right] = (1 - \delta_{ij}) f_{ij} \dot{q}_{ij} \Phi'(q_{ij}, \lambda)$$
$$+ \sum_{k \neq i, j} f_{ik} f_{kj} \left[\Phi(q_{ik}, \lambda) \Phi'(q_{kj}, \lambda) - \Phi'(q_{ik}, \lambda) \Phi(q_{kj}, \lambda) \right]$$

This reduces to the equations of motion, in the case where $f_{ii} = \alpha$, if we use an identity satisfied by the Lamé function:

$$\Phi'(x, \lambda)\Phi(y, \lambda) - \Phi'(y, \lambda)\Phi(x, \lambda) = [\wp(y) - \wp(x)]\Phi(x + y, \lambda) \tag{7.14}$$

which also implies, by taking the limit $y \to -x$:

$$\Phi'(x, \lambda)\Phi(-x, \lambda) - \Phi'(-x, \lambda)\Phi(x, \lambda) = -\wp'(x)$$

To show eq. (7.14) we compare the analyticity and monodromy properties of both sides of the equation in the x variable, using the properties given in Chapter 15. ∎

By a similar method one shows that the Lamé function obeys the identity:

$$\Phi(q, \lambda)\Phi(-q, \lambda) = \wp(\lambda) - \wp(q) \qquad (7.15)$$

hence

$$\frac{1}{2}\text{Tr}\, L^2 = H + \wp(\lambda) \sum_{i<j} f_{ij} f_{ji}$$

where H is the Hamiltonian of the spin Calogero–Moser model, and $\sum_{i \neq j} f_{ij} f_{ji} = \text{Tr}\, F^2 - N\alpha^2$ is an orbit invariant in the centre of the Poisson algebra.

Let us comment on the trigonometric and rational limits of the above formulae. The trigonometric limit is obtained when one of the periods $\omega \to \infty$. We choose the other one as $i\pi$. In this limit the function Φ becomes:

$$\Phi(q, \lambda) \to (\coth q - \coth \lambda)\, e^{q \coth \lambda}$$

The exponential factor in $\Phi(q, \lambda)$ comes from the factor $\exp(\zeta(\lambda)q)$ which is necessary in the elliptic case to ensure the double periodicity of $\Phi(q, \lambda)$ in λ. In the trigonometric case, however, this exponential factor can be eliminated by performing a similarity transformation on $L(\lambda)$ without affecting the periodicity properties of the matrix elements of $L(\lambda)$. So we may define

$$L_{\text{trigo}}(\lambda) = \text{Diag}\left(e^{-q_i \coth \lambda}\right) \lim_{\omega \to \infty} L_{\text{elliptic}}(\lambda)\, \text{Diag}\left(e^{q_i \coth \lambda}\right)$$

$$= \sum_i p_i E_{ii} + \sum_{i \neq j} f_{ij} (\coth q_{ij} - \coth \lambda) E_{ij} \qquad (7.16)$$

The potential $V(q)$ becomes $V(q) = 1/\sinh^2(q)$.

The rational limit is obtained straightforwardly from the trigonometric limit by sending the second period $\omega' \to \infty$. The functions $V(q)$ and $\Phi(q, \lambda)$ become

$$V(q) \to \frac{1}{q^2}, \quad \Phi(q, \lambda) \to \frac{1}{q} - \frac{1}{\lambda},$$

Remark. The diagonal action in eq. (7.7) is equivalent to conjugation of the Lax matrix by a diagonal matrix. The Hamiltonian reduction we have performed is similar to the one appearing in the general rational case, see Chapter 5.

7.3 The r-matrix

In this section we compute the r-matrix associated with the Lax matrix

$$L(\lambda) = \sum_i p_i E_{ii} + \sum_{ij} \Phi(q_{ij}, \lambda)\, f_{ij}\, E_{ij}. \qquad (7.17)$$

One should emphasize that it is only the reduced system ($f_{ii} = \alpha$) which is integrable. The Lax matrix (7.17) is not a function on the reduced phase space and so is not expected to have an r-matrix in the usual sense. This accounts for the extra term in eq. (7.18) below:

Proposition. *The Poisson bracket of the Lax matrix eq. (7.17) is given by:*

$$\{L_1(\lambda), L_2(\mu)\} = [r_{12}(\lambda, \mu), L_1(\lambda)] - [r_{21}(\mu, \lambda), L_2(\mu)] + [\mathcal{D}, r_{12}(\lambda, \mu)] \tag{7.18}$$

where $\mathcal{D} = \sum_i f_{ii} \frac{\partial}{\partial q_i}$ and the r-matrix is expressed as:

$$r_{12}(\lambda, \mu) = a(\lambda, \mu) \sum_i E_{ii} \otimes E_{ii} - \sum_{ij} b(q_{ij}, \lambda, \mu) \, E_{ij} \otimes E_{ji} \tag{7.19}$$

with

$$a(\lambda, \mu) = \zeta(\lambda - \mu) - \zeta(\lambda) + \zeta(\mu)$$

$$b(q, \lambda, \mu) = \frac{\sigma(\lambda - \mu - q)}{\sigma(\lambda - \mu)\sigma(q)} e^{(\zeta(\lambda) - \zeta(\mu))q}$$

<u>Proof.</u> We compute the various terms of eq. (7.18), and collect them according to the number of equal matrix indices. We have (all written indices are different, and sums are implied):

$$\begin{aligned}
\{L_1(\lambda), L_2(\mu)\} = &-\Phi(q_{ij}, \lambda)\Phi(q_{ji}, \mu)(f_{ii} - f_{jj})E_{ij} \otimes E_{ji} \\
&+\Phi'(q_{ij}, \mu)f_{ij}(E_{ii} - E_{jj}) \otimes E_{ij} - \Phi'(q_{ij}, \lambda)f_{ij}E_{ij} \otimes (E_{ii} - E_{jj}) \\
&+\Phi(q_{ij}, \lambda)\Phi(q_{ki}, \mu)f_{kj}E_{ij} \otimes E_{ki} - \Phi(q_{ij}, \lambda)\Phi(q_{jk}, \mu)f_{ik}E_{ij} \otimes E_{jk}
\end{aligned}$$

Similarly one has, using the antisymmetry of the r-matrix:

$$\begin{aligned}
[r_{12}, L_1(\lambda) + L_2(\mu)] = &\\
&\left(a(\lambda, \mu)\Phi(q_{ij}, \mu) + b(q_{ji}, \lambda, \mu)\Phi(q_{ij}, \lambda)\right)f_{ij}(E_{ii} - E_{jj}) \otimes E_{ij} \\
&+\left(a(\lambda, \mu)\Phi(q_{ij}, \lambda) + b(q_{ij}, \lambda, \mu)\Phi(q_{ij}, \mu)\right)f_{ij}E_{ij} \otimes (E_{ii} - E_{jj}) \\
&+\left(b(q_{ij}, \lambda, \mu)\Phi(q_{kj}, \mu) - b(q_{ik}, \lambda, \mu)\Phi(q_{kj}, \lambda)\right)f_{kj}E_{ij} \otimes E_{ki} \\
&+\left(b(q_{kj}, \lambda, \mu)\Phi(q_{ik}, \lambda) - b(q_{ij}, \lambda, \mu)\Phi(q_{ik}, \mu)\right)f_{ik}E_{ij} \otimes E_{jk}
\end{aligned}$$

Finally, we have:

$$[\mathcal{D}, r_{12}] = -b'(q_{ij}, \lambda, \mu)(f_{ii} - f_{jj})E_{ij} \otimes E_{ji}$$

Using the relations $a(\mu, \lambda) = -a(\lambda, \mu)$ and $b(-q, \mu, \lambda) = -b(q, \lambda, \mu)$, we see that eq. (7.18) reduces to the three identities:

$$\Phi'(q, \mu) = a(\lambda, \mu)\Phi(q, \mu) + b(-q, \lambda, \mu)\Phi(q, \lambda)$$
$$\Phi(q, \lambda)\Phi(q', \mu) = b(q, \lambda, \mu)\Phi(q + q', \mu) - b(-q', \lambda, \mu)\Phi(q + q', \lambda)$$
$$b'(q, \lambda, \mu) = \Phi(q, \lambda)\Phi(-q, \mu)$$

The first identity, written in terms of Weierstrass functions, reads:

$$\frac{\sigma(\lambda - q)\sigma(\mu)\sigma(\lambda - \mu + q)}{\sigma(\lambda)\sigma(\mu - q)\sigma(\lambda - \mu)\sigma(q)} = \zeta(\mu - q) + \zeta(q) + \zeta(\lambda - \mu) - \zeta(\lambda)$$

One observes that both sides are elliptic functions of λ (and μ) and have the same poles and residues, hence are equal. The second identity reads:

$$\sigma(q + q')\sigma(\lambda - \mu)\sigma(\lambda - q)\sigma(\mu - q') =$$
$$\sigma(\lambda - \mu - q)\sigma(q')\sigma(\lambda)\sigma(\mu - q - q') + \sigma(\lambda - \mu + q')\sigma(\mu)\sigma(q)\sigma(\lambda - q - q')$$

which is true because both sides vanish at $\lambda = q$ and $\lambda = \mu$, are equal for $\lambda = 0$, and have the same monodromy properties when shifting λ by periods. Finally, the third identity is a consequence of the first one. ∎

Notice that $a(\lambda, \mu)$ and $b(q, \lambda, \mu)$ are true elliptic functions of both λ and μ. Note also that the r-matrix is antisymmetric, i.e. $r_{12}(\lambda, \mu) = -r_{21}(\mu, \lambda)$. The most remarkable feature of this r-matrix is that it depends on the dynamical variables q_{ij}, hence the name dynamical r-matrix.

Remark. Equation (7.18) holds for the non-reduced dynamical system (7.3). We see that $\operatorname{Tr} L^n(\lambda)$ are *not* in involution for this system. Indeed, we have:

$$\{\operatorname{Tr} L^n(\lambda), \operatorname{Tr} L^m(\mu)\}$$
$$= -nm \sum_{\substack{i,j=1 \\ i \neq j}}^{N} \Phi(q_{ij}, \lambda)\Phi(q_{ji}, \mu)(f_{ii} - f_{jj})[L^{n-1}(\lambda)]_{ij}[L^{m-1}(\mu)]_{ji} \quad (7.20)$$

As we have already seen, the f_{ii} are the moments of the diagonal group which acts on $L(\lambda)$ by conjugation. The quantities $\operatorname{Tr} L^n(\lambda)$ are invariant under this action. It follows that one can compute their reduced Poisson bracket on the manifold $(f_{ii} = \alpha)_{i=1,\dots,N}$ by just setting $f_{ii} = \alpha$ in eq. (7.20), see Chapter 14. Therefore $\operatorname{Tr} L^n(\lambda)$ are in involution for the reduced system. This proves integrability of the spin Calogero–Moser model. We will count the number of action variables later on.

Proposition. *The classical r-matrix, eq. (7.19), satisfies the identity* $YB = 0$ *with*

$$YB \equiv -\{L_1, r_{23}\} + \{L_2, r_{13}\} - \{L_3, r_{12}\} + [r_{12}, r_{13}] + [r_{12}, r_{23}] + [r_{13}, r_{23}] \tag{7.21}$$

Proof. It is done by direct calculation. It is more interesting, however, to see how this is related to the Jacobi identity. Let us call

$$Z_{12} = [\mathcal{D}, r_{12}], \quad \mathcal{D} = \sum_i f_{ii} \frac{\partial}{\partial q_i}$$

The Jacobi identity reads

$$[L_1, [r_{12}, r_{23}] + [r_{12}, r_{13}] + [r_{32}, r_{13}] + \{L_2, r_{13}\} - \{L_3, r_{12}\}] + \text{cyclic perm.}$$
$$+ [r_{23}, Z_{12}] + [r_{31}, Z_{23}] + [r_{12}, Z_{31}] - [r_{32}, Z_{13}] - [r_{13}, Z_{21}] - [r_{21}, Z_{32}]$$
$$+ \{L_1, Z_{23}\} + \{L_2, Z_{31}\} + \{L_3, Z_{12}\} = 0$$

Note that the term commuted with L_1 is not quite equal to the left-hand side of eq. (7.21). We want to show that the missing term, $\{L_1, r_{23}\}$, is produced by all the Z_{ij} contributions. Using $r_{12}(\lambda, \mu) = -r_{21}(\mu, \lambda)$, we find

$$[r_{23}, Z_{12}] + [r_{31}, Z_{23}] + [r_{12}, Z_{31}] - [r_{32}, Z_{13}] - [r_{13}, Z_{21}] - [r_{21}, Z_{32}]$$
$$= -[\mathcal{D}, [r_{12}, r_{13}] + [r_{12}, r_{23}] + [r_{13}, r_{23}]]$$

Moreover, we easily compute $\{L_1, \mathcal{D}\} = -\sum_k [L_1, E_{kk}]\partial_{q_k}$ so that:

$$\{L_1, Z_{23}\} = \{L_1, [\mathcal{D}, r_{23}]\} = [\mathcal{D}, \{L_1, r_{23}\}] + [\{L_1, \mathcal{D}\}, r_{23}]$$
$$= [\mathcal{D}, \{L_1, r_{23}\}] - [L_1, \{L_1, r_{23}\}]$$

The Jacobi identity becomes

$$[L_1 + L_2 + L_3, YB] - [\mathcal{D}, YB] = 0$$

where we used that YB is invariant under cyclic permutations of the indices $1, 2, 3$. As a result, the Jacobi identity is satisfied if eq. (7.21) holds, thereby giving a motivation to the computation showing that $YB = 0$. ∎

Remark. Let us comment on the trigonometric limit of these formulae. Using the Lax matrix, eq. (7.16), we find

$$\{L_1(\lambda), L_2(\mu)\} = [r_{12}(\lambda, \mu), L_1(\lambda)] - [r_{21}(\mu, \lambda), L_2(\mu)] + \sum_{\substack{i,j=1 \\ i \neq j}}^{N} \frac{f_{ii} - f_{jj}}{\sinh^2(q_i - q_j)} E_{ij} \otimes E_{ji}$$

where

$$r_{12}(\lambda, \mu) = \coth(\lambda - \mu) C_{12} - \sum_{\substack{i,j=1 \\ i \neq j}}^{N} \coth(q_i - q_j) E_{ij} \otimes E_{ji}$$

and C_{12} is the Casimir element of $sl(N)$: $C_{12} = \sum_{i,j=1}^{N} E_{ij} \otimes E_{ji}$.

7.4 The scalar Calogero–Moser model

The scalar Calogero–Moser model is defined by the Hamiltonian:

$$H_{\text{Cal}} = \frac{1}{2} \sum_{i=1}^{N} p_i^2 + \frac{1}{2} \gamma^2 \sum_{\substack{i,j=1 \\ i \neq j}}^{N} V(q_i - q_j) \tag{7.22}$$

We show here that this model and its r-matrix can be obtained from eq. (7.18) by a Hamiltonian reduction procedure.

Quite generally, we can parametrize the matrix $F = (f_{ij})$ of rank l as follows:

$$f_{ij} = \sum_{r=1}^{l} b_i^r a_j^r \tag{7.23}$$

The l vectors b^r form a basis of the image of F, and the vectors a^r form a basis of a supplementary space of the kernel of F. Moreover, the Poisson bracket, eq. (7.2), is reproduced if we set

$$\{a_i^r, b_j^s\} = -\delta_{rs}\delta_{ij}$$

The equations of motion for the quantities a_i^r, b_i^r read

$$\dot{a}_i^r = \{H, a_i^r\} = -\sum_{k \neq i} V(q_{ik}) f_{ki} a_k^r, \qquad \dot{b}_i^r = \{H, b_i^r\} = \sum_{k \neq i} V(q_{ik}) f_{ik} b_k^r$$

These equations of motion reproduce eq. (7.4) on the reduced manifold $f_{ii} = \alpha$.

To recover the scalar Calogero–Moser model, we choose $l = 1$ in eq. (7.23). So there is only one pair of vectors a and b. We simply denote their components by a_i, b_i. On these variables, the diagonal action eq. (7.7) reads:

$$a_i \longrightarrow d_i \, a_i, \qquad b_i \longrightarrow d_i^{-1} b_i \tag{7.24}$$

The integrable system is obtained by applying the method of Hamiltonian reduction under this group. We fix the moment to $f_{ii} = a_i b_i = \alpha = \sqrt{-1}\gamma$, which removes N degrees of freedom. Then we have to quotient by the isotropy subgroup of the moment, which is again the group of diagonal matrices. At the end the $2N$ degrees of freedom a_i, b_i are eliminated, leaving as reduced Hamiltonian the scalar Calogero–Moser model eq. (7.22). Note that $f_{ij} f_{ji} = \alpha^2 = -\gamma^2$.

In order to perform the reduction at the level of the Lax matrix, we remark that if $g = \text{Diag}\,(a_i^{-1})_{i=1,\dots,N}$, the matrix

$$L^{\text{Cal}}(\lambda) = g L(\lambda)\, g^{-1} = \sum_i p_i\, E_{ii} + \alpha \sum_{ij} \Phi(q_{ij,\lambda})\, E_{ij}$$

is invariant under the diagonal action and so is a function on the reduced phase space. This is the Lax matrix of the scalar Calogero–Moser model.

Proposition. *The Poisson bracket of the Lax matrix $L^{\text{Cal}}(\lambda)$ takes the r-matrix form:*

$$\{L_1^{\text{Cal}}(\lambda), L_2^{\text{Cal}}(\mu)\} = [r_{12}^{\text{Cal}}(\lambda, \mu), L_1^{\text{Cal}}(\lambda)] - [r_{21}^{\text{Cal}}(\mu, \lambda), L_2^{\text{Cal}}(\mu)] \quad (7.25)$$

with

$$r_{12}^{\text{Cal}}(\lambda, \mu) = a(\lambda, \mu) \sum_i E_{ii} \otimes E_{ii} - \sum_{ij} b(q_{ij}, \lambda - \mu) \; E_{ij} \otimes E_{ji}$$

$$- \frac{1}{2} \sum_{ij} \Phi(q_{ij}, \mu) \; (E_{ii} + E_{jj}) \otimes E_{ij}$$

This is a dynamical r-matrix which is no longer antisymmetric.

<u>Proof</u>. Since $L^{\text{Cal}}(\lambda)$ is invariant under the symmetry group, we can compute the Poisson brackets of its matrix elements directly. Using eq. (2.13) in Chapter 2, we get:

$$r_{12}^{\text{Cal}}(\lambda, \mu) = g_1 g_2 \left[r_{12}(\lambda, \mu) + g_1^{-1}\{g_1, L_2(\mu)\} + \frac{1}{2}[u_{12}, L_2(\mu)] \right] g_1^{-1} g_2^{-1}$$

where $u_{12} = g_1^{-1} g_2^{-1}\{g_1, g_2\}$ is here equal to zero. We get

$$r_{12}^{\text{Cal}}(\lambda, \mu) = r_{12}(\lambda, \mu) - \sum_{i \neq j} \Phi(q_{ij}, \mu) E_{ii} \otimes E_{ij}$$

Redefining

$$r_{12}^{\text{Cal}}(\lambda, \mu) \longrightarrow r_{12}^{\text{Cal}}(\lambda, \mu) + \frac{1}{2\alpha} \sum_i \left[E_{ii} \otimes E_{ii}, L_2^{\text{Cal}}(\mu) \right]$$

does not change eq. (7.25) and yields the r-matrix of the scalar Calogero–Moser model. ∎

Finally, we give the formula for the matrix $M^{\text{Cal}} = gMg^{-1} + \dot{g}g^{-1}$. We find $M_{ij}^{\text{Cal}} = -\dot{a}_i/a_i \delta_{ij} - (1 - \delta_{ij})\alpha \Phi'(q_i - q_j, \lambda)$. Using the equation of motion $\dot{a}_i = -\sum_{k \neq i} \alpha V(q_i - q_k) a_i$, we obtain

$$M^{\text{Cal}} = \alpha \delta_{ij} \sum_{k \neq i} V(q_i - q_k) - (1 - \delta_{ij})\alpha \Phi'(q_i - q_j, \lambda)$$

7.5 The spectral curve

The spectral curve of the spin Calogero–Moser model is defined as usual:

$$\Gamma : \Gamma(\lambda, \mu) \equiv \det\left(L(t, \lambda) - \mu I\right) = 0 \tag{7.26}$$

The curve Γ is time-independent due to the Lax equation, eq. (7.11). Note that Γ is invariant under the symmetries eq. (7.7).

Proposition. *The equation of the spectral curve takes the form:*

$$\Gamma(\lambda, \mu) \equiv \sum_{i=0}^{N} r_i(\lambda)\mu^i = 0 \tag{7.27}$$

where the $r_i(\lambda)$ are elliptic functions of λ, independent of t, which can be expanded on the Weierstrass $\wp$ function and its derivatives as:

$$r_i(\lambda) = I_i^0 + \sum_{s=0}^{N-i-2} I_{i,s} \partial_\lambda^s \wp(\lambda) \tag{7.28}$$

In a neighbourhood of $\lambda = 0$, the function $\Gamma(\lambda, \mu)$ can be factorized as:

$$\Gamma(\lambda, \mu) = \prod_{i=1}^{N} (\mu - \mu_i(\lambda)); \quad \mu_i(\lambda) = (\alpha - \nu_i)\lambda^{-1} + h_i(\lambda) \tag{7.29}$$

where $h_i(\lambda)$ are regular functions of λ, and the ν_i are the eigenvalues of the matrix $F = (f_{ij})$, so that $\nu_i = 0$, $i > l$ and $f_{ii} = \alpha$

Proof. The matrix elements $L_{ij}(t, \lambda)$ of the Lax matrix are elliptic functions of the variable λ having an essential singularity at $\lambda = 0$. The functions $r_i(\lambda)$, however, are meromorphic because the essential singularity in $L(\lambda)$ can be gauged away near $\lambda = 0$ since we can write:

$$L(t, \lambda) = G(t, \lambda)\tilde{L}(t, \lambda)G^{-1}(t, \lambda), \quad G_{ij} = \delta_{ij} \exp(\zeta(\lambda)q_i(t)) \tag{7.30}$$

where $\tilde{L}_{ij}(t, \lambda)$ are meromorphic functions of λ in a neighbourhood of the point $\lambda = 0$. Incorporating the constraint $f_{ii} = \alpha$, we have:

$$\tilde{L}(t, \lambda) = \frac{1}{\lambda}(\alpha I - F(t)) + O(\lambda^0) \tag{7.31}$$

where $F(t)$ is the matrix of elements $f_{ij}(t)$. Therefore the elliptic functions $r_i(\lambda)$ have poles of degree at most $N - i$ at the point $\lambda = 0$, so that they can be expanded as linear combinations of the function $\wp(\lambda)$ and its derivatives. We can always factorize the polynomial in μ, $\Gamma(\lambda, \mu)$, around

$\lambda = 0$. The branches $\mu_i(\lambda)$ in eq. (7.29) have simple poles, of the stated form due to eq. (7.31). In particular, since F is of rank l, the eigenvalue $\nu_i = 0$ has multiplicity $N - l$. ∎

The coefficients $I_i^0, I_{i,s}$ of this expansion are the integrals of motion of the spin Calogero–Moser model. They define the moduli of the algebraic curve Γ. We are now in a position to compute the number of action variables.

Proposition. *The number of action variables is half the dimension of the phase space:*

$$Nl - l(l+1)/2 + 1 = \frac{1}{2} \dim \mathcal{M}$$

<u>Proof</u>. The spectral equation $\Gamma(\lambda, \mu) = 0$ depends on $N(N+1)/2$ parameters I_i^0, I_{is}. However, they are not all independent. The constraints come from the conditions $\nu_i = 0$, $i > l$ and ν_i non-dynamical for $1 \le i \le l$ in eq. (7.29) . To see how they translate on the parameters I_i^0, I_{is}, let us introduce the variable $\tilde{\mu} = \mu - \alpha\lambda^{-1}$. Then we have $\Gamma(\lambda, \tilde{\mu} + \alpha\lambda^{-1}) = \tilde{\Gamma}(\lambda, \tilde{\mu})$, which can be expanded as:

$$\tilde{\Gamma}(\lambda, \tilde{\mu}) = \sum_{i=0}^{N} \tilde{\Gamma}_i(\tilde{\mu})\lambda^{-i} + \mathcal{R}(\lambda, \tilde{\mu}), \tag{7.32}$$

where $\tilde{\Gamma}_i(\tilde{\mu})$ are polynomials in $\tilde{\mu}$ and $\mathcal{R}(\lambda, \tilde{\mu}) = O(\lambda)$ is regular at $\lambda = 0$. One can check easily that the degree of $\tilde{\Gamma}_i(\tilde{\mu})$ is $N - i$ and that its coefficients are linear combinations of the parameters I_i^0, I_{is}.

The conditions $\nu_i = 0$, $i > l$, imply that

$$\tilde{\Gamma}_i(\tilde{\mu}) = 0, \quad i > l. \tag{7.33}$$

Altogether this is equivalent to a set of $(N-l)(N-l+1)/2$ linear equations on the parameters I_i^0, I_{is}. The total number of independent parameters is therefore equal to $Nl - l(l-1)/2$.

Next, recall that the expansion around $\lambda = 0$ of the branch μ_i for $i = 1, \ldots, l$ reads $\mu_i = (\alpha - \nu_i)\lambda^{-1} + O(1)$. This yields $l - 1$ additional relations (since the ν_i are constants characterizing the orbit of F, not to be counted as dynamical variables). Note that this gives only $l-1$ constraints, and not l, because we have $N\alpha = \sum \nu_i = \mathrm{Tr}\, F$. This condition also accounts for the fact that the elliptic function $r_{N-1}(\lambda)$ is constant, since it cannot have a single pole of order 1. Finally, the number of independent parameters is equal to $Nl - l(l-1)/2 - (l-1)$ which is exactly half the dimension of the reduced phase space. ∎

We now compute the genus of the spectral curve Γ.

Proposition. *For generic values of the action variables the genus of the spectral curve is given by:*

$$g = Nl - \frac{l(l+1)}{2} + 1 \qquad (7.34)$$

Proof. The idea of the proof is the same as in Chapter 5 and uses the Riemann–Hurwitz theorem. There is a difference, however, because here the base curve is of genus 1. Equation (7.26) presents the compact Riemann surface Γ as an N-sheeted branched covering of the base curve of the variable λ. The sheets are the N roots in μ. By the Riemann–Hurwitz formula we have $2g - 2 = N(2g_0 - 2) + \nu$, where g_0 is the genus of the base curve, $g_0 = 1$, and ν is the number of branch points, i.e. the number of values of λ for which $\Gamma(\lambda, \mu)$ has a double root in μ. This is the number of zeroes of $\partial_\mu \Gamma(\lambda, \mu)$ on the surface $\Gamma(\lambda, \mu) = 0$. But $\partial_\mu \Gamma(\lambda, \mu)$ is a meromorphic function on the surface, hence it has as many zeroes as poles. The poles are located above $\lambda = 0$, or $\mu = \infty$ which is the same, and are easy to count.

Let P_i be the points of Γ lying on the different sheets over the point $\lambda = 0$. In the neighbourhood of P_i the function μ has the expansion $\mu_i = (\alpha - \nu_i)\lambda^{-1} + h_i(\lambda)$. It follows that the function $\partial \Gamma / \partial \mu$ in the neighbourhood of P_i has the form

$$\partial \Gamma / \partial \mu = \prod_{j \neq i} [(\nu_j - \nu_i)\lambda^{-1} - (h_j(\lambda) - h_i(\lambda))]$$

From this, we see that on each of the l sheets $(\lambda, \mu_i(\lambda))$ $(i = 1, \dots, l)$ we have one-pole of order $(N-1)$. On each of the $(N-l)$ sheets $(\lambda, \mu_i(\lambda))$ $(i = l+1, \dots, N)$ we have one-pole of order l. Finally, $\nu = l(N-1) + (N-l)l$. Inserting this value in the Riemann–Hurwitz formula yields the result. ∎

The last two propositions show that the number of independent action variables is exactly the genus of the spectral curve. Among these, one is the total momentum associated with translation invariance and we will have to factorize by this symmetry.

7.6 The eigenvector bundle

As in Chapter 5, we consider at any point $P = (\lambda, \mu)$ of the spectral curve the unique eigenvector $\Psi(0, P)$ of $L(0, \lambda)$ with eigenvalue μ, normalized by $\psi_1(0, P) = 1$. We want to study the analyticity properties in λ of this eigenvector. A first difference with the case of rational Lax matrices is that it has an essential singularity at the points P_i above $\lambda = 0$.

Proposition. *In the neighbourhood of the point P_i the component $\psi_j(0, P)$ has the form*

$$\psi_j(0, P) = \exp\left[\zeta(\lambda)(q_j(0) - q_1(0))\right](c_j^{(i)} + O(\lambda)) \tag{7.35}$$

where $c_j^{(i)}$ are the eigenvectors of the matrix F corresponding to the non-zero eigenvalue ν_i for $i = 1, \ldots, l$:

$$\sum_{j=1}^{N} f_{kj} c_j^{(i)} = \nu_i c_k^{(i)}$$

while for $i > l$ the $c^{(i)}$ form a basis of the kernel of F.

Proof. From equation (7.30), we have $\Psi(0, P) = G(0, \lambda)\tilde{\Psi}(0, P)$, where $\tilde{\Psi}(0, P)$ is an eigenvector of $\tilde{L}(0, \lambda)$. Using eq. (7.31), we have $\tilde{\Psi}(0, P) = c^{(i)} + O(\lambda)$, where $c^{(i)}$ is an eigenvector of F. More precisely, for $i = 1, \ldots, l$, $c^{(i)}$ is the unique eigenvector of F with eigenvalue $\nu_i \neq 0$ normalized by $c_1^{(i)} = 1$, while for $i > l$ it is determined by the limit for $P \to P_i$ of the normalized eigenvector of $\tilde{L}(P)$ corresponding to the eigenvalue $\mu_i(P)$. The degeneracy has been lifted by higher order terms in λ. Therefore we have $\psi_j(0, P) = (c_j^{(i)} + O(\lambda))\exp(\zeta(\lambda)q_j(0))$. Normalizing $\psi_1(0, P) = 1$ yields the result. ∎

We can now compute the number of poles of Ψ on Γ. The result is slightly different from the case of rational Lax matrices, where it was found to be $g + N - 1$.

Proposition. *The number of poles of $\Psi(0, P)$ is:*

$$m = Nl - \frac{l(l+1)}{2} = g - 1 \tag{7.36}$$

Proof. As in the case of rational Lax matrices, let us introduce the function $W(\lambda)$ of the complex variable λ defined by:

$$W(\lambda) = \left(\text{Det}\,|\psi_i(M_j)|\right)^2$$

where the M_j are the N points above λ. It is well-defined on the base curve since the Det^2 does not depend on the order of the M_j.

This function has an essential singularity of the form $e^{2\zeta(\lambda)\sum_i(q_i(0) - q_1(0))}$ at $\lambda = 0$. This does not affect the property that the number of poles of $W(\lambda)$ is equal to the number of its zeroes. This property is obtained by considering the sum of residues of W'/W on the λ-torus, and noting that

$\zeta'(\lambda) = \wp(\lambda) = \frac{1}{\lambda^2} + O(\lambda^2)$ is elliptic and has no residue. Clearly $W(\lambda)$ has a double pole where there exists a point P above λ at which $\Psi(P)$ has a simple pole.

As in the rational case, we show that $W(\lambda)$ has a simple zero for values of λ corresponding to a branch-point of the covering, hence $m = \nu/2 = g-1$, by Riemann–Hurwitz. Here lies the difference with the rational case, because $g_0 = 1$, see Chapter 5.   ■

This result looks different from what we got in the case of rational Lax matrices. There, we had $g + N - 1$ poles for a meromorphic eigenvector at time $t = 0$. However, $N - 1$ poles were located above $\lambda = \infty$ and were not dynamical. Here all poles are dynamical and we have $g - 1$ of them. This is a surprising result. In fact, from eq. (7.35), we see that the components of the eigenvector $\psi_i(0, P)$ at time $t = 0$ are Baker–Akhiezer functions. But generically, such a function has at least g poles, not $g-1$. So, we have to admit that we are not in a generic situation. It was noted in Chapter 3 that when the genus g of the base curve is greater than or equal to 1, the consistency equations of the Lax pair become overdetermined, preventing genericity of the Lax matrix. In fact, this will be a crucial ingredient in the solution of the Calogero–Moser model.

7.7 Time evolution

The next step is to compute the time evolution of the eigenvectors. We let the eigenvector evolve according to the natural equation:

$$\frac{d\Psi}{dt} = M\Psi \tag{7.37}$$

We choose as initial condition the eigenvector $\Psi(0, P)$ normalized with its first component equal to 1. Of course, at subsequent time this normalization will not hold any more. However, we know that, in this setting, the poles of $\Psi(t, P)$ do not evolve with time, see Chapter 5.

Proposition. *The coordinates $\psi_j(t, P)$ of the vector-function $\Psi(t, P)$ are meromorphic functions on Γ except at the points P_i above $\lambda = 0$. Their poles $\gamma_1, \ldots, \gamma_{g-1}$ do not depend on t. In the neighbourhood of P_i they have the form*

$$\psi_j(t, P) = c_j^{(i)}(t, \lambda) \exp\left[\zeta(\lambda)(q_j(t) - q_1(0)) + m_i(\lambda)t\right] \tag{7.38}$$

where $c_j^{(i)}(t, \lambda)$ are regular functions of λ for $\lambda \simeq 0$, $c_j^{(i)}(t, \lambda) = c_j^{(i)}(t) + O(\lambda)$. Here the $c^{(i)}(t)$ are eigenvectors of the matrix $F(t) = (f_{ij}(t))$

corresponding to the eigenvalues ν_i, and:

$$m_i(\lambda) = (-\alpha + \nu_i)\lambda^{-2} - h_i(0)\lambda^{-1} \qquad (7.39)$$

is the singular part of $-\lambda^{-1}\mu_i(\lambda)$ at $\lambda = 0$, see eq. (7.29).

<u>Proof.</u> Let us consider the vector $\tilde{\Psi}(t, P)$ defined as

$$\Psi(t, P) = G(t, \lambda)\tilde{\Psi}(t, P) \qquad (7.40)$$

where $G(t, \lambda)$ is defined in eq. (7.30) and let $\Psi(t, P)$ evolve according to eq. (7.37). The vector $\tilde{\Psi}(t, P)$ is an eigenvector of the matrix $\tilde{L}(t, \lambda)$ and evolves according to the equation

$$(\partial_t - \tilde{M}(t, \lambda))\tilde{\Psi}(t, P) = 0, \quad \tilde{M} = -G^{-1}\partial_t G + G^{-1}MG \qquad (7.41)$$

From eqs. (7.12, 7.13) it follows that:

$$\tilde{M}_{ij} = -\delta_{ij}\zeta(\lambda)\dot{q}_i - (1 - \delta_{ij})f_{ij}\frac{\sigma(\lambda - q_{ij})}{\sigma(\lambda)\sigma(q_{ij})}[\zeta(q_{ij} - \lambda) - \zeta(q_{ij}) + \zeta(\lambda)]$$

so that collecting the coefficient of the $1/\lambda$ term we can write:

$$\tilde{M}(t, \lambda) = -\lambda^{-1}\tilde{L}(t, \lambda) + O(\lambda^0) \qquad (7.42)$$

Hence around P_i we have:

$$\partial_t\tilde{\Psi}(t, \lambda) = \left(-\frac{1}{\lambda}\mu_i(\lambda) + O(1)\right)\tilde{\Psi}(t, P) \qquad (7.43)$$

The quantity:

$$m_i(\lambda) = -(\lambda^{-1}\mu_i(\lambda))_- = (-\alpha + \nu_i)\lambda^{-2} - h_i(0)\lambda^{-1}$$

is independent of time because so is $\mu_i(\lambda)$. Integrating eq. (7.43), multiplying by $G(t, \lambda)$ and normalizing $\psi_1(0, P) = 1$, we get the result. ∎

7.8 Reconstruction formulae

We now reconstruct the original dynamical variables in terms of the Riemann surface Γ and the poles of the eigenvectors. From eq. (7.38) we see that the components $\psi_i(t, P)$ of the eigenvector are Baker–Akhiezer functions. Their behaviour above $\lambda = 0$ is

$$\psi_i(P) = \exp[(q_i(t) - q_1(0))\lambda^{-1} + m_j(\lambda)t](c_i^{(j)}(t) + O(\lambda)), \quad P \to P_j \qquad (7.44)$$

As we already mentioned, there is a paradox with the number of poles of the Baker–Akhiezer function. Generically, such a function has g poles.

The function then exists and is unique up to normalization. In particular, its zeroes are completely determined. One way to construct a Baker–Akhiezer function with $g-1$ poles is to let one of the zeroes of the generic function cancel one of its poles. Clearly, this gives a relation between the parameters defining the function, specifically the $q_i(t)$ defining the essential singularity and the moduli of the curve.

Let $\psi_i(t, P)$ be the Baker–Akhiezer function with the singularities eq. (7.44) at the points P_j above $\lambda = 0$, and a divisor of g poles $(\gamma_0, \gamma_1, \ldots, \gamma_{g-1})$. We will denote by $(\eta_0, \eta_1, \ldots, \eta_{g-1})$ the divisor of its zeroes.

By the general formula eq. (5.48) in Chapter 5, we have:

$$\psi_i(t, P) = d_i(t) e^{[(q_i(t)-q_1(0))\int \Omega_1 + t \int \Omega_2]}$$
$$\times \frac{\theta(\mathcal{A}(P) - U_1(q_i(t) - q_1(0)) - U_2 t - \zeta)}{\theta(\mathcal{A}(P) - \zeta)}$$

where $\mathcal{A}(P)$ is the Abel map, and Ω_1 and Ω_2 are normalized second kind Abelian differentials with singularities $d\lambda^{-1}$ and $dm_j(\lambda)$ at the points P_j above $\lambda = 0$ respectively. Note that the differentials are independent of the index i of the component considered, and so are the vectors U_1 and U_2 of their b-periods. The functions $d_i(t)$ are arbitrary normalizations.

According to Riemann's theorem, the divisors of the poles and zeroes of $\psi_i(t, P)$ satisfy :

$$\mathcal{A}(\eta_0) + \sum_{i=1}^{g-1} \mathcal{A}(\eta_i) - U_1(q_i(t) - q_1(0)) - U_2 t - \zeta = -\mathcal{K}$$

$$\mathcal{A}(\gamma_0) + \sum_{i=1}^{g-1} \mathcal{A}(\gamma_i) - \zeta = -\mathcal{K}$$

Let us assume that the zero η_0 coincides with the pole γ_0. We then get the condition

$$U_1 q_i(t) + U_2 t + V = -\mathcal{K} + \sum_{i=1}^{g-1} \mathcal{A}(\eta_i) \qquad (7.45)$$

where we have denoted $V = -\mathcal{K} - U_1 q_1(0) + \sum_{i=1}^{g-1} \mathcal{A}(\gamma_i)$, a vector in Jac (Γ) *independent* of time. Now the expression in the right-hand side of this equation belongs to the zero divisor of the theta-function (see Chapter 15). This is the constraint we were looking for. As a consequence we have

Proposition. *The quantities $q_i(t)$ satisfying the equation*

$$\theta(U_1 q_i(t) + U_2 t + V) = 0 \qquad (7.46)$$

are the solutions of the equations of motion of the spin Calogero–Moser model. Moreover, the spin variables are reconstructed with the matrix C of eigenvectors of $F = C\nu C^{-1}$ with:

$$c_i^{(j)}(t) = e^{\alpha_j q_i(t)}\,\theta(U_1 q_i(t) + U_2 t + V_j) \qquad (7.47)$$

where V_j is a constant vector.

Proof. Equation (7.46) expresses the fact that the left-hand side of eq. (7.45) belongs to the zero divisor of the theta-function. Note that all $q_i(t)$ satisfy the *same* equation which describes, for all values of time, the intersection of a straight line in the Jacobian torus with the theta divisor.

To proceed to the reconstruction of the spin variables, it is enough to compute the matrix of eigenvectors C. As we have shown, the columns of C are the vectors $c^{(j)}$ whose components are given by the limit when $\lambda \to 0$ of the following expression, taken for $P = (\lambda, \mu_j(\lambda))$:

$$c_i^{(j)}(t) = \lim_{\lambda \to 0} d_i(t) e^{\left[(q_i(t)-q_1(0))\left(\int^P \Omega_1 - \frac{1}{\lambda}\right) + t\left(\int^P \Omega_2 - m_j(\lambda)\right)\right]}$$
$$\times \frac{\theta(\mathcal{A}(P) - U_1(q_i(t) - q_1(0)) - U_2 t - \zeta)}{\theta(\mathcal{A}(P) - \zeta)}$$

The limits $\alpha_j = \lim_{\lambda \to 0}(\int^P \Omega_1 - \frac{1}{\lambda})$ and $\beta_j = \lim_{\lambda \to 0}(\int^P \Omega_2 - m_j(\lambda))$ are well-defined and depend on the point P_j. Note that if $c_i^{(j)} \to d_i c_i^{(j)}$ we have $f_{ij} \to d_i f_{ij} d_j^{-1}$ and we know that we must quotient by this diagonal action. Moreover, if we change the normalization of the vector $c^{(j)}$, i.e. $c_i^{(j)} \to d'_j c_i^{(j)}$, the matrix F is invariant. Hence we can drop all factors that can be absorbed in these invariances, and we can choose $c_i^{(j)}$ as in eq. (7.47) with the constant vector $V_j = \zeta - U_1 q_1(0) - \mathcal{A}(P_j)$. ∎

7.9 Symplectic structure

To compute the symplectic form we need to consider the inverse $\widehat{\Psi}^{-1}(\lambda)$ of the matrix $\widehat{\Psi}(\lambda)$ whose columns are the eigenvectors at the N points above λ. The matrix $\widehat{\Psi}^{-1}(\lambda)$ is built by considering the adjoint system $\Psi^+(P)(L(\lambda) - \mu) = 0$. The row vector $\Psi^+(P)$ is reconstructed from its analytical properties, exactly like $\Psi(P)$. The rows of the matrix $\widehat{\Psi}^{-1}(\lambda)$ are the values of the row vector:

$$\Psi^{(-1)}(P) \equiv \frac{\Psi^+(P)}{\langle \Psi^+(P)\Psi(P) \rangle}$$

at the N points P_j above λ. As before $\langle \Psi^+(P)\Psi(P)\rangle$ is a meromorphic function on Γ with zeroes at the branch points of the covering $\mu \to \lambda$. Note that this function is regular at $\lambda = 0$, in particular the essential singularities cancel. It follows that $\Psi^{(-1)}(P)$ has poles at the branch points, and zeroes at the divisor of poles of Ψ.

We remarked already several times that the components of the eigenvector can be written as ratios of suitable minors of the matrix $L(\lambda) - \mu$. In particular, the poles of $\Psi(P)$ are obtained as the zeroes of the first minor. Note that in this minor, the variables q_1, p_1 have disappeared. This corresponds to a reduction of the system by translational symmetry, which leaves a phase space of dimension $2(g-1)$, hence the $g-1$ poles of $\Psi(P)$. In particular one can choose the origin such that at initial time $q_1(0) = 0$.

Proposition. *The symplectic form of the spin Calogero–Moser model is*

$$\omega = \sum_{i=1}^{g-1} d\lambda_{\gamma_i} \wedge d\mu_{\gamma_i} = \sum_{i=1}^{N} dp_i \wedge dq_i - \omega_K \qquad (7.48)$$

where $(\lambda_{\gamma_i}, \mu_{\gamma_i})$ are the coordinates of the dynamical divisor, and ω_K is the Kirillov symplectic form on the orbit defined by $F = C\nu C^{-1}$, i.e.
$\omega_K = \mathrm{Tr}\left(\nu C^{-1}\delta C \wedge C^{-1}\delta C\right)$.

<u>Proof</u>. As usual, we consider the form $K = K_1 + K_2 + K_3$ with:

$$
\begin{aligned}
K_1 &= \langle \Psi^{(-1)}(P)\delta L(\lambda) \wedge \delta\Psi(P)\rangle \, d\lambda \\
K_2 &= \langle \Psi^{(-1)}(P)\delta\mu \wedge \delta\Psi(P)\rangle \, d\lambda \\
K_3 &= \delta\left(\log \partial_\mu \Gamma\right) \wedge \delta\mu \, d\lambda
\end{aligned}
$$

and write that the sum of its residues vanishes to prove the equality of the two expressions for ω. The poles of K are the poles of $\Psi(P)$, $\{\gamma_i = (\lambda_{\gamma_i}, \mu_{\gamma_i}), \; i = 1, \ldots, g-1\}$, the branch points, and the points P_j above $\lambda = 0$. At the points γ_i, the computation is exactly similar to that in Chapter 5, yielding a sum of residues $2\sum_{i=1}^{g-1} \delta\mu_{\gamma_i} \wedge \delta\lambda_{\gamma_i}$ coming from equal contributions of K_1 and K_2. At the branch points the residues from K_1, K_2, K_3 cancel, as in the rational case. The new features occur above $\lambda = 0$. It is easy to see that K_3 is regular above $\lambda = 0$ because $\delta\mu$ is *regular* at these points, since the coefficients of the polar part of μ_j are non-dynamical. The sum of residues of K_1 and K_2 at the P_j can be written in matrix form as the residue of forms on the base as follows:

$$E_1 \equiv \sum_j \mathrm{Res}_{P_j} K_1 = \mathrm{Res}_{\lambda=0} \mathrm{Tr}\left(\widehat{\Psi}^{-1}\delta L(\lambda) \wedge \delta\widehat{\Psi}\right) d\lambda$$

$$E_2 \equiv \sum_j \mathrm{Res}_{P_j} K_2 = -\mathrm{Res}_{\lambda=0} \mathrm{Tr}\left(\delta\widehat{\Psi} \wedge \delta\widehat{\mu}\widehat{\Psi}^{-1}\right) d\lambda$$

where $\widehat{\mu}$ is the diagonal matrix of the $\mu_j(\lambda)$. Since this is a local computation we first extract the essential singularities by setting $\widehat{\Psi} = G\tilde{\Psi}$ and $L = G\tilde{L}G^{-1}$ with G given in eq. (7.30). We get

$$E_1 = \mathrm{Res}_{\lambda=0} \mathrm{Tr}\left([G^{-1}\delta G, \tilde{L}] + \delta\tilde{L}\right) \wedge \left(G^{-1}\delta G + \delta\tilde{\Psi}\tilde{\Psi}^{-1}\right) d\lambda$$

$$E_2 = -\mathrm{Res}_{\lambda=0} \mathrm{Tr}\left(G^{-1}\delta G \wedge \tilde{\Psi}\delta\widehat{\mu}\tilde{\Psi}^{-1} + \delta\tilde{\Psi} \wedge \delta\widehat{\mu}\tilde{\Psi}^{-1}\right) d\lambda$$

Note that $\mathrm{Tr}\left([G^{-1}\delta G, \tilde{L}] \wedge G^{-1}\delta G\right) = 0$ because G is diagonal, and $\mathrm{Res}_{\lambda=0}\delta\tilde{\Psi} \wedge \delta\widehat{\mu}\tilde{\Psi}^{-1} = 0$ since $\delta\widehat{\mu}$ is regular at $\lambda = 0$. Collecting the remaining terms, we have:

$$E_1 + E_2 = \mathrm{Res}_{\lambda=0} \mathrm{Tr}\left(\delta\tilde{L} \wedge G^{-1}\delta G + \delta\tilde{L} \wedge \delta\tilde{\Psi}\tilde{\Psi}^{-1}\right.$$

$$\left. + [G^{-1}\delta G, \tilde{L}] \wedge \delta\tilde{\Psi}\tilde{\Psi}^{-1} - G^{-1}\delta G \wedge \tilde{\Psi}\delta\widehat{\mu}\tilde{\Psi}^{-1}\right) d\lambda$$

Using $\tilde{L}\tilde{\Psi} = \tilde{\Psi}\widehat{\mu}$ and $\tilde{\Psi}^{-1}\tilde{L} = \widehat{\mu}\tilde{\Psi}^{-1}$, we get

$$\mathrm{Tr}\left([G^{-1}\delta G, \tilde{L}] \wedge \delta\tilde{\Psi}\tilde{\Psi}^{-1}\right) = \mathrm{Tr}\left(G^{-1}\delta G \wedge (\tilde{L}\delta\tilde{\Psi} - \delta\tilde{\Psi}\widehat{\mu})\tilde{\Psi}^{-1}\right)$$

$$= \mathrm{Tr}\left(G^{-1}\delta G \wedge (\tilde{\Psi}\delta\widehat{\mu} - \delta\tilde{L}\tilde{\Psi})\tilde{\Psi}^{-1}\right)$$

where in the last step we have used $\tilde{L}\delta\tilde{\Psi} - \delta\tilde{\Psi}\widehat{\mu} = \tilde{\Psi}\delta\widehat{\mu} - \delta\tilde{L}\tilde{\Psi}$. Hence our expression simplifies to:

$$E_1 + E_2 = \mathrm{Res}_{\lambda=0} \mathrm{Tr}\left(2\delta\tilde{L} \wedge G^{-1}\delta G + \delta\tilde{L} \wedge \delta\tilde{\Psi}\tilde{\Psi}^{-1}\right) d\lambda$$

Note that $(G^{-1}\delta G)_{ij} = \lambda^{-1}\delta q_i \delta_{ij} + O(1)$ so that the first term contributes a residue $2\sum_i \delta p_i \wedge \delta q_i$. For the second term we note that $\tilde{\Psi} = C + O(\lambda)$, in the reduced system where $q_1(0) = 0$, and $\tilde{L} = -\lambda^{-1}(F - \alpha) + O(1)$ so that the residue is $-\mathrm{Tr}\left(\delta F \wedge \delta C C^{-1}\right)$ since $\delta\alpha = 0$. Remembering that $F = C\nu C^{-1}$, this reads $-2\mathrm{Tr}\left(\nu C^{-1}\delta C \wedge C^{-1}\delta C\right)$. We recognize the Kostant–Kirillov form on the coadjoint orbit of ν, see Chapter 3. Writing that the sum of residues vanishes proves the proposition. ∎

This result shows that $\lambda_{\gamma_i}, \mu_{\gamma_i}$ are canonically conjugate variables. Since $(\lambda_{\gamma_i}, \mu_{\gamma_i})$ belong to the spectral curve, they form a set of separated variables.

7.10 Poles systems and double-Bloch condition

In this section we present a natural construction of the Lax pair for the spin Calogero–Moser model by relating it to the matrix KP equation. Before delving into this particular example it is illuminating to present the general context of this sort of connection.

Let us consider a linear differential operator $\mathcal{D}$ in two variables x and t (there could be more than one time) and consider the differential equation $\mathcal{D}\Psi(x,t) = 0$. We will consider as an example the KP operator $\mathcal{D} = \partial_t - \partial_x^2 + u(x,t)$, see eq. (10.10) in Chapter 10. Assume, moreover, that the coefficients of $\mathcal{D}$ are doubly periodic meromorphic functions of the complex variable x with periods $2\omega_i$, $i = 1, 2$. We require nothing concerning the t-dependence. In general we know that for generic simply periodic potentials we can find a solution of the equation $\mathcal{D}\Psi(x,t) = 0$ which is quasi-periodic, i.e. $\Psi(x + 2\omega_1, t) = B_1\Psi(x,t)$. Such solutions are called Floquet or Bloch solutions.

In the case of elliptic potentials, since we have a second period, it is natural to require that Ψ be double Bloch, i.e.

$$\Psi(x + 2\omega_i, t) = B_i\Psi(x,t), \quad i = 1, 2$$

In contrast to the case of the one-period situation it turns out that this is in general *impossible*. Nevertheless, one can find double Bloch solutions for very special potentials, from which the Calogero–Moser model will automatically spring out.

To understand the restrictions coming from the double Bloch condition let us assume that the function $x \to \Psi(x,t)$ is a meromorphic function and has N poles at positions q_i on the torus with periods $2\omega_i$. Applying the Riemann–Roch theorem with $g = 1$ one sees that such functions form a vector space of dimension N. Indeed, for any two such functions Ψ_1, Ψ_2 with the same Bloch multipliers, the quotient Ψ_2/Ψ_1 is a meromorphic function on the torus with N poles (since Ψ_1 has the same number of zeroes and poles, because Ψ_1'/Ψ_1 is elliptic), hence lives in a space of dimension $N - g + 1 = N$. The existence of such functions comes from their explicit construction using the Lamé function. Take any sum of the form:

$$\Psi(x, t, z) = \sum_{i=1}^{N} c_i(t, z)\Phi(x - q_i, z)e^{kx}$$

where $\Phi(q,z)$ is the Lamé function defined in eq. (7.9). Recall that we
have $\Phi(x + 2\omega_i, z) = T_i(z)\Phi(x,z)$, with $T_i(z) = \exp(2\omega_i\zeta(z) - 2\eta_i z)$
(see Chapter 15), so that $\Psi(x + 2\omega_i, t, z) = B_i\Psi(x,t,z)$ with $B_i =$
$T_i(z)\exp(2k\omega_i)$. Given the two Bloch multipliers B_i, we can adjust k
and z to achieve these values. So we have found an explicit form of the
basis of the N-dimensional vector space of double Bloch functions. If we
now require that this function Ψ obeys the equation $\mathcal{D}\Psi = 0$, we impose
in fact more than N conditions. This is because $\mathcal{D}\Psi$ is a double Bloch
function with the same multipliers, but differentiations and multiplication
by potentials can only increase the degree of its divisor of poles. Hence
$\mathcal{D}\Psi$ lives in a space of dimension greater than N, and its vanishing requires more than N linear conditions on the N coefficients c_i. This means
that for a general operator $\mathcal{D}$ with elliptic coefficients there are no double
Bloch solutions.

On the other hand, given a Riemann surface Γ of genus g, one can
construct Baker–Akhiezer functions Ψ on it. It is well known that such
Baker–Akhiezer functions satisfy differential equations of the form $\mathcal{D}\Psi =$
0 for some *specific* operator $\mathcal{D}$, see an example below eq. (7.49). A generic
Baker–Akhiezer function depends on many parameters. It is defined first
by the choice of the Riemann surface Γ, which depends on $3g - 3$ moduli,
i.e. depends on $3g - 3$ complex parameters, and second by the choice of
punctures P_α, local parameters w_α around P_α, and singular parts of order
n_α at P_α. Only the first n_α coefficients in the expansion of w_α are relevant
to the definition of the singular part. Altogether this produces a total of
$(3g - 3) + \sum_\alpha(1 + n_\alpha)$ parameters.

In the case of one puncture, the Baker–Akhiezer function is of the
generic form:

$$\Psi(P,t) = e^{\sum_i t_i \int^P \Omega^{(i)}} \frac{\theta(\mathcal{A}(P) + \sum_i U^{(i)}t_i + \zeta)}{\theta(\mathcal{A}(P) + \zeta)}$$

where $\mathcal{A}(P)$ is the Abel map. We assume that x is t_1, but we could take
for x any combination $x = \sum_i \alpha_i t_i$ of the elementary times t_i of the hierarchy. The condition for such a function Ψ to be double Bloch is that $2\omega_i U_1$
belongs to the lattice of periods of $\mathrm{Jac}\,(\Gamma)$. This means $2g$ conditions on
the parameters of the Baker–Akhiezer functions. The dimension of the
parameter space of Baker–Akhiezer functions is large enough to accomo-
date the $2g$ double Bloch conditions. This provides families of differential
operators possessing double Bloch solutions. In this case the overdetermined linear system on the coefficients $c_i(t,z)$ becomes compatible. The
compatibility conditions eventually take the form of a Lax equation.

Let us apply this strategy to a simple example: we consider a smooth
Riemann surface of genus g and l punctures P_β, $\beta = 1, \ldots, l$ with local

parameters $w_\beta(P)$ around P_β ($w_\beta(P_\beta) = 0$). Fix a divisor of degree $g+l-1$ in general position. Then there exists a unique Baker–Akhiezer function ψ_α having poles at this divisor and behaving in the neighbourhood of each P_β as:

$$\psi_\alpha(x, t, P) = e^{w_\beta^{-1}x + w_\beta^{-2}t} \left(\delta_{\alpha\beta} + \sum_{s=1}^{\infty} \xi_s^{\alpha\beta}(x; t) w_\beta^s \right)$$

In fact, the degree of the divisor of poles being $g+l-1$, we have a vector space of functions of dimension l and we impose a system of l linear inhomogeneous normalization conditions.

Proposition. *Let* $|\Psi(x, t, P)\rangle$ *be the vector with l components* $\psi_\alpha(x, t, P)$. *It satisfies the equation:*

$$(\partial_t - \partial_x^2 + u(x, t))|\Psi(x, t, P)\rangle = 0 \qquad (7.49)$$

where the $l \times l$ matrix u is given by $u^{\alpha\beta}(x, t) = 2\partial_x \xi_1^{\alpha\beta}(x, t)$. Such potentials are called finite-zone potentials.

<u>Proof.</u> In the vicinity of each puncture P_β one can write:

$$(\partial_t - \partial_x^2)\psi_\alpha(x, t, P) = e^{w_\beta^{-1}x + w_\beta^{-2}t} \left(-2\partial_x \xi_1^{\alpha\beta} + O(w_\beta) \right)$$

Since the left-hand side is meromorphic except at the P_j, has the same $g + l - 1$ poles as Ψ, and has an appropriate essential singularity at each puncture, it can be expanded on the ψ_β, so that one can write $(\partial_t - \partial_x^2)\psi_\alpha = -\sum_\beta u^{\alpha\beta}\psi_\beta$, for some $u^{\alpha\beta}(x, t)$ independent of $P \in \Gamma$. Comparing with the right-hand side around P_β we find $u^{\alpha\beta} = 2\partial_x \xi_1^{\alpha\beta}$. ∎

We now express the condition that the potential u is elliptic and obtain its precise form.

Proposition. *If the finite-zone potential u is elliptic, it has necessarily the form:*

$$u(x, t) = \sum_{i=1}^{N} \rho_i(t)\wp(x - q_i(t))$$

where $\rho_i(t)$ is an $l \times l$ matrix of rank 1 of the form $\rho_i = |a_i\rangle\langle b_i|$ with $|a_i\rangle$ an l vector and $\langle b_i|$ an l covector.

<u>Proof.</u> We need the explicit form of $|\Psi\rangle$ as a Baker–Akhiezer function on some curve Γ. Let $P_1, \ldots, P_l$ be the punctures and $\gamma_1, \ldots, \gamma_{g+l-1}$ be the poles of the Baker–Akhiezer function. There exists a unique meromorphic function $h_\alpha(P)$ with poles at the γ_i and such that $h_\alpha(P_\beta) = \delta_{\alpha\beta}$. In particular it has $l - 1$ zeroes at the P_β, $\beta \neq \alpha$, and g other zeroes at a

divisor D_α. Applying Abel's theorem, it follows that $Z_0 + \mathcal{K} \equiv \mathcal{A}(P_\alpha) - \mathcal{A}(D_\alpha) = \sum_\beta \mathcal{A}(P_\beta) - \sum \mathcal{A}(\gamma_i)$ is independent of α. The Baker–Akhiezer function $\psi_\alpha(P)$ reads:

$$\psi_\alpha(P) = h_\alpha(P) \frac{\theta(\mathcal{A}(P) + U_1 x + U_2 t + Z_0 - \mathcal{A}(P_\alpha))\theta(Z_0)}{\theta(\mathcal{A}(P) + Z_0 - \mathcal{A}(P_\alpha))\theta(U_1 x + U_2 t + Z_0)} e^{x \int^P \Omega_1 + t \int^P \Omega_2}$$

In this formula the first theta-function in the denominator cancels the extra g zeroes of $h_\alpha(P)$ at D_α. Then the first theta-function in the numerator is obtained by requiring that there is no monodromy. Finally, the two other theta functions are necessary to ensure the correct normalization of $\psi_\alpha(P)$ at P_α. It is now easy to identify the poles in x of $|\Psi(x, t, P)\rangle$. They occur at positions $x = q_i(t)$ with (compare with eq. (7.46)):

$$\theta(U_1 q_i(t) + U_2 t + Z_0) = 0$$

This defines the functions $q_i(t)$. Consider the residue of $\psi_\alpha(x, t, P)$ when $x = q_i(t)$. As a function of P it is a Baker–Akhiezer function, having poles at the points γ_i. Moreover, in the neighbourhood of each puncture P_β it has the form $\exp\left(w_\beta^{-1} q_i(t) + w_\beta^{-2} t\right) O(w_\beta)$, i.e. the coefficient in front of the exponential *vanishes*. This is because at P_β, $\beta \neq \alpha$ the function h_α vanishes, while at P_α the theta-function $\theta(\mathcal{A}(P) + U_1 q_i(t) + U_2 t + Z_0 - \mathcal{A}(P_\alpha))$ vanishes, due to the definition of $q_i(t)$. In general such a Baker–Akhiezer function vanishes identically, however for the special values of the parameters $(x = q_i(t), t)$ it exists, and is a fortiori unique, up to a normalization constant. We choose some normalization and call it $\sigma_i(t, P)$. This means that:

$$\psi_\alpha(x, t, P) = \frac{a_{i\alpha}(t)\sigma_i(t, P)}{x - q_i(t)} + O(1) \qquad (7.50)$$

Here $a_{i\alpha}(t)$ is the normalization constant independent of P. The potential $u^{\alpha\beta}(x, t)$ is obtained by computing the expansion of $\psi_\alpha(x, t, P)$ around P_β. Writing the expansion of $\sigma_i(t, P)$ as $\sigma_i(t, P) = -\frac{1}{2}(w_\beta b_i^\beta + O(w_\beta^2)) \exp\left(w_\beta^{-1} q_i(t) + w_\beta^{-2} t\right)$, we find around $x = q_i$ and $P = P_\beta$:

$$\xi_1^{\alpha\beta}(x, t) = \frac{a_{i\alpha}(t) b_i^\beta(t)}{x - q_i(t)} + O(1)$$

Finally, $u^{\alpha\beta} = 2\partial_x \xi_1^{\alpha\beta}$ has double poles at $x = q_i$ with coefficient $\rho_i^{\alpha\beta} = a_{i\alpha}(t) b_i^\beta(t)$. If we now impose that u is elliptic, the double pole gives rise to a Weierstrass function $\wp(x - q_i(t))$ with a matrix coefficient ρ_i of rank 1.   ■

We now require that $|\Psi\rangle$ is double Bloch in x, hence can also be written as:

$$|\Psi\rangle = \sum_{i=1}^{N} |s_i(t,\lambda,\mu)\rangle \Phi(x - q_i(t),\lambda)e^{-\frac{\mu}{2}x+\frac{\mu^2}{4}} \qquad (7.51)$$

In this formula λ and μ are free parameters, but the double Bloch condition will turn out to specify the Riemann surface on which (λ,μ) are coordinates. Moreover, μ will become infinite at the punctures, and λ will be the local parameter required to define the essential singularity of the Baker–Akhiezer function. Finally $|s_i\rangle$, in view of eq. (7.50), is proportional to $|a_i(t)\rangle$. We denote by $\psi_i(t,\lambda,\mu)$ the proportionality coefficient $|s_i\rangle = \psi_i(t,\lambda,\mu)|a_i(t)\rangle$ (do not confuse $\psi_i(t,\lambda,\mu)$, $i = 1,\ldots,N$ with $\psi_\alpha(x,t,P)$, $\alpha = 1,\ldots,l$).

Proposition. *The equation*

$$\left(\partial_t - \partial_x^2 + \sum_{i=1}^{N} \rho_i(t)\wp(x - q_i(t)) \right) |\Psi\rangle = 0 \qquad (7.52)$$

has solutions $|\Psi\rangle$ of the form

$$|\Psi\rangle = \sum_{i=1}^{N} \psi_i(t,\lambda,\mu)\Phi(x - q_i(t),\lambda)e^{-\frac{\mu}{2}x+\frac{\mu^2}{4}t}|a_i(t)\rangle \qquad (7.53)$$

if and only if $q_i(t)$ and the quantities $f_{ij} = \langle b_i|a_j\rangle$ satisfy the equations of motion of the spin Calogero–Moser system, eqs. (7.4), and the constraints $f_{ii} = 2$.

<u>Proof</u>. Inserting equation (7.53) into equation (7.52), we find the condition:

$$\mathcal{E} \equiv \sum_{i=1}^{N} \left\{ \Phi(x - q_i,\lambda)\frac{d(\psi_i|a_i\rangle)}{dt} - (\dot{q}_i - \mu)\Phi'(x - q_i,\lambda)\psi_i|a_i\rangle \right.$$
$$\left. - \Phi''(x - q_i,\lambda)\psi_i|a_i\rangle + \sum_{j=1}^{N} f_{ji}\wp(x - q_j)\Phi(x - q_i,\lambda)\psi_i|a_j\rangle \right\} = 0$$

where $\Phi' = \partial_x\Phi$ and so on. The vanishing of the triple pole $(x - q_i)^{-3}$ gives the condition:

$$\langle b_i|a_i\rangle|a_i\rangle = 2|a_i\rangle \qquad (7.54)$$

Using this condition and the Lamé equation (7.10), we can identify the double pole $(x - q_i)^{-2}$. Its vanishing gives the condition:

$$(\dot{q}_i - \mu)\psi_i|a_i\rangle + \sum_{j\neq i} f_{ij}\Phi(q_i - q_j,\lambda)\psi_j|a_i\rangle = 0 \qquad (7.55)$$

We finally identify the residue of the simple pole and obtain the condition:

$$\left(\frac{d}{dt} - \wp(\lambda)\right)\psi_i|a_i\rangle + \sum_{j\neq i} f_{ji}\wp(q_i-q_j)\psi_i|a_j\rangle + \sum_{j\neq i} f_{ij}\Phi'(q_i-q_j, \lambda)\psi_j|a_i\rangle = 0$$

(7.56)

Inserting back eqs. (7.54, 7.55, 7.56) into the expression of $\mathcal{E}$ one sees that $\mathcal{E}$ vanishes identically due to the functional equation (7.14). Hence the vector function $|\Psi\rangle$ given by eq. (7.53) satisfies eq. (7.52) if and only if the conditions (7.54, 7.55, 7.56) are fulfilled.

From eq. (7.54) it follows that the constraints $f_{ii} = 2$ should hold. Equation (7.55) can then be rewritten as a matrix equation for the N-dimensional vector $\Psi = (\psi_i)$ (not to be confused with the l-dimensional object $|\Psi\rangle$):

$$(L(t, \lambda) - \mu I)\Psi = 0$$

(7.57)

where the matrix $L(t, \lambda)$ is given by:

$$L_{ij}(t, \lambda) = \dot{q}_i\delta_{ij} + (1 - \delta_{ij})f_{ij}\Phi(q_i - q_j, \lambda)$$

(7.58)

We recognize the Lax matrix of the Calogero–Moser model, eq. (7.12), so that the N-vector Ψ identifies with the eigenvector considered in section (7.6).

We can rewrite equation (7.56) as:

$$|\dot{a}_i\rangle = -\Lambda_i|a_i\rangle - \sum_{j\neq i} f_{ji}\wp(q_i - q_j)|a_j\rangle$$

(7.59)

where we have defined:

$$\Lambda_i = \frac{\dot{\psi}_i}{\psi_i} - \wp(\lambda) + \sum_{j\neq i} f_{ij}\Phi'(q_i - q_j, \lambda)\frac{\psi_j}{\psi_i}$$

But this last equation can be written:

$$(\partial_t - \wp(\lambda)I - \Lambda - M)\Psi = 0$$

(7.60)

where $\Lambda = \text{Diag}(\Lambda_i)$ and the matrix $M(t, \lambda)$ is given by:

$$M_{ij}(t, \lambda) = -(1 - \delta_{ij})f_{ij}\Phi'(q_i - q_j, \lambda)$$

(7.61)

We recognize the second matrix M of the Lax pair (7.13). The compatibility condition of eq. (7.57, 7.60) reads $\dot{L} = [M + \Lambda + \wp(\lambda)I, L]$. Of course the term $\wp(\lambda)I$ does not contribute to the commutator. Moreover, we can get rid of the diagonal matrix Λ by performing a conjugation by a diagonal matrix on L, which amounts to quotienting out the toral action. In this way we have exactly recovered the Lax pair of the Calogero–Moser model, eq. (7.11), hence the q_i and f_{ij} have to satisfy the Calogero–Moser equations of motion, in order for $|\Psi\rangle$ to be double Bloch. In the course of the proof, λ and μ have been identified as coordinates on the spectral curve of the Calogero–Moser model, i.e. are related by the equation $\det(L(\lambda) - \mu) = 0$. One can see that the punctures P_β are l among the N points above $\lambda = 0$, and λ is a local parameter around each of them. ∎

The outcome of this analysis is that the double Bloch condition singles out very specific finite-zone potentials. It amounts to an overdetermined linear system on the coefficients of the expansion of $|\Psi\rangle$ which is equivalent to the Lax equation. In particular, we have obtained in a simple and natural way the Lax matrices of the Calogero–Moser model in eqs. (7.58, 7.61). The method is clearly general and lends itself to extensions by Baker–Akhiezer functions with different patterns of essential singularities. Note that in our construction, we have considered only two "times", x and t, which parametrize the singularity at each puncture. This provides a whole variety of integrable systems with spectral parameter lying on a genus 1 curve. In view of the counting of parameters in Lax equations in Chapter 3 this is a notable fact.

7.11 Hitchin systems

A remarkable construction, due to Hitchin, provides integrable systems with spectral parameter lying on a curve of arbitrary genus. The Calogero–Moser model can also be seen as a particular case of this construction.

Let Σ be a Riemann surface of genus g, and let G be a complex semisimple Lie group. Let $\mathcal{A}$ be the space of type $(0,1)$ fields on Σ, i.e. fields of the form $A = A_{\bar{z}}(z, \bar{z})d\bar{z}$ for some local coordinate system z, with values in the Lie algebra of G. We define the "gauge group" $\mathcal{G}$ to be the space of maps from Σ to G, so that $h \in \mathcal{G}$ is a function $h(z, \bar{z})$ with values in G. The gauge group acts on $\mathcal{A}$ as follows:

$$A \longrightarrow A^h \equiv h^{-1}Ah + h^{-1}\bar{\partial}h \qquad (7.62)$$

Note that the differences $A - A'$ form a vector space compatible with the gauge group action, so that $\mathcal{A}$ can be seen as an affine space with group action. We call $\mathcal{N} = \mathcal{A}/\mathcal{G}$ the orbit space of $\mathcal{A}$ under $\mathcal{G}$.

A tangent vector at the point $A \in \mathcal{A}$ is of the form $X = X_{\bar{z}}(z, \bar{z})d\bar{z}$ with values in the Lie algebra of G. A covector Φ at the point A is of the form $\Phi = \Phi_z(z, \bar{z})dz$, the pairing between vectors and covectors at the point A being given by:

$$(\Phi, X) = \int_\Sigma \mathrm{Tr}\,(\Phi_z X_{\bar z})dzd\bar z$$

The gauge group acts on vectors and covectors by adjoint action, $X^h = h^{-1}Xh$, $\Phi^h = h^{-1}\Phi h$, and this leaves the pairing invariant.

The starting point of Hitchin construction is the cotangent bundle $T^*\mathcal{A}$ whose points are pairs (A, Φ), where Φ is a cotangent vector at $A \in \mathcal{A}$. The canonical symplectic form on this space reads:

$$\omega = \int_\Sigma \mathrm{Tr}\,(\delta\Phi_z \wedge \delta A_{\bar z})dzd\bar z$$

Note that this symplectic form is invariant under the gauge group action, so we can perform a Hamiltonian reduction by this group. To do that we need to compute the moment μ of this action. In the case of a cotangent bundle it is shown in Chapter 14 that the Hamiltonian generating the infinitesimal group action is given by

$$H_\epsilon \equiv (\mu, \epsilon) = \int \mathrm{Tr}\,(\mu_{z\bar z}\epsilon)dzd\bar z = \alpha(X_\epsilon(A, \Phi)), \quad \epsilon \in \mathrm{Lie}(\mathcal{G})$$

where α is the canonical 1-form $\alpha = \int_\Sigma \mathrm{Tr}\,(\Phi\delta A)$ and $X_\epsilon(A, \Phi)$ is the infinitesimal variation of the point (A, Φ) under gauge group action, namely:

$$X_\epsilon A = \bar\partial_A\epsilon \equiv \bar\partial\epsilon + [A, \epsilon], \quad X_\epsilon\Phi = [\Phi, \epsilon]$$

One gets after an integration by parts:

$$\mu_{z\bar z}dzd\bar z = -\left(\frac{\partial\Phi_z}{\partial\bar z} + [A_{\bar z}, \Phi_z]\right)dzd\bar z$$

$$\mu = \bar\partial_A\Phi \equiv \bar\partial\Phi + A \wedge \Phi + \Phi \wedge A$$

The phase space $\mathcal{P}$ of the Hitchin system is obtained by choosing the moment equal to 0. The stability group of this moment is therefore the whole gauge group $\mathcal{G}$, so that we have:

$$\mathcal{P} = \mu^{-1}(0)/\mathcal{G}$$

Choosing $\mu = 0$ means that $\bar\partial_A\Phi = 0$, and this has a nice geometric interpretation. A cotangent vector at a point $n \in \mathcal{N}$, which is the class of $A \in \mathcal{A}$ under the $\mathcal{G}$ action, may be viewed as a linear form on $T_A\mathcal{A}$ vanishing on vectors tangent to the fibre, that is, such that $(\Phi, \bar\partial_A\epsilon) = 0$ for all ϵ. By integration by parts, this condition is equivalent to $\bar\partial_A\Phi = 0$. This interpretation being covariant under the gauge group action, it follows that $\mathcal{P} = T^*\mathcal{N}$, where $\mathcal{N}$ is the orbit space (avoiding non-generic orbits to obtain a good manifold). By a theorem of Narasimhan and Seshadri, the space $\mathcal{N}$ is known to be isomorphic to the moduli space of (stable) holomorphic G-bundles on Σ, and this implies in particular that it is finite-dimensional.

Theorem. *The phase space* $\mathcal{P} = T^*\mathcal{N}$ *is of finite dimension: for* $g > 1$ *and* G *a semi-simple Lie group, we have*

$$\dim \mathcal{P} = 2 \dim \mathcal{N} = 2\,(g-1)\dim G$$

<u>Proof.</u> Let us sketch some ideas of the proof. The first step is to relate $\mathcal{N}$ to holomorphic G-bundles. Given $A \in \mathcal{A}$ and a sufficiently fine covering U_α of Σ, one solves, for some C^∞ functions $h_\alpha \in G$ defined in U_α, the equation

$$h_\alpha^{-1}\bar{\partial}h_\alpha = A_\alpha \tag{7.63}$$

where $A_\alpha \equiv A|_{U_\alpha}$. We define a principal G-bundle by the transition functions $g_{\alpha\beta} = h_\alpha h_\beta^{-1}$ on $U_\alpha \cap U_\beta$. The action of $\mathcal{G}$ on h_α reads $h_\alpha^h = h_\alpha h$, so that $g_{\alpha\beta}$ is gauge invariant, and the G-bundle is really attached to a point of $\mathcal{N}$. This bundle is holomorphic, $\bar{\partial}g_{\alpha\beta} = 0$, because $g_{\alpha\beta}^{-1}\bar{\partial}g_{\alpha\beta} = h_\beta(A_\alpha - A_\beta)h_\beta^{-1} = 0$ since $A_\alpha = A_\beta$ on $U_\alpha \cap U_\beta$. Of course, h_α is defined up to right multiplication by a holomorphic function f_α, but this yields an equivalent presentation of the same bundle, with transition functions $g'_{\alpha\beta} = f_\alpha g_{\alpha\beta} f_\beta^{-1}$.

Next we remark that the associated determinant bundle has vanishing Chern class. Viewing G as a group of matrices, define the determinant bundle as a line bundle whose transition functions are $\det g_{\alpha\beta}$. By definition of h_α we have $\bar{\partial}\log\det h_\alpha = \operatorname{Tr} A_\alpha = 0$ because we assume G semi-simple. So $\det h_\alpha$ is in fact holomorphic. Then $\det g_{\alpha\beta} = \det h_\alpha/\det h_\beta$ defines a trivial holomorphic line bundle, and its Chern class vanishes.

To compute the dimension of $\mathcal{N}$, one computes the dimension of the cotangent space of $\mathcal{N}$ at a point $n \in \mathcal{N}$. Take a representative element $A \in \mathcal{A}$ of n. As we have seen, the cotangent space at n identifies with the space of forms Φ of type $(1,0)$ satisfying $\bar{\partial}_A\Phi = 0$. Consider the G-bundle attached to n, defined using some choice of functions h_α as in eq. (7.63). Define on each U_α the forms $\widehat{\Phi}_\alpha = h_\alpha \Phi h_\alpha^{-1}$, with values in the Lie algebra of G. By construction we have $\bar{\partial}\widehat{\Phi}_\alpha = 0$ and $\widehat{\Phi}_\alpha = g_{\alpha\beta}\widehat{\Phi}_\beta g_{\alpha\beta}^{-1}$, i.e. the $\widehat{\Phi}_\alpha$ define a *global holomorphic* section with values in 1-forms on Σ of the associated bundle $\operatorname{Ad} P$, the bundle with fibres the Lie algebra of G and adjoint group action. This shows that the cotangent space to $\mathcal{N}$ at n identifies with $T_n^*\mathcal{N} = H^0(\Sigma, \kappa \otimes \operatorname{Ad} P_n)$, where P_n is the principal G-bundle attached to n, and κ is the canonical bundle.

The Riemann–Roch theorem has an extension to vector bundles which reads in our case:

$$\dim H^0(\Sigma, \kappa \otimes \mathrm{Ad}\, P) - \dim H^0(\Sigma, \mathrm{Ad}\, P) = (g-1)\dim G - c(\det \mathrm{Ad}\, P)$$

where the determinant bundle $\det \mathrm{Ad}\, P$ has transition functions $\det g_{\alpha\beta}$, and so has vanishing Chern class $c(\det \mathrm{Ad}\, P) = 0$. We proceed to show that, generically, the vector bundle $\mathrm{Ad}\, P$ has no global holomorphic section, i.e. $\dim H^0(\Sigma, \mathrm{Ad}\, P) = 0$. Indeed, if $\{f_\alpha\}$ with patching condition $f_\alpha = g_{\alpha\beta} f_\beta g_{\alpha\beta}^{-1}$ defines such a section, for any integer m the quantities $\mathrm{Tr}\, f_\alpha^m$ define a global holomorphic function on Σ, hence a constant. This means that the eigenvalues of f_α are constants, independent of α, and one can write $f_\alpha = u_\alpha \Lambda u_\alpha^{-1}$ for some u_α holomorphic on U_α and a constant traceless diagonal matrix Λ. The patching condition reads $[\Lambda, u_\beta^{-1} g_{\alpha\beta} u_\alpha] = 0$. This implies that the matrices $u_\beta^{-1} g_{\alpha\beta} u_\alpha$ are block diagonal (blocks correspond to coincident eigenvalues of Λ). But this would mean that the fibre bundle P is decomposable, since the transition functions $u_\beta^{-1} g_{\alpha\beta} u_\alpha$ provide an equivalent description of our bundle. We exclude this situation because it is not generic. Finally, the dimension of $T_n^* \mathcal{N}$ is $(g-1)\dim G$, and $\dim T^* \mathcal{N} = 2\,(g-1)\dim G$. ∎

Remark. In the case of line bundles, we have $\dim H^0(\Sigma, \mathrm{Ad}\, P) = 1$, since we are here considering global holomorphic *functions* and the same argument shows that $\dim T_n^* \mathcal{N} = g$.

It will be useful to have a heuristic picture of the moduli. First, it is known that all vector bundles on a non-compact Riemann surface are trivial. This implies that one can describe all vector bundles on a compact Riemann surface by using a covering with only two open sets U_0 and U_∞, where U_0 is a small disc around a point, say $z = 0$, and U_∞ is the Riemann surface with the point $z = 0$ removed. The bundles are then described by giving only one transition function $g_{0\infty}$ defined on the annulus $U_0 - \{z = 0\}$. We get an equivalent bundle by changing the transition function

$$g_{0\infty}(z) \rightarrow g'_{0\infty}(z) = f_0(z) g_{0\infty}(z) f_\infty^{-1}(z)$$

where f_0 is analytic non-vanishing on U_0, and f_∞ is analytic non-vanishing on U_∞, i.e. on the whole of $\Sigma - \{z = 0\}$. The moduli space of vector bundles is the space of transition functions $g_{0\infty}$ modulo the above redefinitions. In the case of a line bundle, we can write quite generally

$$g_{0\infty}(z) = z^k e^{\sum_{-\infty}^{\infty} a_n z^n} \tag{7.64}$$

where k is an integer because $g_{0\infty}(z)$ is single valued on the annulus. Consider now on U_∞ the function $f_\infty(z) = \exp\left(\sum_{n=g+1}^\infty a_{-n}\varphi_{-n}(z)\right)$, where $\varphi_{-n}(z)$ is a meromorphic function on Σ such that around $z = 0$ we have $\varphi_{-n}(z) = z^{-n} + O(z^{-g})$ and $\varphi_{-n}(z)$ is regular everywhere else. Notice that such a function exists and is unique for $n \geq g + 1$. Using this f_∞, we can get rid of all the terms $z^{-n}, n \geq g+1$, in eq. (7.64). Then using $f_0(z) = \exp\left(-\sum_{n=0}^\infty a_n z^n\right)$, we can get rid of all the terms $z^n, n \geq 0$, as well. Hence, we are left with

$$g_{0\infty}(z) = z^k e^{\sum_1^g a_{-n} z^{-n}} \tag{7.65}$$

So the line bundles on Σ are holomorphically classified by an integer k, the Chern class of the bundle, and g continuous moduli, which describe in fact the Picard variety of Σ (the Jacobian for $k = 0$).

In the case of higher rank vector bundles, things are much more complicated. If Σ is a sphere, we know by Riemann–Hilbert factorization that we can always decompose a matrix $g_{0\infty}(z)$ as

$$g_{0\infty}(z) = f_0(z)\lambda(z)f_-(z)^{-1} \tag{7.66}$$

where $\lambda(z)$ is a diagonal matrix with diagonal elements z^{k_i}, for some integers k_i, f_0 has an expansion in z, and f_- has an expansion in $1/z$. This exactly means that vector bundles on the sphere are classified by the integers k_i and have no continuous moduli. In other words, on the sphere $\dim \mathcal{N} = 0$. The integers k_i are holomorphic invariants of the bundle. The Chern class of the corresponding determinant bundle is $\sum k_i$.

In the case of a general Riemann surface Σ, we can still use Birkhoff's theorem in the small disc around $z = 0$ to write the transition function on the annulus in the form eq. (7.66). Note that $f_-(z)$ is here only defined on the annulus and cannot in general be extended to $\Sigma - \{0\}$. However one can hope that a similar mechanism as in the case of line bundles allows us to get rid of powers z^{-n} for $n \geq g+1$, so that we can write the transition function as in eq. (7.65), but with k a diagonal matrix with integer entries k_i, and $a_{-1}, \ldots, a_{-g}$ matrices. If all the integers k_i vanish, one can use the freedom of redefinition of the bundle by constant matrices f_0 and f_∞ to diagonalize a_{-1}. We can still quotient away by conjugating by constant diagonal matrices while preserving this form. If $g > 1$, it remains a space of $(g - 1) \dim G$ parameters (in fact $\operatorname{rank} G + [(g - 1) \dim G - \operatorname{rank} G]$) which can be plausibly taken as coordinates on an open set of the moduli space. If $g = 1$, however, we are left with

$$g_{0\infty}(z) = \exp(a_{-1}/z) \tag{7.67}$$

where a_{-1} is diagonal, so that we have rank G parameters, which is known to be the correct dimension of the moduli space in that case. We will content ourselves with this interpretation in the following.

The construction of Hitchin integrable systems on $\mathcal{P}$ will now be done by defining a very simple set of Poisson commuting functions on $T^*\mathcal{A}$, invariant under the gauge group, and by reducing them to $T^*\mathcal{N}$.

Let $P(X)$ be an invariant polynomial on the Lie algebra of G, i.e. $P(h^{-1}Xh) = P(X)$. Recall that the ring of such invariant polynomials is freely generated (Chevalley's theorem) by homogeneous polynomials P_i, $i = 1,\ldots,\text{rank}\,G$, of degrees m_i, the so-called exponents of the Lie algebra. These numbers are such that:

$$\sum_{i=1}^{\text{rank}\,G} (2m_i - 1) = \dim G$$

For any given invariant polynomial P of degree m, e.g. $P(X) = \text{Tr}\,X^m$, consider the function on phase space taking values in differentials of type $(m,0)$ (i.e. of the form $\omega(z,\bar{z})dz^m$):

$$(A,\Phi) \in T^*\mathcal{A} \to P(\Phi)$$

The differential $P(\Phi)$ is holomorphic, since $\bar{\partial}P(\Phi) = m\,\text{Tr}\,(\bar{\partial}\Phi\Phi^{m-1}) = m\,\text{Tr}\,(\bar{\partial}_A\Phi\Phi^{m-1}) = 0$, where we have used cyclicity of trace. Introducing a basis of holomorphic differentials of type $(m,0)$, say $\omega_j^{(m)}$ we can write

$$P(\Phi) = \sum_j H_{P,j}(\Phi)\,\omega_j^{(m)}$$

The functions $H_{P,j}$ on phase space are $\mathcal{G}$-invariant, and define the Hamiltonians of the Hitchin systems. Note that the basis of differentials $\omega_j^{(m)}$ do not contain dynamical variables, since the Riemann surface Σ is a fixed parameter of the construction.

Proposition. *The functions $H_{P_m,j}$ associated with the primitive polynomials P_m, $m = 1,\ldots,\text{rank}\,G$, which generate the ring of invariant polynomials, are in involution. Their number is $(g-1)\dim G$, so they define a Hamiltonian integrable system on $\mathcal{P} = T^*\mathcal{N}$.*

<u>Proof.</u> The functions $H_{P_m,j}$ seen as functions on $T^*\mathcal{A}$ are in involution, $\{H_{P_m,j}, H_{P_n,k}\} = 0$, because they depend only on the momenta Φ. Since the polynomials P are G-invariant, the functions $H_{P,j}$ are gauge invariant. They are thus well-defined on the symplectic quotient and in involution there, because one can compute directly their reduced Poisson brackets,

see Chapter 14. It remains to show that the number of the Hamiltonians is half the dimension of the phase space. Let us count them. We need the number of holomorphic differentials of type $(m, 0)$. By the Riemann–Roch theorem,

$$\dim H^0(\Sigma, \kappa^m) - \dim H^0(\Sigma, \kappa^{1-m}) = c(\kappa^m) + 1 - g$$

where $c(\kappa^m) = m(2g-2)$ and $\dim H^0(\Sigma, \kappa^{1-m}) = 0$ because $1 - m < 0$, so that $\dim H^0(\Sigma, \kappa^m) = (2m - 1)(g - 1)$. The total number of independent Hamiltonians $H_{P_i,j}$ is therefore:

$$\sum_{i=1}^{\mathrm{rank}\,G} \dim H^0(\Sigma, \kappa^{m_i}) = \sum_{i=1}^{\mathrm{rank}\,G} (2m_i - 1)(g - 1) = (g - 1)\dim G$$

This is half the dimension of the phase space. This counting works for $g > 1$. For $g = 0$ there are no regular differentials on the sphere, so that $\dim H^0(\Sigma, \kappa^m) = 0$. The Hitchin construction in this case yields no interesting system. For $g = 1$ one has $\dim H^0(\Sigma, \kappa^m) = 1$, since this space is spanned by dz^m, so that we find $\sum_{i=1}^{\mathrm{rank}\,G} 1 = \mathrm{rank}\,G$ Hamiltonians. ∎

In genus 0 and 1 the Hitchin construction does not provide useful dynamical systems. This is a motivation to generalize it to Riemann surfaces with marked points. Let $z_k \in \Sigma$ be N points on the Riemann surface Σ. At each of these points we associate an element u_k in the dual of the Lie algebra of G, which we shall identify with the Lie algebra itself using the invariant bilinear form.

Instead of choosing the moment in the Hamiltonian reduction equal to zero, we now choose:

$$\mu(A, \Phi) \equiv \bar{\partial}_A \Phi = 2i\pi \sum_k u_k \delta_{z_k} \qquad (7.68)$$

where δ_{z_k} is the Dirac measure at the point z_k, represented locally around z_k by $\delta(z - z_k)dz d\bar{z}$. The stability group of this momentum is the subgroup of gauge transformations leaving u_k invariant: $\mathcal{G}_{z;u} = \{g \in \mathcal{G} \text{ s.t. } h(z_k) \in G_k\}$, where G_k is the stabilizer of u_k. The reduced phase space is as usual:

$$\mathcal{P}_{z;u} \equiv \mu^{-1}(2i\pi \sum_k u_k \, \delta_{z_k})/\mathcal{G}_{z;u}$$

In other words, we have:

$$\mathcal{P}_{z;u} \equiv \{(A, \Phi) | \bar{\partial}_A \Phi = 2i\pi \sum_k u_k \, \delta_{z_k}\}/\mathcal{G}_{z;u} \qquad (7.69)$$

The equation $\bar{\partial}_A \Phi = 2i\pi \sum_k u_k \, \delta_{z_k}$ specifies the behaviour of Φ around the marked points. Indeed, let us parametrize A locally around z_k as $A = g_k^{-1} \bar{\partial} g_k$ with $g_k(z_k) = 1$. Then the condition that (A, Φ) belongs to $\mathcal{P}_{z;u}$ translates locally into the condition $\bar{\partial}(g_k \Phi g_k^{-1}) = 2i\pi u_k \delta_{z_k}$, so that $g_k \Phi g_k^{-1}$ is holomorphic in an open set around z_k excluding z_k, and behaving locally as

$$g_k \Phi g_k^{-1} = \frac{u_k dz}{z - z_k} + O(1) \tag{7.70}$$

(we used $\bar{\partial} \frac{dz}{z - z_0} = 2i\pi \delta_{z_0}$).

The construction of the commuting Hamiltonians then works as above. For P_i an invariant polynomial of degree m_i on the Lie algebra of G, one considers the function

$$(A, \Phi) \in \mathcal{P}_{z;u} \to P_i(\Phi) \tag{7.71}$$

which take values in the space of meromorphic forms of type $(m_i, 0)$ with poles at points z_k of order at most m_i. These functions are in involution and define an integrable system.

7.12 Examples of Hitchin systems

Example 1. Let us illustrate this construction by considering the Riemann sphere with N marked points $z_1, \ldots, z_N$ and fixed parameters u_k in the dual of the Lie algebra of G. We take A of the form

$$A = h^{-1} \bar{\partial} h \tag{7.72}$$

with $h(z, \bar{z}) \in G$ globally defined on the sphere, so that the attached principal bundle is trivial. The condition $\bar{\partial}_A \Phi = 2i\pi \sum_k u_k \delta_{z_k}$ imposes that $h \Phi h^{-1}$ possesses a simple pole at $z = z_k$ with residue $h(z_k) u_k h^{-1}(z_k)$. Since there are no holomorphic differentials on the sphere, the only solution is:

$$\hat{\Phi} = h \Phi h^{-1} = \sum_k \frac{\hat{u}_k \, dz}{z - z_k} \quad \text{with} \quad \hat{u}_k = h(z_k) u_k h^{-1}(z_k) \tag{7.73}$$

In contrast to eq. (7.70), $h(z_k)$ may be different from one. Requiring that $h \Phi h^{-1}$ is regular at infinity yields:

$$\sum_k \hat{u}_k = 0 \tag{7.74}$$

The data (A, Φ) are parametrized by h through the equations (7.72, 7.73), up to left multiplication $h \to lh$ with l holomorphic, hence constant,

and together with the constraint eq. (7.74). Notice that this constraint
is invariant under $h \rightarrow lh$. Gauge transformations act on h as $h \rightarrow hg$,
where g is such that $g(z_k) \in G_k$, the stability group of u_k. Note that $\hat{\Phi}$
and $\hat{u}_k$ are invariant under this action. Quotienting by $\mathcal{G}_{z;u}$ allows us to
gauge h away except at the marked points. At these points only a copy of
G/G_k survives. This is equivalent to the orbit $\mathcal{O}_k$ of u_k under coadjoint
action of G. Noting that $\hat{u}_k$ describes the orbit $\mathcal{O}_k$, we see that:

$$P_{z;u} = \{(\hat{u}_k) \in \prod_k C\mathcal{O}_k \text{ s.t. } \sum_k \hat{u}_k = 0\}/G \qquad (7.75)$$

where G acts on $\hat{u}_k$ by $\hat{u}_k \rightarrow l\hat{u}_k l^{-1}$ for all k.

The reduced symplectic structure is the usual symplectic structure
on coadjoint orbits. Indeed, consider the canonical 1-form on $T^*\mathcal{A}$, i.e.
$2i\pi\alpha = \int_\Sigma \mathrm{Tr}\,(\Phi \delta A)$. Using the parametrization $A = h^{-1}\bar{\partial}h$, one finds
$2i\pi\alpha = -\int_\Sigma \mathrm{Tr}\left[(h^{-1}\delta h)\bar{\partial}_A \Phi\right]$ which, using eq. (7.68), reduces to:

$$\alpha = -\sum_k \mathrm{Tr}\,(u_k h_k^{-1}\delta h_k)$$

We recognize that $\delta\alpha$ is the Kostant–Kirillov symplectic form on the product of coadjoint orbits $\prod_k \mathcal{O}_k$, which means that the Poisson brackets
read:

$$\{\mathrm{Tr}\,(\epsilon\hat{u}_k), \mathrm{Tr}\,(\epsilon'\hat{u}_{k'})\} = \delta_{kk'}\mathrm{Tr}\,([\epsilon, \epsilon']\hat{u}_k)$$

We still have to quotient by the left action of G. Infinitesimal ac-
tion is given by $\delta\hat{u}_k = [\epsilon, \hat{u}_k]$ and is generated by the Hamiltonian
$H_\epsilon = \sum_k \mathrm{Tr}\,(\epsilon\hat{u}_k)$, so the moment is $\mu = \sum_k \hat{u}_k$ and eq. (7.74) shows
that the phase space can be identified with the symplectic quotient
$P_{z;u} = \mu^{-1}(0)/G$.

Choosing the invariant polynomial $P_2(\Phi) = \mathrm{Tr}\,(\Phi^2)$ gives the
Hamiltonian:

$$P_2(\Phi) = \sum_{k,l} \frac{\mathrm{Tr}\,(\hat{u}_k\hat{u}_l)}{(z - z_k)(z - z_l)} = \sum_k \frac{H_k}{z - z_k} \qquad (7.76)$$

with

$$H_k = \sum_{l \neq k} \frac{\mathrm{Tr}\,(\hat{u}_k\hat{u}_l)}{z_k - z_l}$$

The H_k form a family of commuting Hamiltonians usually called the
Gaudin Hamiltonians, which are very closely related to the Neumann
model.

Example 2. Let us explain how one may rederive the elliptic spin Calogero–Moser system from this general construction. Consider the torus $T_\tau \equiv \mathbb{C}/(\mathbb{Z} + \tau\mathbb{Z})$ of periods 1 and τ, $\mathrm{Im}\,\tau > 0$. We shall use a coordinate z on T_τ with the identification $z \sim z + 1$ and $z \sim z + \tau$. We consider the case with one marked point at $z = 0$. Let u be an element of the (dual of the) Lie algebra attached to this marked point. We need a Cartan decomposition of the Lie algebra of G, i.e. consider the basis (H_i, E_α), with H_i in the Cartan subalgebra of the Lie algebra of G and E_α the root generators. We describe any fibre bundle on the torus T_τ using two open sets. U_0, a small disc around $z = 0$, and $U_\infty = T_\tau - \{0\}$. The transition function is $g_{0\infty} = h_\infty h_0^{-1}$, where h_0 and h_∞ are C^∞ functions solving $A = h^{-1}\bar\partial h$ in the respective open sets U_0 and U_∞. It follows from the above discussion, eq. (7.67), that the transition function $g_{0\infty}$ is equivalent to one of the form

$$g_{0\infty} = \exp(q/z), \quad \text{with } q \text{ diagonal}$$

Note that in general h_0 and h_∞ are defined up to left multiplication by holomorphic functions, but the condition $h_\infty h_0^{-1} = g_{0\infty} = \exp(q/z)$ completely determines h_0 and h_∞ up to left multiplication by the same constant diagonal matrix.

The condition $\bar\partial_A \Phi = 2i\pi u \delta_0$ implies, in U_0, that $\widehat\Phi_0 = h_0 \Phi h_0^{-1}$ has a simple pole at $z = 0$ with residue $\widehat{u} = h_0(0) u h_0(0)^{-1}$. Similarly in U_∞, $\widehat\Phi_\infty = h_\infty \Phi h_\infty^{-1}$ is regular. We have $\widehat\Phi_\infty = g_{0\infty} \widehat\Phi_0 g_{0\infty}^{-1}$ so that $\widehat\Phi$ defines a holomorphic section of our holomorphic vector bundle. Writing the patching condition when $z \to 0$ we see that $\widehat\Phi_\infty$ is regular on T_τ except at $z = 0$ where it behaves as (in the case where all holomorphic indices k_i vanish):

$$\widehat\Phi_\infty = e^{q/z} \left(\frac{\widehat{u}}{z} + O(1) \right) e^{-q/z}$$

In the following we drop the subscript ∞ in $\widehat\Phi_\infty$. Introducing the root decompositions:

$$\widehat{u} = \sum_i \widehat{u}_i H_i + \sum_\alpha \widehat{u}_\alpha E_\alpha, \quad \widehat\Phi = \left(\sum_i \widehat\Phi_i H_i + \sum_\alpha \widehat\Phi_\alpha E_\alpha \right) dz$$

we see that $\widehat\Phi_i \sim \frac{\widehat{u}_i}{z}$, and $\widehat\Phi_\alpha \sim \exp\left(\frac{\alpha(q)}{z}\right)\frac{\widehat{u}_\alpha}{z}$. The first condition means that $\widehat\Phi_i$ is an elliptic function with a single pole of order 1, which means that it is a constant. Hence $\widehat{u}_i = 0$ and $\widehat\Phi_i = p_i$, constant. The second condition means that $\widehat\Phi_\alpha$ has an essential singularity of the Baker–Akhiezer type at $z = 0$. The solution is the Lamé function:

$$\widehat{\Phi}_\alpha = -\widehat{u}_\alpha \frac{\sigma(z - \alpha(q))}{\sigma(z)\sigma(\alpha(q))} e^{\alpha(q)\zeta(z)}$$

The section $\widehat{\Phi}$ finally reads:

$$\widehat{\Phi} \equiv \widehat{\Phi}_\infty = \left(\sum_i p_i H_i + \sum_\alpha \widehat{\Phi}_\alpha E_\alpha \right) dz \qquad (7.77)$$

In the $sl(n)$ case this is exactly of the same form as the Lax matrix of the elliptic spin Calogero–Moser model. The variables (p_i, q_i), where $q = \sum_i q_i H_i$, are the momenta and positions of the particles of the Calogero–Moser model, while the $\widehat{u}_\alpha$ are the spin variables. Choosing now $P_2(\Phi) = \mathrm{Tr}\,(\Phi^2)$ and using the property of Lamé functions, eq. (7.15), we can relate $P_2(\Phi)$ to the Hamiltonian of the spin Calogero–Moser model: $P_2(\Phi) = H + \sum_\alpha \mathrm{Tr}(\widehat{u}_\alpha \widehat{u}_{-\alpha}) \wp(z)$ with

$$H = \sum_i p_i^2 - \sum_\alpha \wp(\alpha(q)) \, \mathrm{Tr}(\widehat{u}_\alpha \widehat{u}_{-\alpha})$$

As in the previous example, taking the quotient by the gauge group leaves only the variables q_i and p_i which parametrize respectively the moduli space $\mathcal{N}$ and the Cartan component of $\widehat{\Phi}$, and the spin variables $\widehat{u} = h_0(0) u h_0(0)^{-1}$ describing the G-orbit of u, restricted by $\widehat{u}_i = 0$. Moreover, since h is defined up to left multiplication by a constant diagonal matrix, and this leaves the constraint $\widehat{u}_i = 0$ invariant, the phase space is:

$$\mathcal{P}_u = \{(p, q, \widehat{u} \in \mathcal{O}_u) \text{ s.t. } \widehat{u}_i = 0\}/H$$

where H is the Cartan torus, acting on $\widehat{u}$ by conjugation.

It remains to show that our dynamical variables have the correct Poisson brackets. To do that we compute the canonical 1-form:

$$2i\pi\alpha = \int_\Sigma \mathrm{Tr}\,(\Phi\delta A) = \int_{U_0} \mathrm{Tr}\,(\Phi\delta A) + \int_{\Sigma - U_0} \mathrm{Tr}\,(\Phi\delta A)$$

where we use the description of the bundles on T_τ with two open sets. Writing $A = h_0^{-1}\bar{\partial}h_0 = h_\infty^{-1}\bar{\partial}h_\infty$ in the respective open sets, one computes:

$$\int_{U_0} \mathrm{Tr}\,(\Phi\delta A) = -\int_{U_0} \mathrm{Tr}\,(h_0^{-1}\delta h_0 \bar{\partial}_A\Phi) + \int_{\partial U_0} \mathrm{Tr}\,(\Phi h_0^{-1}\delta h_0)$$

$$\int_{\Sigma - U_0} \mathrm{Tr}\,(\Phi\delta A) = -\int_{\Sigma - U_0} \mathrm{Tr}\,(h_\infty^{-1}\delta h_\infty \bar{\partial}_A\Phi) + \int_{\partial(\Sigma - U_0)} \mathrm{Tr}\,(\Phi h_\infty^{-1}\delta h_\infty)$$

First we have $\int_{\partial(\Sigma-U_0)} = -\int_{\partial U_0}$. Second, since $\bar{\partial}_A \Phi = u\delta_0$, the integral $\int_{\Sigma-U_0}$ vanishes. We get:

$$\alpha = -\mathrm{Tr}\left(uh_0(0)^{-1}\delta h_0(0)\right) + \frac{1}{2i\pi}\int_{\partial U_0} \mathrm{Tr}\left(\Phi(h_0^{-1}\delta h_0 - h_\infty^{-1}\delta h_\infty)\right)$$

The contour integral can be rewritten as $\int_{\partial U_0} \mathrm{Tr}\left(\widehat{\Phi}_\infty(h_\infty h_0^{-1}\delta h_0 h_\infty^{-1} - \delta h_\infty h_\infty^{-1})\right)$. Varying the equation $h_\infty h_0^{-1} = g_{0\infty} = \exp(q/z)$ we have:

$$\frac{\delta q}{z} = \delta h_\infty h_\infty^{-1} - h_\infty h_0^{-1}\delta h_0 h_\infty^{-1}$$

so that we finally get:

$$\alpha = -\mathrm{Tr}\left(uh_0(0)^{-1}\delta h_0(0)\right) - \frac{1}{2i\pi}\int_{\partial U_0} \mathrm{Tr}\left(\widehat{\Phi}_\infty \frac{\delta q}{z}\right)$$
$$= -\sum_i p_i \delta q_i - \mathrm{Tr}\left(uh(0)^{-1}\delta h(0)\right)$$

To evaluate the contour integral, we noticed that δq being diagonal, only the diagonal part of $\widehat{\Phi}$ in eq. (7.77) contributes to the trace. It is clear that the constraint $\widehat{u}_i = 0$ and the quotient by the Cartan torus is a symplectic quotient of the coadjoint orbit, so finally the symplectic form reads:

$$\omega = -\delta\alpha = \sum_i \delta p_i \wedge \delta q_i + \omega_K$$

Remark 1. The solution of the Hitchin systems can be viewed as the projection, under taking the quotient by the gauge group, of a straight line motion on $T^* \mathcal{A}$ since on this space the equations of motion are of the form $\dot{\Phi} = 0$ and $\dot{A} = P'(\Phi) = \mathrm{Const}.$

Remark 2. Note that the constraints $\widehat{u}_i = 0$ are the same as the constraints $f_{ii} = \alpha$ in eq. (7.5), because the Cartan algebra for $sl(N)$ is generated by $E_{ii} - E_{jj}$. Moreover, it is clear that $\widehat{\Phi}$ occuring in eq. (7.77) is the same as the Lax matrix in eq. (7.12). This provides some insight in the true nature of the Lax matrix, which appears as a form-valued holomorphic section of a vector bundle (depending on the dynamical data).

Remark 3. The above construction naturally yields a Lax pair formulation of the equations of motion. Since $\widehat{\Phi} = h\Phi h^{-1}$, we have $d\widehat{\Phi}/dt = [M, \widehat{\Phi}]$ with $M = \dot{h}h^{-1}$. As it was emphasized above, h is determined up to a constant (in z) diagonal matrix, the diagonal action on these data is given by $\widehat{\Phi} \to l\widehat{\Phi}l^{-1}$ and $M \to lMl^{-1} + \dot{l}$.

7.13 The trigonometric Calogero–Moser model

In this section we present the construction, due to Olshanetsky and Perelomov, of the trigonometric spin Calogero–Moser model by Hamiltonian reduction of the geodesic motion on a symmetric space.

The Lax matrix of the elliptic spin Calogero–Moser model, eq. (7.12), reads in the trigonometric limit:

$$L_{ij} = p_i\delta_{ij} + (1 - \delta_{ij})f_{ij}(\coth(q_{ij}) - \coth(\lambda))e^{q_{ij}\coth(\lambda)}$$

In the limit $\lambda \to -\infty$ we get:

$$L_{ij} = p_i\delta_{ij} + (1 - \delta_{ij})\frac{f_{ij}}{\sinh(q_i - q_j)}$$

This Lax matrix, without spectral parameter, is a good Lax matrix in the case of the scalar trigonometric Calogero–Moser model, as it produces N commuting conserved quantities. In the spin case, however, it does not provide enough conserved commuting quantities. We will construct this Lax matrix by a Hamiltonian reduction procedure, starting from a finite-dimensional space.

The construction is similar to that in the case of the open Toda chain, see Chapter 4. One starts from the symplectic space T^*G, where G is a complex Lie group, and reduce under the action on the left by a subgroup H_L and on the right by a subgroup H_R. The general theory, see Chapter 14, shows that the Poisson brackets in the coordinates (g, ξ) are expressed as:

$$\{\xi(X), \xi(Y)\} = \xi([X, Y]), \quad \{\xi(X), g\} = g\, X, \quad \{g, g\} = 0 \qquad (7.78)$$

Recall that here g is the matrix of functions on G whose elements are the matrix elements of g in a faithful representation.

As in the case of the Toda chain we reduce the geodesic motion on G. The Hamiltonian on T^*G reads $H_{\text{geod}} = \frac{1}{2}\,\text{Tr}\,(\xi^2)$. Notice that H_{geod} is bi-invariant, so one can attempt to reduce this dynamical system using Lie subgroups H_L and H_R of G, acting respectively on the left and on the right on T^*G. Recall that the corresponding moments are, see eq. (4.49) in Chapter 4:

$$\mathcal{P}^L(g, \xi) = -P_{\mathcal{H}_L^*}\, g\xi g^{-1}, \quad \mathcal{P}^R(g, \xi) = P_{\mathcal{H}_R^*}\, \xi \qquad (7.79)$$

where we have introduced the projectors on $\mathcal{H}_{L,R}^*$ of elements of $\mathcal{G}^*$ induced by the restriction of these elements (linear forms) to $\mathcal{H}_{L,R}$, the Lie algebras of $H_{L,R}$.

Let us consider an involutive automorphism σ of G and let H be the subgroup of its fixed points. Then H acts on the right on G, defining a principal fibre bundle of total space G and base G/H, which is a symmetric space. Moreover, G acts on the left on G/H and in particular so does H itself. We shall consider the reduction of the geodesic motion on G under the product $H_L \times H_R$ with $H_L = H_R = H$. The reduction is achieved with an adequate choice of the momentum $\mu = (-\mu_L, \mu_R)$ such that $(\mathcal{P}^L, \mathcal{P}^R) = \mu$. We take $\mu_R = 0$ so that the isotropy group of the right action is H_R itself. In this way we are in fact dealing with motions on G/H.

The derivative of σ at the unit element of G is an involutive automorphism of $\mathcal{G}$, also denoted by σ. Let us consider its eigenspaces $\mathcal{H}$ and $\mathcal{K}$ associated with the eigenvalues $+1$ and -1 respectively. We have a decomposition:

$$\mathcal{G} = \mathcal{H} \oplus \mathcal{K} \qquad (7.80)$$

in which $\mathcal{H}$ is the Lie algebra of H. Note that $h\mathcal{K}h^{-1} = \mathcal{K}$, for $h \in \mathcal{H}$.

We shall consider the particular case obtained by taking $G = SL(N, \mathbb{C})$ and $H = SU(N)$. Here, we view $SL(N, \mathbb{C})$ as a *real* Lie algebra. Counting real parameters, $\dim G = 2N^2 - 2$ (note that $\det M = 1$ yields two conditions) and $\dim H = N^2 - 1$. The automorphism σ is $\sigma(g) = (g^*)^{-1}$, where $*$ means transpose and complex conjugate. The set of its fixed points is $SU(N)$. The symmetric space $K = G/H$ may be identified with the space of positive Hermitian matrices. Indeed, for any matrix $M \in SL(N, \mathbb{C})$ we can uniquely write the Cayley decomposition $M = KU$ with $U \in SU(N)$ and K positive Hermitian. This is because MM^* is Hermitian and strictly positive, so we can write $MM^* = VDV^{-1}$ with $V \in SU(N)$ and D diagonal real positive. We define $K = V\sqrt{D}V^{-1}$, where $\sqrt{D}$ is the positive square root. Then we check that $K^{-1}M$ is in $SU(N)$. At the Lie algebra level we have the decomposition (7.80) where $\mathcal{H}$ is the Lie algebra of antihermitian matrices, and $\mathcal{K}$ is the space of Hermitian matrices. Finally, σ reads $X \to -X^*$, and is an automorphism for the structure of a real Lie algebra. The dual of $\mathcal{G}$ is identified with $\mathcal{G}$ using the symmetric real invariant bilinear form $(X, Y) = \mathrm{Tr}\,(XY + X^*Y^*)$.

The choice of the moment μ_L is of course of crucial importance. We will consider μ_L in the (dual of the) Lie algebra of H, i.e. an antihermitian matrix, such that its isotropy subgroup H_μ is a maximal proper Lie subgroup of H_L, so that the phase space of the reduced system is of minimal dimension but non-trivial. As a matter of fact, the dimension of the reduced phase space is $2\dim G - \dim(H \times H) - \dim(H_\mu \times H)$. This is because $\dim T^*G = 2\dim G$, the constraint $(\mathcal{P}^L, \mathcal{P}^R) = (-\mu_L, 0)$ yields $\dim(H \times H)$ equations, and we still have to quotient by $H_\mu \times H$. To analyse the stabilizer of μ we have to solve $h^{-1}\mu h = \mu$. In a basis where μ is diagonal, we see that if μ has $N - l$ equal eigenvalues and the other

l eigenvalues all different, the stabilizer is $U(1)^l \times U(N-l)/U(1)$. This yields a reduced phase space of dimension $2Nl - l(l+1)$. Note that this is the dimension of the phase space of the spin Calogero–Moser model, eq. (7.8), apart from a discrepancy of 2 which corresponds to a reduction by the centre of mass motion, as we shall see.

We now introduce coordinates on the group G. Using the Cayley decomposition, we can write uniquely $g = KU$ with K Hermitian positive and U unitary. Diagonalizing K we write $K = h_L Q h_L^{-1}$ with Q diagonal with real positive entries, and $h_L \in SU(N)$. The columns of h_L form the orthonormal basis of eigenvectors of K, hence are defined up to a phase. So we can write $g = h_L Q h_R^{-1}$, where $h_L, h_R \in SU(N)$ are defined up to right multiplication by the same unitary diagonal matrix.

The set of points of T^*G with given moment $(-\mu_L, 0)$ is defined by the equations:

$$P_{\mathcal{H}}\left(g\xi g^{-1}\right) = P_{\mathcal{H}}\left(h_L Q h_R^{-1} \xi h_R Q^{-1} h_L^{-1}\right) = \mu_L, \quad P_{\mathcal{H}}\left(\xi\right) = 0 \quad (7.81)$$

The second condition means that $\xi \in \mathcal{K}$ is a Hermitian traceless matrix. Setting $L = h_R^{-1}\xi h_R \in \mathcal{K}$, the first condition reads $(h_L Q L Q^{-1} h_L^{-1}, X) = (\mu_L, X), \forall X \in \mathcal{H}$. Equivalently $(QLQ^{-1}, h_L^{-1}X h_L) = (h_L^{-1}\mu_L h_L, h_L^{-1}X h_L)$, yielding:

$$P_{\mathcal{H}}\left(QLQ^{-1}\right) = h_L^{-1}\mu_L h_L$$

Setting $Q = \mathrm{Diag}\left(e^{q_i}\right)$ with $\sum_i q_i = 0$ (hence the reduction by the centre of mass motion), we have $(QLQ^{-1})_{ij} = \exp\left(q_i - q_j\right)L_{ij}$. The projection on $\mathcal{H}$ amounts to take the antiHermitian part of this matrix, giving the equation $\sinh\left(q_{ij}\right)L_{ij} = \widehat{\mu}_{ij}$ with $\widehat{\mu}_{ij} = (h_L^{-1}\mu_L h_L)_{ij}$. The solution of this equation is:

$$L_{ij} = \frac{\widehat{\mu}_{ij}}{\sinh\left(q_{ij}\right)}, \quad i \neq j, \quad L_{ii} = p_i, \quad \sum p_i = 0, \quad \widehat{\mu}_{ii} = 0$$

where the p_i are arbitrary real numbers. We have found:

$$L = \sum_i p_i E_{ii} + \sum_{i \neq j} \frac{\widehat{\mu}_{ij}}{\sinh\left(q_{ij}\right)} E_{ij}$$

We recognize the Lax matrix of the trigonometric spin Calogero–Moser model, with the spin variables $\widehat{\mu}_{ij}$ describing the coadjoint H-orbit of the momentum μ_L. In view of the definition of L, the ambiguity on the definition of h_L, h_R amounts to quotienting by a conjugation by a diagonal unitary matrix.

We still have not explored the implications of the conditions $\widehat{\mu}_{ii} = 0$ on h_L. We do this for the case of the scalar model, i.e. μ_L has $N-1$ equal eigenvalues. This means that μ_L is of the form $\mu_L = i(VV^* - \alpha I)$, where V is an N-vector and α is determined so that μ_L is traceless. We then have $\hat{\mu} = i(WW^* - \alpha I)$, where $W = h_L^{-1}V$. The conditions $\hat{\mu}_{ii} = 0$ read $W_i \bar{W}_i = \alpha$ for all i. Since h_L is defined up to right multiplication by a diagonal unitary matrix, we can always assume that W_i is real positive, so that $W_i = \sqrt{\alpha}$. There always exists some fixed element $h_0^{-1} \in SU(N)$ mapping V to the vector $W = \sqrt{\alpha}(1, \ldots, 1)^t$, and any other solution h_L is of the form $h_L = h_0 h_W$, where h_W runs over the stability group of W, which is exactly $h_0^{-1} H_\mu h_0$, where H_μ is the stabilizer of μ. Here the element h_0 plays essentially no role, and one usually takes $\mu_L = i(WW^* - \alpha I)$. Then all solutions of the constraints $(\mathcal{P}_L, \mathcal{P}_R) = (-\mu_L, 0)$ are of the form $(h_L, h_R).(Q, L)$ with Q, L as above, and $h_L \in H_\mu$, as it should be. Quotienting by $H_\mu \times H$, according to the general procedure, yields the phase space of the Calogero–Moser model. A similar analysis can be performed in the case of the spin Calogero–Moser model.

Finally, it is easy to see that the symplectic structure of the cotangent bundle we started with gives the symplectic structure of the spin Calogero–Moser model under reduction. It is enough to compute the canonical 1-form:

$$\alpha = (\xi, g^{-1}\delta g) = (QLQ^{-1}, h_L^{-1}\delta h_L) + (L, Q^{-1}\delta Q) - (L, h_R^{-1}\delta h_R)$$

Using the constraint eq. (7.81), the first term is $(\mu_L, \delta h_L h_L^{-1})$, that is the Kirillov structure on the spin variables, the second term is $\sum_i p_i \delta q_i$, and the last term vanishes because the constraint $P_\mathcal{H}(\xi) = 0$ also implies $P_\mathcal{H}(L) = 0$.

References

[1] F. Calogero, Exactly solvable one-dimensional many-body systems. *Lett. Nuovo Cimento* **13** (1975) 411–415.

[2] J. Moser, Three integrable Hamiltonian systems connected with isospectral deformations. *Adv. Math.* **16** (1975) 441–416.

[3] M.A. Olshanetsky and A.M. Perelomov, Classical integrable finite-dimensional systems related to Lie algebras. *Physics Reports* **71** (1981).

[4] I.M. Krichever, Elliptic solutions of Kadomtsev–Petviashvilii equation and integrable systems of particles. *Func. Anal. App.* **14** (1980) no. 4, 282–290.

[5] J. Gibbons and T. Hermsen, A generalization of the Calogero–Moser system. *Physica* **11D** (1984) 337.

[6] H. Airault, H. McKean and J. Moser, Rational and elliptic solutions of the KdV equation and related many-body problem. *Comm. Pure Appl. Math.* **30** (1977) 95–125.

[7] N. Hitchin, Stable bundles and integrable systems. *Duke Math. Journ.* **54** (1987) 91.

[8] M.S. Narasimhan and C.S. Seshadri, Holomorphic vector bundles on a compact Riemann surface. *Math. Ann.* **155** (1964) 69–80.

[9] R.C. Gunning, *Lectures on vector bundles over Riemann surfaces*, Princeton University Press (1967).

[10] O. Forster, *Lectures on Riemann Surfaces*, Springer (1981).

8

Isomonodromic deformations

In this chapter, we consider isomonodromic deformations of the first order linear differential operator $\partial_\lambda - M_\lambda(\lambda)$. Here the problem is to determine M_λ such that solutions of this differential equation have a given monodromy. In general, the solution is not unique, and depends on a number of continuous parameters, the so-called isomonodromic deformation parameters. We show that the deformation equations with respect to these parameters are an integrable system. Ordinary integrable systems with a Lax matrix appear as particular cases of such systems, namely when the group generated by the monodromies is finite. However, the new setting is much more general. Just as solutions of Lax equations were written in terms of theta-functions, in the general case, the solutions can be written in terms of new functions called tau-functions. We express the dynamical variables in terms of tau-functions and their so-called Schlesinger transforms. We show that, in terms of tau-functions, the equations of motion take the form of bilinear Hirota equations. Finally, we show that the Painlevé equations can be interpreted as isomonodromic deformation equations.

8.1 Introduction

In Chapter 3, we have seen that the Zakharov–Shabat construction yields hierarchies of integrable equations of the form:

$$\partial_{t_i}\Psi(\lambda) = M_i(\lambda)\Psi(\lambda), \quad M_i(\lambda) = \big(g(\lambda)\xi_i(\lambda)g^{-1}(\lambda)\big)_- \tag{8.1}$$

Here i is a multi-index $i = (k, n, \alpha)$, $\xi_i(\lambda)$ is the diagonal matrix $\xi_i(\lambda) = (\lambda - \lambda_k)^{-n}E_{\alpha\alpha}$, $g(\lambda)$ is a regular matrix, and

$$\Psi(\lambda) = g(\lambda)e^{\sum_i \xi_i(\lambda)t_i}$$

is a matrix of size $N \times N$. The notation $(\ )_-$ means taking the polar part at λ_k. In Chapter 3, the wave function $\Psi(\lambda)$ was defined through the

Lax equation $L(\lambda)\Psi(\lambda) = \Psi(\lambda)\hat{\mu}$, where $\hat{\mu}$ is the diagonal matrix of the eigenvalues of $L(\lambda)$. We have shown that the isospectral flows defined by eq. (8.1) are all commuting and satisfy the zero curvature equations:

$$\partial_{t_i} M_j - \partial_{t_j} M_i - [M_i, M_j] = 0 \qquad (8.2)$$

In this chapter we enlarge the framework by replacing the Lax equation by a linear differential equation in the spectral parameter λ:

$$\partial_\lambda \Psi = M_\lambda \Psi \qquad (8.3)$$

We assume that the entries of the matrix $M(\lambda)$ are rational functions of λ:

$$M_\lambda(\lambda) = \sum_{k=1}^{K} \left(\frac{A_1^{(k)}}{(\lambda - \lambda_k)} + \cdots + \frac{A_{n_k+1}^{(k)}}{(\lambda - \lambda_k)^{n_k+1}} \right) - A_0^{(\infty)} - \cdots - A_{n_\infty-1}^{(\infty)} \lambda^{n_\infty-1}$$

$$(8.4)$$

This is of the form $M_\lambda(\lambda) = \sum_k M_\lambda^{(k)}(\lambda)$, where $M_\lambda^{(k)}$ is the polar part at λ_k (including ∞). The polar parts $M_\lambda^{(k)}$ at λ_k are readily obtained from the expression of M_λ, but some care must be taken concerning the polar part at ∞. To *define* $M_\lambda^{(\infty)}$ one chooses a local parameter $z = 1/\lambda$ and writes the differential equation as $\partial_z \Psi(z) = (M_\lambda^{(\infty)}(z) + \text{regular})\Psi(z)$, so that using eq. (8.4) one gets:

$$M_\lambda^{(\infty)} \equiv -\left(\frac{1}{z^2} M_\lambda(\frac{1}{z}) \right)_- = A_{n_\infty-1}^{(\infty)} z^{-n_\infty-1} + \cdots + A_0^{(\infty)} z^{-2} - \sum_k A_1^{(k)} z^{-1}$$

$$(8.5)$$

The solutions $\Psi(\lambda)$ of the linear differential equation (8.3) have essential singularities at the $\lambda = \lambda_k$ and at $\lambda = \infty$. They are otherwise analytic but have non-trivial monodromies around the singularities. So, they are in fact defined on a Riemann surface with an infinite (in general) number of sheets. This will be understood in the following.

Again we will show that around each pole λ_k of M_λ, the function $\Psi(\lambda)$ admits an expansion of the form

$$\Psi(\lambda) \sim \Psi_{\text{asy}}(\lambda) = g^{(k)}(\lambda)\, e^{\xi^{(k)}(\lambda)}$$

where $g^{(k)}(\lambda) = g_0^{(k)}(1 + g_1^{(k)}(\lambda - \lambda_k) + \cdots)$ is regular at λ_k and

$$\xi^{(k)} = B_0^{(k)} \log(\lambda - \lambda_k) + \sum_{\alpha,n=1}^{n_k} \frac{t_{(k,n,\alpha)}}{(\lambda - \lambda_k)^n} E_{\alpha\alpha}$$

$$\xi^{(\infty)} = B_0^{(\infty)} \log\left(\frac{1}{\lambda}\right) + \sum_{\alpha,n=1}^{n_\infty} t_{(\infty,n,\alpha)} \lambda^n E_{\alpha\alpha} \qquad (8.6)$$

While the equations look the same as in the isospectral context, there are important differences. One of them is that the quantity $\xi^{(k)}$ includes now a logarithmic term, and another one is that the expansions are now only valid in the asymptotic sense (in general), hence the notation $\Psi_{\text{asy}}(\lambda)$. Nevertheless, this is sufficient to show that the polar part $M_\lambda^{(k)}$ of M_λ at λ_k is given by

$$M_\lambda^{(k)} = \left(g^{(k)} \partial_\lambda \xi^{(k)} g^{(k)\,-1} \right)_-$$

The solution $\Psi(\lambda)$ of eq. (8.3) has non-trivial monodromy properties when λ makes a loop around a singularity of $M_\lambda(\lambda)$. We call isomonodromic deformations the deformations of $M_\lambda(\lambda)$ such that the monodromy properties of $\Psi(\lambda)$ are kept fixed. Our aim is to write evolution equations which describe these deformations.

The isomonodromic deformation parameters will include as before the $t_{(k,n,\alpha)}$ occurring in eq. (8.6). Under these time evolutions we have:

$$\partial_{t_i} \Psi = M_i \Psi, \quad M_i = \left(g^{(k)} \partial_{t_i} \xi^{(k)} g^{(k)\,-1} \right)_-, \quad i = (k, n, \alpha)$$

Moreover, this set of commuting flows will be enlarged by varying the positions λ_k of the poles of $M_\lambda(\lambda)$:

$$\partial_{\lambda_k} \Psi = M_{\lambda_k} \Psi, \quad M_{\lambda_k} = \left(g^{(k)} \partial_{\lambda_k} \xi^{(k)} g^{(k)\,-1} \right)_- \tag{8.7}$$

All these flows will be interpreted as isomonodromic deformations of the differential equation $\partial_\lambda \Psi = M_\lambda \Psi$, and will be shown to commute.

8.2 Monodromy data

In this section, we define precisely the monodromy properties of a linear differential equation of the form eq. (8.3). Our purpose is to clarify the relation between the true solution $\Psi(\lambda)$, and local expansions $\Psi_{\text{asy}}(\lambda) = g(\lambda) \exp(\xi(\lambda))$ around each pole λ_k. We start from a wave-function $\Psi(\lambda)$, solution of the linear differential equation eq. (8.3), where $M_\lambda(\lambda)$ is a globally defined rational function of λ. That is, it is a sum of polar parts $M_\lambda^{(k)}(\lambda)$ at given poles λ_k, as in eq. (8.4). Here we consider the parameters $A_j^{(k)}$ as given and our aim is to reconstruct the asymptotic expansions from the differential equation.

We begin by recalling some basic definitions and facts concerning linear differential equations with singular points.

Definition. *The point λ_k is a regular singular point of the linear differential equation eq. (8.3), if $M_\lambda(\lambda)$ has a simple pole at λ_k. When $M_\lambda(\lambda)$*

has a pole of order $n_k + 1 \geq 2$, the point λ_k is called an irregular singular point of rank n_k.

A more refined definition should include the notion of apparent singularities removable through simple changes of variables. Special care must be taken at the point $\lambda = \infty$. To include it into this definition we set $z = 1/\lambda$. The equation becomes:

$$\partial_z \Psi = -1/z^2 M_\lambda(1/z)\Psi \qquad (8.8)$$

When $M_\lambda(\lambda)$ is rational in λ, this equation is of the same form as the original one, so that we can in fact consider that it is defined on the Riemann sphere. Assume first that $M_\lambda(\lambda)$ is a sum of poles at finite distance, i.e. all coefficients $A_j^{(\infty)} = 0$ in eq. (8.4). We then have $M_\lambda(\lambda) = O(1/\lambda)$ at ∞ so that $1/z^2 M_\lambda(1/z)$ has a simple pole at $z = 0$ which is thus a regular singular point. On the other hand, if $A_{n_\infty-1}^{(\infty)} \neq 0$ then $1/z^2 M_\lambda(1/z) \sim 1/z^{n_\infty+1}$ and we have an irregular singularity of rank n_∞ (in particular this is the case if we have only a constant term corresponding to $n_\infty = 1$).

The behaviour of solutions at the two types of singularities is very different. Assume first that eq. (8.3) has a regular singularity at $\lambda = 0$, i.e. one can write it $\lambda \partial_\lambda \Psi(\lambda) = \mathcal{A}(\lambda)\Psi(\lambda)$, with $\mathcal{A}(\lambda)$ analytic around $\lambda = 0$:

$$\mathcal{A}(\lambda) = \mathcal{A}_0 + \mathcal{A}_1 \lambda + \mathcal{A}_2 \lambda^2 + \cdots$$

We also assume that all eigenvalues of $\mathcal{A}_0$ are different and do not differ by integers.

Theorem. *Let $\lambda = 0$ be a regular singularity. There exists a fundamental matrix of solutions in some neighbourhood of $\lambda = 0$ of the form:*

$$\Psi(\lambda) = g(\lambda) e^{B \log \lambda}$$

where $g(\lambda)$ is analytic around $\lambda = 0$, and $B = \text{Diag}\,(\alpha_i)$, where α_i are the eigenvalues of $\mathcal{A}_0$.

The coefficients of the development of $g_{\text{ex}}(\lambda)$ are obtained by plugging a series expansion into the differential equation, and determining the coefficients recursively. Indeed, by a constant similarity transformation, we can assume that $\mathcal{A}_0$ is diagonal. Setting $g = 1 + g_1 \lambda + g_2 \lambda^2 + \cdots$ we get the recursive system:

$$r g_r - [\mathcal{A}_0, g_r] = \sum_{i=1}^{r} \mathcal{A}_{r-i} g_i$$

In components this equation reads $(r - \alpha_i + \alpha_j)(g_r)_{ij} = R_{ij}^{(r-1)}$, where $R^{(r-1)}$ only depends on $g_1, \ldots, g_{r-1}$, so that g_r is uniquely determined if the eigenvalues α_i do not differ by integers. Moreover, it is easy to show that one gets a series with non-vanishing radius of convergence.

If now we have an irregular singularity at $\lambda = 0$, the situation is very different. Let us assume that the equation is of the form

$$\lambda^{n+1} \partial_\lambda \Psi(\lambda) = \mathcal{A}(\lambda) \Psi(\lambda)$$

with $n \geq 1$, and that the most singular term, i.e. $\mathcal{A}_0$, has distinct eigenvalues. One can find a formal expansion $g(\lambda) = g_0 + g_1 \lambda + \cdots$ by plugging it into the differential equation, but one finds a system of linear equations of the form $[g_r, \mathcal{A}_0] = R^{(r-1)}$, where the right-hand side depends on lower order terms. One can obtain formal solutions, but the resulting series is in general *divergent*. The precise meaning of $\Psi \sim \Psi_{\text{asy}}$ is that $\Psi(\lambda) \exp(-\xi(\lambda))$ is asymptotically equal to the formal series $g(\lambda)$.

Theorem. *Let $\lambda = 0$ be an irregular singularity. The differential equation has a fundamental system of* formal *solutions in the neighbourhood of $\lambda = 0$ of the form:*

$$\Psi_{\text{asy}}(\lambda) = g(\lambda) \, e^{\xi(\lambda)} \tag{8.9}$$

where $\xi(\lambda) = B_0 \log \lambda + \cdots$ is a diagonal matrix, and the dots represent a polynomial in $1/\lambda$ with dominant term $\frac{1}{n\lambda^n} \operatorname{Diag}(\alpha_1, \ldots, \alpha_N)$, where the α_i are the eigenvalues of the matrix $\mathcal{A}_0$, and $g(\lambda)$, which is determined up to a right multiplication by a constant diagonal matrix, has a formal expansion of the form $g(\lambda) = g_0 + g_1 \lambda + \cdots$.

In each angular sector *S of angle slightly bigger than π/n with vertex $\lambda = 0$, there exists a unique true solution $\Psi(\lambda)$ which admits, in S, an asymptotic expansion given by the above formal series Ψ_{asy}.*

Proof. We shall not give a complete proof of these theorems here (see the References, particularly Wasow) but sketch a few of the ideas involved. It is simpler to assume that the singularity is at infinity, so we set $z = 1/\lambda$. We consider the equation

$$z^{-q} \Psi'(z) = \mathcal{A}(z) \Psi(z)$$

around the singular point $z = \infty$ of order $(q + 1)$. We assume that $\mathcal{A}(z) = \mathcal{A}_0 + 1/z \mathcal{A}_1 + \cdots$, and make the further simplifying assumption that all eigenvalues of $\mathcal{A}_0$ are different, hence there is no restriction in taking $\mathcal{A}_0$ diagonal. We show below that there is a matrix $P(z)$ such that under the transformation $\Psi(z) = P(z) W(z)$, the differential equation becomes $W'(z) = z^q Q(z) W(z)$ with $Q(z)$ diagonal. This equation is readily solved, yielding:

$$\Psi(z) = P(z) e^{\int^z \zeta^q Q(\zeta) d\zeta} \tag{8.10}$$

The matrix $P(z)$ must obey the equation:

$$z^{-q}P'(z) = -P(z)Q(z) + \mathcal{A}(z)P(z) \tag{8.11}$$

in which $Q(z)$ is diagonal. Of course the transformation $P(z)$ is defined up to multiplication on the right by a diagonal matrix, since this would leave the matrix $Q(z)$ diagonal. We fix this ambiguity by requiring $P_{kk}(z) = 1$. Taking the diagonal element kk of eq. (8.11) one gets $Q_{kk}(z) = (\mathcal{A}(z)P(z))_{kk}$. Plugging this into eq. (8.11) one gets a non-linear system for the $N(N-1)$ non-diagonal elements $P_{ij}(z)$ with $i \neq j$:

$$z^{-q}P'_{ij}(z) = (\mathcal{A}(z)P(z))_{ij} - P_{ij}(\mathcal{A}(z)P(z))_{jj} \tag{8.12}$$

We first show that this equation admits a unique formal solution of the form:

$$P(z) = 1 + \sum_{r \geq 1} \frac{1}{z^r} P_r, \quad (P_r)_{kk} = 0, \quad Q(z) = \mathcal{A}_0 + \sum_{r \geq 1} \frac{1}{z^r} Q_r$$

Inserting into eq. (8.11), the coefficient of $1/z^r$ for $r \geq 1$ gives:

$$\mathcal{A}_0 P_r - P_r \mathcal{A}_0 = Q_r + H_r \tag{8.13}$$

where H_r depends on P_j, Q_j for $j < r$. This is solved recursively uniquely by setting $(Q_r)_{kk} = -(H_r)_{kk}$, which reproduces the solution $Q_{kk} = (AP)_{kk}$, and $(P_r)_{ij} = 1/(\alpha_i - \alpha_j)(H_r)_{ij}$ for $i \neq j$, where we recall that $\mathcal{A}_0 = \text{Diag}(\alpha_1, \ldots, \alpha_N)$. Inserting this expansion for $P(z)$ and $Q(z)$ into eq. (8.10) we get eq. (8.9) with $B_0 = -Q_{q+1}$ and $\xi = -\sum_{i=0}^{q} Q_i z^{q-i+1}/(q-i+1)$. Moreover, $g(z) = P(z) \exp(\sum_{i=q+2}^{\infty} Q_i z^{q-i+1}/(q-i+1))$ is of the form $g = 1 + O(1/z)$ and this uniquely determines this expansion. Note that the most singular term in ξ is indeed of the form $\mathcal{A}_0 z^n/n$ with $n = q+1$ and $\mathcal{A}_0$ diagonal. If $\mathcal{A}_0$ is not diagonal, write $\mathcal{A}_0 = g_0 D g_0^{-1}$. Then $g = g_0(1 + O(1/z))$. Of course g_0 is determined up to a right multiplication by any constant diagonal matrix.

In contrast to the regular singularity case this expansion is, however, only valid in the asymptotic sense.

We then show that there exists a true solution in the sector S having the above formal solution as asymptotics when $z \to \infty$. Write $P = 1 + \tilde{P}$, eq. (8.12) takes the form $z^{-q}\tilde{P}'(z) = f_0(z) + f_1(z, \tilde{P}) + f_2(z, \tilde{P})$, where $f_1(z, \tilde{P})$ is linear in $\tilde{P}$, and $f_2(z, \tilde{P})$ is quadratic in $\tilde{P}$. Moreover, when $z \to \infty$, $f_1(z, \tilde{P}) = [\mathcal{A}_0, \tilde{P}] + \cdots$, hence this linear application is not singular (since $\tilde{P}$ has no diagonal element). Finally, we know this equation has a formal solution $\sum_{r\geq 1} P_r/z^r$. It is a known theorem that there exists, in the interior of any sector of angle less than 2π, an analytic function $\Phi(z)$ with the given asymptotics $\sum_{r\geq 1} P_r/z^r$ when $z \to \infty$. We set $\tilde{P}(z) = U(z) + \Phi(z)$ and get a transformed equation for $U(z)$ of the same form, $z^{-q}U'(z) = g_0(z) + g_1(z, U(z)) + g_2(z, U(z))$, but now when $z \to \infty$, one can show that $g_0(z)$ is asymptotic to 0, and the leading term of $g_1(z, U(z))$ is of the form $[\mathcal{A}_0, U(z)]$. Finally, our equation can be written:

$$z^{-q}U'(z) = [\mathcal{A}_0, U(z)] + R(z, U)$$

where $R(z, U)$ will be treated as a perturbation. The unperturbed equation is readily solved and has the fundamental solution:

$$V(z) = e^{\frac{z^{q+1}}{q+1}\mathcal{A}_0}\, V_0\, e^{-\frac{z^{q+1}}{q+1}\mathcal{A}_0}$$

We now replace the differential equation by a system of coupled integral equations. Written in components, they are of the form:

$$U_{ij}(z) = \int_{\gamma_{ij}} e^{\frac{z^{q+1}-t^{q+1}}{q+1}(\alpha_i-\alpha_j)} t^q R_{ij}(t, U(t))\, dt, \quad i \neq j$$

The integration path γ_{ij} ends at z but its origin may depend on ij. The origins of the paths γ_{ij} represent the $N(N-1)$ integration constants of the problem. So we need to carefully specify the integration paths, and it is here that the sector S appears.

To be able to control the exponentials in the kernel of the integral equation one chooses paths $\gamma_{ij}(t)$ such that $\operatorname{Re}(z^{q+1} - t^{q+1})(\alpha_i - \alpha_j) < 0$. Let us examine the conditions under which such a choice is possible. We use the variables $\zeta = z^{q+1}$ and $\tau = t^{q+1}$. In the τ-plane we draw the lines $\operatorname{Re}\tau(\alpha_i - \alpha_j) = 0$ called Stokes lines. Let Σ be a sector of angle slightly larger than π, such that each Stokes line intersects the interior of Σ on just one half-line. The pre-image of Σ in the z-plane (under $z \to \zeta$) is a sector S:

$$\theta_0 \leq \theta \leq \theta_0 + \pi/n + \delta \tag{8.14}$$

with $n = q+1$, and some small positive δ. The Stokes line $\operatorname{Re}\tau(\alpha_i-\alpha_j) = 0$ divides Σ in two regions where $\operatorname{Re}\tau(\alpha_i - \alpha_j)$ is positive and negative respectively. The path γ_{ij} is taken so that its image in Σ is a straight line from ∞ to ζ with a slope such that $\operatorname{Re}\tau(\alpha_i - \alpha_j) < 0$.

The origins of the paths are chosen at ∞ so that $U_{ij}(z) \to 0$ when $z \to \infty$. The solution $U_{ij}(z)$ is uniquely determined by this requirement, which subsequently yields uniqueness of the solution $\Psi(\lambda)$ having the considered asymptotic expansion in the sector S. With this choice of path, one can check that $U_{ij}(z)$ is asymptotic to 0 when $z \to \infty$, hereby justifying the treatment of $R(z, U)$ as a perturbation. It is clear that for τ on the paths γ_{ij} one can write:

$$\left| e^{\frac{z^{q+1}-t^{q+1}}{q+1}(\alpha_i - \alpha_j)} \right| \leq e^{-\beta |x^{q+1} - t^{q+1}|}$$

for some positive fixed constant β (recall that $\alpha_i \neq \alpha_j$ for $i \neq j$). From this point, one can solve the integral equation by successive approximations and prove that this yields series U a convergent in the sector S and asymptotic to zero. We will not reproduce this analysis here.   ∎

Note that the paths occur in pairs γ_{ij} and γ_{ji} such that their images in Σ are straight lines, one on each side of the line $\mathrm{Re}\,(\zeta - \tau)(\alpha_i - \alpha_j) = 0$. Moreover, we keep the slopes fixed when ζ varies in Σ, and this can be done through the whole sector Σ bounded by some Stokes lines. Note, however, that if ζ goes beyond the boundary, one of the pair of paths has to be modified (see Fig. 8.1) yielding a different value of the integral.

In general the true solution $\Psi_1(\lambda)$ with asymptotic expansion $\Psi_{\mathrm{asy}}(\lambda)$ in the sector S_1 can be analytically continued beyond the sector S_1, but it is very important to realize that its asymptotic expansion will be *different* there. If we consider another sector S_2 adjacent to S_1, there exists another true solution Ψ_2 which has in S_2 the given asymptotic expansion $\Psi_{\mathrm{asy}}(\lambda)$. Since S_1 and S_2 overlap, there is a constant matrix $\mathcal{S}_1$ such that in this overlap $\Psi_2 = \Psi_1 \mathcal{S}_1$. This relation remains true in $S_1 \cup S_2$ by analytic continuation. This phenomenon was first noticed by Stokes and the matrix $\mathcal{S}_1$ is called a Stokes multiplier. Of course, the origin of the Stokes phenomenon is that subdominant exponentials in one sector become dominant in the next sector.

Note that there exists a permutation matrix P such that $P^{-1}\mathcal{S}_1 P$ is triangular with diagonal elements equal to 1. Indeed, let P be a permutation matrix. Then ΨP results from Ψ by some permutation of columns. We have $\Psi P \sim gP \,\exp(P^{-1}\xi P)$, where the matrix $P^{-1}\xi P$ is diagonal and results from ξ by the corresponding permutation of the diagonal elements. In the smallest sector bounded by Stokes rays containing the overlap $S_1 \cap S_2$, one can choose P such that $\mathrm{Re}\,(\alpha_{P(1)}\tau) < \mathrm{Re}\,(\alpha_{P(2)}\tau) < \cdots < \mathrm{Re}\,(\alpha_{P(n)}\tau)$. Since $\Psi_1 P$ and $\Psi_2 P$ have the same asymptotics in the overlap, we have necessarily $\Psi_2 P = \Psi_1 P \mathcal{S}_1'$, where $\mathcal{S}_1'$ is lower triangular. This is because $\exp(-\alpha_i \tau)$ is asymptotically negligible with respect to $\exp(-\alpha_j \tau)$ when $i > j$. Moreover, comparing the asymptotics of $\Psi_1 P$ and $\Psi_2 P$ we see that the diagonal elements of $\mathcal{S}_1'$ are equal to 1. Finally

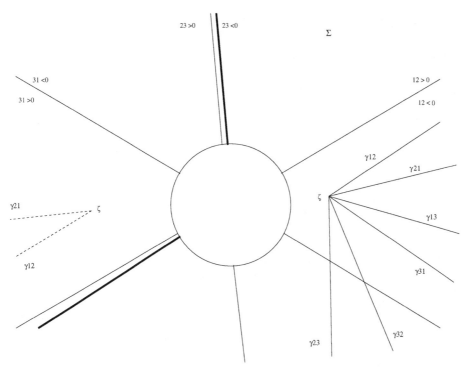

Fig. 8.1. Here the sector Σ is bounded by the bold lines, close to the Stokes rays, i.e. half Stoke lines, labelled 23 and 12. When the point ζ crosses the boundary 12, in the lower left side of the picture, the pair of paths γ_{12} and γ_{21} has to be modified as indicated by the dashed lines. Notice that this new path cannot be continuously deformed into the previous one. The sector Σ is of angle greater than π, so its pre-image is as in eq. (8.14).

$\mathcal{S}_1 = P\mathcal{S}_1'P^{-1}$. Altogether the Stokes matrix $\mathcal{S}_1$ depends on $N(N-1)/2$ continuous parameters.

The monodromy matrix around $\lambda = 0$ is the matrix $\mathcal{M}$ such that

$$\Psi_1(e^{2i\pi}\lambda) = \Psi_1(\lambda)\mathcal{M} \qquad (8.15)$$

where the left-hand side means the analytic continuation of the solution $\Psi_1(\lambda)$, with asymptotics Ψ_{asy} on S_1, around a closed contour around $\lambda = 0$.

We can easily relate the matrix $\mathcal{M}$ to the Stokes multipliers as follows. Let us cover a neighbourhood of the plane at $\lambda = 0$ by $2n$ sectors of angle $\pi/n + \delta$, denoted by $S_1, S_2, \ldots, S_{2n}$. More precisely, the sector S_j is defined by $(j-1)\pi/n \leq \arg(\lambda) \leq j\pi/n + \delta$. First in the sector S_2, we have $\Psi_1(\lambda) = \Psi_2(\lambda)\mathcal{S}_1^{-1}$ (where Ψ_2 has the asymptotic Ψ_{asy} in S_2) since this is true on the overlap. By recursion on the sector S_j we have $\Psi_1(\lambda) = \Psi_j(\lambda)\mathcal{S}_{j-1}^{-1}\cdots\mathcal{S}_1^{-1}$ for $\lambda \in S_j$. Making a complete 2π rotation around $\lambda = 0$, we get a sector S_{2n+1} which projects over S_1, and we have on this sector $\Psi_1(e^{2i\pi}\lambda) = \Psi_{2n+1}(e^{2i\pi}\lambda)\mathcal{S}_{2n}^{-1}\cdots\mathcal{S}_1^{-1}$. But the asymptotic expansion of $\Psi_{2n+1}(e^{2i\pi}\lambda)$ is by definition $\Psi_{\text{asy}}(e^{2i\pi}\lambda) = \Psi_{\text{asy}}(\lambda)\exp(2i\pi B_0)$, so we see that:

$$\Psi_1(e^{2i\pi}\lambda) \simeq \Psi_{\text{asy}}(\lambda)\, e^{2i\pi B_0}\, \mathcal{S}_{2n}^{-1}\cdots\mathcal{S}_1^{-1} \qquad (8.16)$$

By comparing the asymptotic expansions in eqs. (8.15, 8.16) we have:

$$\mathcal{M} = \exp(2i\pi B_0)\, \mathcal{S}_{2n}^{-1}\cdots\mathcal{S}_1^{-1}$$

In the case of regular singularities, there are no Stokes matrices, and the monodromy matrix reduces to $\exp(2i\pi B_0)$.

We now extend these results to the case of several singular points λ_k, including the point ∞. Hence we consider the differential equation eq. (8.3), where $M_\lambda(\lambda)$ is the general rational function of λ given in eq. (8.4). To describe the monodromy data of eq. (8.3), we have to patch the local descriptions around each singularity λ_k. At these points we have formal solutions:

$$\Psi_{\text{asy}}^{(k)}(\lambda) = g^{(k)}(\lambda)\, e^{\xi^{(k)}(\lambda)}, \quad g^{(k)}(\lambda) = g_0^{(k)}\left(1 + g_1^{(k)}(\lambda - \lambda_k) + \cdots\right)$$

Since they obey eq. (8.3) one can readily identify the polar part of M_λ at λ_k. We compute $\partial_\lambda\Psi \cdot \Psi^{-1}$ in a sector so that we can replace Ψ by its asymptotic expansion. We get $\partial_\lambda g g^{-1} + g\partial_\lambda\xi g^{-1} \sim M_\lambda$. Keeping the polar part yields the equality between a finite number of terms:

$$M_\lambda^{(k)} = \left(\Psi_{\text{asy}}^{(k)}(\lambda)\partial_\lambda\xi^{(k)}\Psi_{\text{asy}}^{(k)\,-1}(\lambda)\right)_- \qquad (8.17)$$

Due to our definition of $M_\lambda^{(\infty)}$ we have a similar formula at ∞ using the parameter $z = \lambda^{-1}$:

$$M_\lambda^{(\infty)}(z) = \left(\Psi_{\text{asy}}^{(\infty)}\partial_z\xi^{(\infty)}\Psi_{\text{asy}}^{(\infty)\,-1}\right)_- \qquad (8.18)$$

The global monodromy problem is specified by fixing paths γ_k from a reference point, which we choose to be ∞, to each λ_k. Around λ_k there are sectors $S_j^{(k)}$ and corresponding solutions $\Psi_j^{(k)}$ with the given asymptotics. Starting from the solution $\Psi_1^{(\infty)}$ we want to see how it changes when it is analytically continued around singularities. We can continue $\Psi_1^{(\infty)}$ along the path γ_k and compare the result to $\Psi_1^{(k)}$. This defines matrices $C^{(k)}$

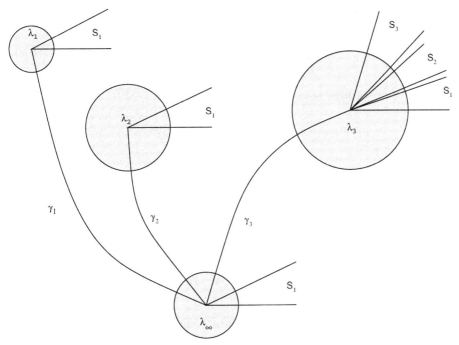

Fig. 8.2. The paths γ_i and the various Stokes sectors at the points λ_k.

such that $\Psi_1^{(\infty)}(\lambda) = \Psi_1^{(k)}(\lambda)\mathcal{C}^{(k)}$. Then the monodromy matrix of $\Psi_1^{(\infty)}$ around λ_k is

$$\mathcal{M}^{(k)} = \mathcal{C}^{(k)-1} e^{2i\pi B_0^{(k)}} \mathcal{S}_{2n}^{(k)-1} \cdots \mathcal{S}_1^{(k)-1} \mathcal{C}^{(k)} \qquad (8.19)$$

Around $\lambda = \infty$ we have the same formula with $\mathcal{C}^{(\infty)} = 1$.

Since the path around all singularities is contractible, these matrices are subjected to the relation:

$$\mathcal{M}^{(\infty)} \cdot \mathcal{M}^{(1)} \cdots \mathcal{M}^{(K)} = 1 \qquad (8.20)$$

where K is the number of singularities at finite distance. The monodromy group is the group generated by the $\mathcal{M}^{(k)}$ subjected to the above relation.

Note that the determinant of each Stokes matrix is equal to 1 since there exists a permutation of the basis in which it is triangular with 1 on the diagonal. It follows that $\det\left(\mathcal{M}^{(k)}\right) = \exp\left(2i\pi \operatorname{Tr} B_0^{(k)}\right)$. Taking the determinant of eq. (8.20) we see that $\sum_k \operatorname{Tr} B_0^{(k)}$ is an integer, where the sum includes ∞. In fact there is a stronger compatibility condition for the existence of $\Psi(\lambda)$, called the Fuchs condition, stating that this integer vanishes.

$$\sum_k \text{Tr}\,(B_0^{(k)}) + \text{Tr}\,(B_0^{(\infty)}) = 0 \qquad (8.21)$$

To show this we first take the trace of eq. (8.18), getting:

$$\text{Tr}\, M_\lambda^{(\infty)} = \left(\text{Tr}\,\partial_z \xi^{(\infty)}\right)_- = \text{Tr}\, B_0^{(\infty)} \frac{1}{z} + \cdots$$

On the other hand, considering the $1/z$ term in eq. (8.5), we get:

$$\text{Tr}\, M_\lambda^{(\infty)} = -\frac{1}{z}\sum_k \text{Tr}\, A_1^{(k)} + \cdots$$

so we get

$$\text{Tr}\, B_0^{(\infty)} = -\sum_k \text{Tr}\, A_1^{(k)} \qquad (8.22)$$

Similarly, at each singularity λ_k at finite distance, we take the trace of eq. (8.17):

$$\text{Tr}\, M_\lambda^{(k)} = \left(\text{Tr}\,\partial_\lambda \xi^{(k)}\right)_- = \text{Tr}\, B_0^{(k)} \frac{1}{\lambda - \lambda_k} + \cdots$$

Considering the $1/(\lambda - \lambda_k)$ term in eq. (8.4), we get:

$$\text{Tr}\, M_\lambda^{(k)} = \frac{1}{\lambda - \lambda_k}\text{Tr}\, A_1^{(k)} + \cdots$$

so that $\text{Tr}\, B_0^{(k)} = \text{Tr}\, A_1^{(k)}$. Inserting this into eq. (8.22), we find the Fuchs relation.

Our next task is to relate the coefficients of $M_\lambda(\lambda)$ and the monodromy data.

Definition. *We define the monodromy data at λ_k to be the $2n_k$ Stokes matrices $S_j^{(k)}$ and the connection matrices $C^{(k)}$.*

We define the singularity data at λ_k to be the coefficients of the singular terms $\xi^{(k)}$. These definitions apply to the point at ∞ as well.

It is important to note that given M_λ, the monodromy data are defined only up to a group of diagonal matrices at each λ_k. In fact we have seen that they are uniquely defined once we have chosen an asymptotic expansion around λ_k. More precisely, once the most singular part of $M_\lambda^{(k)}$

has been diagonalized, there is a canonical asymptotic expansion of the form $\Psi_{\mathrm{asy}}^{(k)}(\lambda) = (1 + O(\lambda - \lambda_k)) \exp{(\xi^{(k)}(\lambda))}$. Conjugating the matrix M_λ by a constant matrix, one can assume that the most singular term at ∞ is diagonal. Then there is a canonical choice of $\Psi_1^{(\infty)}$. At the other singular points, however, one has simply $\Psi_{\mathrm{asy}}^{(k)}(\lambda) = g_0^{(k)}(1 + O(\lambda - \lambda_k)) \exp{(\xi^{(k)}(\lambda))}$, where $g_0^{(k)}$ is the matrix diagonalizing the leading singularity in $M_\lambda^{(k)}$. It is defined only up to multiplication on the right by a diagonal matrix $d^{(k)}$. Hence the $\Psi_1^{(k)}$ are defined only up to right multiplication by $d^{(k)}$. This changes $\xi^{(k)} \to \xi^{(k)}$, $\mathcal{S}_j^{(k)} \to d^{(k)\,-1} \mathcal{S}_j^{(k)} d^{(k)}$, and $\mathcal{C}^{(k)} \to \mathcal{C}^{(k)} d^{(k)}$. Since the monodromy data are significant only up to this diagonal action, we define reduced monodromy data to be the quotient set.

We are now in a position to compare the number of parameters in the matrix M_λ and in the reduced monodromy and singularity data set. Assuming that the leading singular coefficient at ∞ is diagonal, the matrix M_λ depends on $N^2(n_\infty + \sum_{k=1}^K n_k + K - 1) + N$ parameters. On the other hand, each Stokes matrix depends on $N(N-1)/2$ parameters, and there are $2n_k$ such matrices at each singular point including ∞. Similarly each $\xi^{(k)}$ depends on $(n_k + 1)N$ parameters, while the matrix $\mathcal{C}^{(k)}$ (for $k = 1, \ldots, K$) depends on N^2 parameters. Altogether the set of monodromy and singularity data contains $N^2(n_\infty + \sum_{k=1}^K n_k + K) + N(K+1)$ parameters. These parameters are not all independent, since we have to take into account the relation eq. (8.20) between the monodromy matrices which removes N^2 parameters, and we have to quotient by the diagonal group action at each λ_k which removes NK parameters. One then gets exactly the number of free parameters in M_λ.

It is then reasonable to expect that these two sets of data are equivalent, that is to say, for any monodromy and singularity data satisfying the appropriate consistency conditions, one can find a unique differential equation of the form eqs. (8.3, 8.4) with these given monodromy and singularity data. This is called the generalized Riemann problem, which was first studied by Birkhoff and more recently by Malgrange and Sibuya.

The unicity part is easy. Assume that we have two differential equations with the same monodromy and singularity data. Consider corresponding solutions $\Psi(\lambda)$ and $\tilde{\Psi}(\lambda)$, normalized as $(1 + O(\lambda^{-1})) \exp{(\xi^{(\infty)}(\lambda))}$ at ∞. Let us consider $P(\lambda) = \tilde{\Psi}(\lambda)\Psi^{-1}(\lambda)$. We see that $P(\lambda)$ is single valued. The singular parts cancel at λ_k so that $P(\lambda)$ is holomorphic at λ_k and therefore holomorphic everywhere. Since $P(\lambda = \infty) = 1$, we get $P(\lambda) = 1$.

The same argument applied to $\frac{\partial\Psi(\lambda)}{\partial\lambda}\Psi^{-1}(\lambda)$ shows that M_λ is a rational function with a pole of order $n_k + 1$ at λ_k.

The existence part, i.e. the reconstruction of Ψ from its monodromy and singularity data, is much more difficult. We present in the next section a sketch of this construction in the case of regular singularities, by relating it to the Riemann–Hilbert factorization problem. In the following we shall take the general result for granted, and refer to the literature for its proof. However, assuming that Ψ exists, one can write linear differential equations on Ψ which characterize deformations where the *monodromy data are fixed* and we *vary the singularity data*. The compatibility conditions of this system is a set of non-linear differential equations on the coefficients of M_λ. We will directly prove the integrability, in the Frobenius sense, of these equations. Hence there exist locally solutions depending on exactly the number of deformation parameters.

8.3 Isomonodromy and the Riemann–Hilbert problem

In the same way that the Riemann–Hilbert factorization problem is fundamental for the study of isospectral integrable systems, it is also the key ingredient in the construction of Ψ with given monodromy. We shall only consider the restricted Riemann problem, i.e. find a differential equation with first order poles whose solutions have a prescribed monodromy group. Let us fix K points $\lambda_1, \ldots, \lambda_K$ (including $\lambda_K = \infty$) on the Riemann sphere, and K matrices $\mathcal{M}^{(1)}, \ldots, \mathcal{M}^{(K)}$ whose product $\mathcal{M}^{(K)} \cdots \mathcal{M}^{(1)} = 1$. We construct a multivalued function $\Psi(\lambda)$ such that $\Psi(\lambda) \to \Psi(\lambda)\mathcal{M}^{(k)}$, when λ describes a loop around λ_k. Finally, we show that $\det \Psi \neq 0$ and that $\partial_\lambda \Psi \Psi^{-1}$ is rational with simple poles at the λ_k at finite distance.

Following Birkhoff, we draw a simple closed path D visiting all the λ_k, but leaving them outside, and small circles C_k around each λ_k as in Fig. 8.3. We define matrices $M_1, \ldots, M_{K+1}$ with M_1 arbitrary, and $M_{k+1}^{-1} M_k = \mathcal{M}^{(k)}$. Since $\prod \mathcal{M}^{(k)} = 1$ we have $M_{K+1} = M_1$. We then define a C^∞ invertible matrix $M(\lambda)$ on the path D, which is constant equal to M_k between C_k and C_{k+1}, and which inside C_k is a C^∞ interpolation between the two constant values M_k and M_{k+1}. With these data we can consider the following Riemann–Hilbert factorization problem:

$$U_+(\lambda) = U_-(\lambda)M(\lambda) \ \text{ on } D$$

where $U_+(\lambda)$ is analytic inside D and $U_-(\lambda)$ is analytic outside D. As explained in Chapter 3, indices can appear in the solution of the factorization problem, see eq. (3.49). We have absorbed the diagonal matrix of

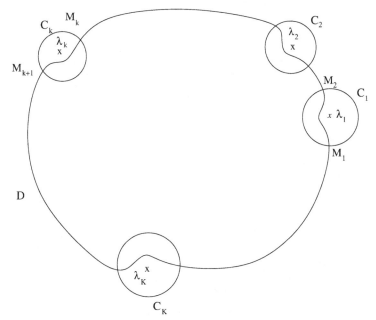

Fig. 8.3. The path D and the small circles C_k around the λ_k.

indices in U_-, which is thus allowed to have a pole or a zero at ∞, otherwise det $U_\pm \neq 0$. Note that U_+ has an analytic continuation outside D, through the segment of D between C_{k-1} and C_k, given by $U_- M_k$. This is because M_k being constant, $U_-(\lambda) M_k$ is analytic outside D and coincides with U_+ on this segment, hence analytically continues $U_+(\lambda)$. Similarly, the analytic continuation of U_- through the segment of D between C_k and C_{k+1} is given by $U_+ M_{k+1}^{-1}$. So, if we perform a loop around λ_k, $U_+(\lambda)$ gets multiplied on the right by $\mathcal{M}^{(k)} = M_{k+1}^{-1} M_k$. We have obtained a multivalued matrix $U(\lambda)$, the analytic continuation of $U_+(\lambda)$, which is analytic outside the C_k, has non-vanishing determinant there, apart possibly for a pole at ∞, and has the given monodromy properties around the C_k.

We now consider a second Riemann–Hilbert factorization problem, relative to the union of the contours C_k. Let $Z_k(\lambda)$ be the matrix defined inside C_k by:

$$Z_k(\lambda) = (\lambda - \lambda_k)^{\frac{1}{2i\pi} \log \mathcal{M}^{(k)}}$$

for some determination of the logarithms. Of course, the matrix $Z_k(\lambda)$ undergoes a right multiplication by $\mathcal{M}^{(k)}$ when λ performs a loop around λ_k. We define the functions $A_k(\lambda)$ on each C_k as $U(\lambda) Z_k^{-1}(\lambda)$. Notice that $A_k(\lambda)$ is univalued and analytic in a vicinity of C_k. Let us denote by $A(\lambda)$ the collection of functions $A_k(\lambda)$ on C_k and consider the factorization problem:

$$V_+(\lambda) = V_-(\lambda)A(\lambda) \quad \text{on } \cup C_k$$

where $V_+(\lambda)$ is a set of invertible matrices $V_{k+}(\lambda)$ analytic inside C_k, V_{K+} having a possible pole at $\lambda_K = \infty$ (we have absorbed the indices here) and $V_-(\lambda)$ is an invertible matrix analytic outside the C_k. Finally, on each C_k we have $V_{k+}(\lambda) = V_-(\lambda)A_k(\lambda)$.

With the solution of these two Riemann–Hilbert problems at hand, we define:

$$\Psi(\lambda) = V_-(\lambda)U(\lambda) \quad \text{outside the } C_k$$

The analytic continuation of this function inside C_k is $V_{k+}(\lambda)Z_k(\lambda)$ by the definition of the second factorization problem. Note that $V_{k+}(\lambda)Z_k(\lambda)$ is analytic inside C_k except at λ_k. Finally, $\Psi(\lambda)$ has the same monodromy around C_k as $Z_k(\lambda)$, hence has the prescribed monodromies $\mathcal{M}^{(k)}$. It is clear that $\partial_\lambda \Psi \Psi^{-1}$ is well-defined and analytic except at the λ_k, where it has a simple pole. This is because inside C_k:

$$\partial_\lambda \Psi \Psi^{-1} = \partial_\lambda V_{k+} V_{k+}^{-1} + V_{k+}(\partial_\lambda Z_k Z_k^{-1})V_{k+}^{-1}$$

and $\partial_\lambda Z_k Z_k^{-1}$ has a simple pole at λ_k. This means that Ψ is a solution of eq. (8.3) with M_λ having only simple poles at the λ_k, thereby solving the restricted Riemann problem. There is still a problem left, that we cannot be sure that the pole at $\lambda_K = \infty$ is simple. When the corresponding monodromy matrix $\mathcal{M}^{(\infty)}$ is diagonalizable, Birkhoff has shown that one can carefully choose Z_{λ_K} so as to achieve a simple pole at ∞. The basic idea is that $\log \mathcal{M}_k$ is only defined up to integers, and one can choose them to remove the left-over poles. It has been found, however, subsequently that the pole may remain of higher order if $\mathcal{M}^{(\infty)}$ is not diagonalizable.

The case of irregular singularities has been treated by similar methods by Birkhoff. Alternatively, one can see higher order poles as obtained from first order poles by confluence of singularities. As an example, consider eq. (8.3) when $M_\lambda(\lambda)$ is a sum of two nearby polar terms:

$$M_\lambda(\lambda) = \frac{1}{2}\frac{A_1 - A_2\epsilon^{-1}}{\lambda + \epsilon} + \frac{1}{2}\frac{A_1 + A_2\epsilon^{-1}}{\lambda - \epsilon} = \frac{A_1}{\lambda} + \frac{A_2}{\lambda^2} + O(\epsilon^2)$$

We see that it is easy to produce poles of order n by letting n simple poles move to a single point.

8.4 Isomonodromic deformations

As we have seen in the previous section, the matrix $\Psi(\lambda)$ is determined by its monodromy data and its singularity data at the λ_k. These are two independent sets of quantities which can be specified independently. We will examine here the matrices $\Psi(\lambda)$, with prescribed monodromy data $S_j^{(k)}, C^{(k)}$, fixed parameters $B_0^{(k)}$, and varying the parameters t_i in $\xi^{(k)}$ and the singularity positions λ_k. Assuming that Ψ exists, we will show that $\Psi(\lambda, \{t_i\}, \{\lambda_k\})$ has to satisfy a hierarchy of equations of the form eq. (8.1) with respect to the times t_i, and new equations, that we call Schlesinger's equations, with respect to the λ_k. Notice that these are all the possible deformation parameters, since once they are fixed, together with the monodromy data, they determine the function $\Psi(\lambda)$ uniquely.

Let us assume that some multivalued function $\Psi(\lambda)$ exists, with given monodromy and singularity data. Specifically, suppose that we are given the positions λ_k of the singularities, some Stokes matrices $S_j^{(k)}$ at these singularities (which satisfy the triangularity condition) and some connection matrices $C^{(k)}$, so that eq. (8.20) is obeyed. By hypothesis, we have an asymptotic expansion of the form $\Psi_{\text{asy}}^{(\infty)}(\lambda, t) = (1 + O(1/\lambda)) \, e^{\xi^{(\infty)}(\lambda, t)}$ in the first sector at ∞, and asymptotic expansions in the sector $S_j^{(k)}$ at λ_k of the form:

$$\Psi(\lambda) \sim \Psi_{\text{asy}}^{(k)}(\lambda) S_j^{(k)\,-1} \cdots S_1^{(k)\,-1} C^{(k)} \tag{8.23}$$

where

$$\Psi_{\text{asy}}^{(k)}(\lambda) = g_0^{(k)} (1 + g_1^{(k)}(\lambda - \lambda_k) + \cdots) \, e^{\xi^{(k)}(\lambda, t)} \tag{8.24}$$

With these assumptions, the matrix $M_\lambda = \partial_\lambda \Psi \Psi^{-1}$ is a rational function of λ, as was noticed above. Hence it is of the form eq. (8.4) and the differential equation $\partial_\lambda \Psi = M_\lambda \Psi$ has a solution $\Psi(\lambda)$ with the above monodromy and singularity data.

Denote by $\{\tau_i\}$ the set of variables $\{t_i\} \cup \{\lambda_k\}$, and by d the exterior differentiation with respect to the parameters τ_i, $d = \sum_i d\tau_i \partial_{\tau_i}$.

Theorem. *The monodromy data for $\Psi(\lambda)$ are independent of the deformation parameters if and only if the function $\Psi(\lambda)$ satisfies differential equations with respect to the deformation parameters:*

$$d\Psi = M\Psi \tag{8.25}$$

where $M = \sum_i M_i d\tau_i$ is a 1-form with coefficients M_i rational functions of λ.

Proof. Let us consider the function $\Psi(\lambda)$ obeying all the above constraints, and consider the 1-form $d\Psi(\lambda)\Psi^{-1}(\lambda)$, as a function of λ. This 1-form is single valued around λ_k because when we turn around λ_k, the matrix $\Psi(\lambda)$ gets multiplied by the monodromy matrix eq. (8.19), which cancels in $d\Psi\,\Psi^{-1}$ because it is assumed to be independent of the deformation parameters. Moreover, in the vicinity of λ_k its asymptotic expansion can be computed in any sector S_j by inserting the asymptotic expansion eq. (8.23), yielding $d\Psi\,\Psi^{-1} \sim d\Psi_{\mathrm{asy}}^{(k)}\,\Psi_{\mathrm{asy}}^{(k)\,-1}$, where we again use the independence of the monodromy data from the deformation parameters, and the known fact that one can differentiate an asymptotic expansion valid in a sector. From the explicit form of $\Psi_{\mathrm{asy}}^{(k)}$ we see that the singularity of the asymptotic expansion of $d\Psi(\lambda)\Psi^{-1}(\lambda)$ is a pole at λ_k. Explicitly, $\partial_{t_i}\Psi(\lambda)\,\Psi^{-1}(\lambda)$ has a pole of order n_k at λ_k, and $\partial_{\lambda_k}\Psi(\lambda)\Psi^{-1}(\lambda)$ has a pole of order n_k+1 at λ_k. Hence $M(\lambda)$ is a rational function of λ.

Conversely, assume that $M_\lambda(\lambda)$ is parametrized by some parameters τ_i and that one can write equations $\partial_{\tau_i}\Psi = M_i(\lambda)\Psi$, where $M_i(\lambda)$ are rational functions of λ. Considering, for example, two adjacent sectors S_1, S_2 at a singularity, the solution $\Psi(\lambda)$ has the asymptotic Ψ_{asy} in S_1 and $\Psi_{\mathrm{asy}}\mathcal{S}$ in S_2. Then $\partial_i\Psi$ has the asymptotic $\partial_i\Psi_{\mathrm{asy}} = M_i\Psi_{\mathrm{asy}}$ in S_1. This relation on Ψ_{asy} remains true in all sectors. In S_2, $\partial_i\Psi$ has the asymptotic $\partial_i\Psi_{\mathrm{asy}}\mathcal{S} + \Psi_{\mathrm{asy}}\partial_i\mathcal{S}$, but this should also be equal to $M_i\Psi$ which has the asymptotic $M_i\Psi_{\mathrm{asy}}\mathcal{S}$. Hence $\partial_i\mathcal{S} = 0$. Similarly, one shows that the connection matrices $\mathcal{C}^{(k)}$ and the $B_0^{(k)}$ are independent of the deformation parameters τ_i.   ∎

We now change our point of view and directly study the system of equations:

$$\partial_\lambda\Psi = M_\lambda\Psi, \quad \partial_{\tau_i}\Psi = M_i(\lambda)\Psi \qquad (8.26)$$

with M_λ of the form eq. (8.4) and M_i rational matrices. This system of equations is compatible if the rational matrices M_λ, M_i obey the zero-curvature conditions:

$$\partial_{\tau_i}M_\lambda = \partial_\lambda M_i + [M_i, M_\lambda] \qquad (8.27)$$

$$\partial_{\tau_i}M_j - \partial_{\tau_j}M_i - [M_i, M_j] = 0 \qquad (8.28)$$

In this case, the Frobenius theorem asserts that one can find solutions $\Psi(\lambda)$ depending on the maximal number of parameters. In our situation the maximal number of isomonodromic parameters τ_i and compatible equations that one can introduce is just the set of times t_i and the λ_k.

We want to prove the compatibility of the system eqs. (8.27, 8.28) directly, for properly defined M_i, yielding the existence of $\Psi(\lambda; \tau_i)$.

We start from the rational matrix M_λ of the form eq. (8.4) and consider the differential equation eq. (8.3). Around λ_k there exists a formal solution

$\Psi_{\mathrm{asy}}^{(k)} = g^{(k)} \exp\left(\xi^{(k)}\right)$. Hence $g^{(k)}$ obeys the equation:

$$\partial_\lambda g^{(k)} = M_\lambda g^{(k)} - g^{(k)} \partial_\lambda \xi^{(k)} \tag{8.29}$$

Next we *define* the rational matrices:

$$M_i(\lambda) = \left(g^{(k)} \partial_{\tau_i} \xi^{(k)} g^{(k)-1}\right)_{-}$$

where τ_i stands for $t_{(k,n,\alpha)}$ and λ_k. Note that the matrices M_i are algebraic functions of the matrix elements of M_λ, since to compute M_i, we need the expansion of $g^{(k)}$ to some finite order, which is obtained algebraically from M_λ by the recurrence relations eq. (8.13).

From M_i, we *define* some vector field X_i acting on M_λ by:

$$X_i M_\lambda = \partial_\lambda M_i + [M_i, M_\lambda] \tag{8.30}$$

It is important to notice that the polar structure of the right-hand side of this equation allows us to consider the flows X_i as acting on the coefficients $A_j^{(k)}$ in eq. (8.4) of M_λ and the λ_k. In particular X_i and ∂_λ commute. To show that, one has to examine the order of the pole at λ_k in both sides of eq. (8.30). Let $i = (k, n, \alpha)$. Around $\lambda_{k'}$, $k' \neq k$, the matrix M_i is regular and eq. (8.30) is compatible with the pole structure of M_λ at $\lambda_{k'}$. Around λ_k, since $\partial_\lambda(A)_- = (\partial_\lambda A)_-$ for any rational matrix $A(\lambda)$, we have

$$\partial_\lambda M_i = (g^{(k)} \partial_\lambda \partial_{\tau_i} \xi^{(k)} g^{(k)-1})_- + ([\partial_\lambda g^{(k)} g^{(k)-1}, g^{(k)} \partial_{\tau_i} \xi^{(k)} g^{(k)-1}])_-$$

In the second term, we can replace $\partial_\lambda g^{(k)} g^{(k)-1} = M_\lambda - g^{(k)} \partial_\lambda \xi^{(k)} g^{(k)-1}$ by M_λ because $g^{(k)} \partial_\lambda \xi^{(k)} g^{(k)-1}$ does not contribute to the commutator since $\xi^{(k)}$ is diagonal. Similarly, writing $g^{(k)} \partial_{\tau_i} \xi^{(k)} g^{(k)-1} = M_i + (g^{(k)} \partial_{\tau_i} \xi^{(k)} g^{(k)-1})_+$ we get

$$\partial_\lambda M_i = (g^{(k)} \partial_\lambda \partial_{\tau_i} \xi^{(k)} g^{(k)-1})_- - [M_i, M_\lambda]_- + [M_\lambda, (g^{(k)} \partial_{\tau_i} \xi^{(k)} g^{(k)-1})_+]_-$$

hence

$$\partial_\lambda M_i + [M_i, M_\lambda] = (g^{(k)} \partial_\lambda \partial_{\tau_i} \xi^{(k)} g^{(k)-1})_- \\ + [M_i, M_\lambda]_+ + [M_\lambda, (g^{(k)} \partial_{\tau_i} \xi^{(k)} g^{(k)-1})_+]_-$$

from which we see that the pole structure is the same as the one of $\partial_{\tau_i} M_\lambda$ (it is the term $\partial_\lambda \partial_{\tau_i} \xi^{(k)}$ which controls this assertion, the action of ∂_{τ_i} on the $A_j^{(k)}$ will be determined later on).

We are now in a position to prove the main theorem of this section:

Theorem. *The flows X_i are all commuting, and we can identify $X_i = \partial_{\tau_i}$.*

<u>Proof</u>. The commutation of the flows X_i, X_j is expressed by $\partial_\lambda F_{ij} - [M_\lambda, F_{ij}] = 0$, where:

$$F_{ij} = X_i M_j - X_j M_i - [M_i, M_j]$$

We show in fact a stronger result, i.e. $F_{ij} = 0$. Our first task is to find the action of the flow X_i on the variables $g^{(k)}$ and $\xi^{(k)}$ around any pole λ_k. To do that we apply X_i to eq. (8.29), getting:

$$\partial_\lambda(X_i g^{(k)} - M_i g^{(k)}) = M_\lambda(X_i g^{(k)} - M_i g^{(k)})$$
$$-(X_i g^{(k)} - M_i g^{(k)})\partial_\lambda \xi^{(k)} - g^{(k)}\partial_\lambda(X_i \xi^{(k)})$$

Writing $(X_i g^{(k)} - M_i g^{(k)}) = g^{(k)} h_i$ and using eq. (8.29), we get for h_i the linear equation:

$$\partial_\lambda h_i - [\partial_\lambda \xi^{(k)}, h_i] = -X_i \partial_\lambda \xi^{(k)}$$

Since $X_i \xi^{(k)}$ is diagonal, a particular solution is $h_i = -X_i \xi^{(k)}$. The general solution is obtained by adding to it $e^{\xi^{(k)}} D_i e^{-\xi^{(k)}}$ for any matrix D_i independent of λ. Through its definition, we see that h_i has at most poles at λ_k, hence D_i must be diagonal, otherwise essential singularities appear in $e^{\xi^{(k)}} D_i e^{-\xi^{(k)}}$. Finally, we get:

$$X_i g^{(k)} = M_i g^{(k)} - g^{(k)} X_i \xi^{(k)} + g^{(k)} D_i$$

Looking at the polar part of this equation we see that $(g^{(k)} X_i \xi^{(k)} g^{(k)-1})_- = (g^{(k)} \partial_{\tau_i} \xi^{(k)} g^{(k)-1})_-$, from which it follows that $X_i \xi^{(k)} = \partial_{\tau_i} \xi^{(k)}$, assuming that $g_0^{(k)}$ is generic. Hence X_i identifies to ∂_{τ_i} on $\xi^{(k)}$.

We can now compute F_{ij}. It is simpler to use a compact notation: let M be the 1-form $\sum M_i d\tau_i$, and let δ be the vector field $\sum_i d\tau_i X_i$ with values in differentials. Note that on $\xi^{(k)}$, δ identifies to d so that $\delta^2 \xi^{(k)} = 0$. The equation on $g^{(k)}$ becomes, with $D = \sum D_i d\tau_i$:

$$\delta g^{(k)} = M g^{(k)} - g^{(k)} \delta \xi^{(k)} + g^{(k)} D \qquad (8.31)$$

The conditions $F_{ij} = 0$ read $\delta M - M \wedge M = 0$. With the help of the above equation one can compute the polar part of δM at λ_k. Since $M^{(k)} = (g^{(k)} \delta \xi^{(k)} g^{(k)-1})_-$, we get:

$$\delta M^{(k)} = (g^{(k)}[D, \delta \xi^{(k)}] g^{(k)-1})_- + [M, g^{(k)} \delta \xi^{(k)} g^{(k)-1}]_-$$

where for $M = \sum M_i d\tau_i$, $N = \sum M_j d\tau_j$, we define

$$[M, N] = \sum_{i \neq j} [M_i, N_j] d\tau_i \wedge d\tau_j$$

The first term vanishes because it involves commutators of diagonal matrices. To evaluate the second term we remark that $M - g^{(k)} \delta \xi^{(k)} g^{(k)-1} = O(1)$, hence, squaring it, we get $M \wedge M - [M, g^{(k)} \delta \xi^{(k)} g^{(k)-1}] = O(1)$ since again the commutator of diagonal matrices vanishes. Taking the polar part of this relation, we conclude that $[M, g^{(k)} \delta \xi^{(k)} g^{(k)-1}]_- = (M \wedge M)_-$, hence $\delta M^{(k)} = (M \wedge M)_-$. Since this is true at each pole λ_k and since M is a rational function of λ, we see that $F_{ij} = 0$ for all i, j.

We have shown that $[X_i, X_j] = 0$. The Frobenius theorem implies that one can simultaneously solve (locally) $\partial_{\tau_i'} M_\lambda = X_i M_\lambda$, thereby obtaining rational matrices M_λ, M_i satisfying eqs. (8.27, 8.28). Since X_i identifies with ∂_{τ_i} on $\xi^{(k)}$, we have, $\partial_{\tau_i'} = \partial_{\tau_i}$ and ∂_{τ_i} identifies to X_i everywhere. $\blacksquare$

Remark 1. The term D in eq. (8.31) appeared because $g^{(k)}$ is defined only up to right multiplication by a diagonal matrix. This gauge transformation did not affect the calculation above because only gauge invariant quantities are considered. For any gauge choice, we have an equation of the form eq. (8.31) for some specific D. Now that δ is identified to $d = \sum d\tau_i \partial_{\tau_i}$, we have $d^2 = 0$, and this implies $dD = 0$. Hence we have $D = dh$, so one can choose a gauge where $D = 0$. In this gauge the evolution equations of $g^{(k)}$ read:

$$dg^{(k)} = M g^{(k)} - g^{(k)} d\xi^{(k)} \tag{8.32}$$

This should be compared to eq. (3.44) in Chapter 3.

Remark 2. The compatibility conditions, eq. (8.27, 8.28) are non-linear differential equations on the coefficients of M_λ. Once we have a complete solution of these equations, one can find Ψ by solving eq. (8.26).

Example. The Schlesinger equations. Consider a differential equation $\partial_\lambda \Psi = M_\lambda \Psi$, where M_λ has only *regular* singularities, i.e.

$$M_\lambda(\lambda) = \sum_k \frac{A^{(k)}}{\lambda - \lambda_k}, \quad A^{(k)} = g_0^{(k)} B_0^{(k)} g_0^{(k)-1} \tag{8.33}$$

We assumed, according to the general analysis, that the matrices $A^{(k)}$ are diagonalizable. Note that there is a hidden regular singularity at ∞. The asymptotic expansions of $\Psi(\lambda)$ at λ_k are easily computed and found to be of the form:

$$\Psi_{\text{asy}}^{(k)}(\lambda) = g_0^{(k)}(1 + O(\lambda - \lambda_k)) e^{\xi^{(k)}(\lambda)}, \quad \xi^{(k)}(\lambda) = B_0^{(k)} \log(\lambda - \lambda_k)$$

This means that the times t_i are all set to 0 and that the only deformation parameters are the positions of the poles λ_k. The deformation equations $\partial_{\lambda_k}\Psi = M_{\lambda_k}\Psi$ are constructed with the help of the general formula:

$$M_{\lambda_k} = \left(\Psi^{(k)}_{asy} \partial_{\lambda_k} \xi^{(k)} \Psi^{(k)}_{asy}{}^{-1} \right)_- = -\frac{A^{(k)}}{\lambda - \lambda_k}$$

The zero curvature conditions read:

$$\partial_{\lambda_l} A^{(k)} = \frac{[A^{(k)}, A^{(l)}]}{\lambda_k - \lambda_l}, \quad l \neq k \tag{8.34}$$

$$\partial_{\lambda_k} A^{(k)} = -\sum_{l \neq k} \frac{[A^{(k)}, A^{(l)}]}{\lambda_k - \lambda_l} \tag{8.35}$$

These equations are called Schlesinger equations. One can check easily that both equations are contained in eq. (8.27), while eq. (8.28) is a direct consequence of them, in agreement with the general theory.

8.5 Schlesinger transformations

We have studied all continuous isomonodromic deformations. They are parametrized by the times t_i and the λ_k. However, there remains discrete isomonodromic deformations. The basic remark is that, although the matrices $B_0^{(k)}$ are not allowed to change continuously, a discrete change $B_0^{(k)} \to B_0^{(k)} + L^{(k)}$, where $L^{(k)}$ is a diagonal matrix with integer entries, does not change the monodromy data. The singularity data are modified by $e^{\xi^{(k)}(\lambda)} \to (\lambda - \lambda_k)^{L^{(k)}} e^{\xi^{(k)}(\lambda)}$, i.e. we add extra zeroes or poles at the singularities. This is a discrete analogue of the t_i deformations. These transformations are called Schlesinger transformations. So, one can consider that $\Psi(\lambda)$ depends not only on the continuous variables t_i and λ_k but also on a set of integers, a diagonal matrix with integer entries above each singularity. For the continuous isomonodromic transformations we have deformation equations $d\Psi = M\Psi$ with M a rational matrix. We show that for Schlesinger transformations we can write analogously difference equations with respect to the integers.

Proposition. *The two wave–functions $\Psi(\lambda)$ associated with the data $B_0^{(k)}$ and $\Psi'(\lambda)$ associated with the data $B_0^{(k)} + L^{(k)}$ are related by:*

$$\Psi'(\lambda) = R(\lambda)\Psi(\lambda) \tag{8.36}$$

where the matrix $R(\lambda)$ is a rational function of λ.

<u>Proof.</u> Consider the matrix $\Psi'(\lambda)\Psi^{-1}(\lambda)$. The essential singularities cancel and it has no monodromy. Hence, it is a rational function. ∎

Note that the integer matrices $L^{(k)}$ have to be restricted by the Fuchs condition, eq. (8.21):

$$\sum_k \text{Tr}\, L^{(k)} + \text{Tr}\, L^{(\infty)} = 0$$

To study Schlesinger's tranformations, it is enough to concentrate on elementary ones:

Definition. *An elementary Schlesinger transformation is associated with the matrices*

$$L^{(l)} \begin{bmatrix} k & k' \\ \alpha & \alpha' \end{bmatrix} = \delta_{kl} E_{\alpha\alpha} - \delta_{k'l} E_{\alpha'\alpha'} \tag{8.37}$$

This shifts the αth diagonal element by $+1$ above λ_k and the α'^{th} diagonal element by -1 above $\lambda_{k'}$ in order to fulfil the Fuchs condition. In the following we restrict ourselves to Schlesinger transformations involving singularities at finite distance.

Proposition. *The matrix $R(\lambda)$ in eq. (8.36) associated with the elementary Schlesinger transformation eq. (8.37) is of the form:*

$$R(\lambda) = 1 - \frac{R_0}{\lambda - \lambda_{k'}}, \quad R^{-1}(\lambda) = 1 + \frac{R_0}{\lambda - \lambda_k} \tag{8.38}$$

where the matrix R_0 is given by:

$$R_0 = \frac{\lambda_k - \lambda_{k'}}{\left(g_0^{(k')-1} g_0^{(k)} \right)_{\alpha'\alpha}} g_0^{(k)} E_{\alpha\alpha'} g_0^{(k')-1}, \quad \text{if } k \neq k' \tag{8.39}$$

$$R_0 = \frac{1}{\left(g_1^{(k)} \right)_{\alpha'\alpha}} g_0^{(k)} E_{\alpha\alpha'} g_0^{(k)-1}, \quad \text{if } k = k' \tag{8.40}$$

The matrices $g_i^{(k)}$ are defined in the asymptotic expansion eq. (8.24). We have $R_0^2 = (\lambda_k - \lambda_{k'})R_0$.

<u>Proof.</u> The conditions determining $R(\lambda)$ are

$$R(\lambda)g^{(l)}(\lambda) = (g')^{(l)}(\lambda)(\lambda - \lambda_l)^{L^{(l)}} \tag{8.41}$$

so that $R(\lambda)$ has a simple pole at $\lambda = \lambda_{k'}$. Similarly, the inverse Schlesinger transform is obtained by changing $L^{(l)}$ to $-L^{(l)}$, so that $R^{-1}(\lambda)$ has a simple pole at λ_k. Asymptotic expansion at ∞ shows that $R(\lambda)$ and $R^{-1}(\lambda)$ tend to 1 at ∞. This motivates eqs. (8.38) which are moreover consistent if $R_0^2 = (\lambda_k - \lambda_{k'})R_0$.

Suppose first $k \neq k'$. Let us write the condition eq. (8.41) in more detail for $l = k', k$. They read

$$\left(1 - \frac{1}{\lambda - \lambda_{k'}} R_0\right) g^{(k')} = (g')^{(k')} \left(1 + \left(\frac{1}{\lambda - \lambda_{k'}} - 1\right) E_{\alpha'\alpha'}\right), \; l = k'$$

$$\left(1 - \frac{1}{\lambda - \lambda_{k'}} R_0\right) g^{(k)} = (g')^{(k)} \left(1 + (\lambda - \lambda_k - 1) E_{\alpha\alpha}\right), \; l = k$$

Looking at the polar terms in the first of these equations, we get $R_0 g_0^{(k')} = -g'_0^{(k')} E_{\alpha'\alpha'}$, or $R_0 = -g'_0^{(k')} E_{\alpha'\alpha'} g_0^{(k')-1}$. The matrix element $\gamma\alpha$ of the right-hand side of the second equation vanishes at $\lambda = \lambda_k$. So we have, using the value of R_0:

$$(g_0^{(k)})_{\gamma\alpha} + \frac{1}{\lambda_k - \lambda_{k'}} (g'_0^{(k')})_{\gamma\alpha'} (g_0^{(k')-1} g_0^{(k)})_{\alpha'\alpha} = 0$$

Solving for $(g'_0^{(k')})_{\gamma\alpha'}$ and inserting back into the formula for R_0 yields eq. (8.39). With this expression one checks immediately that $R_0^2 = (\lambda_k - \lambda_{k'}) R_0$.

Suppose next $k = k'$, which implies $\alpha \neq \alpha'$ in order to get a non-trivial transformation. The equation determining $R(\lambda)$ now reads:

$$\left(1 - \frac{1}{\lambda - \lambda_k} R_0\right) g^{(k)} = (g')^{(k)} \left(\mathrm{Id} + (\lambda - \lambda_k - 1)\left(E_{\alpha\alpha} - \frac{E_{\alpha'\alpha'}}{\lambda - \lambda_k}\right)\right)$$

Comparing the terms of order $(\lambda - \lambda_k)^{-1}$ one gets $R_0 = -g'_0^{(k)} E_{\alpha'\alpha'} g_0^{(k)-1}$ as before. Next the matrix element $\gamma\alpha$ of the right-hand side vanishes at $\lambda = \lambda_k$ so that, considering the terms of order $(\lambda - \lambda_k)^0$ in the left-hand side, one gets (recall that $g^{(k)}(\lambda) = g_0^{(k)}(1 + (\lambda - \lambda_k)g_1^{(k)} + \cdots)$):

$$(g_0^{(k)})_{\gamma\alpha} - (R_0 \, g_0^{(k)} g_1^{(k)})_{\gamma\alpha} = 0$$

This is eq. (8.40). Here one checks that $R_0^2 = 0$. ∎

8.6 Tau-functions

Consider the differential equation eq. (8.3), where M_λ is a rational function of λ depending on isomonodromic deformation parameters. At each singularity λ_k we have asymptotic expansions of the form $\Psi(\lambda) \sim g^{(k)}(\lambda) e^{\xi^{(k)}(\lambda)}$. With any solution of the deformation equations, eqs. (8.25), we can associate a 1-form Υ:

$$\Upsilon = -\sum_k \mathrm{Res}_{\lambda = \lambda_k} \mathrm{Tr}(g^{(k)-1} \partial_\lambda g^{(k)} d\xi^{(k)}) d\lambda \tag{8.42}$$

The sum is over all singularities including ∞.

Theorem. *The deformation equations imply that* Υ *is closed:* $d\Upsilon = 0$.

<u>Proof</u>. We have already proved this equation in a more restricted setting in Chapter 3. Let us repeat the proof of this important result in this more general context. Recall that d is the differential with respect to the isomonodromic deformation parameters t_i and λ_k.

$$d\Upsilon = \sum_k \operatorname{Res}_{\lambda_k} \operatorname{Tr}(g^{(k)-1}dg^{(k)}g^{(k)-1}\partial_\lambda g^{(k)}d\xi^{(k)} - g^{(k)-1}\partial_\lambda dg^{(k)}d\xi^{(k)})d\lambda$$

From the deformation equation eq. (8.32) we get (using that $d\xi^{(k)}\wedge d\xi^{(k)} = 0$ since the matrix $\xi^{(k)}$ is diagonal):

$$d\Upsilon = \sum_k \operatorname{Res}_{\lambda_k} \operatorname{Tr}(d\partial_\lambda \xi^{(k)} \wedge d\xi^{(k)} - \partial_\lambda M \wedge g^{(k)}d\xi^{(k)}g^{(k)-1})d\lambda$$

The first term vanishes because the order of the pole is at least 3. For the same reason

$$\operatorname{Res}_{\lambda_k} \operatorname{Tr}(\partial_\lambda(g^{(k)}d\xi^{(k)}g^{(k)-1}) \wedge (g^{(k)}d\xi^{(k)}g^{(k)-1}))d\lambda$$
$$= \operatorname{Res}_{\lambda_k} \operatorname{Tr}(d\partial_\lambda \xi^{(k)} \wedge d\xi^{(k)})d\lambda = 0 \qquad (8.43)$$

Next we write $g^{(k)}d\xi^{(k)}g^{(k)-1} = M + N^{(k)}$, where $N^{(k)}$, is regular at λ_k. Then eq. (8.43) reads

$$\operatorname{Res}_{\lambda_k} \operatorname{Tr}(\partial_\lambda(M + N^{(k)}) \wedge (M + N^{(k)}))d\lambda = 0$$

Since the residue of a derivative of a function of λ vanishes, we can replace $\partial_\lambda N^{(k)} \wedge M$ by $\partial_\lambda M \wedge N^{(k)}$, getting:

$$\operatorname{Res}_{\lambda_k} \operatorname{Tr}(\partial_\lambda M \wedge N^{(k)})d\lambda = -\frac{1}{2}\operatorname{Res}_{\lambda_k} \operatorname{Tr}(\partial_\lambda M \wedge M)d\lambda$$

It follows that

$$d\Upsilon = -\sum_k \operatorname{Res}_{\lambda_k} \operatorname{Tr}(\partial_\lambda M \wedge (M+N^{(k)}))d\lambda = -\frac{1}{2}\sum_k \operatorname{Res}_{\lambda_k} \operatorname{Tr}(\partial_\lambda M \wedge M)d\lambda$$

But now $\operatorname{Tr}(\partial_\lambda M \wedge M)d\lambda$ is a rational 1-form on the λ Riemann sphere, hence the sum of the residues vanishes. ∎

Example. Let us give the form Υ in the Schlesinger case of regular singularities and deformation parameters λ_k. In that case (see eq. (8.33)):

$$d\xi = -\sum_k \frac{B_0^{(k)}}{\lambda - \lambda_k}d\lambda_k$$

and we have only to keep the constant term in $g^{(k)-1}\partial_\lambda g^{(k)}$, yielding:

$$\Upsilon = \sum_k \mathrm{Tr}(g_1^{(k)} B_0^{(k)}) d\lambda_k$$

Starting from $\partial_\lambda \Psi = \sum_l (g^{(l)} \partial_\lambda \xi^{(l)} g^{(l)-1})_- \Psi$, and expanding

$$\Psi = g_0^{(k)} (1 + (\lambda - \lambda_k) g_1^{(k)} + \cdots) e^{\xi^{(k)}}$$

we get:

$$g_0^{(k)} (g_1^{k} - [B_0^{(k)}, g_1^{(k)}]) g_0^{(k)-1} = \sum_{l \neq k} \frac{g_0^{(l)} B_0^{(l)} g_0^{(l)-1}}{\lambda_k - \lambda_l}$$

so that

$$\Upsilon = \frac{1}{2} \sum_{k \neq l} \mathrm{Tr}(A_k A_l) \frac{d\lambda_k - d\lambda_l}{\lambda_k - \lambda_l}$$

We can verify that this form is closed using the Schlesinger equations eqs. (8.34, 8.35).

By the closedness of Υ, we can introduce a function $\tau(\{t_i\}, \{\lambda_k\})$, defined up to a multiplicative constant, by:

Definition.   *The tau-function is defined by*

$$\Upsilon = d \log \tau \qquad\qquad (8.44)$$

With each solution of the deformation equations eq. (8.25), one can associate a tau-function. Hence, with the Schlesinger transformed solution we can associate a transformed tau-function. There is a simple relation between the original tau-function and its transform by an elementary Schlesinger transformation.

Proposition. *In the gauge eq. (8.32), we have:*

$$\tau \begin{bmatrix} k & k' \\ \alpha & \alpha' \end{bmatrix} (t) = \tau(t) \begin{cases} \frac{1}{\lambda_k - \lambda_{k'}} \left(g_0^{(k')-1} g_0^{(k)} \right)_{\alpha'\alpha} & \text{if } k \neq k' \\ \left(g_1^{(k)} \right)_{\alpha'\alpha} & \text{if } k = k' \end{cases} \qquad (8.45)$$

where the left-hand side denotes the transform of the tau-function under the elementary Schlesinger transformation eq. (8.37).

Proof. Let us denote by Υ' the form associated with the transformed solutions. Using eq. (8.41), we can write:

$$\Upsilon' - \Upsilon = -E_1 - E_2 + E_3$$

where we defined the expressions

$$E_1 = \sum_l \langle R^{-1}\partial_\lambda R g^{(l)} d\xi'^{(l)} g^{(l)-1}\rangle$$

$$E_2 = \sum_l \langle g^{(l)-1}\partial_\lambda g^{(l)} d(\xi'^{(l)} - \xi^{(l)})\rangle$$

$$E_3 = \sum_l \langle L^{(l)}(\lambda - \lambda_l)^{-1} d\xi'^{(l)}\rangle$$

In this section we use the notation $\langle X^{(l)}\rangle = \mathrm{Res}_{\lambda_l} \mathrm{Tr}(X^{(l)})d\lambda$. The term E_3 vanishes because the pole is of order at least 2. Using $\xi'^{(l)} = \xi^{(l)} + L^{(l)} \log(\lambda - \lambda_l)$, the second term E_2 is equal to:

$$E_2 = -\sum_l \mathrm{Tr}(g_1^{(l)} L^{(l)})d\lambda_l = -\left(g_1^{(k)}\right)_{\alpha\alpha} d\lambda_k + \left(g_1^{(k')}\right)_{\alpha'\alpha'} d\lambda_{k'}$$

In the above sum over l, only $l = k, k'$ contribute since otherwise $L^{(l)}$ vanishes. To compute the first term E_1, we split it into two parts: $E_1 = E'_1 + E''_1$

$$E'_1 = \sum_l \langle R^{-1}\partial_\lambda R g^{(l)} d\xi^{(l)} g^{(l)-1}\rangle$$

$$E''_1 = \sum_l \langle R^{-1}\partial_\lambda R g^{(l)} (d\xi'^{(l)} - d\xi^{(l)}) g^{(l)-1}\rangle$$

Using the explicit form for $R(\lambda)$ we can compute (assuming $\lambda_k \neq \lambda_{k'}$)

$$R^{-1}\partial_\lambda R = \frac{R_0}{(\lambda - \lambda_k)(\lambda - \lambda_{k'})} = \frac{R_0}{\lambda_k - \lambda_{k'}}\left(\frac{1}{\lambda - \lambda_k} - \frac{1}{\lambda - \lambda_{k'}}\right)$$

one gets for the first term E'_1:

$$E'_1 = \frac{1}{\lambda_k - \lambda_{k'}}\sum_l \langle \left(\frac{1}{\lambda - \lambda_k} - \frac{1}{\lambda - \lambda_{k'}}\right) R_0 g^{(l)} d\xi^{(l)} g^{(l)-1}\rangle$$

To evaluate this expression, we use the following identity valid for any function $f(\lambda)$ with an expansion around λ_l of the form $f(\lambda) = \sum_{i=-N}^{\infty} f_i(\lambda - \lambda_l)^i$:

$$\mathrm{Res}_{\lambda_l}\left(\frac{1}{\lambda - \lambda_k}f(\lambda)\right) = \begin{cases} -f_-(\lambda)|_{\lambda=\lambda_k} & \text{if } \lambda_l \neq \lambda_k \\ f_0 & \text{if } \lambda_l = \lambda_k \end{cases}$$

We immediately get:

$$E'_1 = \frac{1}{\lambda_k - \lambda_{k'}}\Big\{ -\sum_{l\neq k} \mathrm{Tr}\left(R_0(g^{(l)}d\xi^{(l)}g^{(l)-1})_-\right)|_{\lambda=\lambda_k}$$

$$+ \mathrm{Tr}\left(R_0(g^{(k)}d\xi^{(k)}g^{(k)-1})_0\right) - (k \to k')\Big\}$$

To rewrite these terms, consider the equation of motion eq. (8.32) and expand it around $\lambda = \lambda_k$. We find

$$dg_0^{(k)} g_0^{(k)-1} - g_0^{(k)} g_1^{(k)} g_0^{(k)-1} d\lambda_k = \sum_{l \neq k} M_l|_{\lambda=\lambda_k} + (M^{(k)} - g^{(k)} d\xi^{(k)} g^{(k)-1})|_{\lambda=\lambda_k}$$

Now we have $M_l|_{\lambda=\lambda_k} = (g^{(l)} d\xi^{(l)} g^{(l)-1})_-|_{\lambda=\lambda_k}$ and

$$(M^{(k)} - g^{(k)} d\xi^{(k)} g^{(k)-1})|_{\lambda=\lambda_k} = -(g^{(k)} d\xi^{(k)} g^{(k)-1})_0$$

so that we can rewrite

$$E_1' = \frac{-1}{\lambda_k - \lambda_{k'}} \left\{ \operatorname{Tr} \left(R_0 \left(dg_0^{(k)} g_0^{(k)-1} - g_0^{(k)} g_1^{(k)} g_0^{(k)-1} d\lambda_k \right) - (k \to k') \right) \right\}$$

Similarly, the term E''_1 reads:

$$E''_1 = \frac{1}{\lambda_k - \lambda_{k'}} \Big\{ \sum_{l \neq k} \operatorname{Tr}(R_0 g_0^{(l)} \frac{L^{(l)}}{\lambda_k - \lambda_l} g_0^{(l)-1}) d\lambda_l$$

$$- \operatorname{Tr}(R_0 g_0^{(k)} [g_1^{(k)}, L^{(k)}] g_0^{(k)-1}) d\lambda_k - (k \to k') \Big\}$$

In the right-hand side, only $l = k, k'$ contribute to the sums over l because $L^{(l)}$ vanishes otherwise. After substituting the explicit value of R_0, they produce the contribution

$$\frac{d\lambda_k - d\lambda_{k'}}{\lambda_k - \lambda_{k'}}$$

The terms depending on g_1 in E''_1 give:

$$\left(g_1^{(k)} \right)_{\alpha\alpha} d\lambda_k - \left(g_1^{(k')} \right)_{\alpha'\alpha'} d\lambda_{k'}$$

$$- \frac{\left(g_0^{(k')-1} g_0^{(k)} g_1^{(k)} \right)_{\alpha'\alpha} d\lambda_k - \left(g_1^{(k')} g_0^{(k')-1} g_0^{(k)} \right)_{\alpha'\alpha} d\lambda_{k'}}{\left(g_0^{(k')-1} g_0^{(k)} \right)_{\alpha'\alpha}}$$

they cancel with those coming from E'_1 and E_2. Hence, putting everything together, we get:

$$\Upsilon' - \Upsilon = \operatorname{Tr} \left(R_0 \frac{dg_0^{(k)} g_0^{(k)-1} - dg_0^{(k')} g_0^{(k')-1}}{\lambda_k - \lambda_{k'}} \right) - \frac{d\lambda_k - d\lambda_{k'}}{\lambda_k - \lambda_{k'}}$$

or

$$d\log \frac{\tau'}{\tau} = d\log \frac{\left(g_0^{(k')-1} g_0^{(k)} \right)_{\alpha'\alpha}}{\lambda_k - \lambda_{k'}}$$

Integrating the above equation proves eq. (8.45) for $k \neq k'$. The integration constant has been normalized to 1. The case $k = k'$ is proved similarly. ∎

8.7 Ricatti equation

Notice that the right-hand side of eqs. (8.45) is the product of $\tau(t)$ by the leading term in the expansion of

$$G^{(kk')}(\lambda, \lambda') = \frac{\delta_{kk'}\text{Id} - g^{(k)-1}(\lambda)g^{(k')}(\lambda')}{\lambda - \lambda'} \qquad (8.46)$$

in powers of $z_k = \lambda - \lambda_k$ and $z'_{k'} = \lambda' - \lambda_{k'}$. This double expansion has only *positive* powers of z_k and $z_{k'}$. This is clear when $k \neq k'$, and for $k = k'$ the zero in the denominator is cancelled by a zero in the numerator. The matrix elements of $G^{(kk')}(\lambda, \lambda')$ are algebraic functions of the dynamical variables occuring in M_λ. We can recast the equations of motion of the hierarchy in terms of these new variables. They take a particularly simple Ricatti type form.

Let us consider the generating function for the flows associated with the pole λ_l:

$$\nabla_\alpha^{(l)}(\lambda) = \sum_{n>0}(\lambda - \lambda_l)^{n-1}\frac{\partial}{\partial t_{(l,n,\alpha)}} \qquad (8.47)$$

Strictly speaking, in our formalism there were a finite number of times $t_{(k,n,\alpha)}$ with $n \leq n_k$. We consider now, formally, differential equations $\partial_\lambda\Psi - M_\lambda\Psi = 0$, where M_λ is allowed to have poles of arbitrary order at each λ_k.

Proposition. *The quantity $G^{(kk')}(\lambda, \lambda')$ defined in eq. (8.46) obeys the Ricatti type equation:*

$$\nabla_\alpha^{(l)}(\lambda'')G^{(kk')}(\lambda, \lambda') = G^{(kl)}(\lambda, \lambda'')E_{\alpha\alpha}G^{(lk')}(\lambda'', \lambda') \qquad (8.48)$$
$$+\delta_{kl}E_{\alpha\alpha}\frac{G^{(kk')}(\lambda, \lambda') - G^{(kk')}(\lambda'', \lambda')}{\lambda - \lambda''}$$
$$-\delta_{k'l}\frac{G^{(kk')}(\lambda, \lambda') - G^{(kk')}(\lambda, \lambda'')}{\lambda' - \lambda''}E_{\alpha\alpha}$$

Similarly, the equation of motion relative to the position of the pole λ_l takes the form:

$$\partial_{\lambda_l}G^{(kk')}(\lambda, \lambda') = \text{Res}_{\lambda''=\lambda_l}G^{(kl)}(\lambda, \lambda'')\partial_{\lambda_l}\xi^{(l)}(\lambda'')G^{(lk')}(\lambda'', \lambda')$$
$$+\delta_{kl}\left(\partial_{\lambda_k}\xi^{(k)}(\lambda)G^{(kk')}(\lambda, \lambda')\right)_+ - \delta_{k'l}\left(G^{(kk')}(\lambda, \lambda')\partial_{\lambda_{k'}}\xi^{(k')}(\lambda')\right)_+$$

where $()_+$ means taking the positive power part in the expansions around λ_k and $\lambda_{k'}$ respectively.

<u>Proof.</u> We need the following identities: let $f(\lambda) = \sum_{j=1}^{\infty} f_j \lambda^j$, then we have:

$$\sum_{n=1}^{\infty} \lambda'^{n-1}(\lambda^{-n} f(\lambda))_+ = \frac{f(\lambda) - f(\lambda')}{\lambda - \lambda'}, \quad \sum_{n=1}^{\infty} \lambda'^{n-1}(\lambda^{-n} f(\lambda))_- = \frac{f(\lambda')}{\lambda - \lambda'}$$

Recalling the equations of the hierarchy expressed on the $g^{(k)}$:

$$\frac{\partial}{\partial t_{(l,n,\alpha)}} g^{(k)}(\lambda) = (g^{(l)} E_{\alpha\alpha}(\lambda - \lambda_l)^{-n} g^{(l)-1})_- g^{(k)} - \delta_{kl} g^{(k)} E_{\alpha\alpha}(\lambda - \lambda_k)^{-n}$$

we see that they can be recast in the form:

$$\nabla_\alpha^{(l)}(\lambda') g^{(k)}(\lambda) = -g^{(k)}(\lambda) E_{\alpha\alpha} \frac{\delta_{kl}}{\lambda - \lambda'} + \frac{g^{(l)}(\lambda') E_{\alpha\alpha} g^{(l)-1}(\lambda') g^{(k)}(\lambda)}{\lambda - \lambda'}$$

which proves the first part of the proposition.

Similarly, using the identity for the function $f(\lambda) = \sum_{-\infty}^{\infty} f_n \lambda^n$:

$$f_-(\lambda) = \text{Res}_{\lambda'=0} \left(\frac{1}{\lambda - \lambda'} f(\lambda') \right)$$

we can write the equation of motion for $g^{(k)}$ in the form:

$$\partial_{\lambda_l} g^{(k)}(\lambda)$$
$$= \text{Res}_{\lambda''=\lambda_l} \frac{g^{(l)}(\lambda'') \partial_{\lambda_l} \xi^{(l)}(\lambda'') g^{(l)-1}(\lambda'')}{\lambda - \lambda''} \cdot g^{(k)}(\lambda) - \delta_{kl} g^{(k)}(\lambda) \partial_{\lambda_k} \xi^{(k)}(\lambda)$$

from which the second statement follows.   ∎

We will need, in the next section, the limits of eqs. (8.48) when $l = k, \lambda'' \to \lambda$ and $l = k', \lambda'' \to \lambda'$. We get respectively (if $k \neq k'$):

$$\nabla_\alpha^{(k)}(\lambda) G^{(kk')}(\lambda, \lambda') = G^{(kk)}(\lambda, \lambda) E_{\alpha\alpha} G^{(kk')}(\lambda, \lambda') + E_{\alpha\alpha} \partial_\lambda G^{(kk')}(\lambda, \lambda')$$
$$\tag{8.49}$$

$$\nabla_\alpha^{(k')}(\lambda') G^{(kk')}(\lambda, \lambda') = G^{(kk')}(\lambda, \lambda') E_{\alpha\alpha} G^{(k'k')}(\lambda', \lambda') - \partial_{\lambda'} G^{(kk')}(\lambda, \lambda') E_{\alpha\alpha}$$
$$\tag{8.50}$$

8.8 Sato's formula

It is remarkable that the complete matrix $G^{(kk')}(\lambda, \lambda')$ can be reconstructed from the tau-function and its elementary Schlesinger transforms. As a consequence we can express the matrix elements of $G^{(kk')}(\lambda, \lambda')$ as quotients of tau-functions, as in Sato's formula. We still denote $z_k = \lambda - \lambda_k$ and $z'_{k'} = \lambda' - \lambda_{k'}$ and introduce the notation:

$$t \to t + [z_k]_\alpha \text{ means } t_{(l,n,\gamma)} \to t_{(l,n,\gamma)} + \delta_{kl}\delta_{\gamma\alpha}\frac{z_k^n}{n}$$

Proposition. *Denote by* $G_{\alpha\alpha'}^{(kk')}(\lambda,\lambda')$ *the matrix element* $\alpha\alpha'$ *of* $G^{(kk')}(\lambda,\lambda')$. *We have:*

$$\tau(t)\,G_{\alpha\alpha'}^{(kk')}(\lambda,\lambda') = \tau\begin{bmatrix} k & k' \\ \alpha & \alpha' \end{bmatrix}(t + [z_k]_\alpha - [z'_{k'}]_{\alpha'}), \quad \text{if } (k,\alpha) \neq (k',\alpha')$$

$$\tau(t)\,G_{\alpha\alpha}^{(kk)}(\lambda,\lambda') = \frac{\tau(t) - \tau(t + [z_k]_\alpha - [z'_k]_\alpha)}{\lambda - \lambda'}, \tag{8.51}$$

Proof. This is a generalization of the proof of eq. (3.61) in Chapter 3. From the definition of the tau-function, eq. (8.44), we have:

$$\frac{\partial}{\partial t_{(l,n,\alpha)}} \log \tau = -\text{Res}_{\lambda_l} \text{Tr}\left(g^{(l)-1}(\lambda)\partial_\lambda g^{(l)}(\lambda) E_{\alpha\alpha}(\lambda - \lambda_l)^{-n}\right)$$

Using the identity eq. (3.62) in Chapter 3, we get

$$\nabla_\alpha^{(k)}(\lambda) \log \tau = -\text{Tr}\left(g^{(k)-1}(\lambda)\partial_\lambda g^{(k)}(\lambda) E_{\alpha\alpha}\right) = -G_{\alpha\alpha}^{(kk)}(\lambda,\lambda)$$

From this, it follows, using eqs. (8.49, 8.50) and the definition of $G^{(kk')}(\lambda,\lambda')$, that:

$$\left(\nabla_\alpha^{(k)}(\lambda) - \partial_\lambda\right)\tau(t)G_{\alpha\alpha'}^{(kk')}(\lambda,\lambda') = 0 \tag{8.52}$$

$$\left(\nabla_{\alpha'}^{(k')}(\lambda') + \partial_{\lambda'}\right)\tau(t)G_{\alpha\alpha'}^{(kk')}(\lambda,\lambda') = 0 \tag{8.53}$$

These are differential equations relating the λ-dependence to the time dependence. Their unique solution allows us to express $\tau(t)G^{(kk')}(\lambda,\lambda')$ in the form:

$$\tau(t)G_{\alpha\alpha'}^{(kk')}(\lambda,\lambda') = \tau_{\alpha\alpha'}(t + [z_k]_\alpha - [z'_{k'}]_{\alpha'})$$

To find the functions $\tau_{\alpha\alpha'}$, it is enough to compare the two sides of the equation at $z_k = z'_{k'} = 0$. But there, comparing with eq. (8.45), we find

$$\tau(t)G_{\alpha\alpha'}^{(kk')}(\lambda,\lambda') \to \tau\begin{bmatrix} k & k' \\ \alpha & \alpha' \end{bmatrix}(t)$$

this proves the proposition if $k \neq k'$. If $k = k'$, the right-hand side of eq. (8.49) contains the extra term

$$-\frac{G^{(kk)}(\lambda,\lambda') - G^{(kk)}(\lambda,\lambda)}{\lambda' - \lambda}E_{\alpha\alpha}$$

If $\alpha' \neq \alpha$ this does not affect eq. (8.52), but if $\alpha' = \alpha$ it becomes

$$(\nabla_\alpha^{(k)}(\lambda) - \partial_\lambda)\left(\tau(t)[(\lambda - \lambda')G_{\alpha\alpha}^{(kk)}(\lambda, \lambda') - 1]\right) = 0$$

Using the analogous equation for λ', we deduce that

$$\tau(t)[(\lambda - \lambda')G_{\alpha\alpha}^{(kk)}(\lambda, \lambda') - 1] = \tau_\alpha(t + [z_k]_\alpha - [z'_k]_\alpha)$$

To find the function $\tau_\alpha(t)$ we notice that $\tau(t)[(\lambda - \lambda')G_{\alpha\alpha}^{(kk)}(\lambda, \lambda') - 1] \rightarrow -\tau(t)$ when $\lambda \rightarrow \lambda_k$ and $\lambda' \rightarrow \lambda_k$, hence $\tau_\alpha(t) = -\tau(t)$. This yields the second half of the proposition. ∎

Many remarkable relations can be extracted from this result. In particular, setting $k = k'$, $z_k = 0$, and introducing the matrix $h^{(k)}(\lambda)$ by $g^{(k)}(\lambda) = g_0^{(k)}h^{(k)}(\lambda)$, we find

$$\left(h^{(k)}\right)_{\alpha\alpha'}(\lambda) = (\lambda - \lambda_k)\frac{\tau\begin{bmatrix} k & k \\ \alpha & \alpha' \end{bmatrix}(t - [z_k]_{\alpha'})}{\tau(t)}, \quad \alpha \neq \alpha'$$

$$\left(h^{(k)}\right)_{\alpha\alpha}(\lambda) = \frac{\tau(t - [z_k]_\alpha)}{\tau(t)} \tag{8.54}$$

We have already met these equations in Chapter 3, they are the Sato formulae. We see that we have completely identified the functions $\tau_{\alpha\alpha'}$ occurring in the numerator of eq. (3.61) as the Schlesinger transforms of the tau-function in the denominator.

8.9 The Hirota equations

Hirota noticed that many integrable equations could be recast into a bilinear form in terms of tau-functions. Specifically, introducing the Hirota differential operators D_i with the definition:

$$D_i^n f \cdot g = \left(\frac{\partial}{\partial y_i}\right)^n f(x + y)g(x - y)|_{y=0} \tag{8.55}$$

the equations of motion take the symbolic form: $P(D)\tau \cdot \tau = 0$, where P is a polynomial in D. For instance, the equation

$$(D_1^4 + 3D_2^2 - 4D_1D_3)\tau \cdot \tau = 0 \tag{8.56}$$

is the Hirota form of the KP equation. As a matter of fact, setting

$$u = -2\frac{\partial^2}{\partial t_1^2}\log\tau \tag{8.57}$$

we get the Kadomtsev–Petviashvili equation:

$$3\frac{\partial^2 u}{\partial t_2^2} + \frac{\partial}{\partial t_1}\left(-4\frac{\partial u}{\partial t_3} - 6u\frac{\partial u}{\partial t_1} + \frac{\partial^3 u}{\partial t_1^3}\right) = 0 \tag{8.58}$$

We show in this section that this is a general phenomenon. We have the:

Proposition. *In terms of the tau-function and its elementary Schlesinger transforms, the hierarchy equations take the Hirota bilinear form.*

<u>Proof.</u> The proof is just a rewriting of the Ricatti equation eq. (8.48) in terms of tau-functions. Let us do it in a simple case (the other cases are similar). We assume for simplicity that l, k, k' are all different. Multiplying eq. (8.48) by $\tau^2(t)$ we get:

$$\tau^2(t)\nabla_\beta^{(l)}(\lambda'')\frac{1}{\tau(t)}\tau\begin{bmatrix}k & k'\\ \alpha & \alpha'\end{bmatrix}(t+[z_k]_\alpha - [z'_{k'}]_{\alpha'})$$

$$= \tau\begin{bmatrix}k & l\\ \alpha & \beta\end{bmatrix}(t+[z_k]_\alpha - [z''_l]_\beta)\, \tau\begin{bmatrix}l & k'\\ \beta & \alpha'\end{bmatrix}(t+[z''_l]_\beta - [z'_{k'}]_{\alpha'})$$

The left-hand side can be rewritten in terms of Hirota differential operators, using the identity:

$$f^2\frac{\partial}{\partial t}\left(\frac{g}{f}\right) = -(\dot{f}g - \dot{g}f) = -D_t\, f\cdot g$$

Introducing the generating function for Hirota differential operators:

$$D_\alpha^{(l)}(\lambda'') = \sum_{n>0}(\lambda'' - \lambda_l)^{n-1}D_{t_{(l,n,\alpha)}}$$

and shifting the variables t by $t \to t - [z_k]_\alpha/2 + [z'_{k'}]_{\alpha'}/2$, we get:

$$D_\beta^{(l)}(\lambda'')\tau\left(t - \frac{[z_k]_\alpha}{2} + \frac{[z'_{k'}]_{\alpha'}}{2}\right)\cdot\tau\begin{bmatrix}k & k'\\ \alpha & \alpha'\end{bmatrix}\left(t + \frac{[z_k]_\alpha}{2} - \frac{[z'_{k'}]_{\alpha'}}{2}\right)$$

$$= -\tau\begin{bmatrix}k & l\\ \alpha & \beta\end{bmatrix}\left(t + \frac{[z_k]_\alpha}{2} + \frac{[z'_{k'}]_{\alpha'}}{2} - [z''_l]_\beta\right)$$

$$\times\tau\begin{bmatrix}l & k'\\ \beta & \alpha'\end{bmatrix}\left(t - \frac{[z_k]_\alpha}{2} - \frac{[z'_{k'}]_{\alpha'}}{2} + [z''_l]_\beta\right)$$

We now expand this formula in powers of z_k, $z'_{k'}$ and z''_l using the equation

$$f(t+z)\, g(t-z) = e^{zD_t}\, f\cdot g$$

It is clear that the coefficients in this expansion have the form of Hirota bilinear equations. ∎

Exactly the same method applies to the other Ricatti equations, and shows that they can be recast in Hirota form. This form is very remarkable as it allows for some easy particular solutions and, moreover, lends itself to a beautiful geometric interpretation as Plücker relations in an infinite Grassmannian which will be described in Chapter 9.

8.10 Tau-functions and theta-functions

In this section we show how the Lax matrix approach fits into the isomonodromy approach. We show that it corresponds to very special matrices M_λ. In that case, the tau-functions are essentially Riemann's theta-functions.

We start from a rational Lax matrix $L(\lambda)$ of size $N \times N$, with poles at λ_k. We consider the associated spectral curve, $\Gamma : \det(L(\lambda) - \mu) = 0$, which is a compact Riemann surface presented as an N-sheeted branched covering of the Riemann sphere. For any value of the spectral parameter λ one may consider the (multivalued) $N \times N$ matrix $\widehat{\Psi}(\lambda)$ given by $(\Psi(P_1), \ldots, \Psi(P_N))$, where the P_j are the N points above λ in some order, and $\Psi(P_j)$ is the eigenvector of $L(\lambda)$ for the corresponding eigenvalue $\mu_j(\lambda)$. It has been explained in Chapter 5 that requiring the time evolution equations $\partial_{t_i} \widehat{\Psi} = M_i \widehat{\Psi}$ implies that the components of the eigenvector are Baker–Akhiezer functions on Γ. In general they have essential singularities at all points on Γ above λ_k. Around each puncture λ_k, the matrix $\widehat{\Psi}(\lambda)$ has an expansion of the form $\widehat{\Psi}(\lambda) = g^{(k)}(\lambda) \exp(\xi^{(k)}(\lambda))$ where $g^{(k)}$ is regular at $\lambda = \lambda_k$ and $\xi^{(k)}$ is a diagonal matrix singular at λ_k:

$$\xi^{(k)}(\lambda) = \sum_{n,\alpha} t_{(k,n,\alpha)} \frac{E_{\alpha\alpha}}{(\lambda - \lambda_k)^n}$$

Here $t_{(k,n,\alpha)}$ are the times describing all the integrable flows of the hierarchy. Moreover, $\Psi(P)$ has $g + N - 1$ poles on the Riemann surface, g of them at finite distance being the dynamical divisor D, and the other $N - 1$ being at the points Q_i, $i = 2, \ldots, N$ above $\lambda = \infty$. Note that at λ_k, $\widehat{\Psi}(\lambda)$ has the behaviour considered in this chapter. Hence it is natural to consider the matrix $M_\lambda = \partial_\lambda \widehat{\Psi} \cdot \widehat{\Psi}^{-1}$.

Proposition. *The matrix M_λ is a rational function of λ. It has poles at the λ_k, simple poles at the projections of the branch points of the covering $(\lambda, \mu) \to \lambda$, and at the projections of the poles of $\widehat{\Psi}(\lambda)$.*

<u>Proof</u>. The main point is that while $\widehat{\Psi}(\lambda)$ is multivalued, i.e. its columns undergo permutations when one performs a loop around a branch point, $\partial_\lambda\widehat{\Psi}\cdot\widehat{\Psi}^{-1}$ is independent of the ordering of the columns of $\widehat{\Psi}(\lambda)$, hence it is well-defined as a function of λ. Using the expansions around the punctures λ_k we see that the singular part of M_λ at λ_k is given by $(g^{(k)}\partial_\lambda\xi^{(k)}g^{(k)-1})_-$. It has a finite order pole if there are a finite number of time variables.

Let us consider now a branch point, and for simplicity assume that it is of order 2, that is we assume that the first two columns of $\widehat{\Psi}(\lambda)$ coalesce for $\lambda = \lambda_b$. The corresponding eigenvalues $(\mu_1, \mu_2, \ldots)$ are such that μ_1, μ_2 also coalesce. Locally the equation of Γ is of the form $\lambda-\lambda_b = (\mu-\mu_b)^2$ and the local parameter is $z = \mu - \mu_b = \sqrt{\lambda - \lambda_b}$. The first two eigenvectors are just the evaluation of one meromorphic vector valued function of z on the two sheets above λ. Splitting the even and odd powers of z, we can write the matrix $\widehat{\Psi}(\lambda)$ in the form:

$$\widehat{\Psi}(\lambda) = (\Psi_e(z) + \sqrt{z}\Psi_o(z), \Psi_e(z) - \sqrt{z}\Psi_o(z), \Psi_3(z), \ldots)$$

$$= (\Psi_e(z), \Psi_o(z), \Psi_3(z), \ldots) \begin{pmatrix} 1 & 0 & 0 & \cdots \\ 0 & \sqrt{z} & 0 & \cdots \\ 0 & 0 & 1 & \\ \vdots & \vdots & & \ddots \end{pmatrix} \begin{pmatrix} 1 & 1 & 0 & \cdots \\ 1 & -1 & 0 & \cdots \\ 0 & 0 & 1 & \\ \vdots & \vdots & & \ddots \end{pmatrix}$$

The first matrix $g(z) = (\Psi_e(z), \Psi_o(z), \Psi_3(z), \ldots)$ is regular around $z = 0$. The third matrix is an inessential invertible constant matrix, and the second matrix can be identified with $\exp\xi(\lambda)$ with $\xi(\lambda) = \frac{1}{2}\log(\lambda - \lambda_b)E_{22}$. This produces in M_λ a polar part

$$(g\partial_\lambda\xi g^{-1})_- = \frac{1}{2}\frac{g(\lambda_b)E_{22}g^{-1}(\lambda_b)}{(\lambda - \lambda_b)}$$

At a pole of $\widehat{\Psi}(\lambda)$ (above λ_c) at finite distance, one column (say the jth one) has a pole. Hence one can write, up to right multiplication by a constant invertible matrix, $\widehat{\Psi}(\lambda) = g(\lambda)\exp(\xi(\lambda))$ with $\xi(\lambda) = -\log(\lambda - \lambda_c)E_{jj}$. This again yields a simple pole in M_λ:

$$(g\partial_\lambda\xi g^{-1})_- = -\frac{g(\lambda_c)E_{jj}g^{-1}(\lambda_c)}{(\lambda - \lambda_c)}$$

Finally, above $\lambda = \infty$, in the setup of Chapter 5, we have N points $Q_1, \ldots, Q_N$ on Γ, and $\widehat{\Psi}(\lambda)$ has simple poles at the $N - 1$ points $Q_2, \ldots, Q_N$. When $\lambda \to \infty$, the wave function $\widehat{\Psi}(\lambda)$ has the asymptotic expansion $g_0^{(\infty)}(1 + O(1/\lambda))\exp(B_0^{(\infty)}\log(1/\lambda))C^{(\infty)}$, where $C^{(\infty)}$ is a constant invertible matrix and $B_0^{(\infty)} = -\sum_{i=2}^N E_{ii}$. This implies that around $\lambda = \infty$ the matrix M_λ has the form:

$$(g\partial_\lambda \xi g^{-1})_- = -\frac{g_0^{(\infty)} B_0^{(\infty)} g_0^{(\infty)\,-1}}{\lambda} + O(1/\lambda^2)$$

It is now clear that M_λ is a rational function of λ vanishing at ∞ and is the sum of its polar parts at finite distance. ∎

Remark 1. Note that the last statement of the proof implies that M_λ tends to 0 at ∞, so we must have:

$$M_\lambda = \sum_k \left(g^{(k)} \partial_\lambda \xi^{(k)} g^{(k)\,-1} \right)_- = -\frac{1}{\lambda} g_0^{(\infty)} B_0^{(\infty)} g_0^{(\infty)\,-1} + O\left(\frac{1}{\lambda^2}\right)$$

where the sum over k runs over poles at finite distance. In particular, looking at the $1/\lambda$ terms and taking the trace one gets the Fuchs relation eq. (8.21) again. It is interesting to check it in this context. First we have $\operatorname{Tr} B_0^{(\infty)} = -N+1$. The poles at finite distance are the branch points, each one contributing $1/2$ to the trace, and the g points of D, each one contributing -1. Finally, at the punctures, we have $\operatorname{Tr}(g^{(k)}(\lambda)\partial_\lambda \xi^{(k)} g^{(k)\,-1}(\lambda))_- = \operatorname{Tr} \partial_\lambda \xi^{(k)} = O(1/\lambda^2)$ at ∞, hence they do not contribute to the $1/\lambda$ terms. The Fuchs condition therefore reads $g = \nu/2 - N + 1$. This is just the Riemann–Hurwitz formula.

Remark 2. The matrix $M_\lambda(\lambda)$ has more poles than $L(\lambda)$. However, at the extra poles, namely the poles of $\widehat{\Psi}$ and the branch points, the singularity of M_λ is of regular type. The corresponding singularity data ξ contains only a logarithmic term. At the poles of $L(\lambda)$ we have the whole singularity structure for singularities of irregular type, but without a logarithmic term. The matrix $M_\lambda(\lambda)$ embodies the data allowing us to reconstruct $\widehat{\Psi}$. In particular, it contains data pertaining to the spectral curve, through its branch points, and data pertaining to the eigenvector bundle, through the divisor D.

It follows from the Proposition that the matrix $\widehat{\Psi}(\lambda)$ constructed from the eigenvector bundle satisfies a differential equation $(\partial_\lambda - M_\lambda(\lambda))\widehat{\Psi}(\lambda) = 0$, where M_λ is a rational function of λ of the type studied in this chapter. However, M_λ is a very particular function of λ since the solution $\widehat{\Psi}(\lambda)$ has no monodromy at the punctures, i.e. at the poles of $L(\lambda)$. Its only non–trivial monodromy occurs around the branch points, and acts by permutations of corresponding columns. Globally, the monodromy group is finite, and this is a special feature of the differential equations coming from Lax equations. Finally, there are no Stokes matrices at any singularity. Hence the monodromy data, as defined above, consist only of the connection matrices $\mathcal{C}^{(k)}$, which are time-independent matrices, function of the moduli of the spectral curve.

The general theory nevertheless applies to this very special situation, and in particular the tau-functions can be computed explicitly in terms of theta-functions. We have already met this situation in Chapter 5, but we make here the analysis in a broader context in order to be able to understand the action of Schlesinger transformations. We will see that Schlesinger transformations reduce to very simple translations in the argument of the theta-functions, as we now show.

Let $P_{k\alpha}$ be the N points above λ_k. We take $z_k = \lambda - \lambda_k$ as local parameter around each $P_{k\alpha}$. We introduce the singular parts

$$\xi_{k\alpha}(z_k) = l_{k\alpha} \log(z_k) + \sum_{n \geq 1} t_{(k,n,\alpha)} z_k^{-n} \tag{8.59}$$

We introduced logarithmic terms as in eq. (8.6), but we assume that the coefficients $l_{k\alpha}$ are integers in order to be able to construct Baker–Akhiezer functions. These logarithmic terms will introduce extra zeroes or poles in the Baker–Akhiezer functions at the punctures, and will be useful to help us understand Schlesinger transformations.

Consider Baker–Akhiezer functions with singular parts given by eq. (8.59) at each $P_{k\alpha}$. We introduce poles at the $g + N - 1$ given points $D = (\gamma_1, \ldots, \gamma_g)$ (the dynamical divisor) and $Q_2, \ldots, Q_N$ (above ∞). We choose the numbers $l_{k\alpha}$ such that $\sum_{k\alpha} l_{k\alpha} = 0$ so that the degree of the divisor of prescribed zeroes and poles is still $g + N - 1$. The dimension of the space of such functions is N, by the Riemann–Roch theorem. We fix a particular λ_k and consider the N sheets above it. For each α, one can define a unique Baker–Akhiezer function $\psi_\alpha^{(k)}(P)$ satisfying the N conditions:

$$\psi_\alpha^{(k)}(P) = e^{\xi_{k\alpha}(z_k)}(\delta_{\alpha\beta} + O(z_k)), \quad \text{when } P \to P_{k\beta}$$

We put these N functions in a column vector $\Psi^{(k)}(P)$ and form as usual the $N \times N$ matrix

$$\widehat{\Psi}^{(k)}(\lambda) = (\Psi^{(k)}(P_1), \ldots, \Psi^{(k)}(P_N)) \tag{8.60}$$

where the P_i are the N points above λ. Note that, around our particular λ_k, we have the expansion

$$\widehat{\Psi}^{(k)}(\lambda) = h^{(k)}(\lambda) e^{\xi^{(k)}(z_k)} \tag{8.61}$$

with $h^{(k)}(\lambda) = 1 + O(z_k)$ and $\xi^{(k)}$ being the diagonal matrix $\sum_\alpha \xi_{k\alpha} E_{\alpha\alpha}$, i.e. we have chosen the normalization $g_0^{(k)} = 1$ in the expansion eq. (8.24). We are going to identify the tau-function by comparing the expression of the matrix $\widehat{\Psi}^{(k)}$ in terms of theta-functions with the expression eq. (8.54).

Proposition. *The Baker–Akhiezer function $\psi_\alpha^{(k)}(P)$ can be written in terms of theta-functions as:*

$$\psi_\alpha^{(k)}(P) = e^{\int_{P_{k\alpha}}^{P} \Omega} \cdot \frac{\theta(\mathcal{A}(P) + V - \mathcal{K})\theta(\mathcal{A}(P_{k\alpha}) - \mathcal{A}(D) - \mathcal{K})}{\theta(\mathcal{A}(P_{k\alpha}) + V - \mathcal{K})\theta(\mathcal{A}(P) - \mathcal{A}(D) - \mathcal{K})} \quad (8.62)$$

where the vector V is given by:

$$V = \sum_{kn\alpha} t_{(k,n,\alpha)} U^{(k,n,\alpha)} + \sum_{k'\beta} l_{k'\beta}\mathcal{A}(P_{k'\beta}) - \mathcal{A}(P_{k\alpha}) + \mathcal{A}(Q_1) - \mathcal{A}(D)$$

<u>Proof</u>. Following the general procedure of Chapter 5, we construct

$$\psi_\alpha^{(k)}(P) = C \cdot e^{\int_{P_{k\alpha}}^{P} \Omega} \cdot \frac{\theta(\mathcal{A}(P) + V - \mathcal{K})}{\theta(\mathcal{A}(P) - \mathcal{A}(D) - \mathcal{K})}$$

where C is a normalization constant, Ω is an Abelian differential chosen so that $\psi_\alpha^{(k)}$ has the required properties at the punctures and at the Q_j, the theta-function at the denominator has been introduced to take care of the poles at the dynamical divisor D ($\mathcal{K}$ is the vector of Riemann's constants), and the vector V in the theta-function in the numerator is determined by requiring that the resulting function has no monodromy.

   Let us first determine the form Ω. We write it as a sum of three pieces $\Omega = \Omega^{(t)} + \Omega^{(l)} + \Omega^{(q)}$, where $\Omega^{(t)}$ ensures that $\psi_\alpha^{(k)}(P)$ has the correct essential singularities at the punctures. One can write

$$\Omega^{(t)} = \sum_{k',n,\beta} t_{(k',n,\beta)}\Omega_{P_{k'\beta}}^{(n)} \quad (8.63)$$

where $\Omega_{P_{k'\beta}}^{(n)}$ is the normalized (i.e. the a-periods vanish) second kind Abelian differential with just one singularity at $P_{k'\beta}$ and such that around this point $\Omega_{P_{k'\beta}}^{(n)} = d(z_{k'}^{-n}) + $holomorphic. Note that the integral $\int_{P_{k\alpha}}^{P} \Omega_{P_{k\alpha}}^{(n)}$ is ill-defined. We take as its definition the unique primitive which around $P_{k\alpha}$ behaves as $z_k^{-n} + O(z_k)$ (no constant term). The form $\Omega^{(l)}$ is computed to produce a zero or pole of order $l_{k'\beta}$ at the puncture $P_{k'\beta}$. Noting that $\sum l_{k'\beta} = 0$, it is the unique normalized Abelian differential of the third kind with first order poles at these points with corresponding residues $l_{k'\beta}$. As in the previous case, the integral of this form with origin at $P_{k\alpha}$ is ill-defined. We take it to be the primitive behaving as $l_{k\alpha} \log(z_k) + O(z_k)$, mod $2i\pi$. Finally, the form $\Omega^{(q)}$ is introduced to get $N - 1$ poles at the points $Q_2, \ldots, Q_N$ and $N - 1$ zeroes at the points $P_{k\beta}$ with $\beta \neq \alpha$ (for the special k). It is given by the unique third kind differential with residues -1 at the Q_j and $+1$ at the $P_{k\beta}$.

It is now easy to compute the monodromy of the function $\exp\left(\int_{P_{kq}}^{P}\Omega\right)$. Since the differentials are normalized, there is no monodromy around the a-cycles. Around the cycle b_j the monodromy is given by $\exp\left(\int_{b_j}\Omega\right)$. Using the monodromy property of theta-functions, eq. (15.14) in Chapter 15, we can cancel this monodromy by taking:

$$V_j = \frac{1}{2i\pi}\int_{b_j}\Omega - \mathcal{A}_j(D)$$

Decomposing Ω into its three components, we note that $\int_{b_j}\Omega^{(t)}$ is a linear form in the times with coefficients *independent* of the indices (k,α) since all punctures enter symmetrically in its definition. One can compute more precisely the other two contributions by using Riemann's bilinear identity for third kind differentials. Let Ω_3 be a third kind differential with first order poles at some points P_l with residue r_l ($\sum r_l = 0$). The Riemann bilinear identity reads:

$$\int_{b_j}\Omega_3 = 2i\pi\sum_l r_l\mathcal{A}_j(P_l)$$

Applying this to the form $\Omega^{(l)}$, one gets:

$$\frac{1}{2i\pi}\int_{b_j}\Omega^{(l)} = \sum_{k'\beta} l_{k'\beta}\mathcal{A}_j(P_{k'\beta})$$

where the last sum is over all punctures. Similarly, we find:

$$\frac{1}{2i\pi}\int_{b_j}\Omega^{(q)} = \sum_{\beta\neq\alpha}\mathcal{A}_j(P_{k\beta}) - \sum_{l=2}^{N}\mathcal{A}_j(Q_l) = -\mathcal{A}_j(P_{k\alpha}) + \mathcal{A}_j(Q_1)$$

To get the last equation, note that if $P_1,\ldots,P_N$ are the N points above some λ_0 then the Abel sum $\sum_j \mathcal{A}(P_j)$ is a constant independent of λ_0. This is because the meromorphic function on Γ: $f(P) = (\lambda - \lambda_0)/(\lambda - \lambda_1)$ has zeroes at the points above λ_0 and poles at the points above λ_1, hence these two divisors are mapped to the same point in $\mathrm{Jac}\,(\Gamma)$ due to the Abel theorem. In particular $\sum_\beta \mathcal{A}_j(P_{k\beta}) = \sum_{l=1}^{N}\mathcal{A}_j(Q_l)$.

There remains to compute the normalization constant C, such that $\psi_\alpha^{(k)}(P) = e^{\xi_{k\alpha}(z_k)}(1 + O(z_k))$ when $P \to P_{k\alpha}$. Thanks to our definition of the primitive of the form Ω, we have $\int_{P_{k\alpha}}^{P}\Omega = \xi_{k\alpha}(z_k) + O(z_k)$. Hence we need only to normalize the quotient of theta-functions, and we find the final result, eq. (8.62). ∎

To identify the tau-function, we have to expand this formula for $P \to P_{k\beta}$ (β may be equal to α) in powers of z_k and compare the result with eq. (8.54).

Proposition. *The tau-function associated with the algebro-geometric integrable system is given by:*

$$\tau(t) = e^{\sigma(t,t)+\rho(t)} \, \theta(U \cdot t + W) \tag{8.64}$$

where $\sigma(t,t)$ is bilinear in the times t_i and $\rho(t)$ is linear.

<u>Proof.</u> We first look at $P \to P_{k\alpha}$. Note that, due to the explicit form of V, one can write for P close to $P_{k\alpha}$:

$$\psi_\alpha^{(k)}(z_k) = e^{\xi_{k\alpha}(z_k)} \cdot e^{\int_{P_{k\alpha}}^P (\Omega - d\xi_{k\alpha})} \frac{\theta(\mathcal{A}(P_{k\alpha}) - \mathcal{A}(D) - \mathcal{K})}{\theta(\mathcal{A}(P) - \mathcal{A}(D) - \mathcal{K})}$$
$$\times \frac{\theta(\mathcal{A}(P) - \mathcal{A}(P_{k\alpha}) + U \cdot t + W)}{\theta(U \cdot t + W)}$$

$$W = \sum_{k'\beta} l_{k'\beta} \mathcal{A}(P_{k'\beta}) + \mathcal{A}(Q_1) - \mathcal{A}(D) - \mathcal{K}$$

which compares to $\psi_\alpha^{(k)}(z_k) = e^{\xi_{k\alpha}(z_k)} h_{\alpha\alpha}^{(k)}(z_k)$. The middle term is regular when $z_k \to 0$ and tends to one, so one can write it in the form

$$e^{\int_{P_{k\alpha}}^P (\Omega - d\xi_{k\alpha})} \frac{\theta(\mathcal{A}(P_{k\alpha}) - \mathcal{A}(D) - \mathcal{K})}{\theta(\mathcal{A}(P) - \mathcal{A}(D) - \mathcal{K})} = \exp\left(t.b_\alpha(z_k) + a_\alpha(z_k)\right)$$

where $t.b_\alpha(z_k)$ is an expression linear in the times $t_{(k,n,\alpha)}$, and both $a_\alpha(z_k)$ and $b_\alpha(z_k)$ are of order $O(z_k)$. Considering the last term, note that it can be rewritten as

$$\frac{\theta(\mathcal{A}(P) - \mathcal{A}(P_{k\alpha}) + U \cdot t + W)}{\theta(U \cdot t + W)} = \frac{\theta(U \cdot (t - [z]) + W)}{\theta(U \cdot t + W)}$$

To see that we Taylor expand the Abel map $\mathcal{A}(P) - \mathcal{A}(P_{k\alpha})$ around $P_{k\alpha}$. Writing the first kind differential $\omega_j = \sum_{i=0}^\infty c_i^{(j)} z_k^i dz_k$ one gets, using Riemann's bilinear identities (see eq. (15.9) in Chapter 15):

$$\mathcal{A}_j(P) - \mathcal{A}_j(P_{k\alpha}) = \sum_{n=1}^\infty c_{n-1}^{(j)} \frac{z_k^n}{n} = \frac{-1}{2\pi i} \sum_{n=1}^\infty \frac{z_k^n}{n} \oint_{b_j} \Omega_{P_{k\alpha}}^{(n)} = -\sum_{n=1}^\infty \frac{z_k^n}{n} U_j^{(k,n,\alpha)}$$

This invites us to look for a tau-function of the form:

$$\tau(t) = e^{\rho(t)+\sigma(t,t)} \, \theta\left(\sum_i t_i U^{(i)} + W\right)$$

where $\rho(t)$ is linear in t and $\sigma(t,t)$ is a quadratic form in t. One gets a condition on ρ and σ:

$$t \cdot b_\alpha(z_k) + a_\alpha(z_k) = -\rho([z_k]_\alpha) - 2\sigma(t, [z_k]_\alpha) + \sigma([z_k]_\alpha, [z_k]_\alpha) \quad (8.65)$$

One can always choose ρ and σ satisfying this equation provided that b_α obeys the adequate symmetry property stemming from the fact that the quadratic form σ is symmetric in its arguments. Explicitly, we have the expansion

$$t \cdot b_\alpha(z_k) = \sum b_{(k',n',\alpha'),\alpha}(z_k) t_{(k',n',\alpha')}$$

For $(k', \alpha') \neq (k, \alpha)$ (the equality case was treated in Chapter 5) we have by eq. (8.63):

$$b_{(k',n',\alpha'),\alpha}(z_k) = \int_{P_{k\alpha}}^{P} \Omega_{P_{k'\alpha'}}^{(n')} = \sum_n b_{(k',n',\alpha'),(k,n,\alpha)} z_k^n$$

The condition (8.65) implies $\sigma_{(k',n',\alpha'),(k,n,\alpha)} = \frac{1}{2} n b_{(k',n',\alpha'),(k,n,\alpha)}$. So we must have the relation $n b_{(k',n',\alpha'),(k,n,\alpha)} = n' b_{(k,n,\alpha),(k',n',\alpha')}$ (in this equation, the function $b_{(k,n,\alpha)}(z_{k'})$ is obtained by performing the same construction as above but starting from the privilegied point $P_{k'\alpha'}$). This is a consequence of Riemann's bilinear identities: apply the identity eq. (15.8) in Chapter 15 to the second kind differentials $\Omega_{P_{k\alpha}}^{(n)}$ and $\Omega_{P_{k'\alpha'}}^{(n')}$. Since these differentials are normalized their a-periods vanish and the left-hand side of the identity vanishes. We get $\sum \mathrm{Res}\,(b_{(k',n',\alpha')} db_{(k,n,\alpha)}) = 0$. There are two poles, one at $P_{k\alpha}$ and the other at $P_{k'\alpha'}$. Computing the residues yields the required relation. Altogether this shows that the quadratic form σ exists, is completely determined, and independent of the choice of the particular point $P_{k\alpha}$. The computation of the linear form $\rho(t)$ is then straightforward. ∎

We are now in a position to discuss the effect of a Schlesinger transformation on this tau-function since this only amounts to changing the integers $l_{k\alpha}$. These integers only occur in the contribution $\sum_{k'\beta} l_{k'\beta} \mathcal{A}(P_{k'\beta})$ to the vector W in eq. (8.64), and in a term linear in $l_{k\alpha}$ in $\rho(t)$ (appearing in the exponential prefactor). We see that an elementary Schlesinger transformation (which adds a zero at $P_{k\alpha}$ and a pole at $P_{k'\alpha'}$) is obtained by changing $l_{k\alpha} \to l_{k\alpha} + 1$ and $l_{k'\alpha'} \to l_{k'\alpha'} - 1$. The effect of such a transformation on the theta-function is remarkably simple and amounts to a simple translation of its argument. Up to the exponential prefactor, we have:

$$\tau \begin{bmatrix} k & k \\ \alpha & \alpha' \end{bmatrix}(t) = \theta(U \cdot t + W + \mathcal{A}(P_{k\alpha}) - \mathcal{A}(P_{k'\alpha'}))$$

Remark. This allows us to perform an interesting check of the first of eqs. (8.54). From the definitions, eqs. (8.60, 8.61), the matrix element $h_{\alpha\alpha'}^{(k)}$ is obtained by evaluating $\psi_\alpha^{(k)}(P)$ around the point $P_{k\alpha'}$ in eq. (8.62). The time-dependent theta-function in the numerator of this equation is then exactly what we expect from the numerator in eq. (8.54).

We have found that the tau-function in the algebro-geometric situation is essentially the Riemann theta-function. Moreover, the various matrix elements of $h^{(k)}(t,\lambda)$ are obtained by simple shifts of the arguments of the theta-function. The isomonodromic context of this chapter is a generalization of the algebro-geometric context, so that the tau-functions can be viewed as generalizations of Riemann's theta-functions and should enjoy many of their remarkable properties. In Chapter 9 another framework is proposed allowing us to directly define the tau-function, using the geometry of the infinite Grassmannian.

8.11 The Painlevé equations

An important application of the theory of isomonodromic deformations concerns the Painlevé equations which can be interpreted as isomonodromic deformation equations, but not as isospectral deformations.

The Painlevé property deals with singularities of solutions of differential equations. In this respect there is a striking difference between linear differential equations and non-linear ones. The solutions of linear differential equations have singularities (poles, branch points, essential singularities) only where the coefficients of the equation have singularities, i.e. the position of these singularities are fixed. In contrast, the solutions of non-linear equations can develop singularities at arbitrary points, depending on initial conditions. For example $\dot{y} = y^2$ has the general solution $y = -1/(t - t_0)$. We call these singularities depending on the initial conditions movable singularities. In general one cannot avoid movable poles, however, one can try to find equations whose solutions have no movable singularities other than poles, i.e. the branch points and essential singularities of all solutions are fixed. This is called the Painlevé property. In fact Painlevé and Gambier have classified all differential equations of the form:

$$\frac{d^2 y}{dt^2} = R(t, y, \frac{dy}{dt})$$

where R is a rational function of its arguments, satisfying the Painlevé property. Up to trivial redefinitions they found 50 such equations. Of all these equations only six could not be integrated in terms of already known functions, and are listed below.

(i) $\frac{d^2y}{dt^2} = 6y^2 + t$

(ii) $\frac{d^2y}{dt^2} = 2y^3 + ty + \alpha$

(iii) $\frac{d^2y}{dt^2} = \frac{1}{y}\left(\frac{dy}{dt}\right)^2 - \frac{1}{t}\frac{dy}{dt} + \frac{1}{t}(\alpha y^2 + \beta) + \gamma y^3 + \frac{\delta}{y}$

(iv) $\frac{d^2y}{dt^2} = \frac{1}{2y}\left(\frac{dy}{dt}\right)^2 + \frac{3}{2}y^3 + 4ty^2 + 2(t^2 - \alpha)y + \frac{\beta}{y}$

(v) $\frac{d^2y}{dt^2} = \left\{\frac{1}{2y} + \frac{1}{y-1}\right\}\left(\frac{dy}{dt}\right)^2 - \frac{1}{t}\frac{dy}{dt}$

$\qquad + \frac{(y-1)^2}{t^2}\left\{\alpha y + \frac{\beta}{y}\right\} + \frac{\gamma y}{t} + \frac{\delta y(y+1)}{y-1}$

(vi) $\frac{d^2y}{dt^2} = \frac{1}{2}\left\{\frac{1}{y} + \frac{1}{y-1} + \frac{1}{y-t}\right\}\left(\frac{dy}{dt}\right)^2$

$\qquad - \left\{\frac{1}{t} + \frac{1}{t-1} + \frac{1}{y-t}\right\}\frac{dy}{dt}$

$\qquad + \frac{y(y-1)(y-t)}{t^2(t-1)^2}\left\{\alpha + \frac{\beta t}{y^2} + \frac{\gamma(t-1)}{(y-1)^2} + \frac{\delta t(t-1)}{(y-t)^2}\right\}$

All these equations can be understood in the framework of isomonodromy deformations. The fact that the Painlevé property appears in integrable systems is not an accident. The dynamical variables, i.e. the matrix elements of the $h^{(k)}$ in eq. (8.54), are ratios of tau-functions. In the algebro-geometric case the tau-functions are theta-functions which are *entire* functions, so that we only have movable poles at the zeroes of one theta-function. This remark has been generalized to the isomonodromy case by Malgrange and Miwa.

Here we shall only consider equations (ii) and (vi) in order to illustrate various aspects of the method.

Example 1. The Painlevé (ii) equation. We apply the general construction to the case where all matrices are of size 2×2, with just one singularity at $\lambda = \infty$. Moreover, we require that Ψ belongs to the group $SL(2)$ so that M_λ belongs to the Lie algebra $sl(2)$. We limit ourselves to $n_\infty = 3$, so that $M_\lambda = A_0 + A_1\lambda + A_2\lambda^2$, where A_2 is diagonal and traceless. Altogether there are seven parameters in M_λ. The point at ∞ is an irregular singularity of order 3, so that there are six Stokes sectors, and therefore the Stokes matrices depend on six parameters. The monodromy matrix at ∞ must be equal to 1, yielding three relations, so the monodromy data are expressed by three parameters. Finally, the singularity data depend on four parameters. Since we are looking for an ordinary non-linear differential equation we introduce only one time, t, and assume that the singularity data is given by:

$$\xi(\lambda) = \left(\frac{\lambda^3}{3} + \frac{t\lambda}{2} + \theta \log \left(\frac{1}{\lambda} \right) \right) \sigma_3, \quad \sigma_3 = \begin{pmatrix} 1 & 0 \\ 0 & -1 \end{pmatrix}$$

Clearly one could extend this construction to a whole hierarchy of times. In the following we shall parametrize M_λ by three natural parameters and find their time dependence so that the flow is isomonodromic. We have introduced a branch point parametrized by θ (which will account for the parameter α in Painlevé (ii)), and this is consistent with the Fuchs condition since $\mathrm{Tr}\,(\sigma_3) = 0$. Finally, for irrational θ, any Lax pair interpretation of the differential equation is prohibited since Baker–Akhiezer functions do not have such infinitely branched points.

We set $\Psi(\lambda) = g(\lambda) e^{\xi(\lambda)}$ and take

$$g(\lambda) = 1 + \frac{g_1}{\lambda} + \frac{g_2}{\lambda^2} + \frac{g_3}{\lambda^3} + \cdots$$

There is no restriction in assuming that the leading term is 1. We also impose that $\det g(\lambda) = 1$, since $\Psi(\lambda)$ belongs to $Sl(2)$. The g_i are such that the matrix $\Psi(\lambda)$ satisfies the linear differential equation $\partial_\lambda \Psi = M_\lambda \Psi$ with $M_\lambda = (g \partial_\lambda \xi g^{-1})_-$. Here the symbol $()_-$ means taking the polynomial part of the considered expression. Finally, the isomonodromic deformation equation is $\partial_t \Psi = M_t \Psi$ with $M_t = (g \partial_t \xi g^{-1})_-$. One gets:

$$M_\lambda = \lambda^2 \sigma_3 + \lambda [g_1, \sigma_3] + \frac{t}{2} \sigma_3 + [g_2, \sigma_3] - [g_1, \sigma_3] g_1$$

$$M_t = \frac{1}{2} \lambda \sigma_3 + \frac{1}{2} [g_1, \sigma_3]$$

We see that M_λ depends only on g_1, g_2, and $\partial_\lambda g = M_\lambda g - g \partial_\lambda \xi$ determines the $g_3, g_4, \ldots$ in terms of these two matrices. One finds for $i \geq 3$:

$$[g_i, \sigma_3] = (i - 3) g_{i-3} + \theta g_{i-3} \sigma_3 - \frac{t}{2} [g_{i-2}, \sigma_3]$$

$$+ [g_1, \sigma_3] g_{i-1} + [g_2, \sigma_3] g_{i-2} - [g_1, \sigma_3] g_1 g_{i-2} \qquad (8.66)$$

We parametrize the matrix g_i in the form:

$$g_i = \begin{pmatrix} \Delta_i + a_i & b_i \\ c_i & \Delta_i - a_i \end{pmatrix}$$

where Δ_i is obtained by requiring that $\det g(\lambda) = 1$, yielding $\Delta_1 = 0$, $\Delta_2 = (a_1^2 + b_1 c_1)/2$, $\Delta_3 = a_1 a_2 + (b_1 c_2 + b_2 c_1)/2$, etc. The left-hand side of eq. (8.66) has vanishing diagonal elements. This provides a constraint on the lower order g_i. The off-diagonal elements determine b_i and c_i. In the case $i = 3$, we find one constraint:

$$b_1 c_2 + b_2 c_1 = \frac{\theta}{2}$$

and get the off-diagonal elements:

$$-b_3 = \frac{t}{2} b_1 + a_2 b_1 + a_1 b_2 + b_1 \Delta_2$$

$$c_3 = -\frac{t}{2} c_1 + a_2 c_1 + a_1 c_2 - c_1 \Delta_2$$

Similarly for $i = 4$ we get the constraint:

$$a_1 = -(t + 4\Delta_2) b_1 c_1 + 2 b_2 c_2 + 2 a_1 (b_1 c_2 - b_2 c_1)$$

while a_2 is determined by the equation at order $i = 5$. Finally there are only three free parameters in g_1 and g_2.

The equations of motion $\partial_t g = M_t g - g \partial_t \xi$ read:

$$\dot{g}_1 = -\frac{1}{2} [g_2, \sigma_3] + \frac{1}{2} [g_1, \sigma_3] g_1, \quad \dot{g}_2 = -\frac{1}{2} [g_3, \sigma_3] + \frac{1}{2} [g_1, \sigma_3] g_2$$

Note that since $[g_3, \sigma_3]$ is known in terms of g_1, g_2, these equations close on these two matrices. In components we get:

$$\dot{a}_1 = -b_1 c_1, \quad \dot{b}_1 = b_2 + a_1 b_1, \quad \dot{c}_1 = -c_2 + a_1 c_1$$

$$\dot{b}_2 = -\frac{t}{2} b_1 - a_1 b_2 - 2 b_1 \Delta_2, \quad \dot{c}_2 = \frac{t}{2} c_1 - a_1 c_2 + 2 c_1 \Delta_2$$

It is now natural to introduce the three parameters:

$$x = \frac{b_1}{c_1}, \quad y = a_1 + \frac{b_2}{b_1}, \quad z = b_1 c_1$$

With these parameters one has $a_1 = -tz - 2z^2 - 2zy^2 + \theta y$. Then the equations of motion give:

$$\dot{x} = \frac{1}{2} \theta \frac{x}{z}, \quad \dot{y} = -t/2 - 2z - y^2, \quad \dot{z} = 2zy - \theta/2$$

Eliminating z yields the equation:

$$\frac{d^2 y}{dt^2} = 2y^3 + yt - \frac{1}{2} + \theta$$

One recognizes the Painlevé (ii) equation with $\alpha = \theta - 1/2$. Once the solution $y(t)$ of the Painlevé equation is known, it is easy to reconstruct the matrix $M_\lambda(\lambda; t)$ in terms of $x(t), y(t), z(t)$. The subtle nature of the t-dependence of $M_\lambda(\lambda; t)$ ensuring the isomonodromy property is here particularly striking.

The Schlesinger transformations take a particularly simple form on this example. They just amount to changing $\theta \to \theta_n = \theta + n$, so that the first column of Ψ acquires a zero of order n at $\lambda = \infty$ while the second column acquires a pole of order n. Let $\Psi(n, \lambda)$ be the wave-function constructed with the parameter θ_n, then we have for an elementary Schlesinger transformation $\Psi(n + 1, \lambda) = R(n, \lambda)\Psi(n, \lambda)$, where $R(n, \lambda)$ is a matrix rational in λ. Plugging in the asymptotic expansions at ∞, $\Psi(n, \lambda) \sim g(n, \lambda) \exp(\xi(n, \lambda))$, one gets $R(n, \lambda)g(n, \lambda) = g(n+1, \lambda)(\lambda^{-1}E_{11} + \lambda E_{22})$. Expanding to order λ^{-3}, one finds:

$$R(n, \lambda) = \begin{pmatrix} 0 & b_{1,n+1} \\ -1/b_{1,n+1} & \lambda - y_{n+1} \end{pmatrix}$$

where $b_{1,n+1}$ and y_{n+1} are the parameters entering in $g(n+1, \lambda)$. Moreover, one also gets:

$$a_{1,n+1} = \frac{c_{2,n}}{c_{1,n}}, \quad c_{1,n+1} = c_{3,n} - c_{1,n}(\Delta_{2,n} + a_{2,n}) + c_{2,n}(a_{1,n} - a_{1,n+1})$$

$$\tag{8.67}$$

$$b_{1,n+1} = \frac{1}{c_{1,n}}, \quad b_{2,n+1} = -\frac{a_{1,n}}{c_{1,n}} \tag{8.68}$$

From this we obtain the recursion relations:

$$z_{n+1} = -\frac{t}{2} - y_{n+1}^2 - z_n, \quad y_{n+1} = -y_n + \frac{\theta_n}{2z_n}$$

It is now illuminating to introduce the tau-functions. First we compute the closed form $\Upsilon = -\text{Res}_\infty \text{Tr}(g^{-1}(\lambda)\partial_\lambda d\xi)d\lambda$. Here we simply have $d\xi = \lambda\sigma_3 dt/2$, so that we find:

$$\Upsilon = -\frac{1}{2}\text{Tr}(g_1\sigma_3)dt = -a_1 dt$$

Hence τ_n is defined by $\dot{\tau}_n/\tau_n = -a_{1,n}$. One can express the matrix elements of $g(\lambda)$ in terms of tau-functions according to the general results of the previous sections. Here however, things are so simple that it is even more straightforward to rederive the appropriate results. From eq. (8.67) we have $a_{1,n+1} = -\dot{\tau}_{n+1}/\tau_{n+1} = c_{2,n}/c_{1,n}$. It follows immediately that

$$y_{n+1} = -a_{1,n} + \frac{c_{2,n}}{c_{1,n}} = \frac{\dot{\tau}_n}{\tau_n} - \frac{\dot{\tau}_{n+1}}{\tau_{n+1}} = \frac{d}{dt}\log\left(\frac{\tau_n}{\tau_{n+1}}\right)$$

On the other hand, by the equations of motion we have $y_n = d \log b_{1,n}/dt$. Moreover, by eq. (8.68) we have $c_{1,n} = 1/b_{1,n+1}$. Hence we can take:

$$b_{1,n} = \frac{\tau_{n-1}}{\tau_n}, \quad c_{1,n} = \frac{\tau_{n+1}}{\tau_n}, \quad z_n = \frac{\tau_{n+1}\tau_{n-1}}{\tau_n^2}$$

The Hirota bilinear form of the equations of motion also follows straightforwardly. Let us write the three equations of motion:

$$\dot{a}_{1,n} = -z_n \rightarrow \tau_n \ddot{\tau}_n - \dot{\tau}_n^2 - \tau_{n+1}\tau_{n-1} = 0$$

$$\dot{z}_n = 2y_n z_n - \frac{1}{2}\theta_n \rightarrow \tau_{n+1}\dot{\tau}_{n-1} - \tau_{n-1}\dot{\tau}_{n+1} + \frac{1}{2}\theta_n\tau_n^2 = 0$$

$$\dot{y}_n = -2z_n - y_n^2 - \frac{t}{2} \rightarrow \tau_{n-1}\ddot{\tau}_n + \tau_n\ddot{\tau}_{n-1} - 2\dot{\tau}_n\dot{\tau}_{n-1} + \frac{1}{2}t\tau_n\tau_{n-1} = 0$$

In the last equation we have used the first one to simplify the result.

Example 2. The Painlevé (vi) equation. We consider the case of three regular singularities at finite distance which we put at $\lambda = 0, 1, t$, and we study the isomonodromic deformation equations with respect to the parameter t. We choose the singularity data at these three points as:

$$\xi^{(k)}(\lambda) = \begin{pmatrix} \theta_k & 0 \\ 0 & 0 \end{pmatrix} \log(\lambda - \lambda_k), \quad k = 0, 1, t, \quad \lambda_0 = 0, \ \lambda_1 = 1, \ \lambda_t = t$$

With this we construct the wave-function $\Psi(\lambda)$ having the asymptotic expansion $\Psi(\lambda) \sim g^{(k)}(\lambda) \exp(\xi^{(k)}(\lambda))$, with $k = 0, 1, t$ and satisfying the differential equation $\partial_\lambda \Psi = M_\lambda \Psi$. This implies:

$$M_\lambda(\lambda) = \frac{A^{(0)}}{\lambda} + \frac{A^{(1)}}{\lambda - 1} + \frac{A^{(t)}}{\lambda - t}, \quad A^{(k)} = g_0^{(k)} \begin{pmatrix} \theta_k & 0 \\ 0 & 0 \end{pmatrix} g_0^{(k)\, -1}$$

In particular $A^{(k)}$ is a rank 1 projector, hence $\det A^{(k)} = 0$ and $\operatorname{Tr} A^{(k)} = \theta_k$. The equation of motion reads $\partial_t \Psi = M_t \Psi$ with:

$$M_t = \left(g^{(t)}\partial_t\xi^{(t)}g^{(t)\,-1}\right)_- = -\frac{A^{(t)}}{\lambda - t}$$

We are free to make a global gauge transformation by a matrix which is constant in λ, and we use this freedom to diagonalize M_λ at ∞, so that we assume:

$$M_\lambda(\lambda) = \frac{1}{\lambda}\begin{pmatrix} \kappa_1 & 0 \\ 0 & \kappa_2 \end{pmatrix} + O(\lambda^{-2}), \quad \lambda \to \infty$$

The differential equation $\partial_\lambda \Psi = M_\lambda \Psi$ has a regular singularity at ∞, and the Fuchs condition reads $\kappa_1 + \kappa_2 = \theta_0 + \theta_1 + \theta_t$.

We can write M_λ as:

$$M_\lambda(\lambda) = \begin{pmatrix} m_{11}(\lambda) & m_{12}(\lambda) \\ m_{21}(\lambda) & m_{22}(\lambda) \end{pmatrix} = \frac{1}{\lambda(\lambda-1)(\lambda-t)} A(\lambda)$$

where $A_{ii}(\lambda)$ are second degree polynomials in λ with leading term $\kappa_i \lambda^2$, and $A_{12}(\lambda)$ and $A_{21}(\lambda)$ are degree one polynomials in λ. We set

$$A_{12}(\lambda) = \gamma(\lambda - y)$$

introducing the important parameter y which will end up satisfying the Painlevé (vi) equation. Taking the trace of $M_\lambda(\lambda)$, we have

$$A_{22}(\lambda) = \theta_0(\lambda-1)(\lambda-t) + \theta_1 \lambda(\lambda-t) + \theta_t \lambda(\lambda-1) - A_{11}(\lambda)$$

which eliminates the parameters in $A_{22}(\lambda)$. Moreover, we know that $\det A(\lambda)$ vanishes for $\lambda = 0, 1, t$, so we get three conditions on $A_{21}(\lambda)$:

$$A_{21}(\lambda) = \frac{A_{11}(\lambda)A_{22}(\lambda)}{\gamma(\lambda-y)}, \quad \lambda = 0, 1, t$$

The first two allow us to express $A_{21}(\lambda)$ in terms of $A_{11}(\lambda)$ and $A_{12}(\lambda)$, while the last one provides, since $A_{21}(\lambda)$ is a linear function of λ, a constraint on the parameters in $A_{11}(\lambda)$:

$$A_{21}(t) = (1-t)A_{21}(0) + tA_{21}(1)$$

In order to write this constraint more explicitly, we parametrize $A_{11}(\lambda)$ by the values it takes at $\lambda = t$ and $\lambda = y$, i.e. write the interpolation formula:

$$A_{11}(\lambda) = (\lambda-t)(\lambda-y)\kappa_1 + \frac{(\lambda-y)A_{11}(t) - (\lambda-t)A_{11}(y)}{t-y}$$

Substituting this into the previous equations, one observes that the constraint is quadratic in $A_{11}(y)$ but, unexpectedly, is linear in $A_{11}(t)$. We solve $A_{11}(t)$ in term of $A_{11}(y)$ getting:

$$\begin{aligned}
A_{11}(t) = &-\frac{\kappa_1^2}{\kappa_1 - \kappa_2}(y-t)(y-1+t) \\
&+ \frac{\kappa_1}{\kappa_1 - \kappa_2}\Big((\theta_0 t + \theta_1(t-1))(y-t) + 2A_{11}(y)\Big) \\
&- \frac{A_{11}(y)}{(\kappa_1 - \kappa_2)y(y-1)}\Big(A_{11}(y) + \theta_1 y(t-1) + \theta_0 t(y-1)\Big)
\end{aligned}$$

The dynamical variables are now $y, \gamma, A_{11}(y)$ and we have to write the equations of motion $\partial_t M_\lambda = \partial_\lambda M_t + [M_t, M_\lambda]$. Everything can be expressed in terms of the matrix $A(\lambda)$ since we have

$$M_t(\lambda) = -\frac{1}{t(t-1)(\lambda-t)} A(t)$$

We compute the equation of motion and afterwards we set $\lambda = y$. We get

$$\begin{pmatrix} \dot{m}_{11}(\lambda) & \dot{m}_{12}(\lambda) \\ \dot{m}_{21}(\lambda) & \dot{m}_{22}(\lambda) \end{pmatrix}_{\lambda=y} = \frac{1}{t(t-1)(y-t)^2} \left\{ A(t) - \frac{1}{y(y-1)} [A(t), A(y)] \right\}$$

Taking the matrix element 12 of this equation and evaluating it at $\lambda = y$ gives:

$$\dot{y} = \frac{y(y-1)(y-t)}{t(t-1)} \left[2m_{11}(y) - \frac{\theta_0}{y} - \frac{\theta_1}{y-1} - \frac{\theta_t - 1}{y-t} \right]$$

Taking the matrix element 11 evaluated at $\lambda = y$ gives:

$$\dot{m}_{11}(\lambda)|_{\lambda=y} = \frac{A_{11}(t)}{t(t-1)(y-t)^2} + \frac{\gamma}{t(t-1)} m_{21}(y)$$

We introduce the dynamical variable $z = m_{11}(y)$ and compute

$$\dot{z} = \dot{m}_{11}(\lambda)|_{\lambda=y} + (\partial_\lambda m_{11})|_{\lambda=y} \cdot \dot{y}$$

We find after some algebra:

$$t(t-1)\dot{z} = -(3y^2 - 2(1+t)y + t)z^2$$
$$+ \left(2(\kappa_1 + \kappa_2 - 1)y + 1 - \theta_0 - \theta_t - (\theta_0 + \theta_1)t \right)z + \kappa_1(1 - \kappa_2)$$

Eliminating z between the equations for $\dot{y}$ and $\dot{z}$ yields the Painlevé (*vi*) equation, with the following values of the parameters:

$$\alpha = \frac{1}{2}(\kappa_1 - \kappa_2 + 1)^2, \quad \beta = -\frac{1}{2}\theta_0^2, \quad \gamma = \frac{1}{2}\theta_1^2, \quad \delta = \frac{1}{2}(1 - \theta_t^2)$$

It is known that the other Painlevé equations can be obtained from this one by various limiting procedures.

References

[1] G.D. Birkhoff, *Collected mathematical papers*, Vol. 1. Dover Publications Inc. (1968) 259–306.

[2] E.L. Ince, *Ordinary differential equations*. Dover Publications Inc. (1956).

[3] W. Wasow, *Asymptotic expansions for ordinary differential equations*. Interscience Publishers (1965).

[4] H. Flaschka and A. Newell, Monodromy and Spectrum preserving deformations I. *Commun. Math. Phys.* **76** (1980) 65–166.

[5] M. Jimbo, T. Miwa and K. Ueno, Monodromy preserving deformations of ordinary differential equations with rational coefficients I. *Physica* **2D** (1981) 306–352.

[6] M. Jimbo and T. Miwa, Monodromy preserving deformations of ordinary differential equations with rational coefficients II. *Physica* **2D** (1981) 407–448.

[7] M. Jimbo and T. Miwa, Monodromy preserving deformations of ordinary differential equations with rational coefficients III. *Physica* **4D** (1981) 26–46.

[8] T. Miwa. Painlevé property of monodromy preserving deformation equations and the analyticity of τ–functions. *RIMS* **17** (1981) 703–721.

[9] B. Malgrange, *Sur les déformations isomonodromiques*. Mathematics and Physics (Paris, 1979/1982), Progr. Math. **37**, Birkhäuser, Boston, Mass. (1983).

[10] Y. Sibuya, *Linear differential equations in the complex domain: problems of analytic continuation*. AMS Providence (1990).

[11] D. Anosov and A. Bolibruch, *The Riemann–Hilbert problem*. Aspects of Mathematics Vol. 22 (1994).

9

Grassmannian and integrable hierarchies

We learned in previous chapters that when the Lax matrix has several singularities, the situation mainly reduces to local studies around each singularity. In this chapter we consider the case of one singularity in all its generality, which amounts to the study of the Kadomtsev–Petviashvili equation. This finds a natural presentation in Sato's Grassmannian approach. We give an explicit realization of the Grassmannian in a fermionic Fock space. In this setting we obtain a remarkable formula expressing the τ-function as $\tau(t) = \langle 0|e^{\sum_i H_i t_i} g|0\rangle$. Soliton solutions are then obtained by choosing for g particular elements which have a simple expression in terms of vertex operators. Hirota equations are interpreted as the Plücker equations of an infinite-dimensional Grassmannian, on which the infinite-dimensional group $GL(\infty)$ acts. The time flows are interpreted as the action of an infinite-dimensional natural Abelian subgroup. We also to find particularly simple tau-functions expressed as Schur polynomials. Finally, we show that the full KP hierarchy admits an elegant formulation in terms of pseudo-differential operators.

9.1 Introduction

In Chapter 3 we showed that one can simultaneously solve the hierarchy of equations:

$$\partial_{t_j}\Psi = M_j\Psi, \quad M_j = (g\xi_j g^{-1})_-, \quad \Psi = ge^{\sum \xi_j t_j} \tag{9.1}$$

with $g = 1 + O(1/z)$ regular and ξ_j a diagonal constant matrix singular at the unique puncture that we take at $z = \infty$. Here Ψ is a matrix of fundamental solutions of this system. In this chapter we denote by z the spectral parameter and the singularity is at $z = \infty$. The matrix

299

elements $\Psi_{\alpha\alpha}$ admit a remarkable expression in terms of tau-functions, see eq. (3.61) in Chapter 3.

$$\Psi_{\alpha\alpha}(z; t_1, t_2, \ldots) = \frac{\tau(t - [z^{-1}]\alpha)}{\tau(t)} \exp\left(\sum_{n\geq 1} z^n t_{(n,\alpha)}\right) \qquad (9.2)$$

where $(t - [z^{-1}]\alpha)$ means that times $t_{(n,\alpha)}$ are shifted by $-z^{-n}/n$. The noticable feature of this expression is that the z-dependence is folded into the infinitely many times $t_{(n,\alpha)}$. The local formula eq. (9.2) was also recovered when analyzing isospectral flows in Chapter 5, where we observed that, in this context, the tau-functions are essentially Riemann's theta–functions. In the isomonodromic approach of Chapter 8, it was shown that the notion of tau-function is in fact much more general than that of theta-function.

We have also seen in Chapter 8 that the equations of motion can be recast as bilinear Hirota equations on the tau-function, $P(D)\tau \cdot \tau = 0$. They form an infinite set of bilinear identities which are *equivalent* to the hierarchy equations.

In this chapter we elaborate on the relation between Sato's formula eq. (9.2), Hirota's bilinear equations and representation theory of infinite-dimensional affine Kac–Moody algebras. A motivation for introducing affine Kac–Moody algebras may be seen in the structure of Sato's formula, which, using $\exp(a\partial_t)f(t) = f(t + a)$, can be written as:

$$\Psi(z, t) = \frac{V(z, t)\tau(t)}{\tau(t)}, \quad V(z, t) = e^{\left(\sum_{n\geq 1} z^n t_n\right)} e^{-\left(\sum_{n\geq 1} \frac{z^{-n}}{n} \frac{\partial}{\partial t_n}\right)} \qquad (9.3)$$

The operator $V(z, t)$ is a typical vertex operator which is known to play a central role in the construction of representations of affine Kac–Moody algebras, see Chapter 16. A further motivation can be found in the observation that highest weight representations of affine Kac–Moody algebras by vertex operators naturally produce tau-functions obeying bilinear identities.

We sketch now, in a rather informal way, the ideas allowing us to associate Hirota bilinear equations with vertex operator highest weight representations of affine Kac–Moody algebras. These ideas also underly the fermionic constructions of the following sections.

Let $\mathcal{G}$ be an affine Kac–Moody algebra and $\mathcal{H}$ a Cartan subalgebra of $\mathcal{G}$. We will use freely, in this section, the notion of a group G associated with $\mathcal{G}$. Let X^a be a basis of $\mathcal{G}$. It will often be a Cartan–Weyl basis (H_i, E_α), where H_i is a basis of $\mathcal{H}$, and E_α are the root vectors associated to the roots α, normalized by $(E_\alpha, E_{-\alpha}) = 1$. The tensor Casimir C_{12} of $\mathcal{G}$ is defined by:

$$C_{12} = \sum_a X^a \otimes X_a = \sum_i H^i \otimes H_i + \sum_\alpha E_\alpha \otimes E_{-\alpha}$$

where the indices are raised or lowered with the Killing form. The fundamental property of C_{12} is that for all $g \in G$, we have:

$$C_{12}\, g \otimes g = g \otimes g\, C_{12}$$

Let us give a proof of this property, which will be adapted in the fermionic case. We have to show that

$$\sum_a g^{-1} X^a g \otimes g^{-1} X_a g = \sum_a X^a \otimes X_a$$

but $g^{-1} X^a g = \alpha^a_d X^d$, and $g^{-1} X_b g = \beta^c_b X_c$. The Killing form is such that $(X^a, X_b) = \delta^a_b$ and is invariant under adjoint action. Hence we have $\sum_c \alpha^a_c \beta^c_b = \delta^a_b$, which implies our property.

If $|\Lambda\rangle$ and $|\Lambda'\rangle$ are two highest weight vectors, we have:

$$C_{12}|\Lambda\rangle \otimes |\Lambda'\rangle = (\Lambda, \Lambda')|\Lambda\rangle \otimes |\Lambda'\rangle$$

To see it, we write C_{12} in the Cartan–Weyl basis and use that highest weight vectors $|\Lambda\rangle$ are eigenstates of H_i and are annihilated by the generators E_α for α any positive root. These two relations imply the following identity:

$$C_{12}\, g|\Lambda\rangle \otimes g|\Lambda'\rangle = (\Lambda, \Lambda')\, g|\Lambda\rangle \otimes g|\Lambda'\rangle, \quad \forall\ g \in G \qquad (9.4)$$

Suppose now that the representation of $\mathcal{G}$, with highest weight vector $|\Lambda\rangle$, admits a vertex operator construction in terms of bosonic operators α_n, $n \in \mathbb{Z}$, satisfying $[\alpha_n, \alpha_m] = n\delta_{n+m,0}$, see Chapter 16. The representation space is the bosonic Fock space with vacuum $|\Lambda\rangle$ such that $\alpha_n|\Lambda\rangle = 0$ for $n > 0$. It is generated by acting on $|\Lambda\rangle$ with the operators α_n, $n < 0$. A basis consists of states of the form $\alpha_{-n_1} \cdots \alpha_{-n_k}|\Lambda\rangle$. A vertex operator is an operator acting on the Fock space and has the typical form:

$$V(z) = e^{-\left(\sum_{n<0}(z^{-n}/n)\alpha_n\right)} e^{-\left(\sum_{n>0}(z^{-n}/n)\alpha_n\right)} \qquad (9.5)$$

When we expand $V(z)$ in powers of z, the coefficients are polynomials in the bosonic operators α_n. Together with the α_n themselves, they represent the elements X^a of the Lie algebra.

Note that taking for α_{-n}, $n > 0$ the operator of multiplication by nt_n, and for α_n, $n > 0$ the derivation operator $\partial/\partial t_n$, the vertex operator eq. (9.5) exactly reproduces the vertex operator appearing in eq. (9.3).

Let us introduce the generating function, $H(t) = \sum_{n>0} \alpha_n t_n$, depending on the infinite number of variables t_n. Note that $[H(t), H(t')] = 0$. We also need the dual vacuum $\langle\Lambda|$ such that $\langle\Lambda|\alpha_n = 0$ for $n < 0$. We *define* the tau-function by:

$$\tau_\Lambda^g(t) = \langle\Lambda|e^{H(t)}g|\Lambda\rangle, \quad \text{with} \quad H(t) = \sum_{n>0} \alpha_n t_n, \quad g \in G \qquad (9.6)$$

This is a function on the orbit of $|\Lambda\rangle$ under G.

Since $X^a \in \mathcal{G}$ is represented by some polynomial in the α_n, there exist differential operators $\mathcal{D}_\Lambda^a(t, \partial_t)$ in the variables t_n such that:

$$\langle\Lambda|e^{H(t)}X^a g|\Lambda\rangle = \mathcal{D}_\Lambda^a(t, \partial_t)\langle\Lambda|e^{H(t)}g|\Lambda\rangle \qquad (9.7)$$

This easily follows from:

$$\langle\Lambda|e^{H(t)}\alpha_m g|\Lambda\rangle = \begin{cases} \frac{\partial}{\partial t_m}\langle\Lambda|e^{H(t)}g|\Lambda\rangle & \text{for } m > 0 \\ mt_{-m}\langle\Lambda|e^{H(t)}g|\Lambda\rangle & \text{for } m < 0 \end{cases} \qquad (9.8)$$

The case $m > 0$ is obvious, while for $m < 0$ one uses

$$e^{H(t)}\alpha_m e^{-H(t)} = \alpha_m + mt_{-m}$$

to bring α_m to the left where it is annihilated by the vacuum.

We now multiply the fundamental bilinear relation eq. (9.4) by $e^{H(t')} \otimes e^{H(t'')}$. Let $x_n = \frac{1}{2}(t_n' + t_n'')$, $y_n = \frac{1}{2}(t_n' - t_n'')$, so that

$$e^{H(t')} \otimes e^{H(t'')} = e^{H(y)} \otimes e^{-H(y)} \cdot e^{H(x)} \otimes e^{H(x)}$$

The second factor in the right-hand side of this relation commutes with C_{12}, so that we can push it to its right. Then, taking the scalar product with $\langle\Lambda| \otimes \langle\Lambda'|$, we get

$$\sum_a \langle\Lambda|e^{H(y)}X^a e^{H(x)}g|\Lambda\rangle\langle\Lambda'|e^{-H(y)}X_a e^{H(x)}g|\Lambda'\rangle$$

$$= (\Lambda, \Lambda')\tau_\Lambda^g(x + y)\tau_{\Lambda'}^g(x - y)$$

Using now eq. (9.7) we obtain:

$$\left[\sum_a \mathcal{D}_\Lambda^a(y, \partial_y)\mathcal{D}_{a\Lambda'}(-y, -\partial_y)\right]\tau_\Lambda^g(x + y)\tau_{\Lambda'}^g(x - y)$$

$$= (\Lambda, \Lambda')\tau_\Lambda^g(x + y)\tau_{\Lambda'}^g(x - y)$$

These are Hirota type equations. Expanding in y, we get an infinite number of bilinear Hirota equations which are in fact equations characterizing the orbits of the highest weight vectors $|\Lambda\rangle$ and $|\Lambda'\rangle$.

Clearly this construction can be applied to any vertex operator representation of an affine Kac–Mooody algebra $\mathcal{G}$. The hierarchy of Hirota equations we obtain in this way depends on the affine Kac–Moody algebra but also on the choice of the vertex operator representation. In the next section we make these ideas precise in the case of the group $GL(\infty)$.

9.2 Fermions and $GL(\infty)$

Let $gl(\infty)$ be the Lie algebra of infinite-dimensional band matrices of the form M_{rs} with $r, s \in \mathbb{Z} + \frac{1}{2}$ (the $\frac{1}{2}$ is added for convenience) and such that $M_{rs} = 0$ for $|r - s|$ large. Note that the sum and the product of two such matrices is well-defined and of the same type. Finally, $\widehat{gl}(\infty)$ is a central extension of $gl(\infty)$ that will be described later on, by displaying a representation of this algebra on some Hilbert space.

We introduce fermionic operators β_r, β_r^*, $r \in \mathbb{Z} + \frac{1}{2}$, with the following anticommutation relations:

$$\{\beta_r, \beta_s\} = 0, \quad \{\beta_r^*, \beta_s^*\} = 0, \quad \{\beta_r, \beta_s^*\} = \delta_{r+s,0} \qquad (9.9)$$

where $\{\ ,\ \}$ denotes the anticommutator, ie. $\{a, b\} = ab + ba$ for any operators a and b. By convention, we will say that the fermions β_r have charge $(+1)$ while the fermionic operators β_s^* have charge (-1).

We introduce a vacuum vector $|0\rangle$ such that:

$$\beta_r|0\rangle = \beta_r^*|0\rangle = 0 \qquad \text{for } r \geq \tfrac{1}{2}$$

The Fock space is, by definition, generated by acting successively on the vacuum with the fermionic operators. A basis of the fermionic Fock space is the following set of states:

$$\beta_{-s_n} \cdots \beta_{-s_1} \cdot \beta_{-r_m}^* \cdots \beta_{-r_1}^* |0\rangle \qquad (9.10)$$

with $s_j, r_j \geq \frac{1}{2}$. We introduce a Hermitian structure on the Fock space by defining the dual vacuum vector:

$$\langle 0|\beta_r = \langle 0|\beta_r^* = 0 \qquad \text{for } r \leq -\tfrac{1}{2}$$

such that $\langle 0|0\rangle = 1$, and the adjoint of the fermionic operators:

$$\beta_r^\dagger = \beta_{-r}^*, \quad (\beta_r^*)^\dagger = \beta_{-r}$$

This implies that the states eq. (9.10) form an orthonormal basis of the Fock space. Since the fermions β_s have charge $(+1)$ and the fermions β_r^* charge (-1), the state eq. (9.10) has charge $(n - m)$.

Let $|p\rangle$ be the vector defined by:

$$
\begin{aligned}
|p\rangle &\equiv \beta_{-p+\frac{1}{2}}\cdots\beta_{-\frac{1}{2}}|0\rangle, \quad \text{for} \quad p > 0, & (9.11)\\
&\equiv \beta^*_{p+\frac{1}{2}}\cdots\beta^*_{-\frac{1}{2}}|0\rangle, \quad \text{for} \quad p < 0,
\end{aligned}
$$

These states have charge p and satisfy $\beta_{r-p}|p\rangle = \beta^*_{r+p}|p\rangle = 0$ for $r \geq \frac{1}{2}$. We call them charged vacuum vectors. The states with charge p are obtained by acting on the vacuum $|p\rangle$ with an equal number of β_s and β^*_r. In the following we will mostly restrict ourselves to neutral states which are linear combinations of states of the form eq. (9.10) with the same number of β and β^*.

We will need the notion of normal order of fermionic operators. This order consists in writing all operators β_r, β^*_r with positive indices r on the right and multiplying by the signature of the considered permutation. The important property of normal ordered products is that their vacuum expectation value vanishes.

The relation between a monomial and its normal ordered form is given by Wick's theorem. For two fermions it reads:

$$
\beta_r\beta^*_s = \; :\beta_r\beta^*_s: + \langle 0|\beta_r\beta^*_s|0\rangle, \quad :\beta_r\beta^*_s: \, := \, - :\beta^*_s\beta_r:
$$

This can be seen as follows. To bring the left-hand side to its normal ordered form we have to perform at most one anticommutation, hence adding at most a c-number which can only be the vacuum expectation of the left-hand side since the normal product has vanishing vacuum expectation value. This scalar is called the contraction of the two operators. For more than two operators Wick's theorem is expressed by the induction formula:

$$
\phi_1 : \phi_2 \cdots \phi_n := \; :\phi_1\phi_2\cdots\phi_n: + \sum_{i=2}^{n}(-1)^i\langle 0|\phi_1\phi_i|0\rangle :\phi_2\cdots\widehat{\phi_i}\cdots\phi_n:
$$

where ϕ_j is either some β_r or some β^*_s and the notation $\widehat{\phi_i}$ means omission of the factor ϕ_i. It follows that $\langle 0|\phi_1\phi_2\cdots\phi_n|0\rangle$ is equal to the sum of all possible products of contractions with appropriate signs.

We now construct the algebra $\widehat{gl}(\infty)$. Consider neutral bilinear operators of the form $\beta_r\beta^*_s$. They form a closed algebra under commutation. For example:

$$
[\beta_r\beta^*_s, \beta_n\beta^*_m] = \delta_{s+n,0}(\beta_r\beta^*_m) - \delta_{r+m,0}(\beta_n\beta^*_s)
$$

Let M_{rs} be a band matrix, i.e. an infinite-dimensional matrix such that $M_{rs} = 0$ for $|r - s| > N$ for some given N. We can define the formal operator $X = \sum_{r,s} M_{rs}\beta_r\beta_{-s}^*$. The commutator of two such objects can be computed without ambiguity, due to the band structure of the matrices M and N:

$$[X, Y] = \sum_{rsnm} M_{rs}N_{nm}[\beta_r\beta_{-s}^*, \beta_n\beta_{-m}^*] = \sum_{rm}[M, N]_{rm}\beta_r\beta_{-m}^*$$

We reproduce the Lie algebra of (band) matrices. However, objects like X cannot be represented on the Fock space since their matrix elements on this space may be infinite. For example, $\langle 0|X|0\rangle = \sum_{r>0} M_{rr}$ which can be infinite. To overcome this problem we normal order the bilinear fermionic operator. Then all matrix elements are finite between states with finite number of particles. However, this induces a modification of the commutation rules:

$$[: \beta_r\beta_{-s}^* :, : \beta_n\beta_{-m}^* :] = \delta_{s,n} : \beta_r\beta_{-m}^* : \ - \delta_{r,m} : \beta_n\beta_{-s}^* :$$
$$+\delta_{s,n}\langle 0|\beta_r\beta_{-m}^*|0\rangle - \delta_{r,m}\langle 0|\beta_n\beta_{-s}^*|0\rangle \qquad (9.12)$$

The additional c-number term is the central extension of $gl(\infty)$, and is equal to $\delta_{rm}\delta_{sn}(\theta(m) - \theta(s))$, where $\theta(m) = 1$ for $m > 0$ and vanishes for $m < 0$.

Definition. *The algebra $\widehat{gl}(\infty)$ is the infinite-dimensional Lie algebra of elements of the form:*

$$\widehat{gl}(\infty) = \left\{X = \sum_{r,s} M_{rs} : \beta_r\beta_{-s}^* :, \quad M_{rs} = 0 \ \text{if} \ |r - s| \gg 0\right\} \oplus \mathbb{C}$$

$$(9.13)$$

equipped with the Lie bracket eq. (9.12). Here $|r - s| \gg 0$ means that there exists $N > 0$ such that $M_{rs} = 0$ when $|r - s| > N$.

We will need to consider the group $GL(\infty)$ associated with this Lie algebra. Its precise definition is somewhat tricky. We will adopt here a very naive point of view, and consider group elements of the form:

$$g = e^{X_1}e^{X_2}\cdots e^{X_k}$$

with $X_1, \ldots, X_k \in \widehat{gl}(\infty)$ defined by *finite* sums of bilinears in the fermionic operators. It is then clear that such group elements are well represented in the Fock space. Of course we will quickly encounter the need to extend this setting to infinite sums of bilinears, then one has to be very careful about convergence properties, but we will not attempt to define a general framework and refer instead to the literature.

The group $GL(\infty)$ acts on the fermions by conjugation.

Proposition. *For any $g \in GL(\infty)$ we have*

$$g\beta_s g^{-1} = \sum_r \beta_r a_{rs} \tag{9.14}$$

$$g\beta_{-s}^* g^{-1} = \sum_r \beta_{-r}^* (a^{-1})_{sr} \tag{9.15}$$

for some c-number matrix a_{rs}. With our definition of $GL(\infty)$ the matrices a_{rs} only have a finite but arbitrary large number of non-zero entries.

<u>Proof.</u> We use a matrix notation: β is an ∞ line vector with elements β_s, and β^* is an ∞ column vector with elements β_{-s}^*. We want to show that for a given g,

$$g\beta g^{-1} = \beta \cdot a$$
$$g^{-1}\beta^* g = a \cdot \beta^*$$

with the same matrix a. Note that if $g_1 \beta g_1^{-1} = \beta a_1$ and $g_2 \beta g_2^{-1} = \beta a_2$ then $g_1 g_2 \beta g_2^{-1} g_1^{-1} = \beta a_1 a_2$, and similarly if $g_1^{-1} \beta^* g_1 = b_1 \beta^*$, $g_2^{-1} \beta^* g_2 = b_2 \beta^*$, then $g_2^{-1} g_1^{-1} \beta^* g_1 g_2 = b_1 b_2 \beta^*$. So, it is enough the verify that $a = b$ on simple generators. We take

$$g = e^{\epsilon:\beta_r \beta_s^*:} = e^{-\epsilon\langle 0|\beta_r \beta_s^*|0\rangle} e^{\epsilon\beta_r \beta_s^*}$$

We see that the normal ordering produces simple scalar factors which do not contribute to the adjoint action, and we can omit it. If $(r + s) \neq 0$, we have $g = e^{\epsilon\beta_r\beta_s^*} = 1 + \epsilon\beta_r\beta_s^*$, and therefore,

$$g\beta_k g^{-1} = \beta_k + \epsilon[\beta_r\beta_s^*, \beta_k] = \beta_l(\delta_{kl} + \epsilon\delta_{rl}\delta_{s+k,0})$$
$$g^{-1}\beta_{-l}^* g = \beta_{-l}^* - \epsilon[\beta_r\beta_s^*, \beta_{-l}^*] = (\delta_{kl} + \epsilon\delta_{rl}\delta_{s+k,0})\beta_{-k}^*$$

If $(r + s) = 0$ then $g = e^{\epsilon\beta_r\beta_{-r}^*}$, and

$$g\beta_k g^{-1} = \beta_k e^{\epsilon\delta_{rk}} \quad , \quad g^{-1}\beta_{-k}^* g = \beta_{-k}^* e^{\epsilon\delta_{rk}}$$

This proves the result. ∎

It is convenient to work with generating functions $\beta(z)$ and $\beta^*(z)$, also called fermionic fields, defined by:

$$\beta(z) = \sum_{r\in\mathbb{Z}+\frac{1}{2}} \beta_r z^{-r-1/2} \tag{9.16}$$

$$\beta^*(z) = \sum_{r\in\mathbb{Z}+\frac{1}{2}} \beta_r^* z^{-r-1/2} \tag{9.17}$$

Notice that $\beta(z)$ and $\beta^*(z)$ are single valued, i.e. $\beta(ze^{2i\pi}) = \beta(z)$ and similarly for $\beta^*(z)$.

Proposition. *The fermionic fields satisfy the following "operator product expansion" :*

$$\beta(z)\beta^*(w) =: \beta(z)\beta^*(w): +\frac{1}{z-w}, \quad \text{for} \quad |z| > |w| \quad (9.18)$$

Proof. It is enough to compute the vacuum expectation value of the left-hand side. This is given by $\sum_{r,s} \langle 0|\beta_r\beta_s^*|0\rangle z^{-r-\frac{1}{2}} w^{-s-\frac{1}{2}}$. Since $\langle 0|\beta_r\beta_s^*|0\rangle = 1$ when $r = -s > 0$ and vanishes otherwise, we get $\langle 0|\beta(z)\beta^*(w)|0\rangle = 1/z \sum_{i=0}^\infty (w/z)^i = 1/(z-w)$ when $|z| > |w|$. ∎

By Wick's theorem this can be generalized to any product of fields, yielding:

$$\langle 0| \prod_{j=1}^{N} \beta(z_j) \prod_{j=1}^{N} \beta^*(w_j)|0\rangle = (-1)^{\frac{N(N-1)}{2}} \det\left(\frac{1}{z_i - w_j}\right) \quad (9.19)$$

A direct consequence of eqs. (9.14, 9.15) is that there exists a fermionic analogue of the tensor Casimir. This fact is the crucial point of the fermionic construction of the Hirota equations.

Proposition. *Let S_{12} be defined by:*

$$S_{12} = \sum_{r\in\mathbb{Z}+\frac{1}{2}} \beta_r \otimes \beta_{-r}^* = \oint \frac{dz}{2i\pi} \beta(z) \otimes \beta^*(z) \quad (9.20)$$

Then

$$S_{12}\, g \otimes g = g \otimes g\, S_{12}, \quad \forall g \in GL(\infty) \quad (9.21)$$

The operator $(S^\dagger S)$ is the tensor Casimir for $\widehat{gl}(\infty)$.

Proof. Equation (9.21) is equivalent to :

$$\sum_{r\in\mathbb{Z}+1/2} \beta_r g \otimes \beta_{-r}^* g = \sum_{r\in\mathbb{Z}+1/2} g\beta_r \otimes g\beta_{-r}^*, \quad \forall g \in GL(\infty)$$

From eqs. (9.14, 9.15), we have $\beta_r g = g\beta_s(a^{-1})_{sr}$ and $\beta_{-r}^* g = a_{rs}g\beta_{-s}^*$, so that

$$\sum_r \beta_r g \otimes \beta_{-r}^* g = \sum_{r,s,k} g\beta_s \otimes g\beta_{-k}^*(a^{-1})_{sr}a_{rk} = \sum_s g\beta_s \otimes g\beta_{-s}^*$$

∎

Notice that the definition of the vacuum implies that $\beta_r|0\rangle \otimes \beta^*_{-r}|0\rangle = 0$, $\forall r$, since either $\beta_r|0\rangle = 0$ or $\beta^*_{-r}|0\rangle = 0$. Therefore

$$S_{12}|0\rangle \otimes |0\rangle = \sum_r \beta_r|0\rangle \otimes \beta^*_{-r}|0\rangle = 0 \qquad (9.22)$$

Combining eqs. (9.21, 9.22) we obtain:

$$S_{12}g|0\rangle \otimes g|0\rangle = 0 \qquad (9.23)$$

for any $g \in GL(\infty)$. This relation is the fermionic analogue of the bosonic equation eq. (9.4) and will be very important in the following.

Remark. The transformations eqs. (9.14, 9.15) can alternatively be written in a more compact way by introducing the function $A_n^g(z)$ for $n \in \mathbb{Z}$:

$$A_n^g(z) = \sum_{m\in\mathbb{Z}} z^m a_{m+\frac{1}{2},n+\frac{1}{2}} \qquad (9.24)$$

We then have:

$$(g\beta_s g^{-1}) = \frac{1}{2i\pi} \oint dz \beta(z) A_{s-\frac{1}{2}}^g(z)$$

In eq. (9.24) the sum is finite as long as we use the previous definition for the group $GL(\infty)$. As pointed out above, extending this construction such that the definition of the function $A_n^g(z)$ involves an infinite sum requires choosing an appropriate completion of $GL(\infty)$.

9.3 Boson–fermion correspondence

We now come to a truly remarkable result known as the boson–fermion correspondence. This will be the main technical tool used in the forthcoming sections. We construct bosonic operators from the fermions and conversely.

The bosonic operators H_n are simply bilinear in the fermionic operators.

Proposition. *Define the operators H_n by*

$$H(z) =: \beta(z)\beta^*(z) := \sum_{n\in\mathbb{Z}} z^{-n-1} H_n \qquad (9.25)$$

or equivalently $H_n = \sum_r : \beta_r \beta^*_{-r+n} :$. They obey the following commutation relations:

$$\begin{aligned}
[H_n, \beta(z)] &= z^n \beta(z) \\
[H_n, \beta^*(z)] &= -z^n \beta^*(z) \\
[H_n, H_m] &= n\delta_{n+m,0}
\end{aligned} \qquad (9.26)$$

<u>Proof</u>. The proof follows from standard manipulations using Wick's theorem. By definition, we have:

$$H_n = \oint \frac{dz}{2i\pi} z^n : \beta(z)\beta^*(z) :$$

Using Wick's theorem, we may write the product $H_n \beta(w)$ as:

$$
\begin{aligned}
H_n\beta(w) &= \oint \frac{dz}{2i\pi} z^n : \beta(z)\beta^*(z) : \beta(w) \\
&= \oint \frac{dz}{2i\pi} z^n : \beta(z)\beta^*(z)\beta(w) : - \oint_{|w|<|z|} \frac{dz}{2i\pi} z^n \frac{1}{w-z}\beta(z)
\end{aligned}
$$

Here the integration over z is performed on a circle C_1 around the origin enclosing w so that we can apply eq. (9.18). Similarly,

$$
\begin{aligned}
\beta(w)H_n &= \oint \frac{dz}{2i\pi} z^n \beta(w) : \beta(z)\beta^*(z) : \\
&= \oint \frac{dz}{2i\pi} z^n : \beta(z)\beta^*(z)\beta(w) : - \oint_{|w|>|z|} \frac{dz}{2i\pi} z^n \frac{1}{w-z}\beta(z)
\end{aligned}
$$

The integral is over a circle C_2 around the origin *not* enclosing w. So the commutator $[H_n, \beta(w)]$ is:

$$
\begin{aligned}
[H_n, \beta(w)] &= - \oint_{C_1-C_2} \frac{dz}{2i\pi} z^n \frac{1}{w-z}\beta(z) \qquad (9.27) \\
&= \oint_{C_w} \frac{dz}{2i\pi} z^n \frac{1}{z-w}\beta(z) = w^n \beta(w)
\end{aligned}
$$

where C_w is a small circle around w. The other relations are proved similarly. ∎

From eq. (9.26), we see that H_n obey bosonic commutation relations. Moreover, we have $H_n^\dagger = H_{-n}$. The operator H_0 is both Hermitian and in the centre of the bosonic algebra.

Let us now describe the bosonic Fock space. The vacuum vector $|l\rangle$ is such that:

$$H_n|l\rangle = 0, \quad \text{for} \quad n > 0, \quad H_0|l\rangle = l|l\rangle, \quad l \in \mathbb{Z}$$

Here we choose $l \in \mathbb{Z}$ because in terms of fermionic operators H_0 has integer spectrum. Over each state $|l\rangle$ we construct a bosonic Fock space by acting with the creation operators H_n with $n < 0$. A basis of the bosonic Fock space consists of the states:

$$H_{-n_1} \cdots H_{-n_k}|l\rangle$$

We will say that these states have charge l, so the charge is the H_0-eigenvalue. In particular, the neutral bosonic Fock space is spanned by the states obtained by acting with the H_n on the vacuum $|0\rangle$.

We have constructed bosons from fermions, the unexpected result is that fermions can be reconstructed from bosons. We start with bosonic operators obeying commutation relations $[H_n, H_m] = n\delta_{n+m,0}$. As already noticed, H_0 is in the centre of this algebra, so we call it "momentum" and denote it p. We enlarge the algebra by introducing a "position" operator q such that $[p, q] = -i$. On the vacua $|l\rangle$, the operator e^{iq} acts as a translation operator: $e^{iq}|l\rangle = |l+1\rangle$, compatible with $p|l\rangle = l|l\rangle$.

Proposition. *Let $[p, q] = -i$ and define*

$$\phi(z) = q - ip \log z + i \sum_{n \neq 0} \frac{H_n}{n} z^{-n}$$

Then, the fermionic fields are reconstructed by the formulae:

$$\beta(z) = V_+(z), \quad \beta^*(z) = V_-(z) \tag{9.28}$$

where

$$V_\pm(z) =: \exp(\pm i\phi(z)) :$$

The colons $:$ $:$ *denote the bosonic normal ordering, which consists of writing the operators H_n with $n > 0$ on the right, and the e^{iq} factor on the left. Explicitly, the operators $V_\pm(z)$ read*

$$V_\pm(z) = e^{\pm iq} z^{\pm p} \exp\left(\mp \sum_{n<0} \frac{z^{-n}}{n} H_n\right) \exp\left(\mp \sum_{n>0} \frac{z^{-n}}{n} H_n\right) \tag{9.29}$$

The operators $V_\pm(z)$ are called vertex operators.

<u>Proof.</u> The proof relies on the following relations:

$$V_\pm(z)V_\pm(w) = (z - w) : e^{\pm i\phi(z)} e^{\pm i\phi(w)} : \quad |z| > |w| \tag{9.30}$$

$$V_\pm(z)V_\mp(w) = \frac{1}{z - w} : e^{\pm i\phi(z)} e^{\mp i\phi(w)} : \quad |z| > |w|$$

They follow immediately from the standard Campbell–Hausdorff formula $e^A e^B = e^B e^A e^{[A,B]}$ for two operators A and B whose commutator $[A, B]$ is a c-number. To compute the anticommutator $\{\beta_r, \beta_s^*\}$, we represent the fermionic operators as contour integrals, $\beta_r = \oint dz\, z^{-r-1/2} : \exp(i\phi(z)) :$ and similarly for β_s^*. We then use the above relations and apply contour integral manipulations as in the previous proposition:

$$\{\beta_r, \beta_s^*\} = \left[\oint_{|z|>|w|} - \oint_{|z|<|w|} \right] \frac{dz\,dw}{(2i\pi)^2} \frac{z^{-r-1/2}w^{-s-1/2}}{z-w} : e^{i\phi(z)}e^{-i\phi(w)} :$$

$$= \oint_{C_0} \frac{dw}{2i\pi} \oint_{C_w} \frac{dz}{2i\pi} \frac{z^{-r-1/2}w^{-s-1/2}}{z-w} : e^{i\phi(z)}e^{-i\phi(w)} := \delta_{r+s,0}$$

where C_0 and C_w are small circles around 0 and w respectively. The proof that $\{\beta_r^*, \beta_s^*\} = \{\beta_r, \beta_s\} = 0$ is similar. ∎

When describing the bosonic Fock space, we denoted by $|l\rangle$ states which are eigenstates of the momentum operator, $p|l\rangle = l|l\rangle$, and which satisfy $H_n|l\rangle = 0$ for $n > 0$. Using the bosonic construction of the fermions, one verifies that these states $|l\rangle$ also satisfy : $\beta_{r-l}|l\rangle = \beta_{r+l}^*|l\rangle = 0$ for $r \geq \frac{1}{2}$. Thus they coincide with the states that we defined in eq. (9.11), i.e. fermionic charged vacua of charge l.

The boson–fermion correspondence may also be viewed at the level of vacuum expectation values as a consequence of Cauchy's determinant formula. Wick's theorem allows us to compute expectation values of vertex operators:

$$\langle 0| \prod_{j=1}^{N} : e^{i\phi(z_j)} : \prod_{j=1}^{N} : e^{-i\phi(w_j)} : |0\rangle = \frac{\prod_{i<j}(z_i - z_j)(w_i - w_j)}{\prod_{i,j}(z_i - w_j)} \tag{9.31}$$

Similarly we have:

$$\langle 0| \prod_{j=1}^{N} : e^{i\phi(z_j)} : |-N\rangle = \prod_{i<j}(z_i - z_j) \prod_i z_i^{-N} \tag{9.32}$$

The bosonization formula will follow if the fermionic expectation value eq. (9.19) coincide with the bosonic expectation value eq. (9.31), specifically, if we have:

$$\det\left(\frac{1}{z_i - w_j}\right) = (-1)^{\frac{N(N-1)}{2}} \frac{\prod_{i<j}(z_i - z_j)(w_i - w_j)}{\prod_{i,j}(z_i - w_j)} \tag{9.33}$$

This is nothing but the Cauchy determinant formula.

9.4 Tau-functions and Hirota bilinear identities

The tau-functions are functions on the orbit of the vacuum $|0\rangle$ under the action of the group $GL(\infty)$. To define them as in eq. (9.6), we again introduce the function $H(t)$ by:

$$H(t) = \sum_{n>0} t_n H_n$$

Definition. *We define the tau-function $\tau(t)$ as the vacuum expectation value:*

$$\tau(t;g) = \langle 0|e^{H(t)}g|0\rangle, \quad g \in GL(\infty) \tag{9.34}$$

We wish to prove that the tau-functions obey Hirota equations. We repeat the argument of section (9.1), but using the operator S_{12} in place of C_{12}. From eq. (9.23) we get:

$$S_{12} \cdot g|0\rangle \otimes g|0\rangle = \oint \frac{dz}{2i\pi} \beta(z)g|0\rangle \otimes \beta^*(z)g|0\rangle = 0 \tag{9.35}$$

This is an identity on vectors in Fock space belonging to the orbit of the vacuum vector. It translates readily into bilinear identities on the tau-functions.

Proposition. *Let $\tau(t;g)$ be the tau-function defined in eq. (9.34). It satisfies the identities:*

$$\oint \frac{dz}{2i\pi} \exp\left(\sum_{n>0} 2z^n y_n\right) \exp\left(-\sum_{n>0} \frac{z^{-n}}{n} \frac{\partial}{\partial y_n}\right) \tau(t+y;g)\tau(t-y;g) = 0 \tag{9.36}$$

<u>Proof.</u> Applying $e^{H(t')} \otimes e^{H(t'')}$ to eq. (9.35), and taking the inner product with $\langle +1| \otimes \langle -1|$, we get :

$$\oint \frac{dz}{2i\pi} \langle +1|e^{H(t')}\beta(z)g|0\rangle \langle -1|e^{H(t'')}\beta^*(z)g|0\rangle = 0$$

To compute $\langle +1|e^{H(t')}\beta(z)g|0\rangle$ and $\langle -1|e^{H(t'')}\beta^*(z)g|0\rangle$ we use the bosonization formula. This gives, as we show below:

$$\langle +1|e^{H(t')}\beta(z)g|0\rangle = V_+(z,t')\tau(t';g) \tag{9.37}$$
$$\langle -1|e^{H(t'')}\beta^*(z)g|0\rangle = V_-(z,t'')\tau(t'';g)$$

where $V_\pm(z;t)$ are the vertex operators in their differential operator representation:

$$V_\pm(z;t) = \exp\left(\pm\sum_{n>0} z^n t_n\right) \exp\left(\mp\sum_{n>0} \frac{z^{-n}}{n} \frac{\partial}{\partial t_n}\right)$$

Indeed, inserting eq. (9.28) into the left-hand side of eq. (9.37) we get:

$$\langle +1|e^{H(t')}\beta(z)g|0\rangle$$

$$= \langle +1|e^{H(t')}e^{iq}z^p \exp\left(-\sum_{n<0}\frac{z^{-n}}{n}H_n\right)\exp\left(-\sum_{n>0}\frac{z^{-n}}{n}H_n\right)g|0\rangle$$

Noticing that $\langle +1|e^{iq}z^p = \langle 0|$, and commuting $\exp\left(-\sum_{n<0}\frac{z^{-n}}{n}H_n\right)$ to the left using the Campbell–Hausdorff formula, we get, using $\langle 0|H_n = 0$ for $n < 0$:

$$\langle 0|e^{H(t')}\exp\left(-\sum_{n<0}\frac{z^{-n}}{n}H_n\right) = e^{\xi(t,z)}\langle 0|e^{H(t')}, \quad \xi(t,z) = \sum_{n>0}t_nz^n$$

Hence eq. (9.37) reads:

$$e^{\xi(t',z)}\langle 0|e^{H(t')}\exp\left(-\sum_{n>0}\frac{z^{-n}}{n}H_n\right)g|0\rangle$$

$$= \exp\left(\sum_{n>0}z^n t'_n\right)\exp\left(-\sum_{n>0}\frac{z^{-n}}{n}\frac{\partial}{\partial t'_n}\right)\langle 0|e^{H(t')}g|0\rangle$$

With these formulae, the bilinear identity becomes:

$$\oint\frac{dz}{2i\pi}V_+(z;t')\,V_-(z;t'')\,\tau(t';g)\tau(t'';g) = 0 \tag{9.38}$$

which is eq. (9.36) if we use the variables $t = \frac{1}{2}(t'+t'')$ and $y = \frac{1}{2}(t'-t'')$. ∎

The identity eq. (9.36) contains an infinite set of bilinear Hirota equations. Let us introduce the Hirota operators, see eq. (8.55) in Chapter 8, for the infinite set of variables t_n:

$$D = \{D_1, D_2, \dots, D_n, \dots\}$$

Equation (9.36) can be rewritten in terms of these operators as:

$$\oint\frac{dz}{2i\pi}e^{\left(\sum_{n>0}2z^n y_n\right)}e^{\left(-\sum_{n>0}\frac{z^{-n}}{n}D_n\right)}e^{\left(\sum_{n>0}y_n D_n\right)}\tau(t;g)\cdot\tau(t;g) = 0$$

Using the definition of the elementary Schur polynomials $P_k(t)$:

$$e^{\xi(t,z)} = \exp\left(\sum_{n=1}^{\infty}t_nz^n\right) = \sum_{k=0}^{\infty}P_k(t)z^k \tag{9.39}$$

we can expand in z the quantity $\exp\left(\sum_{n>0} 2z^n y_n\right)$. Integrating over z shows that the bilinear identity eq. (9.36) is equivalent to:

$$\left(\sum_{j=0}^{\infty} P_j(2y)P_{j+1}(-\widehat{D})\exp\sum_{n>0} y_n D_n\right)\tau(t;g)\cdot\tau(t;g) = 0, \quad \forall y_n \qquad (9.40)$$

where $\widehat{D} = \{D_1, \frac{1}{2}D_2, \ldots, \frac{1}{n}D_n, \ldots\}$.

Expanding this equation in powers of y, we get an infinite set of Hirota equations called the KP hierarchy. For instance, the coefficient of y_1^3 gives the equation

$$(D_1^4 + 3D_2^2 - 4D_1D_3)\tau\cdot\tau = 0$$

which is the Hirota form of the KP equation, see eq. (8.56) in Chapter 8.

Remark. Since the states $|l\rangle$ satisfy $\beta_{r-l}|l\rangle = \beta_{r+l}^*|l\rangle = 0$ for $r \geq \frac{1}{2}$, we also have the relation

$$S_{12}\cdot g|l\rangle \otimes g|l\rangle = \sum_r \beta_r g|l\rangle \otimes \beta_{-r}^* g|l\rangle = 0$$

Thus we can similarly prove more general bilinear identities for the τ-functions $\tau_l(t;g) = \langle l|e^{H(t)}g|l\rangle$:

$$\oint \frac{dz}{2i\pi} z^{l-l'}\, V_-(z;t')\, V_+(z;t'')\, \tau_l(t';g)\tau_{l'}(t'';g) = 0, \quad \forall t', t'', l, l',\ l \geq l' \qquad (9.41)$$

9.5 The KP hierarchy and its soliton solutions

To produce a solution of the KP hierarchy, we need to choose an element $g \in GL(\infty)$ and calculate the τ-function. A particularly simple choice is:

$$g_{1-\text{sol}} = e^{a\beta(p)\beta^*(q)} = 1 + a\beta(p)\beta^*(q), \quad \text{with} \quad |p| > |q|$$

This corresponds to the one-soliton solution. In principle one should normal order $\beta(p)\beta^*(q)$ in the exponential, but this would amount to a multiplicative prefactor which is irrelevant in the bilinear Hirota equations. The N-soliton solution is obtained by choosing the group element to be the product of N one-soliton factors.

Proposition. *The N-soliton tau-function of the KP hierarchy is given by:*

$$\tau_N(t;g) = \langle 0|e^{H(t)}g_1 g_2\cdots g_N|0\rangle, \quad \text{with } g_i = 1 + a_i\beta(p_i)\beta^*(q_i)$$

Explicitly, it is equal to:

$$\tau_N(t;g) = 1 + \sum_{r=1}^{N} \sum_{\substack{I \subset \{1,\dots,N\} \\ |I|=r}} \prod_{i<j\in I} \frac{(p_i - p_j)(q_i - q_j)}{(p_i - q_j)(q_i - p_j)} \cdot \prod_{i\in I} e^{\eta_i} \quad (9.42)$$

where $\eta_i = \xi(t, p_i) - \xi(t, q_i) + \log\left(\frac{a_i}{p_i - q_i}\right)$ *with* $\xi(t, p) = \sum_n p^n t_n$.

Proof. Using the commutation relation eq. (9.26) of the fermions with $H(t)$, we get the "evolution equations" of the fermionic fields:

$$e^{H(t)}\beta(z)e^{-H(t)} = e^{\xi(t,z)}\beta(z) \quad (9.43)$$
$$e^{H(t)}\beta^*(z)e^{-H(t)} = e^{-\xi(t,z)}\beta^*(z) \quad (9.44)$$

This allows us to commute $e^{H(t)}$ to the right where it is annihilated by the vacuum, getting:

$$\tau_N(t;g) = \langle 0| \prod_{i=1}^{N} \left(1 + a_i e^{\xi(t,p_i) - \xi(t,q_i)} \beta(p_i)\beta^*(q_i)\right) |0\rangle \quad (9.45)$$

Expanding this formula with the Wick theorem, we get the tau-function as a sum of determinants:

$$\tau_N(t;g) = 1 + \sum_i W_{ii} + \sum_{i<j} \det\,[W_{(ij)}]_{2\times2} + \sum_{i<j<k} \det\,[W_{(ijk)}]_{3\times3} + \cdots$$

$$(9.46)$$

where $[W_{(i_1 i_2 \cdots i_r)}]_{r\times r}$ is the $r \times r$ submatrix of W obtained by selecting out the rows and columns of indices $i_1 i_2 \cdots i_r$, e.g. $\det\,[W_{(ij)}]_{2\times2} = W_{ii}W_{jj} - W_{ij}W_{ji}$. Here W is an $N \times N$ matrix with elements

$$W_{ij} = X_i X_j \langle 0|\beta(p_i)\beta^*(q_j)|0\rangle = \frac{X_i X_j}{p_i - q_j}$$

and $\log\left(\frac{X_i^2}{a_i}\right) = \xi(t, p_i) - \xi(t, q_i)$. Equation (9.42) immediately follows by applying the Cauchy formula eq. (9.33).

$$\tau_N(t;g) = 1 + \sum_i e^{\eta_i} + \sum_{i<j} e^{\eta_i + \eta_j} \frac{(p_i - p_j)(q_i - q_j)}{(p_i - q_j)(q_i - p_j)} + \cdots$$

∎

Equivalently, we can write the tau-function as a determinant:

$$\tau_N(t;g) = \det\left(\delta_{ij} + \frac{X_i X_j}{p_i - q_j}\right)$$

This is proved by comparing the well known expansion formula:

$$\det(1 + W) = 1 + \sum_i W_{ii} + \sum_{i<j} \det\left[W_{(ij)}\right]_{2\times 2} + \cdots \tag{9.47}$$

with eq. (9.46).

9.6 Fermions and Grassmannians

The previous construction can be given a more geometrical interpretation by relating it to Grassmannians. We first explain this notion in the finite-dimensional case. We then extend it to the infinite-dimensional case and make contact with the above fermionic description. The Hirota equations can be interpreted as the Plücker equations of the Grassmannian, and the time flows are induced by one-parameter subgroups of $GL(\infty)$.

Let V be a vector space of dimension N. The Grassmannian $Gr(M, V)$, also denoted by $Gr(M, N)$, is the space of M-dimensional hyperplanes in V. Since a hyperplane is uniquely determined by an M-dimensional frame in V modulo the linear transformations $GL(M)$, $Gr(M, V)$ may be defined as:

$$Gr(M, V) \equiv \{M - \text{frames in } V\}/GL(M) \tag{9.48}$$

Grassmannians are projective algebraic varieties. If $P(\Lambda^M V)$ denotes the projective space of dimension $\binom{N}{M} - 1$ on the Mth exterior power of V, one can embed them into $P(\Lambda^M V)$ using the Plücker coordinates and Plücker relations, as we now explain.

We define the application Θ from $Gr(M, V)$ to $P(\Lambda^M V)$, by associating an M-dimensional frame $(x_{(1)}, \ldots, x_{(M)})$ with their exterior product:

$$\begin{aligned}\Theta \; : \quad &Gr(M, V) \longrightarrow P(\Lambda^M V)\\ &(x_{(1)}, \ldots, x_{(M)}) \longrightarrow \hat{x} = x_{(1)} \wedge \cdots \wedge x_{(M)}\end{aligned} \tag{9.49}$$

A change of basis vectors $x_{(k)}$ by an arbitrary element g of $GL(M)$ multiplies $\hat{x}$ by the scalar $\det g \neq 0$, so the right-hand side is invariant in $P(\Lambda^M V)$.

The group $GL(N)$ acts transitively on the set of M-hyperplanes. If $(e_1, \ldots, e_N)$ is a basis of V, we can take a reference M-hyperplane spanned by $e_1, \ldots, e_M$. Any other basis of any other M-hyperplane can be written

$ge_1, \ldots, ge_M$ for some $g \in GL(N)$. Under the application Θ this yields a point $ge_1 \wedge \cdots \wedge ge_M$. In other words the Grassmannian is the orbit of any of its points under $GL(N)$:

$$Gr(M, V) \sim \{ge_1 \wedge \cdots \wedge ge_M | g \in GL(N)\} \tag{9.50}$$

Writing $x_{(j)} = \sum_i e_i\, x^i{}_j$, we have

$$\widehat{x} = \sum_{h_1, \ldots, h_M} X^{h_1 \cdots h_M} e_{h_1} \wedge e_{h_2} \cdots \wedge e_{h_M} \tag{9.51}$$

The Plücker coordinates of an M-hyperplane in $P(\Lambda^M V)$ are

$$X^{h_1 \cdots h_M} = \det \left(x^{h_i}{}_j \right)_{i,j=1,\ldots,M}$$

The Grassmannian has an image in $P(\Lambda^M V)$ characterized by algebraic relations which are called the Plücker relations.

Proposition. *The image, in the projective space, of the Grassmannian is the set of points whose coordinates $X^{k_1 \cdots k_M}$ satisfy the bilinear Plücker relations, for all $k_1, \ldots, k_{M-1}$ and $h_1, \ldots, h_{M+1}$:*

$$\sum_{j=1}^{M+1} X^{k_1 \cdots h_j \cdots k_{M-1}} X^{h_1 \cdots \widehat{h}_j \cdots h_{M+1}} = 0 \tag{9.52}$$

where the index $\widehat{h}_j$ has to be omitted.

Proof. In order to prove these Plücker relations, we need a:

Lemma. *Let $\widehat{x} \in \Lambda^M V$, and define*

$$V' = \{v^* \in V^* | i(v^*)\widehat{x} = 0\}$$

and $W = (V')^\perp \subset V$. Then W is the smallest subspace of V such that $\widehat{x} \in \Lambda^M V$. Moreover, $W = \{i(\xi)\widehat{x} | \xi \in \Lambda^{M-1} V^*\}$.

Proof. Let $w_1, \ldots, w_l$ be a basis of W completed by $u_{l+1}, \ldots, u_N$ to a basis of V. Let $w_1^*, \ldots, w_l^*, u_{l+1}^*, \ldots, u_N^*$ be the dual basis. Then V' is just the span of $u_{l+1}^*, \ldots, u_N^*$. In this basis one can write $\widehat{x} = \widehat{x}_1 + \widehat{x}_2 + \cdots$, where $\widehat{x}_1 \in \Lambda^M W$, while the other terms can be expanded with increasing number of u_k. For instance $\widehat{x}_2 = \sum_j u_j \wedge \widetilde{v}_j$, with $\widetilde{v}_j \in \Lambda^{M-1} W$. By definition of V', we have $i(u_j^*)\widehat{x} = 0$, and this implies that all $\widehat{x}_j = 0$ for $j \geq 2$. Hence $\widehat{x} \in\!\in \Lambda^M W$. To show the second part of the Lemma, we use the formula

$$\langle i(\xi)\widehat{x}, v^* \rangle = \pm \langle i(v^*)\widehat{x}, \xi \rangle$$

so that $v^* \in V'$ if and only if $i(\xi)\widehat{x} \in (V')^{\perp} = W$.   ∎

We now return to the proof of the Proposition. We have to characterize the elements $\widehat{x}$ which are factorizable, that is

$$\widehat{x} = w_1 \wedge \cdots \wedge w_M$$

This is equivalent to the fact that W is the span of $w_1, \ldots, w_M$, and so is of dimension M. Another way of saying this is

$$w \wedge \widehat{x} = 0, \quad \forall\, w \in W$$

Taking into account the second part of the Lemma, this is equivalent to

$$\widehat{x} \text{ factorizable iff } i(\xi)\widehat{x} \wedge \widehat{x} = 0, \; \forall \xi \in \Lambda^{M-1}V^*$$

To express this condition, it is sufficient to vary ξ over a basis of $\Lambda^{M-1}V^*$. One takes $\xi = e^*_{k_1} \wedge \cdots \wedge e^*_{k_{M-1}}$ with $k1 < \cdots < k_{M-1}$. Using $\widehat{x}$ in the form eq. (9.51), we get

$$i(\xi)\widehat{x} = \sum_h X^{k_1 \cdots k_{M-1}h} e_h$$

Similarly, writing $\widehat{x} = \sum X^{h_1 \cdots \widehat{h_j} \cdots h_{M+1}} e_{h_1} \wedge \cdots \wedge e_{h_{M+1}}$ we get

$$i(\xi)\widehat{x} \wedge \widehat{x} = \sum_j X^{k_1 \cdots h_j \cdots k_{M-1}} X^{h_1 \cdots \widehat{h_j} \cdots h_{M+1}} e_{h_1} \wedge \cdots \wedge e_{h_j} \wedge \cdots \wedge e_{h_{M+1}}$$

yielding eq. (9.52).   ∎

Remark. It is easy to see that the Plücker relations are necessary relations. Let us demand that a vector of the form $z = \sum_m z^{(m)} x_{(m)}$ belongs to the hyperplane spanned by the $x_{(j)}$. This means

$$0 = z \wedge \widehat{x} = \sum_{h_1 < \cdots < h_{M+1}} Z^{h_1 \cdots h_{M+1}} \, e_{h_1} \wedge \cdots \wedge e_{h_{M+1}}$$

with

$$Z^{h_1 \cdots h_{M+1}} = \sum_{j=1}^{M+1} \sum_{m=1}^{M} (-)^{j-1} x^{h_j}_m z^{(m)} X^{h_1 \cdots \widehat{h_j} \cdots h_{M+1}} \tag{9.53}$$

We next choose $z^{(m)}$ as the minor of the following determinant:

$$X^{k_1 \cdots k_{M-1}h_j} = \det \begin{pmatrix} x^{k_1}_1 & \cdots & x^{k_1}_M \\ \vdots & & \vdots \\ x^{k_{M-1}}_1 & \cdots & x^{k_{M-1}}_M \\ x^{h_j}_1 & \cdots & x^{h_j}_M \end{pmatrix} = \sum_m z^{(m)} x^{h_j}_m$$

Expanding this determinant with respect to the last row shows that the coefficient $z^{(m)}$ only depends on the indices $k_1, \ldots, k_{M-1}$ and not on the index h_j. Inserting this formula into eq. (9.53) gives the Plücker relations.

To introduce the infinite-dimensional Grassmannian, we first need to consider the formal limit of the above constructions to infinite-dimensional spaces V_∞. A basis of V_∞ will be denoted by e_s with $s \in \mathbb{Z} + \frac{1}{2}$. We equip V_∞ with the scalar product $(e_s, e_r) = \delta_{r+s,0}$. To make sense of exterior products in this context, consider formally the vacuum vector $|0\rangle$ which is the semi-infinite wedge vector:

$$|0\rangle = e_{\frac{1}{2}} \wedge e_{\frac{3}{2}} \wedge e_{\frac{5}{2}} \wedge \cdots \qquad (9.54)$$

and consider elements which are finite perturbations of this state in the following sense:

$$|Y\rangle = e_{-s_n} \wedge \cdots \wedge e_{-s_1} \wedge e_{\frac{1}{2}} \wedge \cdots \wedge \widehat{e}_{r_1} \wedge \cdots \wedge \widehat{e}_{r_m} \wedge \cdots \qquad (9.55)$$

for $0 < s_1 < \cdots < s_n$ and $0 < r_1 < \cdots < r_m$, where $\widehat{e}_{r_j}$ have been omitted in the infinite wedge product. We say that the state $|Y\rangle$ contains a finite number of n particles and a finite number m of holes. Note that the tail of these wedge products is of the form $\wedge e_s \wedge e_{s+1} \wedge \cdots$ for sufficiently large positive s, so these states are finite perturbations of the vacuum. We will work in the space of linear combinations of states $|Y\rangle$.

We now recast the fermionic constructions of the previous sections into the infinite Grassmannian language. The basis for this interpretation is to write the fermionic vacuum $|0\rangle$ as a Dirac sea:

$$|0\rangle = \beta_{\frac{1}{2}} \beta_{\frac{3}{2}} \cdots | - \infty\rangle \qquad (9.56)$$

where $| - \infty\rangle$ is the charge $-\infty$ fermionic vacuum.

The dictionary between the semi-infinite wedge product and the fermionic language is as follows. We identify the fermionic state eq. (9.56) with the vacuum vector eq. (9.54). Then we identify the fermionic operators with the exterior and inner products:

$$\beta_r = e_r \wedge, \qquad \beta_s^* = e_s \neg \qquad (9.57)$$

where $e_r \wedge$ means taking the wedge product with e_r, and $e_s \neg$ means taking the interior product with e_s, using the scalar product $(e_r, e_s) = \delta_{r+s,0}$, which also means removal of e_{-s}.

It is a direct consequence of exterior calculus that the definition eq. (9.57) forms a representation of the anticommutation relations eq. (9.9).

We define the infinite Grassmannian, or rather its image by Θ, as the orbit of the vacuum under $GL(\infty)$, as in eq. (9.50):

$$g|0\rangle = (g \cdot e_{\frac{1}{2}}) \wedge (g \cdot e_{\frac{3}{2}}) \wedge (g \cdot e_{\frac{5}{2}}) \wedge \cdots \qquad (9.58)$$

We have denoted $g \cdot e_s = \sum_r e_r a_{rs}$, where a_{rs} is related to g as in eq. (9.14). Since a_{rs} is a band matrix with $a_{rr} \neq 0$ this produces a combination of finite perturbations of the vacuum state, hence belonging to the space of semi-infinite wedge products. The expression of $g|0\rangle$ is factorized, so naturally belongs to the infinite Grassmannian.

In the fermionic language, we have:

$$g|0\rangle = g\beta_{\frac{1}{2}}g^{-1} \cdot g\beta_{\frac{3}{2}}g^{-1} \cdots | -\infty\rangle$$

i.e. g acts on the β_r by conjugation. Note that one has naturally $g|-\infty\rangle = |-\infty\rangle$. To see it, we write $g = e^X$ and verify that $X|-\infty\rangle = 0$. This is because X is a combination of elements $\beta_r \beta_s^*$ and $|-\infty\rangle = \cdots \beta_{-\frac{3}{2}}^* \beta_{-\frac{1}{2}}^* |0\rangle$ by eq. (9.11). When $s < 0$ we get $\beta_s^{*2} = 0$ and if $s > 0$ we can push β_s^* to the right where it annihilates the vacuum. This may serve as a motivation for our definition of the action of $GL(\infty)$ in eq. (9.58).

Note that $GL(\infty)$ acts transitively on the Grassmannian but the vacuum vector $|0\rangle$ has a stability subgroup. In particular, elements g which correspond to upper triangular matrices a act trivially. From eq. (9.58) we see that when a is upper triangular, i.e. $a_{rs} = 0$ for $r < s$, the vacuum vector gets multiplied by the scalar $\prod_r a_{rr} \neq 0$, so is invariant in $P(\Lambda V)$.

In $GL(\infty)$ we can factorize any element $g = g_-^{-1} g_+$ so that $g_\pm$ correspond respectively to upper and lower triangular band matrices, and only g_- acts effectively on the vacuum.

Finally, we write the Plücker equations of the Grassmannian and identify them with eq. (9.35). Formally an element of the infinite Grassmannian is of the form:

$$|X\rangle = \sum_{k_1 k_2 \cdots k_m \cdots} X^{k_1 k_2 \cdots k_m \cdots} e_{k_1} \wedge e_{k_2} \wedge \cdots \wedge e_{k_m} \wedge \cdots$$

where $e_{k_1} \wedge \cdots$ is a semi-infinite wedge product, i.e. there is no hole after some k_m. The straight extension of the Plücker relation eq. (9.52) to this setting reads:

$$\sum_j X^{k_1 k_2 \cdots h_j \cdots} X^{h_1 h_2 \cdots \widehat{h_j} \cdots} = 0$$

The sum over j is finite since we cannot add a particle at position h_j after k_m. This can be rewritten using eqs. (9.57) in the elegant form:

$$\sum_j \beta_j \otimes \beta_{-j}^* |X\rangle \otimes |X\rangle = 0$$

which is exactly eq. (9.35). Of course this simple way of stating the Plücker relation relies on the infinite number of indices. As we have already shown, the condition eq. (9.35) translates into Hirota's equations on the tau-function which are therefore equivalent to the equations of the Grassmannian in the infinite wedge space. This is a beautiful discovery of Sato.

We can now identify the flows of the KP hierarchy as the action of an Abelian subgroup of $GL(\infty)$ in $\mathrm{Gr}(V_\infty)$. The KP time evolutions are defined in eq. (9.34) by the action on $g|0\rangle$ of $e^{\sum_{n>0} H_n t_n}$. We want to compute the action of this operator on semi-infinite wedge products.

Proposition. *Let $\mathcal{S}$ be the shift operator on V_∞ defined by:*

$$\mathcal{S} : e_s \quad \to \quad e_{s+1}$$

We extend $\mathcal{S}$ to semi-infinite wedge products by:

$$\mathcal{S}\; e_{i_1} \wedge e_{i_2} \wedge \cdots = (\mathcal{S}e_{i_1}) \wedge e_{i_2} \wedge \cdots \; + \; e_{i_1} \wedge (\mathcal{S}e_{i_2}) \wedge \cdots \; + \; \cdots$$

The action of H_n on the states eq. (9.55) is given by the shift operator $\mathcal{S}^n$.

Proof. In the fermionic language, a semi-infinite wedge product is of the form $\beta_{i_1} \cdots \beta_{i_m}|l\rangle$, where $|l\rangle$ is the vacuum of charge l and the indices $i_1 < i_2 < \cdots < i_m < l$ may be negative. The operator $e^{H(t)}$ acts by conjugation on the β_{i_j} and leaves the vacuum $|l\rangle$ invariant. At the infinitesimal level, this yields the action of H_n on this state as a finite sum of commutators $\sum_j \beta_{i_1} \cdots [H_n, \beta_{i_j}] \cdots \beta_{i_m}|l\rangle$. This translates exactly to the action of $\mathcal{S}^n$ on the infinite wedge product because $[H_n, \beta_{i_j}] = \beta_{i_j+n}$. ∎

One realization of the infinite-dimensional space V_∞ is the set of functions on the circle $|z| = 1$. A basis is given by the powers z^n for $n \in \mathbb{Z}$. We identify the abstract basis elements e_s with the powers $z^{s-\frac{1}{2}}$. On this basis, eq. (9.24) shows that the group $GL(\infty)$ acts as:

$$z^n \to A_n^g(z) = \sum_m a_{m+\frac{1}{2},n+\frac{1}{2}} z^m$$

In particular, when a_{rs} is lower triangular with diagonal elements equal to 1, we have:

$$z^n \quad \to \quad A_n^g(z) = \sum_{m \leq n} a_{m+\frac{1}{2},n+\frac{1}{2}}\; z^m = z^n + a_{n-\frac{1}{2},n+\frac{1}{2}} z^{n-1} + \cdots \tag{9.59}$$

This representation will be studied in detail below. For the time being we look at the image under Θ. The vacuum vector becomes:

$$|0\rangle = z^0 \wedge z^1 \wedge z^2 \wedge \cdots$$

The group $GL(\infty)$ acts on the vacuum as in eq. (9.58):

$$g|0\rangle = A_0^g(z) \wedge A_1^g(z) \wedge A_2^g(z) \wedge \cdots \qquad (9.60)$$

In this setting we can give the expression of the τ-function:

Proposition. *The tau-function may be written as an infinite-dimensional determinant:*

$$\tau(t;g) = \det \left[\oint \frac{dz}{2i\pi} e^{\xi(t,z)} z^{-m-1} A_n^g(z) \right]_{m,n\geq0} \qquad (9.61)$$

with $\xi(t,z) = \sum_n z^n t_n$.

Proof. Recall the definition of the tau-function as $\tau(t;g) = \langle 0|e^{H(t)}g|0\rangle$. The fermionic operators satisfy the "evolution equations", eqs. (9.43, 9.44). The time evolution of $g|0\rangle$ under $e^{H(t)}$ amounts to the substitution $z^m \to z^m e^{\xi(t,z)}$ in eq. (9.60), thus

$$e^{H(t)}g|0\rangle = \left(e^{\xi(t,z)}A_0^g(z)\right) \wedge \left(e^{\xi(t,z)}A_1^g(z)\right) \wedge \left(e^{\xi(t,z)}A_2^g(z)\right) \wedge \cdots$$

To compute the expectation value $\langle 0|e^{H(t)}g|0\rangle$ we have to extract the component of $e^{H(t)}g|0\rangle$ on $|0\rangle$. This is given by the determinant eq. (9.61). ∎

9.7 Schur polynomials

In this section we show that a basis of semi-infinite wedge products is indexed by Young diagrams. This allows us to investigate the structure of tau-functions. In particular, we get polynomial tau-functions which turn out to be the Schur polynomials. We use the properties of Schur polynomials to prove that knowledge of the tau-function allows us to reconstruct the corresponding element of the Grassmannian, see eqs. (9.67, 9.68).

In the finite-dimensional case, a basis of $\Lambda^M V$ consists of the vectors $|Y\rangle = e_{k_1} \wedge \cdots \wedge e_{k_M}$. The notation $|Y\rangle$ refers to Young diagrams, which are defined as follows. Recall that, for any integer $p > 0$, a partition of p is defined as a collection of integers n_j, $j = 1, \ldots, M$ such that $p = \sum_{j=1}^M n_j$ and $1 \leq n_1 \leq n_2 \leq \cdots \leq n_M$. With such a partition one associates a diagram of boxes with M lines, and n_M boxes in the first line, n_{M-1} boxes in the second line, up to n_1 boxes in the last line. This is called a Young diagram with p boxes.

With the set of number $(k_1, \ldots, k_M)$ with $1 \leq k_1 < \cdots < k_M \leq N$ (note that $k_i \neq k_j$ for $i \neq j$) we associate the Young diagram $Y = [n_j]$

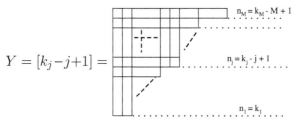

Fig. 9.1. A Young diagram

with the jth line of length $n_j = k_j - j + 1$ (now $1 \le n_1 \le \cdots \le n_M \le N$).
These Young diagrams have at most M lines and N columns.

Consider variables $u_1, u_2, \ldots, u_k$, where k is arbitrarily large. Symmetric polynomials of degree d in these variables may be expressed on several bases. To construct one of them consider the elementary basic symmetric polynomials h_j of degree j defined by:

$$\prod_{i=1}^{k} \frac{1}{1 - u_i s} = \sum_{j=0}^{\infty} h_j(u) s^j$$

With any partition of d into at most k parts, namely ($\lambda_1 \ge \lambda_2 \ge \cdots \lambda_k \ge 0$) such that $d = \sum_i \lambda_i$, we associate the Young diagram Y with at most k lines, whose i^{th} line has λ_i boxes, and the symmetric polynomial of degree d:

$$h_Y = \prod_i h_{\lambda_i}$$

It is known that the h_Y form an integral basis of the space of symmetric polynomials of degree d. Any symmetric polynomial with integer coefficients can be expanded on this basis with integer coefficients.

With the same Young diagram Y one can associate another symmetric polynomial, known as the Schur polynomial:

$$S_Y(u) = \frac{\det u_j^{\lambda_i + j - i}}{\Delta(u)}, \quad \Delta(u) = \prod_{i<j}(u_i - u_j) \tag{9.62}$$

where $\Delta(u) = \det u_j^{j-i}$ is the Vandermonde determinant. Note that the numerator and the denominator are antisymmetric polynomials, and that the denominator divides the numerator so that S_Y is a symmetric polynomial of degree d. It can also be shown that the Schur polynomials form an integral basis of symmetric polynomials of degree d indexed by the

Young diagrams. Hence we can express S_Y in terms of the $h_{Y'}$, and this relation is given by the formula

$$S_Y = \det\left(h_{\lambda_i+j-i}\right)$$

We will prove this formula below using the boson–fermion correspondence, see eq. (9.65).

We can also relate these definitions to the definition of "elementary Schur polynomials" introduced in eq. (9.39). We have

$$h_i(u) = P_i(t), \quad \text{for } t_n = \frac{1}{n}\sum_j u_j^n$$

This may be seen by writing:

$$\sum_i h_i(u)s^i = e^{-\sum_j \log(1-u_j s)} = e^{\sum_{i=1}^{\infty}(\frac{1}{i}\sum_j u_j^i)s^j} = \sum_i P_i(t)s^i$$

Any polynomial $f(t_1, t_2, \ldots)$ can be seen as a symmetric polynomial of the variables u and conversely. This is because symmetric polynomials in u_j can be expressed on the Newton sums t_n. In particular, we shall denote by $\chi_Y(t)$ the Schur polynomial $S_Y(u)$ expressed in terms of the variables t_n.

$$\chi_Y(t) = S_Y(u), \quad t_n = \frac{1}{n}\sum_j u_j^n$$

The next proposition computes the expectation value $\langle 0|e^{H(t)}|Y\rangle$, using the boson–fermion correspondence.

Proposition. Let $|Y\rangle = \beta_{-s_m}\cdots\beta_{-s_1}\cdot\beta^*_{-r_1}\cdots\beta^*_{-r_m}|0\rangle$ be a neutral state with $s_m > \cdots > s_1$ and $r_m > \cdots > r_1$, then

$$\langle 0|e^{H(t)}|Y\rangle = \det\left(P_{n_a+a-b}(t)\right) \tag{9.63}$$

with n_a the number of boxes in the ath line, counted from the bottom, for a Young diagram $Y = [n_a]$ with $L = (r_m + \frac{1}{2})$ lines and $C = (s_m + \frac{1}{2})$ columns, enclosed in a hook[1] of width m. The first m lines have length $(s_{m-k+1} + k - \frac{1}{2})$, and the first m columns have length $(r_{m-k+1} + k - \frac{1}{2})$ for $k = 1, \ldots, m$. See Fig. (9.2).

Proof. Let us give the details of the proof of eq. (9.63) in the simplest case $m = 1$. Thus we consider the state $|Y\rangle = \beta_{-s}\beta^*_{-r}|0\rangle$. Using the definition eq. (9.11) for the charged vacuum $|-r-\frac{1}{2}\rangle$ we can write:

$$|Y\rangle = \beta_{-s}\beta_{\frac{1}{2}}\cdots\beta_{r-1}|-r-\frac{1}{2}\rangle$$

---

[1] This means that there is no box in the rectangle of size $(L-m)\times(C-m)$ in the lower right corner.

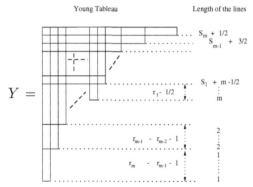

Fig. 9.2. The Young diagram associated with the state with m particules and holes.

Expressing the fermionic operators as contour integrals, $\beta_k = \oint \frac{dz}{2i\pi} z^{k-\frac{1}{2}} \beta(z)$, we obtain

$$\langle 0|e^{H(t)}|Y\rangle = \oint \prod_{a=1}^{r+\frac{1}{2}} z_a^{r-m_a-\frac{1}{2}} \langle 0|e^{H(t)}\beta(z_{r+1/2})\cdots\beta(z_1)|-r-\tfrac{1}{2}\rangle$$

with $(m_1,\ldots,m_{r+1/2}) = (1,2,\ldots,r-\tfrac{1}{2},r+s)$. Using the evolution equations eq. (9.43) and the fact that $H(t)$ annihilates $|-r-\tfrac{1}{2}\rangle$, we find:

$$\langle 0|e^{H(t)}|Y\rangle = \oint \prod_{a=1}^{r+\frac{1}{2}} z_a^{r-m_a-\frac{1}{2}} e^{\xi(t,z_a)} \langle 0|\beta(z_{r+1/2})\cdots\beta(z_1)|-r-\tfrac{1}{2}\rangle$$

The expectation value can be evaluated using the bosonization formulae eqs. (9.28, 9.30):

$$\langle 0|\beta(z_{r+1/2})\cdots\beta(z_1)|-r-\tfrac{1}{2}\rangle = \prod_{a<b}(z_a - z_b) \prod_a z_a^{-r-1/2}$$

Thus, because the product $\prod_{a<b}(z_a - z_b)$ is the Vandermonde determinant $\det(z_a^{b-1})$, we get using eq. (9.39):

$$\langle 0|e^{H(t)}|Y\rangle = \oint \prod_{a=1}^{r+\frac{1}{2}} z_a^{-m_a-1} e^{\xi(t,z_a)} \prod_{a<b}(z_a - z_b) \qquad (9.64)$$

$$= \det\left(\oint dz\, z^{b-m_a-2}\, e^{\xi(t,z)}\right) = \det\left(P_{m_a-b+1}(t)\right)$$

Since $m_a = n_a + a - 1$ this proves the result eq. (9.63) for $m = 1$. The general case is proved similarly starting from the representation of the state $|Y\rangle = \beta_{-s_1} \cdots \beta_{-s_m} \cdot \beta^*_{-r_1} \cdots \beta^*_{-r_m}|0\rangle$ as

$$|Y\rangle = \beta_{-s_1} \cdots \beta_{-s_m} \cdot \beta_{\frac{1}{2}} \cdots \widehat{\beta}_{r_k} \cdots \beta_{r_m-1}\widehat{\beta}_{r_m}| - r_m - \tfrac{1}{2}\rangle$$

where the operators $\widehat{\beta}_{r_k}$ have been omitted. ∎

This proposition allows us to associate a Young diagram Y with the state $|Y\rangle \equiv \beta_{-s_m} \cdots \beta_{-s_1} \cdot \beta^*_{-r_1} \cdots \beta^*_{-r_m}|0\rangle$ in the following way:

For example, in the simple case $m = 1$ which has been considered in the proof of the proposition, Y is a hook diagram of width one, with the first line containing $s + 1/2$ boxes, followed by $r - 1/2$ lines with one box. In the next proposition, we compute $\langle 0|e^{H(t)}|Y\rangle$ in another way and get the Jacobi–Trudy identity:

Proposition. *The Schur polynomial S_Y associated with the Young diagram Y, eq. (9.62), can be expressed in terms of the variables t_n as:*

$$\chi_Y(t) = \det\left(P_{n_a+a-b}(t)\right) \tag{9.65}$$

Proof. We use the same technique as before, and limit ourselves to the case $m = 1$. In the expression eq. (9.64) of $\langle 0|e^{H(t)}|Y\rangle$, we insert $\exp \xi(t, z) = \prod(1 - zu_i)^{-1}$. Next we use the Cauchy identity in the form:

$$\prod_{i,a} \frac{1}{1 - u_i z_a} = \frac{1}{\Delta(u)\Delta(z)}\det\left(\frac{1}{1 - z_a u_i}\right) \tag{9.66}$$

where $\Delta(u) = \prod_{a<b}(u_a - u_b)$ is the Vandermonde determinant. We obtain:

$$\langle 0|e^{H(t)}|Y\rangle = \frac{1}{\Delta(u)}\det\left(\oint dz \frac{z^{-m_a-1}}{1 - zu_i}\right) = \frac{\det(u_i^{m_a})}{\Delta(u)} = S_Y(u)$$

with $m_a = n_a + a - 1$. Setting $u = u(t)$ and using eq. (9.63) yields the result. ∎

Remark. The polynomials $S_Y(u)$ introduced in the previous proposition are the characters of the group of linear transformations in the representation ρ_Y corresponding to the Young diagram Y. More precisely, the character $\mathrm{Tr}\,(\rho_Y(g))$ for $g \in GL(k)$ is invariant under conjugation and only depends on the eigenvalues of g. One has $\mathrm{Tr}\,(\rho_Y(g)) = S_Y(u)$, where $u_1, \ldots, u_k$ are the eigenvalues of g.

Any symmetric polynomial f of the variables u_i can be expanded on the basis $S_Y(u)$. Viewing u_i as functions of t_n one can expand it as $f(t) = \sum_Y \zeta_Y \chi_Y(t)$.

Proposition. *The coefficients ζ_Y of the expansion $f(t) = \sum_Y \zeta_Y \chi_Y(t)$ can be recovered by:*

$$\zeta_Y = \chi_Y(\widehat{\partial_t}) \cdot f(t)|_{t=0}$$

where $\chi_Y(\widehat{\partial_t})$ are the differential operators obtained by replacing t_n by $\frac{1}{n}\frac{\partial}{\partial t_n}$ in the definition eq. (9.63) of $\chi_Y(t_n)$.

Proof. We first rewrite the Cauchy identity eq. (9.66) in the form:

$$\prod_{i,j} \frac{1}{1 - u_i y_j} = \sum_Y S_Y(u) S_Y(y)$$

where the sum runs over all Young diagrams. This is shown by expanding each matrix entry $1/(1 - u_i y_j)$ in powers of $u_i y_j$. The jth column can be written as $\sum_{k_j} y_j^{k_j} u_i^{k_j}$, so using the multilinearity of the determinant, one gets:

$$\det\left(\frac{1}{1 - u_i y_j}\right) = \sum_{k_1, k_2, \dots} \prod_j y_j^{k_j} \det(u_i^{k_j})$$

Since $\det(u_i^{k_j})$ is antisymmetric one can replace the sum over the unrestricted k_i by a sum over $k_1 > k_2 > \cdots$. The coefficient of $\det(u_i^{k_j})$ becomes $\det(y_i^{k_j})$ in this restricted sum. Dividing by $\Delta(u)\Delta(y)$ one gets the result. Writing now $1/(1 - u_i y_j) = \exp(-\log(1 - u_i y_j))$ and expanding the logarithm, we find:

$$\sum_Y S_Y(u) S_Y(y) = e^{\sum_n \frac{1}{n}(\sum_i u_i^n)(\sum_j y_j^n)} = e^{\sum_n n t_n \tilde{H}_n}$$

where $t_n = 1/n \sum u_i^n$, and $\tilde{H}_n = 1/n \sum y_i^n$. Thus we have the algebraic identity $\exp\left(\sum_n n t_n \tilde{H}_n\right) = \sum_Y \chi_Y(t) \chi_Y(\tilde{H})$. Replacing the variables $\tilde{H}_n$ by the commuting operators $1/n H_n$ we finally obtain:

$$e^{H(t)} = \sum_Y \chi_Y(\tilde{H}) \chi_Y(t), \quad \tilde{H}_n = \frac{1}{n} H_n$$

Taking the matrix element of this identity between $\langle 0|$ and $|Y\rangle$ and remembering eqs. (9.63, 9.65), we get:

$$\chi_Y(t) = \langle 0|e^{H(t)}|Y\rangle = \sum_{Y'} \chi_{Y'}(t)\langle 0|\chi_{Y'}(\tilde{H})|Y\rangle$$

yielding $\langle 0 | \chi_{Y'}(\tilde{H}) | Y \rangle = \delta_{YY'}$. We now compute:

$$\chi_Y(\widehat{\partial}_t) \cdot f(t) = \chi_Y(\widehat{\partial}_t) \sum_{Y'} \zeta_{Y'} \chi_{Y'}(t) = \chi_Y(\widehat{\partial}_t) \sum_{Y'} \langle 0 | e^{H(t)} | Y' \rangle \zeta_{Y'}$$

$$= \sum_{Y'} \langle 0 | \chi_Y(\tilde{H}) e^{H(t)} | Y' \rangle \zeta_{Y'}$$

Letting $t \to 0$ yields the result.   ∎

In particular, for a general tau-function of the form $\tau(t, g) = \langle 0 | e^{H(t)} g | 0 \rangle$ we insert the completeness relation $1 = \sum_Y |Y\rangle\langle Y|$ to get:

$$\tau(t; g) = \sum_Y \zeta_Y(g) \; \chi_Y(t) \tag{9.67}$$

with $\zeta_Y(g) = \langle Y | g | 0 \rangle$. The sum is over all Young diagrams. The previous proposition shows that conversely:

$$\zeta_Y(g) = \chi_Y(\widehat{\partial}_t) \cdot \tau(t; g)|_{t=0} \tag{9.68}$$

This allows us to reconstruct the components of $g|0\rangle$ on the basis $|Y\rangle$ knowing the tau-function. We see that general tau-functions corresponding to any element of $GL(\infty)$ can be expressed on Schur polynomials. There is just one Schur polynomial when g is chosen so that $g|0\rangle = |Y\rangle$.

9.8 From fermions to pseudo-differential operators

We show that the time evolution of the wave function of the KP hierarchy can be written very concisely in terms of pseudo-differential operators. This will be the basis for the study of the KP hierarchy in the next chapter.

We start from eq. (9.61) for the tau-function. We use the fundamental differential property of the elementary Schur polynomials which follows directly from their definition, eq. (9.39):

$$\frac{\partial}{\partial t_n} P_k(t) = \partial^n P_k(t) = P_{k-n}(t) \tag{9.69}$$

where $\partial = \frac{\partial}{\partial t_1}$. In this section we will freely use the notion of pseudo-differential operators explained in Chapter 10.

Recalling that $\xi(t, z) = \sum_n t_n z^n$, we have $z^{-n} e^{\xi(t,z)} = \partial^{-n} e^{\xi(t,z)}$, and eq. (9.61) for the tau-function can be written as:

$$\tau(t; g) = \det \left[F_{mn}^g(t) \right]_{m,n \geq 1} \tag{9.70}$$

where the functions $F_{mn}^g(t)$ are given by:

$$F_{mn}^g(t) = \partial^{-n} F_{m0}^g(t) = \oint dz e^{\xi(z,t)} z^{-n-1} A_m^g(z)$$

We introduce now a pseudo-differential operator, Φ, by adding one line and one column to the determinant in eq. (9.70).

$$\Phi \cdot f \equiv \frac{1}{\tau(t;g)} \det \begin{pmatrix} \partial^{-n} f \\ \partial^{-n} F_{m;0}^g(t) \end{pmatrix}_{\substack{m \geq 1 \\ n \geq 0}} = \left[1 + \sum_{n>0} w_n(t) \partial^{-n}\right] \cdot f \quad (9.71)$$

From this, we define a function $\Psi(t, z)$ by the action of Φ on $e^{\xi(z,t)}$, specifically:

$$\Psi(t, z) = \Phi e^{\xi(z,t)} = \widehat{w}(z,t) e^{\xi(z,t)}; \quad \widehat{w}(z,t) \equiv 1 + \sum_{n>0} w_n(t) z^{-n} \quad (9.72)$$

We show that $\Psi(t, z)$ is the Baker–Akhiezer function. For this, it is sufficient to prove that it is expressed by Sato's formula, eq. (9.77), in terms of the tau-function.

Proposition. *Let $\tau(t;g)$ be defined by eq. (9.70), and let $\Psi(t, z)$ be the function defined in eq. (9.72). We have:*

$$\Psi(z, t) = \frac{\tau(t - [z^{-1}]; g)}{\tau(t;g)} e^{\xi(z,t)} = \frac{\langle 1 | e^{H(t)} \beta(z) g | 0 \rangle}{\langle 0 | e^{H(t)} g | 0 \rangle} \quad (9.73)$$

The second expression is just a rewriting of the first in terms of fermions.

<u>Proof.</u> By definition, $\tau(t;g)\widehat{w}(z,t)$ is the following determinant:

$$\tau(t;g)\widehat{w}(z,t) = \det \begin{pmatrix} 1 & z^{-1} & z^{-2} & \cdots \\ F_{1,0}^g & \partial^{-1} F_{1,0}^g & \partial^{-2} F_{1,0}^g & \cdots \\ F_{2,0}^g & \partial^{-1} F_{2,0}^g & \partial^{-2} F_{2,0}^g & \cdots \\ \vdots & \vdots & \vdots & \cdots \end{pmatrix}$$

Subtracting z^{-1} times the jth column from the $(j+1)^{\text{th}}$ column reduces the first line to $(1, 0, 0, \ldots)$. Expanding the determinant with respect to this first line gives:

$$\tau(t;g)\widehat{w}(z,t) = \det \left[(1 - z^{-1}\partial) \partial^{-n} F_{m,0}^g \right]_{n,m \geq 1}$$

On the other hand, since $e^{\xi(\zeta, t - [z^{-1}])} = (1 - \zeta/z) e^{\xi(\zeta,t)}$ for $|\zeta| < |z|$, we have:

$$F_{m,n}^g(t - [z^{-1}]) = (1 - z^{-1}\partial) F_{m,n}^g(t)$$

hence $\tau(t;g)\widehat{w} = \tau(t - [z^{-1}]; g)$ which proves eq. (9.73). ∎

The pseudo-differential operator Φ has coefficients depending on the times of the KP hierarchy. One can express its time-evolution simply:

Proposition. *We have:*

$$\frac{\partial}{\partial t_n}\Phi = (\Phi\partial^n\Phi^{-1})_+\Phi - \Phi\partial^n = -(\Phi\partial^n\Phi^{-1})_-\Phi \qquad (9.74)$$

where the subscript $+$ refers to the projection on the differential part of the operator, and the subscript $-$ refers to the projection on the negative powers of ∂.

<u>Proof.</u> We introduce differential operators $D_N = \partial^N + \cdots$ which are finite order approximations of Φ:

$$\Phi = \lim_{N\to\infty} D_N\partial^{-N}$$

The operator D_N is defined by truncating the determinant eq. (9.71) to a finite-dimensional determinant:

$$D_N \cdot f = \frac{1}{\widehat{\tau}_N(t;g)}\det\begin{pmatrix} \partial^N f & \partial^{N-1} f & \cdots & f \\ F^g_{1,0} & \partial^{-1}F^g_{1,0} & \cdots & \partial^{-N}F^g_{1,0} \\ \vdots & \vdots & & \vdots \\ F^g_{N,0} & \partial^{-1}F^g_{N,0} & \cdots & \partial^{-N}F^g_{N,0} \end{pmatrix}$$

with $\widehat{\tau}_N(t;g) = \det\left(\partial^{-n}F^g_m\right)_{n,m=1,\ldots,N}$. The time evolution of D_N is easy to find:

$$\frac{\partial}{\partial t_n}D_N = (D_N\partial^n D_N^{-1})_+ D_N - D_N\partial^n$$

Since both side of this equation are differential operators of order $(N-1)$, to prove this equality it is enough to check it on N linearly independent functions. We choose them to be $F^g_{p,N} = \partial^{-N}F^g_{p,0}$, for $p = 1,\ldots,N$, which span the kernel of D_N. We have:

$$\frac{\partial D_N}{\partial t_n} \cdot F^g_{p,N} = -D_N \cdot \frac{\partial F^g_{p,N}}{\partial t_n} = \left(-D_N\partial^n + (D_N\partial^n D_N^{-1})_+ D_N\right) \cdot F^g_{p,N}$$

The first equality follows by applying $\frac{\partial}{\partial t_n}$ to $D_N F^g_{p,N} = 0$. To prove the second equality we substitute $\partial_{t_n}F^g_{p,N} = \partial^n F^g_{p,N}$ and we use again the fact that $F^g_{p,N}$ is in the kernel of D_N to add the second term which vanishes on $F^g_{p,N}$. This term is chosen so that, combined with the first one, we get a differential operator of degree $N-1$. This proves the evolution equation for D_N. Taking the limit $N \to \infty$, we get the evolution equation of Φ. ∎

Note that eq. (9.74) has exactly the form of eq. (3.45) in Chapter 3, but the element of the loop group, $g^{(k)}(\lambda)$, is here replaced by the pseudo-differential operator $\Phi(\partial^{-1})$. This formulation of the KP hierarchy will be studied in detail in Chapter 10.

9.9 The Segal–Wilson approach

Up to now, we have used the description of the Grassmannian embedded into projective space by Plücker coordinates, using the fermionic language. This has the advantage of providing a well-defined computational framework by regularizing potential infinities using normal ordering. In this section we shall look at the Grassmannian as the space of suitable subspaces of a Hilbert space, selected by imposing appropriate functional constraints.

We start again from the space V_∞ with basis $z^n, n \in \mathbb{Z}$ which can be seen as the space of functions on the unit circle $|z| = 1$. With a function on the circle we associate its Fourier expansion $f(z) = \sum_n a_n z^n$. The space V_∞ is turned into a Hilbert space H by introducing the L^2 norm $\int |f|^2$, or what amounts to the same, $\sum_n |a_n|^2$. We can decompose H as a direct sum of subspaces $H^\pm$, where H^+ is generated by $z^n, n \geq 0$ and H^- by $z^n, n < 0$. We consider the set of subspaces W which are comparable to H^+ in the following sense:

$$W \in \mathrm{Gr} \quad \text{iff} \quad \begin{cases} \mathrm{pr}_+ : W \to H^+ \text{ is Fredholm} \\ \mathrm{pr}_- : W \to H^- \text{ is compact} \end{cases}$$

The fact that the projection pr_- is compact means that it is a norm limit of operators with finite-dimensional images. In our fermionic language these finite-dimensional images correspond to states with a finite number of particles. The fact that the projection pr_+ is Fredholm means that the kernel of pr_+ is finite-dimensional, and its image is closed and of finite codimension. In the fermionic language this means a finite number of holes.

Let us illustrate these conditions by an example: assume that W is spanned by $e_{-2}, e_1 + e_{-1}, e_3, e_4, \ldots$. Then $\mathrm{pr}_+ W$ is spanned by $e_1, e_3, e_4, \ldots$ and is of codimension 2 in H^+ since e_0 and e_2 are missing. Its kernel is spanned by e_{-2} and is of dimension 1. Similarly $\mathrm{pr}_- W$ is spanned by e_{-1}, e_{-2} and is of finite dimension 2. Under the Plücker embedding W goes to $e_{-2} \wedge (e_1 + e_{-1}) \wedge e_3 \wedge e_4 \wedge \cdots$ which expands on two semi-infinite wedge products. These two terms have the property Index $(\mathrm{pr}_+) = -1$ where the index of the Fredholm operator pr_+ is defined by

$$\text{Index } (\mathrm{pr}_+) \equiv \dim \text{Ker } \mathrm{pr}_+ - \text{codim Im } \mathrm{pr}_+$$
$$= \text{no. of particles} - \text{no. of holes}$$

Notice that this is also the common fermionic charge of all the above states. In general we have Ker $\mathrm{pr}_+ \subset$ Im pr_-, but the first one is of finite dimension while the second one may become infinite-dimensional under the limiting procedure yielding a general compact operator.

We recall some properties of compact and Fredholm operators. If u is compact and v is continuous (i.e. bounded) then uv and vu are compact. The product of a Fredholm operator and a compact operator is compact. Finally, the sum of a Fredholm operator and a compact operator is Fredholm, and the product of two Fredholm operators is Fredholm. This allows us to consider the group $GL(\infty)$ of matrices having the following block structure on the decomposition $H = H^+ \oplus H^-$:

$$h = \begin{pmatrix} a & b \\ c & d \end{pmatrix} \qquad a, d \text{ Fredholm and } b, c \text{ compact}$$

The product of two such elements is of the same form. Moreover, a group element h acts on the Grassmannian, moving W to hW, where the projections for hW are given by:

$$\begin{pmatrix} \mathrm{pr}'_+ \\ \mathrm{pr}'_- \end{pmatrix} = \begin{pmatrix} a & b \\ c & d \end{pmatrix} \begin{pmatrix} \mathrm{pr}_+ \\ \mathrm{pr}_- \end{pmatrix}$$

and have therefore the required properties.

A subgroup denoted by Γ^+ is of particular interest. An element of Γ^+ is given by the multiplication by an L^2 non-vanishing function h on the unit circle, extending to a non-vanishing analytic function h in the unit disc and normalized by $h(0) = 1$. In particular the expansion $h(z)$ has only positive powers of z. It can be represented in the block form as:

$$h \in \Gamma^+, \qquad h = \begin{pmatrix} a & b \\ 0 & d \end{pmatrix}$$

where a and d are invertible, hence Fredholm, and b is compact. Indeed, let us write $h(z) = 1 + a_1 z + a_2 z^2 + \cdots$ and consider its action by multiplication on $g(z) = \sum_k b_k z^k$. If g has only positive powers, hg has only positive powers, moreover, $1/h$ can be expanded on positive powers, so that a is invertible. It is continuous by the Schwartz inequality and so is Fredholm. If g has only negative powers of z, we remark that

$$h(z)z^{-n} = z^{-n} + a_1 z^{-n+1} + \cdots + a_{n-1} z^{-1} + \sum_{j \geq 0} a_{n+j} z^j$$

We see that the H^- part induces a triangular system with 1 on the diagonal, hence is invertible, so that d is Fredholm. To show that b is compact, consider the truncation $h_N = \sum_{j=0}^{N} a_j z^j$ which is such that $\dim\{\mathrm{pr}_+(h_N g)|g \in H^-\} < N$, i.e. the corresponding b_N is of finite rank. We have $||h - h_N|| \to 0$ when $N \to \infty$ because $||(h - h_N)g|| \leq ||g|| \sum_{j=N}^{\infty} |a_j|^2$ by the Schwartz inequality.

In the following we restrict ourselves to spaces W which are transversal to H^-, i.e. such that $\mathrm{pr}_+ : W \to H^+$ is an isomorphism. Such W have charge 0, and are "small" deformations of the vacuum $|0\rangle$. In this case one can define an operator $A : H^+ \to H^-$ as follows. With any $f \in H^+$ one can associate a unique element $w \in W$ such that $f = \mathrm{pr}_+(w)$. We denote $w = \mathrm{pr}_+^{-1}(f)$ and form

$$f \in H^+ \to A(f) = \mathrm{pr}_-(w), \quad w = \mathrm{pr}_+^{-1}(f)$$

The operator A is compact, and we have

$$(\mathrm{pr}_+|_W)^{-1}(f) = w = f + Af$$

with $Af \in H^-$.

We now want to understand the tau-function in this context. Let us fix an element W of the Grassmannian and let $h \in \Gamma^+$ be given by $h = \exp(\sum_{i>0} t_i z^i)$. This introduces the times of the KP hierarchy. We denote

$$h^{-1} = \begin{pmatrix} a & b \\ 0 & d \end{pmatrix}$$

Definition. *Assuming that W is transverse to H^- so that $(\mathrm{pr}_+|_W)^{-1}$ exists, we set:*

$$\tau_W(h) = \det\left(h \, \mathrm{pr}_+ \, h^{-1} (\mathrm{pr}_+|_W)^{-1}\right) = \det\left(1 + a^{-1}bA\right) \qquad (9.75)$$

To understand the second formula, consider $f \in H^+$. Then we have $\mathrm{pr}_+ h^{-1}(\mathrm{pr}_+|_W)^{-1}(f) = af + bAf \in H^+$ and $h \, \mathrm{pr}_+ \, h^{-1}(\mathrm{pr}_+|_W)^{-1}(f) = (1 + a^{-1}bA)f$. We have here an operator from H^+ to H^+ of the form $1 + $ compact and we want to take its determinant. This is possible if the operator $a^{-1}bA$ is of trace class, which is ensured if h is sufficiently regular on the unit circle.

Note that the operator $a : H^+ \to H^+$ is triangular with 1 on the diagonal, hence one can set $\det a = 1$. If it were allowed to write $\det MN = \det M \cdot \det N$ for infinite-dimensional matrices, we could content ourselves with defining $\tau_W = \det(a + bA)$. The definition eq. (9.75) performs a regularization of this too naive expression.

It is important to show that this definition agrees with the previous construction in eq. (9.34). Recall the expression eq. (9.60) of the Plücker embedding of some element W of the Grassmannian. The two definitions agree if we identify

$$A_s^g(z) = (\mathrm{pr}_+|_W)^{-1}(z^s)$$

Hence $A_0(z) \wedge A_1(z) \wedge \cdots$ represents the Plücker embedding of W. By definition the action of h^{-1} on W is $A_s^g(z) \to e^{\xi(t,z)}A_s^g(z)$, which under the Plücker embedding is represented by multiplication by $e^{H(t)}$ due to eq. (9.43). Next the projection P_+ becomes $|0\rangle\langle 0|$ and the multiplication by h is achieved by multiplying by $e^{-H(t)}$, so that:

$$h\,\mathrm{pr}_+\,h^{-1}(\mathrm{pr}_+|_W)^{-1}\,(H^+) \xrightarrow{\text{Plücker}} e^{-H(t)}|0\rangle\langle 0|e^{H(t)}A_0(z) \wedge A_1(z) \wedge \cdots$$

Taking the scalar product with $\langle 0|$ produces the determinant of the operator $h\,\mathrm{pr}_+\,h^{-1}(\mathrm{pr}_+|_W)^{-1}$. Since $\langle 0|e^{-H(t)}|0\rangle = 1$ due to normal ordering of H, we reproduce exactly eq. (9.61).

When M, N are of the form $1 + m, 1 + n$ with m and n of trace class, one is allowed to write $\det MN = \det M \det N$. For $h \in \Gamma^+$ one can show that $a^{-1}bA$ is of trace class. In particular one obtains:

$$\tau_W(h_1 h_2) = \tau_W(h_1)\tau_{h_1^{-1}W}(h_2) \tag{9.76}$$

Indeed,

$$\tau_W(h_1)\tau_{h_1^{-1}W}(h_2)$$
$$= \det(h_2\mathrm{pr}_+h_2^{-1}(\mathrm{pr}_+|_{h_1^{-1}W})^{-1}) \cdot \det(\mathrm{pr}_+h_1^{-1}(\mathrm{pr}_+|_W)^{-1}h_1)$$
$$= \det(h_1 h_2\mathrm{pr}_+h_2^{-1}(\mathrm{pr}_+|_{h_1^{-1}W})^{-1}\mathrm{pr}_+h_1^{-1}(\mathrm{pr}_+|_W)^{-1}) = \tau_W(h_1 h_2)$$

This is because if $f \in H^+$ then $w = (\mathrm{pr}_+|_W)^{-1}f \in W$ so that $h_1^{-1}w \in h_1^{-1}W$. Under pr_+ this gives some $g \in H^+$ which under $(\mathrm{pr}_+|_{h_1^{-1}W})^{-1}$ reproduces $h_1^{-1}w = h_1^{-1}(\mathrm{pr}_+|_W)^{-1}f$.

To define Baker–Akhiezer functions, we assume that $h \in \Gamma^+$ and is such that $h^{-1}W$ is transverse to H^-.

Definition. *Let* $h(z;t) = \exp(\sum_{i>0} t_i z^i) = e^{\xi(z,t)} \in \Gamma^+$. *The Baker–Akhiezer function* $\Psi_W(h, z)$ *is the unique function such that* $h^{-1}\Psi_W(h, z)$ *is the inverse image under* $\mathrm{pr}_+|_{h^{-1}W}$ *of the constant function* $1 \in H^+$.

This means that $h^{-1}\Psi_W(h, z) \in h^{-1}W$ and can be written as $1 + \sum_{i=1}^{\infty} a_i(h)z^{-i}$. That is to say, the Baker–Akhiezer function is the unique function $\Psi_W(h, z) \in W$ having the form:

$$\Psi_W(h, z) = h(z;t)\left(1 + \sum_{i=1}^{\infty} a_i(h)z^{-i}\right)$$

We see that Ψ_W has an essential singularity at $z = \infty$. This is one of the essential feature of the Baker–Akhiezer functions. The Baker–Akhiezer function Ψ_W becomes a function of t and z and can be explicitly expressed in terms of the tau-function by means of Sato's formula:

Proposition. *We have*

$$\Psi_W(t, z) = \frac{\tau_W(t - [z^{-1}])}{\tau_W(t)} e^{\xi(t,z)} \qquad (9.77)$$

where the notation $t - [z^{-1}]$ *refers to the substitution of* t_n *by* $t_n - z^{-n}/n$.

<u>Proof.</u> To avoid notational conflicts, in this proof we denote by ζ the current variable on the circle (previously denoted by z). The group element $h(\zeta; t - [z^{-1}])$ is given by $\exp(\sum_{n>0}(t_n - z^{-n}/n)\zeta^n) = h(\zeta; t)q_z$ where $q_z = 1 - \zeta/z \in \Gamma^+$. Equation (9.77) is equivalent to

$$e^{-\xi(t,z)}\Psi_W(t, z) = \frac{\tau_W(hq_z)}{\tau_W(h)}$$

Since $\tau_W(hq_z) = \tau_W(h)\tau_{h^{-1}W}(q_z)$, we have to show that $e^{-\xi(z,t)}\Psi_W(t, z) = \tau_{h^{-1}W}(q_z)$. By definition, $e^{-\xi(z,t)}\Psi_W(t, z) = (\mathrm{pr}_+|_{h^{-1}W})^{-1}(1)$. Replacing $h^{-1}W$ by W, we have to show the equality of the two functions of z: $(\mathrm{pr}_+|_W)^{-1}(1) = \tau_W(q_z) = \det(1 + a^{-1}bA)$, where the operators a and b are the ones appearing in the block representation of q_z^{-1} and A is the operator induced by W. Since $q_z^{-1}(\zeta) = \sum_{n\geq 0} \zeta^n/z^n$, we have:

$$b(\zeta^{-n}) = \mathrm{pr}_+\left(\frac{\zeta^{-n}}{1 - \zeta/z}\right) = z^{-n}q_z^{-1}(\zeta)$$

while the action of a^{-1} on this element of H^+ is simply represented by the multiplication by $q_z(\zeta)$. Hence for any element $g \in H^-$ we have $a^{-1}bg = g(z) \cdot 1 \in H^+$, i.e. the operator is of rank 1, so that $\det(1 + a^{-1}bA) = 1 + \mathrm{Tr}(a^{-1}bA)$. Since the image is spanned by the function 1, it is enough to compute the action of $a^{-1}bA$ on 1 to get the trace. But by definition A sends the basis element 1 on the negative power part of $(\mathrm{pr}_+|_W)^{-1}(1)$ denoted by $f(\zeta)$, so that $1 + \mathrm{Tr}(a^{-1}bA) = 1 + f(z)$.   ∎

Let us explain how the algebro-geometric solutions of KP fit into this setting (see Chapter 10 for a description of these solutions). Let Γ be a compact Riemann surface of genus g and $\mathcal{L}$ be a line bundle on Γ, see Chapter 15. Fix a puncture x_∞ on Γ and let z^{-1} be a local parameter around x_∞. Let D_∞ be a small disc around x_∞. Sections of $\mathcal{L}$ locally appear as functions on D_∞. Consider also the open set Γ_0 which is the complement of the disc D_∞. The two open sets D_∞ and Γ_0 cover Γ.

With this set of data one associates an element W of the Grassmannian such that $w \in W$ if w is the boundary value on the circle of a holomorphic section of the restriction of $\mathcal{L}$ on Γ_0.

One can show that pr_- is compact. To show that pr_+ is Fredholm one first shows that $\mathrm{Ker}(\mathrm{pr}_+ : W \to H^+) = H^0(\Gamma, \mathcal{L}_\infty)$, where $\mathcal{L}_\infty = \mathcal{L} - [x_\infty]$ is the difference of the line bundle $\mathcal{L}$ and the point bundle at x_∞, i.e. a section of $\mathcal{L}_\infty$ arises from a section of $\mathcal{L}$ vanishing at x_∞. Recall that $H^0(\Gamma, \mathcal{L}_\infty)$ is the set of global holomorphic sections of $\mathcal{L}_\infty$. A function belongs to $\mathrm{Ker}(\mathrm{pr}_+ : W \to H^+)$ if and only if it has only strictly negative powers of z, hence extends to the interior of D_∞ and vanishes at x_∞. By definition it extends to Γ_0 so providing a global section of $\mathcal{L}_\infty$.

Next we show that $H^+/\mathrm{pr}_+ W = H^1(\Gamma, \mathcal{L}_\infty)$. It is a well-known fact that the *non-compact* Riemann surface Γ_0 has no sheaf cohomology $H^1(\Gamma_0, \mathcal{L}_\infty) = 0$ and it is obviously the same for D_∞. Hence any non-vanishing element in $H^1(\Gamma, \mathcal{L}_\infty)$ comes from some analytic function $\phi_{0\infty}$ on the annulus $\Gamma_0 \cap D_\infty$ which cannot be written as $\phi_0 - \phi_\infty$, where ϕ_0 extends to an analytic section of $\mathcal{L}$ on Γ_0 and ϕ_∞ is an analytic function on D_∞ vanishing at x_∞. But the part of $\phi_{0\infty}$ with strictly negative powers of z extends uniquely to a function ϕ_∞ vanishing at x_∞. So $H^1(\Gamma, \mathcal{L}_\infty)$ is isomorphic to the set of analytic functions on S^1 with only non-negative powers of z modulo those which extend to sections of $\mathcal{L}$. This is precisely the definition of $H^+/\mathrm{pr}_+ W$.

Thus we arrived at the conclusion that the Fredholm index of pr_+ is given by the Riemann–Roch theorem:

$$\mathrm{Index}\,(W) = \dim H^0(\mathcal{L}_\infty) - \dim H^1(\mathcal{L}_\infty) = 1 - g + c(\mathcal{L}_\infty) = c(\mathcal{L}) - g$$

where $c(\mathcal{L})$ is the Chern class of $\mathcal{L}$ and g is the genus of the Riemann surface. Recall that $c(\mathcal{L}_\infty) = c(\mathcal{L}) - 1$ (see Chapter 15). In particular, in the interesting case where W is transverse to H^- the index of W vanishes, which needs $c(\mathcal{L}) = g$.

In practice one takes a puncture x_∞ and a set of g points in generic position and one considers meromorphic functions with poles at these g points and an essential singularity at x_∞. These data define a unique W transverse to H^- in the Grassmannian. Elements of W are boundary values on the circle of meromorphic functions on Γ_0 with poles only at these g points. The Baker–Akhiezer function specified by W has an essential singularity at x_∞ and g poles. It identifies with those defined in Chapter 5.

References

[1] R. Hirota, Exact solution of the Korteweg–de Vries equation for multiple collisions of solitons. *Phys. Rev. Lett.* **27** (1971) 1192.

[2] M. Sato and Y. Sato, Solitons equations as dynamical systems on infinite-dimensional Grassmann manifolds. *Lect. Notes in Num. Appl. Anal.* **5** (1982) 259, or *Proc. U.S.–Japan Seminar Nonlinear PDE in Applied Science, Tokyo 1982*, Ed. Lax, Fujita, 259–271, North Holland/Kinokuniya.

[3] E. Date, M. Kashiwara, M. Jimbo and T. Miwa, Transformation groups for soliton equations. in *Proceedings of RIMS symposium, Kyoto 1981*. World Scientific (1983) 39–119.

[4] M. Jimbo and T. Miwa, Solitons and infinite dimensional Lie algebras. *RIMS* **19** (1983) 943–1001.

[5] G. Segal and G. Wilson, Loop groups and equations of KdV type. *Publ. Math. I.H.E.S.* **61** (1985) 5–65.

[6] V. Kac, *Infinite-dimensional Lie algebras*. Cambridge University Press (1985).

[7] V. Kac and A. Raina, *Bombay lectures on highest weight representations of infinite-dimensional Lie algebras*. World Scientific (1987).

[8] L. Dickey, On the tau-function of matrix hierarchies of integrable equations. *J. Math. Phys.* **32** (1991) 2996–3002.

[9] W. Fulton and J. Harris, *Representation theory*. Springer (1991).

[10] C. Itzykson and J.-B. Zuber, Combinatorics of the modular group II: the Kontsevich integrals. *Int. J. Mod. Phys.* **A7** (1992) 5661–5705.

10

The KP hierarchy

In the previous chapter we showed that the equations of the KP hierarchy can be written as:

$$\partial_{t_n}\Phi = -(\Phi\partial^n\Phi^{-1})_-\Phi$$

where Φ is a pseudo-differential operator. This is identical to the standard form of the equations of an integrable hierarchy, but we are dealing here with the algebra of pseudo-differential operators, instead of a loop algebra. In this chapter, we explain this setting and investigate the corresponding hierarchy. We show that the general solution can be expressed with the Grassmannian tau-function. With any Riemann surface one can associate particular finite-zone solutions. More generally, we construct solutions corresponding to slow modulations of the algebro-geometric solutions following the Whitham procedure. We also present the reduction of this hierarchy to the generalized KdV equations, and discuss their Poisson structures. We show that these Poisson stuctures can be obtained by Hamiltonian reduction from the Kostant–Kirillov bracket on a Kac–Moody algebra.

10.1 The algebra of pseudo-differential operators

We briefly expose the theory of pseudo-differential operators first introduced in this context by Gelfand and Dickey.

The algebra of differential operators is the algebra generated by $\mathbb{C}$-valued functions of one variable x and the derivation symbol ∂, with the usual Leibnitz rule, $\partial.a = a.\partial + (\partial a)$, where (∂a) means the derivative $(\partial_x a)(x)$ of the function $a(x)$. This defines the multiplication law between the symbol ∂ and the functions. An element in this algebra is a finite sum $A = \sum_{i=0}^{N} a_i\partial^i$, with N finite but arbitrary. The coefficients a_i are functions of x. To define the algebra of pseudo-differential operators, we extend the algebra of differential operators by introducing the "integration" symbol, ∂^{-1} and its powers, with the following algebraic rules:

$$\partial^{-1}\partial = \partial\partial^{-1} = 1$$

$$\partial^{-1}a = \sum_{i=0}^{\infty}(-1)^i(\partial^i a)\partial^{-i-1} \tag{10.1}$$

The algebra of pseudo-differential operators consists of elements which are semi-infinite sums of the form $A = \sum_{-\infty}^{N} a_i\partial^i$. Equations (10.1) define the multiplication by ∂^{-1} in the pseudo-differential algebra since they allow us to push all ∂^{-1} symbols to the right. Here we don't have to deal with the convergence questions involved in this reshuffling, since only a finite number of terms appear at each order in ∂^{-i}. This rule is motivated by the integration by parts formula. Symbolically we have:

$$\partial^{-1}(a.u) = \int dx\ au = \int dx\ a\partial(\partial^{-1}u) = a\partial^{-1}u - \int dx\ (\partial a)\partial^{-1}u$$

$$= a\partial^{-1}u - (\partial a)\partial^{-2}u + \int dx\ (\partial^2 a)\partial^{-2}u$$

and so on. The algebra of pseudo-differential operators is an associative algebra with a unit. It possesses a natural anti-homomorphism, that is $(AB)^* = B^*A^*$, which we call the formal adjoint, defined by:

$$(a\ \partial^i)^* \equiv (-\partial)^i\ a \tag{10.2}$$

for any function a. We summarize these facts in the following definitions:

Let $\mathcal{P} = \left\{A = \sum_{-\infty}^{N} a_i\partial^i\right\}$ be the set of formal pseudo-differential operators in one variable. We denote by $\mathcal{P}_+ = \left\{A = \sum_{i=0}^{N} a_i\partial^i\right\}$ the subalgebra of differential operators, and by $\mathcal{P}_- = \left\{A = \sum_{-\infty}^{-1} a_i\partial^i\right\}$ the subalgebra of integral operators. We have the direct sum decomposition of $\mathcal{P}$ as a vector space:

$$\mathcal{P} = \mathcal{P}_+ \oplus \mathcal{P}_-$$

Notice that $\mathcal{P}$ is naturally a Lie algebra. $\mathcal{P}_+$ and $\mathcal{P}_-$ are Lie subalgebras, but $\mathcal{P}_+$ and $\mathcal{P}_-$ do not commute.

For $A \in \mathcal{P}$, we define its residue, denoted by $\mathrm{Res}_\partial A$, as the coefficient of ∂^{-1} in A,

$$\mathrm{Res}_\partial A \equiv a_{-1}(x) \tag{10.3}$$

On $\mathcal{P}$ there exists a natural linear form:

Proposition. *The algebra $\mathcal{P}$ is equipped with a linear form, denoted by $\langle\ \rangle$, called the Adler trace. It is defined for any element $A = \sum_{-\infty}^{N} a_i \partial^i$ by:*

$$\langle A \rangle = \int dx\, \mathrm{Res}_\partial A = \int dx\, a_{-1}(x) \qquad (10.4)$$

This linear form satisfies the fundamental trace property $\langle AB \rangle = \langle BA \rangle$, hence defines an ad*-invariant non-degenerate scalar product on $\mathcal{P}$ by: $(A, B) = \langle AB \rangle$. Using this bilinear form we have the duality: $\mathcal{P}_+^* = \mathcal{P}_-$.*

<u>Proof.</u> There is a unique way of writing a pseudo-differential operator in the form $A = \sum_{-\infty}^{N} a_i \partial^i$, i.e. with all ∂^i on the right. Let us prove the trace property. It is sufficient to verify it on operators of the form $A = a\ \partial^k$ and $B = b\ \partial^j$. If k and j are both positive or both strictly negative, we clearly have $\langle AB \rangle = \langle BA \rangle = 0$. Thus we take $A = a\ \partial^k$ and $B = b\ \partial^{-j-1}$ with $k,\ j \geq 0$. Using the relation:

$$\partial^{-j-1} a = \sum_{v=0}^{\infty} (-1)^v \binom{j+v}{v} (\partial^v a)\, \partial^{-j-1-v} \qquad (10.5)$$

which is shown by induction, starting from the definition relation eq. (10.1), and the identity between binomial coefficients

$$\binom{j+1+\mu}{\mu} = \sum_{\nu=0}^{\mu} \binom{j+\nu}{\nu}$$

we get:

$$\langle BA \rangle = (-1)^{k-j} \binom{k}{j} \int dx\ \left(b\ \partial^{k-j} a \right) = \binom{k}{j} \int dx\ \left(a\ \partial^{k-j} b \right)$$

Similarly, using the Leibnitz rule, $AB = \sum_{v=0}^{\infty} \binom{k}{v} (a\ \partial^v b)\, \partial^{k-j-v-1}$, we find that $\langle AB \rangle$ is given by:

$$\langle AB \rangle = \binom{k}{j} \int dx\ \left(a\ \partial^{k-j} b \right)$$

Clearly, $\langle AB \rangle$ and $\langle BA \rangle$ coincide.

The invariance of the scalar product means $(A, [B, C]) = ([A, B], C)$. It follows from the trace property.

We already noticed that $\mathcal{P}_+$ and $\mathcal{P}_-$ are isotropic with respect to the trace $(\mathcal{P}_\pm, \mathcal{P}_\pm) = 0$. To check that $\mathcal{P}_+^* = \mathcal{P}_-$, we consider in $\mathcal{P}_+$ elements

$a_i \partial^i$, $i = 0, 1, \ldots, \infty$. They are paired with elements $\partial^{-i-1} b_i$ in $\mathcal{P}_-$ since:

$$\langle \partial^{-i-1} b_i, a_j \partial^j \rangle = \delta_{ij} \int dx \, (a_i b_i)$$

Choosing the coefficients a_i and b_i in an orthonormal basis under $\int dx \, ab$, we get dual bases of $\mathcal{P}_+$ and $\mathcal{P}_-$. ∎

10.2 The KP hierarchy

We introduce the KP flows by applying the Adler–Kostant–Symes construction to the Lie algebra of pseudo-differential operators.

Consider the formal group $G = \exp(\mathcal{P}_-)$, called the Volterra group. We have $G \sim 1 + \mathcal{P}_-$ because powers of elements in $\mathcal{P}_-$ are in $\mathcal{P}_-$. Let Φ be an element of G:

$$\Phi = 1 + \sum_{i=1}^{\infty} w_i \partial^{-i} \in (1 + \mathcal{P}_-) \tag{10.6}$$

The element Φ has an inverse because, writing $\Phi^{-1} = 1 + \sum_1^\infty w_i' \partial^{-i}$ and demanding $\Phi^{-1} \Phi = 1$, one recursively computes the coefficients w_i'. One finds that they are of the form $w_i' = -w_i + p_i(w_1, \ldots, w_{i-1})$, where p_i is a polynomial in its arguments and their derivatives (up to order $i - 2$). For example:

$$w_1' = -w_1, \quad w_2' = -w_2 + w_1^2$$
$$w_3' = -w_3 - w_1(\partial w_1) + 2 w_1 w_2 - w_1^3$$

The left and right inverses are identical.

In the Adler–Kostant–Symes scheme we consider the decomposition of the Lie algebra $\mathcal{P} = \mathcal{P}_+ + \mathcal{P}_-$ and introduce a second Lie algebra structure on the underlying vector space, called $\mathcal{P}_R$. In $\mathcal{P}_R$, $\mathcal{P}_+$ and $\mathcal{P}_-$ commute, see eq. (4.13) in Chapter 4. The Lie algebra $\mathcal{P}_R$ acts on the Volterra group by

$$\delta_A \Phi = (\Phi A \Phi^{-1})_+ \Phi - \Phi A_+ = -(\Phi A \Phi^{-1})_- \Phi + \Phi A_-$$

for any $A = A_+ + A_- \in \mathcal{P}_R$. We have $[\delta_A, \delta_B] = \delta_{[A,B]_R}$. See eq. (14.35) in Chapter 14. The signs are slightly different from those in that chapter because we use here the decomposition $A = A_+ + A_-$ instead of $A = A_+ - A_-$ in order to conform ourselves with the usual conventions in KP theory.

If $A \in A_+$, the formula simplifies to $\delta_A \Phi = -(\Phi A \Phi^{-1})_- \Phi$. The KP flows are defined by taking A in a the *Abelian* subalgebra ∂^k for $k > 0$.

Definition. *For any* $\Phi \in (1 + \mathcal{P}_-)$, *define the* k^{th} *KP flow by:*

$$\partial_{t_k} \Phi = - \left(\Phi \cdot \partial^k \cdot \Phi^{-1} \right)_- \Phi \qquad (10.7)$$

These flows coincide with eq. (9.74) in Chapter 9. By construction they commute, but it is instructive to check this essential commutativity property directly.

Proposition. *The KP flows* ∂_{t_k} *all commute.*

Proof. Consider the pseudo-differential operators $\Theta^k = \Phi \partial^k \Phi^{-1}$. One has:

$$\partial_{t_k} \partial_{t_l} \Phi - \partial_{t_l} \partial_{t_k} \Phi = \left([\Theta^l_-, \Theta^k] - [\Theta^k_-, \Theta^l] + [\Theta^k_-, \Theta^l_-] \right)_- \Phi$$

Replace $\Theta^k = \Theta^k_- + \Theta^k_+$ and similarly for Θ^l. One gets:

$$[\partial_{t_k}, \partial_{t_l}]\Phi = \left([\Theta^l_+, \Theta^k_-]_- + [\Theta^l_-, \Theta^k_+]_- + [\Theta^l_-, \Theta^k_-]_- \right)\Phi = [\Theta^l, \Theta^k]_- \Phi$$

The result follows because $[\Theta^l, \Theta^k] = 0$. ∎

From Φ, we construct the pseudo-differential operator:

$$Q = \Phi \cdot \partial \cdot \Phi^{-1}, \quad Q = \partial + \sum_{i=1}^{\infty} q_{-i} \, \partial^{-i}$$

It is easy to check that there is no ∂^0 term. Given Q one can reconstruct Φ up to some constants:

Proposition. *The pseudo-differential operator* Φ *is determined by* Q *up to the transformation* $\Phi \to \Phi C$, *where* $C = 1 + \sum_{i=1}^{\infty} c_i \partial^{-i}$ *and the* c_i *are constants independent of* x.

Proof. Obviously $\Phi \to \Phi C$, with C independent of x, leaves Q invariant. Using the expression of Φ^{-1}, we find immediately $q_{-i} = -(\partial w_i) + h_i(w_1, \ldots, w_{i-1})$, where h_i is a differential polynomial in its arguments. The derivatives are at most of order $(i - 1)$. Conversely, w_i is recursively determined by q_{-i} as $w_i = \int^x (h_i - q_{-i}) dx$, up to an integration constant. These constants can be absorbed in C. ∎

On the pseudo-differential Q the KP evolution equations take the Lax form:

Proposition. *The time evolutions of Q are given by:*

$$\partial_{t_k} Q = [\, B_k \,,\, Q \,] \quad \text{with} \quad B_k = \left(Q^k\right)_+ \tag{10.8}$$

Moreover, the differential operators B_k satisfy the zero curvature condition:

$$\partial_{t_k} B_l - \partial_{t_l} B_k - [B_k, B_l] = 0$$

<u>Proof.</u> We have to compute $\partial_{t_k} Q$. Using the definition eq. (10.7) written as $\partial_{t_k} \Phi = -(Q^k)_- \Phi$, we find:

$$\partial_{t_k} Q = \partial_{t_k}(\Phi \partial \Phi^{-1}) = -\left[\left(Q^k\right)_- , Q\right] = \left[\left(Q^k\right)_+ , Q\right]$$

The zero-curvature condition follows from a direct computation:

$$\partial_{t_k} B_l - \partial_{t_l} B_k = \partial_{t_k}\left(Q^l\right)_+ - \partial_{t_l}\left(Q^k\right)_+ = \left([B_k, Q^l] - [B_l, Q^k]\right)_+$$

Therefore, using the decomposition $Q^k = \left(Q^k\right)_+ + \left(Q^k\right)_-$, we obtain:

$$\partial_{t_k} B_l - \partial_{t_l} B_k - [B_k, B_l] = \left([B_k, Q^l] - [B_l, Q^k] - [B_k, B_l]\right)_+$$

$$= -[B_k - Q^k, B_l - Q^l]_+ = -\left[\left(Q^k\right)_- , \left(Q^l\right)_-\right]_+ = 0$$

∎

The Lax equation eq. (10.8) shows that Q is the analogue of the Lax matrix L, and the differential operator B_k is the analogue of M_k. The pseudo-differential operator Φ is the analogue of $g(\lambda)$ in the loop–algebra situation of Chapter 3. In particular the evolution equation eq. (10.7) is analogous to eq. (3.50) for g. We will keep, however, the traditional notations in this chapter. This analogy can be pursued to get conserved quantities as traces of powers of the Lax matrix if we replace the ordinary trace by the Adler trace.

Proposition. *The quantities $H_k = \langle Q^k \rangle$ are conserved.*

<u>Proof.</u> Using eq. (10.8) we have $\partial_{t_l} H_k = \langle [B_l, Q^k] \rangle$ which vanishes due to the cyclicity of Adler's trace. ∎

Remark 1. The equations (10.8) are consistent in the sense that $[B_k, Q] \in \mathcal{P}_-$. This is because $[B_k, Q] = [(Q^k)_+, Q] = [Q^k - (Q^k)_-, Q] = -[(Q^k)_-, Q]$ which expands

on the negative powers ∂^{-j}, $j \geq 1$. Hence the Lax equations (10.8) produce non-linear equations of motion for the coefficients of Q, i.e. for the functions $\{q_{-i}\}$.

Remark 2. The first KP-flow ∂_1 is identified with ∂, because we have $Q_+ = \partial$ so that the first flow reads:

$$\partial_1 Q = [(Q)_+, Q] = [\partial, Q] = \sum_{i=1}^{\infty} (\partial q_{-i}) \partial^{-i}$$

Therefore, the KP-time t_1 is naturally identified with the variable x introduced in the definition of the algebra $\mathcal{P}$.

Example. To illustrate these formulae we compute the first few equations of motion. First we have:

$$(Q^2)_+ = \partial^2 + 2q_{-1}, \quad (Q^3)_+ = \partial^3 + 3q_{-1}\partial + 3(\partial q_{-1}) + 3q_{-2}$$

The time evolution with respect to t_2 reads:

$$\partial_{t_2} q_{-1} = \partial^2 q_{-1} + 2\partial q_{-2}$$
$$\partial_{t_2} q_{-2} = \partial^2 q_{-2} + 2\partial q_{-3} + 2q_{-1}\partial q_{-1}$$

$$\vdots$$

Similarly, the time evolution with respect to t_3 of q_{-1} is given by:

$$\partial_{t_3} q_{-1} = \partial^3 q_{-1} + 3\partial^2 q_{-2} + 3\partial q_{-3} + 6q_{-1}\partial q_{-1}$$

Eliminating q_{-2} and q_{-3} between these equations and renaming $u = -2q_{-1}$, one gets:

$$3\partial_{t_2}^2 u = \partial(4\partial_{t_3} u + 6u\partial u - \partial^3 u)$$

This is the KP equation, see eq. (8.58) in Chapter 8. It is the first of an infinite hierarchy of non-linear partial differential equations for $q_{-1}, q_{-2}, \ldots$.

10.3 The Baker–Akhiezer function of KP

By analogy with the Lax situation, we consider eigenvectors of the operator Q, together with their time evolutions under the KP flows, that is we look for an eigenfunction $\Psi(t, z)$ of Q such that:

$$(Q - z)\Psi = 0, \quad Q = \Phi \partial \Phi^{-1}$$
$$(\partial_{t_m} - B_m)\Psi = 0, \quad B_m = (Q^m)_+ \tag{10.9}$$

For $m = 2$, we find

$$(\partial_{t_2} - \partial^2 + u)\Psi = 0 \tag{10.10}$$

In the algebro-geometric case such eigenfunctions were shown to be Baker–Akhiezer functions and we will continue to call them by this name. In order to make connection with previous expressions of the Baker–Akhiezer function, we define the action of pseudo-differential operators on exponentials:

$$\partial^i e^{zx} = z^i e^{zx}, \quad \text{for all } i \in \mathbb{Z}$$

This extends to the action of any pseudo-differential operator by writing it first in normal form with all ∂^i on the right. This definition is compatible with the algebra structure, in particular $(\Phi_1\Phi_2)\, e^{zx} = \Phi_1\left((\Phi_2)\, e^{zx}\right)$.

Proposition. *The Baker–Akhiezer function $\Psi(t,z)$ obeying eqs. (10.9) can be written as:*

$$\Psi(t,z) = \Phi e^{\xi(t,z)} = (1 + w_1 z^{-1} + w_2 z^{-2} + \cdots)\, e^{\xi(t,z)} \equiv \widehat{w}(t,z)\, e^{\xi(t,z)}$$
$$(10.11)$$

where $\xi(t,z) = \sum_{i=1}^{\infty} t_i z^i$. This defines

$$\widehat{w}(t,z) = 1 + w_1 z^{-1} + w_2 z^{-2} + \cdots$$

Proof. The expansion $\widehat{w}(t,z)$ results clearly from the definition of the action of Φ on $\exp \xi(t,z)$, noting that $t_1 = x$. Then $Q\Psi = (\Phi\partial\Phi^{-1})\Phi e^{\xi(t,z)} = z\Psi$. Similarly, using

$$\partial_{t_m} e^{\xi(t,z)} = \partial^m e^{\xi(t,z)} = z^m e^{\xi(t,z)}$$

the evolution of Ψ with respect to t_m is given by:

$$\partial_{t_m}\Psi = \partial_{t_m}(\Phi e^{\xi(t,z)}) = (\partial_{t_m}\Phi)e^{\xi(t,z)} + \Phi\left(\partial_{t_m}e^{\xi(t,z)}\right)$$
$$= -(Q^m)_- \Phi e^{\xi(t,z)} + \Phi\left(\partial^m e^{\xi(t,z)}\right) = \left(-(Q^m)_- + Q^m\right)\Phi e^{\xi(t,z)}$$
$$= (Q^m)_+ \Psi$$

In the last equality, we have written $\Phi\partial^m e^\xi = (\Phi\partial^m\Phi^{-1})\Phi e^\xi = Q^m\Psi$ and we used the decomposition $Q^m = (Q^m)_+ + (Q^m)_-$. ∎

It is useful to introduce the adjoint Baker–Akhiezer function by:

$$\Psi^* = (\Phi^*)^{-1}\, e^{-\xi(t,z)}$$

where Φ^* is the formal adjoint of the pseudo-differential operator Φ, defined in eq. (10.2). This adjoint function satisfies the adjoint system:

Proposition. *The adjoint Baker–Akhiezer function obeys:*

$$(Q^* - z)\Psi^* = 0, \quad Q^* = -(\Phi^*)^{-1}\partial\phi^*$$
$$(\partial_{t_m} + B_m^*)\Psi^* = 0, \quad B_m^* = (Q^m)_+^*$$

<u>Proof</u>. Since the formal adjoint is an antihomomorphism we have $Q^* = \left(\Phi \cdot \partial \cdot \Phi^{-1} \right)^* = -(\Phi^*)^{-1} \cdot \partial \cdot \Phi^*$. As a consequence, we have:

$$Q^* \ \Psi^* = -(\Phi^*)^{-1} \ \partial \ e^{-\xi(t,z)} = z \ \Psi^*$$

Similarly, we have:

$$\begin{aligned} \partial_{t_m} \Psi^* &= \left(\partial_{t_m} (\Phi^*)^{-1} \right) \ e^{-\xi(t,z)} + (\Phi^*)^{-1} \ \partial_{t_m} \ e^{-\xi(t,z)} \\ &= \left((Q^m)^*_- - (Q^m)^* \right) \ \Psi^* = -(Q^m)^*_+ \ \Psi^* \end{aligned}$$

∎

Note that, compared to the algebro-geometric Baker–Akhiezer functions, the puncture is at $z = \infty$ where the exponential factor $\exp \xi(t,z)$ has an essential singularity, while $\hat{w}(x, z)$ is formally regular.

The fact that the Baker–Akhiezer functions are solutions of the linear system eq. (10.9) implies that they satisfy an important bilinear identity:

Theorem. *The following bilinear identity holds for all* $(i_1, \dots, i_m)$, $i_j \geq 0$:

$$\oint \frac{dz}{2i\pi} (\partial^{i_1}_{t_1} \cdots \partial^{i_m}_{t_m} \Psi(t, z)) \cdot \Psi^*(t, z) = 0 \tag{10.12}$$

It can be rewritten more compactly as:

$$\oint \frac{dz}{2i\pi} \Psi(t, z) \cdot \Psi^*(t', z) = 0, \quad \forall t, t' \tag{10.13}$$

The integrals over z are residues around $z = \infty$, i.e. integrals on big circles around $z = \infty$.

<u>Proof</u>. Notice that the integrands are meromorphic functions around ∞ because the essential singularities cancel. We first need a formula expressing the Adler residue, eq. (10.3), of a product of two pseudo-differential operators $D = \sum_i d_i \partial^i$ and $F = \sum_i f_i \partial^i$:

Lemma.

$$\oint \frac{dz}{2i\pi} (De^{zx})(Fe^{-zx}) = \mathrm{Res}_\partial (DF^*) \tag{10.14}$$

<u>Proof</u>. The left-hand side is the coefficient of z^{-1} in the integrand:

$$\oint \frac{dz}{2i\pi} (De^{zx})(Fe^{-zx}) = \oint \frac{dz}{2i\pi} \sum_i d_i z^i \sum_j f_j (-z)^j = \sum_j (-1)^j d_{-j-1} f_j$$

Similarly, we compute the Adler residue, i.e. the coefficient of ∂^{-1} in (DF^*), using $F^* = \sum_i (-\partial)^i f_i$:

$$\text{Res}_\partial (DF^*) = \text{Res}_\partial \left(\sum_{i,j} d_i \partial^i (-\partial)^j f_j\right) = \sum_j (-1)^j d_{-j-1} f_j$$

∎

We can now prove the identity eq. (10.12). Since $\partial_{t_m} \Psi = B_m \Psi$, where B_m is a polynomial in ∂ (only a finite number of positive powers appear), it is sufficient to prove this equality for $(i, 0, \ldots, 0)$, $i \geq 0$. In this case:

$$\oint \frac{dz}{2i\pi} (\partial^i \Psi) \cdot \Psi^* = \oint \frac{dz}{2i\pi} (\partial^i \Phi e^{\xi(t,z)}) \cdot ((\Phi^*)^{-1} e^{-\xi(t,z)})$$

$$= \oint \frac{dz}{2i\pi} (\partial^i \Phi e^{zx}) \cdot ((\Phi^*)^{-1} e^{-zx})$$

$$= \text{Res}_\partial (\partial^i \Phi \cdot \Phi^{-1}) = \text{Res}_\partial (\partial^i) = 0$$

The compact expression eq. (10.13) is obtained formally by Taylor expanding around $t = t'$. ∎

In the previous proposition, we proved that the KP equations imply that the Baker–Akhiezer functions satisfy the bilinear identities eq. (10.13). We now establish the converse statement, meaning that the whole KP hierarchy is equivalent to these bilinear identities.

Proposition. *Consider two formal series:*

$$\Psi = \Phi e^{\xi(t,z)}, \quad \Psi^* = \Phi^* e^{-\xi(t,z)}$$

where Φ and $\Phi^\star$ are two pseudo-differential operators of the form:

$$\Phi = 1 + \sum_{i=1}^{\infty} w_i \partial^{-i}, \quad \Phi^\star = 1 + \sum_{i=1}^{\infty} w_i^\star (-\partial)^{-i}$$

where $\{w_i, w_i^\star\}$ are functions of the variables $\{t_i\}$. Let us assume that the bilinear identity eq. (10.13) is satisfied. Then one has

$$\Phi^\star = (\Phi^*)^{-1} \tag{10.15}$$

Moreover, defining $Q = \Phi \partial \Phi^{-1}$, we have $\partial_{t_m} \Phi = -(Q^m)_- \Phi$. Hence Ψ is the Baker–Akhiezer function of the KP hierarchy.

Proof. We have used the notation $\Phi^\star$ (with a different star) to avoid introducing a new letter, but at this stage it is an independent pseudo-differential operator. By definition the functions Ψ and Ψ^* are:

$$\Psi = \left(1 + \sum_{i=1}^{\infty} w_i z^{-i}\right) e^{\xi(t,z)}, \quad \Psi^* = \left(1 + \sum_{i=1}^{\infty} w_i^\star z^{-i}\right) e^{-\xi(t,z)}$$

and we assume that $\oint \frac{dz}{2i\pi} \partial^i \Psi \cdot \Psi^* = 0$ for any $i \geq 0$. We first prove that this implies that Φ^* and Φ^* are inverse to each other. Indeed, using eq. (10.14), we have for any $i \geq 0$:

$$\text{Res}_\partial \left(\partial^i \Phi \, (\Phi^*)^* \right) = \oint \frac{dz}{2i\pi} (\partial^i \Phi e^{\xi(t,z)})(\Phi^* e^{-\xi(t,z)}) = \oint \frac{dz}{2i\pi} \partial^i \Psi \cdot \Psi^* = 0$$

where the hypothesis is used in the last step. But by construction $\Phi(\Phi^*)^* = 1 + X$, with $X \in \mathcal{P}_-$, therefore the above equation implies $\text{Res}\,(\partial^i X) = 0$, for all $i \geq 0$ and thus $X = 0$, so that $\Phi^* = (\Phi^*)^{-1}$. Now let $Q = \Phi \partial \Phi^{-1}$, and $B_m = (Q^m)_+$. We show that $\partial_{t_m} \Phi = -(Q^m)_- \Phi$. First, observe that using $\partial_{t_m} e^{\xi(t,z)} = \partial^m e^{\xi(t,z)}$ and $\Phi \partial^m = Q^m \Phi$, we have:

$$\left((\partial_{t_m} \Phi) + (Q^m)_- \Phi \right) e^{\xi(t,z)} = \partial_{t_m} \left(\Phi e^{\xi(t,z)} \right) - \left(\Phi \partial_{t_m} - (Q^m)_- \Phi \right) e^{\xi(t,z)}$$

$$= \partial_{t_m} \left(\Phi e^{\xi(t,z)} \right) - \left(\Phi \partial^m - (Q^m)_- \Phi \right) e^{\xi(t,z)}$$

$$= \left(\partial_{t_m} - Q^m + (Q^m)_- \right) \Phi \; e^{\xi(t,z)}$$

$$= \left(\partial_{t_m} - (Q^m)_+ \right) \Phi \; e^{\xi(t,z)}$$

By hypothesis, Ψ and Ψ^* satisfy the bilinear identity eq. (10.13). Therefore, since $(Q^m)_+$ is a differential *polynomial* we have, for any $i \geq 0$:

$$0 = \oint \frac{dz}{2i\pi} \partial^i (\partial_{t_m} - (Q^m)_+) \Phi e^{\xi(t,z)} \cdot \Phi^* e^{-\xi(t,z)}$$

$$= \oint \frac{dz}{2i\pi} \partial^i \left((\partial_{t_m} \Phi) + (Q^m)_- \Phi \right) e^{\xi(t,z)} \cdot \Phi^* e^{-\xi(t,z)}$$

Equivalently, one can write:

$$0 = \text{Res}_\partial \left(\partial^i \left((\partial_{t_m} \Phi) + (Q^m)_- \Phi \right) (\Phi^*)^* \right)$$

$$= \text{Res}_\partial \left(\partial^i \left((\partial_{t_m} \Phi) + (Q^m)_- \Phi \right) \Phi^{-1} \right)$$

Since this is true for any $i \geq 0$, it implies: $((\partial_{t_m} \Phi) + (Q^m)_- \Phi)\Phi^{-1} = 0$. Multiplying on the right by Φ proves the result. ∎

10.4 Algebro-geometric solutions of KP

It is quite a remarkable fact that with any Riemann surface of genus g one can associate a solution of the KP hierarchy. We explain this construction in this section.

Let Γ be a smooth algebraic curve of genus g. Fix a point P_∞ on Γ and a local coordinate $w(P) = z^{-1}$ in a neighbourhood of the puncture

$P_\infty(\ z=\infty)$. Then for each set of g points $\gamma_1,\ldots,\gamma_g$ in a general position there exists a unique function $\Psi(t,P)$ of the variable $P \in \Gamma$ which is meromorphic outside P_∞ and has at most simple poles at the points γ_s, and in the neighbourhood of the puncture P_∞ one requires:

$$\Psi(t,P) = e^{\xi(t,z)}\left(1 + \sum_{s=1}^{\infty} w_s(t)z^{-s}\right) \tag{10.16}$$

We now recall the fundamental formula expressing the Baker–Akhiezer functions in terms of Riemann theta functions (see Chapter 5). Let $\Omega^{(i)}$ be the unique normalized meromorphic differential with a pole at P_∞, of the form

$$\Omega^{(i)} = d(z^i + O(z^{-1}))$$

and holomorphic everywhere else. The normalization condition is that all its a-periods vanish,

$$\oint_{a_k} \Omega^{(i)} = 0$$

With it, we define a vector $U^{(i)}$ with coordinates

$$U_k^{(i)} = \frac{1}{2\pi i} \oint_{b_k} \Omega^{(i)} \tag{10.17}$$

The Baker–Akhiezer function $\Psi(t,P)$ is equal to

$$\Psi(t,P) = \frac{\theta(\mathcal{A}(P) + U^{(1)}x + U^{(2)}t - \zeta)\theta(\zeta)}{\theta(\mathcal{A}(P) - \zeta)\theta(U^{(1)}x + U^{(2)}t - \zeta)} e^{(\sum_i t_i \int_{P_\infty}^P \Omega^{(i)})} \tag{10.18}$$

where $\int_{P_\infty}^P \Omega^{(i)}$ is the unique primitive of $\Omega^{(i)}$ behaving as $z^i + O(z^{-1})$ modulo periods in the vicinity of P_∞. The vector ζ is equal to $\zeta = \mathcal{A}(D) + \mathcal{K}$ with $D = \gamma_1 + \cdots + \gamma_g$ and $\mathcal{K}$ is the vector of Riemann constants. Finally, $\mathcal{A}(P)$ is the Abel map with origin at P_∞.

Remark. The Baker–Akhiezer function is intrinsically defined by its analyticity properties. In the above formula, the choice of a-cycles and b-cycles and the normalization of the differentials is at our disposal. Another more canonical normalization is obtained by requiring that the forms $\Omega'^{(j)}$ have pure imaginary periods on any cycle. They are obtained by the transformation

$$\Omega^{(j)} = \Omega'^{(j)} + \sum_{i=1}^{g} \alpha_i^{(j)}\omega_i \tag{10.19}$$

where the ω_i are the g normalized holomorphic differentials. Writing these normalization conditions gives $2g$ real conditions on the g complex parameters $\alpha_i^{(j)}$ which can be solved. Indeed, taking the integral of this formula over the cycle a_i, we get

$\alpha_i^{(j)} = -\oint_{a_i} \Omega'^{(j)}$ is pure imaginary. Taking the integral over the cycle b_i, we get the unique solution

$$\alpha_i^{(j)} = 2i\pi (\text{Im } \mathcal{B})_{ik}^{-1} \text{Im } U_k^{(j)}$$

where $\mathcal{B}$ is matrix of b-periods of the ω_i. With such $\Omega'^{(j)}$, we can write

$$\Psi(t, P) = e^{\left(\sum_j t_j \int_{P_\infty}^P \Omega'^{(j)}\right)} \varphi\left(\sum t_j U'^{(j)}, P\right) \qquad (10.20)$$

where the vector $U'^{(j)}$ has $2g$ components

$$U_i'^{(j)} = -\frac{1}{2i\pi} \oint_{a_i} \Omega'^{(j)}, \quad U_{g+i}'^{(j)} = \frac{1}{2i\pi} \oint_{b_i} \Omega'^{(j)}, \quad j = 1, \ldots, g$$

The function $\varphi(z, P)$ is equal to

$$\varphi(z, P) = e^{2i\pi\left(\sum_{i=1}^g z_i \mathcal{A}_i(P)\right)} \frac{\theta(\mathcal{A}(P) + \sum_i (z_{i+g} I_i + z_i \mathcal{B}_i) - \zeta)\theta(\zeta)}{\theta(\mathcal{A}(P) - \zeta)\theta(\sum_i (z_{i+g} I_i + z_i \mathcal{B}_i) - \zeta)}$$

The vectors I_i and $\mathcal{B}_i$ are such that $(I_i)_j = \delta_{ij}$ and $(\mathcal{B}_i)_j = \mathcal{B}_{ij}$. This is obtained by plugging eq. (10.19) in eq. (10.18).

Note that the function $\varphi(z, P)$ is periodic with period 1 in each of the $2g$ variables z_i. This form will be useful in the considerations of the last sections on Whitham equations.

The Baker–Akhiezer function automatically produces solutions of the KP hierarchy as follows. Consider eq. (10.16) for the asymptotic form of the Baker–Akhiezer function around P_∞. Let us rewrite it as

$$\Psi(t, P) = \Phi e^{\xi(t,z)}, \quad \Phi = 1 + \sum_{s=1}^{\infty} w_s(t)\partial^{-s}$$

where $\partial = \partial_{t_1} = \partial_x$ and $\xi(t, z) = \sum_i t_i z^i$. This defines the pseudo-differential operator Φ. From it we *define*

$$Q = \Phi \partial \Phi^{-1}$$

Then we have

Proposition. *Let $\Psi(t, P)$ be the above Baker–Akhiezer function. Then it satisfies the equations of motion of the KP hierarchy*

$$(Q - z)\Psi = 0$$

$$(\partial_{t_i} - (Q^i)_+)\Psi = 0$$

<u>Proof.</u> The first equation has a meaning as an expansion around P_∞ and directly follows from the definition of Q. To prove the second equation,

consider the function $(\partial_{t_i} - (Q^i)_+)\Psi$ on Γ. It has the same analyticity properties as Ψ, apart from the behavior around P_∞ where we have

$$(\partial_{t_i} - (Q^i)_+)\Psi = (\partial_{t_i} - Q^i + (Q^i)_-)\Psi = O(z^{-1})e^{\xi(t,z)}$$

We used that $\Phi\partial_{t_i}e^{\xi(t,z)} = Q^i\Phi e^{\xi(t,z)}$. Hence the expression in the left-hand side identically vanishes by the unicity of the Baker–Akhiezer function. ∎

Remark. We stress that this construction associates solutions of the KP hierarchy with any Riemann surfaces. Special curves may lead to additional interesting structures, as we have seen in Chapter 7 on Calogero–Moser systems.

We now give the global definition of the adjoint Baker–Akhiezer function. For any set of g points in general position there exists a unique meromorphic differential Ω with a double pole at P_∞:

$$\Omega = dz(1 + O(z^{-2})) \qquad (10.21)$$

and zeroes at the points γ_s:

$$\Omega(\gamma_s) = 0, \quad s = 1, \dots, g \qquad (10.22)$$

Besides γ_s this differential has g other zeroes that we denote by γ_s^*.

The adjoint Baker–Akhiezer function is the unique function $\Psi^*(t, P)$ of the variable $P \in \Gamma$ which is meromorphic outside P_∞, has at most simple poles at the points γ_s^* (if all of them are distinct), and behaves in the neighbourhood of the puncture P_∞ as

$$\Psi^*(t, P) = e^{-\xi(t,z)}\left(1 + \sum_{s=1}^{\infty} w_s^*(t)z^{-s}\right)$$

The adjoint Baker–Akhiezer function $\Psi^*(t, P)$ is equal to

$$\Psi^*(t, P) = \frac{\theta(\mathcal{A}(P) - U^{(1)}x - U^{(2)}t - \zeta^*)\theta(\zeta^*)}{\theta(\mathcal{A}(P) - \zeta^*)\theta(U^{(1)}x + U^{(2)}t + \zeta^*)}e^{-(\sum_i t_i \int_{P_\infty}^P \Omega^i)}$$

where

$$\zeta^* = \mathcal{A}(D^*) + \mathcal{K}$$

where $D^* = \gamma_1^* + \cdots + \gamma_g^*$.

Proposition. *The adjoint Baker–Akhiezer function satisfies the equations:*

$$(Q^* - z)\Psi^* = 0$$

$$(\partial_{t_i} + (Q^i)^*_+)\Psi^* = 0 \qquad\qquad (10.23)$$

where Q^* is the formal adjoint of Q.

<u>Proof.</u> Consider, for any positive integer i, the form $(\partial^i \Psi)\Psi^*\Omega$, where Ω is defined by eqs. (10.21, 10.22). This is a meromorphic 1-form on Γ with a *unique* pole at P_∞ of order $2+i$ because the poles of Ψ and Ψ^* are cancelled by the zeroes of Ω. Moreover, the essential singularities of Ψ and Ψ^* at P_∞ cancel. Around P_∞, we have $\Psi = \Phi e^{\xi}(t,z)$ and $\Psi^* = \Phi^* e^{-\xi(t,z)}$, where $\Phi^\star$ is defined from the expansion of Ψ^*. So we have

$$(\partial^i \Phi e^{\xi})\Phi^\star e^{-\xi}\Omega = z^i dz\,(1 + O(1/z))$$

Since the sum of residues of any meromorphic 1-form must vanish, we get:

$$\oint (\partial^i \Psi) \cdot \Psi^*\Omega = 0, \quad \forall i \geq 0$$

where the integral is taken on a small circle around P_∞. This means that Ψ and $\Psi^*\Omega$ satisfy the bilinear identities eq. (10.13), and therefore by eq. (10.15) we have $\Phi^\star\Omega = (\Phi^*)^{-1}dz$. The adjoint equations of motion follow because Ω is independent of t. ∎

We have shown that the adjoint Baker–Akhiezer function is equal to the formal adjoint of Ψ, up to the factor Ω/dz. This also shows that Baker–Akhiezer functions constructed from Riemann surfaces automatically satisfy the fundamental bilinear identities eq. (10.13).

10.5 The tau-function of KP

The bilinear identity eq. (10.13) allows us to express the Baker–Akhiezer function in terms of a tau-function.

Proposition. *Assume that Ψ and Ψ^* are Baker–Akhiezer functions of the KP hierarchy satisfying eq. (10.13). Then there exists a function τ such that*

$$\Psi(t,z) = \frac{\tau(t - [z^{-1}])}{\tau(t)}\,e^{\xi(t,z)}, \quad \Psi^*(t,z) = \frac{\tau(t + [z^{-1}])}{\tau(t)}\,e^{-\xi(t,z)} \quad (10.24)$$

where $[z^{-1}] = \left(\frac{1}{z}, \frac{1}{2z^2}, \frac{1}{3z^3}, \cdots\right)$ and $\xi(t,z) = \sum_{i=1}^{\infty} t_i z^i$.

<u>Proof.</u> Note that for $f(z) = 1 + \sum_{i=1}^{\infty} f_i z^{-i}$ we have, by the residue theorem:

$$\oint \frac{dz}{2i\pi} \frac{f(z)}{1 - \frac{z}{z'}} = z'(f(z') - 1) \tag{10.25}$$

where z' is big enough to be inside the integration contour around ∞. The two terms correspond to the poles at $z = z'$ and $z = \infty$. The bilinear identity applied to t and $t' = t - [z_1^{-1}]$ yields:

$$0 = \oint \frac{dz}{2i\pi} \Psi(t, z) \Psi^*(t - [z_1^{-1}], z) = \oint \frac{dz}{2i\pi} \frac{\widehat{w}(t, z)\widehat{w}^*(t - [z_1^{-1}], z)}{1 - z/z_1}$$

where $\widehat{w}(t, z)$ is defined in eq. (10.11). To show the second equality, we used that

$$e^{-\xi(t - [z_1^{-1}], z)} = e^{-\xi(t, z)} \frac{1}{1 - z/z_1}$$

Up to now the arguments are similar to the proof of eq. (9.77) in Chapter 9. Notice that the Cauchy kernel in the right-hand side is produced by the very specific choice of shift we consider. Applying the residue formula eq. (10.25), we get $\widehat{w}(t, z_1)\widehat{w}^*(t - [z_1^{-1}], z_1) = 1$, or:

$$\widehat{w}^*(t - [z_1^{-1}], z_1) = \frac{1}{\widehat{w}(t, z_1)} \tag{10.26}$$

Similarly, applying the bilinear identity to t and $t' = t - [z_1^{-1}] - [z_2^{-1}]$, we see that the following quantity vanishes:

$$\oint \frac{dz}{2i\pi} \Psi(t, z) \Psi^*(t - [z_1^{-1}] - [z_2^{-1}], z) = \oint \frac{dz}{2i\pi} \frac{\widehat{w}(t, z)\widehat{w}^*(t - [z_1^{-1}] - [z_2^{-1}], z)}{(1 - z/z_1)(1 - z/z_2)}$$

Since there is no residue at ∞ we get $\widehat{w}(t, z_1)\widehat{w}^*(t - [z_1^{-1}] - [z_2^{-1}], z_1) = \widehat{w}(t, z_2)\widehat{w}^*(t - [z_1^{-1}] - [z_2^{-1}], z_2)$. Eliminating $\widehat{w}^*$ using eq. (10.26), we obtain the functional equation:

$$\frac{\widehat{w}(t - [z_2^{-1}], z_1)}{\widehat{w}(t, z_1)} = \frac{\widehat{w}(t - [z_1^{-1}], z_2)}{\widehat{w}(t, z_2)}$$

We want to show that this equation implies:

$$\widehat{w}(t, z) = \frac{\tau(t - [z^{-1}])}{\tau(t)} \tag{10.27}$$

It is trivial to verify that this solves the equation. We now proceed to show that this is the general solution. Taking the logarithm of the functional equation, we are led to study an equation of the form

$$f(t - [u^{-1}], v) - f(t, v) = f(t - [v^{-1}], u) - f(t, u) \tag{10.28}$$

where the function f is:

$$f(t,v) = \log \widehat{w}(t,v) = \frac{1}{v}w_1(t) + \frac{1}{v^2}(w_2(t) - \frac{1}{2}w_1^2(t)) + \cdots$$

Introducing the generating function of time derivatives:

$$\nabla_v = \sum_{i=1}^{\infty} v^{-i-1}\partial_{t_i}$$

we remark that $(\partial_v - \nabla_v)\phi(t - [v^{-1}]) = 0$ for any function $\phi(t)$. Applying $(\partial_v - \nabla_v)$ to eq. (10.28), we get

$$(\partial_v - \nabla_v)f(t - [u^{-1}], v) - (\partial_v - \nabla_v)f(t, v) = -(\partial_v - \nabla_v)f(t, u)$$

Expanding in v^{-1} and setting $(\partial_v - \nabla_v)f(t,v) = \sum_{i=1}^{\infty} \gamma_i(t)v^{-i-1}$, this reads:

$$\gamma_i(t - [u^{-1}]) - \gamma_i(t) = \partial_{t_i}f(t, u) \qquad (10.29)$$

Considering $F_{ij}(t) = \partial_{t_i}\gamma_j(t) - \partial_{t_j}\gamma_i(t)$ we get the condition $F_{ij}(t - [u^{-1}]) = F_{ij}(t)$. Expanding in powers of u^{-1} one sees that $F_{ij}(t)$ is independent of all the time variables $t_1, t_2, \ldots$. But by construction, F_{ij} is a local differential polynomial in $w_1(t), w_2(t), \ldots$ (for example $F_{12} = \partial_{t_2}w_1 - 2\partial_{t_1}w_2 + 2w_1\partial_{t_1}w_1 - \partial_1^2 w_1$, etc....). Using the equations of motion of the KP hierarchy we can replace all the $\partial_{t_k}w_l$ for $k \geq 2$ by higher derivatives of w_l with respect to $t_1 = x$. Hence we can write F_{ij} as a polynomial in the $\partial^k w_l$. But the monomials in F_{ij} are independent, and we know that F_{ij} is constant, hence it reduces to its constant term. Since it vanishes for the particular solution $w = 0$, we see that $F_{ij} = 0$. So we can write $\gamma_i(t) = \partial_{t_i} \log \tau(t)$. Finally, inserting this into eq. (10.29) we get eq. (10.27).

The formula for $\Psi^*(t, z)$ is then a straightforward consequence of eq. (10.26) where one substitutes $t \to t + [z_1^{-1}]$. ∎

Equation (10.24) is Sato's formula. Using it, the bilinear identity eq. (10.13) can be rewritten as bilinear identities for the tau-functions. Of course they coincide with the Hirota bilinear identities we obtained using vertex operators, eq. (9.38) in Chapter 9.

Remark 1. We recall that in terms of the fermionic description of the previous chapter, the tau-function and the Baker–Akhiezer function have a particularly elegant formulation:

$$\tau(t; g) = \langle 0|e^{H(t)}g|0\rangle$$

and

$$\Psi(t, z) = \frac{\langle 1|e^{H(t)}\beta(z)g|0\rangle}{\tau(t; g)}, \quad \Psi^*(t, z) = \frac{\langle -1|e^{H(t)}\beta^*(z)g|0\rangle}{\tau(t; g)}$$

where g is an element of the group $GL(\infty)$, and $\beta(z)$ and $\beta^*(z)$ are the fermionic operators.

Remark 2. The Grassmannian formulation also shows that $\tau(t)$ is given by an infinite determinant. In particular interesting cases, it degenerates to a finite determinant.

10.6 The generalized KdV equations

The KP hierarchy is a system of evolution equations for the infinite set of functions $w_i(t)$ or equivalently the coefficients $q_{-i}(t)$ appearing in Q. To reduce the sytem to a finite number of coefficients, we remark that since Q obeys the Lax equations $\partial_{t_k}Q = [(Q^k)_+, Q]$, any power Q^{n+1} also obeys the same equation. The main remark is that one can impose consistently that Q^{n+1} is a differential operator, i.e. $(Q^{n+1})_- = 0$. This is because one then has $[(Q^k)_+, Q^{n+1}]_- = 0$, and so $\partial_{t_k}Q^{n+1}$ is a differential operator. Moreover, the two sides of the Lax equation are differential operators of the same order because one can also write $\partial_{t_k}Q^{n+1} = [-(Q^k)_-, Q^{n+1}]$, which is in fact of order ∂^{n-1}. It follows that one can further impose that the coefficient of ∂^n in Q^{n+1} vanishes.

To summarize, we impose that $Q^{n+1} = L$ is a *differential* operator:

$$L = \partial^{n+1} - \sum_{i=0}^{n-1} u_i \partial^i \qquad (10.30)$$

With this L one can write Lax equations which define flows on the finite number of functions u_i. These flows close on the u_i because given L one can reconstruct Q such that $Q^{n+1} = L$, so that $(Q^k)_+$ may be viewed as a function of the u_i.

Proposition. *Let L be the differential operator eq. (10.30). There exists a unique pseudo-differential operator $Q = \partial + q_{-1}\partial^{-1} + \cdots$ such that $Q^{n+1} = L$. We will denote it by $Q = L^{\frac{1}{n+1}}$.*

Proof. If $Q = \partial + q_0 + q_{-1}\partial^{-1} + \cdots$, one first sees that $Q^{n+1} = \partial^{n+1} + (n+1)q_0\partial^n + \cdots$. Since there is no term ∂^n in L one has $q_0 = 0$. Then by induction one shows that:

$$(n+1)q_{-1} = -u_{n-1}, \quad (n+1)q_{-2} = -u_{n-2} - \frac{n(n+1)}{2}\partial q_{-1}$$

$$(n+1)q_{-i} = -u_{n-i} + p_i(q_{-1}, \ldots, q_{-i+1})$$

where p_i is a differential polynomial in its arguments. Knowing the u_i, this system uniquely determines the q_i recursively. ∎

We can rewrite the reduced KP flows directly in terms of L. These systems are called the generalized KdV hierarchies. The KdV hierarchy corresponds to $n = 1$, and the generalized ones to $n = 2, 3, \ldots$. It is worth writing these equations once more:

Proposition. *Let L be the differential operator eq. (10.30). Then the Lax equations*

$$\partial_{t_k} L = \left[\left(L^{\frac{k}{n+1}} \right)_+, L \right] \qquad (10.31)$$

are consistent for all $k \in \mathbb{N}$.

<u>Proof.</u> We introduce the pseudo-differential operator $Q = L^{\frac{1}{n+1}}$. Notice that Q^k, $\forall k \in \mathbb{N}$, commutes with L since $LQ^k = Q^{n+1+k} = Q^k L$. Then we have:

$$\left[\left(Q^k \right)_+, L \right] = \left[Q^k, L \right] - \left[\left(Q^k \right)_-, L \right] = - \left[\left(Q^k \right)_-, L \right]$$

From the last equality, it follows that the differential operator $\left[\left(Q^k \right)_+, L \right]$ is of order less or equal to $n - 1$, so that the Lax equation eq. (10.31) is an equation on the coefficients of L.   ∎

Example. Let us consider the KdV case $n = 1$. The operator L is the second order differential operator

$$L = \partial^2 - u$$

We first find Q such that $Q^2 = L$. One has $Q^2 = \partial^2 + 2q_{-1} + (2q_{-2} + \partial q_{-1})\partial^{-1} + \cdots$ so that $q_{-1} = -\frac{1}{2}u$, $q_{-2} = \frac{1}{4}\partial u$, etc. . . . .

$$Q = \partial - \frac{1}{2}u\partial^{-1} + \frac{1}{4}(\partial u)\partial^{-2} + \cdots$$

We again check on this simple example that all the q_{-j} are recursively determined in terms of u by requiring that no ∂^{-j} terms occur in Q^2. To obtain the KdV flows, we only have to compute $(Q^k)_+$, $k = 1, 2, \ldots$. For $k = 1$, we have $(Q)_+ = \partial$, and $\partial_1 L = [\partial, L]$. This reduces to the identification $\partial_{t_1} = \partial$. For $k = 2$, we have $(Q^2)_+ = L$, and we get the trivial equation $\partial_{t_2} L = 0$. The first non-trivial case is $k = 3$. We have:

$$(Q^3)_+ = \partial^3 - \frac{3}{2}u\partial - \frac{3}{4}(\partial u)$$

so the Lax equation reads $\partial_{t_3} u = [(Q^3)_+, \partial^2 - u]$. This is the Korteweg–de Vries equation:

$$4\partial_{t_3} u = \partial^3 u - 6u(\partial u)$$

This is the first of a hierarchy of equations obtained by taking $k = 3, 5, 7, \ldots$, called the KdV hierarchy (note that for k even we get trivial equations), which will be studied in detail in Chapter 11.

We now show that the generalized KdV equations are Hamiltonian systems. The differential operator L is an element of $\mathcal{P}_+$. So we have to specify a Poisson structure on the space $\mathcal{F}(\mathcal{P}_+)$ of functions on $\mathcal{P}_+$. If we view $\mathcal{P}_+$ as the dual of the Lie algebra $\mathcal{P}_-$ through the Adler trace, there is a natural Poisson bracket on $\mathcal{P}_+$: the Kostant–Kirillov bracket. For any functions f and g on $\mathcal{P}_+$, it is defined as usual by:

$$\{f, g\}_1(L) = \langle L, [df, dg]\rangle \quad \forall L \in \mathcal{P}_+ \qquad (10.32)$$

where we understand that $df, dg \in \mathcal{P}_-$. In particular, if for any $X = \sum_{j=0}^{\infty} \partial^{-j-1} x_j \in \mathcal{P}_-$, we define the linear function $f_X(L)$ by:

$$f_X(L) = \langle L, X\rangle \qquad (10.33)$$

and we have $df_X = X \in \mathcal{P}_-$. Therefore $\{f_X, f_Y\}_1 = f_{[X,Y]} = \langle L, [X, Y]\rangle$ for any $X, Y \in \mathcal{P}_-$.

Proposition. *Let $L \in \mathcal{P}_+$ be the differential operator of order $n + 1$ as in eq. (10.30). Define the functions of L by*

$$H_k(L) = \left(\frac{n+1}{n+k+1}\right) \langle L^{\frac{k}{n+1}+1}\rangle$$

They are the conserved quantities of the generalized KP hierarchy. Then:
(i) The quantities H_k are the Hamiltonians of the generalized KdV flows under the bracket eq. (10.32):

$$\dot{L} = \{H_k, L\}_1 = \left[\left(L^{\frac{k}{n+1}}\right)_+, L\right] \qquad (10.34)$$

(ii) The functions $H_k(L)$ are in involution with respect to this bracket.

<u>Proof.</u> Recall that $Q = L^{\frac{1}{n+1}}$ so that H_k is proportional to $\langle Q^{k+n+1}\rangle$, which are the conserved quantities of the KP hierarchy, and so are conserved also in the generalized KdV hierarchies. We first need to compute the differential of the Hamiltonian H_k. Let L and δL be differential operators of the form eq. (10.30). One has, using the cyclicity of Adler's trace:

$$\langle (L + \delta L)^{\nu}\rangle = \langle L^{\nu}\rangle + \nu\langle L^{\nu-1}\delta L\rangle + \cdots$$

which implies $d\langle L^{\nu}\rangle = \nu(L^{\nu-1})_{(-n)}$, where the notation $(\ )_{(-n)}$ means projection on $\mathcal{P}_-$ truncated at the first n terms. This projection appears because $\delta L = -\delta u_{n-1}\partial^{n-1} - \cdots - \delta u_0$, which is dual to elements of the form $\partial^{-1}x_0 + \cdots + \partial^{-n}x_n$ under the Adler trace. Hence:

$$dH_k(L) = \left(L^{\frac{k}{n+1}}\right)_{(-n)} = \left(Q^k\right)_{(-n)} \quad \in \mathcal{P}_- \tag{10.35}$$

We define $\theta^{(k)}_{-(n+1)}$ as the terms left over in the truncation:

$$\left(L^{\frac{k}{n+1}}\right)_- = dH_k + \theta^{(k)}_{-(n+1)} \tag{10.36}$$

We now prove eq. (10.34). Consider the function $f_X(L) = \langle LX \rangle$, then

$$\dot{f}_X = \{H_k, f_X\}(L) = \langle L, [dH_k, df_X] \rangle = \langle [L, dH_k], X \rangle$$

where we used the invariance of the Adler trace. Since $X \in \mathcal{P}_-$, only $[L, dH_k]_+$ contributes to this expression. But

$$[L, dH_k]_+ = \left[L, \left(L^{\frac{k}{n+1}}\right)_-\right]_+ - \left[L, \theta^{(k)}_{-(n+1)}\right]_+ = \left[\left(L^{\frac{k}{n+1}}\right)_+, L\right]$$

where we have used $[L^{\frac{k}{n+1}}, L] = 0$, and the fact that $[L, \theta^{(k)}_{-(n+1)}]_+ = 0$. So $[L, dH_k]_+$ is a differential operator of order at most $n-1$, and this is enough to prove eq. (10.34).

Next we show that the Hamiltonians H_k are in involution. We have:

$$\{H_k, H_l\}_1(L) = \langle L, [dH_k, dH_l] \rangle = \langle [L, dH_k]_+, dH_l \rangle$$

$$= \langle \left[\left(L^{\frac{k}{n+1}}\right)_+, L\right], dH_l \rangle$$

Using again the fact that $[L, dH_k]_+$ is of order at most $n-1$, we can replace dH_l by $\left(L^{\frac{l}{n+1}}\right)_-$, and get:

$$\{H_k, H_l\}_1(L) = \langle \left[\left(L^{\frac{k}{n+1}}\right)_+, L\right]\left(L^{\frac{l}{n+1}}\right)_- \rangle = \langle \left[\left(L^{\frac{k}{n+1}}\right)_+, L\right] L^{\frac{l}{n+1}} \rangle$$

In the last step we used that $\langle \mathcal{P}_+, \mathcal{P}_+ \rangle = 0$ in order to replace $\left(L^{\frac{l}{n+1}}\right)_-$ by $L^{\frac{l}{n+1}}$. Finally, from the invariance of the trace, we obtain:

$$\{H_k, H_l\}_1(L) = \langle \left(L^{\frac{k}{n+1}}\right)_+ \left[L, L^{\frac{l}{n+1}}\right] \rangle = 0$$

$\blacksquare$

This proposition shows that the generalized KdV hierarchies are Hamiltonian systems. In the next section we show that there exists in fact another local Hamiltonian structure for the same hierarchy.

10.7 KdV Hamiltonian structures

We will establish recursion relations between the equations of motion written with Hamiltonians H_k and H_{k+n+1}. These relations are called Lenard recursion relations. They suggest introducing a second Poisson bracket on $\mathcal{F}(\mathcal{P}_+)$ such that the generalized KdV equations can be writtens as Hamilton equations with respect to both Poisson brackets; such systems are called bihamiltonian systems.

Proposition. *Let L be the differential operator eq. (10.30). Let us introduce two operators $\mathcal{D}_1$ and $\mathcal{D}_2$ acting on any $X \in \mathcal{P}_-$ by:*

$$\mathcal{D}_1(X) = [L, X]_+$$

$$\mathcal{D}_2(X) = (LX)_+ L - L(XL)_+ - \frac{1}{n+1}[L, (\partial^{-1}[X, L]_{-1})]$$

Then, the functions $H_k(L)$ satisfy the following recursion relation:

$$\mathcal{D}_1(dH_{k+n+1}) = \mathcal{D}_2(dH_k) \tag{10.37}$$

<u>Proof.</u> Note that for any $X \in \mathcal{P}_-$ we have $0 = \langle [X, L] \rangle = \int dx\, [X, L]_{-1}$ so that $[X, L]_{-1}$ is a total derivative and the object $(\partial^{-1}[X, L]_{-1})$ appearing in the definition of $\mathcal{D}_2$ is by definition a primitive of this total differential, i.e. it is local. For example, we have $[\partial^2 - u, \sum_{i=1}^{\infty} x_i \partial^{-i}]_{-1} = \partial(\partial x_1 + 2x_2)$. Using eq. (10.36) for dH_{k+n+1}, we have:

$$\mathcal{D}_1(dH_{k+n+1}) = [L, dH_{k+n+1}]_+ = [L, Q_-^{k+n+1}]_+ - [L, \theta_{-(n+1)}^{(k+n+1)}]_+$$
$$= -[L, Q_+^{k+n+1}]$$

We used $[L, \theta_{-(n+1)}^{(k+n+1)}]_+ = 0$ and the fact that $Q^{k+n+1} = Q_+^{k+n+1} + Q_-^{k+n+1}$ commutes with L.

Now, the simple recursion relation: $Q^k L = Q^{k+n+1}$ implies:

$$\left(Q^{k+n+1}\right)_+ = \left(Q^k L\right)_+ = \left(Q_+^k L\right)_+ + \left(Q_-^k L\right)_+ = Q_+^k L + \left(Q_-^k L\right)_+$$

Thus we obtain:

$$\mathcal{D}_1(dH_{k+n+1}) = -\left[L, \left(Q^{k+n+1}\right)_+\right] = -\left[L, Q_+^k\right]L + \left[L, \left(Q_-^k L\right)_+\right]$$

$$= [L, dH_k]_+ L - [L, (dH_k L)_+] - [L, (\theta_{-(n+1)}^{(k)} L)_+] \tag{10.38}$$

where we used again the decomposition eq. (10.36) for dH_k. The remarkable fact is that the last term involving $\theta_{-(n+1)}^{(k)}$ can also be expressed in terms of dH_k. Indeed, defining v_0 by $\theta^{(k)}_{-(n+1)} = v_0\partial^{-(n+1)} + \cdots$, we have $(\theta^{(k)}_{-(n+1)}L)_+ = v_0$. Also, $[(L^{\frac{k}{n+1}})_-, L]_- = -[(L^{\frac{k}{n+1}})_+, L]_- = 0$, and looking at the Adler residue gives:

$$0 = [(L^{\frac{k}{n+1}})_-, L]_{-1} = [dH_k, L]_{-1} + [\theta^{(k)}_{-(n+1)}, L]_{-1} = [dH_k, L]_{-1} - (n+1)\partial v_0$$

Because the residue of a commutator is a total derivative, we can integrate this equation, and write consistently that:

$$v_0 = \frac{1}{n+1}\left(\partial^{-1}[dH_k, L]_{-1}\right)$$

Inserting this result into eq. (10.38), we get:

$$[L, dH_{k+n+1}]_+ = (LdH_k)_+L - L(dH_kL)_+ - \frac{1}{n+1}[L, (\partial^{-1}[dH_k, L]_{-1})]$$

which is the claimed statement.   ∎

This recursion relation led Adler to conjecture that the operators $\mathcal{D}_1$ and $\mathcal{D}_2$ define two Poisson brackets on $\mathcal{F}(\mathcal{P}_+)$ denoted by $\{\ ,\ \}_1$ and $\{\ ,\ \}_2$ through the equations:

$$\{f, g\}_1 = \langle \mathcal{D}_1(df) \cdot dg \rangle$$
$$\{f, g\}_2 = \langle \mathcal{D}_2(df) \cdot dg \rangle$$

More precisely, let us define as in eq. (10.33) the functions $f_X(L) = \langle LX \rangle$ such that $df_X = X$. The explicit expressions for the two brackets are then:

$$\{f_X, f_Y\}_1(L) = \langle LXY \rangle - \langle LYX \rangle \tag{10.39}$$
$$\{f_X, f_Y\}_2(L) = \langle (LX)_+(LY)_- \rangle - \langle (XL)_+(YL)_- \rangle$$
$$-\frac{1}{n+1}\int dx \left((\partial^{-1}[L, X]_{-1})[L, Y]_{-1}\right)$$

The first bracket is the Kostant–Kirillov bracket, eq. (10.32).

We now prove that $\{\ ,\ \}_2$ is a Poisson bracket. The antisymmetry can easily be checked using the cyclicity of Adler trace and isotropy of $\mathcal{P}_\pm$. To check the Jacobi identity, we change variables, from the coefficients u_i of L, see eq. (10.30), to new variables p_j in which the Jacobi identity is obvious. This change of variables is called the Miura transformation. The Poisson bracket $\{\ ,\ \}_2$ becomes very simple in the variables p_j. This is the content of the Kupershmidt–Wilson theorem.

Theorem. *Let us write $L = L_{n+1}L_n \cdots L_1$, with $L_i = \partial - p_i$:*

$$L = (\partial - p_{n+1}) \cdot (\partial - p_n) \cdots (\partial - p_1), \quad \sum_{i=1}^{n+1} p_i = 0 \qquad (10.40)$$

Let us define a Poisson bracket on the functions $p_k(x)$ by:

$$\{p_k(x), p_l(y)\} = \left(\delta_{kl} - \frac{1}{n+1}\right)\delta'(x - y) \qquad (10.41)$$

Then we have $\{f_X, f_Y\}(L) = \{f_X, f_Y\}_2(L)$. This shows that $\{\ ,\ \}_2$ satisfies the Jacobi identity.

<u>Proof.</u> Viewing $f_X(L)$ as a function of p_k, we have:

$$\{f_X, f_Y\}(L) = \sum_{k,l} \int dx \, dy \frac{\delta f_X}{\delta p_k(x)}\{p_k(x), p_l(y)\}\frac{\delta f_Y}{\delta p_l(y)}$$

We now check that inserting eq. (10.41) into this formula reproduces the second Poisson structure. Define the operators:

$$L_{ij} = L_i L_{i-1} \cdots L_j, \quad \text{for } i \geq j, \quad L_{01} = 1, \ L_{n+1,n+2} = 1 \qquad (10.42)$$

From the expression of f_X, we have:

$$\delta f_X/\delta p_k(x) = -(L_{k-1,1}XL_{n+1,k+1})_{-1}(x)$$

Thus,

$$\{f_X, f_Y\}(L) = \sum_{k,l} \int dx \, dy (L_{k-1,1}XL_{n+1,k+1})_{-1}(x)$$

$$\times \{p_k(x), p_l(y)\} \times (L_{l-1,1}YL_{n+1,l+1})_{-1}(y)$$

Using the Poisson bracket of the p_k, this is rewritten as:

$$\{f_X, f_Y\}(L) = E_1 - E_2 \qquad (10.43)$$

where E_1 is produced by the Kronecker delta in eq. (10.41) while E_2 is produced by the $1/(n+1)$ term. We have:

$$E_1 = \sum_{k=1}^{n+1} \int dx (\partial(L_{k-1,1}XL_{n+1,k+1})_{-1})(L_{k-1,1}YL_{n+1,k+1})_{-1}$$

$$E_2 = \frac{1}{n+1}\sum_{k,l=1}^{n+1} \int dx (\partial(L_{k-1,1}XL_{n+1,k+1})_{-1})(L_{l-1,1}YL_{n+1,l+1})_{-1}$$

We can write $\partial(L_{k-1,1}XL_{n+1,k+1})_{-1} = ([L_k, L_{k-1,1}XL_{n+1,k+1}])_{-1}$. This is because $L_k = \partial - p_k$ and one checks that for any pseudo-differential $f = \sum_n f_n \partial^n$ one has $[\partial, f]_{-1} = (\partial f_{-1})$ and $[p_k, f]_{-1} = 0$. Thus:

$$E_1 = \sum_{k=1}^{n+1} \int dx \Big[(L_{k1}XL_{n+1,k+1})_{-1}(L_{k-1,1}YL_{n+1,k+1})_{-1} \tag{10.44}$$

$$-(L_{k-1,1}XL_{n+1,k})_{-1}(L_{k-1,1}YL_{n+1,k+1})_{-1} \Big]$$

$$E_2 = \frac{1}{n+1} \sum_{k,l=1}^{n+1} \int dx \Big[(L_{k1}XL_{n+1,k+1})_{-1}(L_{l-1,1}YL_{n+1,l+1})_{-1}$$

$$-(L_{k-1,1}XL_{n+1,k})_{-1}(L_{l-1,1}YL_{n+1,l+1})_{-1} \Big]$$

We use the identity, true for all p,

$$\int dx (U)_{-1}(V)_{-1} = \langle U_-(\partial - p)V_- \rangle = \langle (\partial - p)U_-V_- \rangle \tag{10.45}$$

to rewrite the expression E_1 as:

$$E_1 = \sum_{k=1}^{n+1} \Big(\langle (L_{k1}XL_{n+1,k+1})_- L_k (L_{k-1,1}YL_{n+1,k+1})_- \rangle$$

$$- \langle L_k(L_{k-1,1}XL_{n+1,k})_- (L_{k-1,1}YL_{n+1,k+1})_- \rangle \Big)$$

Next replace all the terms U_- by $U - U_+$, to get

$$E_1 = \sum_{k=1}^{n+1} \Big(- \langle (L_{k1}XL_{n+1,k+1})_+ L_{k1}YL_{n+1,k+1} \rangle$$

$$+ \langle (L_{k-1,1}XL_{n+1,k})_+ L_{k-1,1}YL_{n+1,k} \rangle \Big)$$

In the sum, the terms cancel two by two, and we are left with

$$E_1 = \langle (XL)_+ YL \rangle - \langle (LX)_+ LY \rangle$$

We now turn to the summation over k in E_2. Again, terms cancel two by two and we are left with:

$$E_2 = \frac{1}{n+1} \sum_{l=1}^{n+1} \int dx \, [X, L]_{-1}(L_{l-1,1}YL_{n+1,l+1})_{-1}$$

$$= \frac{1}{n+1} \sum_{l=1}^{n+1} \langle [X, L]_{-1} L_{l-1,1} Y L_{n+1,l+1} \rangle$$

$$= \frac{1}{n+1} \sum_{l=1}^{n+1} \langle L_{n+1,l+1}[X, L]_{-1} L_{l-1,1} Y \rangle$$

But we have $[L, f] = \sum_{l=1}^{n+1} L_{n+1,l+1} \partial f L_{l-1,1}$ for any function f. Thus the sum can finally be written as:

$$E_2 = \frac{1}{n+1} \int dx \ (\partial^{-1}([X, L]_{-1})[Y, L]_{-1}$$

This completes the proof that $\{ \ , \ \} = \{ \ , \ \}_2$. Since $\{ \ , \ \}$ obviously satisfies the Jacobi identity, so does $\{ \ , \ \}_2$. ∎

Proposition. *The Hamiltonians H_k are in involution with respect to both Poisson brackets.*

Proof. We already know that $\{H_k, H_l\}_1 = 0$. Using the recursion relation eq. (10.37), we have:

$$\{H_k, H_l\}_2 = \langle \mathcal{D}_2 dH_k, dH_l \rangle = \langle \mathcal{D}_1 dH_{n+1+k}, dH_l \rangle = \{H_{n+1+k}, H_l\}_1 = 0$$

∎

Example. For $n = 1$, we have $L = \partial^2 - u$ with $u = p^2 + p'$. The Poisson bracket eq. (10.41) becomes $\{p(x), p(y)\} = \frac{1}{2}\delta'(x - y)$. This implies

$$\{u(x), u(y)\}_2 = [u(x)\partial_x + \partial_x u(x) - \frac{1}{2}\partial_x^3]\delta(x - y)$$

We recognize the Virasoro algebra (see Chapter 11).

10.8 Bihamiltonian structure

The two Poisson brackets $\{ \ , \ \}_1$ and $\{ \ , \ \}_2$ have a remarkable compatibility property, called the Magri compatibility condition.

Proposition. *The two Poisson structures $\{ \ , \ \}_1$ and $\{ \ , \ \}_2$ are compatible, in the sense that for any λ_1, λ_2 the application $(f, g) \to \lambda_1\{f, g\}_1 + \lambda_2\{f, g\}_2$ is a Poisson bracket.*

Proof. The condition that the sum of two Poisson brackets $\{f, g\}_1 = \sum_{ij} P_{ij}^{(1)} \partial_i f \partial_j g$ and $\{f, g\}_2 = \sum_{ij} P_{ij}^{(2)} \partial_i f \partial_j g$ satisfies the Jacobi identity reads:

$$\left(P_{il}^{(1)} \partial_l P_{jk}^{(2)} + P_{il}^{(2)} \partial_l P_{jk}^{(1)} \right) \partial_i f \partial_j g \partial_k h + \text{ cyc. perm.} = 0$$

Since no second order derivatives occur, it is sufficient to check it on the linear functions $f = f_X$, $g = f_Y$, and $h = f_Z$. The condition becomes:

$$\langle \mathcal{D}_1(X), d\langle \mathcal{D}_2(Y), Z \rangle \rangle + \langle \mathcal{D}_2(X), d\langle \mathcal{D}_1(Y), Z \rangle \rangle + \text{ cyc. perm.} = 0$$

where d is the differential on phase space. We have:

$$d\langle \mathcal{D}_1(X), Y \rangle = [X, Y]$$

$$d\langle \mathcal{D}_2(X), Y \rangle = X(LY)_- - (YL)_- X + (XL)_- Y - Y(LX)_-$$

$$+ (YLX - XLY)_- + \frac{1}{n+1} \left\{ [(\partial^{-1}[Y, L]_{-1}), X] - [(\partial^{-1}[X, L]_{-1}), Y] \right\}$$

Inserting this into the above condition, one verifies that it is indeed satisfied.   ∎

One of the main advantages of bihamiltonian structures is that they automatically produce commuting Hamiltonians, as we now explain.

Let $\{\ ,\ \}_1$ and $\{\ ,\ \}_2$ be two compatible Poisson brackets, and consider the linear combination

$$\{\ ,\ \}_\lambda = \{\ ,\ \}_1 - \lambda \{\ ,\ \}_2$$

Let us assume the existence of H_λ, a Casimir function of the Poisson bracket $\{\ ,\ \}_\lambda$. This means

$$\{H_\lambda, f\}_\lambda = 0 \quad \forall f \tag{10.46}$$

Suppose that we can expand $H_\lambda = \sum_{n=0}^{\infty} H_n \lambda^n$. Then the above relation gives

$$\{H_0, f\}_1 = 0, \quad \{H_n, f\}_1 = \{H_{n-1}, f\}_2, \quad \forall f \tag{10.47}$$

This shows that the flows generated by the H_n are related by recursion relations of the Lenard type. Moreover, the H_n are in involution with respect to both Poisson brackets $\{\ ,\ \}_{1,2}$. The proof is done by induction. First, by eq. (10.47), one has $\{H_0, H_n\}_1 = 0, \forall n$. Suppose that we have shown that $\{H_m, H_n\}_1 = 0, \forall n$, then we have $\{H_{m+1}, H_n\}_1 = 0, \forall n$. This is because, using the recursion relations, we have

$$\{H_{m+1}, H_n\}_1 = \{H_m, H_n\}_2 = -\{H_n, H_m\}_2 = -\{H_{n+1}, H_m\}_1 = 0$$

Hence the H_n are in involution with respect to the first Poisson bracket. But by the recursion relations, we have $\{H_m, H_n\}_2 = \{H_{m+1}, H_n\}_1 = 0$, so they are also in involution with respect to the second bracket.

All this means that bihamiltonian structures are consubstantial to integrable systems. The unique feature of the present situation is that both Poisson brackets are local.

10.9 The Drinfeld–Sokolov reduction

The two compatible Poisson brackets $\{\ ,\ \}_1$ and $\{\ ,\ \}_2$ have a nice Lie algebraic interpretation, which we now explain. They can be obtained, through Hamiltonian reduction, from the Kostant–Kirillov bracket on coadjoint orbits of central extensions of loop algebras.

Consider the loop algebra of traceless $(n+1) \times (n+1)$ matrices $U(x)$, with matrix elements functions of x. Let $\mathcal{G}$ be the central extension of this loop algebra. It consists of pairs $(U(x), u)$ also denoted by $U(x) + u\,K$, where u is a number, and K is called the central element. The commutator on $\mathcal{G}$ is defined as:

$$[(U(x), u), (V(x), v)] \equiv (\,[U(x), V(x)]\,,\, \omega(U, V)\,)$$

where in the right-hand side, $[U(x), V(x)]$ is the loop algebra commutator and $\omega(U, V)$ is a bilinear antisymmetric form. The central element $K = (0, 1)$ commutes with everything. The Jacobi identity for this bracket reduces to the cocycle condition on ω:

$$\omega([U, V], W) + \omega([W, U], V) + \omega([V, W], U) = 0$$

Note that this is a linear condition on ω, so the sum of two cocycles is a cocycle. Trivial cocycles are given by $\Sigma([U, V])$ where Σ is a linear form on the loop algebra. Such a linear form can be written as

$$\Sigma(U) = \int dx\, \mathrm{Tr}\,(\Sigma(x)U(x))$$

where we used the natural invariant scalar product on the loop algebra:

$$(U, V) = \int dx\, \mathrm{Tr}\,(U(x)V(x))$$

The standard non-trivial cocycle is:

$$\omega_0(U, V) = \int dx\, \mathrm{Tr}\,(U(x)\partial_x V(x))$$

and we can take for ω any linear combination

$$\omega(U, V) = \omega_0(U, V) + \Sigma([U, V])$$

The dual $\mathcal{G}^*$ of $\mathcal{G}$ can be identified with $\mathcal{G}$ using the non-degenerate bilinear form $(U + u\,K, V + v\,K) = (U, V) + uv$. Then the coadjoint action of $\mathcal{G}$ on $\mathcal{G}^*$ reads:

$$\mathrm{ad}^*_{(V,v)}(U, u) = (u\partial_x V - [U + \Sigma, V], 0) \qquad (10.48)$$

To see it, we apply the definition

$$\left(\mathrm{ad}^*_{(V,v)}(U, u)\right)(W, w) = -\Big((U, u), [(V, v), (W, w)]\Big)$$
$$= -(U, [V, W]) - u\omega(V, W)$$
$$= \int dx\, \mathrm{Tr}\,\Big((-[U, V] + u\partial V - [\Sigma, V])W\Big)$$

We see that u is invariant by the coadjoint action eq. (10.48), so in the following we fix it to the value $u = 1$.

The coadjoint action of (V, v) becomes a gauge transformation on the operator $\partial - U - \Sigma$, namely:

$$\mathrm{ad}^*_{(V,v)}(U, 1) = (U', 0), \quad \text{with } U' = \partial V - [U + \Sigma, V] \qquad (10.49)$$

By construction any orbit of the gauge action in $\mathcal{G}^*$ is equipped with an invariant symplectic form, the Kostant–Kirillov form. Explicitly, at the point U, the induced Poisson bracket reads:

$$\{f, g\}(U) = \left(\mathrm{ad}^*_{df}(U) \right)(dg) = \left(\partial df - [U, df], dg \right) - \left(\Sigma, [df, dg] \right) \quad (10.50)$$

where the differentials df and dg of functions on $\mathcal{G}^*$ are viewed as elements of $\mathcal{G}$. The two terms in the right-hand side of eq. (10.50) obviously satisfy the Jacobi identity separately, and so does their sum, hence they define two compatible Poisson brackets. We show below that, with a proper choice of Σ and an appropriate symplectic reduction, these two Poisson brackets reduce to the brackets $\{\,,\,\}_1$ and $\{\,,\,\}_2$ considered in the previous section, which are then compatible by construction.

We choose to reduce by a subgroup of the loop group, namely the loop group N_- of lower triangular matrices with 1 on the diagonal. The dual of its Lie algebra can be identified with the loop algebra $\mathcal{N}_+$ of strictly upper triangular matrices.

The Hamiltonian which generates the gauge action by (V, v) is simply $H_{(V,v)}(U, u) = (U, V)$ as in the general theory, see Chapter 14. Alternatively, this follows from eq. (10.50) with $f_V(U) = (U, V)$ and $df_V = V$. The moment at the point $(U, 1)$ is $\mathcal{P}(U) = P_{\mathcal{N}_+} U$.

To perform the Hamiltonian reduction, we must set $\mathcal{P}(U)$ to a fixed value $\mu \in \mathcal{N}_+$, which determines the nature of the reduced symplectic manifold. We take:

$$\mu = \sum_{i=1}^{n} E_{i,i+1} = \mathcal{E}_+ \in \mathcal{N}_+$$

where $\mathcal{E}_+$ is the sum of simple root vectors of the Lie algebra $sl(n+1)$ in the vector representation.

So far, we did not specify the form $\Sigma(U)$ in the central extension we considered. We require now that $\Sigma \in \mathcal{N}_-$, and more precisely:

$$\Sigma = \alpha \Sigma_0, \quad \text{with } \Sigma_0 = E_{n+1,1} \in \mathcal{N}_- \qquad (10.51)$$

This choice will lead to the Poisson bracket $\{\,,\,\}_1$. It has the important property that $[\Sigma, V] = 0$ for any $V \in \mathcal{N}_-$, so that the coadjoint action of

N_- reduces to a gauge action on $\partial - U$. Moreover, the group of stability of μ is the whole group N_-, as we now show. The variation of the moment under the coadjoint action of N_- is given by

$$\delta_V \mu = P_{\mathcal{N}_+} (\partial V - [\mu + \Sigma, V])$$

where $V \in \mathcal{N}_-$. Due to the specific form of the momentum μ, the commutator $[\mu, V]$ cannot have matrix elements above the diagonal and is killed by the projection on $\mathcal{N}_+$. Similarly, ∂V is lower triangular and does not survives the projection. Finally, we recall that $[\Sigma, V] = 0$ for the specific choice of Σ we made. Altogether $\delta_V \mu = 0$.

The matrices $U(x)$ such that $P_{\mathcal{N}_+} U = \mu$ have the form $U(x) = B(x) + \mu$, where $B(x)$ is a lower triangular matrix, including the diagonal. The reduced phase space is obtained by quotienting by the group N_- with group action $U \to U' = \partial V + [U, V]$ with $V \in \mathcal{N}_-$. Note that this leaves the form of U invariant.

Alternatively, the reduced phase space can be identified with the set of differential operators $\partial - B - \mu$ quotiented by the group N_- acting by gauge transformations.

One can use this gauge action to bring $\partial - B - \mu$ to either one of the two forms (note that B has $(n+1)$ more parameters than N_-):

$$D_p = \partial - \begin{pmatrix} p_1 & 1 & 0 & \cdots \\ 0 & p_2 & 1 & \cdots \\ \vdots & & \ddots & \\ 0 & \cdots & & 1 \\ 0 & \cdots & \cdots & 0 & p_{n+1} \end{pmatrix}, \quad D_u = \partial - \begin{pmatrix} 0 & 1 & 0 & \cdots \\ 0 & 0 & 1 & \cdots \\ \vdots & & \ddots & \\ 0 & \cdots & & 0 & 1 \\ u_0 & u_1 & \cdots & & u_n \end{pmatrix}$$

$$(10.52)$$

Since we consider the loop algebra of traceless matrices we have $\sum_i p_i = 0$ and $u_n = 0$. The sets $(p_1, \dots, p_{n+1})$ and $(u_0, \dots, u_n)$ constitute two different coordinate systems on the reduced phase space $F_\mu = \mathcal{M}_\mu / N_-$.

Note that with any point in $\mathcal{M}_\mu$, i.e. with any matrix differential operator $D = \partial - B - \mu$, one can associate a scalar differential operator of order $(n+1)$. To do that we consider the matrix differential equation $D\Psi = 0$ and write the differential equation of order $(n+1)$ induced on the first component ψ_1 of Ψ, which is of the form $L\psi_1 = 0$. Since the group action of N_- on Ψ leaves ψ_1 invariant, this differential equation is *invariant* under gauge transformations, so the coefficients of the equation are invariant functions under N_-. For the two particular forms D_p, D_u, we get:

$$L = (\partial - p_{n+1}) \cdots (\partial - p_1), \quad L = \partial^{n+1} - u_{n-1}\partial^{n-1} - \cdots - u_1\partial - u_0$$

It remains to express the reduced Poisson bracket in terms of the invariant operator L. We recall that the reduced bracket of invariant functions can be computed straightforwardly using the Poisson bracket on the unreduced phase space, see Chapter 14. We take as invariant functions the functions $f_X = \langle LX \rangle$, where X is any pseudo-differential operator on $\mathcal{P}_-$, and $\langle \ \rangle$ is the Adler trace. Hence we have at the point $(p + \mu)$ (where p is diagonal)

$$\{f_X, f_Y\}_{\text{reduced}} = (\text{ad}^*_{df_X}(p + \mu), df_Y) \tag{10.53}$$

Separating the ω_0 and Σ parts in the cocycle definition, we can write also

$$\{f_X, f_Y\}_{\text{reduced}} = \{f_X, f_Y\}_{\omega_0} + \{f_X, f_Y\}_{\Sigma}$$

with

$$\{f_X, f_Y\}_{\omega_0} = (\partial df_X - [p + \mu, df_X], df_Y), \quad \{f_X, f_Y\}_{\Sigma} = -([\Sigma, df_X], df_Y)$$

To compute df_X and df_Y, we first need to compute the variation of L when $p + \mu \to p + \mu + b$, where b is a small lower triangular matrix (including diagonal). Writing the system $(\partial - p - b - \mu)\Psi = 0$ in terms of ψ_1 and keeping only terms of first order in b we find, with the notations of eq. (10.42):

$$\delta L = - \sum_{i \geq j} L_{n+1,i+1} b_{ij} L_{j-1,1}$$

The differential df_X is defined by the relation $(df_X, b) = \langle \delta L X \rangle$ so that df_X is the upper triangular traceless matrix:

$$(df_X)_{ji} = -\left(L_{j-1,1} X L_{n+1,i+1}\right)_{-1} + \frac{1}{n+1}\delta_{ij} \sum_k \left(L_{k-1,1} X L_{n+1,k+1}\right)_{-1} \tag{10.54}$$

We are now in a position to prove the main result of this section:

Proposition. *One has*

$$\{f_X, f_Y\}_{\omega_0} = \{f_X, f_Y\}_2, \quad \{f_X, f_Y\}_{\Sigma} = \{f_X, f_Y\}_1$$

Proof. We start with the coadjoint action:

$$\text{ad}^*_{df_X}(p + \mu) = \partial df_X - [p + \mu + \Sigma, df_X]$$

Noting that df_Y is upper triangular in eq. (10.53), we need only keep the lower triangular part in $\text{ad}^*_{df_X}(p + \mu)$. Using eq. (10.49) (with $V = df_X$ and $U = p + \mu$), we remark that the Σ *independent* term in this expression is upper triangular, so that we need only keep the diagonal part in this term, that is $(df_X)_{kk}$. We get:

$$\{f_X, f_Y\}_{\text{reduced}} = \sum_k (\partial(df_X)_{kk})(df_Y)_{kk} - ([\Sigma, df_X], df_Y)$$

The first term is just $\{f_X, f_Y\}_{\omega_0}$. Substituting the expressions of df_X and df_Y it immediately yields eq. (10.43), which shows that it coincides with the bracket $\{\ ,\ \}_2$. This also shows that the Poisson bracket of the diagonal coordinates $p_i(x)$ is given by eq. (10.41).

We now look at the Σ dependent term which is $\{f_X, f_Y\}_\Sigma$ and show that it reproduces the bracket $\{\ ,\ \}_1$. Due to the choice $\Sigma = \alpha\Sigma_0$ we have to compute:

$$([\Sigma_0, df_X], df_Y) = \sum_{j=1}^{n+1} (df_X)_{1,j}(df_Y)_{j,n+1} - (df_Y)_{1,j}(df_X)_{j,n+1}$$

Inserting the expressions eq. (10.54) for df_X and df_Y, and noting that the terms proportional to the Kroneker deltas do not contribute, we get:

$$([\Sigma_0, df_X], df_Y) = \sum_{j=1}^{n+1} \int dx \, (XL_{n+1,j+1})_{-1}(L_{j-1,1}Y)_{-1}$$
$$-(YL_{n+1,j+1})_{-1}(L_{j-1,1}X)_{-1}$$

Using eq. (10.45), this can be rewritten in terms of Adler traces:

$$([\Sigma_0, df_X], df_Y) = \sum_{j=1}^{n+1} \langle (XL_{n+1,j+1})_- L_j(L_{j-1,1}Y)_- \rangle$$
$$-\langle (YL_{n+1,j+1})_- L_j(L_{j-1,1}X)_- \rangle$$

Substituting everywhere $U_- = U - U_+$ and using the isotropy of $\mathcal{P}_\pm$ so that $\langle UV_+ \rangle = \langle U_-V \rangle$, this reads:

$$([\Sigma_0, df_X], df_Y) = \sum_{j=1}^{n+1} \langle XLY \rangle - \langle (XL_{n+1,j})_- L_{j-1,1}Y \rangle$$
$$-\langle (XL_{n+1,j+1})_+ L_{j,1}Y \rangle \ - \ (X \leftrightarrow Y)$$

In the sum the first term yields $(n+1)\langle XLY \rangle$, while the second and third terms regroup themselves as

$$-\sum_{j=2}^{n+1} \langle \left((XL_{n+1,j})_- + (XL_{n+1,j})_+ \right) L_{j-1,1}Y \rangle = -n\langle XLY \rangle$$

and we finally get:

$$-([\Sigma_0, df_X], df_Y) = -\langle XLY \rangle + \langle YLX \rangle = \langle L[X,Y] \rangle = \{f_X, f_Y\}_1$$

∎

This construction, due to Drinfeld and Sokolov, can be generalized to
Lie algebras other than $sl(n+1)$, replacing $\mathcal{E}_+$ by the sum of simple root
vectors and Σ_0 by the root vector $E_{-\alpha}$, where α is the longest root. This
also shows the nice interplay between different Lie algebra structures (the
one induced by the algebra of pseudo-differential operators, and the Kac–
Moody one) producing the same Kostant–Kirillov Poisson brackets, after
suitable Hamiltonian reduction.

10.10 Whitham equations

In many cases, solutions of non-linear partial differential equations take
the form of modulated wavetrains, i.e. at small scale they look like sinu-
soidal solutions, but at large scale the parameters of the sinusoid slowly
evolve. Whitham equations describe the slow variations of these parameters. It turns out that algebro-geometric solutions of KP are particularly
well suited examples of Whitham analysis.

In the algebro-geometric solutions, the field $u = -2q_{-1}$ of the KP hierarchy is of the form:

$$u(t) = u_0 \left(\sum_i t_i U^{(i)}(m), m \right)$$

where t_i are the KP time variables and m denotes the moduli of the
Riemann surface, Γ, used to build the solution. The quantities $U^{(i)}(m)$,
defined in eq. (10.17), are functions of the moduli only. The vector

$$V = \sum_i t_i U^{(i)}(m) \tag{10.55}$$

lives on the Jacobian of Γ and $u(t)$ is a pseudo-periodic function of each
time t_i.

We now look for solutions of KP, close to these algebro-geometric solutions, but where the moduli m slowly evolve. To describe the slow modulation, we introduce the large scale variables $T_i = \epsilon t_i$, and we express the
idea of a modulated wavetrain by searching for $u(t)$ in the form

$$u(t) = u_0(\epsilon^{-1} S(T), m(T)) + \epsilon u_1(T) + \cdots, \qquad T_i = \epsilon t_i \tag{10.56}$$

Our purpose is to find the equations for $S(T)$ and $m(T)$ such that $u(t)$ in
eq. (10.56) is a solution of the KP equation to first order in ϵ, valid over
a time scale $t \sim \epsilon^{-1}$. This means that the first order term must remain
uniformly bounded over a period of time ϵ^{-1}.

To this aim, we take advantage that time dependence in the algebro-
geometric solution is entirely contained in the variable V, eq. (10.55), so

that we can write it explicitly as a function of V and m. Once this is done, we consider V and m as independent variables. We *postulate* the equations of motion for $V = \epsilon^{-1}S$:

$$\partial_{T_i} S = U^{(i)}(m(T)) \tag{10.57}$$

These equations are obviously satisfied for the modulated solutions. As a consequence, we can write the time derivatives of $u(t)$ in eq. (10.56) to order ϵ as

$$\partial_{t_i} u(t) = U^{(i)} \cdot \partial_V u_0 + \epsilon(\partial_{T_i} u_0 + \partial_{t_i} u_1(t)) \tag{10.58}$$

where the slow time derivatives are defined by:

$$\partial_{T_i} u_0 = \sum_j (\partial_{T_i} m_j) \partial_{m_j} u_0$$

They come from the variation of the moduli only.

Note that eq. (10.57) already imposes constraints on the time evolution of the moduli. Specifically, integrability conditions of eq. (10.57) imply

$$\partial_{T_i} U^{(j)} = \partial_{T_j} U^{(i)} \tag{10.59}$$

The slow modulation equations will have to be compatible with these constraints.

The equations for the time evolution of the moduli we are aiming at are the Whitham equations, eqs. (10.77). The main idea of the derivation, which is rather long, is to average over fast oscillations and retain terms involving only slow modulations. In the algebro-geometric setting, the specific feature we use is that, the time flow being a linear motion on the Jacobian torus, by the ergodicity theorem we can replace the fast time average by average over the torus.

Let us now start from the linear system satisfied by the Baker–Akhiezer function, and limit ourselves to the first three times:

$$\partial_{t_2} \Psi = (Q^2)_+ \Psi, \quad \partial_{t_3} \Psi = (Q^3)_+ \Psi$$

or

$$(\partial_{t_2} - L)\Psi = 0, \quad L \equiv \partial_x^2 - u \tag{10.60}$$

$$(\partial_{t_3} - A)\Psi = 0, \quad A \equiv \partial_x^3 - \frac{3}{2}u\partial_x - v \tag{10.61}$$

where we set $v = -\frac{3}{2}\partial_x u - 3q_{-2}$, and we identify $t_1 = x$, $T_1 = X$. The compatibility condition of this system

$$F \equiv \partial_{t_2} A - \partial_{t_3} L - [L, A] = 0$$

is the KP equation.

Proposition. *Let us denote by $L = L_0 + \epsilon L_1 + \cdots$ and $A = A_0 + \epsilon A_1 + \cdots$ the operators corresponding to the small perturbation eq. (10.56) of a finite-zone solution with L_1, A_1 linear in u_1. To order ϵ the zero curvature condition reduces to:*

$$F_1 + \partial_{T_2} A_0 - \partial_{T_3} L_0 + L_0^{(1)} \partial_X A_0 - A_0^{(1)} \partial_X L_0 = 0 \qquad (10.62)$$

with $L_0^{(1)} = -2\partial_x$, $A_0^{(1)} = -3\partial_x^2 + \frac{3}{2} u_0$ and

$$F_1 = \partial_{t_2} A_1 - \partial_{t_3} L_1 - [L_1, A_0] - [L_0, A_1]$$

<u>Proof</u>. We suppose that the perturbed operators satisfy the zero curvature equation. Since $L = L_0 + \epsilon L_1 + \cdots$ and $A = A_0 + \epsilon A_1 + \cdots$, one has

$$F = F_0 + \epsilon F_1 + \cdots \qquad (10.63)$$

It is important to realize, however, that the leading term F_0 also produces a correction of order ϵ due to the deformation of ∂_{t_i}, see eq. (10.58). To extract this term, let us write $L_0 = \sum_i l_i \partial^i$, $A_0 = \sum_j a_j \partial^j$, with l_i, a_j functions of u_0 and its derivatives. The product is

$$L_0 A_0 = \sum_{i,j,k} \binom{i}{k} l_i (\partial^k a_j) \partial^{i+j-k}$$

The term of order ϵ induced by eq. (10.58) in $(\partial^k a_j)$ is $\epsilon k \partial_X \partial^{k-1} a_j$. Hence the first order term is equal to $-\epsilon L_0^{(1)} \partial_X A_0$, where $L_0^{(1)} = -\sum_{i>0} i l_i \partial^{i-1}$. One gets the similar contribution $-\epsilon A_0^{(1)} \partial_X L_0$ from the product $A_0 L_0$ with a similar definition of $A_0^{(1)}$. ∎

We will get rid of F_1 in eq. (10.63) by an averaging procedure, leading to a direct determination of the variation of the moduli in terms of the slow variables.

To work conveniently with these averages, we introduce a definition. For D a differential operator, we define the differential operators $D^{(j)}$ by

$$(D^* f) g = \sum_{j \geq 0} \partial^j (f D^{(j)} g) \qquad (10.64)$$

To show how this definition works, consider $D = a \partial^i$, so that $(D^* f) \cdot g = ((-)^i \partial^i (af)) \cdot g$. Let us write $\partial = \partial_1 + \partial_2$, where $\partial_{1,2}$ act respectively on the first and second factor around the dot:

$$\partial(f \cdot g) = \partial_1 f \cdot g + f \cdot \partial_2 g \equiv \partial f \cdot g + f \cdot \partial g$$

This is just a way to encode the Leibnitz rule. Then

$$(D^*f) \cdot g = (\partial_2 - \partial)^i af \cdot g = \sum_{j \geq 0} (-1)^j \binom{i}{j} \partial^j (fa\partial^{i-j}g)$$

So, for $D = a\partial^i$, we get $D^{(j)} = (-1)^j \binom{i}{j} a\partial^{i-j}$. In particular, by linearity, we get for any differential operator $D = \sum_i a_i \partial^i$

$$D^{(0)} = D, \quad D^{(1)} = -\sum_{i>0} ia_i\partial^{i-1}$$

Note that the notation $D^{(1)}$ is consistent with the notation $L_0^{(1)}$, $A_0^{(1)}$ introduced earlier. We also have the identity:

$$(D_1 D_2)^{(j)} = \sum_{k=0}^{j} D_1^{(k)} D_2^{(j-k)} \tag{10.65}$$

This follows from the binomial identity

$$\sum_p \binom{n}{j-p}\binom{m-k}{p} = \binom{m}{k}\binom{n+m-k}{j}$$

Let Ψ_0 and Ψ_0^* be the Baker–Akhiezer functions corresponding to the exact algebro-geometric solution u_0. Recall that Ψ_0 and Ψ_0^* can be written in the form (see eq. (10.20))

$$\Psi_0 = e^{\sum_i t_i \mathcal{P}^{(i)}(P)}\varphi(P, V, m), \quad \Psi_0^* = e^{-\sum_i t_i \mathcal{P}^{(i)}(P)}\varphi^*(P, V, m) \tag{10.66}$$

where $\varphi(P, V, m), \varphi(P, V, m)^*$ are periodic, of period 1 in each component of V, and so bounded in V. We have introduced the notation

$$\mathcal{P}^{(i)}(P) = \int_{P_\infty}^{P} \Omega^{(i)}$$

where the forms $\Omega^{(i)}$ have purely imaginary periods over any cycle.

In these Baker–Akhiezer functions, we make the substitution $V \to S/\epsilon$. They satisfy the equations $(\partial_{t_2} - L_0)\Psi_0 = O(\epsilon)$ and $(\partial_{t_3} - A_0)\Psi_0 = O(\epsilon)$. The right-hand sides are not zero because in L_0 and A_0 we made the substitution $V \to S/\epsilon$.

Proposition. *We have the identity*

$$\partial_{t_2}(\Psi_0^* A_1 \Psi_0) - \partial_{t_3}(\Psi_0^* L_1 \Psi_0)$$
$$= \Psi_0^* F_1 \Psi_0 + \sum_{j \geq 1} \partial^j \left(\Psi_0^*(A_0^{(j)} L_1 - L_0^{(j)} A_1)\Psi_0\right) + O(\epsilon) \tag{10.67}$$

<u>Proof.</u> From eq. (10.23), and the renaming of the operators Q^i as in eqs. (10.60, 10.61), we have

$$\partial_{t_2}(\Psi_0^* A_1 \Psi_0) = -(L_0^* \Psi_0^*) A_1 \Psi_0 + \Psi_0^* \partial_{t_2} A_1 \Psi_0 + \Psi_0^* A_1 L_0 \Psi_0 + O(\epsilon)$$

$$= \Psi_0^* (\partial_{t_2} A_1 + [A_1, L_0]) \Psi_0 - \sum_{j \geq 1} \partial^j (\Psi_0^* L_0^{(j)} A_1 \Psi_0) + O(\epsilon)$$

Writing the second term $\partial_{t_3}(\Psi_0^* L_1 \Psi_0)$ in a similar way yields the result. ∎

We now take the average of eq. (10.67) over the times t_1, t_2, t_3. This average is taken over a time scale ℓ which is large compared to 1 but small compared to $1/\epsilon$. For the quantity $\mathcal{O}$, we denote this average by

$$\langle\langle\mathcal{O}\rangle\rangle_{t_i} = \frac{1}{2\ell} \int_{-\ell}^{\ell} \mathcal{O} \, dt_i$$

Over this time scale, the point V describes in general an almost dense trajectory on the torus, so that the average can also be interpreted as an average on the torus. The time scales are chosen so that the moduli can be considered as constant in the averaging.

In agreement with our hypothesis, we assume that L_1 and A_1 remain bounded when the t_i evolve in an interval of order $O(\epsilon^{-1})$. Note that in eq. (10.67), the exponential factors cancel between Ψ_0 and Ψ_0^*. Since the average of derivatives of bounded functions vanishes, only one term survives in the averaging of eq. (10.67) and we get:

$$\langle\langle\Psi_0^* F_1 \Psi_0\rangle\rangle_{t_1,t_2,t_3} = 0$$

Hence, by averaging eq. (10.62), we get an equation, valid at order ϵ:

$$\langle\langle\Psi_0^*(\partial_{T_2} A_0 - \partial_{T_3} L_0 + L_0^{(1)} \partial_X A_0 - A_0^{(1)} \partial_X L_0)\Psi_0\rangle\rangle_{t_1,t_2,t_3} = 0 \quad (10.68)$$

In this equation all quantities are computed with the exact algebro-geometric solution u_0. In the following we shall drop the suffix 0.

The next two propositions are devoted to the computation of the various terms in this equation.

Proposition. *With the parametrization eq. (10.66) of the Baker–Akhiezer function, we have:*

$$\langle\langle\Psi^* \partial_{T_3} L\Psi\rangle\rangle = \partial_{T_3}\mathcal{P}^{(2)}\langle\langle\varphi^*\varphi\rangle\rangle + \partial_{T_3}U_j^{(2)}\langle\langle\varphi^*\varphi_j\rangle\rangle$$
$$+ \partial_{T_3}U_j^{(1)}\langle\langle\varphi^*\widehat{L}^{(1)}\varphi_j\rangle\rangle \quad (10.69)$$

$$\langle\langle\Psi^* \partial_{T_2} A\Psi\rangle\rangle = \partial_{T_2}\mathcal{P}^{(3)}\langle\langle\varphi^*\varphi\rangle\rangle + \partial_{T_2}U_j^{(3)}\langle\langle\varphi^*\varphi_j\rangle\rangle$$
$$+ \partial_{T_2}U_j^{(1)}\langle\langle\varphi^*\widehat{A}^{(1)}\varphi_j\rangle\rangle \quad (10.70)$$

and

$$\langle\langle \Psi^*(L^{(1)}\partial_X A - A^{(1)}\partial_X L)\Psi\rangle\rangle = \partial_X \mathcal{P}^{(3)}\langle\langle\Psi^* L^{(1)}\Psi\rangle\rangle$$
$$- \partial_X \mathcal{P}^{(2)}\langle\langle\Psi^* A^{(1)}\Psi\rangle\rangle + \partial_X U_j^{(3)}\langle\langle\varphi^* \widehat{L}^{(1)}\varphi_j\rangle\rangle$$
$$- \partial_X U_j^{(2)}\langle\langle\varphi^* \widehat{A}^{(1)}\varphi_j\rangle\rangle \qquad (10.71)$$

where $\varphi_j = \partial_{V_j}\varphi$ *and* $\widehat{L}^{(1)} = e^{-\sum_i t_i \mathcal{P}^{(i)}} L^{(1)} e^{\sum_i t_i \mathcal{P}^{(i)}}$ *and similarly for* $\widehat{A}^{(1)}$.

<u>Proof.</u> Let us choose two Riemann surfaces $\Gamma(m)$ and $\Gamma' \simeq \Gamma(m + \delta m)$. Comparison of functions defined on different Riemann surfaces requires a "connection". This is achieved by choosing a meromorphic function on each Riemann surface and keeping it fixed. We choose to keep $\mathcal{P}^{(1)}$ fixed. Let Ψ and Ψ' be corresponding Baker–Akhiezer functions on Γ and Γ'. Consider the expression

$$\partial_{t_2}(\Psi^*\Psi') = -(L^*\Psi^*)\Psi + \Psi^* L'\Psi'$$
$$= \Psi^* L'\Psi' - \sum_{j\geq 0}\partial^j(\Psi^* L^{(j)}\Psi')$$
$$= \Psi^*(L' - L)\Psi' - \sum_{j\geq 1}\partial^j(\Psi^* L^{(j)}\Psi')$$

Subtracting the same equation for $\Psi' = \Psi$, we get

$$\partial_{t_2}(\Psi^*(\Psi' - \Psi)) = \Psi^*(L' - L)\Psi' - \sum_{j\geq 1}\partial^j(\Psi^* L^{(j)}(\Psi' - \Psi))$$

If $\Psi' = \Psi + \delta\Psi$, this gives:

$$\partial_{t_2}(\Psi^*\delta\Psi) = (\Psi^*\delta L\Psi) - \sum_{j\geq 1}\partial^j(\Psi^* L^{(j)}\delta\Psi) \qquad (10.72)$$

Now from eq. (10.66) we have

$$\delta\Psi = e^{\sum_i t_i \mathcal{P}^{(i)}}\left(\delta m\,\partial_m\varphi + \sum_{i,j} t_i \delta U_j^{(i)}\varphi_j + \sum_i t_i \delta\mathcal{P}^{(i)}\varphi\right) \qquad (10.73)$$

where we recall that $\varphi_j = \partial_{V_j}\varphi$.

We now average eq. (10.72). In the left-hand side, in the average over t_2, the terms which do not contain an explicit factor t_2 vanish because they are the average of a derivative of a bounded function. The terms containing an explicit factor t_2 are treated by first averaging over t_2 on the interval $[-\ell, \ell]$:

$$\langle\langle \partial_{t_2}(t_2 f(t_1, t_2, t_3))\rangle\rangle_{t_1, t_2, t_3} = \frac{1}{2\ell}\langle\langle \ell f(t_1, \ell, t_3) + \ell f(t_1, -\ell, t_3)\rangle\rangle_{t_1, t_3}$$
$$= \langle\langle f \rangle\rangle$$

where $\langle\langle f \rangle\rangle$ means average on the torus. We treat similarly the average over $t_1 = x$ in the right-hand side. Note that since we have kept $\mathcal{P}^{(1)}$ fixed, there is no $\delta\mathcal{P}^{(1)}$ contribution.

Interpreting δ as a small variation of the moduli m in the direction T_3, we arrive at

$$\langle\langle \Psi^* \partial_{T_3} L\Psi \rangle\rangle = \partial_{T_3}\mathcal{P}^{(2)}\langle\langle\varphi\varphi^*\rangle\rangle + \partial_{T_3}U_j^{(2)}\langle\langle\varphi_j\varphi^*\rangle\rangle$$
$$+\partial_{T_3}U_j^{(1)}\langle\langle\varphi^* \widehat{L}^{(1)}\varphi_j\rangle\rangle$$

where the last term comes from the term $j = 1$ in eq. (10.72). In the same way, we get

$$\langle\langle \Psi^* \partial_{T_2} A\Psi \rangle\rangle = \partial_{T_2}\mathcal{P}^{(3)}\langle\langle\varphi\varphi^*\rangle\rangle + \partial_{T_2}U_j^{(3)}\langle\langle\varphi_j\varphi^*\rangle\rangle$$
$$+\partial_{T_2}U_j^{(1)}\langle\langle\varphi^* \widehat{A}^{(1)}\varphi_j\rangle\rangle$$

This proves eqs. (10.69, 10.70).

To prove eq. (10.71), note that the vanishing of the curvature, $F = 0$, implies $0 = (F^*g)f = \sum_j \partial^j(gF^{(j)}f)$, and we deduce that $F^{(j)} = 0, \forall j$. By eq. (10.65), this can be written as

$$\partial_{t_3}L^{(j)} - \partial_{t_2}A^{(j)} + \sum_{k=0}^{j}[L^{(k)}, A^{(j-k)}] = 0$$

This relation implies, by performing the time derivatives with eqs. (10.60, 10.61), the identity

$$\sum_{j\geq 1}\partial^{j-1}\left(\partial_{t_3}(\Psi^* L^{(j)}\Psi') - \partial_{t_2}(\Psi^* A^{(j)}\Psi')\right)$$
$$= \sum_{j\geq 1}\partial^{j-1}\left(\Psi^* L^{(j)}(A' - A)\Psi' - \Psi^* A^{(j)}(L' - L)\Psi'\right)$$

Averaging this equation with $\Psi' = \Psi + \delta\Psi$ we obtain:

$$\langle\langle \partial_{t_3}(\Psi^* L^{(1)}\delta\Psi) - \partial_{t_2}(\Psi^* A^{(1)}\delta\Psi)\rangle\rangle = \langle\langle \Psi^* L^{(1)}\delta A\Psi - \Psi^* A^{(1)}\delta L\Psi\rangle\rangle$$

Indeed, the order zero term, i.e. $\Psi' = \Psi$, gives vanishing averages because it is always a derivative of a bounded function. The first order term (in $\delta\Psi$), produces potentially dangerous terms linear in the time variables. However, the averages vanish when $j \geq 2$ because we have at least two derivatives. The average finally reduces to eq. (10.71) when we interpret $\delta = \partial_X$. ∎

Using the results of this proposition, eq. (10.68) becomes:

$$0 = \left(\partial_{T_2}\mathcal{P}^{(3)} - \partial_{T_3}\mathcal{P}^{(2)}\right)\langle\langle\Psi^*\Psi\rangle\rangle$$
$$+\partial_X\mathcal{P}^{(3)}\langle\langle\Psi^*L^{(1)}\Psi\rangle\rangle - \partial_X\mathcal{P}^{(2)}\langle\langle\Psi^*A^{(1)}\Psi\rangle\rangle \quad (10.74)$$

The terms $\langle\langle\varphi^*\varphi_j\rangle\rangle$ cancel because we assumed $U^{(i)} = \partial_{T_i}S$, and so the compatibility condition eq. (10.59) holds. For the same reason the terms $\langle\langle\varphi^*\widehat{L}^{(1)}\varphi_j\rangle\rangle$ and $\langle\langle\varphi^*\widehat{A}^{(1)}\varphi_j\rangle\rangle$ also cancel.

The last step in our derivation of the Whitham equations consists of evaluating the averages $\langle\langle\Psi^*L^{(1)}\Psi\rangle\rangle$ and $\langle\langle\Psi^*A^{(1)}\Psi\rangle\rangle$.

Proposition. *Let $\Omega^{(i)}$ be the second kind Abelian differentials with pure imaginary periods used to construct the Baker–Akhiezer function on Γ. We have:*

$$\langle\langle\Psi^*L^{(1)}\Psi\rangle\rangle\Omega^{(1)} = -\langle\langle\Psi^*\Psi\rangle\rangle\Omega^{(2)} \quad (10.75)$$
$$\langle\langle\Psi^*A^{(1)}\Psi\rangle\rangle\Omega^{(1)} = -\langle\langle\Psi^*\Psi\rangle\rangle\Omega^{(3)} \quad (10.76)$$

<u>Proof</u>. Consider eq. (10.72) with $\delta = d$ now representing the differential on the curve Γ. Since $\delta L = 0$, it reduces to:

$$\partial_{t_2}(\Psi^*d\Psi) = -\partial_x(\Psi^*L^{(1)}d\Psi) - \sum_{j\geq 2}\partial^j(\Psi^*L^{(j)}d\Psi)$$

We have, recalling that $d\mathcal{P}^{(i)} = \Omega^{(i)}$,

$$d\Psi = e^{\sum_i t_i\mathcal{P}^{(i)}}\left(\sum_i t_i\Omega^{(i)}\varphi + d\varphi\right)$$

By averaging, the terms with ∂^j, $j \geq 2$, all vanish. Treating carefuly the terms linear in the times as in the proof of the previous proposition, we get:

$$\Omega^{(2)}\langle\langle\Psi^*\Psi\rangle\rangle = -\Omega^{(1)}\langle\langle\Psi^*L^{(1)}\Psi\rangle\rangle$$

The other formula is proved similarly. ∎

Inserting these formulae into eq. (10.74) we get our final result:

Proposition. *The slow modulations obey the Whitham equations*

$$\left(\partial_{T_2} - \frac{\Omega^{(2)}}{\Omega^{(1)}}\partial_X\right)\mathcal{P}^{(3)} = \left(\partial_{T_3} - \frac{\Omega^{(3)}}{\Omega^{(1)}}\partial_X\right)\mathcal{P}^{(2)} \qquad (10.77)$$

Had we kept fixed any meromorphic function on the Riemann surfaces instead of $\mathcal{P}^{(1)}$, the Whitham equation would have taken the more symmetric form

$$\Omega^{(1)}\left(\partial_{T_2}\mathcal{P}^{(3)} - \partial_{T_3}\mathcal{P}^{(2)}\right)$$
$$+\Omega^{(2)}\left(\partial_{T_3}\mathcal{P}^{(1)} - \partial_{T_1}\mathcal{P}^{(3)}\right)$$
$$+\Omega^{(3)}\left(\partial_{T_1}\mathcal{P}^{(2)} - \partial_{T_2}\mathcal{P}^{(1)}\right) = 0$$

It is important to check the consistency equations, eqs. (10.59). When the point P on Γ describes a non-trivial cycle, the forms $\Omega^{(i)}(P)$ do not change, but the functions $\mathcal{P}^{(i)}(P)$ change by a period $U^{(i)}$. Hence the above equation implies

$$\Omega^{(1)}(P)\left(\partial_{T_2}U^{(3)} - \partial_{T_3}U^{(2)}\right) + \text{cyclic perm.} = 0$$

which implies eqs. (10.59) because the $\Omega^{(i)}$ are linearly independent.

In the KdV case, there is no time T_2. Keeping $\int_{P_\infty}^{P}\Omega^{(2)}$ fixed, we get

$$\partial_{T_3}\mathcal{P}^{(1)} - \partial_{T_1}\mathcal{P}^{(3)} = 0$$

Differentiating with respect to the point P on the Riemann surface, we get the Whitham equations in their usual form:

$$\partial_{T_3}\Omega^{(1)} - \partial_{T_1}\Omega^{(3)} = 0 \qquad (10.78)$$

We will recover this equation in Chapter 11 where other proofs are available. In the KP case, however, the above derivation of Whitham equations, which is due to Krichever, is the only one known.

Remark. Assuming that the Riemann surface Γ is generic, the forms $\Omega^{(1)}$ and $\Omega^{(2)}$ have no common zero. Let us assume that $\langle\langle\Psi^*\Psi\rangle\rangle$ and $\langle\langle\Psi^*L^{(1)}\Psi\rangle\rangle$ are meromorphic functions. They have respectively $2g$ and $2g + 1$ poles, hence $2g$ and $2g + 1$ zeroes. Looking at eq. (10.75), we see that the zeroes of $\langle\langle\Psi^*\Psi\rangle\rangle$ are the $2g$ zeroes of $\Omega^{(1)}$. The form

$$\Omega = \frac{\Omega^{(1)}}{\langle\langle\Psi^*\Psi\rangle\rangle} \qquad (10.79)$$

has a double pole at ∞, is otherwise regular and has zeroes at the poles of Ψ and Ψ^*. It coincides with the form defined in eq. (10.21).

10.11 Solution of the Whitham equations

There is a simple method to find explicit solutions to the Whitham equations. Let us present it in the simple case of hyperelliptic curves, which is appropriate to the KdV equation. In this case the curve is of the form

$$\mu^2 = R(\lambda) = \prod_{i=1}^{2g+1} (\lambda - \lambda_i)$$

where the λ_i are slowly modulated. Recall that in KdV, only the odd times survive. The forms $\Omega^{(2i-1)}$ are given by

$$\Omega^{(2i-1)} = \left(\frac{2i-1}{2}\right) \frac{\lambda^{g+i-1} + P^{(2i-1)}(\lambda)}{\sqrt{R(\lambda)}} d\lambda \qquad (10.80)$$

where the polynomial $P^{(2i-1)}(\lambda)$ is of degree $g-1$ and chosen so that all the periods of $\Omega^{(2i-1)}$ are pure imaginary. At infinity

$$\Omega^{(2i-1)} = d(z^{2i-1} + O(z^{-1})), \quad z = \sqrt{\lambda}$$

Let us introduce the normalized form

$$\mathcal{S} = \sum_i T_{2i-1} \Omega^{(2i-1)} + \Omega^{(n)}$$

where n is chosen at will and is a free parameter of the solution.

Proposition. *Let us assume that for each branch point λ_j, either λ_j is independent of the times T_{2i-1} or $\mathcal{S}$ vanishes at λ_j. This is a system of $2g+1$ equations for the $2g+1$ quantities λ_j, which allows us to express them in terms of the T_{2i-1}. Then*

$$\partial_{T_{2i-1}} \mathcal{S} = \Omega^{(2i-1)}$$

It follows that the Whitham equations, eqs. (10.78), are satisfied. We have more generally:

$$\partial_{T_{2i-1}} \Omega^{(2j-1)} = \partial_{T_{2j-1}} \Omega^{(2i-1)} \qquad (10.81)$$

$$U_j^{(2i-1)} = \partial_{T_{2i-1}} S_j, \quad S_j = \oint_{c_j} \mathcal{S}$$

where c_j is a basis of non-trivial cycles on Γ.

<u>Proof.</u> Let us consider the analyticity properties of $\partial_{T_{2i-1}} \mathcal{S}$. First, at infinity we have

$$\mathcal{S} = d\left(\sum_i T_{2i-1} z^{2i-1} + z^{2n-1} + O(z^{-1})\right)$$

hence $\partial_{T_{2i-1}}\mathcal{S} = d(z^{2i-1} + O(z^{-1}))$. At finite distance, we have

$$\partial_{T_{2i-1}}\mathcal{S} = \Omega^{(2i-1)} + \frac{1}{2}\left(\sum_k \frac{\partial_{T_{2i-1}}\lambda_k}{\lambda - \lambda_k}\right)\mathcal{S} + \sum_k c_k^{(i)}\omega_k$$

The second term in the right-hand side comes from the derivation of the factor $1/\sqrt{R(\lambda)}$ in eq. (10.80), while the last term comes from differentiating the polynomials $P^{(2i-1)}(\lambda)$ and $P^{(n)}$. The ω_k are the holomorphic differentials. The right-hand side is regular at finite distance since either $\partial_{T_{2i-1}}\lambda_k = 0$ or $\mathcal{S}|_{\lambda_k} = 0$. Finally, all periods of $\partial_{T_{2i-1}}\mathcal{S}$ are purely imaginary for T_{2i-1} real. Hence we have $\partial_{T_{2i-1}}\mathcal{S} = \Omega^{(2i-1)}$. This in turn implies eq. (10.81). Moreover, we have

$$U_j^{(2i-1)} = \oint_{c_j} \Omega^{(2i-1)} = \partial_{T_{2i-1}} \oint_{c_j} \mathcal{S} = \partial_{T_{2i-1}}\mathcal{S}_j$$

So we have solved both eq. (10.57) and the Whitham equations, eq. (10.78). ∎

References

[1] G.B. Whitham, *Linear and Nonlinear Waves*. Wiley (1974).

[2] I.M. Gelfand and L.A. Dickey, Fractional powers of operators and Hamiltonian sytems. *Funkz. Anal. Priloz.* **10** (1976) 13–29.

[3] F. Magri, A simple model of integrable Hamiltonian equation. *J. Math. Phys.* **19** (1978) 1156–1162.

[4] M. Adler, On a trace functional for formal pseudo-differential operators and the symplectic structure of the Korteweg–de Vries equations. *Invent. Math.* **50** (1979) 219–248.

[5] D.R. Lebedev and Yu.I. Manin, Hamiltonian Gelfand–Dickey operator and coadjoint representation of the Volterra group. *Funkz. Analys. Priloz.* **13** (1979) 40–46.

[6] H. Flaschka, M.G. Forest and D.W. McLaughlin, Korteweg–de Vries equation. *Comm. Pure Appl. Math.* **33** (1980) 739–784.

[7] A.G. Reyman and M.A. Semenov-Tian-Shansky, Family of Hamiltonian structures, hierarchy of Hamiltonians, and reduction for matrix first order differential operators. *Funkz. Analys. Priloz.* **14** (1980) 77–78.

[8] V.G. Drinfeld and V.V. Sokolov, Equations of the Korteweg–de Vries type and simple Lie algebras. *Doklady AN SSSR* **258** (1981) 11–16.

[9] M. Jimbo and T. Miwa, Solitons and infinite-dimensional Lie algebras. *RIMS* **19** (1983) 943–1001.

[10] I. Krichever, Method of averaging for two dimensional integrable equations. *Funkz. Analys. Priloz.* **22** (1988) 37–52.

[11] L.A. Dickey, *Soliton Equations and Hamiltonian Systems.* World Scientific (1991).

11

The KdV hierarchy

In this chapter we study the Korteweg–de Vries equation, which occupies a central place in the modern theory of integrable systems. All the aspects of integrable systems discussed so far converge in this chapter to draw a particularly rich landscape. In particular, the methods of pseudo-differential operators allow us to easily discuss the formal aspects, the tau-functions yield soliton solutions, and the algebro-geometric methods yield finite-zone solutions. The soliton solutions which we obtained in the Grassmannian setting by using vertex operators are also degenerate cases of these finite-zone solutions. Finally, we use a fermionic fomalism to analyse the structure of the local fields and show that the equations of the hierarchy can be recast in a very compact form. This is used to give a new derivation of the Whitham equations in the KdV case.

11.1 The KdV equation

The Korteweg–de Vries (KdV) equation was introduced historically as an approximation of the equations of hydrodynamics, describing unidimensional long waves in shallow water. In their pioneering work, Gardner, Greene, Kruskal and Miura found an unexpected connection with the inverse scattering problem of the Schroedinger equation. More recently, the Hamiltonian aspects of KdV theory connected it to conformal field theory. The KdV equation reads:

$$4\partial_t u = -6u\partial_x u + \partial_x^3 u \tag{11.1}$$

The numerical factors in front of each term in eq. (11.1) can be modified by rescaling u, x and t. The KdV equation can be written as the zero curvature condition

$$F_{xt} \equiv \partial_x A_t - \partial_t A_x - [A_x, A_t] = 0$$

with the connection A_x, A_t, depending on a spectral parameter λ:

$$A_x = \begin{pmatrix} 0 & 1 \\ \lambda + u & 0 \end{pmatrix}, \quad A_t = \frac{1}{4} \begin{pmatrix} \partial_x u & 4\lambda - 2u \\ 4\lambda^2 + 2\lambda u + \partial_x^2 u - 2u^2 & -\partial_x u \end{pmatrix}$$

(11.2)

Note that $\partial_x - A_x = D_u - \lambda \Sigma_0$ with the notations of eqs. (10.51, 10.52) in Chapter 10.

Alternatively, one can recast the KdV equation in the Lax form $\partial_t L = [M, L]$, where L and M are the following *differential* operators:

$$L = \partial^2 - u \tag{11.3}$$

$$M = \frac{1}{4}(4\partial^3 - 3u\partial - 3\partial u) = \frac{1}{4}(4\partial^3 - 6u\partial - 3(\partial_x u)) = (L^{\frac{3}{2}})_+$$

The operator ∂ acts as ∂_x, and the notation $(L^{\frac{3}{2}})_+$ refers to the pseudo-differential operator formalism introduced in Chapter 10. In the Lax equation $[M, L]$ is the commutator of differential operators.

Of course these two descriptions are not independent. To relate them, consider the linear system:

$$(\partial_x - A_x) \begin{pmatrix} \Psi \\ \chi \end{pmatrix} = 0, \quad (\partial_t - A_t) \begin{pmatrix} \Psi \\ \chi \end{pmatrix} = 0 \tag{11.4}$$

The x-equation yields $\chi = \partial_x \Psi$ and

$$(L - \lambda)\Psi = 0 \quad \text{with} \quad L = \partial_x^2 - u \tag{11.5}$$

The time evolution of Ψ is given by $4\partial_t \Psi = \partial_x u \cdot \Psi + (4\lambda - 2u)\partial_x \Psi$. Using eq. (11.5), this may be rewritten as:

$$(\partial_t - M)\Psi = 0 \quad \text{with} \quad M = \frac{1}{4}(4\partial^3 - 3u\partial - 3\partial u) \tag{11.6}$$

The compatibility condition of eqs. (11.5, 11.6) is the Lax equation $\partial_t L = [M, L]$, which is equivalent to the KdV equation.

Equation (11.5) is the Schroedinger equation with potential u. The parameter λ gets an interpretation as a point of the spectrum of this operator. This is the origin of the terminology "spectral parameter".

As explained in Chapter 3, a general consequence of the zero curvature condition is the existence of non-trivial conserved quantities. This requires, however, imposing appropriate boundary conditions. We shall consider here for definiteness either potentials $u(x)$ fast decreasing at $x \to \pm\infty$, or potentials periodic under $x \to x + \ell$. Let us assume for definiteness that $u(x)$ is periodic. Since A_x and A_t are local in u, they are also periodic. In this case conserved quantities are generated by $\text{Tr}\, T_\ell(\lambda)$, where $T_\ell(\lambda)$ is the monodromy matrix associated with the linear system eq. (11.4):

$$T_\ell(\lambda) = \overleftarrow{\exp}\left(\int_0^\ell dx A_x(x,t,\lambda)\right) \qquad (11.7)$$

To compute the trace $\operatorname{Tr} T_\ell(\lambda)$, we remark that it is invariant under *periodic* gauge transformations. We build a gauge in which the connection is diagonal, making the calculation of the monodromy matrix and its trace simple.

We shall present this computation for general connections whose components A_x and A_t belong to the $sl(2)$ algebra, so that it can be applied to a wider class of systems. The commutation relations of the $sl(2)$ algebra are:

$$[H, E_\pm] = \pm 2 E_\pm, \quad [E_+, E_-] = H$$

Its fundamental representation is given by:

$$H = \begin{pmatrix} 1 & 0 \\ 0 & -1 \end{pmatrix}, \quad E_+ = \begin{pmatrix} 0 & 1 \\ 0 & 0 \end{pmatrix}, \quad E_- = \begin{pmatrix} 0 & 0 \\ 1 & 0 \end{pmatrix}$$

Proposition. *Let $A_x = A_h H + A_- E_- + A_+ E_+$, where $A_h(x,\lambda)$, $A_\pm(x,\lambda)$ are periodic functions of x. There exists a periodic gauge transformation $g(x,\lambda)$: $A_x \to {}^g A_x \equiv g^{-1} A_x g - g^{-1} \partial_x g$ such that ${}^g A_- = {}^g A_+ = 0$, and*

$$^g A_h = \frac{1}{\ell} \mathcal{P}_\ell(\lambda) H$$

where $\mathcal{P}_\ell(\lambda)$, independent of x, is given by: $\mathcal{P}_\ell(\lambda) = \int_0^\ell dx\, v(x,\lambda)$. The function $v(x,\lambda)$ is a solution of the Ricatti equation:

$$v' + v^2 = V \quad \text{with} \quad V = \mathcal{A}' + \mathcal{A}^2 + A_- A_+, \quad \mathcal{A} = A_h - \frac{1}{2}\frac{A'_+}{A_+} \qquad (11.8)$$

<u>Proof.</u> The proof consists of performing successively three gauge transformations: the first one annihilates the component along E_-, the second one annihilates the component along E_+ and the third one is chosen to ensure that the component along H is constant.

Let us perform first a gauge transformation with $g = g_1 = \exp(f_- E_-)$, then

$$^g A_x = (A_h + A_+ f_-)H - (f'_- + 2A_h f_- + A_+ f_-^2 - A_-)E_- + A_+ E_+$$

where prime ($'$) means derivative with respect to x. The coefficient of E_- vanishes if f_- is a solution of the Ricatti equation

$$f'_- + 2A_h f_- + A_+ f_-^2 - A_- = 0$$

If one sets $f_- = \frac{1}{A_+}(v - \mathcal{A})$ and then $\mathcal{A} = A_h - \frac{1}{2}\frac{A'_+}{A_+}$, this equation becomes the Ricatti equation (11.8). As usual the substitution $v = y'/y$ linearizes the equation which becomes the Schroedinger equation:

$$y'' - Vy = 0 \qquad (11.9)$$

The potential $V(x, \lambda)$ being periodic, one can take for $y(x, \lambda)$ any one of the two quasi-periodic Bloch waves (Floquet solutions), $y_\pm(x, \lambda)$:

$$y_\pm(x + \ell, \lambda) = \exp(\pm \mathcal{P}_\ell(\lambda))\, y_\pm(x, \lambda) \qquad (11.10)$$

$\mathcal{P}_\ell(\lambda)$ is called the quasi-momentum. For definiteness, we shall take $v = y'_+/y_+$ which is periodic. We shall, moreover, assume that the Wronskian of y_+ and y_- is normalized by $y_+ y'_- - y'_+ y_- = 1$. Notice that

$$\mathcal{P}_\ell(\lambda) = \ln\left(\frac{y_+(\ell, \lambda)}{y_+(0, \lambda)}\right) = \int_0^\ell dx\, v(x, \lambda) \qquad (11.11)$$

Similarly, let us define $g_2 = \exp(f_+ E_+)$ and compute the matrix $^g A_x$ with $g = g_1 g_2$:

$$^g A_x = (A_h + A_+ f_-)H + (-f'_+ + 2(A_h + A_+ f_-)f_+ + A_+)E_+$$

We reduce $^g A_x$ to the diagonal form if we choose for f_+ the periodic solution of the equation $-f'_+ + 2(A_h + A_+ f_-)f_+ + A_+ = 0$. This solution is $f_+ = A_+ y_+ y_-$ which is also periodic.

Finally, taking $g_3 = \exp(hH)$, the gauge transformed matrix $^g A_x$ with $g = g_1 g_2 g_3$ reads:

$$^g A_x = \left(-h' + A_h + A_+ f_-\right)H = \left(-h' + \frac{y'_+}{y_+} + \frac{1}{2}\frac{A'_+}{A_+}\right)H$$

Choosing for h the periodic function $h = \frac{1}{2}\ln(A_+ y_+^2 e^{-2\mathcal{P}_\ell(\lambda)x/\ell})$ reduces the coefficient of H to the constant $\frac{1}{\ell}\mathcal{P}_\ell(\lambda)H$. ∎

Conserved quantities are obtained by looking at the trace of the monodromy matrix $\mathrm{Tr}\,(T_\ell(\lambda))$. Once the connection has been diagonalized this trace is easy to compute. It follows from the previous proposition that the two eigenvalues of $T_\ell(\lambda)$ are $\exp(\pm \mathcal{P}_\ell(\lambda))$, hence:

$$\mathrm{Tr}\, T_\ell(\lambda) = 2\cosh \mathcal{P}_\ell(\lambda), \quad \mathcal{P}_\ell(\lambda) = \int_0^\ell dx\, v(x, \lambda)$$

The function $\mathcal{P}_\ell(\lambda)$ can serve, as well as $\mathrm{Tr}\, T_\ell(\lambda)$, as a generating function for the integrals of motion. To construct them we only have to solve the Ricatti equation (11.8) for $v(x, \lambda)$.

Let us apply the above proposition to the KdV equation. In view of the expression of the KdV connection (11.2), we have $A_h = 0$, $A_- = \lambda + u$ and $A_+ = 1$. Thus $V = \lambda + u$ and the Ricatti equation reads:

$$v' + v^2 = \lambda + u \tag{11.12}$$

The Schroedinger equation associated with the Ricatti eq. (11.12) coincides with eq. (11.5). The quantity $\mathcal{P}_\ell(\lambda)$ is the quasi-momentum of the Bloch eigenfunctions of the differential operator $L = \partial^2 - u$ with periodic potential u and eigenvalue λ.

To obtain *local* conserved quantities we expand $\mathcal{P}_\ell(\lambda)$ around $\lambda = \infty$. When $\lambda \to \infty$, the solution of the Ricatti equation admits the asymptotic expansion:

$$v = \sqrt{\lambda} + \sum_{n \geq 0} \frac{(-1)^n v_n}{(\sqrt{\lambda})^n}, \quad 2v_{n+1} = v_n' + \sum_{p=0}^{n} v_p v_{n-p}, \quad v_0 = 0, \; 2v_1 = -u$$

This gives a recursion relation for computing the coefficients v_n. Since its solution does not require any integration it leads to local integrals of motion. The first few coefficients are:

$$v_1 = -\tfrac{1}{2}u, \quad v_2 = -\tfrac{1}{4}u', \quad v_3 = \tfrac{1}{8}(u^2 - u'')$$

$$v_4 = \tfrac{1}{2}v_3' + \tfrac{1}{8}uu', \quad v_5 = \tfrac{1}{2}v_4' + \tfrac{1}{32}u'^2 + \tfrac{1}{16}uu'' - \tfrac{1}{16}u^3$$

The conserved quantities are given by the integral:

$$\mathcal{P}_\ell(\lambda) = \int_0^\ell v \, dx = \sqrt{\lambda}\ell + \frac{1}{2\sqrt{\lambda}} \left(\int_0^\ell u \, dx \right) - \frac{1}{(2\sqrt{\lambda})^3} \left(\int_0^\ell u^2 \, dx \right)$$

$$+ \frac{1}{(2\sqrt{\lambda})^5} \left(\int_0^\ell (u'^2 + 2u^3) dx \right) - \cdots \tag{11.13}$$

11.2 The KdV hierarchy

We now particularize the formalism of pseudo-differential operators studied in Chapter 10 to the KdV situation. This amounts to studying the implications of the condition that $L = Q^2$ is a second order differential operator:

$$Q^2 = \Phi \partial^2 \Phi^{-1} = L = \partial^2 - u, \quad \Phi = 1 + \sum_1^\infty w_i \partial^{-i}$$

A first consequence is that only odd times survive in the KdV hierarchy. Indeed, recalling the equations of motion of the KP hierarchy,

$$\partial_{t_i} \Phi = -(\Phi \partial^i \Phi^{-1})_- \Phi$$

we see that for $i = 2j$, $(\Phi \partial^{2j} \Phi^{-1}) = L^j$ is a differential operator, so that its projection $(\ )_-$ vanishes.

Recall that we have defined two formal Baker–Akhiezer functions, see eq. (10.11) in Chapter 10:

$$\Psi(t, z) = \Phi e^{\xi(t,z)}, \quad \Psi^*(t, z) = (\Phi^*)^{-1} e^{-\xi(t,z)}, \quad \xi(t, z) = \sum_{i=1}^{\infty} t_{2i-1} z^{2i-1}$$

where $t_1 = x$. The function $\Psi(t, z)$ is an eigenfunction of L with the eigenvalue

$$\lambda = z^2$$

and $\Psi^*(t, z)$ is its formal adjoint (see eq. (10.2) in Chapter 10). Since L is obviously formally self-adjoint, $\Psi^*(t, z)$ is also an eigenfunction of L with the same eigenvalue,

$$(\partial_x^2 - u)\Psi(t, z) = \lambda \Psi(t, z), \quad (\partial_x^2 - u)\Psi^*(t, z) = \lambda \Psi^*(t, z) \qquad (11.14)$$

The Wronskian of these two solutions is a constant that we now compute.

Proposition. *The Wronskian of the two Baker–Akhiezer functions Ψ and Ψ^* is given by:*

$$W(\Psi, \Psi^*) \equiv \Psi'(t, z)\Psi^*(t, z) - \Psi^{*'}(t, z)\Psi(t, z) = 2z \qquad (11.15)$$

where we have denoted $\Psi' \equiv \partial_x \Psi$.

Proof. From the definition of the Baker–Akhiezer functions we see that the essential singularities cancel in $W(\Psi, \Psi^*)$ and that W admits a power series expansion in z around ∞ of the form $W(z) = 2z + \alpha + \beta/z + \cdots$. We prove that only the first term is present by showing that the residue of $W(z)z^i$ vanishes for $i \geq -1$.

$$\oint \frac{dz}{2i\pi} z^i W(z)$$

$$= \oint \frac{dz}{2i\pi} \left(\partial \Phi \partial^i e^{zx} \right) \left((\Phi^*)^{-1} e^{-zx} \right) - \left(\Phi e^{zx} \right) \left(\partial (\Phi^*)^{-1} (-\partial)^i e^{-zx} \right)$$

Note that the terms involving the times $t_3, t_5, \ldots$ in $\xi(t, z)$ cancel because there are only derivatives with respect to $x = t_1$. Using eq. (10.14) in Chapter 10, we can rewrite this as an Adler residue in the pseudo-differential algebra:

$$\oint \frac{dz}{2i\pi} z^i W(z) = \mathrm{Res}_\partial \left(\partial \Phi \partial^i \Phi^{-1} + \Phi \partial^i \Phi^{-1} \partial \right) = \mathrm{Res}_\partial \left(\partial L^{\frac{i}{2}} + L^{\frac{i}{2}} \partial \right)$$

For i even, this vanishes because $L^{\frac{i}{2}}$ is a differential operator. For i odd we are going to show that:

$$L^{\frac{i}{2}*} = -L^{\frac{i}{2}}, \quad i = \text{odd}$$

so that $\partial L^{\frac{i}{2}} + L^{\frac{i}{2}}\partial$ is formally self-adjoint and so cannot have a residue. It is sufficient to show that $L^{\frac{1}{2}*} = -L^{\frac{1}{2}}$. But $L^{\frac{1}{2}} = \partial - \frac{1}{2}u\partial^{-1} + \frac{1}{4}u'\partial^{-2} + \cdots$ is the unique solution of the equation $(L^{\frac{1}{2}})^2 = L$ with leading term ∂. Similarly, $-L^{\frac{1}{2}}$ is the unique solution of the same equation with leading term $-\partial$. Since $(L^{\frac{1}{2}*})^2 = L^* = L$ and $L^{\frac{1}{2}*} = -\partial + \cdots$ the result follows. ∎

We introduce the quantity $S(t, \lambda)$ which will be useful in expanding the compact pseudo-differential expressions. It will also play an important role in the last two sections on Whitham theory. It is defined by

$$S(t, \lambda) \equiv \Psi^*(t, z)\Psi(t, z) = 1 + \sum_{j=1}^{\infty} \lambda^{-j} S_{2j}(t)$$

Note that the essential singularities cancel in $S(t, \lambda)$ and that $S(t, \lambda)$ is a function of $\lambda = z^2$. This is because $\Psi^*(t, z)$ and $\Psi(t, -z)$ are solutions of the same eq. (11.14), and have the same behaviour at $z \to \infty$. So we have

$$\Psi^*(t, z) = c(z, t_3, \ldots)\Psi(t, -z) \tag{11.16}$$

Inserting this into eq. (11.15) we see that $c(z, t)$ is even in z, and so is $\Psi^*(t, z)\Psi(t, z)$. The aim of the following propositions is to show that the whole KdV hierarchy can be written in terms of $S(t, \lambda)$.

Proposition. *The Baker–Akhiezer functions can be expressed as:*

$$\Psi(t, z) = S^{1/2}(t, \lambda)e^{X(t,z)}, \quad \Psi^*(t, z) = S^{1/2}(t, \lambda)e^{-X(t,z)} \tag{11.17}$$

where
$$\partial_x X(t, z) = \frac{z}{S(t, \lambda)}, \quad X(t, z) = \xi(t, z) + O(1/z)$$

<u>Proof.</u> This parametrization obviously satisfies $\Psi^*(t, z)\Psi(t, z) = S(t, \lambda)$ and this defines $X(t, z)$. Inserting it into eq. (11.15) yields the equation relating $X(t, z)$ and $S(t, \lambda)$. The asymptotic form of $X(t, z)$ when $z \to \infty$ follows by comparing the asymptotic expansions of $\Psi(t, z)$. ∎

As a consequence, eq. (11.14) translates into an equation on $S(t, \lambda)$. It is convenient to write it in the form of the Ricatti equation (11.12) with:

$$v(t, z) = \frac{\partial_x \Psi(t, z)}{\Psi(t, z)} = \frac{1}{2}\partial_x \log S(t, \lambda) + \frac{z}{S(t, \lambda)} \tag{11.18}$$

There is a simple expression of the coefficients S_{2j} as residues of fractional powers of L.

Proposition. *The coefficients S_{2j} are the local densities of the conserved quantities of the KdV hierarchy, as computed in Chapter 10.*

$$S_{2j} = \text{Res}_\partial \left(L^{\frac{2j-1}{2}} \right) \tag{11.19}$$

As a consequence, the Hamiltonians of the KdV hierarchy are given by:

$$H_{2j-1}(L) = \frac{2}{2j+1} \int dx\, S_{2j+2}(x) \tag{11.20}$$

Proof.

$$\begin{aligned}
S_{2j} &= \oint \frac{dz}{2i\pi} z^{2j-1} \Psi^*(t,z)\Psi(t,z) = \oint \frac{dz}{2i\pi} ((\Phi^*)^{-1} e^{-zx}) \cdot (\Phi \partial^{2j-1} e^{zx}) \\
&= \text{Res}_\partial (\Phi \partial^{2j-1} \Phi^{-1}) = \text{Res}_\partial \left(L^{\frac{2j-1}{2}} \right)
\end{aligned}$$

where we have used eq. (10.14) in Chapter 10. ∎

Due to eq. (11.16), replacing $\Psi(t,z)$ by $\Psi^*(t,z)$ in eq. (11.18) amounts to changing $z \to -z$. In particular this shows that the coefficients v_{2n} are derivatives with respect to x of local densities.

One can compute the coefficients S_{2j} by induction as follows. Since Ψ and Ψ^* obey eq. (11.14), their product $S(t,\lambda) = \Psi^*(t,z)\Psi(t,z)$ obeys the third order equation:

$$\left(\frac{1}{4}\partial^3 - \frac{1}{2}u' - u\partial - \lambda\partial\right) S(t,\lambda) = 0$$

Expanding in z one gets $\partial S_{2j+2} = (\frac{1}{4}\partial^3 - \frac{1}{2}u' - u\partial)S_{2j}$. This recursion relation can also be understood as the Lenard recursion relation, eq. (10.37) in Chapter 10. In fact, since

$$H_k(L) = \frac{2}{k+2}\langle L^{\frac{k+2}{2}}\rangle = \frac{2}{k+2} \int dx\, \text{Res}_\partial L^{\frac{k}{2}+1} = \frac{2}{k+2} \int dx\, S_{k+3} \tag{11.21}$$

we get, using eq. (10.35) with $n = 1$, $dH_k = S_{k+1}\partial^{-1}$. The Lenard relation becomes $\mathcal{D}_1(S_{2j+2}\partial^{-1}) = \mathcal{D}_2(S_{2j}\partial^{-1})$ which is exactly the previous recursion relation. The first few values of S_{2j} are:

$$S_2 = -\frac{1}{2}u, \quad S_4 = -\frac{1}{8}u'' + \frac{3}{8}u^2, \quad S_6 = -\frac{5}{16}u^3 - \frac{5}{32}u'^2 + \frac{5}{16}(uu')' - \frac{1}{32}u^{(iv)}$$

One can recast the equations of motion of the KdV hierarchy as equations on $S(t, \lambda)$. It is convenient to introduce a generating function for all the time derivatives:

$$\nabla(\lambda) = \sum_{j \geq 1} \lambda^{-j} \partial_{2j-1} \tag{11.22}$$

Then we have:

Proposition. *The equations of the KdV hierarchy are equivalent to:*

$$\nabla(\lambda)S(t, \lambda') = \frac{S(t, \lambda) \cdot \partial_x S(t, \lambda') - S(t, \lambda') \cdot \partial_x S(t, \lambda)}{\lambda - \lambda'}, \quad |\lambda| > |\lambda'| \tag{11.23}$$

<u>Proof.</u> In these equations, we have $\lambda = z^2, \lambda' = z'^2$. We shall first prove a similar equation on $\Psi(t, z)$:

$$\nabla(\lambda)\Psi(t, z') = \frac{2S(t, \lambda)\partial_x - \partial_x S(t, \lambda)}{2(\lambda - \lambda')}\Psi(t, z'), \quad \text{for} \quad |\lambda| > |\lambda'| \tag{11.24}$$

Recall that, from eq. (10.9) in Chapter 10, the equation of motion for Ψ is

$$\partial_{2j-1}\Psi(t, z) = (L^{(2j-1)/2})_+ \Psi(t, z) \tag{11.25}$$

Now, there is a simple recursion relation:

$$\left(L^{\frac{2j+1}{2}}\right)_+ = \left(L^{\frac{2j-1}{2}}\right)_+ L + S_{2j}\partial_x - \frac{1}{2}\partial_x S_{2j} \tag{11.26}$$

Indeed, L being a differential operator, we have:

$$\left(L^{\frac{2j+1}{2}}\right)_+ = \left(L^{\frac{2j-1}{2}}L\right)_+ = \left(L^{\frac{2j-1}{2}}\right)_+ L + \left(\left(L^{\frac{2j-1}{2}}\right)_- L\right)_+$$

To compute the plus part in the last term, we only have to keep the first two terms in the expansion of $(L^{(2j-1)/2})_-$ because L is a second order differential operator. We have:

$$\left(L^{\frac{2j-1}{2}}\right)_- = S_{2j}\partial^{-1} - \frac{1}{2}S'_{2j}\partial^{-2} + \cdots$$

where the first coefficient (which is the residue of the considered pseudo-differential operator) is determined by eq. (11.19), and the second coefficient is then fixed by the fact that the left-hand side is formally anti self-adjoint. The recursion relation eq. (11.26) follows immediately. It implies in turn:

$$\left(L^{\frac{2j-1}{2}}\right)_+ = \sum_{i=0}^{j-1}(S_{2i}\partial - \frac{1}{2}S'_{2i})L^{j-i-1} \tag{11.27}$$

Using $L\Psi(t, z') = \lambda'\Psi(t, z')$, we have:

$$\partial_{2j-1}\Psi(t, z') = \sum_{i=0}^{j-1}(S_{2i}\partial - \frac{1}{2}S'_{2i})\lambda'^{j-i-1}\Psi(t, z')$$

The computation of $\nabla(\lambda)\Psi(t, z')$ is then straightforward, yielding eq. (11.24). Changing $z' \to -z'$, we see that the Baker–Akhiezer function $\Psi^*(t, z')$ also obeys eq. (11.24), and eq. (11.23) follows immediately for $S = \Psi^*\Psi$. ∎

It is worth noticing that eq. (11.23) can be rewritten as a local conservation law:

$$\nabla(\lambda)\left(\frac{1}{S(\lambda')}\right) = \partial_x\left(\frac{1}{\lambda - \lambda'}\frac{S(\lambda)}{S(\lambda')}\right)$$

Using this formalism we now show that the conserved quantities given in eq. (11.13) are the same as the ones in eq. (11.21). This amounts to showing that $v(t, z)$ and $S(t, \lambda)$ differ by the derivative in x of a local function.

Proposition. *We have the relation:*

$$\frac{dv(t, z)}{dz} = S(t, \lambda) + \frac{1}{2}\partial_x\left(2z\frac{d}{d\lambda} + \nabla(\lambda)\right)\log S(t, \lambda) \qquad (11.28)$$

where $v = \partial_x\Psi/\Psi$, $\lambda = z^2$.

Proof. From eq. (11.23) we have:

$$\nabla(\lambda)\log S(t, \lambda') = \frac{S(\lambda)}{\lambda - \lambda'}\left(\partial_x\log S(t, \lambda') - \partial_x\log S(t, \lambda)\right)$$

We substitute eq. (11.18) in this equation and take the limit $\lambda' \to \lambda$. One gets:

$$\nabla(\lambda)\log S(t, \lambda) = -\frac{S(t, \lambda)}{z}\frac{dv(t, z)}{dz} + \frac{1}{z} - 2z\frac{d}{d\lambda}\log S(t, \lambda)$$

We differentiate this expression with respect to x and substitute $\partial_x v = \lambda + u - v^2$ to get the result. ∎

Remark 1. As for the KP hierarchy, there exists a tau-function $\tau(t)$ such that:

$$\Psi(t, z) = e^{\xi(t,z)}\frac{\tau(t - [z^{-1}])}{\tau(t)}, \quad \Psi^*(t, z) = e^{-\xi(t,z)}\frac{\tau(t + [z^{-1}])}{\tau(t)}$$

with $[z^{-1}] = (\dots, \frac{z^{-2j+1}}{2j-1}, \dots)$. It is related to the generating function $S(t, \lambda)$ by:

$$S(t, \lambda) = \frac{\tau(t + [z^{-1}])\tau(t - [z^{-1}])}{\tau^2(t)} = 1 + \partial_x\nabla(\lambda)\log\tau(t) \qquad (11.29)$$

This last formula follows by inserting the parametrization of Ψ, Ψ^* in terms of tau-functions in eq. (11.28).

Remark 2. The Schroedinger equations, eq. (11.14), can also be translated into an equation on $S(t, \lambda)$. Using the value of the Wronskian, it is straightforward to show that

$$2\partial_x^2 \log S(t, \lambda) + (\partial_x \log S(t, \lambda))^2 + 4\lambda S^{-2}(t, \lambda) = 4(u + \lambda) \tag{11.30}$$

Remark 3. We can give more information on the decomposition eq. (11.17). Using eq. (11.24), we have:

$$\nabla(\lambda)X(t, z') = \frac{z'}{\lambda - \lambda'} \left(\frac{S(t, \lambda)}{S(t, \lambda')} \right) \tag{11.31}$$

$$= \frac{z'}{\lambda - \lambda'} \left(\frac{S(t, \lambda) - S(t, \lambda')}{S(t, \lambda')} \right) + \frac{z'}{\lambda - \lambda'}, \quad \text{for } |\lambda| > |\lambda'|$$

Notice that we can expand $X(t, z)$ as:

$$X(t, z) = \xi(t, z) + \tilde{X}(t, z), \quad \text{with} \quad \xi(t, z) = \sum_{j \geq 1} z^{2j-1} t_{2j-1} \tag{11.32}$$

with $\tilde{X}(t, z)$ regular at $z = \infty$. This decomposition follows from the fact that $\nabla(\lambda)\xi(z', t) = \frac{z'}{\lambda - \lambda'}$ for $|\lambda| > |\lambda'|$.

11.3 Hamiltonian structures and Virasoro algebra

Recall eq. (10.39) in Chapter 10 which defined two Poisson structures:

$$\{f_X, f_Y\}_1(L) = \langle LXY \rangle - \langle LYX \rangle$$
$$\{f_X, f_Y\}_2(L) = \langle (LX)_+(LY)_- \rangle - \langle (XL)_+(YL)_- \rangle$$
$$- \frac{1}{2} \int dx \left((\partial^{-1}[L, X]_{-1}) [L, Y]_{-1} \right)$$

Here $L = \partial^2 - u$, $X = X_{-1}\partial^{-1} + X_{-2}\partial^{-2} + \cdots$ and $f_X(L) = \langle LX \rangle$, where $\langle \rangle$ denotes the Adler trace. To compute the Poisson brackets of u it is enough to take $X = X(x)\partial^{-1}$ so that $f_X(L) = -\int dx u(x)X(x)$. The two Poisson brackets become:

$$\left\{ \int dx u X, \int dx u Y \right\}_1 = \int dx (X'Y - XY') \tag{11.33}$$

$$\left\{ \int dx u X, \int dx u Y \right\}_2 = -\int dx u (X'Y - XY') - \frac{1}{2} \int dx X Y''' \tag{11.34}$$

Equivalently, we can write:

$$\{u(x), u(y)\}_1 = -(\partial_x - \partial_y)\delta(x - y) \tag{11.35}$$

$$\{u(x), u(y)\}_2 = \frac{1}{2}(u(x) + u(y))(\partial_x - \partial_y)\delta(x - y) - \frac{1}{4}(\partial_x^3 - \partial_y^3)\delta(x - y) \tag{11.36}$$

A noticeable feature of these two Poisson brackets is that they are both local in x. Alternatively, we can expand the field $u(x)$ in Fourier series: $u(x) = \sum_k u_k e^{ikx}$ (we chose $\ell = 2\pi$). Taking $X = e^{-inx}$ and $Y = e^{-imx}$, eqs. (11.33, 11.34) become respectively:

$$\{u_n, u_m\}_1 = -\frac{i}{\pi} n \delta_{n+m,0} \tag{11.37}$$

$$\{u_n, u_m\}_2 = -\frac{1}{2i\pi}\left((n - m)u_{n+m} - \frac{n^3}{2}\delta_{n+m,0}\right) \tag{11.38}$$

The bracket $\{ , \}_1$ is called the Gardner–Faddeev–Zakharov bracket, while the bracket $\{ , \}_2$ is called the Magri–Virasoro bracket. It coincides with the Kostant–Kirillov bracket associated with the Virasoro algebra.

The Lax operator L can also be written in factorized form $L = (\partial + p)(\partial - p)$ so that $u = p' + p^2$. This is the Miura transformation. The second Poisson bracket has a simple expression in this parametrization:

$$\{p(x), p(y)\}_2 = \frac{1}{4}(\partial_x - \partial_y)\delta(x - y) \tag{11.39}$$

or in Fourier modes $p(x) = \sum_k p_k e^{ikx}$:

$$\{p_n, p_m\}_2 = \frac{in}{4\pi}\delta_{n+m,0}$$

It was shown in Chapter 10 that these Poisson brackets are compatible in the sense that their sum is again a Poisson bracket. Moreover, the Hamiltonians H_n of the KdV hierarchy are in involution with respect to both Poisson brackets.

Let us give an example of the Hamiltonian equations of motion. Taking

$$H_1 = \frac{1}{4}\int dx\, u^2, \quad H_3 = -\frac{1}{16}\int dx(u'^2 + 2u^3)$$

one gets:

$$\partial_{t_3} u = \{H_3, u\}_1 = \{H_1, u\}_2 = \frac{1}{4}(-6uu' + u''')$$

illustrating the fact that one finds the same equations of motion with the two Poisson brackets, but with different Hamiltonians.

11.4 Soliton solutions

Considering the KP hierarchy with $Q = L^{\frac{1}{2}}$, we see that the equations of motion $\partial_{t_i}\Phi = -(L^{\frac{i}{2}})_-\Phi$ imply that Φ is stationary with respect to the even times t_{2j}. Conversely, any solution of the KP hierarchy which is stationary for any even time is such that $(Q^{2j})_- = 0$, hence, in particular, $Q^2 = L$ is a differential operator. This solution is thus a solution of the KdV hierarchy. At the tau-function level, to obtain the decoupling of the even time variables it is sufficient to have:

$$\tau_{KP}(t_{\text{even}}, t_{\text{odd}}) = e^{\sum_{n \text{ even}} c_n t_n} \tau_{KdV}(t_{\text{odd}}) \tag{11.40}$$

This is because the action of a Hirota differential operator with respect to even time variables on such a KP tau-function vanishes:

$$D_{t_{2k}}^m e^{c_{2k} t_{2k}} \tau_{KdV}(t_{\text{odd}}) \cdot e^{c_{2k} t_{2k}} \tau_{KdV}(t_{\text{odd}})$$
$$= \partial_y^m e^{c_{2k}(t_{2k}+y)} \tau_{KdV}(t_{\text{odd}}) e^{c_{2k}(t_{2k}-y)} \tau_{KdV}(t_{\text{odd}})|_{y=0} = 0$$

The Hirota equations of the KdV hierarchy are thus obtained from the Hirota equations of the KP hierarchy by simply erasing the even times. We get, for instance, the Hirota form of the KdV equation:

$$(D_1^4 - 4D_1D_3)\tau \cdot \tau = 0 \tag{11.41}$$

(compare with eq. (8.56) in Chapter 8). Setting

$$u = -2\frac{\partial^2}{\partial t_1^2} \log \tau = -\frac{D_1^2 \tau \cdot \tau}{\tau^2} \tag{11.42}$$

we recover the KdV equation on u. Indeed, one has:

$$-u'' + 3u^2 = \frac{D_1^4 \tau \cdot \tau}{\tau^2}, \qquad D_1^4 \tau \cdot \tau = 2\tau^{(iv)}\tau - 8\tau'''\tau' + 6(\tau'')^2$$
$$-\dot{u} = \partial_x \frac{D_1 D_3 \tau \cdot \tau}{\tau^2}, \qquad D_1 D_3 \tau \cdot \tau = 2(\dot{\tau}'\tau - \tau'\dot{\tau})$$

Combining these expressions we see that the KdV equation is equivalent to the Hirota equation (11.41).

Recall that the KP tau-functions are constructed by choosing an element $g \in GL(\infty)$ (see eq. (9.34) in Chapter 10):

$$\tau_{KP}(t) = \langle 0|e^{H(t)}g|0\rangle$$

with $H(t) = \sum_n H_n t_n$ and where H_n are bosonic oscillators, (not to be confused with the Hamiltonians). See Chapter 9. This tau-function satisfies the bilinear identity (9.36), which reduces to the KdV Hirota bilinear identity whenever $\tau_{KP}(t)$ is of the form eq. (11.40). The main problem is to find the group elements g such that this property holds.

Recall that the Lie algebra of $GL(\infty)$ consists of fermionic bilinears of the form $X = \sum_{rs} M_{rs} : \beta_r \beta^*_{-s} :$ (see eq. (9.13) in Chapter 9). If g commutes with the H_{2k}, one can push the term $\exp\left(\sum_k t_{2k} H_{2k}\right)$ in τ_{KP} to the right, where it hits the vacuum and disappears since $H_{2k}|0\rangle = 0$. In fact commutation up to a central element is sufficient since a central term would produce an exponential of a linear combination of the even times. Using eq. (9.26) in Chapter 9, we have:

$$[H_n, \beta_r] = \beta_{r+n}, \quad [H_n, \beta^*_s] = -\beta^*_{s+n}$$

so that the H_{2k} commute with X, up to central terms, if $M_{rs} = M_{r+2, s-2}$. This means that the infinite band matrix M_{rs} is made of diagonals whose elements reproduces themselves with period 2. This characterizes the Lie algebra $\widehat{sl(2)} \subset \widehat{gl}(\infty)$, see Chapter 16. We have found:

Proposition. *The tau-function of the KdV hierarchy is given by:*

$$\tau_{KdV}(t) = \langle 0|e^{H(t)}g|0\rangle, \quad \text{with} \quad H(t) = \sum_{k>0} t_{2k-1} H_{2k-1} \text{ and } g \in \widehat{SL(2)}$$

As an application, we can find the KdV soliton solutions. Recall that the KP soliton solutions are obtained for $g = \prod g_i$ with $g_i = 1 + a_i \beta(p_i)\beta^*(q_i)$, where $\beta(p)$ and $\beta^*(q)$ are the fermionic fields. Since we have:

$$[H_n, \beta(z)\beta^*(w)] = (z^n - w^n) : \beta(z)\beta^*(w) : + \frac{z^n - w^n}{z - w}$$

we see that g_i commutes with H_{2k} if $p_i = -q_i$. Hence we have:

Proposition. *The n-soliton tau-functions of the KdV hierarchy are:*

$$\tau_n(t, g) = \langle 0|e^{H(t)} \prod_{i=1}^{n} (1 + a_i \beta(p_i)\beta^*(-p_i)) |0\rangle \qquad (11.43)$$

Explicitly, they are equal to:

$$\tau_n(X|p) = 1 + \sum_{p=1}^{n} \sum_{\substack{I \subset \{1,\ldots,n\} \\ |I|=p}} \prod_{i<j\in I} \left(\frac{p_i - p_j}{p_i + p_j}\right)^2 \cdot \prod_{i\in I} X_i \qquad (11.44)$$

$$= 1 + \sum_i X_i + \sum_{i<j} X_i X_j \left(\frac{p_i - p_j}{p_i + p_j}\right)^2 + \cdots$$

where $\xi(p,t) = \sum_{n \text{ odd}} p^n t_n$, and $X_i = \frac{a_i}{2p_i} e^{2\xi(p_i,t)}$. This can be written in compact form as:

$$\tau_n(X|p) = \det(1 + W), \quad \text{with} \quad W_{ij} = \sqrt{X_i} \frac{\sqrt{4p_i p_j}}{(p_i + p_j)} \sqrt{X_j} \quad (11.45)$$

<u>Proof.</u> We have $e^{H(t)}\beta(p_i)e^{-H(t)} = e^{\xi(t,p_i)}\beta(p_i)$ and $e^{H(t)}\beta^*(-p_i)e^{-H(t)} = e^{\xi(t,p_i)}\beta^*(-p_i)$ so that pushing $e^{H(t)}$ to the right amounts to replacing $a_i \to (2p_i)X_i$. The expression (11.44) is obtained by applying Wick's theorem.

$$\langle 0| \prod_i (1 + 2p_i X_i \beta(p_i)\beta^*(-p_i))|0\rangle = 1 + \sum_i 2p_i X_i \langle 0|\beta(p_i)\beta^*(-p_i)|0\rangle$$

$$+ \sum_{i<j} 4p_i p_j X_i X_j \langle 0|\beta(p_i)\beta^*(-p_i)\beta(p_j)\beta^*(-p_j)|0\rangle + \cdots$$

Each of these vacuum expectation values is a determinant (see eq. (9.19) in Chapter 9). We get:

$$\langle 0| \prod_i (1 + 2p_i X_i \beta(p_i)\beta^*(-p_i))|0\rangle$$

$$= 1 + \sum_i X_i + \sum_{i<j} X_i X_j \det\left(\frac{1}{p_i + p_j}\right) + \cdots \quad (11.46)$$

By the Cauchy formula, eq. (9.33), these determinants can be rewritten as in eq. (11.44). On the other hand, eq. (11.45) reduces to eq. (11.46) by virtue of the expansion formula, eq. (9.47) in Chapter 9. ∎

Remark 1. Using the bosonization formula of Chapter 9, one recognizes that the operator

$$V(p) = p\,\beta(p)\beta^*(-p)$$

coincides with the vertex operator defining the level one vertex representation of the affine $\widehat{sl}(2)$ algebra, (see Chapter 16). In the bosonic representation, the group elements g_i can be written:

$$g_i \equiv 1 + a_i\beta(p_i)\beta^*(-p_i) = 1 + \frac{a_i}{p_i}V(p_i)$$

Hence the soliton solutions of KdV are directly related to vacuum expectation values of vertex operators.

The one-soliton solution is obtained when $\tau = 1 + X$, with $X = 1 + e^{(2p(x-x_0)+2p^3 t)}$. One gets:

$$u(x,t) = -\frac{2p^2}{\cosh^2(p(x - x_0) + p^3 t)}$$

It corresponds to a bump (or rather a dip) of height $2p^2$ propagating with velocity $-p^2$. In sharp contrast with the case of linear partial differential equations where wave packets spread out in time, here the bump preserves its shape for all times. Note that the centre of the bump is located at $X(x,t) = 1$.

Consider now the general n-soliton solution, eq. (11.44). We can analyse its shape asymptotically when $t \to \pm\infty$. Assume that $p_1 > p_2 > \cdots > p_n > 0$. We want to show that we have asymptotically n solitons moving from right to left with velocities $-(p_1)^2, \ldots, -(p_n)^2$. Let us assume $t \to -\infty$ and consider what happens around $X_i = 1$, i.e. $x_i(t) = -p_i^2 t$. The values of the other X_j when $x \sim x_i(t)$ are: $X_j \sim (a_j / 2p_j) \exp(2p_j(p_j^2 - p_i^2)t)$. Hence for large negative time, if $p_j^2 < p_i^2$ then $X_j(x_i(t), t)$ is very large, while if $p_j^2 > p_i^2$, then $X_j(x_i(t), t)$ is very small. So we can split the indices j into two subsets, relative to the index i. One subset $\mathcal{I}_+$ is such that $p_j^2 < p_i^2$ and corresponds to X_j very large. The other one $\mathcal{I}_-$ is such that $p_j^2 > p_i^2$ and corresponds to X_j very small. To evaluate the tau functions, eq. (11.44), when $x \sim x_i(t)$, one has to keep the terms containing the maximum number of X_j, $j \in \mathcal{I}_+$. There are two such terms, yielding:

$$\tau(x,t)|_{x \sim x_i(t)} \sim \prod_{j \in \mathcal{I}_+} X_j \prod_{j < k \in \mathcal{I}_+} \frac{(p_j - p_k)^2}{(p_j + p_k)^2} \left(1 + X_i \prod_{j \in \mathcal{I}_+} \frac{(p_i - p_j)^2}{(p_i + p_j)^2}\right)$$

where we have to keep the second term because $X_i \sim 1$. When we compute the KdV field using eq. (11.42), the prefactor disappears because it involves an exponential linear in x, t. Hence the tau-function reduces to one-soliton tau-functions, but with parameter

$$a_i \to a_i^{\mathrm{in}} = a_i \prod_{j \in \mathcal{I}_+} \frac{(p_i - p_j)^2}{(p_i + p_j)^2}$$

So around $x = x_i(t)$ the KdV field $u(x,t)$ looks like a one-soliton field, and the n-soliton solution appears as n widely separated solitons travelling to the left with velocities $-p_i^2$. At some finite times these solitons will collide and the above analysis becomes invalid. However, for $t \to +\infty$ the solitons separate again, and we get the same picture with the roles of $\mathcal{I}_+$ and $\mathcal{I}_-$ reversed, so that:

$$a_i \to a_i^{\mathrm{out}} = a_i \prod_{j \in \mathcal{I}_-} \frac{(p_i - p_j)^2}{(p_i + p_j)^2}$$

It follows that the interaction introduces a phase shift:

$$\delta_i = \log \frac{a_i^{\text{out}}}{a_i^{\text{in}}} = \log \prod_{j<i} \frac{(p_i - p_j)^2}{(p_i + p_j)^2} - \log \prod_{j>i} \frac{(p_i - p_j)^2}{(p_i + p_j)^2}$$

This formula shows that the delay is the sum of delays introduced by pairwise interactions. Moreover, each individual soliton keeps its form and its velocity after interactions.

Knowing the picture of the n-soliton solutions in the regime $t \to \pm\infty$ as a superposition of n one-soliton solutions, we can easily compute the conserved quantities. Since the quantities are conserved one can evaluate them at $t \to -\infty$. From eq. (11.29) we can write $S(x,t,\lambda) = 1 + \sum_i \partial_x \nabla(\lambda) \log \tau_i(x,t)$, where τ_i is the one-soliton tau-function with parameters p_i, a_i^{in}, since for any given value of x only one soliton contributes. We have $\tau_i = 1 + a_i^{\text{in}} \exp(2\xi(t,p_i))$, so that

$$\partial_{t_{2j-1}} \log \tau_i = 2p_i^{2j-1} \frac{a_i^{\text{in}} e^{2\xi(t,p_i)}}{1 + a_i^{\text{in}} e^{2\xi(t,p_i)}}$$

yielding

$$H(\lambda) = \int_{-\ell}^{\ell} dx\, S(x,t,\lambda) = 2\ell + 2 \sum_{j>1} \lambda^{-j} \sum_i p_i^{2j-1}$$

So the complete set of conserved quantities is provided by:

$$H_{2j-1} = \frac{4}{2j+1} \sum_{i=1}^{n} p_i^{2j+1} \tag{11.47}$$

Because the Hamiltonians H_{2j-1} are conserved for any j, it follows that in the scattering process of solitons, only permutations of the p_i can occur. The scattering of solitons is completely described by this permutation and the time delays δ_i.

11.5 Algebro-geometric solutions

We wish to apply the analytical methods of Chapter 5 to construct solutions of the KdV equation. As explained in that chapter, one way to get a Lax matrix compatible with the equations of the KdV hierarchy is to seek for stationary solutions with respect to some higher time t_j. Then the zero-curvature condition, $\partial_i A_j - \partial_j A_i - [A_i, A_j] = 0$, becomes a Lax equation since the stationarity condition with respect to time t_j means $\partial_j A_i = 0$, for all i. The Lax matrix is A_j and its associated spectral curve is independent of all times t_i. A very simple example of this situation occurs when u is stationary with respect to $t_3 = t$. In that case the Lax matrix is A_t given in eq. (11.2). The associated spectral curve is:

$$\Gamma: \quad \det(A_t - \mu) = \mu^2 - \frac{1}{4}\lambda^3 + \frac{1}{4}(3u^2 - u'')\lambda + \frac{1}{16}(2uu'' - u'^2 - 4u^3) = 0$$

The zero-curvature condition becomes the Lax equation $\partial_x A_t = [A_x, A_t]$ and reduces to the stationary KdV equation $6uu' - u''' = 0$. Integrating, one gets $3u^2 - u'' = C_1$ and $2u^3 - u'^2 = 2C_1 u + C_2$ for some constants C_1, C_2. So the spectral curve reads $\mu^2 = \lambda^3/4 - C_1\lambda/4 - C_2/16$, and is independent of x as it should be. This is a genus 1 curve, so that u is given by an elliptic integral.

More interesting solutions will be obtained by assuming u to be stationary with respect to some higher time t_{2j-1}. Let us compute the matrices $A_{t_{2j-1}}$. We start from the equations of the hierarchy, eq. (11.25). Since $L^{\frac{2j-1}{2}}$ is anti self-adjoint, for either one of the two solutions Ψ and Ψ^* of the KdV hierarchy we have:

$$\partial_{t_{2j-1}}\begin{pmatrix} \Psi \\ \partial_x \Psi \end{pmatrix} = \begin{pmatrix} (L^{\frac{2j-1}{2}})_+ \Psi \\ \partial_x(L^{\frac{2j-1}{2}})_+ \Psi \end{pmatrix} = A_{t_{2j-1}}\begin{pmatrix} \Psi \\ \partial_x \Psi \end{pmatrix}$$

Using the identity (11.27) and $\partial_x^2 \Psi = (\lambda - u)\Psi$ one gets:

$$A_{t_{2j-1}} = \sum_{i=0}^{j-1} \lambda^{j-i-1}\begin{pmatrix} -\frac{1}{2}S'_{2i} & S_{2i} \\ (\lambda - u)S_{2i} - \frac{1}{2}S''_{2i} & \frac{1}{2}S'_{2i} \end{pmatrix}$$

Notice that $A_{t_{2j-1}}$ depends only on λ, in agreement with the fact that Ψ and Ψ^* play the same role. In particular, for $j = 1$ one finds A_x, and for $j = 2$ one finds A_t. Writing $(L^{\frac{2j-1}{2}})_+ = (L^{\frac{2j-1}{2}}) - (L^{\frac{2j-1}{2}})_-$, we see that:

$$A_{t_{2j-1}}\begin{pmatrix} \Psi \\ \partial_x \Psi \end{pmatrix} = \lambda^{\frac{2j-1}{2}}\begin{pmatrix} \Psi \\ \partial_x \Psi \end{pmatrix} + O(1)$$

Hence we have identified, asymptotically for $\lambda \to \infty$, the eigenvectors of $A_{t_{2j-1}}$ and the eigenvalues:

$$\mu = \lambda^{\frac{2j-1}{2}} + O(1) \tag{11.48}$$

The matrix $A_{t_{2j-1}}$ being traceless 2×2, its associated spectral curve is a hyperelliptic curve of genus $(j-1)$ of the form $\mu^2 = R(\lambda)$ with $R(\lambda) = \prod_{j=0}^{2j-1}(\lambda - \lambda_j) = \det A_{t_{2j-1}}$. This curve is not a general hyperelliptic

Riemann surface of genus $(j-1)$. In fact, since $\mu = \sqrt{R(\lambda)}$ is an eigenvalue of $A_{t_{2j-1}}(\lambda)$, it has to be of the specific form eq. (11.48), showing that $R(\lambda)$ has the very special form $R(\lambda) = \lambda^{2j-1} + C_1\lambda^{j-1} + C_2\lambda^{j-2} + \cdots$. To overcome this peculiarity, we notice that the stationarity condition can be generalized by imposing the condition:

$$\sum_j c_j \partial_{t_{2j-1}} u = 0$$

for some constant coefficients c_j. Then the corresponding Lax matrix becomes $\sum_j c_j A_{t_{2j-1}}$. For any time t_{2i-1}, the zero curvature condition implies the Lax equation (because the $A_{t_{2j-1}}$ depend only on u):

$$\partial_{t_{2i-1}}\left(\sum_j c_j A_{t_{2j-1}}\right) = \left[A_{t_{2i-1}}, \left(\sum_j c_j A_{t_{2j-1}}\right)\right]$$

In the following we consider the hyperelliptic curve Γ constructed from such a Lax matrix. It is of the generic form:

$$\Gamma: \quad \mu^2 = R(\lambda) = \prod_{i=1}^{2g+1}(\lambda - \lambda_i) \tag{11.49}$$

The point at ∞ is a branch point, and a local parameter around that point is $z = \sqrt{\lambda}$.

We want to construct a section Ψ of the eigenvector bundle on Γ, obeying

$$(\partial_{t_{2i-1}} - A_{t_{2i-1}})\begin{pmatrix}\Psi \\ \partial_x\Psi\end{pmatrix} = 0, \quad \forall i$$

The choice $\partial_x\Psi$ for the second component is dictated by the equation for $i = 1$ which then reduces to:

$$(\partial_x^2 - u)\Psi = \lambda\Psi$$

So it is enough to consider Ψ. A consequence of eq. (11.48) is that Ψ has the asymptotic behaviour at infinity $\Psi = e^{\xi(t,z)}(1 + O(z^{-1}))$, where $z = \sqrt{\lambda}$.

We know (see Chapter 5) that Ψ has $g + N - 1 = g + 1$ poles on Γ (N is the size of the Lax matrix). Here one of the poles is at ∞ because $\partial_x\Psi \sim z\Psi$ and $z = \sqrt{\lambda}$ is the local parameter at ∞. Hence we require that Ψ has g poles $(\gamma_1, \ldots, \gamma_g)$ at finite distance. Recall that the positions of these poles are independent of all the times t_{2j-1}. With these data we construct the Baker–Akhiezer function on Γ which is the *unique* function with the following analytical properties:

- It has an essential singularity at the point P at infinity:

$$\Psi(t,z) = e^{\xi(t,z)}\left(1 + \frac{\alpha(t)}{z} + O(1/z^2)\right) \tag{11.50}$$

where $z = \sqrt{\lambda}$ and $\xi(t,z) = \sum_{i\geq 1} z^{2i-1} t_{2i-1}$.

- It has g simple poles, independent of all times. The divisor of these poles is $D = (\gamma_1, \ldots, \gamma_g)$.

This Baker–Akhiezer function solves the KdV hierarchy equations as the following two propositions show.

Proposition. *There exists a function $u(x,t)$ such that*

$$(\partial_x^2 - u)\,\Psi = \lambda\Psi \tag{11.51}$$

<u>Proof.</u> Consider on Γ the function $\partial_x^2\Psi - \lambda\Psi$. To define this object as a function on the curve, λ is viewed as a meromorphic function on Γ. Note that λ has only a double pole at ∞ and such a function exists only if Γ is hyperelliptic. We see that $\partial_x^2\Psi - \lambda\Psi$ has the same analytical properties as Ψ itself at finite distance on Γ. At infinity we have by eq. (11.50):

$$\partial_x^2\Psi - \lambda\Psi = e^{\xi(t,z)}(2\partial_x\alpha + O(1/z)), \quad z = \sqrt{\lambda}$$

So it is a Baker–Akhiezer function, but with a normalization $2\partial_x\alpha$ instead of 1 at infinity. By the uniqueness theorem of such functions, we have:

$$\partial_x^2\Psi - \lambda\Psi = u\Psi, \quad u = 2\partial_x\alpha \tag{11.52}$$

                                                               ■

    Having found the potential u, we construct the differential operator $L = \partial^2 - u$ and show that the Baker–Akhiezer function Ψ obeys all the equations of the associated KdV hierarchy.

Proposition. *The evolution of Ψ is given by:*

$$\partial_{t_{2i-1}}\Psi = (L^{\frac{2i-1}{2}})_+\Psi$$

where $L = \partial^2 - u$ is the KdV operator constructed above.

<u>Proof.</u> Consider the function $\partial_{t_{2i-1}}\Psi - (L^{\frac{2i-1}{2}})_+\Psi$. It has the same analytical properties as Ψ at finite distance on Γ. At infinity we have

$\partial_{t_{2i-1}}\Psi = z^{2i-1}\Psi + e^{\xi}O(1/z)$ and $(L^{\frac{2i-1}{2}})_{+}\Psi = L^{\frac{2i-1}{2}}\Psi - (L^{\frac{2i-1}{2}})_{-}\Psi = z^{2i-1}\Psi + e^{\xi}O(1/z)$, where we have used $L\Psi = z^2\Psi$. Summarizing, we get:

$$\partial_{t_{2i-1}}\Psi - (L^{\frac{2i-1}{2}})_{+}\Psi = e^{\xi(t,z)}O(z^{-1}), \quad z \to \infty$$

By unicity, this Baker–Akhiezer function which vanishes at ∞ vanishes identically.   ∎

Remark. Because the Schroedinger operator in eq. (11.51) is self-adjoint, the adjoint Baker–Akhiezer function $\Psi^*(P)$ is easily related to $\Psi(P)$. In fact one can choose

$$\Psi^*(P) = \Psi(\sigma(P))$$

where σ is the hyperelliptic involution on Γ. Note, however, that this choice does not correspond to the normalization selected by the definition of $\Psi^*(P)$ in terms of pseudo-differential operators.

The Baker–Akhiezer function $\Psi(t, z)$ can be written explicitly in terms of theta-functions, see Chapter 5. Let $\Omega^{(2j-1)}$ be the unique normalized second kind differential (all the a-periods vanish) with a pole of order $2j$ at infinity, such that:

$$\Omega^{(2j-1)} = d\left(z^{2j-1} + \text{regular}\right), \quad \text{for} \quad z \to \infty$$

Let $U_k^{(2j-1)}$ be its b-periods:

$$U_k^{(2j-1)} = \frac{1}{2i\pi} \oint_{b_k} \Omega^{(2j-1)}$$

In terms of these data we have:

Proposition. *The Baker–Akhiezer function with the divisor of poles $D = (\gamma_1, \ldots, \gamma_g)$ can be expressed as:*

$$\Psi(t, P) = e^{\int_{\infty}^{P} \sum_j t_{2j-1}\Omega^{(2j-1)}} \frac{\theta(\mathcal{A}(P) + \sum_j t_{2j-1}U^{(2j-1)} - \zeta)\,\theta(\zeta)}{\theta(\mathcal{A}(P) - \zeta)\,\theta(\sum_j t_{2j-1}U^{(2j-1)} - \zeta)}$$
$$(11.53)$$

where $\mathcal{A}(P)$ is the Abel map on Γ with base point ∞, and $\zeta = \mathcal{A}(D) + \mathcal{K}$ with $\mathcal{K}$ the Riemann's constant vector. The KdV field, u, is given by the Its–Matveev formula:

$$u(x, t) = -2\partial_x^2 \log\theta\left(\sum_j t_{2j-1}U^{(2j-1)} - \zeta\right) + \text{const.} \quad (11.54)$$

<u>Proof.</u> In eq. (11.53) the integral $\int_\infty^P$ has to be understood in the following sense: for z in a vicinity of ∞, one *defines* $\int_\infty^P \Omega^{(2j-1)}$ as the unique primitive of $\Omega^{(2j-1)}$ which behaves as $z^{2j-1} + O(1/z)$ (no constant term). Of course, when this is analytically continued on the Riemann surface, b-periods will appear. However, they will cancel out in eq. (11.53) due to the monodromy properties of theta-functions, leaving us with a well-defined normalized Baker–Akhiezer function. The formula for the KdV field is found by using:

$$\lambda + u = (\partial_x^2 \log \Psi) + (\partial_x \log \Psi)^2$$

Setting $\Omega^{(1)}(z) = d(z + \frac{\beta}{z} + O(z^{-2}))$, where β does not depend on times, we have:

$$\partial_x \log \Psi = z + \partial_x \log \theta\Big(\mathcal{A}(P) + \sum_j t_{2j-1} U^{(2j-1)} - \zeta\Big)$$

$$- \partial_x \log \theta\Big(\sum_j t_{2j-1} U^{(2j-1)} - \zeta\Big) + \frac{\beta}{z} + O(z^{-2})$$

We evaluate this expression when $z \to \infty$. Using Riemann's bilinear identities, we can expand the Abel map $\mathcal{A}(P)$ around ∞ (see eq. (5.56) in Chapter 5), and we have:

$$\theta\Big(\mathcal{A}(P) + \sum_j t_{2j-1} U^{(2j-1)} - \zeta\Big) = \theta\Big(\sum_j (t_{2j-1} - \frac{z^{-2j+1}}{2j-1}) U^{(2j-1)} - \zeta\Big)$$

Keeping the $1/z$ terms, we obtain:

$$\partial_x \log \Psi = z - \frac{1}{z} \partial_x^2 \log \theta\Big(\sum_j t_{2j-1} U^{(2j-1)} - \zeta\Big) + \frac{\beta}{z} + O(\frac{1}{z^2})$$

Differentiating once more with respect to x, we also get $\partial_x^2 \log \Psi = O(1/z)$. It follows that $z^2 + u = z^2 - 2\partial_x^2 \log \theta + 2\beta + O(1/z)$, proving the result. ∎

On a hyperelliptic curve Γ, one can easily express the Baker–Akhiezer function Ψ knowing the divisor of its zeroes. We concentrate first on the x dependence. The higher times t_{2j-1} will be considered next. Let $D(x)$ be the divisor of the zeroes of the Baker–Akhiezer function Ψ. It is of degree g, and coincides with the divisor of the poles, D, for $x = 0$:

$$D(x) \equiv \{\gamma_1(x), \ldots, \gamma_g(x)\} \tag{11.55}$$

The points $\gamma_i(x)$ have coordinates $\left(\lambda_{\gamma_i(x)}, \mu_{\gamma_i(x)} = \sqrt{R(\lambda_{\gamma_i(x)})}\right)$. In this formula, the expression $\sqrt{R(\lambda_{\gamma_i(x)})}$ refers to the determination of the square root corresponding to the sheet to which the point $\gamma_i(x)$ belongs, while $-\sqrt{R(\lambda_{\gamma_i(x)})}$ corresponds to the other sheet. Let us define the polynomial in λ:

$$B(\lambda, x) = \prod_{i=1}^{g}(\lambda - \lambda_{\gamma_i(x)})$$

In terms of these data, we have:

Proposition. *The Baker–Akhiezer function is equal to:*

$$\frac{\Psi(x, \lambda)}{\Psi(x_0, \lambda)} = \sqrt{\frac{B(\lambda, x)}{B(\lambda, x_0)}} \exp\left(\int_{x_0}^{x} \frac{\sqrt{R(\lambda)}}{B(\lambda, x)} dx\right) \qquad (11.56)$$

The KdV potential u is expressed as:

$$u = 2\sum_{i=1}^{g} \lambda_{\gamma_i(x)} - \sum_{i=1}^{2g+1} \lambda_j \qquad (11.57)$$

<u>Proof.</u> We need to find the equations governing the x dependence of the divisor $D(x)$. To this end we consider the function $\partial_x \Psi/\Psi$. It is a meromorphic function on Γ, has poles at the points $\gamma_i(x)$ and behaves like $z + O(1/z)$ at infinity. Hence we can write

$$\frac{\partial_x \Psi}{\Psi} = \frac{\sqrt{R(\lambda)} + Q(\lambda, x)}{\prod_{i=1}^{g}(\lambda - \lambda_{\gamma_i(x)})} \qquad (11.58)$$

where $Q(\lambda, x)$ is a polynomial of degree $g - 1$ in λ. We determine $Q(\lambda, x)$ by requiring that $\frac{\partial_x \Psi}{\Psi}$ has a pole above $\lambda = \lambda_{\gamma_i(x)}$ on the sheet $\mu_{\gamma_i(x)}$ and not on $-\mu_{\gamma_i(x)}$. Thus we find the g conditions $Q(\lambda_{\gamma_i(x)}, x) = \mu_{\gamma_i(x)}$ which completely determine the polynomial:

$$Q(\lambda, x) = \sum_i \mu_{\gamma_i(x)} \frac{\prod_{j\neq i}(\lambda - \lambda_{\gamma_j(x)})}{\prod_{j\neq i}(\lambda_{\gamma_i(x)} - \lambda_{\gamma_j(x)})}$$

On the other hand, in the vicinity of $\lambda_{\gamma_i(x)}$, we have:

$$\frac{\partial_x \Psi}{\Psi} = -\frac{\partial_x \lambda_{\gamma_i(x)}}{\lambda - \lambda_{\gamma_i(x)}} + O(1) \qquad (11.59)$$

because in the vicinity of the zero λ_{γ_i}, we have $\Psi(x, \lambda) \sim (\lambda - \lambda_{\gamma_i})\tilde{\Psi}$. Comparing the residues of the poles at $\lambda = \lambda_{\gamma_i}$ in eq. (11.58) and in eq. (11.59), we get the equations of motion for the divisor $D(x)$:

$$\partial_x \lambda_{\gamma_i(x)} = -2\frac{\mu_{\gamma_i}(x)}{\prod_{j \neq i}(\lambda_{\gamma_i}(x) - \lambda_{\gamma_j}(x))} \tag{11.60}$$

One can now reconstruct the Baker–Akhiezer function itself. Indeed, inserting eq. (11.60) into eq. (11.58), we get:

$$\frac{\partial_x \Psi}{\Psi} = \frac{\sqrt{R(\lambda)}}{B(\lambda, x)} - \frac{1}{2}\sum_i \frac{1}{\lambda - \lambda_{\gamma_i}(x)} \partial_x \lambda_{\gamma_i(x)}$$

Integrating this formula from x_0 to x gives eq. (11.56). To compute the potential $u(x)$, we insert eq. (11.56) into $(\partial_x^2 - u)\Psi = \lambda\Psi$. We get the polynomial identity

$$R = -\frac{1}{2}BB'' + \frac{1}{4}B'^2 + (u + \lambda)B^2$$

where $' = \partial_x$. Comparing the terms in λ^{2g} we obtain eq. (11.57).   ■

One can find the generalization of eq. (11.60) for any flow of the KdV hierarchy.

Proposition. *On the coordinates λ_{γ_i} of the divisor $D(x)$ the equations of motion with respect to the time t_{2j-1} read:*

$$\partial_{t_{2j-1}}\lambda_{\gamma_i} = -2\frac{P_j(\lambda_{\gamma_i})\mu_{\gamma_i}}{\prod_{l \neq i}(\lambda_{\gamma_i} - \lambda_{\gamma_l})}, \quad P_j(\lambda) = \left(\frac{\prod_{i=1}^{g}(1 - \lambda_{\gamma_i}\lambda^{-1})}{\sqrt{\prod_{i=1}^{2g+1}(1 - \lambda_i\lambda^{-1})}}\lambda^{j-1}\right)_+$$
$$\tag{11.61}$$

where the $(\ )_+$ *means taking the polynomial part in the expansion at* $\lambda = \infty$.

<u>Proof.</u> The only difference from the previous discussion is that, in the derivation of eq. (11.58), the behaviour at ∞ is replaced by:

$$\partial_{t_{2j-1}}\Psi/\Psi = z^{2j-1} + O(1/z) \tag{11.62}$$

Following the same reasoning as before, we can write:

$$\frac{\partial_{t_{2j-1}}\Psi}{\Psi} = \frac{P_j(\lambda)\sqrt{R(\lambda)} + Q_j(\lambda)}{B(\lambda)}$$

where $P_j(\lambda)$ is a polynomial in λ of degree $(j-1)$ and $Q_j(\lambda)$ is a polynomial in λ of degree $(g-1)$. The polynomial $P_j(\lambda) = \lambda^{j-1} + \cdots$ is uniquely determined by imposing the asymptotic eq. (11.62), which gives $(j-1)$ linear conditions. The solution is given by eq. (11.61). The polynomial Q_j is determined as above, yielding $Q_j(\lambda_{\gamma_i}) = \mu_{\gamma_i} P_j(\lambda_{\gamma_i})$. One gets the equation of motion for the divisor $D(x)$:

$$\partial_{t_{2j-1}}\lambda_{\gamma_i} = -2\frac{P_j(\lambda_{\gamma_i})\mu_{\gamma_i}}{\prod_{l\neq i}(\lambda_{\gamma_i} - \lambda_{\gamma_l})}$$

∎

This can be used to give a direct proof of the linearization of the flows on the Jacobian of the hyperelliptic curve Γ. For this purpose, it is sufficient to consider the time evolutions of the Abel sums

$$\frac{\partial}{\partial_{t_{2j-1}}}\sum_i \int^{\lambda_{\gamma_i}} \frac{\lambda^k}{\sqrt{R(\lambda)}}d\lambda = \sum_i \frac{\lambda_{\gamma_i}^k}{\sqrt{R(\lambda_{\gamma_i})}}\partial_{t_{2j-1}}\lambda_{\gamma_i}$$

$$= -2\sum_i \frac{\lambda_{\gamma_i}^k P_j(\lambda_{\gamma_i})}{\prod_{l\neq i}(\lambda_{\gamma_i} - \lambda_{\gamma_l})}$$

The right-hand side is equal to the integral:

$$\frac{-2}{2i\pi}\int_\Upsilon d\lambda \frac{\lambda^k P_j(\lambda)}{B(\lambda)}, \quad B(\lambda) = \prod_l (\lambda - \lambda_{\gamma_l})$$

where Υ is a loop surrounding all the λ_{γ_i}. We deform the contour to a loop around ∞ so that:

$$-2\sum_i \frac{\lambda_{\gamma_i}^k P_j(\lambda_{\gamma_i})}{\prod_{l\neq i}(\lambda_{\gamma_i} - \lambda_{\gamma_l})} = 2\mathrm{Res}_{\lambda=\infty}\frac{P_j(\lambda)}{B(\lambda)}\lambda^k d\lambda$$

To compute this residue at $\lambda = \infty$, we write:

$$2\mathrm{Res}_{\lambda=\infty}\frac{P_j(\lambda)}{B(\lambda)}\lambda^k d\lambda = \mathrm{Res}_\infty \frac{P_j(\lambda)\sqrt{R(\lambda)}}{B(\lambda)}\frac{\lambda^k d\lambda}{\sqrt{R(\lambda)}}$$

where the right-hand side is a residue computed on the curve Γ. The factor 2 appears because ∞ is a branch point of the covering $z \to \lambda = z^2$ around that point, so that the residue on Γ has to be computed with $1/z$ as local parameter. The first factor behaves as $z^{2j-1} + O(1/z)$ by construction. So we have:

$$\frac{\partial}{\partial_{t_{2j-1}}}\sum_i \int^{\gamma_i} \omega_k = \mathrm{Res}_{z'=0}\left(z'^{-2j+1} + O(z')\right)\omega_k$$

where $z' = 1/z$ is the local parameter at ∞, and $\omega_k = \lambda^k d\lambda/\sqrt{R(\lambda)}$ is an unnormalized Abelian differential. This shows that the flow linearizes on the Jacobian because the right-hand side is a constant. Since this equation is linear in ω_k it remains true for normalized Abelian differentials. One can then use Riemann's bilinear identities to evaluate the residue further. The time derivatives of the Abel map are given by:

$$\frac{\partial}{\partial t_{2j-1}} A(D) = \mathrm{Res}_\infty \left(z'^{-2j+1} + O(z') \right) \omega_k = -\frac{1}{2i\pi} \oint_{b_k} \Omega^{(2j-1)} = -U^{(2j-1)}$$

We recover exactly the expected slope of the linear flow on the Jacobian.

Remark. The equation of motion of the divisor D can be recast in compact form by introducing the generating function of time derivatives, eq. (11.22). Remembering that for a function $f(\lambda) = \sum_{n=0}^{\infty} f_n \lambda^{-n}$ we have:

$$\sum_{j=1}^{\infty} \lambda'^{-j} (f(\lambda)\lambda^{j-1})_+ = \frac{f(\lambda')}{\lambda' - \lambda}$$

we get:

$$\nabla(\lambda)\lambda_{\gamma_i} = -2 \frac{\sqrt{\lambda}}{\lambda - \lambda_{\gamma_i}} \frac{B(\lambda)}{\sqrt{R(\lambda)}} \frac{\sqrt{R(\lambda_{\gamma_i})}}{\prod_{l \neq i}(\lambda_{\gamma_i} - \lambda_{\gamma_l})} \tag{11.63}$$

Finally, we show that eq. (11.57) can be generalized to a whole set of so-called trace identities.

Proposition. *The generating function $S(t, \lambda)$ of the local quantities S_{2n} has a simple expression in terms of the divisor $D(x)$.*

$$S(t, \lambda) = \sqrt{\lambda} \frac{B(\lambda, x)}{\sqrt{R(\lambda)}} = \frac{\prod_{i=1}^{g}(1 - \lambda_{\gamma_i(t)}\lambda^{-1})}{\sqrt{\prod_{j=1}^{2g+1}(1 - \lambda_j \lambda^{-1})}} \tag{11.64}$$

<u>Proof.</u> Recall that we have defined $S(t, \lambda) = \Psi\Psi^*$, where Ψ and Ψ^* are normalized such that their Wronskian is equal to $2z$. We can evaluate $S(t, \lambda)$ using eq. (11.56) for Ψ and a similar equation for Ψ^* with the sign of the exponential reversed, provided that we normalize these expressions to have the correct Wronskian. Using eq. (11.56), one gets $W(\Psi, \Psi^*) = 2\Psi(x_0, \lambda)\Psi^*(x_0, \lambda)\sqrt{R(\lambda)}/B(\lambda, x_0)$, and the expression of $S(t, \lambda)$ follows. ∎

Taking the logarithm of eq. (11.64), we get:

$$\sum_{n=0}^{\infty} \lambda^{-n} \frac{1}{n} \left(\sum_i \lambda_{\gamma_i}^n \right) = \frac{1}{2} \sum_{n=0}^{\infty} \lambda^{-n} \frac{1}{n} \left(\sum_i \lambda_i^n \right) - \log S(\lambda)$$

Identifying the powers of λ^{-n}, we find:

$$\sum_i \lambda_{\gamma_i} = \frac{1}{2}\sum \lambda_i + \frac{u}{2}, \quad \sum_i \lambda_{\gamma_i}^2 = \frac{1}{2}\sum \lambda_i^2 + \frac{1}{4}u'' - \frac{1}{2}u^2$$

The first equation is eq. (11.57). We see that all the symmetric functions of the λ_{γ_i} have a local expression in terms of the potential u.

11.6 Finite-zone solutions

The Its–Matveev formula, eq. (11.54), shows that, as a function of the *real* variable x, the potential $u(x,t)$ is almost periodic. Specifically, the argument of the theta-function is a straight line, $Y(x) = U^{(1)}x + Y_0$, which wraps densely around the Jacobian torus, and for ℓ sufficiently large $Y(x+\ell)$ returns arbitrarily close to $Y(x)$. This means that $Y(x+\ell) \simeq Y(x)+n+\mathcal{B}m$, where $n, m \in \mathbb{Z}^g$ and $\mathcal{B}$ is the matrix of b-periods. The effect of the translation by $n + \mathcal{B}m$ does not affect the potential because of the second order derivative in front of the logarithm. Hence $u(x+\ell) \simeq u(x)$. One can choose the moduli of the curve Γ so that the potential is exactly periodic. This amounts to the condition:

$$\ell U^{(1)} = n + \mathcal{B}m \tag{11.65}$$

which gives $2g$ real conditions on the $2g + 1$ complex parameters λ_i, the branch points of the hyperelliptic curve Γ:

$$\Gamma: \quad \mu^2 = R(\lambda) = \prod_{i=1}^{2g+1} (\lambda - \lambda_i)$$

If, however, the parameters λ_i are real (as will be the case in the following), these conditions quantize all the moduli of Γ (up to a translation $\lambda_i \to \lambda_i + \alpha$ which does not change the periods $U^{(1)}$). Under these conditions the potential $u(x)$ given by eq. (11.57) is periodic, and eq. (11.53) shows that the Baker–Akhiezer function $\Psi(x, \lambda)$ is a Bloch wave:

$$\Psi(x + \ell, \lambda) = e^{\mathcal{P}(\lambda)}\Psi(x, \lambda) \tag{11.66}$$

Up to now, the potential $u(x)$ was complex. We determine below the conditions on branch points λ_i and the divisor $D(x)$ ensuring that $u(x)$ is real and periodic. To do this, we first derive general properties of Bloch waves for generic real periodic potentials. We will then particularize these properties to the algebro-geometric solutions. We will get in this way the very special finite-zone potentials.

Consider the Schroedinger equation $\Psi'' - u\Psi = \lambda\Psi$ with a real periodic potential $u(x)$ with period ℓ. The space of solutions is spanned by the two solutions $y_1(x, \lambda)$ and $y_2(x, \lambda)$ with initial conditions:

$$y_1(0, \lambda) = 1, \quad y_1'(0, \lambda) = 0, \quad y_2(0, \lambda) = 0, \quad y_2'(0, \lambda) = 1$$

Because the initial conditions are independent of λ, y_1 and y_2 are entire functions of λ. Since $u(x)$ is periodic, $y_1(x + \ell, \lambda)$ and $y_2(x + \ell, \lambda)$ are two other solutions and we can write:

$$\begin{pmatrix} y_1(x + \ell, \lambda) \\ y_2(x + \ell, \lambda) \end{pmatrix} = T(\lambda) \begin{pmatrix} y_1(x, \lambda) \\ y_2(x, \lambda) \end{pmatrix}$$

where $T(\lambda)$ is a 2×2 matrix, the monodromy matrix. Because the Wronskian $W(y_1, y_2) = 1$, we have $\det T(\lambda) = 1$. Notice that for real λ, y_i are real so that $T(\lambda)$ is real, and in general $\overline{T(\lambda)} = T(\bar\lambda)$. The Bloch waves are the two solutions which diagonalize the monodromy matrix. Denote by $e^{\pm\mathcal{P}(\lambda)}$ its eigenvalues (with product 1). The two Bloch waves are such that $\Psi_\pm(x + \ell, \lambda) = e^{\pm\mathcal{P}(\lambda)}\Psi_\pm(x, \lambda)$, and we choose to normalize them by $\Psi_\pm(0, \lambda) = 1$. The quantity $\mathcal{P}(\lambda)$ is called the quasi-momentum. Writing $\Psi_\pm(x, \lambda) = y_1(x, \lambda) + \beta y_2(x, \lambda)$, the Bloch condition on $\Psi_\pm(x, \lambda)$ and $\partial_x \Psi_\pm(x, \lambda)$ gives:

$$\Psi_\pm(x, \lambda) = y_1(x, \lambda) + \frac{e^{\pm\mathcal{P}(\lambda)} - y_1(\ell, \lambda)}{y_2(\ell, \lambda)} y_2(x, \lambda) \tag{11.67}$$

where $\mathcal{P}(\lambda)$ is obtained by solving the equation:

$$e^{\mathcal{P}(\lambda)} + e^{-\mathcal{P}(\lambda)} = y_1(\ell, \lambda) + y_2'(\ell, \lambda) \equiv t(\lambda) \tag{11.68}$$

This shows that $t(\lambda) = \operatorname{Tr} T(\lambda)$ is an entire function of λ, hence $e^{\mathcal{P}(\lambda)}$ has no pole or zero at finite distance. One can discuss the nature of the solutions of eq. (11.68) when λ is real, which also implies that $t(\lambda)$ is real. When $\Delta(\lambda) \equiv t^2(\lambda) - 4$ is negative, the quasi-momentum $\mathcal{P}(\lambda)$ is pure imaginary. In this case $\Psi(x, \lambda)$ has an oscillatory behaviour in x at large scale corresponding to propagation of waves. This defines what is called the allowed zones in the spectrum. When $\Delta(\lambda)$ is positive, $\Psi(x, \lambda)$ has an exponential behaviour at large scale, so waves cannot propagate. This defines the forbidden zones in the spectrum. Finally, when $\Delta(\lambda) = 0$, we have $t(\lambda) = \pm 2$ and $e^{\mathcal{P}(\lambda)} = \pm 1$ (same sign). This means that $\Psi(x + \ell, \lambda) = \pm\Psi(x, \lambda)$ so that $\Psi(x, \lambda)$ is periodic or antiperiodic. This means that the periodic and antiperiodic levels are the boundaries of allowed and forbidden zones.

When $\lambda \to \infty$ one can as a first approximation neglect the potential u. Then for $\lambda \to +\infty$ we have $\Psi(x, \lambda) \sim e^{\pm\sqrt{\lambda}x}$ (forbidden zone) and

$t(\lambda) \sim 2\cosh\sqrt{\lambda}\ell$. For $\lambda \to -\infty$ we have similarly $\Psi(x,\lambda) \sim e^{\pm i\sqrt{-\lambda}x}$ (allowed zone) and $t(\lambda) \sim 2\cos\sqrt{-\lambda}\ell$. In fact, when $u \neq 0$ one can show that for generic potentials there are an infinite number of forbidden zones of exponentially decreasing sizes extending in the region $\lambda \to -\infty$.

Finite-zone potentials are such that this phenomenon does not occur, i.e. there are a finite number of forbidden zones.

The classical theory of Sturm–Liouville equations gives a rather detailed information on the poles of the Bloch waves and the boundaries of the zones. We recall the main facts. Consider a differential equation $y'' - uy = \lambda y$ with a real periodic potential u. We have:

$$\int_0^\ell f\partial^2 g = \int_0^\ell g\partial^2 f + [fg' - f'g]_0^\ell$$

so that we get a self-adjoint problem when the boundary conditions are such that the term $[fg' - f'g]_0^\ell$ vanishes. In this case the spectrum is real.

Proposition. *The poles of the Bloch wave $\Psi(x,\lambda)$ and the periodic and antiperiodic levels are all real. The periodic and antiperiodic levels form a sequence $\beta_1 > \beta_2 \geq \beta_3 > \beta_4 \geq \beta_5 > \cdots$ such that β_1 is a periodic level, β_2, β_3 are antiperiodic levels, β_4, β_5 are periodic, and so on. The allowed zones are the intervals $[\beta_{2j}, \beta_{2j-1}]$. The forbidden zones are the intervals $[\beta_1, \infty]$ and the $[\beta_{2j+1}, \beta_{2j}]$. There is at least one-pole of $\Psi(x,\lambda)$ in each forbidden zone, except $[\beta_1, +\infty]$.*

<u>Proof.</u> It follows from eq. (11.67) that the poles in λ of $\Psi_\pm(x,\lambda)$ are located where $y_2(\ell,\lambda) = 0$. For these values of λ, $y_2(x,\lambda)$ is solution of the Sturm–Liouville problem with boundary conditions $y(0,\lambda) = y(\ell,\lambda) = 0$. This problem is self-adjoint, so that these values of λ are real. Similarly, the periodic and antiperiodic levels correspond to the boundary conditions (valid in the case of a periodic potential) $y(\ell,\lambda) = \pm y(0,\lambda)$ and $y'(\ell,\lambda) = \pm y'(0,\lambda)$, which also lead to a self-adjoint problem. These levels are therefore also real. We have shown that all the roots of $\Delta(\lambda)$ are real. We now show that $\partial_\lambda t(\lambda)$ has a definite sign in the allowed zones. We have $\partial_\lambda t(\lambda) = \partial_\lambda y_1(\ell,\lambda) + \partial_\lambda y_2'(\ell,\lambda)$. Since $v = \partial_\lambda y_j$ obeys the differential equation $v'' - (u+\lambda)v = y$ with initial conditions $v(0) = v'(0) = 0$, one has:

$$\partial_\lambda y_j(x,\lambda) = \int_0^x (y_1(x)y_2(\xi) - y_1(\xi)y_2(x))y_j(\xi)d\xi, \quad j = 1,2$$

$$\partial_\lambda y_j'(x,\lambda) = \int_0^x (y_1'(x)y_2(\xi) - y_1(\xi)y_2'(x))y_j(\xi)d\xi$$

so that:

$$\partial_\lambda t(\lambda) = \int_0^\ell (A(\lambda)y_2^2(\xi) + B(\lambda)y_1(\xi)y_2(\xi) + C(\lambda)y_1^2(\xi))d\xi$$

where $A(\lambda) = y_1'(\ell, \lambda)$, $B(\lambda) = y_1(\ell, \lambda) - y_2'(\ell, \lambda)$, $C(\lambda) = -y_2(\ell, \lambda)$. The quadratic form appearing in the integrand has discriminant

$$B^2 - 4AC = (y_1 - y_2')^2 + 4y_2y_1' = t^2(\lambda) - 4 = \Delta(\lambda) < 0$$

so it is of fixed sign in the whole domain of integration. It follows that $\partial_\lambda t(\lambda) \neq 0$ in an allowed zone, and that the sign of $\partial_\lambda t(\lambda)$ is the same as the sign of $C(\lambda) = -y_2(\ell, \lambda)$. In particular $y_2(\ell, \lambda)$ cannot vanish in an allowed zone. We see that $t(\lambda)$ either crosses the lines $t = \pm 2$ or is tangent to them, but cannot have an extremum in the region $|t(\lambda)| < 2$. From this, it follows that the periodic and antiperiodic levels are distributed as indicated in the proposition. Obviously, the sign of $\partial_\lambda t(\lambda)$ changes when one goes from one allowed zone to the next one, so that $y_2(\ell, \lambda)$ has at least one zero in each forbidden zone $[\beta_{2j+1}, \beta_{2j}]$ with $\beta_{2j+1} \neq \beta_{2j}$ and $j \geq 1$. ■

We now return to the case where the periodic potential $u(x)$ is produced by the algebro-geometric construction on the hyperelliptic curve

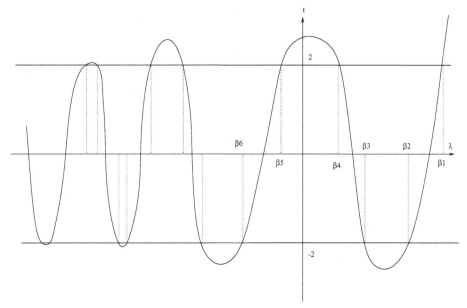

Fig. 11.1. The graph of the function $t(\lambda)$ for real λ.

$\Gamma(\lambda, \mu)$. As we have seen, the two Bloch waves $\Psi_\pm(x, \lambda)$ are the two values of the Baker–Akhiezer function $\Psi(x, (\lambda, \mu))$ at the two points $(\lambda, \pm\mu)$ above λ. At a branch point, we have $\Psi_+(x, \lambda) = \Psi_-(x, \lambda)$ so that the quasi-momentum satisfies $e^{\mathcal{P}(\lambda)} = \pm 1$, i.e. the branch points are zone boundaries and therefore real. Since the region $\lambda \to \infty$ is a forbidden zone and corresponds to $R(\lambda) > 0$, we see, following sign changes, that allowed zones correspond to $R(\lambda) < 0$ and forbidden zones to $R(\lambda) > 0$. In particular, the branch points form a sequence $\lambda_1 > \lambda_2 > \lambda_3 > \cdots > \lambda_{2g+1}$ with $\lambda_1 = \beta_1$ and $\{\lambda_i\} \subset \{\beta_i\}$. There may be degenerate forbidden zones with $\beta_{2j+1} = \beta_{2j}$ which do not correspond to branch points of the spectral curve. To compare eq. (11.56) with this discussion, we set $x_0 = 0$, and we see that for $x = 0$, the divisor $D(0)$ of zeroes of $\Psi(x, \lambda)$ coincides with the divisor of its poles. By eq. (11.67) and the discussion which follows it, we see that the elements of $D(0)$ are all real, and lie in forbidden zones. We know that the divisor $D(0)$ is of degree g if Γ is of genus g, so we put one-pole of $\Psi(x, \lambda)$ in each forbidden zone $[\lambda_{2j+1}, \lambda_{2j}]$, for $j = 1, \ldots, g$, and no pole in $[\lambda_1, +\infty]$ to get a periodic motion. The equation of motion, eq. (11.60), are thus regular and show that all points of $D(x)$ stay real for all x, since $R(\lambda_{\gamma_i(x)}) > 0$, and remain in forbidden zones. Conversely, if all λ_j and $\lambda_{\gamma_i(x)}$ are real, eq. (11.57) shows that u is real.

We have also to choose the curve so that the potential is periodic. This will be the case if the points of $D(x)$ describe cyclic motions on g cycles belonging to the real slice of Γ (i.e. λ and μ real and so $R(\lambda) > 0$) and if this motion is periodic of period ℓ (so that $u(x)$ has period ℓ). As we already remarked, this requires that the moduli be quantized.

It is worth mentioning that, once a genus g Riemann surface Γ is chosen with moduli quantized to produce a periodic finite-zone real potential, the KdV flows with respect to higher times preserve these conditions, and automatically produce a family of continuous deformations parameters.

To express the periodicity conditions further, we view the curve Γ as a two-sheeted covering of the λ plane with cuts on the forbidden zones.

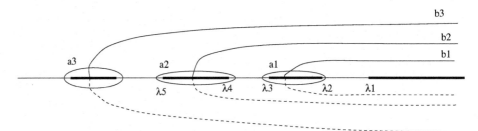

Fig. 11.2. The curve $\Gamma(\lambda, \mu)$ seen as a two sheeted cover of the λ plane with cuts along the forbidden zones.

We choose for the a-cycles loops around the compact cuts on the upper sheet. Then the b-cycles start from ∞ on the upper sheet, cross the a-cycles once, go to the lower sheet through the cut and return to ∞. The regular Abelian differentials are of the form $\lambda^j d\lambda/s'$ for $j = 0, \dots, g-1$. Their a-periods are real and their b-periods are pure imaginary. Hence normalized regular Abelian differentials are real combinations of the above, so the matrix of b-periods is pure imaginary. Moreover, the second kind Abelian differential $\Omega^{(1)}$ is of the form $\lambda^g d\lambda/s'$ plus a combination of first kind differentials with real coefficients, so that $U^{(1)}$ is real. More generally, all $\Omega^{(i)}$ will have all their periods pure imaginary. They are thus the forms considered in eq. (10.19) in Chapter 10.

The periodicity condition eq. (11.65) can only be realized with $m = 0$ and becomes

$$\ell U^{(1)} = \ell \frac{1}{2i\pi} \int_{b_j} \Omega^{(1)} = n_j$$

The Bloch momentum $\mathcal{P}$ is naturally defined only up to $2i\pi\mathbb{Z}$. This is consistent with the following:

Proposition. *The Bloch quasi-momentum $\mathcal{P}(\lambda)$ is the primitive to the form $\Omega^{(1)}$:*

$$d\mathcal{P} = \ell\Omega^{(1)} \tag{11.69}$$

<u>Proof.</u> Recall that $\mu = e^{\mathcal{P}(\lambda)}$ satisfies $\mu + \mu^{-1} = t(\lambda)$, where $t(\lambda)$ is an entire function of λ. Hence μ has no pole or zero at finite distance. Moreover, for periodic motions of the divisor $D(x)$, μ is a well-defined function on the curve Γ (minus the point at ∞) since, by eq. (11.66), we can write it as the quotient of two Baker–Akhiezer functions:

$$\mu(P) = \frac{\Psi(P, x + \ell)}{\Psi(P, x)}$$

We also deduce from this formula the behaviour at ∞: $\mu \sim e^{\ell\sqrt{\lambda}}$. Hence $d\mathcal{P} = d\mu/\mu$ is an Abelian differential on Γ, without singularities at finite distance, and having a double pole at ∞, i.e. $d\mu/\mu \sim \ell dz$ where $z = \sqrt{\lambda}$ is the local parameter. For real λ, μ is real of fixed sign on a forbidden zone, so that the a-periods of $d\mathcal{P} = d\mu/\mu$ vanish. This characterizes $d\mathcal{P} = \ell\Omega^{(1)}$. On the other hand in an allowed zone, μ is of modulus one, so the b-periods of $d\mu/\mu$ are pure imaginary, and of the form $2i\pi n_j$ since μ is well-defined on the curve, in agreement with the periodicity condition. ∎

From eq. (11.56), we can get still another expression for the quasi-momentum. In the case of a periodic motion of the divisor $D(x)$, it is given by:

$$P(\lambda) = \int_0^\ell \frac{\sqrt{R(\lambda)}}{B(\lambda, x)} dx \tag{11.70}$$

At a point β_j which is a boundary of a non-degenerate forbidden zone, $P(\lambda)$ changes from real to pure imaginary, which implies that $R(\lambda)$ changes sign. Hence all such zone boundaries appear among the branch points of Γ. All other periodic or antiperiodic levels are therefore points where $t(\lambda)$ is tangent to $t = \pm 2$, and there are an infinite number of them.

These remarks also allow us to compare the spectral curve Γ_T of the monodromy matrix $T(\lambda)$ corresponding to a finite-zone potential u with the finite genus curve Γ used to built the algebro-geometric solution.

By definition, the "curve" Γ_T is given by the equation $\mu^2 - t(\lambda)\mu + 1 = 0$. Setting $s = 2\mu - t(\lambda)$, we get the hyperelliptic type equation $s^2 = \Delta(\lambda)$, where $\Delta(\lambda) = t^2(\lambda) - 4$. The entire function $\Delta(\lambda)$ is not a polynomial, but admits an infinite product representation:

$$\Delta(\lambda) = \prod_{i=1}^{2g+1} \left(1 - \frac{\lambda}{\lambda_i}\right) \prod_{j=1}^\infty \left(1 - \frac{\lambda}{\beta_j}\right)^2$$

where the β_j are the points where $t(\lambda)$ is tangent to the lines ± 2, i.e. correspond to the degenerate forbidden zones, and the λ_i are the points where $t(\lambda)$ crosses the lines ± 2, i.e. correspond to the branch points of Γ. This infinite product is convergent because $t(\lambda) \sim 2\cos\sqrt{-\lambda}\ell$ when λ is large, or else $\beta_j \sim -j^2\pi^2/\ell$ when j is large. Note that the potential u is determined by Γ, so the β_j are really functions of the moduli λ_i of Γ.

The bianalytic transformation $s = s' \prod_{j=1}^\infty (1 - \frac{\lambda}{\beta_j})$ transforms the equation of Γ_T into the equation $s'^2 = \prod_{i=1}^{2g+1}(1 - \frac{\lambda}{\lambda_i})$, that is the equation of Γ. All points $(s = 0, \lambda = \beta_j)$ are singular points of Γ_T and Γ is the desingularization of Γ_T. Conversely, Γ_T is obtained from Γ by identifying pairs of points $(\lambda = \beta_j, s' = \pm s'(\beta_j))$ which accumulate at ∞.

11.7 Action-angle variables

We want to express the restriction of the symplectic forms corresponding to the two Poisson brackets $\{\ \}_{1,2}$ on the finite-zone solutions of the KdV hierarchy.

We begin with the first Poisson bracket (11.35). This Poisson bracket is degenerate, and the kernel is $\int_0^\ell u(x)dx$. So we consider a symplectic leaf where this integral is kept constant. On such a leaf, the associated symplectic form reads:

$$\omega_1 = \frac{1}{4} \int_0^\ell dx \, \delta u(x) \wedge \int_0^x dy \, \delta u(y)$$

To build finite-zone solutions one chooses a genus g hyperelliptic Riemann surface, and a divisor $D = (\gamma_1, \ldots, \gamma_g)$ of degree g on it. These data determine the potential u. The variations δu are expressed through the variations of the moduli of the curve, and the variations of the divisor, D, of the poles of the Baker–Akhiezer function. When all the times t_{2j-1} are set to zero, the Baker–Akhiezer function is equal to one, and its zeroes coincide with its poles. In agrement with eq. (11.55) we denote $\gamma_i = (\lambda_{\gamma_i}, \mu_{\gamma_i})$, where $\mu_{\gamma_i} = \sqrt{R(\lambda_{\gamma_i})}$.

Proposition. *The restriction of the symplectic form ω_1 on the finite-zone solution constructed from the curve*

$$\Gamma: \ \mu^2 = R(\lambda) = \prod_{i=1}^{2g+1} (\lambda - \lambda_i)$$

is expressed in terms of the divisor of poles $\gamma_i = (\lambda_{\gamma_i}, \mu_{\gamma_i})$ of the Baker–Akhiezer function as:

$$\omega_1 = \sum_{i=1}^{g} \delta \mathcal{P}(\gamma_i) \wedge \delta \lambda_{\gamma_i}$$

where $\mathcal{P}$ is the Bloch momentum such that $d\mathcal{P} = \ell \Omega^{(1)}$ and $\mathcal{P} = z\ell + O(1/z)$ at infinity.

Proof. By analogy with the discussion in Chapter 5, we introduce a 1-form on the curve Γ with values in 2-forms on phase space:

$$K = \langle \Psi^* \delta L \wedge \delta \Psi \rangle \, \Omega$$

where Ω is the form defined in eq. (10.21) in Chapter 10. There are several important differences coming from the field theoretical context. First, the notation $\langle \, \rangle$ means:

$$\langle f(x) \rangle = \lim_{\ell \to \infty} \frac{1}{\ell} \int_0^\ell f(x) dx \tag{11.71}$$

This is a Whitham average and is also an average on the Liouville torus. Second, the variations $\delta \Psi$ have to be defined by keeping the primitive $k = \int^P \Omega^{(1)}$ (behaving as $k = z + O(z^{-1})$ at ∞) fixed instead of keeping λ

fixed. Otherwise, eq. (10.66) in Chapter 10 shows that a term linear in x occurs in $\delta\Psi$ and the average is not defined. Comparing with eq. (10.79) in the same chapter, we can write:

$$K = \langle\Psi^*\delta L \wedge \delta\Psi\rangle\frac{dk}{\langle\Psi^*\Psi\rangle}, \quad \Omega = \frac{dk}{\langle\Psi^*\Psi\rangle} \tag{11.72}$$

Just as taking the variation δ by keeping λ fixed introduces poles at the branch points, that is where $d\lambda = 0$, taking the variation δ by keeping k fixed introduces poles at the points where $dk = \Omega^{(1)} = 0$.

We write as usual that the sum of residues of K on Γ vanishes. The residue of K at ∞ is precisely the form ω_1. Indeed, using k as a local parameter, we have $\Psi = e^{kx}(1 + \alpha/k + \cdots)$ and $\Psi^* = e^{-kx}(1 + \cdots)$. Since we vary while keeping k fixed, we have $\delta\Psi = e^{kx}(\delta\alpha/k + O(1/k^2))$. Remembering that $\delta L = -\delta u$, we get

$$\text{Res}_\infty K = \langle\delta u \wedge \delta\alpha\rangle$$

There is an extra minus sign because the local parameter is really $1/k$. To relate α to the potential u, notice that by definition $\Omega^{(1)} = d(z + O(z^{-1}))$ at ∞, so that

$$\lambda = k^2 + c + O(k^{-1})$$

for some constant c. Reproducing the reasoning leading to eq. (11.52), we find $u = 2\partial\alpha - c$. When the curve Γ is such that the potential u is periodic, we have $\Psi(x + \ell) = e^{k\ell}\Psi(x)$, so that α is periodic. Hence $\langle u\rangle = -c$. Recalling that we keep $\langle u\rangle$ fixed, we have $\delta c = 0$, and so $\delta\alpha = \frac{1}{2}\int^x \delta u$, giving

$$\text{Res}_\infty K = \frac{2}{\ell}\omega_1$$

The form K has poles at finite distance at the poles of Ψ (the poles of Ψ^* are cancelled by Ω) and at the zeroes of $\Omega^{(1)}$, coming from $\delta\Psi$.

At a pole γ_i of Ψ we have $\delta\Psi = (\delta k_{\gamma_i}/(k - k_{\gamma_i}))(\Psi + O(1))$, since we keep k constant in the variation. We get a contribution:

$$\text{Res}_{\gamma_i} K = \text{Res}_{\gamma_i}\frac{\langle\Psi^*\delta L\Psi\rangle}{\langle\Psi^*\Psi\rangle} \wedge \frac{\delta k_{\gamma_i}}{k - k_{\gamma_i}}dk$$

Varying $L\Psi = \lambda\Psi$, we have $\langle\Psi^*\delta L\Psi\rangle = -\langle\Psi^*(L - \lambda)\delta\Psi\rangle + \delta\lambda\langle\Psi^*\Psi\rangle$. Integrating by parts, and using $(L - \lambda)\Psi^* = 0$, we get:

$$\langle\Psi^*(L - \lambda)\delta\Psi\rangle = \frac{1}{\ell}\left[\Psi^*\partial_x\delta\Psi - \partial_x\Psi^*\delta\Psi\right]_0^\ell = \frac{1}{\ell}W(\delta\Psi, \Psi^*)\Big|_0^\ell$$

Taking ℓ large but close enough to an almost period of u, we write:

$$\Psi(x + \ell) = e^{\ell k}\Psi(x), \quad \Psi^*(x + \ell) = e^{-\ell k}\Psi^*(x)$$

so that this quantity vanishes. Finally, $\langle \Psi^* \delta L \Psi \rangle = \delta \lambda \langle \Psi^* \Psi \rangle$ and the contribution of the pole γ_i is:

$$\mathrm{Res}_{\gamma_i} K = \delta \lambda_{\gamma_i} \wedge \delta k_{\gamma_i}$$

We now analyze what happens at zeroes s_i of dk. We have:

$$\delta \Psi = \frac{d\Psi}{d\lambda} \delta \lambda + \sum_j \frac{d\Psi}{dm_j} \delta m_j$$

where m_j are the moduli of Γ. The polar part at s_i comes from the first term and we can replace $\delta \Psi \to \frac{d\Psi}{d\lambda} \delta \lambda$. Moreover, δL is the multiplication operator by $-\delta u$, so

$$\langle \Psi^* \delta L \frac{d\Psi}{d\lambda} \rangle \wedge \delta \lambda \, \Omega = \langle (\delta L - \delta \lambda) \Psi^* \frac{d\Psi}{d\lambda} \rangle \wedge \delta \lambda \, \Omega$$

where we added the $\delta \lambda$ term which cancels in the wedge product. Varying $(L - \lambda)\Psi^* = 0$, we can write

$$\mathrm{Res}_{s_i} K = -\mathrm{Res}_{s_i} \langle (L - \lambda) \delta \Psi^* \frac{d\Psi}{d\lambda} \rangle \wedge \delta \lambda \, \Omega$$

Integrating by parts gives

$$\mathrm{Res}_{s_i} K = -\mathrm{Res}_{s_i} \left[\langle \delta \Psi^* (L - \lambda) \frac{d\Psi}{d\lambda} \rangle + \frac{1}{\ell} W(\delta \Psi^*, \frac{d\Psi}{d\lambda}) \Big|_0^\ell \right] \wedge \delta \lambda \, \Omega$$

Differentiating $(L - \lambda)\Psi = 0$ with respect to λ, we have $(L - \lambda)\frac{d\Psi}{d\lambda} = \Psi$. Using that Ψ and Ψ^* are Bloch waves and the fact that $\delta k = 0$, we find

$$\frac{1}{\ell} W(\delta \Psi^*, \frac{d\Psi}{d\lambda}) \Big|_0^\ell = \frac{dk}{d\lambda} W(\delta \Psi^*, \Psi) \Big|_0$$

Since dk vanishes at s_i, to get a residue for this term one has to take the polar contribution in $\delta \Psi^*$, proportional to $\delta \lambda$, which disappears due to the wedge product. We get finally

$$\mathrm{Res}_{s_i} K = -\mathrm{Res}_{s_i} K_1, \quad K_1 = \langle \delta \Psi^* \Psi \rangle \wedge \delta \lambda \, \Omega$$

To evaluate the sum of residues of K_1 at s_i, we write that the sum of residues of K_1 vanishes. Poles of K_1 are at the s_i, the poles γ_i^* of Ψ^*, and at infinity. At infinity, Ω has a double pole but $\delta \lambda$ and $\langle \delta \Psi^* \Psi \rangle$ have a simple zero so that there is no residue. At γ_i^*, we write $\delta \Psi^* = \frac{\delta k_{\gamma_i^*}}{k - k_{\gamma_i^*}} (\Psi^* + \cdots)$, and using eq. (11.72) for Ω, we find

$$\sum_{s_i} \mathrm{Res}_{s_i} K_1 = -\sum_{\gamma_i^*} \delta k_{\gamma_i^*} \wedge \delta \lambda_{\gamma_i^*}$$

Recalling that $\gamma_i^* = \sigma(\gamma_i)$, where σ is the hyperelliptic involution $(\lambda, k) \to (\lambda, -k)$, we get $\lambda_{\gamma_i^*} = \lambda_{\gamma_i}$ and $k_{\gamma_i^*} = -k_{\gamma_i}$. Hence the result

$$\sum_{s_i} \mathrm{Res}_{s_i} K = -\sum_{\gamma_i} \delta k_{\gamma_i} \wedge \delta \lambda_{\gamma_i}$$

from which the proposition follows.   ∎

We can analyse the second symplectic structure in a similar way. The symplectic form ω_2 is expressed most simply in terms of the Miura variable p such that $u = p' + p^2$. Since the second Poisson bracket is given by eq. (11.39), it has a kernel $\int_0^\ell p(x)dx$ in the variable p. So this quantity has to be fixed. The symplectic form in the variable p reads:

$$\omega_2 = -\int_0^\ell dx\, \delta p(x) \wedge \int_0^x dy\, \delta p(y)$$

With the same notations as in the previous proposition:

Proposition. *The restriction of the symplectic form ω_2 on the finite-zone solution is expressed as:*

$$\omega_2 = 2 \sum_{i=1}^{g} \delta \mathcal{P}(\gamma_i) \wedge \frac{\delta \lambda_{\gamma_i}}{\lambda_{\gamma_i}}$$

<u>Proof.</u> Now we introduce the 1-form K on the curve Γ, with values in 2-forms on phase space:

$$K = \langle \Psi^* \delta L \wedge \delta \Psi \rangle \frac{\Omega}{\lambda}$$

Again, the variations are done by keeping the meromorphic function k fixed. In contrast to the previous case, there is no pole at ∞, but a new pole appears at $\lambda = 0$. Let us compute the residue of K at $\lambda = 0$. We will show that

$$\mathrm{Res}_{\lambda=0} K = \frac{1}{\ell} \omega_2$$

The Miura variable p is related to Ψ by

$$p = \frac{\partial_x \Psi}{\Psi}\bigg|_{\lambda=0}$$

and we can express $\Psi(x) \equiv \Psi(\lambda = 0, x)$ in terms of p as:

$$\Psi(x) = e^{\int_0^x p(y)dy}$$

When u is periodic of period ℓ, we can choose p periodic and Ψ is a Bloch wave with Bloch momentum given by $\mathcal{P}(\lambda = 0) = \int_0^\ell p(y)dy$, hence in the kernel of ω_2. The residue of K at $\lambda = 0$ is:

$$\langle \Psi^* \delta L \wedge \delta \Psi \rangle = -\langle \Psi^* \Psi (\delta p' + 2p\delta p) \wedge \int_0^x \delta p \rangle$$

$$= -\langle \Psi^* \Psi \, 2p\delta p \wedge \int_0^x \delta p \rangle + \langle \delta p \wedge \partial_x \left(\Psi^* \Psi \int_0^x \delta p \right) \rangle$$

$$- \frac{1}{\ell} \left[\Psi^* \Psi \delta p \wedge \int_0^x \delta p \right]_0^\ell = \frac{W(\Psi, \Psi^*)}{\ell} \int_0^\ell \delta p \wedge \int_0^x \delta p$$

where we have used $(\Psi^* \Psi)' = -W(\Psi, \Psi^*) + 2p\Psi^*\Psi$ and the fact that $\Psi^*\Psi$ and p are periodic, so the last term is proportional to $\delta \int_0^\ell p$, hence vanishes. Now the residue is easily computed once we notice that

$$\Omega = \frac{d\lambda}{W(\Psi, \Psi^*)}$$

This is because the right-hand side has zeroes at the poles of Ψ and Ψ^*. Moreover, the Wronskian vanishes at the branch points, thus cancelling the zeroes of $d\lambda$. Finally, at ∞ we can normalize Ψ and Ψ^* such that $W(\Psi, \Psi^*) \simeq 2z$, so that the form behaves as dz ($\lambda = z^2$). This uniquely identifies Ω.

The analysis of the residues of K at the poles of Ψ and Ψ^* is exactly the same as in the previous case, replacing Ω by Ω/λ everywhere, which accounts for the result $\delta \mathcal{P}(\gamma_i) \wedge \delta\lambda_{\gamma_i}/\lambda_{\gamma_i}$. Finally, at the points s_i, the analysis is also similar but the form K_1 involves Ω/λ and so has no pole at ∞ but a possible pole at $\lambda = 0$. In fact, there is no pole at $\lambda = 0$ because $\delta\lambda$ vanishes there: the function $k(\lambda)$ has an expansion of the form

$$k(\lambda) = \frac{1}{\ell} \int_0^\ell p(y)dy + a\lambda + \cdots$$

where the first term is the Bloch momentum at $\lambda = 0$. Variations are done by keeping k fixed and $\int_0^\ell p(y)dy$ fixed. So $\delta\lambda = -(\delta \log a)\lambda + \cdots$ vanishes at $\lambda = 0$. ∎

11.8 Analytical description of solitons

The soliton solutions can be viewed as a singular limit of the finite-zone solutions in which the branch points λ_j coincide in pairs. On these degenerate Riemann surfaces, everything becomes rational and all calculations can be performed to the end.

Consider the curve $s^2 = \lambda \prod_{i=1}^{n} (\lambda - \lambda_i)^2$, where we set $\lambda_i = p_i^2$ with $0 < p_1 < p_2 < \cdots$. This singular curve is desingularized by setting $s = z \prod(\lambda - \lambda_i)$, which leads to $\lambda = z^2$. Hence the desingularized curve is the Riemann sphere, which we identify to the complex z plane. The singular curve is obtained from the z plane by identifying the points $z = \pm p_i$ which are both mapped to the point $(s = 0, \lambda = \lambda_i)$. A general Baker–Akhiezer function $\Psi(t, z)$ on the Riemann sphere is given by:

$$\frac{\Psi(t, z)}{\Psi(0, z)} = \prod_{i=1}^{n} \frac{z - z_i(t)}{z - z_i(0)} e^{\xi(z,t)} \tag{11.73}$$

If such a function comes from a function defined on the singular curve, and this is the case if it is a singular limit of a finite-zone solution, it must take the same value at identified points, so we must have $\Psi(t, p_i) = \Psi(t, -p_i)$, or

$$\prod_{j=1}^{n} \frac{p_i + z_j(t)}{p_i - z_j(t)} = a_i e^{2\xi(p_i,t)}, \quad i = 1, \ldots, n \tag{11.74}$$

where a_i are time-independent constants. Equation (11.74) is a linear system for the symmetric functions of the $z_j(t)$. It determines the time evolution of the divisor of zeroes of Ψ.

On the other hand, the soliton solution is well known in terms of tau-functions, see eq. (11.44). From this we can construct the Baker–Akhiezer function through the Sato formula:

$$\frac{\Psi(t, z)}{\Psi(0, z)} = \frac{\tau(t - [z^{-1}])}{\tau(t)} \frac{\tau(t)|_{t=0}}{\tau(t - [z^{-1}])|_{t=0}} e^{\xi(z,t)} \tag{11.75}$$

Recall that $\tau(t)$ depends on times only through the variables

$$X_i(t) = X_i(0) e^{2\xi(p_i,t)}$$

so that:

$$X_i(t - [z^{-1}]) = \frac{z - p_i}{z + p_i} X_i(t)$$

Plugging this into eq. (11.44) for $\tau(t)$, we find a formula of the type:

$$\frac{\tau(t - [z^{-1}])}{\tau(t)} = \frac{\prod(z - z_i(t))}{\prod(z + p_i)} \tag{11.76}$$

This shows that the right-hand side of eq. (11.75) is exactly of the form eq. (11.73). We shall now prove that the $z_i(t)$ defined by eq. (11.74) and eq. (11.76) are the same. For this it is sufficient to show that the $z_i(t)$ from eq. (11.76) satisfy eq. (11.74). We note the relation

$$\tau_n(X) = \tau_{n-1}(X) + X_i\, \tau_{n-1}\left(\frac{(p_i - p_k)^2}{(p_i + p_k)^2}X_k\right)$$

where $\tau_{n-1}(X)$ is given by the formula for $(n-1)$ solitons (X_i removed), and in the coefficient of X_i all arguments X_k in $\tau_{n-1}(X)$ are replaced as indicated. This formula is obvious from eq. (11.44). With its help, we can compare the residue of the pole $z = -p_i$ on both sides of eq. (11.76). We find:

$$-\frac{2p_i}{z+p_i}X_i\frac{\tau_{n-1}\left(\frac{p_i-p_j}{p_i+p_j}X_j\right)}{\tau_n(X)} = -\frac{1}{z+p_i}\frac{\prod_j(p_i + z_j(t))}{\prod_{j\neq i}(p_i - p_j)}$$

Similarly, comparing the two sides of eq. (11.76) for $z = p_i$, we find:

$$\frac{\tau_{n-1}\left(\frac{p_i-p_j}{p_i+p_j}X_j\right)}{\tau_n(X)} = \frac{1}{2p_i}\frac{\prod_j(p_i - z_j(t))}{\prod_{j\neq i}(p_i + p_j)}$$

Combining the two equations yields

$$\prod_j \frac{p_i + z_j(t)}{p_i - z_j(t)} = \prod_{j\neq i}\frac{p_i - p_j}{p_i + p_j}X_i(t) \tag{11.77}$$

This is exactly eq. (11.74). We also found the value of the constants a_i in that equation:

$$a_i = \prod_{j\neq i}\frac{p_i - p_j}{p_i + p_j}X_i(0)$$

The equations of motion for the divisor $D(t) = \{z_i(t)\}$ are obtained by taking the limit of eq. (11.63) in which we have to replace $\lambda_{\gamma_i(t)} = z_i^2(t)$. They read:

$$\nabla(z^2)z_i = -\frac{\prod_j(z_i^2 - p_j^2)\prod_{j\neq i}(z^2 - z_j^2)}{\prod_j(z^2 - p_j^2)\prod_{j\neq i}(z_i^2 - z_j^2)}$$

In particular, taking the coefficient of z^{-2} we find:

$$\partial_x z_i = -\frac{\prod_j(z_i^2 - p_j^2)}{\prod_{j\neq i}(z_i^2 - z_j^2)} \tag{11.78}$$

The solution of these equations is of course given by solving the algebraic system eq. (11.74).

We can also compute the degenerate limit of the Baker–Akhiezer function starting from eq. (11.56). It suffices to notice that:

$$\frac{\sqrt{R(\lambda)}}{B(\lambda, x)}\bigg|_{\lambda=z^2} = z - \frac{1}{2}\sum_i \left(\frac{1}{z - z_i} + \frac{1}{z + z_i}\right)\partial_x z_i \qquad (11.79)$$

so that performing the integral in eq. (11.56) we get eq. (11.73).

One can use the rationality of the spectral curve in this degenerate case to express the tau-function as a rational functions of the divisor $D(t)$.

Proposition. *The tau-function admits a simple expression in terms of the divisor $D(t)$.*

$$\tau(t) = (-1)^{\frac{n(n-1)}{2}} 2^n \left(\prod_{j=1}^n p_j\right) \frac{\prod_{i<j}(z_i + z_j)\prod_{i<j}(p_i + p_j)}{\prod_{i,j}(p_i - z_j)}$$

<u>Proof.</u> We start from the logarithm of eq. (11.76) and apply the operator $\partial_z - \nabla(z^2)$ which vanishes on $\log \tau(t - [z])$. We get:

$$\nabla(z^2)\log \tau(t) = \sum_i \frac{1}{z - z_i} - \sum_i \frac{1}{z + p_i} + \sum_i \frac{1}{z - z_i}\nabla(z^2)z_i$$

Since the tau-function depends on times only through the $z_i(t)$, we have $\nabla(z^2)\log \tau(t) = \sum_i \partial_{z_i} \log \tau(t)\nabla(z^2)z_i$. Identifying the residues of the poles in z at $z = \pm p_l$ and $z = z_l$, one finds the n conditions:

$$\sum_i \partial_{z_i} \log \tau(t)\frac{1}{p_l^2 - z_i^2}\partial_x z_i = \sum_i \frac{1}{p_l - z_i}\frac{1}{p_l^2 - z_i^2}\partial_x z_i, \quad l = 1, \dots, n$$

This is a linear system for the quantities $\partial_{z_i} \log \tau(t)\partial_x z_i$, the solution of which requires the inversion of the Cauchy matrix $M_{ij} = 1/(p_i^2 - z_j^2)$. Recalling eq. (9.33) in Chapter 9 we find:

$$(M^{-1})_{ij} = (-1)^{n-1}\frac{\prod_{k\neq i}(p_j^2 - z_k^2)\prod_k(p_k^2 - z_i^2)}{\prod_{k\neq j}(p_j^2 - p_k^2)\prod_{k\neq i}(z_i^2 - z_k^2)} = \partial_x z_i \frac{\prod_{k\neq i}(p_j^2 - z_k^2)}{\prod_{k\neq j}(p_j^2 - p_k^2)}$$

This gives:

$$\partial_{z_i} \log \tau = (-1)^{n-1}\sum_{j,l} \frac{\prod_{k\neq i}(p_j^2 - z_k^2)\prod_{k\neq j}(p_k^2 - z_l^2)}{\prod_{k\neq j}(p_j^2 - p_k^2)\prod_{k\neq l}(z_l^2 - z_k^2)}\frac{1}{p_j - z_l}$$

We consider this expression as a rational function of z_i. It has poles at $z_i = p_j$ for any j and at $z_i = \pm z_j$ for $j \neq i$, and goes to 0 at $\lambda_i = \infty$.

One sees easily that the residue at $z_i = p_j$ is equal to -1, the residue at $z_i = z_j$ vanishes. We finally get:

$$\partial_{z_i} \log \tau = \sum_j \frac{1}{p_j - z_i} + \sum_{l \neq i} \frac{r_l}{z_i + z_l}$$

with

$$r_l = \sum_j \frac{\prod_{k \neq j}(z_l^2 - p_k^2) \prod_{k \neq i,l}(p_j^2 - z_k^2)}{\prod_{k \neq j}(p_j^2 - p_k^2) \prod_{k \neq l,i}(z_l^2 - z_k^2)}$$

In fact we have $r_l = 1$. This is equivalent to the identity:

$$\sum_j \frac{\prod_{k \neq i,l}(p_j^2 - z_k^2)}{\prod_{k \neq j}(p_j^2 - p_k^2)} \frac{1}{z_l^2 - p_j^2} = \frac{\prod_{k \neq i,l}(z_l^2 - z_k^2)}{\prod_k(z_l^2 - p_k^2)}$$

which is easily checked by comparing the residues in z_l^2 on both sides. This gives the final simple result:

$$\partial_{z_i} \log \tau = \sum_j \frac{1}{p_j - z_i} + \sum_{l \neq i} \frac{1}{z_i + z_l}$$

Integrating this formula, we find:

$$\tau = C \frac{\prod_{i<j}(z_i + z_j)}{\prod_{i,j}(p_i - z_j)} \tag{11.80}$$

where the constant C does not depend on the z_i hence is independent of times. To find it, note that when the times are set to $-\infty$ we have $\tau = 1$. Equation (11.76) implies that the $z_i(-\infty)$ are equal to the $-p_i$ up to ordering. Inserting these values into eq. (11.80) gives

$$C = (-1)^{\frac{n(n-1)}{2}} 2^n \prod_j p_j \prod_{i<j}(p_i + p_j)$$

■

To end this section, we would like to discuss the Poisson structures of the KdV equation in these variables. As we know, we have a whole hierarchy of these structures. Let ω_k, $k = 1, 2$ be the restriction of the kth symplectic form on the manifold of finite-zone solutions. We have seen that:

$$\omega_k = c_k \sum_{i=1}^{n} \delta \mathcal{P}(\lambda_{\gamma_i}) \wedge \frac{\delta \lambda_{\gamma_i}}{\lambda_{\gamma_i}^{k-1}} \tag{11.81}$$

where c_k is a normalization constant and the Bloch momentum $\mathcal{P}(\lambda)$ is defined as:

$$\mathcal{P}(\lambda) = \frac{1}{2} \log \frac{\Psi(\lambda, x = \ell)}{\Psi(\lambda, x = -\ell)}$$

To compute this Bloch momentum in the soliton limit, we use eq. (11.73) and we send $\ell \to \infty$. Equation (11.74) shows that, up to ordering, we have $z_i(\ell) \to p_i$, $z_i(-\ell) \to -p_i$. Using eq. (11.70) and eq. (11.79), which is symmetric in the z_i, we get:

$$\mathcal{P}(z) = \ell z - \frac{1}{2} \log \prod_j \left(\frac{z + p_j}{z - p_j} \right) \quad \mathrm{mod} \;\; i\pi$$

Hence, recalling that $\lambda_{\gamma_i} = z_i^2$, we find:

$$\omega_k = c_k \sum_{i=1}^{n} \delta \log \prod_j \left(\frac{z_i - p_j}{z_i + p_j} \right) \wedge \frac{\delta z_i}{z_i^{2k-3}} = 2c_k \sum_{ij} \frac{\delta z_i \wedge \delta p_j}{(z_i^2 - p_j^2) z_i^{2k-4}}$$

The form ω_2, normalized with $c_2 = 2$, corresponds to the Magri–Virasoro bracket:

$$\omega_2 = 4 \sum_{ij} \frac{\delta z_i \wedge \delta p_j}{(z_i^2 - p_j^2)} = 4 \sum_j \left(\sum_i \frac{\delta z_i}{z_i^2 - p_j^2} \right) \wedge \delta p_j \tag{11.82}$$

In the coordinates $(X_i(0), p_i)$, it reads

$$\omega_2 = 2 \sum_i \frac{\delta p_i}{p_i} \wedge \frac{\delta X_i}{X_i} - 8 \sum_{i<j} \frac{\delta p_i \wedge \delta p_j}{p_i^2 - p_j^2} \tag{11.83}$$

To see that, we differentiate eq. (11.77), getting

$$2p_i \sum_j \frac{\delta z_j}{z_j^2 - p_i^2} = -\frac{\delta X_i}{X_i} + 2p_i \sum_{j \neq i} \frac{\delta p_j}{p_i^2 - p_j^2} + \Lambda_i \delta p_i$$

where Λ_i is a coefficient whose value does not matter. Inserting into eq. (11.82), we arrive at eq. (11.83). We can write the Poisson brackets

$$\{p_i, p_j\}_2 = 0, \quad \{p_i, X_j\}_2 = \frac{1}{2} p_i X_j \delta_{ij}, \quad \{X_i, X_j\}_2 = 2 \frac{p_i p_j}{p_i^2 - p_j^2} X_i X_j$$

As a simple consistency check, we use the Hamiltonians, eq. (11.47), and compute

$$\partial_{t_{2j+1}} X_i = \{H_{2j-1}, X_i\}_2 = 2p_i^{2j+1} X_i$$

in agreement with the fact that H_{2j-1} generate the flow t_{2j+1} with $\{ , \}_2$.

11.9 Local fields

In this and the next section, we study the Whitham average of local fields in the KdV hierarchy. By definition, the space of local fields is the vector space generated by monomials of the form

$$\mathcal{O}(u, u', u'', \dots)$$

where the prime denotes ∂_x. In particular, we wish to know when the Whitham average of a local field vanishes. One such a case is when the local field we consider is a derivative with respect of any time of the hierarchy of another local field. This is a motivation for presenting local fields in the form:

$$\mathcal{O}(u, u', u'', \dots) = \sum_{|\nu| \geq 0} \partial^\nu E_{\mathcal{O}, \nu}(S_2, S_4, \dots) \qquad (11.84)$$

where $\nu = (i_1, i_3, \dots)$ is a multi-index, $\partial^\nu = \partial_{t_1}^{i_1} \partial_{t_3}^{i_3} \cdots$, and $|\nu| = i_1 + 3i_3 + \cdots$. This is not all the story, however, because some linear combinations in the right-hand side of eq. (11.84) identically vanish by the equations of motion of the KdV hierarchy. We define a *null vector* as such a combination which vanishes when we take the equations of motion into account. Our first task is to describe the null vectors. We will analyse the Whitham average in the next section.

The possibility of going from one presentation to the other in eq. (11.84) relies on the possibility of replacing the odd derivatives $\partial_x^{2j-1} u$ by the higher times derivatives $\partial_{t_{2j-1}}$, according to the equations of motion

$$\partial_{t_{2j-1}} u = [(L^{\frac{2j-1}{2}})_+, L] = \frac{1}{2^{2j-1}} u^{(2j-1)} + \cdots$$

Similarly, the even derivatives $\partial_x^{2j} u$ can be replaced by the densities of the integrals of motion S_{2j}:

$$S_{2j} = \mathrm{Res}_\partial L^{\frac{2j-1}{2}} = -\frac{1}{2^{2j-1}} u^{(2j-2)} + \cdots$$

Let us compare the dimensions of the space $\mathcal{L}$ and the space $\mathcal{E}$ generated by elements of the form $\partial^\nu E(S_2, S_4, \dots)$, where $E(S_2, S_4, \dots)$ is a monomial of $S_2, S_4, \dots$.

Since both spaces are infinite, we introduce a grading by attributing to ∂_x weight 1 and to u weight 2. As a consequence, $\partial_{t_{2j-1}}$ has weight $2j-1$ and S_{2j} has weight $2j$. This makes the two spaces $\mathcal{L}$ and $\mathcal{E}$ graded vector spaces:

$$\mathcal{L} = \bigoplus_{n=0}^{\infty} \mathcal{L}_n, \quad \mathcal{E} = \bigoplus_{n=0}^{\infty} \mathcal{E}_n$$

At each grade n, the vector spaces $\mathcal{L}_n$ and $\mathcal{E}_n$ are finite-dimensional. We define the characters:

$$\chi(\mathcal{L}) = \sum_{n=0}^{\infty} q^n \dim \mathcal{L}_n, \quad \chi(\mathcal{E}) = \sum_{n=0}^{\infty} q^n \dim \mathcal{E}_n$$

The space $\mathcal{L}_n$ is made of monomials in $u, u', u'', \dots$ of weight n. It is quite clear that

$$\chi(\mathcal{L}) = \prod_{j \geq 2} \frac{1}{1 - q^j} = (1 - q) \prod_{j \geq 1} \frac{1}{1 - q^j} = 1 + q^2 + q^3 + 2q^4 + 2q^5 + \cdots$$

because the number of local fields of weight k is the number of ways to write $2n_1 + 3n_2 + \cdots = k$, so that $\chi(\mathcal{L}) = \left(\sum_{n_1} q^{2n_1} \right) \left(\sum_{n_2} q^{3n_2} \right) \cdots$. Similarly, the character of the vector space $\mathcal{E}$ can be easily computed and is found to be:

$$\chi(\mathcal{E}) = \prod_{j \geq 1} \frac{1}{1 - q^{2j-1}} \prod_{j \geq 1} \frac{1}{1 - q^{2j}} = \prod_{j \geq 1} \frac{1}{1 - q^j}$$
$$= 1 + q + 2q^2 + 3q^3 + 5q^4 + 7q^5 + \cdots$$

The first infinite product counts the factors ∂^ν and the second product counts the factors $E_{\mathcal{O},\nu}$ in eq. (11.84).

Note that we have $\chi(\mathcal{E}) > \chi(\mathcal{L})$, meaning that for each n dim $\mathcal{E}_n \geq$ dim $\mathcal{L}_n$. However, the equations of motion of the KdV hierarchy imply that many expressions of the form of those in the left-hand side of eq. (11.84) vanish. So the equality in eq. (11.84) is meant modulo the equations of motion of the KdV hierarchy. Let us give some examples of null vectors:

$$\text{level } 1 : \partial_{t_1} \cdot 1 = 0,$$
$$\text{level } 2 : \partial_{t_1}^2 \cdot 1 = 0,$$
$$\text{level } 3 : \partial_{t_1}^3 \cdot 1 = 0, \quad \partial_{t_3} \cdot 1 = 0$$
$$\text{level } 4 : \partial_{t_1}^4 \cdot 1 = 0, \quad \partial_{t_1} \partial_{t_3} \cdot 1 = 0, \quad (\partial_{t_1}^2 S_2 - 4S_4 + 6S_2^2) = 0,$$
$$\text{level } 5 : \partial_{t_1}^5 \cdot 1 = 0, \quad \partial_{t_1}^2 \partial_{t_3} \cdot 1 = 0, \quad \partial_{t_5} \cdot 1 = 0,$$
$$\partial_{t_1}((\partial_{t_1}^2 S_2 - 4S_4 + 6S_2^2) = 0, \quad (\partial_{t_3} S_2 - \partial_{t_1} S_4) = 0$$

We have written all the null vectors explicitly to show that their numbers exactly match the character formula. The non-trivial null vector at level 4 expresses S_4 in terms of u: $8S_4 = -u'' + 3u^2$. With this identification, the non-trivial null vector at level 5, $\partial_{t_3} S_2 - \partial_{t_1} S_4 = 0$, gives the KdV equation eq. (11.1).

The goal of our study is to find all the null vectors. They will have a simple description, given in eqs. (11.90, 11.97) below, at the price of introducing a fermionic language.

A generating function for the monomials spanning the vector space $\mathcal{E}$ is:

$$\mathcal{E}(u, v) \equiv e^{\sum_j u_{2j-1} \partial_{t_{2j-1}}} e^{\sum_n v_{2n} J_{2n}} \qquad (11.85)$$

where the coefficients J_{2n} are defined by

$$\log S(\lambda) \equiv -\sum_{n>0} \frac{1}{n} J_{2n} \lambda^{-n} \qquad (11.86)$$

One can express any monomial in the S_{2n} in terms of the J_{2n} and vice versa.

To describe the null vectors, we introduce fermionic and bosonic fields as in Chapter 9:

$$\beta(\lambda) = \sum_{r \in \mathbb{Z}+\frac{1}{2}} \beta_r \lambda^{-r-1/2}, \quad \beta^*(\lambda) = \sum_{r \in \mathbb{Z}+\frac{1}{2}} \beta_r^* \lambda^{-r-1/2}, \qquad (11.87)$$

$$H(\lambda) = \sum_{n \in \mathbb{Z}} H_n \lambda^{-n-1} =: \beta(\lambda)\beta^*(\lambda) :$$

The vacuum $|0\rangle$ is characterized by $\beta_r|0\rangle = 0$, $\beta_r^*|0\rangle = 0$, for $r > 0$.

By the Campbell–Hausdorff formula we have:

$$e^{\sum_n v_{2n} J_{2n}} = \langle 0| e^{-\sum_{n>0} v_{2n} H_n} e^{-\sum_{n>0} \frac{1}{n} J_{2n} H_{-n}} |0\rangle$$

Hence the monomials in J_{2n} are in one to one correspondence with the components of the vector $\mathcal{Z}|0\rangle$ in Fock space, where:

$$\mathcal{Z} \equiv e^{-\sum_{n>0} \frac{1}{n} J_{2n} H_{-n}} = e^{\int \frac{d\lambda}{2i\pi} \log S(\lambda) H(\lambda)} \qquad (11.88)$$

The full set of elements of the space $\mathcal{E}$ is obtained by acting on this vector with the time derivatives $\partial_{t_{2j-1}}$. Null vectors admit a particularly simple description in this setting.

Proposition. *Let* $\nabla(\lambda) = \sum_{j \geq 1} \lambda^{-j} \partial_{t_{2j-1}}$ *be the operator defined in eq. (11.22), and $\mathcal{Q}$ be the fermionic operator:*

$$\mathcal{Q} = \int \frac{d\lambda}{2i\pi} \beta(\lambda) \nabla(\lambda) \qquad (11.89)$$

Then the equations of motion of the KdV hierarchy imply:

$$\mathcal{Q} \, \mathcal{Z}|0\rangle = 0 \qquad (11.90)$$

<u>Proof.</u> We need the two formulae

$$\nabla(\lambda)\mathcal{Z} = \mathcal{Z} \int \frac{d\lambda'}{2i\pi} S^{-1}(\lambda')\nabla(\lambda)S(\lambda')H(\lambda') \tag{11.91}$$

and

$$\beta(\lambda)\mathcal{Z} = S^{-1}(\lambda)\ \mathcal{Z}\ \beta(\lambda) \tag{11.92}$$

These formulae are straightforward consequences of the fact that all H_n, $n > 0$, mutually commute, and eq. (9.43) in Chapter 9 applies. So we can write

$$\mathcal{Q}\ \mathcal{Z}|0\rangle = \mathcal{Z} \int \frac{d\lambda}{2i\pi}\frac{d\lambda'}{2i\pi} S^{-1}(\lambda)S^{-1}(\lambda')\nabla(\lambda)S(\lambda')\beta(\lambda)H(\lambda')|0\rangle$$

$$= \mathcal{Z} \int \frac{d\lambda}{2i\pi}\frac{d\lambda'}{2i\pi}\frac{\partial_x \log S(\lambda') - \partial_x \log S(\lambda)}{\lambda - \lambda'}\beta(\lambda)H(\lambda')|0\rangle \tag{11.93}$$

In the last step we used eq. (11.23) valid for $|\lambda| > |\lambda'|$. We examine separately the $\partial_x \log S(\lambda')$ and the $\partial_x \log S(\lambda)$ terms. The $\partial_x \log S(\lambda')$ term reads:

$$\int_{|\lambda|>|\lambda'|} \frac{d\lambda}{2i\pi}\frac{d\lambda'}{2i\pi}\frac{1}{\lambda - \lambda'}\partial_x \log S(\lambda')\beta(\lambda)H(\lambda')|0\rangle$$

One can do the integral over λ. Poles can occur at $\lambda = 0$ and $\lambda = \lambda'$. At $\lambda = 0$, the integrand is actually regular. In fact, potentially dangerous terms come from $\beta(\lambda)H(\lambda')|0\rangle$. But this is regular at $\lambda = 0$ because we have

$$\beta(\lambda)H(\lambda') =: \beta(\lambda)\beta(\lambda')\beta^*(\lambda') : -\frac{1}{\lambda - \lambda'}\beta(\lambda'),\quad |\lambda| > |\lambda'|$$

and by definition of the vacuum and normal ordered product, the term $: \beta(\lambda)H(\lambda') : |0\rangle$ is regular at $\lambda = 0$. The same formula is used to analyse the poles at $\lambda = \lambda'$. One has two terms. The first one is

$$\int \frac{d\lambda}{2i\pi}\frac{1}{\lambda - \lambda'} : \beta(\lambda)\beta(\lambda')\beta^*(\lambda') : |0\rangle$$

which is zero because at $\lambda = \lambda'$ we get the product of two fermionic fields at the same point inside the normal product, and this vanishes. The second term is equal to

$$-\int \frac{d\lambda}{2i\pi}\frac{1}{(\lambda - \lambda')^2}\beta(\lambda')|0\rangle$$

and this obviously vanishes. Next, we examine the $\partial_x \log S(\lambda)$ term:

$$\int_{|\lambda|>|\lambda'|} \frac{d\lambda \, d\lambda'}{2i\pi \, 2i\pi} \frac{1}{\lambda - \lambda'} \partial_x \log S(\lambda) \beta(\lambda) H(\lambda')|0\rangle$$

This time we can do the λ' integral. But, by the properties of the vacuum vector, the integrand is regular at $\lambda' = 0$ because $H(\lambda')|0\rangle = O(1)$ and the integral vanishes. We have shown that $\mathcal{Q} \, \mathcal{Z} \, |0\rangle = 0$. ∎

Equation (11.90) is not sufficient to characterize null vectors. We exhibit now a second equation which, together with the first one, will provide a complete set of constraints.

Proposition. *Let $\mathcal{C}$ be the fermionic operator:*

$$\mathcal{C} = \mathcal{C}_0 + \mathcal{C}_1 \tag{11.94}$$

where

$$\mathcal{C}_0 = \int \frac{d\lambda}{2i\pi} \beta(\lambda) \, \lambda \frac{d}{d\lambda} \beta(\lambda) \tag{11.95}$$

$$\mathcal{C}_1 = \int_{|\lambda_1|<|\lambda_2|} \frac{d\lambda_1 \, d\lambda_2}{2i\pi \, 2i\pi} \log\left(1 - \frac{\lambda_1}{\lambda_2}\right) \beta(\lambda_1)\beta(\lambda_2)\nabla(\lambda_1)\nabla(\lambda_2) \tag{11.96}$$

Then the equations of motion of the KdV hierarchy imply:

$$\mathcal{C} \, \mathcal{Z}|0\rangle = 0 \tag{11.97}$$

<u>Proof.</u> We first calculate $\mathcal{C}_0\mathcal{Z}|0\rangle$. Since $\beta(\lambda)^2 = 0$, we immediately get:

$$\mathcal{C}_0\mathcal{Z}|0\rangle = \mathcal{Z} \int \frac{d\lambda}{2i\pi} S^{-2}(\lambda)\beta(\lambda)\lambda\frac{d}{d\lambda}\beta(\lambda)|0\rangle$$

Next we calculate $\mathcal{C}_1\mathcal{Z}|0\rangle$. We have

$$\mathcal{C}_1\mathcal{Z}|0\rangle = -\int_\Upsilon \frac{d\lambda_2}{2i\pi}\beta(\lambda_2)\nabla(\lambda_2) \int \frac{d\lambda_1}{2i\pi} \log\left(1 - \frac{\lambda_1}{\lambda_2}\right)\beta(\lambda_1)\nabla(\lambda_1)\mathcal{Z}|0\rangle$$

where Υ is the contour $|\lambda_1| < |\lambda_2|$. We use eqs. (11.91, 11.92) to write

$$\beta(\lambda_1)\nabla(\lambda_1)\mathcal{Z} = \mathcal{Z} \int \frac{d\lambda}{2i\pi} S^{-1}(\lambda)S^{-1}(\lambda_1)\nabla(\lambda_1)S(\lambda)\beta(\lambda_1)H(\lambda)$$

$$= \mathcal{Z} \int_{|\lambda_1|>|\lambda|} \frac{d\lambda}{2i\pi} \frac{1}{\lambda_1 - \lambda} \partial_x \log\left(\frac{S(\lambda)}{S(\lambda_1)}\right)\beta(\lambda_1)H(\lambda)$$

So, one has to evaluate the integral

$$\int_{|\lambda_1|>|\lambda|} \frac{d\lambda_1 \, d\lambda}{2i\pi \, 2i\pi} \log\left(1 - \frac{\lambda_1}{\lambda_2}\right) \frac{\partial_x \log S(\lambda) - \partial_x \log S(\lambda_1)}{\lambda_1 - \lambda}\beta(\lambda_1)H(\lambda)|0\rangle$$

This is exactly the same type of integral we met in eq. (11.93). Again, we see that the term $\partial_x \log S(\lambda_1)$ vanishes because in the λ integral, the integrand is regular at $\lambda = 0$. Only the term $\partial_x \log S(\lambda)$ contributes. This time, however, the double pole gives a non-vanishing contribution:

$$\int_{|\lambda_1|>|\lambda|} \frac{d\lambda_1}{2i\pi} \log\left(1 - \frac{\lambda_1}{\lambda_2}\right) \frac{1}{(\lambda_1 - \lambda)^2} = -\frac{1}{\lambda_2 - \lambda}$$

Hence, we find

$$\mathcal{C}_1 \mathcal{Z}|0\rangle = -\int \frac{d\lambda_2}{2i\pi} \beta(\lambda_2)\nabla(\lambda_2)\mathcal{Z} \int_{|\lambda|<|\lambda_2|} \frac{d\lambda}{2i\pi} \frac{\partial_x \log S(\lambda)}{\lambda_2 - \lambda}\beta(\lambda)|0\rangle$$

In this formula, $\nabla(\lambda_2)$ acts on $\mathcal{Z}$ and $\partial_x \log S(\lambda)$. In the terms coming from the action of $\nabla(\lambda_2)$ on $\mathcal{Z}$, we have three integrals. One of them can be evaluated because it does not involve any $S(\lambda)$. We finally get a total vanishing contribution. The terms coming from the action of $\nabla(\lambda_2)$ on $\partial_x \log S(\lambda)$ can be put in the form

$$-\mathcal{Z}\int_{|\lambda_2|>|\lambda|} \frac{d\lambda\, d\lambda_2}{2i\pi\, 2i\pi} \left(\frac{\partial_x \log S(\lambda_2)\partial_x \log S(\lambda)}{(\lambda_2 - \lambda)^2}\right.$$
$$\left. + \frac{\partial_x^2 \log S(\lambda) - \partial_x^2 \log S(\lambda_2) - (\partial_x \log S(\lambda_2))^2}{(\lambda_2 - \lambda)^2}\right) : \beta(\lambda_2)\beta(\lambda) : |0\rangle$$

In the second term, one can always perform one of the integrals. In the first term, we take the half sum with λ and λ_2 exchanged, getting an integral which localizes at $\lambda = \lambda_2$. Putting everything together, we get

$$\mathcal{C}_1 \mathcal{Z}|0\rangle = \frac{1}{4}\mathcal{Z}\int \frac{d\lambda}{2i\pi\lambda}[2\partial_x^2 \log S(\lambda) + (\partial_x \log S(\lambda))^2]\beta(\lambda)\lambda\frac{d}{d\lambda}\beta(\lambda)|0\rangle$$

Remembering eq. (11.30), we finally obtain:

$$(\mathcal{C}_0 + \mathcal{C}_1)\mathcal{Z}|0\rangle = \int \frac{d\lambda}{2i\pi\lambda}(u + \lambda)\beta(\lambda)\lambda\frac{d}{d\lambda}\beta(\lambda)|0\rangle$$

This integral vanishes because $\beta(\lambda)\lambda\frac{d}{d\lambda}\beta(\lambda)|0\rangle = O(\lambda)$.   ∎

Remark. It is useful to write the operators $\mathcal{Q}$ and $\mathcal{C}$ explicitly. For $\mathcal{Q}$ we find

$$\mathcal{Q} = \sum_{j\geq 1} \beta_{-j+\frac{1}{2}}\partial_{t_{2j-1}}$$

From this expression it is quite clear that

$$\mathcal{Q}^2 = 0$$

Similarly, we find

$$\mathcal{C} = \sum_{r \geq \frac{1}{2}} \beta_{-r} \left(-2r\beta_r + \sum_{j_1 \geq 1} \sum_{j_2=1}^{r-\frac{1}{2}} \frac{1}{r - j_2 + \frac{1}{2}} \beta_{r+1-j_1-j_2} \partial_{t_{2j_1-1}} \partial_{t_{2j_2-1}} \right)$$

Note that $\mathcal{Q}$ and $\mathcal{C}$ commute.

We now show that eqs. (11.90, 11.97), contain all the information about the KdV hierarchy by enumerating all null vectors and comparing characters.

Let $\mathcal{F}^*_{-n}$ be the Fock space consisting of elements of the form

$$\langle 0 | \beta_{s_1} \cdots \beta_{s_m} \beta^*_{r_1} \cdots \beta^*_{r_{m+n}}$$

with $s_i, r_i \geq \frac{1}{2}$ and all different. Note that we have $n + m$ operators $\beta^*_{r_i}$ of charge -1, and m operators β_{s_i} of charge $+1$, so that the total charge of the state is $-n$. Attributing a weight $2s$ at β_s and $2r$ at β^*_r turns the dual Fock space $\mathcal{F}^*_{-n}$ of charge $-n$ into a graded vector space.

Introducing a parameter q to count the weight and x to count the charge, the character of $\mathcal{F}^*_{-n}$ is easily calculated:

$$\chi(\mathcal{F}^*_{-n}) = \int \frac{dx}{2i\pi x} x^n \prod_{j \geq 1} (1 + q^{2j-1}x)(1 + q^{2j-1}x^{-1})$$

Changing $x = q^2 x$ in the above integral, we find the relation $\chi(\mathcal{F}^*_{-n}) = q^{2n-1}\chi(\mathcal{F}^*_{-n+1})$, so that $\chi(\mathcal{F}^*_{-n}) = q^{n^2}\chi(\mathcal{F}^*_0)$. On the other hand, $\mathcal{F}^*_0$ is isomorphic to the bosonic Fock space generated by the H_n, which has weight $2n$. Its character is therefore $\prod_{j \geq 1}(1 - q^{2j})^{-1}$. Hence we have shown

$$\chi(\mathcal{F}^*_{-n}) = q^{n^2} \prod_{j \geq 1} \frac{1}{1 - q^{2j}} \qquad (11.98)$$

Proposition. *Let $\mathcal{F}^*_{-2}$ be the dual Fock space of charge -2. The application $\mathcal{C} : \mathcal{F}^*_{-2} \to \mathcal{F}^*_0$ is injective.*

Proof. Introduce a linear transformation on the space of fermions $\tilde{\beta}_r = \sum_{r'} N_{rr'} \beta_{r'}$, where the matrix $N_{rr'}$ is defined by

$$\tilde{\beta}_r = \beta_r, \quad r \leq -\frac{1}{2}$$

$$\tilde{\beta}_r = 2r\beta_r - \sum_{j_1 \geq 1} \sum_{j_2=1}^{r-\frac{1}{2}} \frac{1}{r - j_2 + \frac{1}{2}} \beta_{r+1-j_1-j_2} \partial_{t_{2j_1-1}} \partial_{t_{2j_2-1}}, \quad r \geq \frac{1}{2}$$

The dual fermions are $\tilde{\beta}^*_{-r} = ({}^tN^{-1})_{r,r'}\beta^*_{-r'}$ so that $\tilde{\beta}, \tilde{\beta}^*$ satisfy the canonical anticommutation relations. Because the transformation N is triangular and leaves β_{-r}, β^*_r, $r \geq \frac{1}{2}$ invariant, the vacuum is invariant and the Fock spaces are the same. In the $\tilde{\beta}_r$ basis we can write

$$\mathcal{C} = \sum_{r \geq \frac{1}{2}} \tilde{\beta}_{-r}\tilde{\beta}_r$$

Let us call $X_+ \equiv \mathcal{C}$ and introduce $X_- = \sum_{r \geq \frac{1}{2}} \tilde{\beta}^*_{-r}\tilde{\beta}^*_r$. We have the sl_2 algebra

$$[X_+, X_-] = H_0, \quad [H_0, X_\pm] = \pm 2X_\pm$$

where $H_0 = \sum_r : \tilde{\beta}_r\tilde{\beta}^*_{-r} := \sum_r : \beta_r\beta^*_{-r} :$ is the charge operator The spaces $\mathcal{F}^*_{-n}$ are eigenspaces of H_0. So we have a representation of the sl_2 algebra on the full Fock space $\mathcal{F}^* = \bigoplus_n \mathcal{F}^*_n$.

Note, moreover, that $X_\pm, H_0$ are of grade zero, so their action on $\mathcal{F}^*_{-n}$ preserves the gradation of these spaces, and we can restrict them to subspaces of given grade. Subspaces of given grade are finite-dimensional. For each grade, we can decompose $\mathcal{F}^*$ into a direct sum of irreducible finite-dimensional representations of sl_2. Finally, on a finite-dimensional irreducible representation, $X_+ : \mathcal{F}^*_{n-2} \to \mathcal{F}^*_n$ is injective for $n \leq 0$.   ∎

The weights were so chosen that, taking into account the weights of $\partial_{t_{2j-1}}$ defined above, the operators $\mathcal{Q}$ and $\mathcal{C}$ have weight zero. Hence they preserve the gradings of the spaces on which they act. This is an important observation in the proof of the following:

Proposition. *The character of the space of vectors in $\mathcal{E}$, solutions of the equations eq. (11.90) and eq. (11.97), is*

$$\chi_1 = \prod_{j \geq 2} \frac{1}{1 - q^j}$$

It is equal to the character of the space of local fields. As a consequence eqs. (11.90, 11.97) capture the complete information about the KdV hierarchy.

<u>Proof.</u> The character we are looking for is of the form

$$\chi_1 = \prod_{j \geq 1} \frac{1}{1 - q^{2j-1}} \cdot \chi$$

where the first factor comes from the u-exponential in eq. (11.85), while the second factor, χ, comes from the v-exponential, subjected to the conditions eqs. (11.90, 11.97). To compute χ, we count the dimension of the dual spaces. The $\mathcal{C}$ conditions are taken into account by considering the space $\mathcal{F}_0^*/\mathcal{F}_{-2}^*\mathcal{C}$. Due to the previous proposition, the character of this space is $\chi(\mathcal{F}_0^*) - \chi(\mathcal{F}_{-2}^*)$. Next, we have to take into account the $\mathcal{Q}$ conditions. Because $\mathcal{Q}$ and $\mathcal{C}$ commute, this is achieved by replacing the spaces $\mathcal{F}_{-n}^*$ by the spaces $\mathcal{H}_{-n}^* = \mathcal{F}_{-n}^*/\mathcal{F}_{-n-1}^*\mathcal{Q}$. It follows that:

$$\chi = \chi(\mathcal{H}_0^*) - \chi(\mathcal{H}_{-2}^*) \tag{11.99}$$

To compute the characters $\chi(\mathcal{H}_{-n}^*)$, we take into account that $\mathcal{Q}$ is a nilpotent operator with trivial cohomology. In fact, we can construct a homotopy:

$$\mathcal{Q}^* = \sum_{j \geq 1} \beta_{j-\frac{1}{2}}^* \partial_{t_{2j-1}}^{-1}$$

The operator $\mathcal{Q}\mathcal{Q}^* + \mathcal{Q}^*\mathcal{Q}$ acting on any vector reproduces it up to a constant. Hence

$$\mathrm{Ker}\,(\mathcal{Q} : \mathcal{F}_{-n-1}^* \to \mathcal{F}_{-n}^*) = \mathrm{Im}\,(\mathcal{Q} : \mathcal{F}_{-n-2}^* \to \mathcal{F}_{-n-1}^*)$$

Summing over this complex, we get, using eq. (11.98):

$$\chi(\mathcal{H}_{-n}^*) = \left(q^{n^2} - q^{(n+1)^2} + q^{(n+2)^2} + \cdots \right) \prod_{j \geq 1} \frac{1}{1 - q^{2j}}$$

Inserting into eq. (11.99), we get $\chi = (1-q)\prod_{j \geq 1}(1-q^{2j})^{-1}$. ∎

Equations (11.90, 11.97), which code for the non-linear KdV hierarchy, are actuatly linear in $\mathcal{Z}$. The non-linearity comes from the explicit form of $\mathcal{Z}$ as an exponential, as in eq. (11.88).

11.10 Whitham's equations

We study eqs. (11.90, 11.97) in the case of finite-zone solutions. They acquire a beautiful geometrical meaning when we consider their average over the Liouville torus in the spirit of the Whitham theory (see Chapter 10). Specifically, $\mathcal{C}$ reflects the Riemann bilinear identities and $\mathcal{Q}$ gives rise to the Whitham equations directly.

For any local quantity $\mathcal{O}(u, u', \ldots)$, the Whitham average is defined by

$$\langle\langle \mathcal{O} \rangle\rangle = \lim_{\ell \to \infty} \frac{1}{2\ell} \int_{-\ell}^{\ell} \mathcal{O}(u(x,t),\ldots) dx$$
$$= \int_0^{2\pi} \frac{d\theta_1}{2\pi} \cdots \int_0^{2\pi} \frac{d\theta_g}{2\pi} \mathcal{O}(u(\theta, \{\lambda_i\}), \ldots)$$

where, in the second expression, we have replaced the average over x by an average over the Liouville torus, which is justified for almost all trajectories. Let us remark that if the expression, $\mathcal{O}$, is the x-derivative of a bounded function, its Whitham average vanishes. However, we will show below that there exist more subtle vanishing conditions.

We consider finite-zone solutions constructed with the hyperelliptic curve eq. (11.49) and the dynamical divisor $D(x) = \{\gamma_i = (\lambda_{\gamma_i}, \mu_{\gamma_i})\}$, defined in eq. (11.55). With these data, we can write:

$$\langle\langle\mathcal{O}\rangle\rangle = \mathcal{N}^{-1} \int_{a_1} d\lambda_{\gamma_1} \cdots \int_{a_n} d\lambda_{\gamma_g} \mathcal{O}(u(\{\lambda_{\gamma_i}\},\{\lambda_i\}),\ldots) \left|\frac{\partial\theta}{\partial\gamma}\right|$$

where the normalization factor $\mathcal{N}$ is defined so that $\langle\langle 1\rangle\rangle = 1$. The Jacobian determinant $|\partial\theta/\partial\gamma|$ is easily computed.

Proposition. *We have*

$$\mathcal{N}^{-1}\left|\frac{\partial\theta}{\partial\gamma}\right| = \Delta^{-1}\frac{\prod_{i<j}(\lambda_{\gamma_i} - \lambda_{\gamma_j})}{\prod_i \sqrt{R(\lambda_{\gamma_i})}} \tag{11.100}$$

where

$$\Delta = \det\left(\oint_{a_i} \frac{\lambda^{j-1}d\lambda}{\sqrt{R(\lambda)}}\right)_{i,j=1,\ldots,g}$$

<u>Proof.</u> The angles on the torus are given by $\theta_k = \sum_i \int^{\gamma_i} \omega_k$, where ω_k are the normalized Abelian differentials. So

$$d\theta_k = \sum_i \omega_k(\gamma_i)$$

But $\omega_k = \sum_{j=1}^g c_{kj}\frac{\lambda^{j-1}}{\sqrt{R(\lambda)}}d\lambda$, where the coefficients c_{kj} are determined by normalizing the a-periods. Hence

$$\left|\frac{\partial\theta}{\partial\gamma}\right| = \det c \det \frac{\lambda_{\gamma_i}^{j-1}}{\sqrt{R(\lambda_{\gamma_i})}} = \det c \frac{\prod_{i<j}(\lambda_{\gamma_i} - \lambda_{\gamma_j})}{\prod_i \sqrt{R(\lambda_{\gamma_i})}}$$

The factor $\det c$ cancels out in the normalization factor, which is found by imposing $\langle\langle 1\rangle\rangle = 1$ giving $\mathcal{N} = \det c\, \Delta$. ∎

By eqs. (11.84, 11.64), with every local field $\mathcal{O}$ we can associate a symmetric function of the points γ_i, $L_{\mathcal{O}}(\lambda_{\gamma_1},\ldots,\lambda_{\gamma_g}) = E_{\mathcal{O},0}(S_2,S_4,\ldots)$. This function depends on the moduli λ_i and the λ_{γ_i}. In the Whitham average, all terms with $\nu > 0$ in eq. (11.84) vanish because they are exact derivatives. We have

$$\langle\langle\mathcal{O}\rangle\rangle = \Delta^{-1} \int_{a_1} \frac{d\lambda_{\gamma_1}}{\sqrt{R(\lambda_{\gamma_1})}} \cdots \int_{a_n} \frac{d\lambda_{\gamma_g}}{\sqrt{R(\lambda_{\gamma_g})}} L_{\mathcal{O}}(\lambda_{\gamma_1},\ldots,\lambda_{\gamma_g}) \prod_{i<j}(\lambda_{\gamma_i}-\lambda_{\gamma_j})$$

(11.101)

We can expand the symmetric function $L_{\mathcal{O}}(\lambda_{\gamma_1},\ldots,\lambda_{\gamma_g})$ on the Schur polynomials introduced in eq. (9.62) in Chapter 9:

$$L_{\mathcal{O}} = \sum_Y L_{\mathcal{O}}^Y S_Y$$

Using the determinant formula for S_Y associated with the Young diagram $Y = [n_i]$, we see that averaging reduces to computing

$$\langle\langle\mathcal{O}\rangle\rangle = \Delta^{-1} \sum_Y L_{\mathcal{O}}^Y \det\left(\int_{a_j} \frac{d\lambda}{\sqrt{R(\lambda)}}\lambda^{n_i+j-i}\right)$$

These formulae are particularly useful for describing the Whitham average because the variables λ_{γ_i} are separated and we have only one dimensional integrals to compute.

By eq. (11.64), the value of the coefficients J_{2n} defined in eq. (11.86) are:

$$J_{2n} = \sum_{i=1}^{g} \lambda_{\gamma_i}^n - \frac{1}{2}\sum_{i=1}^{2g+1} \lambda_i^n$$

Using the boson–fermion correspondance of Chapter 9, we can write $\mathcal{Z}$ in eq. (11.88) as:

$$\mathcal{Z}|0\rangle = G_\Gamma \frac{\prod_{i=1}^{g} \lambda_{\gamma_i}^g}{\prod_{i<j}(\lambda_{\gamma_i}-\lambda_{\gamma_j})} \beta^*(\lambda_{\gamma_1})\beta^*(\lambda_{\gamma_2})\cdots\beta^*(\lambda_{\gamma_g})|g\rangle \qquad (11.102)$$

where G_Γ depends only on the curve Γ:

$$G_\Gamma = \exp\left(\frac{1}{2}\sum_{n\geq 1}\frac{1}{n}\left(\sum_{i=1}^{2g+1}\lambda_i^n\right)H_{-n}\right)$$

In the Whitham theory, we assume that the moduli, λ_i, become functions of the slow modulation times $T_{2j-1} = \epsilon t_{2j-1}$.

$$\partial_{t_{2j-1}}\mathcal{O} = \partial_{t_{2j-1}}\mathcal{O}|_{\lambda_i} + \sum_i \frac{\partial\lambda_i}{\partial_{t_{2j-1}}}\partial_{\lambda_i}\mathcal{O}$$

Upon averaging, the first term drops out and we are left with

$$\partial_{t_{2j-1}}\langle\langle\mathcal{O}\rangle\rangle = \sum_i \frac{\partial\lambda_i}{\partial t_{2j-1}}\partial_{\lambda_i}\langle\langle\mathcal{O}\rangle\rangle \equiv \epsilon\partial_{T_{2j-1}}\langle\langle\mathcal{O}\rangle\rangle$$

The modulation equations are obtained by keeping the leading terms in ϵ in eqs. (11.90, 11.97). They become

$$\mathcal{Q}_0\langle\langle\mathcal{Z}\rangle\rangle|0\rangle = 0, \quad \mathcal{C}_0\langle\langle\mathcal{Z}\rangle\rangle|0\rangle = 0 \tag{11.103}$$

with

$$\mathcal{Q}_0 = \sum_{j\geq 1}\beta_{-j+\frac{1}{2}}\partial_{T_{2j-1}} \tag{11.104}$$

and

$$\mathcal{C}_0 = \int\frac{d\lambda}{2i\pi}\beta(\lambda)\,\lambda\frac{d}{d\lambda}\,\beta(\lambda) \tag{11.105}$$

The rest of this section is devoted to the analysis of these equations. We need some preparation.

On a Riemann surface, there is a natural pairing between meromorphic differentials. If Ω_1 and Ω_2 are two such differentials on Γ, we define

$$(\Omega_1 \bullet \Omega_2) = \sum_{i=1}^g\left(\int_{a_j}\Omega_1\int_{b_j}\Omega_2 - \int_{a_j}\Omega_2\int_{b_j}\Omega_1\right) \tag{11.106}$$

The Riemann bilinear identities express this quantity in terms of residues (see Chapter 15).

We also have a pairing between cycles. If C_1 and C_2 are two cycles, the pairing is simply

$$C_1 \circ C_2 = \sum_{j=1}^g(n_j^1 m_j^2 - m_j^1 n_j^2)$$

where $C_i = \sum_{j=1}^g(n_j^i a_j + m_j^i b_j)$. We can write this intersection number in a way similar to eq. (11.106).

Proposition. *Let ω_i be the normalized holomorphic differentials. Let η_i, $i = 1, \ldots, g$, be the second kind differentials dual to the ω_i, normalized by*

$$(\omega_i \bullet \eta_j) = \delta_{ij}, \quad (\eta_i \bullet \eta_j) = 0, \quad (\omega_i \bullet \omega_j) = 0$$

then

$$C_1 \circ C_2 = \sum_{j=1}^g\left(\int_{C_1}\omega_j\int_{C_2}\eta_j - \int_{C_2}\omega_j\int_{C_1}\eta_j\right) \tag{11.107}$$

<u>Proof</u>. The normalization conditions of ω_j and η_j mean that the matrix P, defined by

$$P = \begin{pmatrix} \int_{a_j} \omega_i & \int_{b_j} \omega_i \\ \int_{a_j} \eta_i & \int_{b_j} \eta_i \end{pmatrix}$$

is a symplectic matrix:

$$PJ\,{}^tP = J, \quad J = \begin{pmatrix} 0 & \mathrm{Id} \\ -\mathrm{Id} & 0 \end{pmatrix}$$

Since $J^2 = -\mathrm{Id}$, the right inverse of P is $-J\,{}^tPJ$. Using the fact that the right inverse and the left inverse are the same, we deduce that ${}^tPJP = J$. Now let $\langle C_i| = (n_j^i, m_j^i)$. We can rewrite the intersection number as $C_1 \circ C_2 = \langle C_1|J|C_2\rangle$. Using the relation ${}^tPJP = J$, this is equal to $C_1 \circ C_2 = \langle C_1|\,{}^tPJP|C_2\rangle$. This is equivalent to eq. (11.107) because $\langle C_1|\,{}^tP$ is the vector of periods of the forms ω_j, η_j along the cycle C_1, and similarly for $P|C_2\rangle$. ∎

For a hyperelliptic curve, one has the explicit formula:

Proposition. *The intersection number is given by*

$$C_1 \circ C_2 = \frac{1}{4i\pi} \int_{C_1} \frac{d\lambda_1}{\sqrt{R(\lambda_1)}} \int_{C_2} \frac{d\lambda_2}{\sqrt{R(\lambda_2)}} C(\lambda_1, \lambda_2) \qquad (11.108)$$

where the antisymmetric polynomial $C(\lambda_1, \lambda_2)$ is defined by

$$C(\lambda_1, \lambda_2) = \sqrt{R(\lambda_1)}\frac{\partial}{\partial\lambda_1}\left(\frac{\sqrt{R(\lambda_1)}}{\lambda_1 - \lambda_2}\right) - (\lambda_1 \leftrightarrow \lambda_2)$$

<u>Proof</u>. The first term in the expression of $C_1 \circ C_2$ reads

$$\frac{1}{4i\pi} \int_{C_2} \frac{d\lambda_2}{\sqrt{R(\lambda_2)}} \int_{C_1} d\lambda_1 \frac{d}{d\lambda_1}\left(\frac{\sqrt{R(\lambda_1)}}{\lambda_1 - \lambda_2}\right) \qquad (11.109)$$

The integral over λ_1 can be performed and gets contributions only at intersection points of C_1 and C_2. Let the curves have a positive intersection at $\lambda = \lambda_0$. We get a contribution

$$\sqrt{R(\lambda_0)}\left(-\frac{1}{\lambda_0 + i\epsilon - \lambda_2} + \frac{1}{\lambda_0 - i\epsilon - \lambda_2}\right) = 2i\pi\sqrt{R(\lambda_0)}\delta(\lambda_0 - \lambda_2)$$

The integral over λ_2 now gives $1/2$. The second term is treated similarly and also gives $1/2$. ∎

It is important to realize that the average eq. (11.101) can vanish for particular antisymmetric polynomials:

$$M_{\mathcal{O}}(\lambda_{\gamma_1}, \ldots, \lambda_{\gamma_g}) \equiv \prod_{i<j}(\lambda_{\gamma_i} - \lambda_{\gamma_j}) L_{\mathcal{O}}(\lambda_{\gamma_1}, \ldots, \lambda_{\gamma_g})$$

There are two general reasons why such an integral can vanish:

- The first one is when $M_{\mathcal{O}}$ is "an exact form"

$$M_{\mathcal{O}}(\lambda_{\gamma_1}, \ldots, \lambda_{\gamma_g}) = \sum_i (-1)^i M(\lambda_{\gamma_1}, \ldots, \widehat{\lambda_{\gamma_i}}, \ldots, \lambda_{\gamma_g}) P(\lambda_{\gamma_i})$$

where $P(\lambda)$ is a polynomial such that $P(\lambda)/\sqrt{R(\lambda)}$ has vanishing a-periods. In particular, if $\deg P \geq 2g$, we can write

$$\frac{P}{\sqrt{R}} = \frac{S}{\sqrt{R}} + \frac{d}{d\lambda}(Q\sqrt{R})$$

with $\deg S \leq 2g-1$ and $\deg Q = \deg P - 2g$. The exact derivative term has vanishing periods.

- The second one, less trivial, is when

$$M_{\mathcal{O}}(\lambda_{\gamma_1}, \ldots, \lambda_{\gamma_g})$$
$$= \sum_{i,j} (-1)^{i+j} M(\lambda_{\gamma_1}, \ldots, \widehat{\lambda_{\gamma_i}}, \ldots, \widehat{\lambda_{\gamma_j}}, \ldots, \lambda_{\gamma_g}) C(\lambda_{\gamma_i}, \lambda_{\gamma_j}) \quad (11.110)$$

since we are integrating over non-intersecting a-cycles.

This second condition, which is a direct consequence of Riemann bilinear identities, ensures that the second of eqs. (11.103) is automatically satisfied.

Proposition. *The equation*

$$\mathcal{C}_0\langle\langle \mathcal{Z} \rangle\rangle |0\rangle = 0 \qquad\qquad (11.111)$$

follows from eq. (11.110).

Proof. We have to evaluate

$$\mathcal{C}_0\langle\langle \mathcal{Z} \rangle\rangle |0\rangle = \Delta^{-1} \prod_{i=1}^{g} \int_{a_i} \frac{\lambda_{\gamma_i}^g \, d\lambda_{\gamma_i}}{\sqrt{R(\lambda_{\gamma_i})}} \, \mathcal{C}_0 \, G_\Gamma \, \beta^*(\lambda_{\gamma_1}) \cdots \beta^*(\lambda_{\gamma_g}) |g\rangle$$

The commutation relation

$$\beta(\lambda) G_\Gamma = \lambda^{-g-\frac{1}{2}} \sqrt{R(\lambda)} \, G_\Gamma \, \beta(\lambda)$$

implies

$$\mathcal{C}_0 \, G_\Gamma = G_\Gamma \int \frac{d\lambda}{2i\pi} \lambda^{-2g-1} R(\lambda)\beta(\lambda)\lambda \frac{d}{d\lambda}\beta(\lambda)$$

So, we have to compute

$$\int \frac{d\lambda}{2i\pi} \lambda^{-2g-1} R(\lambda)\beta(\lambda)\lambda \frac{d}{d\lambda}\beta(\lambda)\beta^*(\lambda_{\gamma_1})\cdots\beta^*(\lambda_{\gamma_g})|g\rangle$$

This is done using Wick's theorem. Since we are using the charged vacuum, the contraction is

$$\langle g|\beta(z)\beta^*(w)|g\rangle = \frac{z^g w^{-g}}{z - w}, \quad |z| > |w|$$

We have three terms corresponding to zero, one and two contractions respectively. The term with zero contraction reads

$$E_0 = \int \frac{d\lambda}{2i\pi} \lambda^{-2g} R(\lambda) : \beta(\lambda)\frac{d}{d\lambda}\beta(\lambda)\beta^*(\lambda_{\gamma_1})\cdots\beta^*(\lambda_{\gamma_g}) : |g\rangle$$

This term vanishes because $\beta(\lambda)\frac{d}{d\lambda}\beta(\lambda)|g\rangle = \lambda^{2g}|g+2\rangle + O(\lambda^{2g+1})$. The term with one contraction can be written as

$$E_1 = 2\sum_{i=1}^{g}(-1)^i \sum_{n\geq 0} \int \frac{d\lambda}{2i\pi}\frac{\lambda_{\gamma_i}^{-g}}{\lambda - \lambda_{\gamma_i}}\sqrt{R(\lambda)}\frac{d}{d\lambda}\left(\sqrt{R(\lambda)}\lambda^n\right)$$

$$: \beta_{-g-n-\frac{1}{2}}\beta^*(\lambda_{\gamma_1})\cdots\widehat{\beta^*(\lambda_{\gamma_i})}\cdots\beta^*(\lambda_{\gamma_g}) : |g\rangle$$

When we perform the λ integral, we get a contribution at the pole λ_{γ_i} which produces an exact form and therefore does not contribute in the Whitham average. The point $\lambda = 0$ does not contribute because the integrand is regular there. The term with two contractions reads

$$E_2 = \sum_{ij}(-1)^{i+j} \int \frac{d\lambda}{2i\pi}\lambda_{\gamma_i}^{-g}\lambda_{\gamma_j}^{-g}\left(\frac{\sqrt{R(\lambda)}}{\lambda - \lambda_{\gamma_i}}\frac{d}{d\lambda}\frac{\sqrt{R(\lambda)}}{\lambda - \lambda_{\gamma_j}} - i \leftrightarrow j\right)$$

$$: \beta^*(\lambda_{\gamma_1})\cdots\widehat{\beta^*(\lambda_{\gamma_i})}\cdots\widehat{\beta^*(\lambda_{\gamma_j})}\cdots\beta^*(\lambda_{\gamma_g}) : |g\rangle$$

where the hat means that the corresponding quantity is omitted. Performing the λ integral, we get an expression of the form eq. (11.110) vanishing under the Whitham average. ∎

Proposition. *The equation*

$$\mathcal{Q}_0 \langle\langle \mathcal{Z} \rangle\rangle |0\rangle = 0 \tag{11.112}$$

implies the modulation equations

$$\frac{\partial}{\partial T_{2p-1}} \Omega^{(2q-1)} = \frac{\partial}{\partial T_{2q-1}} \Omega^{(2p-1)}$$

where $\Omega^{(2p-1)}$ are the normalized second kind differentials with a pole at ∞ such that $\Omega^{(2p-1)} = d(z^{2p-1} + O(z^{-1}))$ at ∞.

Proof. To extract this particular modulation equation, one has to extract a particular component in eq. (11.112). Consider the co-vector

$$\lambda^{-1/2} d\lambda \langle 0| : \beta^*_{p-\frac{1}{2}} \beta^*_{q-\frac{1}{2}} \lambda^{\frac{1}{2}} \frac{d}{d\lambda} \lambda^{\frac{1}{2}} \beta(\lambda) :$$

where the factor $\lambda^{-1/2} d\lambda = 2dz$ is introduced to get a 1-form on Γ. Applying the Wick theorem, with contraction $\langle 0|\beta^*_{p-\frac{1}{2}} \beta_{-j+\frac{1}{2}}|0\rangle = \delta_{pj}$, we have

$$\langle 0| \lambda^{\frac{1}{2}} \frac{d}{d\lambda} \lambda^{\frac{1}{2}} \beta(\lambda) \beta^*_{p-\frac{1}{2}} \beta^*_{q-\frac{1}{2}} \mathcal{Q}_0$$
$$= \langle 0| \lambda^{\frac{1}{2}} \frac{d}{d\lambda} \lambda^{\frac{1}{2}} \beta(\lambda) \beta^*_{p-\frac{1}{2}} \partial_{T_{2q-1}} - \langle 0| \lambda^{\frac{1}{2}} \frac{d}{d\lambda} \lambda^{\frac{1}{2}} \beta(\lambda) \beta^*_{q-\frac{1}{2}} \partial_{T_{2p-1}}$$

On the other hand, we easily get

$$\langle 0| \beta^*(\lambda) \beta^*_{p-\frac{1}{2}} \mathcal{C}_0 = (p - \frac{1}{2}) \langle 0| \beta^*(\lambda) \beta_{p-\frac{1}{2}} - \langle 0| \lambda^{\frac{1}{2}} \frac{d}{d\lambda} \lambda^{\frac{1}{2}} \beta(\lambda) \beta^*_{p-\frac{1}{2}}$$

hence, having in mind eq. (11.111), we can write

$$\langle 0| \lambda^{\frac{1}{2}} \frac{d}{d\lambda} \lambda^{\frac{1}{2}} \beta(\lambda) \beta^*_{p-\frac{1}{2}} \beta^*_{q-\frac{1}{2}} \mathcal{Q}_0$$
$$= (p - \frac{1}{2}) \langle 0| \beta^*(\lambda) \beta_{p-\frac{1}{2}} \partial_{T_{2q-1}} - (q - \frac{1}{2}) \langle 0| \beta^*(\lambda) \beta_{q-\frac{1}{2}} \partial_{T_{2p-1}}$$

This yields the equation

$$(2q-1) \partial_{T_{2p-1}} \langle 0| : \beta_{q-\frac{1}{2}} \beta^*(\lambda) : \langle\langle \mathcal{Z} \rangle\rangle |0\rangle$$
$$= (2p-1) \partial_{T_{2q-1}} \langle 0| : \beta_{p-\frac{1}{2}} \beta^*(\lambda) : \langle\langle \mathcal{Z} \rangle\rangle |0\rangle$$

It remains to evaluate

$$\langle 0| \beta^*(\lambda) \beta_{p-\frac{1}{2}} \langle\langle \mathcal{Z} \rangle\rangle |0\rangle$$
$$= \Delta^{-1} \prod_{i=1}^{g} \int_{a_i} \frac{\lambda^g_{\gamma_i} d\lambda_{\gamma_i}}{\sqrt{R(\lambda_{\gamma_i})}} \langle 0| \beta^*(\lambda) \beta_{p-\frac{1}{2}} \, G_\Gamma \, \beta^*(\lambda_{\gamma_1}) \cdots \beta^*(\lambda_{\gamma_g}) |g\rangle$$

Pushing G_Γ to the left and writing $\beta_{p-\frac{1}{2}} = \int \frac{d\lambda_1}{2i\pi} \lambda_1^{p-1} \beta(\lambda_1)$, we arrive at

$$\langle 0|\beta^*(\lambda)\beta_{p-\frac{1}{2}} \, G_\Gamma \, \beta^*(\lambda_{\gamma_1}) \cdots \beta^*(\lambda_{\gamma_g})|g\rangle = \frac{\lambda^{g+\frac{1}{2}}}{\sqrt{R(\lambda)}}$$

$$\int \frac{d\lambda_1}{2i\pi} \lambda_1^{p-g-\frac{3}{2}} \sqrt{R(\lambda_1)} \langle 0|\beta^*(\lambda)\beta(\lambda_1)\beta^*(\lambda_{\gamma_1}) \cdots \beta^*(\lambda_{\gamma_g})\beta_{-g+\frac{1}{2}} \cdots \beta_{-\frac{1}{2}}|0\rangle$$

The last vacuum expectation value is just the determinant of a $(g+1) \times (g+1)$ matrix

$$\det \begin{pmatrix} \frac{1}{\lambda - \lambda_1} & \lambda^{-j+1} \\ \frac{1}{\lambda_{\gamma_i} - \lambda_1} & \lambda_{\gamma_i}^{-j+1} \end{pmatrix}_{i,j=1,\dots,g}$$

The λ_1 integral can be evaluated:

$$\frac{\lambda^{g+\frac{1}{2}}}{\sqrt{R(\lambda)}} \int_{|\lambda_1|>|\lambda|} \frac{d\lambda_1}{2i\pi} \sqrt{R(\lambda_1)} \frac{1}{\lambda - \lambda_1} = \frac{\lambda^{g+\frac{1}{2}}}{\sqrt{R(\lambda)}} \left(\lambda^{p-g-\frac{3}{2}} \sqrt{R(\lambda)} \right)_+$$

where $()_+$ means the polynomial part at ∞. Putting back the factor $\lambda^{-1/2}d\lambda$, we finally get:

$$\lambda^{-1/2}d\lambda \, \langle 0|\beta^*(\lambda)\beta_{p-\frac{1}{2}}\langle\langle \mathcal{Z}\rangle\rangle|0\rangle$$

$$= \Delta^{-1} \det \begin{pmatrix} \frac{\lambda^g}{\sqrt{R(\lambda)}} \left(\lambda^{p-g-\frac{3}{2}} \sqrt{R(\lambda)} \right)_+ d\lambda & \frac{\lambda^{g-j+1}}{\sqrt{R(\lambda)}} d\lambda \\ \int_{a_i} \frac{d\lambda_{\gamma_i}}{2i\pi} \frac{\lambda_{\gamma_i}^g}{\sqrt{R(\lambda_{\gamma_i})}} \left(\lambda_{\gamma_i}^{p-g-\frac{3}{2}} \sqrt{R(\lambda_{\gamma_i})} \right)_+ & \int_{a_i} \frac{d\lambda_{\gamma_i}}{2i\pi} \frac{\lambda_{\gamma_i}^{g-j+1}}{\sqrt{R(\lambda_{\gamma_i})}} \end{pmatrix}$$

Here the indices $i,j = 1,\dots,g$, so that the matrix inside the determinant is $(g+1) \times (g+1)$. Note that this is a second kind differential. At infinity it behaves as $(z^{2p-2} + O(1))dz$. Moreover, it is evident that its a-periods all vanish. It is equal to $\frac{1}{2p-1}\Omega^{(2p-1)}$. $\blacksquare$

References

[1] D.J. Korteweg and G. de Vries, On the change of form of long waves advancing in a rectangular channel and on a new type of long stationary wave. *Philos. Mag.* **39** (1895) 422–443.

[2] C.S. Gardner, J.M. Greene, M.D. Kruskal and R.M. Miura, Method for solving the Korteweg–de Vries equation. *Phys. Rev. Lett.* **19** (1967) 1095.

[3] P.D. Lax, Integrals of nonlinear equations of evolution and solitary waves. *Comm. Pure Appl. Math.* **21** (1968) 467–490.

[4] V.E. Zakharov and L.D. Faddeev, Korteweg–de Vries equation: a completely integrable Hamiltonian system. *Funct. Anal. Appl.* **8** 4 (1971) 280–287.

[5] S. Novikov, S.V. Manakov, L.P. Pitaevskii and V.E. Zakharov, *Theory of solitons. The inverse scattering method.* Consultants Bureau (1984).

[6] L.D. Faddeev and L.A. Takhtajan, *Hamiltonian Methods in the Theory of Solitons.* Springer (1986).

[7] L.A. Dickey, *Soliton Equations and Hamiltonian Systems.* World Scientific (1991).

[8] O. Babelon, D. Bernard and F. Smirnov, Null-vectors in integrable field theory. *Commun. Math. Phys.* **186** (1997) 601–648.

[9] O. Babelon, D. Bernard and F. Smirnov, Form factors, KdV and deformed hyperelliptic curves. *Nucl. Phys. B (Proc. Suppl.)* **58** (1997) 21–33.

[10] I. Krichever and D.H. Phong, Symplectic forms in the theory of solitons. *Surveys in Differential Geometry* **4** (1998) 239–313, International Press.

12

The Toda field theories

In this chapter, we study Toda field theories, which are generalizations of the Liouville equation. The equations of motion are both integrable, i.e. they admit a zero curvature representation, and conformally invariant. They allow us to see the interplay between conformal symmetry and the classical Yang–Baxter equation. The sine-Gordon theory is a Toda field theory associated with the affine Lie algebra $\widehat{sl}_2$. In particular, soliton solutions are constructed by the action of very special elements of the dressing group on the vacuum solution. Compared to the KP situation, we have here an example with two singularities, each one accomodating one of the two relativistic light-cone variables $x \pm t$, and their corresponding hierarchy of times. We study soliton and finite-zone solutions of the sine-Gordon equation. In the next chapter we will apply the inverse scattering method to discuss further solutions.

12.1 The Liouville equation

In his studies on surfaces with constant curvature, Liouville introduced the equation

$$(\partial_t^2 - \partial_x^2)\varphi = -4e^{2\varphi} \tag{12.1}$$

In the light-cone coordinates $x_\pm = x \pm t$, $\partial_{x_\pm} = \frac{1}{2}(\partial_x \pm \partial_t)$, the equation reads $\partial_{x_+}\partial_{x_-}\varphi = e^{2\varphi}$. Remarkably, Liouville was able to give the general solution of this non-linear partial differential equation:

$$e^{2\varphi} = -\frac{\partial F(x_+)\partial G(x_-)}{(F(x_+) - G(x_-))^2} \tag{12.2}$$

where F and G are arbitrary functions of the single variables x_+ and x_- respectively and $\partial F(x_+) = \partial_{x_+}F(x_+)$, $\partial G(x_-) = \partial_{x_-}G(x_-)$. Such functions are called chiral functions. A very important property of this equation is its invariance under changes of coordinates:

$$x_+ = f(x'_+), \quad x_- = g(x'_-), \quad \varphi = \varphi' - \frac{1}{2} \log \left(\partial f \partial g \right)$$

We see that the equation for the primed variables is the same as for the unprimed ones. We call this invariance the conformal invariance of the Liouville equation.

There is an important connection between the Liouville equation and the Schroedinger equation, eq. (11.5) in Chapter 11. The field $e^{-\varphi}$ satisfies two chiral equations:

$$(\partial_{x_-}^2 - u(x_-))e^{-\varphi} = 0, \quad u = (\partial_{x_-} \varphi)^2 - \partial_{x_-}^2 \varphi, \quad \partial_{x_+} u = 0$$
$$(\partial_{x_+}^2 - \bar{u}(x_+))e^{-\varphi} = 0, \quad \bar{u} = (\partial_{x_+} \varphi)^2 - \partial_{x_+}^2 \varphi, \quad \partial_{x_-} \bar{u} = 0$$

Indeed, the two Schroedinger equations are obtained readily by computing $\partial_{x_\pm}^2 e^{-\varphi}$ and the chirality of u and $\bar{u}$ is then proved using the Liouville equation.

Conversely, starting from two arbitrary chiral potentials $u(x_-)$ and $\bar{u}(x_+)$, we construct solutions $\xi_i(x_-)$ and $\bar{\xi}_i(x_+)$, $i = 1, 2$, of the two Schroedinger equations, normalized such that their Wronskians are:

$$W(\xi_1, \xi_2) = 1, \quad W(\bar{\xi}_1, \bar{\xi}_2) = -1$$

where the Wronskian is defined as $W(f, g) = f'g - fg'$. It is easy to check that $\varphi(x_+, x_-)$, given by the formula,

$$e^{-\varphi(x_+, x_-)} = \xi_1(x_-)\bar{\xi}_1(x_+) + \xi_2(x_-)\bar{\xi}_2(x_+) \tag{12.3}$$

satisfies the Liouville equation. One can relate this solution to Liouville's solution eq. (12.2) by writing it as:

$$e^{-\varphi} = \frac{F - G}{\sqrt{-F'G'}} = \frac{F}{\sqrt{F'}} \frac{1}{\sqrt{-G'}} - \frac{1}{\sqrt{F'}} \frac{G}{\sqrt{-G'}}$$

This leads us to set:

$$\bar{\xi}_1 = \frac{F}{\sqrt{F'}}, \quad \bar{\xi}_2 = -\frac{1}{\sqrt{F'}}, \quad \xi_1 = \frac{1}{\sqrt{-G'}}, \quad \xi_2 = \frac{G}{\sqrt{-G'}}$$

The Wronskian conditions $W(\xi_1, \xi_2) = 1$ and $W(\bar{\xi}_1, \bar{\xi}_2) = -1$ are automatically satisfied. It follows that $F = -\bar{\xi}_1/\bar{\xi}_2$ and $G = \xi_2/\xi_1$. Finally, one can use this identification to compute the potentials u and $\bar{u}$ in terms of F and G.

$$\bar{u} = -\frac{1}{2}S(F), \quad u = -\frac{1}{2}S(G), \quad S(f) = \frac{f'''}{f'} - \frac{3}{2}\frac{f''^2}{f'^2}$$

where $S(f)$ is the so-called Schwarzian derivative of f.

Note that the two solutions ξ_1, ξ_2 are defined up to a linear transformation of determinant 1. This translates into a homographic transformation for G which leaves the potential u invariant, and similarly for F and $\bar{u}$. When the two homographic transformations of F and G are inverse to each other, the Liouville field φ is invariant.

Invariance of the Liouville equation under change of coordinates reflects itself into covariance properties of the Schroedinger equation. Changing $x = f(x')$, we have:

$$\xi'(x') = \frac{\xi(x)}{\sqrt{\partial f}}, \quad u'(x') = u(x)(\partial f)^2 + S(f), \quad (\partial'^2 - u')\xi' = (\partial f)^{\frac{3}{2}}(\partial^2 - u)\xi$$

These equations exhibit the covariance properties of the objects involved. Specifically, ξ is a differential of weight $-\frac{1}{2}$, u is a Schwartzian connection, and $(\partial^2 - u)\xi$ is a differential of weight $\frac{3}{2}$.

In the next section, we generalize this setup to a large class of two-dimensional field theories, called Toda field theories, based on Lie algebras.

12.2 The Toda systems and their zero-curvature representations

Toda field theories are two-dimensional generalizations of the Toda chains studied in Chapter 4.

We use the same notations for Lie algebras as in that chapter. Let $\mathcal{G}$ be a simple Lie algebra of rank r, and consider a Cartan decomposition:

$$\mathcal{G} = \mathcal{N}_- \oplus \mathcal{H} \oplus \mathcal{N}_+$$

Let $\Phi(x,t)$ be the Toda field taking values in the Cartan subalgebra

$$\Phi(x,t) = \sum_{i=1}^{r} \Phi_i(x,t)H_i$$

where H_i form an orthonormal basis of the Cartan algebra. By a straightforward generalization of eq. (4.24), we define the Toda field theories by their equations of motion:

$$(\partial_t^2 - \partial_x^2)\Phi = -2 \sum_{\alpha \text{ simple}} H_\alpha \exp\left(2\alpha(\Phi)\right) \tag{12.4}$$

To write these equations in components, we introduce $\varphi^i = \Lambda^{(i)}(\Phi)$ for $i = 1, \ldots, r$, where $\Lambda^{(i)}$ are the r fundamental weights of $\mathcal{G}$. Since we have $\Lambda^{(i)}(H_{\alpha_j}) = \frac{(\alpha_i, \alpha_i)}{2}\delta_{ij}$, and $\alpha_i(\Phi) = \sum_j \Lambda^{(j)}(\Phi)\, a_{ji}$, where a_{ji} are the elements of the Cartan matrix, the Toda equations of motion can be written as:

$$(\partial_t^2 - \partial_x^2)\varphi^i = -\alpha_i^2\, e^{\,2\sum_j \varphi^j\, a_{ji}} \tag{12.5}$$

In particular in the sl_2 case, the Toda field has only one component $\varphi^1 = \Lambda(\Phi)$, and setting $\varphi^1 = \frac{1}{2}\varphi$, we recover eq. (12.1).

All these equations share with the Liouville equation the important property of being invariant under a change of coordinates:

Proposition. *The Toda field equations are invariant under the transformations $x_\pm \to x'_\pm$ with $x_+ = f(x'_+)$, $x_- = g(x'_-)$, and*

$$\Phi'(x'_+, x'_-) = \Phi(f(x'_+), g(x'_-)) + \frac{1}{2}H_\rho \ln(\partial f(x'_+)\partial g(x'_-))$$

Here $H_\rho \in \mathcal{H}$ is the Weyl vector characterized by the property $\alpha_j(H_\rho) = 1$ for all simple roots α_j.

Proof. The proof is obvious once we write the field equations in the light-cone coordinates:

$$\partial_{x_+}\partial_{x_-}\Phi = \frac{1}{2} \sum_{\alpha \text{ simple}} H_\alpha e^{2\alpha(\Phi)}$$

∎

We introduce now a zero curvature representation for the Toda field equations.

Proposition. *Let*

$$A_x = \partial_t\Phi + \sum_{\alpha \text{ simple}} n_\alpha e^{\alpha(\Phi)}(E_\alpha + E_{-\alpha}) \tag{12.6}$$

$$A_t = \partial_x\Phi - \sum_{\alpha \text{ simple}} n_\alpha e^{\alpha(\Phi)}(E_\alpha - E_{-\alpha}) \tag{12.7}$$

Then the Toda field equations eq. (12.4) can be rewritten as the zero curvature equation $\partial_x A_t - \partial_t A_x - [A_x, A_t] = 0$. The constants n_α are such that $n_\alpha^2\,(E_\alpha, E_{-\alpha}) = 1$.

Proof. The proof is exactly the same as for the open Toda chain, and we do not repeat it. ∎

It will be often convenient to work with the light-cone coordinates. In these coordinates, $A_{x_\pm} = \frac{1}{2}(A_x \pm A_t)$. Define the elements of $\mathcal{G}$:

$$\mathcal{E}_\pm = \sum_{\alpha_j \text{ simple}} n_{\alpha_j} E_{\pm\alpha_j}$$

then

$$A_{x_+} = \partial_{x_+}\Phi + e^{-\text{ad}\,\Phi}\mathcal{E}_-, \quad A_{x_-} = -\partial_{x_-}\Phi + e^{\text{ad}\,\Phi}\mathcal{E}_+ \qquad (12.8)$$

Example. Let us give the example associated with the Lie algebra $\mathcal{G} = sl_2$. Let E_+, E_-, H, be its generators with commutation relations

$$[H, E_\pm] = \pm 2E_\pm, \quad [E_+, E_-] = H$$

We have $\Phi = \frac{1}{2}\varphi H$, and the Lax connection reads, in the fundamental representation:

$$A_{x_+} = \begin{pmatrix} \frac{1}{2}\partial_{x_+}\varphi & 0 \\ e^\varphi & -\frac{1}{2}\partial_{x_+}\varphi \end{pmatrix} \quad \text{and} \quad A_{x_-} = \begin{pmatrix} -\frac{1}{2}\partial_{x_-}\varphi & e^\varphi \\ 0 & \frac{1}{2}\partial_{x_-}\varphi \end{pmatrix}$$

The zero curvature condition $\partial_{x_+}A_{x_-} - \partial_{x_-}A_{x_+} - [A_{x_+}, A_{x_-}]$ is equivalent to the Liouville equation (12.1).

12.3 Solution of the Toda field equations

The Toda field equations being conformally invariant, one can solve them by splitting the chiralities, as in the Liouville case. To do that, we use the zero-curvature representation. When Φ is a solution of the Toda field equations, we have $F_{x_+x_-} = 0$ (and conversely), so we can solve the linear system $(\partial_{x_\pm} - A_{x_\pm})\Psi = 0$. Equivalently, we can write $A_{x_\pm}$ as a pure gauge

$$A_{x_\pm} = (\partial_{x_\pm}\Psi) \cdot \Psi^{-1} \qquad (12.9)$$

We denote by $\mathcal{B}_\pm = \mathcal{H} \oplus \mathcal{N}_\pm$ the two Borel subalgebras of $\mathcal{G}$.

Proposition. *Let $Q_\pm \in \exp\mathcal{B}_\pm$ be defined by two decompositions of Ψ as:*

$$\Psi = e^{\pm\Phi} N_\mp Q_\pm \quad \text{with} \quad N_\mp \in \exp\mathcal{N}_\mp, \quad Q_\pm \in \exp\mathcal{B}_\pm \qquad (12.10)$$

then $Q_\pm$ satisfy the following equations:

$$\partial_{x_-} Q_- Q_-^{-1} = 0, \qquad \partial_{x_+} Q_- Q_-^{-1} = -\overline{P} + \mathcal{E}_- \qquad (12.11)$$

$$\partial_{x_+} Q_+ Q_+^{-1} = 0, \qquad \partial_{x_-} Q_+ \cdot Q_+^{-1} = P + \mathcal{E}_+ \qquad (12.12)$$

where $P(x_-)$ and $\overline{P}(x_+)$ are chiral fields with values in the Cartan subalgebra.

Proof. We write Ψ in two different ways: $\Psi = e^{-\Phi} G_1 = e^{\Phi} G_2$. Plugging this into eq. (12.9) we get:

$$\partial_{x_-} G_1 \cdot G_1^{-1} = e^{2\mathrm{ad}\Phi}\mathcal{E}_+, \partial_{x_-} G_2 \cdot G_2^{-1} = -2\partial_{x_-}\Phi + \mathcal{E}_+ \qquad (12.13)$$

$$\partial_{x_+} G_1 \cdot G_1^{-1} = 2\partial_{x_+}\Phi + \mathcal{E}_-, \partial_{x_+} G_2 \cdot G_2^{-1} = e^{-2\mathrm{ad}\Phi}\mathcal{E}_- \qquad (12.14)$$

Let us prove eqs. (12.11). Using the Gauss decomposition $G_1 = N_+ Q_-$ with $N_+ \in \exp \mathcal{N}_+$ and $Q_- \in \exp \mathcal{B}_-$, we obtain:

$$N_+ (\partial_{x_-} Q_- Q_-^{-1}) N_+^{-1} + \partial_{x_-} N_+ N_+^{-1} = e^{2\mathrm{ad}\Phi}\mathcal{E}_+$$

or, multiplying on the right by N_+ and on the left by N_+^{-1},

$$\partial_{x_-} Q_- Q_-^{-1} = -N_+^{-1}\partial_{x_-} N_+ + N_+^{-1} e^{2\mathrm{ad}\Phi}\mathcal{E}_+ N_+$$

Since the left-hand side is in $\mathcal{B}_-$ and the right-hand side is in $\mathcal{N}_+$, they both vanish, so that $\partial_{x_-} Q_- Q_-^{-1} = 0$ and $\partial_{x_-} N_+ N_+^{-1} = e^{2\mathrm{ad}\Phi}\mathcal{E}_+$. This proves that Q_- only depends on x_+. Next, using eq. (12.14) and again the decomposition $G_1 = N_+ Q_-$, we obtain:

$$N_+ (\partial_{x_+} Q_- Q_-^{-1}) N_+^{-1} + \partial_{x_+} N_+ N_+^{-1} = 2\partial_{x_+}\Phi + \mathcal{E}_- \qquad (12.15)$$

The right-hand side has lowest height -1 given by the $\mathcal{E}_-$ term. So the lowest height term in $\partial_{x_+} Q_- Q_-^{-1}$ is also equal to $\mathcal{E}_-$. Since $\partial_{x_+} Q_- Q_-^{-1} \in \mathcal{B}_-$, it is necessarily of the form $-\overline{P} + \mathcal{E}_-$, with $\overline{P} \in \mathcal{H}$ only depending on x_+. The equations for Q_+ are proved similarly. ■

One can reconstruct the Toda field, Φ, from the knowledge of $Q_\pm$. Let $|\Lambda^{(i)}\rangle$, $i = 1, \ldots, r = \mathrm{rank}\,\mathcal{G}$ be highest weight vectors for the fundamental representations of $\mathcal{G}$. Recall the main properties of $|\Lambda^{(i)}\rangle$:

$$H|\Lambda^{(i)}\rangle = \Lambda^{(i)}(H)|\Lambda^{(i)}\rangle, \qquad E_\alpha|\Lambda^{(i)}\rangle = 0 \quad \text{for} \quad \alpha > 0$$

We denote by $\langle\Lambda^{(i)}|$ the conjugate highest weight which satisfies (see Chapter 16):

$$\langle\Lambda^{(i)}| = \Lambda^{(i)}(H)\langle\Lambda^{(i)}|, \qquad \langle\Lambda^{(i)}|E_{-\alpha} = 0 \quad \text{for} \quad \alpha > 0$$

We can compute any scalar product of the form $\langle \Lambda^{(i)} | \mathcal{X} | \Lambda^{(i)} \rangle$, where $\mathcal{X}$ is any element of the universal enveloping algebra, by pushing all the E_α to the right and all the $E_{-\alpha}$ to the left using the commutation relations, and the fact that $|\Lambda^{(i)}\rangle$ is a common eigenvector of all the H. Finally, we normalize the scalar product by:

$$\langle \Lambda^{(i)} | \Lambda^{(i)} \rangle = 1$$

With these definitions at hand, we have:

Proposition. *For any fundamental representation with highest weight* $\Lambda^{(i)}$*, define:*

$$\xi^{(i)} = \langle \Lambda^{(i)} | e^{-\Phi} \Psi = \langle \Lambda^{(i)} | Q_+$$
$$\overline{\xi}^{(i)} = \Psi^{-1} e^{-\Phi} | \Lambda^{(i)} \rangle = Q_-^{-1} | \Lambda^{(i)} \rangle \qquad (12.16)$$

The vectors $\xi^{(i)}$ *and* $\overline{\xi}^{(i)}$ *are chiral:* $\partial_{x_+} \xi^{(i)} = 0$ *and* $\partial_{x_-} \overline{\xi}^{(i)} = 0$.
The Toda field Φ *can be reconstructed by the formula:*

$$e^{-2\Lambda^{(i)}(\Phi)} = \xi^{(i)} \cdot \overline{\xi}^{(i)} = \langle \Lambda^{(i)} | Q_+ Q_-^{-1} | \Lambda^{(i)} \rangle \qquad (12.17)$$

<u>Proof.</u> First $e^{-\Phi} \Psi = N_- Q_+$ and $\langle \Lambda^{(i)} | N_- = \langle \Lambda^{(i)} |$ by the highest weight condition. So $\langle \Lambda^{(i)} | e^{-\Phi} \Psi = \langle \Lambda^{(i)} | Q_+$ depends only on x_+. Similarly $\Psi^{-1} e^{-\Phi} | \Lambda^{(i)} \rangle = Q_-^{-1} | \Lambda^{(i)} \rangle$ depends only on x_-. By the definition of $\xi^{(i)}$ and $\overline{\xi}^{(i)}$, we see that Ψ and Ψ^{-1} cancel in the scalar product $\xi^{(i)} \cdot \overline{\xi}^{(i)}$, leaving $\langle \Lambda^{(i)} | e^{-2\Phi} | \Lambda^{(i)} \rangle = e^{-2\Lambda^{(i)}(\Phi)}$. The knowledge of these quantities for $i = 1, \ldots, r$ completely characterizes Φ. ∎

Equations (12.12) determine $Q_\pm$ in terms of the two chiral fields P and $\overline{P}$ with values in $\mathcal{H}$. For example, consider the equation $\partial_{x_-} Q_+ = (P + \mathcal{E}_+) Q_+$ in the Liouville case, $\mathcal{G} = sl_2$. We have:

$$Q_+ = \begin{pmatrix} q_{11} & q_{12} \\ 0 & q_{22} \end{pmatrix}, \quad (P + \mathcal{E}_+) = \begin{pmatrix} p & 1 \\ 0 & -p \end{pmatrix}$$

The vector ξ reads $\xi = (1,0)Q_+ = (q_{11}, q_{12})$ and the first order differential equation for Q_+ yields:

$$(\partial_{x_-}^2 - u)\xi = 0, \quad u = p' + p^2$$

The relation between u and p is called the Miura transformation. The first order linear system for Q_+ is a matrix version of the Schroedinger equation, which is recovered as an equation for the first row of the matrix Q_+.

The chiral fields P and $\bar{P}$ are the two arbitrary functions parametrizing the general solution of Toda field equations.

Note that the splitting of chiralities in Toda field theories brings us back to the Drinfeld–Sokolov linear systems of Chapter 10.

We describe more explicitly the case of $sl(n+1)$. The n fundamental representations are the vector representation and its wedge products. The vector representation acts on $\mathbb{C}^{n+1}$ with basis $|\epsilon_j\rangle$, $j = 1, \ldots, (n+1)$. The highest weight vector is $|\epsilon_1\rangle$. The elements of the Cartan algebra are the traceless diagonal matrices, and $\mathcal{E}_{\pm} = \sum E_{i,i\pm 1}$, where $E_{ij} = |\epsilon_i\rangle\langle\epsilon_j|$ are the canonical matrices acting on $\mathbb{C}^{n+1}$. In the vector representation the chiral fields are $\xi(x) = \langle\epsilon_1|Q_+$ and $\bar{\xi}(x) = Q_-^{-1}|\epsilon_1\rangle$. Let us decompose the fields P and $\bar{P}$ as $P = \sum_j P_j E_{jj}$ and $\bar{P} = \sum_j \bar{P}_j E_{jj}$, with $\sum P_j = \sum \bar{P}_j = 0$. Denote by ξ_j and $\bar{\xi}_j$ the components of the chiral fields: $\xi(x) = \sum_j \langle\epsilon_j| \xi_j$, and $\bar{\xi}(x) = \sum_j \bar{\xi}_j|\epsilon_j\rangle$. The functions ξ_j and $\bar{\xi}_j$ satisfy the differential equations of order $(n+1)$:

$$(\partial_{x_+} - \bar{P}_{n+1})\cdots(\partial_{x_+} - \bar{P}_1)\,\bar{\xi}_j = 0 \qquad (12.18)$$
$$(\partial_{x_-} - P_{n+1})\cdots(\partial_{x_-} - P_1)\,\xi_j = 0 \qquad (12.19)$$

Indeed, the first row of Q_+ is $(\xi_1, \ldots, \xi_{n+1})$ and the explicit form of $\mathcal{E}_+$ immediately yields eqs. (12.19).

Proposition. *The components of the Toda field along the fundamental weights $\Lambda^{(p)}$ are given by:*

$$e^{-2\Lambda^{(p)}(\Phi)} = \det\left(\partial_{x_-}^i \xi \cdot \partial_{x_+}^j \bar{\xi}\right), \quad i, j = 0, \ldots, p-1 \qquad (12.20)$$

<u>Proof.</u> The highest weight vector of the $\Lambda^{(p)}$ is the wedge product $|\Lambda^{(p)}\rangle = |\epsilon_1\rangle \wedge \cdots \wedge |\epsilon_p\rangle$, and we have $\mathcal{E}_-^{p-1}|\epsilon_1\rangle = |\epsilon_p\rangle$ for $p = 1, \ldots, n$. Thus,

$$|\Lambda^{(p)}\rangle = |\epsilon_1\rangle \wedge \mathcal{E}_-|\epsilon_1\rangle \wedge \cdots \wedge \mathcal{E}_-^{p-1}|\epsilon_1\rangle$$

Acting with Q_-^{-1} we obtain an expression for the chiral field $\bar{\xi}^{(p)}$:

$$\bar{\xi}^{(p)} = Q_-^{-1}|\Lambda^{(p)}\rangle = Q_-^{-1}|\epsilon_1\rangle \wedge Q_-^{-1}\mathcal{E}_-|\epsilon_1\rangle \wedge \cdots \wedge Q_-^{-1}\mathcal{E}_-^{p-1}|\epsilon_1\rangle \quad (12.21)$$

We now use the equation of motion (12.12) to express $Q_-^{-1}\mathcal{E}_-^j|\epsilon_1\rangle$ in terms of the derivatives $\partial_{x_+}^j\bar{\xi}$. Using eq. (12.12) and differentiating it, we get:

$$Q_-^{-1}\mathcal{E}_-|\epsilon_1\rangle = -\partial_{x_+}Q_-^{-1}|\epsilon_1\rangle + Q_-^{-1}\bar{P}|\epsilon_1\rangle$$
$$Q_-^{-1}\mathcal{E}_-^2|\epsilon_1\rangle = \partial_{x_+}^2 Q_-^{-1}|\epsilon_1\rangle + Q_-^{-1}\left(\bar{P}\mathcal{E}_- + \mathcal{E}_-\bar{P} - \bar{P}^2 - \partial_{x_+}\bar{P}\right)|\epsilon_1\rangle$$
$$Q_-^{-1}\mathcal{E}_-^j|\epsilon_1\rangle = (-1)^j\partial_{x_+}^j Q_-^{-1}|\epsilon_1\rangle + \cdots$$

The extra terms cancel in the wedge product (12.21). Therefore,

$$\bar{\xi}^{(p)} = (-1)^{\frac{p(p-1)}{2}} Q_-^{-1}|\epsilon_1\rangle \wedge (\partial_{x_+} Q_-^{-1})|\epsilon_1\rangle \wedge \cdots \wedge (\partial_{x_+}^{p-1} Q_-^{-1})|\epsilon_1\rangle$$

$$= (-1)^{\frac{p(p-1)}{2}} \bar{\xi} \wedge \partial_{x_+} \bar{\xi} \wedge \cdots \wedge \partial_{x_+}^{p-1} \bar{\xi}$$

$$= (-1)^{\frac{p(p-1)}{2}} \sum_{j_1 < \cdots < j_p} |\epsilon_{j_1}\rangle \wedge \cdots \wedge |\epsilon_{j_p}\rangle \, \det \begin{pmatrix} \bar{\xi}_{j_1} & \cdots & \bar{\xi}_{j_p} \\ \partial_{x_+}\bar{\xi}_{j_1} & \cdots & \partial_{x_+}\bar{\xi}_{j_p} \\ \vdots & & \vdots \\ \partial_{x_+}^{p-1}\bar{\xi}_{j_1} & \cdots & \partial_{x_+}^{p-1}\bar{\xi}_{j_p} \end{pmatrix}$$

A similar expression is obtained for the chiral fields:

$$\xi^{(p)} = \langle \Lambda^{(p)}|Q_+ = (-1)^{\frac{p(p-1)}{2}} \xi \wedge \partial_{x_-}\xi \wedge \cdots \partial_{x_-}^{p-1}\xi$$

Equation (12.20) follows from $\exp(-2\Lambda^{(p)}(\Phi)) = \xi^{(p)} \cdot \bar{\xi}^{(p)}$. Noting that $\langle \epsilon_{j_1}| \wedge \cdots \wedge \langle \epsilon_{j_p}|\epsilon_{k_1}\rangle \wedge \cdots \wedge |\epsilon_{k_p}\rangle = \epsilon_{j_1 \cdots j_p}^{k_1 \cdots k_p}$, we can also write

$$e^{-2\Lambda^{(p)}(\Phi)}$$

$$= \sum_{j_1 < \cdots < j_p} \det \begin{pmatrix} \xi_{j_1} & \cdots & \xi_{j_p} \\ \partial_{x_-}\xi_{j_1} & \cdots & \partial_{x_-}\xi_{j_p} \\ \vdots & & \vdots \\ \partial_{x_-}^{p-1}\xi_{j_1} & \cdots & \partial_{x_-}^{p-1}\xi_{j_p} \end{pmatrix} \det \begin{pmatrix} \bar{\xi}_{j_1} & \cdots & \bar{\xi}_{j_p} \\ \partial_{x_+}\bar{\xi}_{j_1} & \cdots & \partial_{x_+}\bar{\xi}_{j_p} \\ \vdots & & \vdots \\ \partial_{x_+}^{p-1}\bar{\xi}_{j_1} & \cdots & \partial_{x_+}^{p-1}\bar{\xi}_{j_p} \end{pmatrix}$$

This is the sl_{n+1} generalization of the Liouville solution eq. (12.3). ■

It is interesting to give a direct and algebraic proof that the formula we just obtained for Φ does indeed satisfy the Toda field equations. We first rewrite the field equations directly in terms of the tau-functions:

$$\tau_i(x_+, x_-) = \exp\left(-2\Lambda^{(i)}(\Phi)\right) \tag{12.22}$$

Proposition. *In terms of the functions τ_i, the field equations become:*

$$\tau_i(\partial_{x_-}\partial_{x_+}\tau_i) - (\partial_{x_+}\tau_i)(\partial_{x_-}\tau_i) = -\frac{(\alpha_i, \alpha_i)}{2} \prod_{j \neq i} \tau_j^{-a_{ji}} \tag{12.23}$$

where a_{ij} is the Cartan matrix of $\mathcal{G}$.

Proof. Recall the Toda field equations, eqs. (12.5), for the components $\varphi^i = \Lambda^{(i)}(\Phi) = -\frac{1}{2}\log\tau_i$. We have $\exp\left(2\sum_j \varphi^j a_{ji}\right) = \prod_j \tau_j^{-a_{ji}}$. Equation (12.23) for the tau-functions then follows using $a_{ii} = 2$ and

$$\partial_{x_+}\partial_{x_-}\log\tau_i = \frac{1}{\tau_i^2}\left(\tau_i \partial_{x_-}\partial_{x_+}\tau_i - \partial_{x_+}\tau_i \partial_{x_-}\tau_i\right).$$

■

Remark. Using Hirota's differential operators we can rewrite eq. (12.23) as:

$$D_+ D_- \tau_i \cdot \tau_i = -(\alpha_i, \alpha_i) \prod_{j \neq i} \tau_j^{-a_{ji}}$$

In general $0 \leq a_{ij} a_{ji} \leq 4$ and $a_{ij} \leq 0$ for $j \neq i$, so the right-hand side of this equation may be a polynomial of degree greater than 2 in the τ_j. However, in the sl_{n+1} case, we have $a_{ij} = -1$ for $j = i \pm 1$ (the other a_{ij} vanish) so this equation takes a bilinear form:

$$D_+ D_- \tau_i \cdot \tau_i = -2\tau_{i-1}\tau_{i+1}$$

We now show that eqs. (12.23) can be derived directly from the chiral equations, eqs. (12.12, 12.11). The tau-functions, eq. (12.17), are directly expressed in terms of $G = Q_+ Q_-^{-1}$:

$$\tau_i = \langle \Lambda^{(i)} | G | \Lambda^{(i)} \rangle$$

The left-hand side of eq. (12.23) can be written as:

$$
\begin{aligned}
\tau_i(\partial_{x_-}\partial_{x_+}\tau_i) - (\partial_{x_+}\tau_i)(\partial_{x_-}\tau_i) = \ &\langle \Lambda^{(i)} | \otimes \langle \Lambda^{(i)} | G \otimes \partial_{x_-}\partial_{x_+} G \\
&- \partial_{x_+} G \otimes \partial_{x_-} G | \Lambda^{(i)} \rangle \otimes | \Lambda^{(i)} \rangle
\end{aligned}
$$

Using the equations of motion of $Q_\pm$, we see that the fields P and $\bar{P}$ drop out because they always hit the highest weight vector giving the scalar factors $\Lambda^{(i)}(P)$ or $\Lambda^{(i)}(\overline{P})$. We get:

$$
\begin{aligned}
\tau_i(\partial_{x_-}\partial_{x_+}\tau_i) - (\partial_{x_+}\tau_i)(\partial_{x_-}\tau_i) = \ &- \langle \Lambda^{(i)} | G | \Lambda^{(i)} \rangle \langle \Lambda^{(i)} | \mathcal{E}_+ G \mathcal{E}_- | \Lambda^{(i)} \rangle \\
&+ \langle \Lambda^{(i)} | \mathcal{E}_+ G | \Lambda^{(i)} \rangle \langle \Lambda^{(i)} | G \mathcal{E}_- | \Lambda^{(i)} \rangle
\end{aligned}
$$

This can be rewritten as

$$\tau_i(\partial_{x_-}\partial_{x_+}\tau_i) - (\partial_{x_+}\tau_i)(\partial_{x_-}\tau_i) = -\tfrac{1}{2}\langle \Xi^{(i)} | \, G \otimes G \, | \Xi^{(i)} \rangle \qquad (12.24)$$

where

$$| \Xi^{(i)} \rangle = | \Lambda^{(i)} \rangle \otimes \mathcal{E}_- | \Lambda^{(i)} \rangle - \mathcal{E}_- | \Lambda^{(i)} \rangle \otimes | \Lambda^{(i)} \rangle = | \Lambda^{(i)} \rangle \wedge \mathcal{E}_- | \Lambda^{(i)} \rangle$$

The proof of the equations of motion, eq. (12.23), now consists of comparing the right-hand sides of eq. (12.23) and eq. (12.24). We need the following lemma:

Lemma. *In the tensor product of two copies of the fundamental representation with highest weight $\Lambda^{(i)}$, $|\Xi^{(i)}\rangle$ is a highest weight vector with weight $-\sum_{j \neq i} a_{ji}\Lambda^{(j)}$ and norm $\langle \Xi^{(i)} | \Xi^{(i)} \rangle = (\alpha_i, \alpha_i)$.*

<u>Proof.</u> We must show that E_α kills $|\Xi^{(i)}\rangle$ for all positive roots α. It is enough to consider α simple. In the tensor product, E_α is represented by $E_\alpha \otimes I + I \otimes E_\alpha$. Hence we have to show that:

$$(E_\alpha \otimes I + I \otimes E_\alpha)|\Xi^{(i)}\rangle = |\Lambda^{(i)}\rangle \otimes E_\alpha \mathcal{E}_-|\Lambda^{(i)}\rangle - E_\alpha \mathcal{E}_-|\Lambda^{(i)}\rangle \otimes |\Lambda^{(i)}\rangle = 0$$

But for α a simple positive root,

$$E_\alpha \mathcal{E}_-|\Lambda^{(i)}\rangle = \sum_{\beta \text{ simple}} [E_\alpha, E_{-\beta}]|\Lambda^{(i)}\rangle = H_\alpha|\Lambda^{(i)}\rangle = \Lambda^{(i)}(H_\alpha)|\Lambda^{(i)}\rangle$$

where we have used that for α and β simple, $[E_\alpha, E_{-\beta}] = 0$ for $\beta \neq \alpha$. The first part of the lemma follows. For the second part, we must prove that $|\Xi^{(i)}\rangle$ is an eigenstate of $(H_\alpha \otimes I + I \otimes H_\alpha)$ and find the eigenvalue. Let us compute:

$$(H_\alpha \otimes I + I \otimes H_\alpha)|\Xi^{(i)}\rangle = H_\alpha|\Lambda^{(i)}\rangle \otimes \mathcal{E}_-|\Lambda^{(i)}\rangle - H_\alpha \mathcal{E}_-|\Lambda^{(i)}\rangle \otimes |\Lambda^{(i)}\rangle$$
$$+|\Lambda^{(i)}\rangle \otimes H_\alpha \mathcal{E}_-|\Lambda^{(i)}\rangle - \mathcal{E}_-|\Lambda^{(i)}\rangle \otimes H_\alpha|\Lambda^{(i)}\rangle$$

Next we notice that $\mathcal{E}_-|\Lambda^{(i)}\rangle = E_{-\alpha_i}|\Lambda^{(i)}\rangle$, because $E_{-\beta}|\Lambda^{(i)}\rangle$ vanishes unless $\beta = \alpha_i$. To see it, we compute the norm of this vector:

$$\langle\Lambda^{(i)}|E_\beta E_{-\beta}|\Lambda^{(i)}\rangle = \Lambda^{(i)}(H_\beta)$$

This vanishes if $\beta \neq \alpha_i$ by definition of the fundamental weights. Hence $H_\alpha \mathcal{E}_-|\Lambda^{(i)}\rangle = \left(-\alpha_i(H_\alpha) + \Lambda^{(i)}(H_\alpha)\right)E_{-\alpha_i}|\Lambda^{(i)}\rangle$, which gives:

$$(H_\alpha \otimes I + I \otimes H_\alpha)|\Xi^{(i)}\rangle = [2\Lambda^{(i)}(H_\alpha) - \alpha_i(H_\alpha)]|\Xi^{(i)}\rangle$$

Thus the weight of $|\Xi^{(i)}\rangle$ is $2\Lambda^{(i)} - \alpha_i = -\sum_{j\neq i} a_{ji}\Lambda^{(j)}$ because $\alpha_i = \sum_j \Lambda^{(j)}a_{ji}$. Finally, we have: $\langle\Xi^{(i)}|\Xi^{(i)}\rangle = \Lambda^{(i)}(H_{\alpha_i}) = (\alpha_i, \alpha_i)$.   ∎

We now finish the proof of the equations of motion. We have to compute $\langle\Xi^{(i)}| \, G \otimes G \, |\Xi^{(i)}\rangle$. But $G \otimes G$ is the representation of G in the tensor product representation. Since this scalar product only depends on the values of the weights and the Lie algebra structure, we can compute it once we know the highest weight vector and its normalization:

$$|\Xi^{(i)}\rangle = (\alpha_i, \alpha_i)^{\frac{1}{2}} \bigotimes_{j\neq i} \left(|\Lambda^{(j)}\rangle\right)^{\otimes(-a_{ij})}$$

This yields eq. (12.22).

12.4 Hamiltonian formalism

The Toda field equation eq. (12.4) can be derived from the Lagrangian density:

$$\mathcal{L} = \frac{1}{2}(\partial_t\Phi, \partial_t\Phi) - \frac{1}{2}(\partial_x\Phi, \partial_x\Phi) - \sum_{\alpha_i \text{ simple}} \exp(2\alpha(\Phi))$$

As usual we can go to the Hamiltonian formalism by introducing the conjugate momentum $\Pi = \partial_t\Phi$. We expand it on an orthonormal basis of the Cartan subalgebra as $\Pi(x) = \sum_i \Pi_i(x)H_i$. The canonical equal-time Poisson bracket between Φ_i and its conjugate momentum Π_i is

$$\{\Pi_i(x), \Phi_j(y)\} = \frac{1}{2}\delta_{ij}\delta(x - y)$$

Since we compute equal time Poisson brackets, we will often assume in the following that $t = 0$, and we do not write the time dependence explicitly. Replacing $\partial_t\Phi(x)$ by $\Pi(x)$ in the expression of A_x, we get:

$$A(x) \equiv A_x(x) = \Pi + \sum_{\alpha \text{ simple}} e^{\alpha(\Phi)}n_\alpha(E_\alpha + E_{-\alpha})$$

The Poisson brackets of A takes the familiar r-matrix form. In exactly the same way as for the Toda chain in Chapter 4, we have:

Proposition. *There exists a matrix $r_{12} \in \mathcal{G}\otimes\mathcal{G}$, independent of the fields Π and Φ, such that:*

$$\{A_1(x), A_2(y)\} = [r_{12}, A_1(x) + A_2(y)]\delta(x - y) \qquad (12.25)$$

We have

$$r_{12} = \frac{1}{2}\sum_{\alpha \text{ positive}} \frac{E_\alpha \otimes E_{-\alpha} - E_{-\alpha} \otimes E_\alpha}{(E_\alpha, E_{-\alpha})} + \lambda \cdot C_{12} \qquad (12.26)$$

where C_{12} the tensor Casimir:

$$C_{12} = \sum_i H_i \otimes H_i + \sum_{\alpha \text{ positive}} \frac{E_\alpha \otimes E_{-\alpha} + E_{-\alpha} \otimes E_\alpha}{(E_\alpha, E_{-\alpha})} \qquad (12.27)$$

As explained in Chapter 4, the two values $\lambda = \pm\frac{1}{2}$ of the parameter multiplying the tensor Casimir yield matrices $r_{12}^\pm$ which are solutions of the classical Yang–Baxter equation

$$[r_{12}^\pm, r_{13}^\pm] + [r_{12}^\pm, r_{23}^\pm] + [r_{13}^\pm, r_{23}^\pm] = 0$$

These two solutions, which play an important role below, are:

$$r_{12}^+ = \frac{1}{2} \sum_i H_i \otimes H_i + \sum_{\alpha \text{ positive}} \frac{E_\alpha \otimes E_{-\alpha}}{(E_\alpha, E_{-\alpha})} \qquad (12.28)$$

$$r_{12}^- = -\frac{1}{2} \sum_i H_i \otimes H_i - \sum_{\alpha \text{ positive}} \frac{E_{-\alpha} \otimes E_\alpha}{(E_\alpha, E_{-\alpha})} \qquad (12.29)$$

Recall that the wave function $\Psi(x)$ is the solution of the linear system $(\partial_x - A_x)\Psi(x) = 0$, normalized with the boundary condition $\Psi(0) = 1$. As explained in Chapter 3, the Poisson brackets of the matrix elements of $\Psi(x)$ can be computed explicitly and take the following form:

Proposition. *The wave function $\Psi(x)$ satisfies the quadratic bracket relation:*

$$\{\Psi_1(x), \Psi_2(x)\} = [r_{12}^\pm, \Psi_1(x)\Psi_2(x)] \qquad (12.30)$$

We can use either r_{12}^+ or r_{12}^-, since the difference is the tensor Casimir C_{12} which commutes with $\Psi(x) \otimes \Psi(x)$.

With each highest weight vector $|\Lambda^{(r)}\rangle$ we associate the two chiral vectors $\xi^{(r)}$ and $\overline{\xi}^{(r)}$ by eqs. (12.16). The Poisson brackets between these chiral vectors can be computed and obey what is called an exchange algebra. Since we compute the Poisson brackets at equal time, $t = 0$, we do not distinguish the notations $x_+ = x_- = x$.

Proposition. *The exchange algebra of the chiral fields reads:*

$$\{\xi_1^{(r)}(x), \xi_2^{(r')}(y)\} = -\xi_1^{(r)}(x)\xi_2^{(r')}(y)\, r_{12}^\pm \qquad \text{for } x \gtrless y$$
$$\{\overline{\xi}_1^{(r)}(x), \overline{\xi}_2^{(r')}(y)\} = -r_{12}^\mp\, \overline{\xi}_1^{(r)}(x)\overline{\xi}_2^{(r')}(y) \qquad \text{for } x \gtrless y \qquad (12.31)$$
$$\{\xi_1^{(r)}(x), \overline{\xi}_2^{(r')}(y)\} = \xi_1^{(r)}(x) \cdot r_{12}^- \cdot \overline{\xi}_2^{(r')}(y) \qquad \text{for } x \neq y$$
$$\{\overline{\xi}_1^{(r)}(x), \xi_2^{(r')}(y)\} = \xi_2^{(r')}(y) \cdot r_{12}^+ \cdot \overline{\xi}_1^{(r)}(x) \qquad \text{for } x \neq y$$

<u>Proof.</u> We have:

$$\{\xi_1^{(r)}(x), \xi_2^{(r')}(y)\} = \langle\Lambda^{(r)}| \otimes \langle\Lambda^{(r')}|\{e^{-\Phi_1(x)}\Psi_1(x), e^{-\Phi_2(y)}\Psi_2(y)\}$$
$$\langle\Lambda^{(r)}| \otimes \langle\Lambda^{(r')}| \exp[-\Phi_1(x)]\exp[-\Phi_2(y)]$$
$$\cdot \Big[\{\Psi_1(x), \Psi_2(y)\} - \{\Phi_1(x), \Psi_2(y)\}\Psi_1(x) - \{\Psi_1(x), \Phi_2(y)\}\Psi_2(y)\Big]$$

Thus we need to evaluate the Poisson brackets of the wave function at different points and the Poisson brackets between the wave function $\Psi(x)$

and the Toda field $\Phi(y)$. The latter is equal to:

$$\{\Phi_1(x), \Psi_2(y)\} = -\theta(y - x)\, \Psi_2(y)\Psi_2^{-1}(x)\, C_{12}^0\, \Psi_2(x) \qquad (12.32)$$

where

$$C_{12}^0 = \tfrac{1}{2}\sum_i H_i \otimes H_i$$

and $\theta(x - y)$ is the step function. To show it, we take the Poisson bracket of $(\partial_y - A(y))\Psi(y) = 0$ with $\Phi(x)$, getting:

$$[\partial_y - A_2(y)]\{\Phi_1(x), \Psi_2(y)\} = \{\Phi_1(x), A_2(y)\}\Psi_2(y)$$

The right-hand side is known: $\{\Phi_1(x), A_2(y)\} = -C_{12}^0\, \delta(x - y)$. The result follows by integrating this differential equation and imposing the boundary condition expressing the ultralocality of the model $\{\Phi_1(x), \Psi_2(y)\} = 0$ if $x > y$.

Let us now compute the Poisson brackets of the wave function at different points. For $x > y$ we can write $\Psi(x) = \Psi(x, y)\Psi(y)$, where $\Psi(x, y)$ is the transport matrix from y to x. Using the ultralocality property of the Poisson bracket, we have $\{\Psi_1(x), \Psi_2(y)\} = \Psi_1(x, y)\{\Psi_1(y), \Psi_2(y)\}$. Equation (12.30) then yields:

$$\{\Psi_1(x), \Psi_2(y)\} = \Psi_1(x)\Psi_2(y)[-r_{12} + \Psi_1^{-1}(y)\Psi_2^{-1}(y)r_{12}\Psi_1(y)\Psi_2(y)] \qquad (12.33)$$

Combining this with eq. (12.32), we finally evaluate the Poisson brackets between components of the chiral field $\xi(x)$ as:

$$\{\xi_1^{(r)}(x), \xi_2^{(r')}(y)\} = \langle\Lambda^{(r)}| \otimes \langle\Lambda^{(r')}| \exp[-\Phi_1(x)]\exp[-\Phi_2(y)]$$
$$\cdot[-\Psi_1(x)\Psi_2(y)\cdot r_{12} + \Psi_1(x)\Psi_1^{-1}(y)\cdot[r_{12} - C_{12}^0].\Psi_1(y)\Psi_2(y)]$$

In this formula, one can take $r_{12} = r_{12}^\pm$. Choosing $r_{12} = r_{12}^+$, we have $1 \otimes \langle\Lambda^{(r')}|r_{12}^+ = 1 \otimes \langle\Lambda^{(r')}|C_{12}^0$, and we see that the last term vanishes. This proves the first of eqs. (12.31) for $x > y$. The other cases are proved similarly. ∎

12.5 Conformal structure

The solutions of the Toda field theory were parametrized by two chiral fields P and $\overline{P}$. The transformation which relates the original fields to these chiral fields is highly non-local. There exist, however, remarkable quantities which are local in terms of both sets of fields. The purpose of this section is to describe them and to find their Poisson bracket algebra. As we will see, the Virasoro algebra appears as a subalgebra, so that we are in fact dealing with the conformal symmetry algebra of the theory.

We consider only one chirality, and recall the linear systems eq. (12.12) and eq. (12.13):

$$[\partial_{x_-} - P - \mathcal{E}_+]Q_+ = 0 \tag{12.34}$$

$$[\partial_{x_-} + 2\partial_{x_-}\Phi - \mathcal{E}_+]N_-Q_+ = 0 \tag{12.35}$$

These equations have exactly the same structure, but one is expressed in terms of P while the other is expressed in terms of $\partial_{x_-}\Phi$, the derivative of the original field of the theory. The relation between the two equations is a gauge transformation $N_- \in \mathcal{N}_-$.

Notice that the fields $\xi^{(r)}(x)$ are invariant under this gauge transformation because $\langle \Lambda^{(r)}|N_- = \langle \Lambda^{(r)}|$. The vector $\xi^{(r)}(x)$ is just the first row of Q_+ or N_-Q_+.

For finite-dimensional highest weight representations of $\mathcal{G}$, one can deduce from the first order linear systems eq. (12.34) and eq. (12.35) a single differential equation of higher order for the components of vector $\xi^{(r)}(x)$,

$$L \cdot \xi = 0 \tag{12.36}$$

Since the components of $\xi^{(r)}(x)$ are invariant under the gauge transformation N_-, the coefficients of this equation will also be invariant. Moreover, these coefficients are local differential polynomials in terms of P or Φ. These are the local quantities we were looking for.

In the following, we shall restrict ourselves to the $sl(n+1)$ case in the fundamental representation. In this representation, the vector $\xi(x)$ has $(n+1)$ components, $\xi(x) = (\xi_i(x))$, $i = 1, \ldots, (n+1)$ and $\mathcal{E}_+ = \sum_i E_{i,i+1}$, where E_{ij} is the canonical basis of $(n+1) \times (n+1)$ matrices. Finally $P = \sum_i P_i E_{ii}$ with $\sum_i P_i = 0$.

Using eq. (12.34) or eq. (12.35) one can write the operator L in eq. (12.36) explicitly, cf. eq. (12.19):

$$L = \partial_{x_-}^{n+1} - \sum_{i=0}^{n-1} u_i \partial_{x_-}^i$$

with

$$L = (\partial_{x_-} - P_{n+1}) \cdots (\partial_{x_-} - P_1), \quad \sum_i P_i = 0 \tag{12.37}$$

From eq. (12.35) L admits a similar expression but with $P \to -2\partial_{x_-}\Phi$. It is easy to compute u_{n-1}:

$$u_{n-1} = \frac{1}{2}(P, P) + (H_\rho, \partial_{x_-} P) = 2(\partial_{x_-}\Phi, \partial_{x_-}\Phi) - 2(H_\rho, \partial_{x_-}^2\Phi) \tag{12.38}$$

where H_ρ is the element in $\mathcal{H}$ such that $[H_\rho, \mathcal{E}_+] = \mathcal{E}_+$, namely in our case $H_\rho = \mathrm{Diag}(\frac{n}{2}, \frac{n}{2} - 1, \ldots, -\frac{n}{2})$, yielding

$$(H_\rho, \partial_{x_-} P) = n\partial_{x_-} P_1 + (n-1)\partial_{x_-} P_2 + \cdots + \partial_{x_-} P_n$$

Using the Toda equations of motion to eliminate the higher order time derivatives of Φ, we also have $u_{n-1} = \mathcal{H} - \mathcal{P}$, where:

$$\mathcal{H} = \frac{1}{2}(\Pi, \Pi) + \frac{1}{2}(\Phi_x, \Phi_x) + \sum_{\alpha \text{ simple}} e^{2\alpha(\Phi)} - (H_\rho, \Phi_{xx})$$

$$\mathcal{P} = (\Pi, \Phi_x) - (H_\rho, \Pi_x)$$

These are the energy and momentum densities respectively.

The function u_{n-1} in eq. (12.38) is the generalization of the potential u introduced in the Liouville theory. We now compute its Poisson brackets. As before we set $t = 0$ and identify $x_- = x$.

Proposition. *Let* $C_{12}^0 = \frac{1}{2} \sum_i H_i \otimes H_i$. *Then*

$$\{P_1(x), P_2(y)\} = (\partial_x - \partial_y)\delta(x - y)C_{12}^0 \tag{12.39}$$
$$\{u_{n-1}(x), u_{n-1}(y)\} = \left(u_{n-1}\partial_x + \partial_x u_{n-1} - (H_\rho, H_\rho)\partial_x^3\right)\delta(x - y)$$

We recognize the standard Virasoro algebra, see Chapter 11. The value of the central charge is $(H_\rho, H_\rho) = n(n+1)(n+2)/12$.

Proof. We have to compute the Poisson brackets of P. To do this we start from the exchange algebra, and remark that the scalar product $\xi^{(r)}(x)|\Lambda^{(r)}\rangle$ satisfies:

$$\left\{\log\left(\xi^{(r)}(x)|\Lambda^{(r)}\rangle\right), \log\left(\xi^{(r')}(y)|\Lambda^{(r')}\rangle\right)\right\}$$
$$= -\frac{1}{2}\sum_i \Lambda^{(r)}(H_i)\Lambda^{(r')}(H_i)\epsilon(x - y) \tag{12.40}$$

where $\epsilon(x)$ is the sign of x. Note also that eq. (12.34) implies that

$$\xi^{(r)}(x)|\Lambda^{(r)}\rangle = e^{\int^x \Lambda^{(r)}(P)}\theta$$

where θ is some integration constant. Equation (12.39) follows by differentiating eq. (12.40) with respect to x and y and using the independence of the fundamental weights. The Poisson bracket for u_{n-1} then follows readily, either from the expression of u_{n-1} in terms of P or in terms of Φ and Π.   ∎

We now explore the Poisson algebra of the other invariant coefficients u_i in eq. (12.37). Since the quantity u_{n-1} is the generator of the conformal symmetry underlying the theory, we call this Poisson bracket algebra an extended conformal algebra. Comparing with Chapter 10, we know that the Poisson brackets of the u_i coincides with the second Hamiltonian structure of the generalized KdV equation. We are going to rederive this result starting from the exchange algebra eq. (12.31). In the course of this computation we will exhibit the extended conformal properties of the fields ξ_i.

Any component $\xi_i(x)$, $i = 1, \ldots, n+1$, of the vector $\xi(x)$ obeys the differential equation $L \cdot \xi_i = 0$ with L a differential operator of order $n+1$. Once we know the functions $\xi_i(x)$, the operator L can be reconstructed as:

$$L \cdot \xi = \det \begin{pmatrix} \xi & \xi_1 & \cdots & \xi_{n+1} \\ \xi' & \xi_1' & \cdots & \xi_{n+1}' \\ \vdots & \vdots & & \vdots \\ \xi^{(n+1)} & \xi_1^{(n+1)} & \cdots & \xi_{n+1}^{(n+1)} \end{pmatrix} = 0 \qquad (12.41)$$

It follows that the coefficients u_i of L are given by Wronskian type expressions of the ξ_i, and we can directly calculate their Poisson brackets knowing those of the ξ_i. To express the result conveniently we need to recall some notations about pseudo-differential operators, see Chapter 10. We have introduced the derivation symbol $\partial = \partial_x$ (here identified with ∂_{x_-}), with the usual Leibnitz rule $\partial.a = a.\partial + (\partial a)$, where (∂a) means $\partial_x a(x)$, and the integration symbol ∂^{-1} with the following computational rules:

$$\partial^{-1}\partial = \partial\partial^{-1} = 1, \quad \partial^{-i-1}f = \sum_{v=0}^{\infty}(-1)^v \binom{i+v}{v}(\partial^v f)\partial^{-i-1-v} \quad (12.42)$$

The elements $A = \sum_{-\infty}^{N} a_i \partial^i$ form an associative algebra with unit, called the algebra of formal pseudo-differential operators in one variable. It is equipped with a linear form satisfying the fundamental trace property $\langle AB \rangle = \langle BA \rangle$, called the Adler trace, and defined by:

$$\langle A \rangle = \int dx \, a_{-1}$$

With these notations we have:

Proposition. Let $X = \sum_i \partial^{-i-1}X_i$ and let $f_X(L) = \langle LX \rangle$. Then one has:

$$\{f_X(L), \xi_i(x)\} = \left((XL)_+ + \frac{1}{n+1}\partial^{-1}([X, L]_{-1}) \right) \xi_i(x) \quad (12.43)$$

<u>Proof.</u> Using eq. (12.42) we find:

$$(XL)_+ = \sum_{k>i\geq 0} \sum_{s=0}^{k-i-1} (-1)^{k-i-s} \binom{k-1-s}{i} \partial^{k-i-s-1}(X_i u_k) \partial^s$$

and

$$[X,L]_{-1} = -\sum_{k>i\geq 0} (-1)^{k+i} \binom{k}{i} \partial^{k-i}(X_i u_k)$$

So we have to prove that:

$$\{f_X(L), \xi_q(x)\} = -\frac{1}{(n+1)} \sum_{k>i\geq 0} (-1)^{k+i} \binom{k}{i} \partial^{k-i-1}(X_i u_k)\xi_q(x)$$

$$+ \sum_{k>i\geq 0} \sum_{s=0}^{k-i-1} (-1)^{k+i+s} \binom{k-s-1}{i} \partial^{k-i-s-1}(X_i u_k)\partial^s \xi_q \qquad (12.44)$$

From eq. (12.41), the coefficients u_i in the operator L are given by $u_i = (-1)^{i+1} \det M_i$, where M_i is the matrix of elements

$$(M_i)_{ab} = \partial^a \xi_b \qquad \text{with} \qquad \begin{cases} a = 0, 1, ..., n+1, \ a \neq i \\ b = 1, ..., n+1 \end{cases}$$

Let $[\Delta_i(x)]_{kl}$ with $k = 0, \ldots, n+1$, $k \neq i$ and $l = 1, \ldots, n+1$ be the minors of the matrix M_i. By using the Leibnitz rules for the Poisson brackets, we obtain

$$\{f_X(L), \xi_q(x)\} = -\int dz X_i(z)(-1)^i[\Delta_i(z)]_{kl}\partial_z^k\{\xi_l(z), \xi_q(x)\} \quad (12.45)$$

where all repeated indices are summed over. The Poisson brackets of the chiral fields were obtained in eq. (12.31):

$$\{\xi_1(z), \xi_2(x)\} = -\xi_1(z)\xi_2(x)r_{12}^- - \theta(z-x)\xi_1(z)\xi_2(x)(r_{12}^+ - r_{12}^-)$$

We have thus two contributions to eq. (12.45). The contribution of the term independent of $\theta(z-x)$ is

$$\int dz X_i(z)(-1)^i[\Delta_i(z)]_{kl}\partial^k\xi_m(z)\xi_n(x)r_{mn,lq}^-$$

Remembering that $\sum_{k\neq i}[\Delta_i(z)]_{kl}\partial^k\xi_m(z) = (\det M_i)\,\delta_{lm}$, we see that the above expression vanishes since the r-matrices are such that: $\mathrm{Tr}_1 r_{12}^- = \mathrm{Tr}_2 r_{12}^- = \mathrm{Tr}_{12} r_{12}^- = 0$.

Calculating the $\theta(z-x)$ dependent term, we get

$$\{f_X(L), \xi_q(x)\} = \int dz \sum_{a\neq k} \binom{k}{a} X_i(z)(-1)^i[\Delta_i(z)]_{kl}\partial_z^a\xi_m(z)\xi_n(x)$$

$$\cdot(r^+ - r^-)_{mn,lq}\partial_z^{k-a}\theta(z-x)$$

The term $a = k$ in the sum may be excluded, again due to the trace properties of the matrices $r_{12}^\pm$. To evaluate this expression further we remark that $r_{12}^+ - r_{12}^- = C_{12}$, where C_{12} is the Casimir element. To evaluate C_{12} for two vector representations of sl_{n+1} we first compute it in gl_{n+1} according to eq. (12.27). Here the E_α are the E_{ab} for $a < b$, and $E_{-\alpha} = E_{ba}$, while $H_i = E_{ii}$, so we get for C_{12} the permutation operator $P_{12} = \sum E_{ab}\otimes E_{ba}$, or $P_{ij,kl} = \delta_{il}\delta_{jk}$. We restrict this to sl_{n+1} by requiring that the partial traces of C_{12} vanish, yielding $C_{12} = -\frac{1}{n+1}1 + P_{12}$. Using this result, we finally write the Poisson brackets (12.45) as:

$$\{f_X(A), \xi_q(x)\} = \frac{1}{(n+1)}U_q(x) - V_q(x)$$

where $U_q(x)$ comes from the identity factor in C_{12} and $V_q(x)$ from the factor P_{12}.

$$U_q(x) = \int dz X_i(z)(-1)^i[\Delta_i(z)]_{kl}\sum_{a\neq k}\binom{k}{a}\partial_z^a\xi_l(z)\xi_q(x)\partial_z^{k-a-1}\delta(z-x)$$

$$V_q(x) = \int dz X_i(z)(-1)^i[\Delta_i(z)]_{kl}\sum_{a\neq k}\binom{k}{a}\partial_z^a\xi_q(z)\xi_l(x)\partial_z^{k-a-1}\delta(z-x)$$

The $U_q(x)$ term is easy to deal with. Noticing that, for $k \neq i$ and $a \leq n+1$, we have

$$\sum_l[\Delta_i(z)]_{kl}\partial^a\xi_l(z) = (\det M_i)\,\delta_{ka} + (-1)^{k+i+1}(\det M_k)\,\delta_{ia} \quad (12.46)$$

we get

$$U_q(x) = -\sum_{k>i}(-1)^{k+i}\binom{k}{i}\partial^{k-i-1}(X_iu_k)\xi_q(x) \quad (12.47)$$

After integrating over z the expression of $V_q(x)$ becomes

$$V_q(x) = \sum_{a\neq k}(-1)^{i+k+a+1}\binom{k}{a}\partial^{k-a-1}(X_i[\Delta_i]_{kl}\partial^a\xi_q)\xi_l(x)$$

The identity $(\partial^c A)B = \sum_{b=0}^{c}(-1)^b \binom{c}{d} \partial^{c-b}(A\partial^b B)$ allows us to rewrite this as

$$V_q(x) = \sum_{a \neq k} \sum_{b=0}^{k-a-1} (-1)^{k+a+i+b+1} \binom{k}{a} \binom{k-a-1}{b}$$

$$\times \partial^{k-a-b-1}(X_i[\Delta_i]_{kl}\partial^a \xi_q \partial^b \xi_l)$$

Suming over l using eq. (12.46) and performing the derivatives, we obtain

$$V_q(x) = -\sum_{s \geq 0}(-1)^{k+i} \binom{k-1-s}{i}$$

$$\times \left\{ \sum_{a<k}(-1)^a \binom{k}{a} \binom{k-1-a}{k-1-s} \right\} \partial^{k-i-1-s}(X_i u_k)\partial^s \xi_q$$

The sum over a in the curly bracket is equal to $(-1)^s$, so that finally

$$V_q(x) = -\sum_{s \geq 0}(-1)^{k+i+s} \binom{k-s-1}{i} \partial^{k-i-s-1}(X_i u_k)\partial^s \xi_q$$

which ends the proof.   ∎

   It is now very easy to derive the Poisson bracket algebra of the coefficient u_k:

Proposition. *The Poisson brackets of the u_k reproduce the second Poisson bracket structure of the KdV hierarchy:*

$$\{f_X(L), f_Y(L)\} = \langle (LX)_+(LY)_- \rangle - \langle (XL)_+(YL)_- \rangle \qquad (12.48)$$

$$-\frac{1}{n+1}\int dx(\partial^{-1}[X,L]_{-1})[Y,L]_{-1}$$

Proof. We start from $L\xi_i = 0$, so that $\{f_X(L),L\}\xi_i + L\{f_X(L),\xi_i\} = 0$, or equivalently

$$\{f_X(L),L\}\xi_i + L\left((XL)_+ + \frac{1}{n+1}\partial^{-1}[X,L]_{-1}\right)\xi_i = 0$$

Next, using again the differential equation for ξ, we rewrite this equation as:

$$\left\{ \{f_X(L),L\} + \left(L(XL)_+ - (LX)_+L \right. \right. \qquad (12.49)$$

$$\left. \left. + \frac{1}{n+1}(L\partial^{-1}[X,L]_{-1} - \partial^{-1}[X,L]_{-1}L)\right)\right\}\xi_i = 0$$

Because $L(XL)_+ - (LX)_+L = (LX)_-L - L(XL)_-$, the differential operator in the left-hand side of eq. (12.49) is of order at most n. Since it annihilates the $(n+1)$ linearly independent functions ξ_i, it is identically zero. The result then follows by multiplying eq. (12.49) by Y and taking the Adler trace. ∎

Remark. Equation (12.43) describes the conformal properties of the field ξ_i. To extract it, let us apply it with $X = \partial^{-n}X_{n-1}$, so that $f_X(L) = -\int_0^{2\pi} dx\, X_{n-1}(x)u_{n-1}(x)$. We then get

$$\{f_X(L), \xi_i(x)\} = \left(X_{n-1}\,\partial - \Delta^{(1)}\,X'_{n+1}\right)\xi_i(x)$$

with $\Delta^{(1)} = \frac{n}{2} = \Lambda^{(1)}(H_\rho)$, where $\Lambda^{(1)}$ is the highest weight of the vector representation. This shows that $\xi_i(x)$ is a differential of weight $-n/2$. Using $\xi^{(r)} = \xi^{(1)}\wedge\cdots\wedge\partial^{(r-1)}\xi^{(1)}$, we see that $\xi^{(r)}$ is a differential of weight $-\Delta^{(r)} = -\Lambda^{(r)}(H_\rho) = -(r-1)(n-r)/2$. In fact, each $\xi^{(1)}$ has conformal weight $-n/2$ and each ∂ has weight 1.

12.6 Dressing transformations

We now describe the group of dressing transformations in Toda field theories. In Chapter 3, dressing transformations were introduced as special gauge transformations preserving the form of the Lax connection. On the wave function they read (see eq. (3.98) in Chapter 3):

$$\Psi \to \Psi^g = \Theta_+\Psi g_+^{-1} = \Theta_-\Psi g_-^{-1}, \qquad \Theta_-^{-1}\Theta_+ = \Psi g_-^{-1}g_+\Psi^{-1} \quad (12.50)$$

Here g is an element of $G = \exp\mathcal{G}$, and $g_\pm$, $\Theta_\pm$ are obtained by solving a suitable factorization problem in G which we now explain. The last condition in eq. (12.50) ensures that the two gauge transformations by $\Theta_\pm$ produce the same gauge transformed connection. We use this condition to determine $\Theta_\pm$ by a factorization problem in G. The right factors, $g_\pm$, play no role in the transformation of the connection, but are essential for understanding the structure of the dressing group.

Proposition. *Let $g \in G = \exp\mathcal{G}$. Define a factorization problem*

$$g = g_-^{-1}g_+ \tag{12.51}$$

by requiring that $g_\pm \in \exp\mathcal{B}_\pm$, and that the components of g_+ and g_- on $\exp(\mathcal{H})$ are inverse to each other. Then the gauge transformations $\Theta_\pm$ in eq. (12.50) preserve the form of the Toda connection.

<u>Proof.</u> We want the gauge transformed connection $A_{x_\pm}^g = \left(\partial_{x_\pm}\Psi^g\right)\Psi^{g\,-1}$ to be of the form:

$$A_{x_\pm}^g = \pm\partial_{x_\pm}\Phi^g + e^{\mp\mathrm{ad}\,\Phi^g}\mathcal{E}_\mp$$

for some field $\Phi^g \in \mathcal{H}$. If this is the case, the gauge transformation acts on the Toda field Φ. We have:

$$A_\mu^g = \partial_\mu \Theta_\pm \Theta_\pm^{-1} + \Theta_\pm A_\mu \Theta_\pm^{-1} \quad \text{with} \quad A_{x_\pm} = \pm \partial_{x_\pm} \Phi + e^{\mp \text{ad}\,\Phi} \mathcal{E}_\mp$$

First, one has to see that $A_{x_-}^g$ decomposes on elements of height $(0, 1)$ of the Lie algebra $\mathcal{G}$, and A_{x_+} on elements of height $(0, -1)$. The choice $\Theta_\pm \in \exp \mathcal{B}_\pm$ precisely ensures this condition. To show it, we use the fact that one can perform the gauge transformation using either Θ_+ or Θ_-. Using Θ_+ we find that $A_{x_-}^g$ contains only positive heights, while using Θ_- we see that its maximal height is 1, so finally $A_{x_-}^g$ decomposes on elements of height $(0, 1)$. Similarly, $A_{x_+}^g$ decomposes on elements of height $(0, -1)$. Let us write

$$\Theta_+ = e^{\Delta_+} M_+, \qquad \Delta_+ \in \mathcal{H}, \quad M_+ \in \exp \mathcal{N}_+$$
$$\Theta_- = e^{\Delta_-} M_-, \qquad \Delta_- \in \mathcal{H}, \quad M_- \in \exp \mathcal{N}_-$$

Performing the gauge transformation Θ_+ we have:

$$A_{x_-}^g = \partial_{x_-}(e^{\Delta_+} M_+) M_+^{-1} e^{-\Delta_+} + e^{\Delta_+} M_+ \left(-\partial_{x_-} \Phi + e^{\text{ad}\,\Phi} \mathcal{E}_+\right) M_+^{-1} e^{-\Delta_+}$$

Projecting this equation on $\mathcal{H}$, we obtain:

$$\partial_{x_-} \Phi^g = \partial_{x_-} \Phi - \partial_{x_-} \Delta_+$$

Now performing the gauge transformation with Θ_-, we get:

$$A_{x_-}^g = \partial_{x_-}(e^{\Delta_-} M_-) M_-^{-1} e^{-\Delta_-} + e^{\Delta_-} M_- \left(-\partial_{x_-} \Phi + e^{\text{ad}\,\Phi} \mathcal{E}_+\right) M_-^{-1} e^{-\Delta_-}$$

Comparing the terms of height $+1$ we get $e^{\text{ad}\,\Phi^g} \mathcal{E}_+ = e^{\text{ad}\,(\Phi+\Delta_-)} \mathcal{E}_+$, so we must have $\alpha_i(\Phi^g) = \alpha_i(\Phi + \Delta_-)$ for all α_i simple, which implies

$$\Phi^g = \Phi + \Delta_- \tag{12.52}$$

The same analysis performed with A_{x_+} gives

$$\partial_{x_+} \Phi^g = \partial_{x_+} \Phi + \partial_{x_+} \Delta_-$$

and $\Phi^g = \Phi - \Delta_+$. We see that these four equations are compatible if and only if

$$\Delta_- = -\Delta_+$$

i.e if Θ_+ and Θ_- have opposite components on $\exp \mathcal{H}$. ∎

Proposition. *The induced action on the chiral fields $\xi(x_+)$ and $\bar{\xi}(x_-)$ is:*

$$\xi^g(x_+) = \xi(x_+)\, g_-^{-1}, \quad \bar{\xi}^g(x_-) = g_+\, \bar{\xi}(x_-) \tag{12.53}$$

Note that the factors $g_\pm$ of g act separately on the two chiral sectors. The dressed Toda field is:

$$e^{-2\Lambda^{(r)}(\Phi^g(x,t))} = \xi^{(r)}(x_-)\, g_-^{-1} \cdot g_+\, \bar{\xi}^{(r)}(x_+)$$

Proof. We have

$$\begin{aligned}
\xi^g &= \langle\Lambda|(e^{-\Phi^g}\Psi^g) = e^{-\Lambda(\Phi^g)}\langle\Lambda|\,(\Theta_-\Psi)\,g_-^{-1} \\
&= e^{-\Lambda(\Phi^g - \Phi - \Delta_-)}\langle\Lambda|\,(e^{-\Phi}\Psi)\,g_-^{-1} = \xi \cdot g_-^{-1} \tag{12.54}
\end{aligned}$$

In eq. (12.54) we used eq. (12.52) and the properties of the highest weight vector $\langle\Lambda|$. The transformation law for $\bar{\xi}$ is found in a similar way. ■

Proposition. *Let $g = g_-^{-1}g_+$ and $h = h_-^{-1}h_+$ in G. Their multiplication in the dressing group is:*

$$(g_+, g_-) \bullet (h_+, h_-) = (g_+h_+, g_-h_-) \tag{12.55}$$

In particular, the plus and minus components commute.

Proof. This follows from the general statement in Chapter 3 but is also obvious from the action on the chiral fields $\xi, \bar{\xi}$. ■

We denote by G^* the new group equipped with this multiplication law. For infinitesimal transformations, $g \simeq 1 + X$, and $g_\pm \simeq 1 + X_\pm$ with $X = X_+ - X_-$, the composition law corresponds to:

$$[X, Y]_R = [X_+, Y_+] - [X_-, Y_-] \tag{12.56}$$

This defines a new Lie algebra $\mathcal{G}^*$ which is the Lie algebra of G^*. As in Chapter 4, we introduce the operators $R^\pm$ such that $X_\pm = R^\pm X$. Then we have:

$$R^+X = \mathrm{Tr}_2\,(r_{12}^+X_2), \quad R^-X = \mathrm{Tr}_2\,(r_{12}^-X_2) \tag{12.57}$$

where $r_{12}^\pm$ are the matrices in eqs. (12.28, 12.29). This exhibits the deep relation between the group of dressing transformations and the r-matrix formalism.

Using the action of dressing transformations on the chiral fields $\xi(x_+)$ and $\bar{\xi}(x_-)$, it is easy to understand the Poisson–Lie group nature of the group of dressing transformations, see Chapter 14. Indeed, the exchange algebra

$$\{\xi_1(x), \xi_2(y)\} = -\xi_1(x)\xi_2(y)r_{12}^{\pm} \qquad \text{for } x \gtrless y$$

is not invariant by $\xi \to \xi g_-^{-1}$ since $r_{12}^{\pm}$ does not commute with $g_-^{-1} \otimes g_-^{-1}$. However, if we introduce non-trivial Poisson brackets for the g_-^{-1} variables, the covariance of the Poisson brackets for ξ can be recovered. More generally, we have:

Proposition. *Let G^* be equipped with the Poisson brackets:*

$$\{(g_+)_1, (g_+)_2\}_{G^*} = -[r_{12}^{\pm}, (g_+)_1(g_+)_2]$$
$$\{(g_-)_1, (g_-)_2\}_{G^*} = -[r_{12}^{\mp}, (g_-)_1(g_-)_2]$$
$$\{(g_-)_1, (g_+)_2\}_{G^*} = -[r_{12}^{-}, (g_-)_1(g_+)_2]$$
$$\{(g_+)_1, (g_-)_2\}_{G^*} = -[r_{12}^{+}, (g_+)_1(g_-)_2]$$

then the action $\xi \to \xi g_-^{-1} \; \bar{\xi} \to g_+\bar{\xi}$ leaves the exchange algebra (12.31) invariant.

Proof. Consider the action of g_- on ξ: $\xi^g = \xi g_-^{-1}$. Introduce the Poisson brackets on g_-^{-1}:

$$\{(g_-^{-1})_1, (g_-^{-1})_2\} = [-r_{12}^{\pm}, (g_-^{-1})_1(g_-^{-1})_2]$$

We view ξ^g as a function on the product of the group G^* by the phase space M equipped with the product Poisson structure, i.e.

$$\{\xi_1(x), (g_-^{-1})_2\} = 0$$

Then we have:

$$\{\xi_1^g(x), \xi_2^g(y)\}_{G^* \times M} = \{\xi_1(x), \xi_2(y)\}_M (g_-^{-1})_1(g_-^{-1})_2$$
$$+\xi_1(x)\xi_2(y)\{(g_-^{-1})_1, (g_-^{-1})_2\}_{G^*}$$
$$= \xi_1(x)\xi_2(y)\left[r_{12}^{\pm}(g_-^{-1})_1(g_-^{-1})_2 + \{(g_-^{-1})_1, (g_-^{-1})_2\}_{G^*}\right]$$
$$= \xi_1^g(x)\xi_2^g(y)r_{12}^{\pm}$$

The other cases are treated similarly.   ∎

Note that for the factorized element, $g = g_-^{-1}g_+$, the Poisson bracket on G^* reads:

$$\{g_1, g_2\}_{G^*} = -g_1 r_{12}^{+} g_2 - g_2 r_{12}^{-} g_1 + g_1 g_2 r_{12}^{\pm} + r_{12}^{\mp} g_1 g_2$$

Since dressing transformations are not symplectic transformations, they are not generated by scalar Hamiltonians. However, since they are Lie–Poisson actions, there exists a non-Abelian analogue of the moment map.

In other words, there exists a non-Abelian Hamiltonian generating the dressing transformations.

If we consider the system on the finite interval of length ℓ with periodic boundary conditions, then $\Psi(x+\ell) = \Psi(x) \cdot T$, where $T \equiv \Psi(\ell)$ is the monodromy matrix. The Poisson brackets of T are: $\{T_1, T_2\} = [r_{12}^{\pm}, T_1 T_2]$. The non-Abelian Hamiltonian of dressing transformations turns out to be the monodromy matrix T. This means that the variations of the chiral fields under infinitesimal dressing transformations are given by the following Poisson brackets:

Proposition. *The monodromy matrix generates dressing transformations:*

$$\delta_X \xi(x) = -\xi(x) X_- = \mathrm{Tr}_2 \left(X_2 T_2^{-1} \{\xi_1(x), T_2\} \right)$$
$$\delta_X \bar{\xi}(x) = X_+ \bar{\xi}(x) = \mathrm{Tr}_2 \left(X_2 T_2^{-1} \{\bar{\xi}_1(x), T_2\} \right) \qquad (12.58)$$

<u>Proof</u>. We have to compute the Poisson bracket of the monodromy matrix and the chiral fields. Using eqs. (12.32, 12.33) we have:

$$\{T_1, e^{-\Phi_2(x)} \Psi_2(x)\} = -T_1 e^{-\Phi_2(x)} \Psi_2(x) r_{12}^{\pm}$$
$$+ T_1 \Psi_1^{-1}(x) e^{-\Phi_2(x)} (r_{12}^{\pm} - C_{12}^0) \Psi_1(x) \Psi_2(x)$$

Projecting on $1 \otimes \langle \Lambda^{(r)} |$ and choosing the plus sign in $r_{12}^{\pm}$, we see that the second term cancels, yielding $\{T_1, \xi_2(x)\} = -T_1 \xi_2 r_{12}^+$, or equivalently:

$$\{\xi_1(x), T_2\} = -\xi_1(x) T_2 \, r_{12}^-$$

The first part of the proposition then follows from eq. (12.57). The second part is proved similarly. $\blacksquare$

We see that Toda field theories provide a particularly simple example of the action of dressing transformations and non-Abelian Hamiltonians.

12.7 The affine sinh-Gordon model

We now show that the sinh-Gordon theory belongs to the general class of Toda field theories. It is associated with the affine Lie algebra $\widehat{sl_2}$. The usual formula for the Lax connection is written in the loop representation of this algebra, with vanishing central charge. In the following we will consider non-zero central charge. The benefit will be that all the algebraic structure of Toda theories then applies to the sinh-Gordon theory as well. In particular, the existence of highest weight representations allows us to disentangle the action of the group of dressing transformations. As we will see, this leads directly to the N-soliton formula in terms of vertex operators.

The affine algebra $\widehat{sl_2}$ admits the Cartan decomposition: $\widehat{sl_2} = \widehat{\mathcal{N}}_- \oplus \widehat{\mathcal{H}} \oplus \widehat{\mathcal{N}}_+$. In the principal gradation, a basis of $\widehat{sl_2}$ is given by (see Chapter 16):

$$\widehat{\mathcal{H}} = \{H, d, K\}$$
$$\widehat{\mathcal{N}}_+ = \{E_+^{(2n-1)} = \lambda^{2n-1}E_+,\ E_-^{(2n-1)} = \lambda^{2n-1}E_-,\ H^{(2n)} = \lambda^{2n}H, n > 0\}$$
$$\widehat{\mathcal{N}}_- = \{E_+^{(2n+1)} = \lambda^{2n+1}E_+,\ E_-^{(2n+1)} = \lambda^{2n+1}E_-,\ H^{(2n)} = \lambda^{2n}H, n < 0\}$$

$$(12.59)$$

In particular, the simple root vectors can be taken as $E_{\pm\alpha_1} = \lambda^{\pm 1}E_\pm$ and $E_{\pm\alpha_2} = \lambda^{\pm 1}E_\mp$. The commutation relations are:

$$\left[H^{(r)}, H^{(s)}\right] = Kr\,\delta_{r+s,0}$$
$$\left[H^{(r)}, E_\pm^{(s)}\right] = \pm 2E_\pm^{(r+s)}$$
$$\left[E_+^{(r)}, E_-^{(s)}\right] = H^{(r+s)} + \frac{K}{2}r\delta_{r+s,0}$$

Following the general construction of Toda field theories, we define

$$\mathcal{E}_\pm = E_{\pm\alpha_1} + E_{\pm\alpha_2} = \lambda^{\pm 1}(E_+ + E_-)$$

The connection $A_{x_\pm}$ is given by

$$A_{x_\pm} = \pm\partial_{x_\pm}\Phi + me^{\mp\mathrm{ad}\Phi}\mathcal{E}_\mp \qquad (12.60)$$

where the field Φ takes values in the Cartan subalgebra $\widehat{\mathcal{H}}$ of $\widehat{sl_2}$. Let us decompose it on the generators H, d and K of $\widehat{\mathcal{H}}$:

$$\Phi = \frac{1}{2}H\,\varphi + d\,\eta + \frac{1}{4}K\,\zeta$$

The zero curvature condition, $\partial_{x_+}A_{x_-} - \partial_{x_-}A_{x_+} - [A_{x_+}, A_{x_-}] = 0$, can be worked out using only the Lie algebra structure of $\widehat{sl_2}$. It gives

$$\partial_{x_+}\partial_{x_-}\varphi = m^2 e^{2\eta}(e^{2\varphi} - e^{-2\varphi}) \qquad (12.61)$$
$$\partial_{x_+}\partial_{x_-}\eta = 0 \qquad (12.62)$$
$$\partial_{x_+}\partial_{x_-}\zeta = m^2 e^{2\eta}(e^{2\varphi} + e^{-2\varphi}) \qquad (12.63)$$

Thanks to the field η, the above equations are conformally invariant. This is in contrast with the sinh-Gordon equation which is not conformally invariant. In fact, performing a change of coordinates $x_+ = f(x'_+)$ and $x_- = g(x'_-)$, the equations are invariant if we redefine the fields by:

$$\varphi'(x'_+, x'_-) = \varphi(f(x'_+), g(x'_-))$$
$$\zeta'(x'_+, x'_-) = \zeta(f(x'_+), g(x'_-))$$
$$\eta'(x'_+, x'_-) = \eta(f(x'_+), g(x'_-)) + \log(\partial f \partial g)$$

There are two real forms of eqs. (12.61–12.63). One is when φ, ζ and η are all real, and the other one is when φ is pure imaginary and ζ, η are real. These two forms correspond to the sinh-Gordon and sine-Gordon case respectively (when one sets $\eta = 0$, which can be done consistently and decouples the equation for φ).

In a loop representation, $K = 0$, the field ζ and its equation of motion, eq. (12.63), disappear. Setting $\eta = 0$, we are left with the standard sinh-Gordon equation. Let us write the Lax connection in this case:

$$A_x = \begin{pmatrix} \frac{1}{2}\partial_t \varphi & m(\lambda e^\varphi + \lambda^{-1} e^{-\varphi}) \\ m(\lambda e^{-\varphi} + \lambda^{-1} e^\varphi) & -\frac{1}{2}\partial_t \varphi \end{pmatrix} \tag{12.64}$$

$$A_t = \begin{pmatrix} \frac{1}{2}\partial_x \varphi & -m(\lambda e^\varphi - \lambda^{-1} e^{-\varphi}) \\ -m(\lambda e^{-\varphi} - \lambda^{-1} e^\varphi) & -\frac{1}{2}\partial_x \varphi \end{pmatrix} \tag{12.65}$$

The benefit of having extended the loop algebra to the full affine algebra is that we have now at our disposal highest weight representations and the general structures of Toda field theories can be applied straightforwardly, provided that the action of "group" elements on the highest weight vector is defined. In the following we shall restrict ourselves to integrable highest weights and work freely with formal expressions. Final formulae will be evaluated explicitly in the level 1 representations of $\widehat{sl_2}$, where they will be seen to make sense.

As usual with Toda field theories, with a highest weight vector $|\Lambda\rangle$ one associates two sets of fields $\xi(x, t)$ and $\bar{\xi}(x, t)$ defined by :

$$\xi(x, t) = \langle \Lambda | e^{-\Phi} \Psi(x, t), \quad \bar{\xi}(x, t) = \Psi^{-1}(x, t) e^{-\Phi} |\Lambda\rangle \tag{12.66}$$

These fields are chiral: $\partial_{x_+} \xi = 0$ and $\partial_{x_-} \bar{\xi} = 0$. For any highest weight Λ we can reconstruct $\Lambda(\Phi)$ by the formula:

$$\exp(-2\Lambda(\Phi)) = \xi(x_-) \cdot \bar{\xi}(x_+)$$

The affine $\widehat{sl_2}$ algebra has two fundamental highest weights, which we shall denote by Λ^- and Λ^+, see Chapter 16. They are characterized by $\Lambda^\pm(H) = \pm\frac{1}{2}$, $\Lambda^\pm(K) = 1$ and $\Lambda^\pm(d) = 0$, so that we have $\Lambda^\pm(\Phi) = \frac{1}{4}(\pm\varphi + \zeta)$. This is enough to reconstruct the fields φ and ζ. The field

η cannot be obtained by these highest weight projections. This is not a problem if we are interested in the sinh-Gordon model, since then the free field η is equal to 0. Defining the tau-functions:

$$\tau_\pm = \exp\left(-2\Lambda^\pm(\Phi)\right)$$

we have:

$$e^{-\varphi} = \frac{\tau_+}{\tau_-} \quad \text{and} \quad e^{-\zeta} = \tau_+\tau_- \tag{12.67}$$

In terms of the tau-functions, the equations of motion take the form:

$$\tau_\pm(\partial_{x_-}\partial_{x_+}\tau_\pm) - (\partial_{x_+}\tau_\pm)(\partial_{x_-}\tau_\pm) = -m^2\, e^{2\eta}\, \tau_\mp^2 \tag{12.68}$$

When $\eta = 0$, this is just the Hirota bilinear form of the sinh-Gordon equation.

We now describe the chiral fields $\xi(x_-)$ and $\bar\xi(x_+)$ for the simplest solution, the vacuum solution of eq. (12.61):

$$\varphi_{\text{vac}} = 0, \quad \eta_{\text{vac}} = 0, \quad \zeta_{\text{vac}} = 2m^2 x_+ x_-$$

One can insert this solution into the linear system and compute the vacuum wave function $\Psi_{\text{vac}}(x,t)$. We have $\Phi_{\text{vac}} = \frac{1}{2}m^2 x_+ x_- K$ and the linear system becomes:

$$\left(\partial_{x_+} - \frac{1}{2}m^2 x_- K - m\mathcal{E}_-\right)\Psi = 0, \quad \left(\partial_{x_-} + \frac{1}{2}m^2 x_+ K - m\mathcal{E}_+\right)\Psi = 0$$

The first equation is readily solved by $\Psi = e^{\frac{1}{2}m^2 x_+ x_- K}e^{mx_+\mathcal{E}_-}\tilde\Psi(x_-)$. Inserting this into the second equation, and using $[\mathcal{E}_+,\mathcal{E}_-] = K$, which implies $\exp(-mx_+\,\text{ad}\mathcal{E}_-)\mathcal{E}_+ = \mathcal{E}_+ + mx_+ K$, one gets $(\partial_{x_-} - m\mathcal{E}_+)\tilde\Psi = 0$. We finally obtain:

$$\Psi_{\text{vac}}(x,t) = e^{\frac{m^2}{2}x_+ x_- K}e^{mx_+\mathcal{E}_-}e^{mx_-\mathcal{E}_+} = e^{-\frac{m^2}{2}x_+ x_- K}e^{mx_-\mathcal{E}_+}e^{mx_+\mathcal{E}_-} \tag{12.69}$$

The two expressions are equal thanks to the Campbell–Haussdorf formula. One can then compute the chiral fields:

$$\xi_{\text{vac}}(x_-) = \langle\Lambda|e^{mx_-\mathcal{E}_+}, \quad \bar\xi_{\text{vac}}(x_+) = e^{-mx_+\mathcal{E}_-}|\Lambda\rangle \tag{12.70}$$

The reconstruction formula for Φ reads

$$\exp(-2\Lambda(\Phi)) = \xi_{\text{vac}}(x_-)\cdot\bar\xi_{\text{vac}}(x_+) = \exp\left(-m^2 x_+ x_-\Lambda(K)\right)$$

as it should be. The tau-functions of the vacuum solution are:

$$\tau_+ = \tau_- = \tau_0 = \exp[-m^2 x_+ x_-] \tag{12.71}$$

Remark. We can embed the vacuum equations of motion $\partial_{x_+}\partial_{x_-}\zeta_{\text{vac}} = 2m^2$ into a larger hierarchy. Introduce the variables $x_\pm^{(r)}$ for r odd, and consider the following connection, generalizing eqs. (12.60) evaluated at Φ_{vac},

$$A_{x_\pm^{(r)}}^{\text{vac}} = \pm\frac{1}{4}\partial_{x_\pm^{(r)}}\zeta\, K + m\mathcal{E}_\mp^{(r)}$$

where $\mathcal{E}_\pm^{(r)} = \lambda^{\pm r}(E_+ + E_-)$, with r odd. Since $[\mathcal{E}_+^{(r)}, \mathcal{E}_-^{(s)}] = Kr\delta_{rs}$, the zero curvature condition reduces to $\partial_{x_+^{(r)}}\partial_{x_-^{(s)}}\zeta_{\text{vac}} = 2m^2 r\delta_{rs}$. A solution is $\zeta_{\text{vac}} = 2m^2\sum_r r\, x_+^{(r)}x_-^{(r)}$. In the same way as before, we can calculate $\Psi_{\text{vac}}(x_+^{(r)}, x_-^{(r)})$

$$\begin{aligned}
\Psi_{\text{vac}}(x_+^{(r)}, x_-^{(r)}) &= e^{-\frac{m^2}{2}\left(\sum_r r x_+^{(r)} x_-^{(r)}\right) K}\, e^{m\sum_r x_-^{(r)}\mathcal{E}_+^{(r)}}\, e^{m\sum_r x_+^{(r)}\mathcal{E}_-^{(r)}} \\
&= e^{\frac{m^2}{2}\left(\sum_r r x_+^{(r)} x_-^{(r)}\right) K}\, e^{m\sum_r x_+^{(r)}\mathcal{E}_-^{(r)}}\, e^{m\sum_r x_-^{(r)}\mathcal{E}_+^{(r)}}
\end{aligned}$$

The vacuum chiral fields are thus:

$$\xi_{\text{vac}}(x_-^{(r)}) = \langle\Lambda| e^{m\sum_r x_-^{(r)}\mathcal{E}_+^{(r)}}, \quad \bar{\xi}_{\text{vac}}(x_+^{(r)}) = e^{-m\sum_r x_+^{(r)}\mathcal{E}_-^{(r)}}|\Lambda\rangle \qquad (12.72)$$

The $x_\pm^{(r)}$ are the collection of elementary times defining the sinh-Gordon hierarchy, see Chapter 3. Each chirality is attached separately to the poles at $\lambda = 0, \infty$ in the loop representation.

12.8 Dressing transformations and soliton solutions

As shown in eq. (12.53), dressing transformation act on the chiral fields ξ and $\bar{\xi}$ by:

$$\xi^g(x_-) = \xi(x_-) \cdot g_-^{-1}, \quad \bar{\xi}^g(x_+) = g_+ \cdot \bar{\xi}(x_+)$$

In our case, we start from an element $g \in \exp \widehat{sl_2}$. The dressing group element (g_+, g_-) is defined by the factorization problem

$$g = g_-^{-1}\, g_+ \quad \text{with} \quad g_\pm \in B_\pm = (\exp\widehat{\mathcal{H}})(\exp\widehat{\mathcal{N}}_\pm)$$

and, moreover, we require that g_- and g_+ have inverse components on the Cartan torus.

When we dress the vacuum solution with an element $g = g_-^{-1}g_+$, we get a new solution Φ^g such that:

$$e^{-2\Lambda(\Phi^g)} = \langle\Lambda| e^{mx_-\mathcal{E}_+}\, g\, e^{-mx_+\mathcal{E}_-}|\Lambda\rangle$$

To construct new solutions of the sinh-Gordon equation, one has to choose particular elements g of the affine group.

Remarkable elements can be constructed with vertex operators in the two level one representations of the affine $\widehat{sl_2}$ algebra. Let us recall here the main facts, and refer to Chapter 16 for more details. One introduces bosonic oscillators p_n for n odd, such that $[p_m, p_n] = m\delta_{n+m,0}$ and $p_n^\dagger = p_{-n}$. They generate a Fock space over the vacuum state $|0\rangle$ which is specified by $p_n|0\rangle = 0$ for $n > 0$. Let $Z(\lambda)$ be the generating function:

$$Z(\lambda) = -i\sqrt{2} \sum_{n \text{ odd}} p_{-n} \frac{\lambda^n}{n}$$

The level one vertex operator representations with highest weight $\Lambda_\pm$ of the Lie algebra $\widehat{sl_2}$ are then obtained by setting:

$$\sum_{n \text{ odd}} \lambda^{-n}(E_+^n + E_-^n) = \frac{i}{\sqrt{2}}\lambda\frac{d}{d\lambda}Z(\lambda)$$

$$\sum_{n \text{ even}} \lambda^{-n} H^n + \sum_{n \text{ odd}} \lambda^{-n}(E_+^n - E_-^n) = \pm V(\lambda) \qquad (12.73)$$

where $V(\lambda)$ denotes the vertex operator:

$$V(\lambda) = \tfrac{1}{2} : e^{-i\sqrt{2}Z(\lambda)} := \tfrac{1}{2} : e^{i\sqrt{2}Z(-\lambda)} :$$

The double dots means that the expressions are normal ordered, i.e. we write all p_n with $n > 0$ to the right. The representations with highest weight $\Lambda^\pm$ correspond to the plus and minus sign respectively, in eq. (12.73).

Recall the formula, (see eq. (16.25)):

$$V(\lambda)V(\mu) = \left(\frac{\lambda - \mu}{\lambda + \mu}\right)^2 : V(\lambda)V(\mu) : \qquad |\lambda| > |\mu|$$

This means that inside expectation values $V^2(\mu) = 0$. So there is an element $g \in \exp \widehat{sl_2}$ such that $\rho_{\Lambda^+}(g) = \rho_{\Lambda^+}(\exp aV(\mu)) = 1 + aV(\mu)$. In the representation ρ_{Λ^-} we have $\rho_{\Lambda^-}(g) = 1 - aV(\mu)$ due to the sign change in eq. (12.73). More generally, we consider the product of such elements $g = g_1 g_2 \cdots g_N$. We have:

$$\rho_{\Lambda^\pm}(g) = (1 \pm 2a_1V(\mu_1))(1 \pm 2a_2V(\mu_2)) \cdots (1 \pm 2a_NV(\mu_N)) \qquad (12.74)$$

Proposition. *Let $\tau_\pm^{(N)}$ be the tau-functions:*

$$\tau_\pm^{(N)}(x_\pm) = \langle \Lambda^\pm | e^{mx_- \mathcal{E}_+} \prod_{i=1}^N (1 \pm 2a_iV(\mu_i)) e^{-mx_+ \mathcal{E}_-} |\Lambda^\pm\rangle$$

Then we have:

$$\frac{\tau_{\pm}^{(N)}}{\tau_0} = 1 + \sum_{p=1}^{N} (\pm)^p \sum_{k_1 < k_2 < \cdots < k_p} X_{k_1} \cdots X_{k_p} \prod_{k_i < k_j} \left(\frac{\mu_{k_i} - \mu_{k_j}}{\mu_{k_i} + \mu_{k_j}} \right)^2 \quad (12.75)$$

with

$$X_i = a_i \exp\left(-2m(\mu_i x_- + \mu_i^{-1} x_+)\right)$$

and $\tau_0 = \exp\left(-m^2 x_+ x_-\right)$. *The parameters* μ_i *are interpreted as the rapidities of the solitons and* a_i *are related to their positions.*

<u>Proof.</u> The commutation relations between $\mathcal{E}_{\pm}$ and $V(\lambda)$ are

$$[\mathcal{E}_{\pm}, V(\lambda)] = -2\lambda^{\pm 1} V(\lambda)$$

see eq. (16.29) in Chapter 16. They imply that

$$e^{mx_{\pm} \mathcal{E}_{\mp}} V(\mu_i) e^{-mx_{\pm} \mathcal{E}_{\mp}} = e^{-2m\mu_i^{\mp 1} x_{\pm}} V(\mu_i)$$

Commuting $e^{mx_- \mathcal{E}_+}$ to the right, and $e^{-mx_+ \mathcal{E}_-}$ to the left amounts to the replacement $a_i \rightarrow X_i$ and, moreover, produces the factor τ_0 since $[\mathcal{E}_+, \mathcal{E}_-] = K$, so that we get:

$$\tau_{\pm}^{(N)} = \tau_0 \langle \Lambda^{\pm} | \prod_{i=1}^{N} (1 + 2X_i V(\mu_i)) | \Lambda^{\pm} \rangle$$

The explicit formula (12.75) then follows from Wick's theorem applied to the vertex operators $V(\mu_i)$.   ∎

Remark. One can incorporate the whole hierarchy of times, as in eq. (12.72), to this expression. Using the commutation relations $[E_+^n + E_-^n, V(\lambda)] = -2\lambda^n V(\lambda)$, X_i is replaced by

$$X_i = a_i \exp\left(-2m \sum_s (\mu_i^s x_-^{(s)} + \mu_i^{-s} x_+^{(s)})\right)$$

and τ_0 by $\exp\left(-m^2 \sum_s s x_+^{(s)} x_-^{(s)}\right)$.

The N-soliton tau-functions can also be written as the determinant of an $N \times N$ matrix:

$$\frac{\tau_{\pm}^{(N)}}{\tau_0} = \det(1 \pm \mathcal{V}) \quad \text{with} \quad \mathcal{V}_{ij} = 2 \frac{\sqrt{\mu_i \mu_j}}{\mu_i + \mu_j} \sqrt{X_i X_j} \quad (12.76)$$

This is proved by using the Cauchy identity:

$$\det\left(2 \frac{\sqrt{\mu_i \mu_j}}{\mu_i + \mu_j}\right) = \prod_{i<j} \left(\frac{\mu_i - \mu_j}{\mu_i + \mu_j}\right)^2$$

and the expansion formula eq. (9.47) in Chapter 9.

For the sine-Gordon equation, the parameters a_i, μ_i in eq. (12.74) satisfy specific reality conditions. Replacing φ by $i\varphi$ in eq. (12.67), the one-soliton solution reads

$$e^{-i\varphi} = \frac{\tau_+}{\tau_-} = \frac{1+X}{1-X}, \qquad X = ae^{-2m((\mu+\mu^{-1})x-(\mu-\mu^{-1})t)}$$

For this to be of modulus 1, one takes $a = i\epsilon e^{\gamma}$ with γ real and $\epsilon = \pm 1$. At fixed t, when x goes from $-\infty$ to $+\infty$, φ goes from $-\epsilon\pi$ to 0. The topological charge is defined as

$$Q = \frac{1}{\pi}(\varphi(+\infty) - \varphi(-\infty))$$

Solitons, with charge $+1$, correspond to $\epsilon = +1$ and antisolitons, with charge -1, to $\epsilon = -1$.

The "breathers" correspond to pairs of complex conjugated rapidities $(\mu_1 = \mu, \mu_2 = \bar{\mu})$ and positions $(X_1 = X, X_2 = -\bar{X})$. In this case we have:

$$e^{-i\varphi} = \frac{1 + X_1 + X_2 + \left(\frac{\mu_1-\mu_2}{\mu_1+\mu_2}\right)^2 X_1 X_2}{1 - X_1 - X_2 + \left(\frac{\mu_1-\mu_2}{\mu_1+\mu_2}\right)^2 X_1 X_2} = \frac{1 + 2i\,\mathrm{Im}\,X + \left(\frac{\mathrm{Im}\,\mu}{\mathrm{Re}\,\mu}\right)^2 |X|^2}{1 - 2i\,\mathrm{Im}\,X + \left(\frac{\mathrm{Im}\,\mu}{\mathrm{Re}\,\mu}\right)^2 |X|^2}$$

This shows that $\varphi(x = -\infty) = \varphi(+\infty) = 0$, and the topological charge is 0. Moreover, if we perform a Lorentz boost to go to the rest frame where $|\mu| = 1$, we see that φ is periodic in time, hence the denomination of this solution.

12.9 N-soliton dynamics

The space of N-soliton solutions forms a finite-dimensional submanifold in the infinite-dimensional phase space of sine-Gordon, on which the symplectic form of sine-Gordon can be restricted. This yields a finite-dimensional symplectic manifold. The soliton solutions then become a relativistically invariant N-body problem first introduced by Ruijsenaars, which is a relativistic generalization of the Calogero model.

Consider the *sine*–Gordon model with equation of motion:

$$\partial_{x_+}\partial_{x_-}\varphi = 2m^2 \sin(2\varphi) \tag{12.77}$$

The natural symplectic form is the canonical one:

$$\Omega_{SG} = \int_{-\infty}^{+\infty} dx\, \delta\pi(x) \wedge \delta\varphi(x) \tag{12.78}$$

where $\pi(x) = \partial_t \varphi(x)$ is the momentum conjugate to the field $\varphi(x)$, and δ denotes the differential on the phase space. The N-soliton solution is given by $e^{-i\varphi} = \frac{\tau_+}{\tau_-}$, where $\tau_\pm$ are given by eq. (12.75). These solutions depend on $2N$ parameters a_i, μ_i which provide coordinates on the N-soliton submanifold. We wish to compute the restriction of the symplectic form eq. (12.78) on this $2N$-dimensional manifold. We present the proofs in the *soliton* case. This means that the μ_i are real, and the a_i are pure imaginary. The analysis for the antisolitons is the same and for breathers some technicalities must be added. We now express the restriction of Ω_{SG} on the N-soliton submanifold.

Proposition. *In the coordinates a_i and μ_i, the restriction of Ω_{SG} is given by:*

$$\omega = 2 \sum_{i=1}^N \frac{\delta a_i}{a_i} \wedge \frac{\delta \mu_i}{\mu_i} + 2 \sum_{i<j} \left(\frac{4\mu_i \mu_j}{\mu_i^2 - \mu_j^2} \right) \frac{\delta \mu_i}{\mu_i} \wedge \frac{\delta \mu_j}{\mu_j} \qquad (12.79)$$

<u>Proof.</u> Consider first the one-soliton solution. In this case, the computation can be done directly using the formula $e^{-i\varphi} = (1 + X)/(1 - X)$, which gives $\pi = 2i\dot{X}/(1 - X^2)$. One finds:

$$\Omega_{SG}\Big|_{\text{restricted}} = \left[-2 \int_{-\infty}^{+\infty} dx \partial_x \left(\frac{1}{1 - X^2} \right) \right] \cdot \frac{\delta \mu}{\mu} \wedge \frac{\delta a}{a} \qquad (12.80)$$

Since X is pure imaginary, and $X(+\infty) = 0$ and $X(-\infty) = \infty$, the integral evaluates to the numerical constant -2. So the formula is proved for one soliton. Consider next the general case with an arbitrary number of solitons with parameters (μ_i, a_i). Since the symplectic form can be computed at any time, we evaluate it at $t \to \pm\infty$. In these limits the sine-Gordon field becomes asymptotically equal to the sum of one-soliton solutions with parameters $(\mu_i^{\text{in}}, a_i^{\text{in}})$ and $(\mu_i^{\text{out}}, a_i^{\text{out}})$:

$$\mu_i^{\text{in}} = \mu_i^{\text{out}} = \mu_i$$

$$a_i^{\text{out}} \overset{\text{in}}{=} a_i \prod_{|\mu_j| \lessgtr |\mu_i|} \left(\frac{\mu_i - \mu_j}{\mu_i + \mu_j} \right)^2$$

These formulae are obtained as follows. Let us look at the quantities $X_i = a_i \exp(-2m(\mu_i x_- + \mu_i^{-1} x_+))$ entering the definition of the N-soliton solution. When t is large, X_i is of order 1 when x is located at $x \sim x_i = \frac{\mu_i - \mu_i^{-1}}{\mu_i + \mu_i^{-1}} t$. The values of the other X_j when $x \sim x_i$ are $X_j \sim a_j e^{4m(\mu_j^2 - \mu_i^2)t/\mu_i \mu_j(\mu_i + \mu_i^{-1})}$. Hence for large positive time, if $\mu_j^2 > \mu_i^2$,

$X_j(x_i, t)$ is very large, while if $\mu_j^2 < \mu_i^2$, $X_j(x_i, t)$ is very small. So we can split the N solitons into two subsets. One, $\mathcal{I}_+$, is such that $\mu_j^2 > \mu_i^2$, and the other one, $\mathcal{I}_-$, is such that $\mu_j^2 < \mu_i^2$. We can evaluate the tau-functions, eq. (12.76), when $x \sim x_i$. One has to keep the terms containing the maximum number of X_j, $j \in \mathcal{I}_+$. These terms read:

$$\tau_\pm(x_i, t) \sim \prod_{j \in \mathcal{I}_+} X_j \prod_{k_j < k_i} \left(\frac{\mu_{k_i} - \mu_{k_j}}{\mu_{k_i} + \mu_{k_j}} \right)^2 \left[1 \pm X_i \prod_{j \in \mathcal{I}_+} \left(\frac{\mu_i - \mu_j}{\mu_i + \mu_j} \right)^2 \right]$$

where we need to keep the second term because $X_i \sim 1$. So we have

$$e^{-i\varphi} = \frac{\tau_+}{\tau_-} = \frac{1 + X_i \prod_{j \in \mathcal{I}_+} \left(\frac{\mu_i - \mu_j}{\mu_i + \mu_j} \right)^2}{1 - X_i \prod_{j \in \mathcal{I}_+} \left(\frac{\mu_i - \mu_j}{\mu_i + \mu_j} \right)^2}$$

which is indeed a one-soliton formula, but with modified parameter a^{out}.

Since asymptotically the solitons decouple, in the symplectic form the crossed terms vanish (the overlap integrals are zero). Therefore the symplectic form reduces to the sum of the one-soliton expressions, but with the modified in and out parameters:

$$\omega = 2 \sum_{i=1}^{N} \frac{\delta a_i^{\text{in}}}{a_i^{\text{in}}} \wedge \frac{\delta \mu_i}{\mu_i} = 2 \sum_{i=1}^{N} \frac{\delta a_i^{\text{out}}}{a_i^{\text{out}}} \wedge \frac{\delta \mu_i}{\mu_i}$$

Both forms are equal to eq. (12.79), as can be checked easily.   ∎

As a byproduct, we see that the transformation from the in-variables $(\mu_i^{\text{in}}, a_i^{\text{in}})$ to the out-variables $(\mu_i^{\text{out}}, a_i^{\text{out}})$ is symplectic, as it should be.

The 2-form (12.79) is non-degenerate. It therefore defines a symplectic structure on the restricted phase space. The corresponding Poisson brackets are found by inverting the symplectic form (12.79):

$$\{\mu_i, \mu_j\} = 0, \quad \{\mu_i, a_j\} = -\frac{1}{2} \mu_i a_j \delta_{ij}, \quad \{a_i, a_j\} = -\frac{1}{2} \left(\frac{4\mu_i \mu_j}{\mu_i^2 - \mu_j^2} \right) a_i a_j$$

$$\tag{12.81}$$

Knowing the Poisson brackets between the parameters (μ_i, a_i), we can compute the Poisson brackets between the elements of the matrix $\mathcal{V}$ in eq. (12.76). Remarkably, they can be written in the r-matrix form:

$$\{\mathcal{V}_1, \mathcal{V}_2\} = [r_{12}, \mathcal{V}_1] - [r_{21}, \mathcal{V}_2] \tag{12.82}$$

where $r_{12} = \sum_{ij;kl} r_{ij;kl} E_{ij} \otimes E_{kl}$ and

$$r_{ij;kl} = -\frac{1}{16} \left(\frac{\mu_i + \mu_j}{\mu_i - \mu_j} \right) (\mathcal{V}_{jk}\delta_{il} + \mathcal{V}_{jl}\delta_{ik} + \mathcal{V}_{ik}\delta_{jl} + \mathcal{V}_{il}\delta_{jk})$$

Equation (12.82) can be checked by straightforward computation.

From the definition of the sine-Gordon field in terms of the tau-functions, we see that it can be expressed in terms of only the eigenvalues Q_i of the matrix $\mathcal{V}$:

$$e^{-i\varphi} = \frac{\det(1 + \mathcal{V})}{\det(1 - \mathcal{V})} = \prod_{i=1}^{N} \left(\frac{1 + Q_i}{1 - Q_i} \right) \tag{12.83}$$

Recall that, in the N-soliton case, we can assume $a_j = ie^{\gamma_j}$ and $\mu_j > 0$ so that the matrix $\mathcal{V}$ is of the form $i\tilde{\mathcal{V}}$, where $\tilde{\mathcal{V}}$ is a real symmetric matrix. The matrix $\tilde{\mathcal{V}}$ can be diagonalized with a real orthogonal matrix U. Moreover, its eigenvalues $\tilde{Q}_j$ are real positive. To show the positivity assertion, we show that $\tilde{\mathcal{V}}$ is positive. For any real vector y we have

$$(y\tilde{\mathcal{V}}y) = 2 \sum_{ij} \frac{x_i x_j}{\mu_i + \mu_j}$$

with $x_i = \sqrt{\mu_i e^{\gamma_i}} e^{-m(\mu_i x_- + \mu_i^{-1} x_+)} y_i$. But since all μ_i are assumed positive, we have for real x_i:

$$\sum_{ij} \frac{x_i x_j}{\mu_i + \mu_j} = \int_0^\infty dz \sum_{ij} e^{-z(\mu_i+\mu_j)} x_i x_j = \int_0^\infty dz (\sum_j e^{-z\mu_j} x_j)^2 \geq 0$$

So the eigenvalues Q_j of $\mathcal{V}$ are of the form $i\times$real positive number, or $Q_j = ie^{q_j}$. Finally, note that eq. (12.82) implies $\{Q_i, Q_j\} = 0$.

We want to express the evolution of solitons in terms of the variables Q_i. It is convenient to consider the evolution in the light-cone coordinates separately, and we take for definiteness the evolution with respect to $x_- = x - t$. Let U be the real orthogonal matrix which diagonalizes the symmetric matrix $\mathcal{V}$ and *define* a symmetric matrix L by:

$$L = -2m\, U\mu U^{-1}, \quad \mathcal{V} = U^{-1} Q U \tag{12.84}$$

where Q denotes the diagonal matrix diag(Q_i) and μ the diagonal matrix diag(μ_i) with μ_i the rapidities of the solitons. The matrix L plays the role of a Lax operator. Obviously, the quantities $\mathrm{tr}(L^n) = (-2m)^n \sum_{i=1}^N \mu_i^n$ are conserved during the evolution of the solitons. They are in involution under the Poisson bracket (12.81). The time evolution of L takes the Lax form, and we have:

Proposition. *The x_- evolution of L is given by a Lax equation*

$$\dot{L} = [M, L] \quad \text{with} \quad M = \dot{U}U^{-1}$$

where the dot means $\frac{\partial}{\partial x_-}$. Moreover, L and M can be expressed in terms of the quantities Q_i and $\dot{Q}_i$:

$$L_{ij} = 2\frac{\sqrt{\dot{Q}_i \dot{Q}_j}}{Q_i + Q_j}, \quad \text{and} \quad M_{ij} = \frac{\sqrt{\dot{Q}_i \dot{Q}_j}}{Q_i - Q_j}(1 - \delta_{ij}) \qquad (12.85)$$

If we introduce the coordinates $Q_j = ie^{q_j}$, the equations of motion read:

$$\ddot{q}_i = \sum_{k \neq i} \frac{2\dot{q}_i \dot{q}_k}{\sinh(q_i - q_k)} \qquad (12.86)$$

Proof. The Lax form of the equation of motion follows from the definition of L, eq. (12.84), and the fact that μ is conserved. Let us prove eq. (12.85). We have to be careful about what we mean by the square roots. Denote, as before, $V = i\tilde{V}$, $Q = i\tilde{Q}$, $X = i\tilde{X}$. The tilde quantities are all real positive definite, and we can take their positive square root. We start from the relation $\mu\tilde{V} + \tilde{V}\mu = 2|e\rangle\langle e|$, where $|e\rangle$ is the column vector with components $e_i = \sqrt{\tilde{X}_i \mu_i}$. Multiplying on the left by U and on the right by U^{-1} we get, using that U is orthogonal, $\tilde{Q}L + L\tilde{Q} = -4m|\tilde{e}\rangle\langle\tilde{e}|$ with $|\tilde{e}\rangle = U|e\rangle$. Since $\tilde{Q}$ is diagonal, we obtain from this relation:

$$L_{ij} = -4m \frac{\tilde{e}_i \tilde{e}_j}{\tilde{Q}_i + \tilde{Q}_j}$$

Next, since $\sqrt{\dot{\tilde{X}}_i} = -m\mu_i \sqrt{\tilde{X}_i}$, we have

$$\dot{\tilde{V}} = -m(\mu\tilde{V} + \tilde{V}\mu) = -2m|e\rangle\langle e| = U^{-1}\left(\dot{\tilde{Q}} + [\tilde{Q}, M]\right)U$$

which implies $-2m|\tilde{e}\rangle\langle\tilde{e}| = \dot{\tilde{Q}} + [\tilde{Q}, M]$. In components, this reads

$$-2m\tilde{e}_i\tilde{e}_j = \dot{\tilde{Q}}_i\delta_{ij} + (\tilde{Q}_i - \tilde{Q}_j)M_{ij}$$

If $i = j$, we find $\tilde{e}_i = \sqrt{-\dot{\tilde{Q}}_i/2m}$, and if $i \neq j$, we find the value of M_{ij} in terms of $\tilde{Q}_i$, $\dot{\tilde{Q}}_i$. This completely determines the antisymmetric matrix M. The equations of motions, eq. (12.86), are obtained by writing the Lax equation. ∎

Instead of the variables $\dot{Q}_i$, let us introduce the variables $\rho_i = \dot{Q}_i/Q_i$. The Lax matrix becomes

$$L_{ij} = 2\frac{\sqrt{Q_i Q_j}}{Q_i + Q_j}\sqrt{\rho_i \rho_j}$$

Comparing the matrices $\mathcal{V}$ and L, one sees that L has exactly the same form as $\mathcal{V}$ with the change of variables $(\mu_i, a_i) \to (Q_i, \rho_i)$. This symmetry extends at the level of the symplectic structure.

Proposition. *The transformation $(\mu_i, a_i) \to (Q_i^{-1}, \rho_i)$ is a symplectic transformation.*

<u>Proof.</u> Let us consider

$$\omega_1 = 2\sum_{i=1}^{N}\frac{\delta a_i}{a_i}\wedge\frac{\delta\mu_i}{\mu_i} + 2\sum_{i<j}\frac{4\mu_i\mu_j}{\mu_i^2 - \mu_j^2}\frac{\delta\mu_i}{\mu_i}\wedge\frac{\delta\mu_j}{\mu_j}$$

$$\omega_2 = 2\sum_{i=1}^{N}\frac{\delta\rho_i}{\rho_i}\wedge\frac{\delta Q_i}{Q_i} + 2\sum_{i<j}\frac{4Q_iQ_j}{Q_i^2 - Q_j^2}\frac{\delta Q_i}{Q_i}\wedge\frac{\delta Q_j}{Q_j}$$

We first express ω_1 using the independent variables μ_i, Q_i. We get $\omega_1 = 2\sum_{ij}A_{ij}\frac{\delta Q_i}{Q_i}\wedge\frac{\delta\mu_j}{\mu_j} + 2\sum_{i<j}B_{ij}\frac{\delta\mu_i}{\mu_i}\wedge\frac{\delta\mu_j}{\mu_j}$. Recall that using the symplectic form ω_1 we have shown that $\mathcal{V}$ obeys an r-matrix relation, hence its eigenvalues Poisson commute. Noting that

$$\begin{pmatrix} 0 & A \\ -A^t & B \end{pmatrix}^{-1} = \begin{pmatrix} (A^t)^{-1}BA^{-1} & -(A^t)^{-1} \\ A^{-1} & 0 \end{pmatrix}$$

we see that the vanishing of the Poisson brackets $\{Q_i, Q_j\}$ requires $B_{ij} = 0$. To compute $A_{ij} = \frac{Q_i}{a_j}\frac{\partial a_j}{\partial Q_i}$ (at μ_k fixed), we vary the matrix $\mathcal{V}$ in eq. (12.76) (at μ fixed), getting $d\mathcal{V} = \frac{1}{2}(\mathcal{V}a^{-1}da + a^{-1}da\mathcal{V})$, where $a = \text{Diag}(a_i)$. Similarly, varying $\mathcal{V} = U^{-1}QU$, we get $d\mathcal{V} = U^{-1}(dQ + [Q, dUU^{-1}])U$ so that:

$$dQ + [Q, dUU^{-1}] = \frac{1}{2}\left(Ua^{-1}daU^{-1}Q + QUa^{-1}daU^{-1}\right)$$

Looking at the diagonal part of this equation, we get

$$\frac{dQ_i}{Q_i} = \sum_j U_{ij}U_{ji}^{-1}\frac{da_j}{a_j} = \sum_{j,k}U_{ij}U_{ji}^{-1}\frac{Q_k}{a_j}\frac{\partial a_j}{\partial Q_k}\frac{dQ_k}{Q_k}$$

Setting $\mathcal{M}_{ij} = U_{ji}^{-1}U_{ij}$, this means $A\mathcal{M} = 1$ or $A = \mathcal{M}^{-1}$.

We now express the form ω_2 in terms of the variables (Q_i, μ_i) exactly in the same way. In particular, starting from ω_2 one may deduce an r-matrix relation for L implying that $\{\mu_i, \mu_j\} = 0$, so that $\omega_2 = 2\sum_{ij} C_{ij} \frac{\delta Q_i}{Q_i} \wedge \frac{\delta \mu_j}{\mu_j}$ with $C_{ij} = -\frac{\mu_j}{\rho_i} \frac{\partial \rho_i}{\partial \mu_j}$ (at Q_k fixed). The computation of C_{ij} is the same as for A_{ij} except that U is replaced by U^{-1}, which leads to $\mathcal{M}C = -1$ so that $C = -A$ and $\omega_2 = -\omega_1$. Changing Q_i to Q_i^{-1} changes ω_2 to $-\omega_2$, thereby proving the symplecticity of the transformation. ∎

We introduce a new set of variables p_i by:

$$\rho_i = \exp(p_i) \prod_{k \neq i} \left(\frac{Q_k + Q_i}{Q_k - Q_i} \right) = e^{p_i} \prod_{k \neq i} \coth \left(\frac{q_i - q_k}{2} \right) \qquad (12.87)$$

It is easy to check that the variables p_i, q_i are canonically conjugated:

$$\{q_i, q_j\} = 0, \quad \{p_i, p_j\} = 0, \quad \{p_i, q_j\} = \delta_{ij}$$

We have already noticed that L obeys an r-matrix relation, which implies that the function $\mathcal{T}(z) = \det(1 + zL)$ is a generating function of commuting Hamiltonians (which we already knew by construction of L):

$$\{\mathcal{T}(z), \mathcal{T}(z')\} = 0$$

Expanding $\mathcal{T}(z)$ in power of z, we find:

$$\mathcal{T}(z) = 1 + \sum_{p=1}^{N} z^p H_p = 1 + \sum_{p=1}^{N} z^p \sum_{k_1 < \cdots < k_p} \rho_{\rho_{k_1}} \cdots \rho_{k_p} \prod_{k_i < k_j} \left(\frac{Q_{k_i} - Q_{k_j}}{Q_{k_i} + Q_{k_j}} \right)^2$$

The Hamiltonian generating the evolution in the light-cone coordinate, x_-, is $H_- = \mathrm{Tr}(L)$. The evolution in the other light-cone coordinate, x_+, is generated by $H_+ = \mathrm{Tr}(L^{-1}) = H_{N-1}/\det(L)$:

$$H_\pm = \sum_j e^{\mp p_j} \prod_{k \neq j} \coth \left(\frac{q_j - q_k}{2} \right)$$

These are the Hamiltonians written by Ruijsenaars and Schneider. By construction, their corresponding flows are linearized in the variables (μ_i, X_i).

We end this section by presenting the reality conditions corresponding to solitons, antisolitons and breathers. For the sine-Gordon field eq. (12.83) and the Hamiltonians $H_\pm$ to be real, the coordinates (Q_i, p_i) have to come in pairs $(j, \bar{j})$, with $Q_{\bar{j}} = -\bar{Q}_j$ and $p_{\bar{j}} = \bar{p}_j$. The case $j = \bar{j}$

corresponds to a soliton or an antisoliton, and the case $j \neq \bar{j}$ to a breather. Therefore, in the coordinates q_j such that $Q_j = ie^{q_j}$:

$$\mathrm{Im}\ q_s = 0 \quad \text{for } s \text{ a soliton}$$
$$\mathrm{Im}\ q_{\bar{s}} = \pi \quad \text{for } \bar{s} \text{ an antisoliton}$$
$$q_{\bar{b}} = \bar{q}_b \quad \text{for } b \text{ a breather}$$

Similarly, the momenta p_s and $p_{\bar{s}}$ are real and p_b is complex with $p_{\bar{b}} = \bar{p}_b$.

12.10 Finite-zone solutions

The simplest example of finite-zone solutions is the one-zone solution. We study it in detail. The multi-zone solutions are then simple generalizations.

The one-zone solution is obtained by looking at solutions $\varphi = \varphi(z)$ of the sinh-Gordon equation depending only on the variable $z = m(\mu x_- + \mu^{-1}x_+)$. Then $\partial_{x_+}\partial_{x_-}\varphi = 2m^2 \sinh(2\varphi)$ becomes $\varphi'' = e^{2\varphi} - e^{-2\varphi}$, where $'$ means the derivative with respect to z. This integrates to $\varphi'^2 = e^{2\varphi} + e^{-2\varphi} - 2E$, where E is a constant. Setting $e^{-2\varphi} = t$, this equation takes the form:

$$t'^2 = 4t(t^2 - 2Et + 1)$$

For $E = \pm 1$, the right-hand side is a perfect square and the equation is readily solved. For $E = +1$ one finds the soliton and antisoliton solutions. For general E, the equation can be compared with the equation for the Weierstrass $\wp$ function:

$$\wp'(z)^2 = 4\wp^3(z) - g_2\wp(z) - g_3$$

and we have:

$$e^{-2\varphi} = t(z) = \wp(z - z_0) + \frac{2E}{3}, \quad g_2 = \frac{16}{3}E^2 - 4, \quad g_3 = \frac{8}{3}E\left(\frac{8}{9}E^2 - 1\right)$$

and z_0 is an arbitrary constant. Note that the discriminant $g_2^3 - 27g_3^2 = 64(E^2 - 1)$ vanishes only for the cases $E = \pm 1$.

To see how this simple one-zone solution fits into the general formalism of finite-zone solutions, we need to construct a Lax matrix, L, satisfying the compatibility conditions

$$[\partial_{x_+} - A_{x_+}, L] = 0, \quad [\partial_{x_-} - A_{x_-}, L] = 0$$

As explained in Chapter 5, the general technique is to impose a stationarity condition with respect to some time. The one-zone solution only depends on $z = m(\mu x_- + \mu^{-1} x_+)$, hence satisfies the stationarity condition:

$$(\mu^{-1}\partial_{x_-} - \mu\partial_{x_+})\varphi = 0 \qquad (12.88)$$

So we choose as Lax matrix $L(\lambda) = \frac{1}{m}(\mu^{-1}A_{x_-} - \mu A_{x_+})$. From eqs. (12.64, 12.65) we find

$$A_{x_+} = \begin{pmatrix} \frac{1}{2}\partial_{x_+}\varphi & m\lambda^{-1}e^{-\varphi} \\ m\lambda^{-1}e^{\varphi} & -\frac{1}{2}\partial_{x_+}\varphi \end{pmatrix}, \quad A_{x_-} = \begin{pmatrix} -\frac{1}{2}\partial_{x_-}\varphi & m\lambda e^{\varphi} \\ m\lambda e^{-\varphi} & \frac{1}{2}\partial_{x_-}\varphi \end{pmatrix}$$

and the Lax matrix is:

$$L(\lambda) = \begin{pmatrix} -\frac{1}{2m}(\mu^{-1}\partial_{x_-}\varphi + \mu\partial_{x_+}\varphi) & \frac{\lambda}{\mu}e^{\varphi} - \frac{\mu}{\lambda}e^{-\varphi} \\ \frac{\lambda}{\mu}e^{-\varphi} - \frac{\mu}{\lambda}e^{\varphi} & \frac{1}{2m}(\mu^{-1}\partial_{x_-}\varphi + \mu\partial_{x_+}\varphi) \end{pmatrix}$$

The spectral curve, $\Gamma : \det(L(\lambda) - \Lambda) = 0$, reads

$$\Lambda^2 = \frac{1}{4m^2}(\mu^{-1}\partial_{x_+}\varphi + \mu\partial_{x_-}\varphi)^2 - e^{2\varphi} - e^{-2\varphi} + \frac{\lambda^2}{\mu^2} + \frac{\mu^2}{\lambda^2}$$

Hence, for these special one-zone solutions, the quantity $\frac{1}{4m^2}(\mu^{-1}\partial_{x_+}\varphi + \mu\partial_{x_-}\varphi)^2 - e^{2\varphi} - e^{-2\varphi}$ is a constant independent of $x_\pm$. Indeed, noting that $\mu^{-1}\partial_{x_+}\varphi + \mu\partial_{x_-}\varphi = 2m\varphi'$, we find that this constant is $-2E$. As an equation in the (Λ, λ) plane, the curve finally reads:

$$\Gamma : \quad \Lambda^2 = \frac{\lambda^2}{\mu^2} + \frac{\mu^2}{\lambda^2} - 2E$$

Besides the hyperelliptic involution, $\Lambda \to -\Lambda$, this curve has another involution $\lambda \to -\lambda$. Let Γ' be the curve obtained from Γ by quotienting by this involution. This is done by setting

$$Y' = \mu\lambda^2\Lambda, \quad \lambda' = \lambda^2 \quad \text{or} \quad \Lambda = \frac{Y'}{\mu\lambda'}, \quad \lambda = \sqrt{\lambda'}$$

The transformation $\Gamma' \to \Gamma$ is one to two. The curve Γ is a two-sheeted unbranched covering of Γ'. The equation of the curve Γ' reads

$$\Gamma' : \quad Y'^2 = \lambda'(\lambda'^2 - 2E\mu^2\lambda' + \mu^4)$$

The points $P_+(\lambda' = 0)$ and $P_-(\lambda' = \infty)$ are branch points on Γ'. Their pre-images on Γ are two singular points which blow up to four points on its desingularization. This makes the situation more complicated on Γ than on Γ', and this is why we usually work on Γ', and eventualy lift the results to Γ if needed.

Next, we describe the singularity structure of the wave function Ψ in this one-zone case. The wave function satisfies the equations

$$(\partial_{x_\pm} - A_{x_\pm})\Psi = 0, \quad (L(\lambda) - \Lambda)\Psi = 0 \tag{12.89}$$

These equations are naturally written on Γ and we need to transport them to Γ'. This is achieved by performing a gauge transformation:

$$\Psi' = g\Psi = \begin{pmatrix} e^{\frac{\varphi}{2}} & 0 \\ 0 & \frac{1}{\lambda}e^{-\frac{\varphi}{2}} \end{pmatrix}\Psi$$

where the important factor is $1/\lambda$ and the factors $e^{\pm\varphi/2}$ have been introduced for later convenience. The gauge transformed connection $A'_{x_\pm}$ and the new Lax matrix $L' = gLg^{-1}$ still obey the Lax equations $[\partial_{x_\pm} - A'_{x_\pm}, L'] = 0$, but are now rational functions of $\lambda' = \lambda^2$. Explicitly, the transformed connection is:

$$A'_{x_-} = \begin{pmatrix} 0 & m\lambda'e^{2\varphi} \\ me^{-2\varphi} & 0 \end{pmatrix}, \quad A'_{x_+} = \begin{pmatrix} \partial_{x_+}\varphi & m \\ m\lambda'^{-1} & -\partial_{x_+}\varphi \end{pmatrix} \tag{12.90}$$

The transformed L' is still equal to $\frac{1}{m}(\mu^{-1}A'_{x_-} - \mu A'_{x_+})$ thanks to the stationarity condition, eq. (12.88). It is now straightforward to extract the singular parts of Ψ' at the branch points $\lambda' = 0$ and $\lambda' = \infty$.

In the neighbourhood of the point $\lambda' = \infty$ of Γ', we can use the local parameter $\sqrt{\lambda'}$. The eigenvalue of L' is $\sqrt{\lambda'}/\mu$. The eigenvector is such that $\psi'_1/\psi'_2 \sim \exp(2\varphi)\sqrt{\lambda'}$. Since Ψ' is an eigenvector of L' and $L' = \frac{1}{m}(\mu^{-1}A'_{x_-} - \mu A'_{x_+})$, we have $A'_{x_-}\Psi' = m\sqrt{\lambda'}\Psi' + O(1)$ around ∞. So we get

$$\Psi'_\infty \sim e^{m\sqrt{\lambda'}x_-}\begin{pmatrix} \chi_1 \\ \frac{1}{\sqrt{\lambda'}}\chi_2 \end{pmatrix}, \quad \frac{\chi_1}{\chi_2} = e^{2\varphi} \tag{12.91}$$

Similarly, at the point $\lambda' = 0$ the eigenvalue of L' is $\mu/\sqrt{\lambda'} + O(\sqrt{\lambda'})$ and $\psi'_1/\psi'_2 \sim -\sqrt{\lambda'}$. We have $A'_{x_+}\Psi' = -m/\sqrt{\lambda'}\Psi' + O(1)$. To get the precise form of Ψ' at $\lambda' = 0$, write $\Psi'_0 = \exp(-\frac{m}{\sqrt{\lambda'}}x_+)(f, -g/\sqrt{\lambda'})$ with $g = f + O(\sqrt{\lambda'})$ and plug into the equations $(\partial_{x_\pm} - A'_{x_\pm})\Psi' = 0$. A careful analysis of these equations shows that f is a constant, independent of $x_\pm$, and so can be normalized to 1. We finally get:

$$\Psi'_0 \sim e^{-\frac{m}{\sqrt{\lambda'}}x_+}\begin{pmatrix} 1 \\ -\frac{1}{\sqrt{\lambda'}} \end{pmatrix} \tag{12.92}$$

We see that Ψ' has a pole at $\lambda' = 0$ on Γ'.

The one-zone analysis can be generalized immediately to provide finite-zone solutions to the sinh-Gordon equation. Consider an arbitrary hyperelliptic curve Γ', of genus g, having $\lambda' = 0$ and $\lambda' = \infty$ as branch points:

$$Y'^2 = \lambda'P(\lambda'), \quad \deg P = 2g$$

On this curve we construct Baker–Akhiezer vectors Ψ' with $g + N - 1 = g + 1$ poles ($N = 2$). We set one of these poles at $\lambda' = 0$. The other g poles at finite distance are independent of $x_\pm$. The essential singularities are set at $P_+(\lambda' = 0)$ and $P_-(\lambda' = \infty)$ with behaviour specified as in eqs. (12.91, 12.92), so ψ_2' has a pole at P_+ and a zero at P_-. This characterise Ψ' uniquely.

Proposition. *We have:*

$$\left[\partial_{x_-} - \begin{pmatrix} 0 & m\lambda' \frac{\chi_1}{\chi_2} \\ m\frac{\chi_2}{\chi_1} & 0 \end{pmatrix} \right] \begin{pmatrix} \psi_1' \\ \psi_2' \end{pmatrix} = 0$$

$$\left[\partial_{x_+} - \begin{pmatrix} \partial_{x_+} \log \chi_1 & m \\ \frac{m}{\lambda'} & \partial_{x_+} \log \chi_2 \end{pmatrix} \right] \begin{pmatrix} \psi_1' \\ \psi_2' \end{pmatrix} = 0 \qquad (12.93)$$

Proof. Consider $\partial_- \psi_1' - m \frac{\chi_1}{\chi_2} \lambda' \psi_2'$. It is a Baker–Akhiezer function. At P_-, it behaves like $e^{m\sqrt{\lambda'}x} - O(1)$, while at P_+ it behaves like $e^{\frac{m}{\sqrt{\lambda'}}x} + O(\sqrt{\lambda'})$ and therefore has a zero at this point. It also has g poles at finite distance inherited from Ψ'. So, it is a Baker–Akhiezer function with g poles and a prescribed zero at $\lambda' = 0$ and therefore vanishes. We prove similarly that $\partial_{x_-} \psi_2' - \frac{\chi_2}{\chi_1} \psi_1' = 0$ and that $\partial_{x_+} \psi_1' - m\psi_2 - \frac{\partial_{x_+} \chi_1}{\chi_1} \psi_1' = 0$. Finally, considering $\partial_{x_+} \psi_2' - \frac{m}{\lambda'} \psi_1' - \frac{\partial_{x_+} \chi_2}{\chi_2} \psi_2'$, we see that it has a simple pole at $\lambda' = 0$ but a double zero at $\lambda' = \infty$, hence also vanishes. ∎

We set $\chi_1/\chi_2 = \exp(2\varphi)$ and $\chi_1\chi_2 = \exp(2\sigma)$ and compute the zero curvature condition for the system (12.93). We get $\partial_{x_+} \partial_{x_-} \sigma = 0$ and $\partial_{x_+} \partial_{x_-} \varphi = 2m^2 \sinh(2\varphi)$. The free field σ decouples and can be set equal to 0. Then the connection appearing in eq. (12.93) is exactly the same as $A'_{x_\pm}$ in eq. (12.90).

We see that, starting from any hyperelliptic curve with branch points at 0 and ∞, and imposing the essential singularities eqs. (12.91, 12.92) at these two punctures yields readily a finite-zone solution of the sinh-Gordon equation. Using two punctures allows one to get independent equations of motion with respect to x_+ and x_-, so producing relativistic generalizations of the finite-zone solutions for the KdV equation.

References

[1] A.C. Scott, F.Y.F. Chu and D.W. McLaughlin, The soliton: a new concept in applied science. *Proc. IEEE* **61** (1973) 1443.

[2] R.F. Dashen, B. Hasslascher and A. Neveu, Particle spectrum in model field theories from semiclassical functional integral techniques. *Phys. Rev.* **D11** (1975) 3424.

[3] A.N. Leznov and M.V. Saveliev, Representation of zero curvature for the system of non-linear partial differential equations $x_{\alpha,z\bar{z}} = \exp(kx)_\alpha$ and its integrability. *Lett. Math. Phys.* **3** (1979) 489–494.

[4] A.B. Zamolodchikov and Al.B. Zamolodchikov, Factorized S-matrices in two dimensions as the exact solutions of certain relativistic quantum field theory models. *Ann. Phys.* **120** (1979) 253.

[5] J.L. Gervais and A. Neveu, Novel triangle relation and absence of tachyons in Liouville string field theory. *Nucl. Phys.* **B238** (1984) 125.

[6] A.A. Belavin, A.M. Polyakov and A.B. Zamolodchikov, Infinite conformal symmetry in two-dimensional quantum field theory. *Nucl. Phys.* **B241** (1984) 333.

[7] M. Semenov-Tian-Shansky. Dressing transformations and Poisson group actions. *Publ. RIMS* **21** (1985) 1237.

[8] L.D. Faddeev and L.A. Takhtajan, *Hamiltonian Methods in the Theory of Solitons.* Springer (1986).

[9] S.N.M. Ruijsenaars H. Schneider, A new class of integrable systems and its relation to solitons. *Ann. Phys.* **170** (1986) 370–405.

[10] O. Babelon, Extended conformal algebra and the Yang–Baxter equation. *Phys. Lett.* **B215** (1988) 523.

[11] O. Babelon D. Bernard, Affine Solitons: a relation between tau functions, dressing and Bäcklund transformations. *Int. J. Mod. Phys.* **A8** (1993) 507–543.

13
Classical inverse scattering method

We introduce the basic tools of the classical inverse scattering method created by Gardner, Green, Kruskal and Miura. This very ingenious method was first exploited to solve the KdV equation. Here we apply it to the sine-Gordon equation since it is rewarding to show how the method applies to a more involved situation. An advantage of the inverse scattering method is that it allows us to construct action–angle variables for a full infinite-dimensional phase space of the field theory corresponding to fields decreasing rapidly at infinity. The idea of this approach is to associate a scattering problem with the field configuration. The key point is that the scattering data have a very simple time evolution, and that the field can be reconstructed from these data. Of course particular solutions like soliton solutions, which depend on a finite number of parameters, are also recovered this way.

13.1 The sine-Gordon equation

Let us consider the sine-Gordon equation

$$\partial_t^2 \varphi - \partial_x^2 \varphi = -\frac{8m^2}{\beta} \sin(2\beta\varphi) \qquad (13.1)$$

with $\varphi \equiv \varphi(x,t)$ a real scalar field. One can write this equation as the compatibility condition of a linear system

$$(\partial_x - A_x)\,\Psi = 0 \qquad (13.2)$$
$$(\partial_t - A_t)\,\Psi = 0 \qquad (13.3)$$

with:

$$A_x = i \begin{pmatrix} \frac{\beta}{2}\partial_t\varphi & m(\lambda e^{i\beta\varphi} - \lambda^{-1}e^{-i\beta\varphi}) \\ m(\lambda e^{-i\beta\varphi} - \lambda^{-1}e^{i\beta\varphi}) & -\frac{\beta}{2}\partial_t\varphi \end{pmatrix} \qquad (13.4)$$

486

$$A_t = i \begin{pmatrix} \frac{\beta}{2}\partial_x\varphi & -m(\lambda e^{i\beta\varphi} + \lambda^{-1}e^{-i\beta\varphi}) \\ -m(\lambda e^{-i\beta\varphi} + \lambda^{-1}e^{i\beta\varphi}) & -\frac{\beta}{2}\partial_x\varphi \end{pmatrix} \quad (13.5)$$

As compared to eqs. (12.64, 12.65) in Chapter 12, we have changed $\varphi \to i\varphi$ to get the sine-Gordon equation, and $\lambda \to i\lambda$ so that $i(\partial_x - A_x)$ and $i(\partial_t - A_t)$ are formally self-adjoint for real λ.

In this chapter, we will study the solutions of the sine-Gordon equation on the line with boundary conditions:

$$\lim_{x\to-\infty}\varphi(x) = 0, \quad \lim_{x\to+\infty}\varphi(x) = \frac{Q\pi}{\beta}, \quad Q \in \mathbb{Z}$$

The number Q is the topological charge of the field configuration. These asymptotic conditions are manifestly compatible with the equations of motion and correspond to a minimum of the sine-Gordon potential. We will assume that they are reached rapidly enough.

To solve a non-linear partial differential equation by the classical inverse scattering method, we proceed in three steps.

- Given the initial data, i.e. $\varphi(x,0)$ for the sine-Gordon case, we solve the direct scattering problem:

$$[\partial_x - A_x(\lambda, \varphi(x,0))]\Psi = 0$$

We thus determine some scattering data $a(\lambda,0)$, $b(\lambda,0)$ for the continuous spectrum and discrete spectum.

- We determine the time evolution of the scattering data by using the second equation:

$$[\partial_t - A_t(\lambda, \varphi)]\Psi = 0 \quad (13.6)$$

In the regions $x = \pm\infty$ where the asymptotics of the field $\varphi(x)$ are known, it allows us to get $a(\lambda,t)$, $b(\lambda,t)$ ....

- Knowing the scattering data at time t, we reconstruct the matrix $A_x(\lambda, t)$ (and thus $\varphi(x,t)$) by solving the Gelfand–Levitan–Marchenko equation, which is a linear integral equation.

In the rest of this chapter, we apply this strategy to solve the sine-Gordon equation.

13.2 The Jost solutions

In our case, the direct scattering problem is defined by the equation

$$\left[\frac{\partial}{\partial x} - i \begin{pmatrix} \frac{\beta}{2}\partial_t\varphi & m(\lambda e^{i\beta\varphi} - \lambda^{-1}e^{-i\beta\varphi}) \\ m(\lambda e^{-i\beta\varphi} - \lambda^{-1}e^{i\beta\varphi}) & -\frac{\beta}{2}\partial_t\varphi \end{pmatrix} \right]\Psi = 0$$
$$(13.7)$$

This linear system possesses a few simple properties:

Let * denote complex conjugation. For λ real, we have $A_x^* = \sigma_2 A_x \sigma_2$, (and more generally for complex λ we have $(A_x(\lambda, x))^* = \sigma_2 A_x(\lambda^*, x)\sigma_2$). Therefore, if Ψ is a solution of eq. (13.7), so is $\overline{\Psi} = \sigma_2 \Psi^*$. This defines a conjugate solution:

$$\overline{\Psi} = \begin{pmatrix} \overline{\psi}_1 \\ \overline{\psi}_2 \end{pmatrix} = \begin{pmatrix} -i\psi_2^* \\ i\psi_1^* \end{pmatrix} \tag{13.8}$$

Similarly, we have $A_x(-\lambda) = \sigma_3 A_x(\lambda)\sigma_3$, so that $\sigma_3 \Psi(x, -\lambda)$ is also a solution.

For two solutions $\xi = \begin{pmatrix} \xi_1 \\ \xi_2 \end{pmatrix}$ and $\Psi = \begin{pmatrix} \psi_1 \\ \psi_2 \end{pmatrix}$, we may define the Wronskian $W(\xi, \psi)$ by:

$$W(\xi, \psi) = \xi_1 \psi_2 - \xi_2 \psi_1$$

Since $\mathrm{Tr}\,(A_x) = 0$, we have $\frac{d}{dx}W(\xi, \psi) = 0$, and the Wronskian is x-independent.

Assume that $\varphi \to 0$ for $x \to -\infty$, and $\varphi \to \frac{\pi Q}{\beta}$ for $x \to +\infty$. Then the linear system eq. (13.7) becomes respectively

$$\left[\frac{\partial}{\partial x} - ik(\lambda)\sigma_1\right]\Psi = 0, x \to -\infty; \quad \left[\frac{\partial}{\partial x} - ie^{Qi\pi}k(\lambda)\sigma_1\right]\Psi = 0, x \to +\infty$$

where $k(\lambda) = m(\lambda - \lambda^{-1})$. It follows that the solutions behave asymptotically, for $x \to \pm\infty$, as

$$\Psi|_{x \to -\infty} \sim c_1 \begin{pmatrix} 1 \\ 1 \end{pmatrix} e^{ik(\lambda)x} + c_2 \begin{pmatrix} 1 \\ -1 \end{pmatrix} e^{-ik(\lambda)x}$$

$$\Psi|_{x \to +\infty} \sim c_1' \begin{pmatrix} 1 \\ e^{i\pi Q} \end{pmatrix} e^{ik(\lambda)x} + c_2' \begin{pmatrix} 1 \\ -e^{i\pi Q} \end{pmatrix} e^{-ik(\lambda)x}$$

The Jost solutions are solutions of the linear system with specific behaviour at infinity. Noting that $\mathrm{Im}\,k(\lambda) = (1 + |\lambda|^{-2})\mathrm{Im}\,\lambda$, we set the definition:

Definition. *The Jost solutions f_1 and f_2 are defined by the following asymptotic behavior:*

$$f_1 \sim \begin{pmatrix} 1 \\ e^{i\pi Q} \end{pmatrix} e^{ik(\lambda)x}, \quad x \to +\infty; \quad f_2 \sim \begin{pmatrix} 1 \\ -1 \end{pmatrix} e^{-ik(\lambda)x}, \quad x \to -\infty$$

$$\tag{13.9}$$

These solutions are chosen in such way that the asymptotics remain bounded if we give a positive imaginary part to λ.

Let us examine in detail the asymptotic form of our Jost solutions when $|x| \to \infty$.

	$x = -\infty$	$x = +\infty$
f_1	$c_{11}(\lambda)\begin{pmatrix}1\\1\end{pmatrix}e^{ik(\lambda)x} + c_{12}(\lambda)\begin{pmatrix}1\\-1\end{pmatrix}e^{-ik(\lambda)x}$	$\begin{pmatrix}1\\e^{iQ\pi}\end{pmatrix}e^{ik(\lambda)x}$

	$x = -\infty$	$x = +\infty$
f_2	$\begin{pmatrix}1\\-1\end{pmatrix}e^{-ik(\lambda)x}$	$c_{21}(\lambda)\begin{pmatrix}1\\e^{iQ\pi}\end{pmatrix}e^{ik(\lambda)x} + c_{22}(\lambda)\begin{pmatrix}1\\-e^{iQ\pi}\end{pmatrix}e^{-ik(\lambda)x}$

We get some relations between the coefficients $c_{ij}(\lambda)$ by calculating the Wronskians at $x = +\infty$ and $x = -\infty$:

	$x = -\infty$	$x = +\infty$
$W(f_1, f_2)$	$-2c_{11}(\lambda)$	$-2c_{22}(\lambda)e^{iQ\pi}$
$W(f_1, \overline{f}_2)$	$2ic_{12}(\lambda)$	$2ic_{21}^*(\lambda)$
$W(f_1, \overline{f}_1)$	$2i\left(|c_{11}|^2 + |c_{12}|^2\right)$	$2i$

Since the Wronskians are constant, we deduce that $c_{11}(\lambda) = c_{22}(\lambda)e^{iQ\pi}$ and $c_{12}(\lambda) = c_{21}^*(\lambda)$. We set $c_{11}(\lambda) = c_{22}(\lambda)e^{iQ\pi} = a(\lambda)$ and $c_{12}(\lambda) = c_{21}^*(\lambda) = -b(\lambda)$. We have

$$|a(\lambda)|^2 + |b(\lambda)|^2 = 1 \tag{13.10}$$

The asymptotic behavior of the Jost solutions can be rewritten as:

	$x = -\infty$	$x = +\infty$
f_1	$a(\lambda)\begin{pmatrix}1\\1\end{pmatrix}e^{ik(\lambda)x} - b(\lambda)\begin{pmatrix}1\\-1\end{pmatrix}e^{-ik(\lambda)x}$	$\begin{pmatrix}1\\e^{iQ\pi}\end{pmatrix}e^{ik(\lambda)x}$

	$x = -\infty$	$x = +\infty$
f_2	$\begin{pmatrix}1\\-1\end{pmatrix}e^{-ik(\lambda)x}$	$-b^*(\lambda)\begin{pmatrix}1\\e^{iQ\pi}\end{pmatrix}e^{ik(\lambda)x} + a(\lambda)\begin{pmatrix}e^{iQ\pi}\\-1\end{pmatrix}e^{-ik(\lambda)x}$

Comparing the asymptotic expansions of $\bar{f}_i$ and $\sigma_3 f_i(-\lambda)$ we get, for real λ:

$$\bar{f}_1 = -i\frac{b^*}{a}f_1 - \frac{i}{a}f_2, \quad \bar{f}_2 = \frac{i}{a}f_1 + \frac{ib}{a}f_2 \qquad (13.11)$$

$$\sigma_3 f_1(-\lambda) = e^{-i\pi Q}\frac{b^*}{a}f_1 + e^{-i\pi Q}\frac{1}{a}f_2, \quad \sigma_3 f_2(-\lambda) = \frac{1}{a}f_1 + \frac{b}{a}f_2 \quad (13.12)$$

From the last relation one gets the symmetry properties

$$a(-\lambda) = e^{-i\pi Q}a^*(\lambda), \quad b(-\lambda) = -e^{-i\pi Q}b^*(\lambda)$$

The main property that we shall prove below is that f_1, f_2 and $a(\lambda)$ can be analytically continued to the upper half-plane . Assuming this result for the time being, we can give a more precise definition of the scattering data.

For λ in the upper half-plane , $-\lambda^*$ is also in the upper half-plane (it is the symmetric of λ with respect to the imaginary axis) and the relation $a(-\lambda) = e^{-i\pi Q}a^*(\lambda)$ for λ real extends to $a(-\lambda^*) = e^{-i\pi Q}(a(\lambda))^*$. Note also that f_1 and f_2 obey the symmetry property:

$$f_1(x, -\lambda^*) = e^{i\pi Q}\sigma_1(f_1(x, \lambda))^*, \quad f_2(x, -\lambda^*) = -\sigma_1(f_2(x, \lambda))^* \quad (13.13)$$

This is because $A_x(x, -\lambda^*) = \sigma_1(A_x(x, \lambda))^*\sigma_1$ so that $f(x, -\lambda^*)$ obeys the same equation as $\sigma_1(f(x, \lambda))^*$, and one can then compare the asymptotics at $x = \pm\infty$. Suppose that $a(\lambda)$ has some complex zeros λ_n, $\mathrm{Im}\,\lambda_n > 0$. From the symmetry property of $a(\lambda)$, these zeroes are either purely imaginary or occur in pairs symmetric with respect to the imaginary axis. Because $2a(\lambda) = -W(f_1, f_2)$, the two Jost solutions f_1 and f_2 become linearly dependent when $\lambda = \lambda_n$. Looking at the asymptotics, we see that this solution is normalizable. It corresponds to a bound state for the linear scattering problem (13.7). At λ_n, the solutions f_1, f_2 are proportional:

$$f_2(x, \lambda_n) = c_n f_1(x, \lambda_n), \quad a(\lambda_n) = 0 \text{ and } \mathrm{Im}(\lambda_n) > 0 \qquad (13.14)$$

Equation (13.13) implies that, for λ_n pure imaginary, $c_n^* = -e^{i\pi Q}c_n$, and a similar relation when we have a pair of symmetric roots. The functions $a(\lambda)$ and $b(\lambda)$ and the numbers λ_n and c_n constitute the scattering data. The function $a(\lambda)$ is called the Jost function.

Proposition. *The Jost solutions $f_1(x, \lambda)$ and $f_2(x, \lambda)$ are analytic functions of λ in the upper half-plane $\mathrm{Im}(\lambda) > 0$.*

<u>Proof.</u> Denote $A_{\pm\infty} = \lim_{x\to\pm\infty} A_x$, i.e. $A_{-\infty} = ik(\lambda)\sigma_1$ and $A_\infty = ie^{i\pi Q}k(\lambda)\sigma_1$. We can write the linear systems for $f_{1,2}$ as $(\partial_x - A_{\pm\infty})f_{1,2} =$

$(A_x - A_{\pm\infty})f_{1,2}$. These equations, plus the boundary conditions, are equivalent to the integral equations

$$f_1(x, \lambda) = f_1^0(x, \lambda) + \int_{-\infty}^{+\infty} dy\, G_1(x, y) V_1(y, \lambda) f_1(y, \lambda) \quad (13.15)$$

$$f_2(x, \lambda) = f_2^0(x, \lambda) + \int_{-\infty}^{+\infty} dy\, G_2(x, y) V_2(y, \lambda) f_2(y, \lambda) \quad (13.16)$$

where we have set $V_{1,2}(x, \lambda) = A_x(x, \lambda) - A_{\pm\infty}(x, \lambda)$ and

$$f_1^0 = \begin{pmatrix} 1 \\ e^{i\pi Q} \end{pmatrix} e^{ik(\lambda)x}, \quad f_2^0 = \begin{pmatrix} 1 \\ -1 \end{pmatrix} e^{-ik(\lambda)x}$$

The kernels G_1 and G_2 are:

$$G_1(x, y) = -\frac{1}{2}\theta(y - x)\left[\begin{pmatrix} 1 & e^{i\pi Q} \\ e^{i\pi Q} & 1 \end{pmatrix} e^{ik(\lambda)(x-y)}\right.$$

$$\left. + \begin{pmatrix} 1 & -e^{i\pi Q} \\ -e^{i\pi Q} & 1 \end{pmatrix} e^{-ik(\lambda)(x-y)}\right]$$

$$G_2(x, y) = \frac{1}{2}\theta(x - y)\left[\begin{pmatrix} 1 & 1 \\ 1 & 1 \end{pmatrix} e^{ik(\lambda)(x-y)} + \begin{pmatrix} 1 & -1 \\ -1 & 1 \end{pmatrix} e^{-ik(\lambda)(x-y)}\right]$$

where $\theta(x)$ is the step function. To analyse the integral equation for f_1, we set $f_1 = e^{ik(\lambda)x}\tilde{f}_1$. Using the properties of the step function, we get a Volterra type integral equation:

$$\tilde{f}_1(x, \lambda) = \tilde{f}_1^0 - \int_x^\infty dy K(x, y; \lambda) \tilde{f}_1(y, \lambda) \quad (13.17)$$

with

$$K(x, y; \lambda) = \frac{1}{2}\left[\begin{pmatrix} 1 & e^{i\pi Q} \\ e^{i\pi Q} & 1 \end{pmatrix} + \begin{pmatrix} 1 & -e^{i\pi Q} \\ -e^{i\pi Q} & 1 \end{pmatrix} e^{2ik(\lambda)(y-x)}\right] V_1(y, \lambda)$$

By iteration, we find the unique solution:

$$\tilde{f}_1 = (1 - K_1 + K_1^2 + \cdots)\tilde{f}_1^0$$

We show that this series converges uniformly, thereby proving the existence of the Jost solution f_1. In the upper half-plane, we can bound the exponential $|\exp(2ik(\lambda)(y - x))| \le 1$ so that $|K(x, y; \lambda)| \le M(y; \lambda)$, where $M(y; \lambda) > 0$ is such that $\int^\infty dy M(y; \lambda)$ converges. This allows us to bound the nth iterate in the following way:

$$|(K^n \tilde{f}_1^0)(x)| = |\int_x^\infty dx_1 K(x, x_1) \cdots \int_{x_{n-1}}^\infty K(x_{n-1}, x_n)\tilde{f}_1^0| \le$$

$$|\tilde{f}_1^0| \int_{x \le x_1 \le \cdots \le x_n} dx_1 \cdots dx_n M(x_1) \cdots M(x_n) = \frac{1}{n!}\left(\int_x^\infty dy M(y)\right)^n |\tilde{f}_1^0|$$

Because of the $n!$ in the denominator, the series is absolutely and uniformly bounded by $\exp\left(\int_x^\infty dy\, M(y)\right)$. This is the main observation in the theory of Volterra integral equations. Each term in the expansion is analytic in λ for $\mathrm{Im}\,\lambda > 0$, hence the uniform limit is also analytic. Note, however, that for $\lambda = 0$, $V(y,\lambda)$ has a pole so that singularities are expected for f_1 and f_2. The equation for f_2 is treated in the same way. ∎

Remark 1. If we assume, in the previous proof, that the potential φ is such that V_1, V_2 have compact support (i.e. the potential reaches its limiting values $0, Q\pi/\beta$ at finite distance), then the integration domain in the integral equation is finite, and one can bound $e^{2ik(\lambda)(y-x)}$ even for $\mathrm{Im}(k) < 0$. It follows that $f_1(x,\lambda)$ and $f_2(x,\lambda)$ are analytic functions of λ in the whole λ-plane except at $\lambda = 0$ and $\lambda = \infty$. In this case, the scattering data $a(\lambda)$, $b(\lambda)$ can be analytically continued in the λ-plane. This remains true if V_1, V_2 vanish at $\pm\infty$ rapidly enough to compensate for the growth of $e^{2ik(\lambda)(y-x)}$ when $\mathrm{Im}(k) < 0$. In particular, choosing λ to be a zero λ_n of $a(\lambda)$ in the upper half-plane , and looking at the asymptotic expansions of f_1, f_2 when $x = -\infty$, we see that $c_n = -1/b(\lambda_n)$ in eq. (13.14).

The next task is to compute the asymptotics of the Jost solutions when $\lambda \to 0$ and $\lambda \to \infty$.

Proposition. *The Jost solutions f_1, f_2 have the following asymptotics in λ, valid for any x:*

$$f_1 = e^{ik(\lambda)x}\left(e^{\frac{i\beta\varphi}{2}\sigma_3}\begin{pmatrix} e^{-\frac{i\pi Q}{2}} \\ e^{-\frac{i\pi Q}{2}} \end{pmatrix} + O(\frac{1}{|\lambda|})\right), \quad |\lambda| \to \infty$$

$$f_2 = e^{-ik(\lambda)x}\left(e^{\frac{i\beta\varphi}{2}\sigma_3}\begin{pmatrix} 1 \\ -1 \end{pmatrix} + O(\frac{1}{|\lambda|})\right), \quad |\lambda| \to \infty$$

$$f_1 = e^{ik(\lambda)x}\left(e^{-\frac{i\beta\varphi}{2}\sigma_3}\begin{pmatrix} e^{\frac{i\pi Q}{2}} \\ e^{\frac{i\pi Q}{2}} \end{pmatrix} + O(|\lambda|)\right), \quad |\lambda| \to 0$$

$$f_2 = e^{-ik(\lambda)x}\left(e^{-\frac{i\beta\varphi}{2}\sigma_3}\begin{pmatrix} 1 \\ -1 \end{pmatrix} + O(|\lambda|)\right), \quad |\lambda| \to 0 \qquad (13.18)$$

<u>Proof.</u> It is easily checked that the factors $e^{\pm\frac{i\beta\varphi}{2}\sigma_3}$ are such that the right-hand sides of these equations have the correct asymptotics for f_1, f_2 when $x \to \pm\infty$ respectively, see eq. (13.9). Let us consider the case $\lambda \to \infty$ and analyse the asymptotics of f_1 for definiteness, the other cases being similar. We introduce $\widehat{f_1}$ by performing the gauge transformation:

$$f_1 = e^{ik(\lambda)x}e^{i\frac{\beta\varphi}{2}\sigma_3}\widehat{f_1}$$

It obeys the transformed equation $[\partial_x + ik(\lambda)(1 - \sigma_1) - \widehat{A}_x]\widehat{f}_1 = 0$ with:

$$\widehat{A}_x = i \begin{pmatrix} \frac{\beta}{2}(\dot\varphi - \varphi') & m\lambda^{-1}(1 - e^{-2i\beta\varphi}) \\ m\lambda^{-1}(1 - e^{2i\beta\varphi}) & -\frac{\beta}{2}(\dot\varphi - \varphi') \end{pmatrix}$$

Notice that the boundary conditions on φ are such that $\widehat{A}_x$ rapidly vanishes when $x \to \pm\infty$. Equivalently, $\widehat{f}_1$ is a solution of the integral equation:

$$\widehat{f}_1(x, \lambda) = \widehat{f}_1^0 - \int_x^\infty dy\, e^{ik(\lambda)(y-x)(1-\sigma_1)} \widehat{A}_x(y, \lambda) \widehat{f}_1(y, \lambda), \quad \widehat{f}_1^0 = e^{-i\frac{\pi Q}{2}} \begin{pmatrix} 1 \\ 1 \end{pmatrix}$$

We decompose this equation on the eigenvectors of the matrix $(1 - \sigma_1)$ corresponding to the eigenvalues $2, 0$ respectively:

$$\widehat{f}_1(x, \lambda) = F(x, \lambda) \begin{pmatrix} 1 \\ -1 \end{pmatrix} + G(x, \lambda) \begin{pmatrix} 1 \\ 1 \end{pmatrix}$$

This yields the two coupled scalar integral equations:

$$F(x, \lambda) = -i \int_x^\infty dy\, e^{2ik(\lambda)(y-x)} \Big\{ \frac{\beta}{2}(\dot\varphi - \varphi')G(y, \lambda)$$
$$- m\lambda^{-1}\Big((1 - \cos(2\beta\varphi))F(y, \lambda) - i\sin(2\beta\varphi)G(y, \lambda)\Big)\Big\}$$
$$G(x, \lambda) = e^{-i\frac{\pi Q}{2}} - i \int_x^\infty dy \Big\{ \frac{\beta}{2}(\dot\varphi - \varphi')F(y, \lambda)$$
$$+ m\lambda^{-1}\Big((1 - \cos(2\beta\varphi))G(y, \lambda) - i\sin(2\beta\varphi)F(y, \lambda)\Big)\Big\}$$

This is a Volterra system, so the iteration procedure converges, exactly as above. Note that there is a Fourier exponential in the first equation, but not in the second one. Now for $h(x)$ rapidly decreasing at $+\infty$ we have:

$$\int_x^\infty dy\, e^{i\lambda y} h(y) = O(\lambda^{-1})$$

It follows that the iteration procedure yields $G(x, \lambda) = e^{-i\frac{\pi Q}{2}} + O(\lambda^{-1})$ and $F(x, \lambda) = O(\lambda^{-1})$. ∎

Proposition. *The Jost function is analytic in the upper half-plane* $\operatorname{Im}(\lambda) > 0$. *Furthermore,*

$$a(\lambda) = e^{-\frac{i\pi Q}{2}} + O\left(\frac{1}{|\lambda|}\right), \quad |\lambda| \to \infty; \quad a(\lambda) = e^{\frac{i\pi Q}{2}} + O\left(|\lambda|\right), \quad |\lambda| \to 0$$

Recall also that $a(-\lambda) = e^{-i\pi Q} a^*(\lambda)$ *for real* λ.

<u>Proof.</u> This follows from the relation $a(\lambda) = -\frac{1}{2}W(f_1, f_2)$, and from the fact that f_1, f_2 are analytic in the upper half-plane. ∎

Remark 2. When $x \to -\infty$, we see that

$$\widehat{f}_1 \sim a(\lambda) \begin{pmatrix} 1 \\ 1 \end{pmatrix} - b(\lambda) \begin{pmatrix} 1 \\ -1 \end{pmatrix} e^{-2ik(\lambda)x}$$

Letting $x \to -\infty$ in the above Volterra system for F, G, we can directly identify:

$$a(\lambda) = e^{-i\frac{\pi Q}{2}} - i \int_{-\infty}^{\infty} dy \left\{ \frac{\beta}{2}(\dot{\varphi} - \varphi')F(y, \lambda) \right.$$

$$\left. + m\lambda^{-1}\left((1 - \cos(2\beta\varphi))G(y, \lambda) - i\sin(2\beta\varphi)F(y, \lambda) \right) \right\}$$

$$b(\lambda) = -i \int_{-\infty}^{\infty} dy\, e^{2ik(\lambda)y} \left\{ \frac{\beta}{2}(\dot{\varphi} - \varphi')G(y, \lambda) \right.$$

$$\left. - m\lambda^{-1}\left((1 - \cos(2\beta\varphi))F(y, \lambda) - i\sin(2\beta\varphi)G(y, \lambda) \right) \right\}$$

We see in the expression for $a(\lambda)$ that it can readily be extended in the upper half-plane, since F and G have such an extension. By contrast, since $b(\lambda)$ is a Fourier transform on the whole real axis, it cannot be extended outside the real λ axis in general. However, if the field φ attains its asymptotic values at finite distance, the integral is over a finite interval, and $b(\lambda)$ admits an analytic continuation in the whole λ-plane. Similarly, if we assume that the field $\varphi(x)$ is C^{∞}, the functions F, G, solutions of the integral equation, will also be C^{∞} in x, so that $\lambda^n b(\lambda) \to 0$ for $\lambda \to \infty$ for any $n > 0$, and $\lambda^{-n}b(\lambda) \to 0$ for $\lambda \to 0$. This is because, integrating by parts,

$$\int_{-\infty}^{\infty} dx\, e^{i\lambda x} h(x) = -\frac{1}{i\lambda} \int_{-\infty}^{\infty} dx\, e^{i\lambda x} h'(x)$$

where the second integral exists by hypothesis. Iterating this formula, we see that the Fourier transform of a C^{∞} function $h(x)$ goes to 0 faster than any power of λ when $\lambda \to \infty$. Noting that $|a|^2 + |b|^2 = 1$, we see that the modulus of a tends rapidly to 1 for $\lambda \to 0, \infty$ and so all information is contained in its phase.

We will need alternative representations of the Jost solutions f_1 and f_2, in which the λ dependence is explicit. It is convenient to first get rid of the explicit φ dependence in the asymptotic expansions eq. (13.18), by defining:

$$f_1 = e^{\frac{i}{2}(\beta\varphi - Q\pi)\sigma_3} \widehat{f}_1, \quad f_2 = e^{\frac{i}{2}\beta\varphi\sigma_3} \widehat{f}_2$$

Proposition. *These functions admit the following Fourier representations:*

$$\widehat{f}_1(x, \lambda) = e^{ik(\lambda)x} \widehat{f}_1^0 + \int_x^{\infty} dy \left(\widehat{U}_1(x, y) + \lambda^{-1}\widehat{W}_1(x, y) \right) e^{ik(\lambda)y} \quad (13.19)$$

$$\widehat{f}_2(x, \lambda) = e^{-ik(\lambda)x} \widehat{f}_2^0 + \int_{-\infty}^x dy \left(\widehat{U}_2(x, y) + \lambda^{-1}\widehat{W}_2(x, y) \right) e^{-ik(\lambda)y}$$

where $\widehat{U}_i(x,y), \widehat{W}_i(x,y)$ *are two component vectors, and*

$$\widehat{f}_1^0 = \begin{pmatrix} 1 \\ e^{i\pi Q} \end{pmatrix}, \quad \widehat{f}_2^0 = \begin{pmatrix} 1 \\ -1 \end{pmatrix}$$

<u>Proof.</u> Consider a function $f(\lambda)$ analytic in the upper half-plane . Since $k(\lambda) = m(\lambda - \lambda^{-1})$, the map $\lambda \to k$ covers twice the upper half k-plane, the two values λ and $-1/\lambda$ map to the same value of k. So the function $f(\lambda)$ can be written:

$$f(\lambda) = \frac{1}{2}\left(f(\lambda) + f(-\frac{1}{\lambda})\right) + \frac{1}{2}\left(f(\lambda) - f(-\frac{1}{\lambda})\right) = g_1(k) + \left(\lambda + \frac{1}{\lambda}\right)g_2(k)$$

where g_1 and g_2 are analytic functions of k. If f is bounded at $\lambda = 0$ and $\lambda = \infty$, this implies that $g_1(k)$ is bounded and $g_2(k)$ tends to zero at $k \to \infty$. Alternatively, since $\lambda + \frac{1}{\lambda} = 2\lambda^{-1} + \frac{1}{m}k$, we can write $f(\lambda) = h_1(k) + \lambda^{-1}h_2(k)$, and the function h_1 is bounded when $k \to \infty$, while h_2 tends to zero. By the classical Paley–Wiener theorem, one can represent the functions h_i in the form:

$$h_i(k) = c_i + \int_0^\infty u_i(y)e^{iky}dy \qquad (13.20)$$

where the functions $u_i(y)$ are sufficiently regular so that the Fourier integral tends to zero when $k \to \infty$, and c_i is then the limiting value of h_i. The analyticity of h_i in the upper half-plane is accounted for by the support of the Fourier transform on the positive half–line. To apply this to the Jost solutions, we note that $e^{-ikx}\widehat{f}_1$ is analytic in the upper half-plane. Moreover, $e^{-ikx}\widehat{f}_1$ is bounded at $\lambda \to 0, \infty$ by eq. (13.18), hence one can represent it as in eq. (13.20). Multiplying by e^{ikx} and changing variables $y + x \to y$ in the integral (13.20), we arrive at eq. (13.19). The other equation is obtained similarly. ∎

The kernels $\widehat{U}_i, \widehat{W}_i$ are determined by eq. (13.7) as follows. The function $\widehat{f}_1$ obeys the gauge transformed equation $(\partial_x - \widehat{A}_x)\widehat{f}_1 = 0$ with the connection:

$$\widehat{A}_x = i\begin{pmatrix} \frac{\beta}{2}(\dot{\varphi} - \varphi') & me^{i\pi Q}(\lambda - \lambda^{-1}e^{-2i\beta\varphi}) \\ me^{i\pi Q}(\lambda - \lambda^{-1}e^{2i\beta\varphi}) & -\frac{\beta}{2}(\dot{\varphi} - \varphi') \end{pmatrix}$$

We will use the notation

$$\widehat{A}_x = \lambda\widehat{A}_1 + \widehat{A}_0 + \lambda^{-1}\widehat{A}_{-1}$$

When we insert the representation eq. (13.19) into that equation, terms in λ and λ^{-2} appear. To rewrite them, we use:

$$\lambda^{-2}e^{ik(\lambda)y} = e^{ik(\lambda)y} - \frac{1}{im\lambda}\partial_y e^{ik(\lambda)y}, \quad \lambda e^{ik(\lambda)y} = \frac{1}{im}\partial_y e^{ik(\lambda)y} + \frac{1}{\lambda}e^{ik(\lambda)y}$$

and perform integration by parts. The differential equation then translates into the following relations on the kernels $\widehat{U}_1(x,y)$ and $\widehat{W}_1(x,y)$:

$$\left(\partial_x + \frac{1}{im}\widehat{A}_1(x)\partial_y\right)\widehat{U}_1 = \widehat{A}_0(x)\widehat{U}_1 + \left(\widehat{A}_1(x) + \widehat{A}_{-1}(x)\right)\widehat{W}_1$$

$$\left(\partial_x - \frac{1}{im}\widehat{A}_{-1}(x)\partial_y\right)\widehat{W}_1 = \widehat{A}_0(x)\widehat{W}_1 + \left(\widehat{A}_1(x) + \widehat{A}_{-1}(x)\right)\widehat{U}_1$$

and the boundary conditions:

$$\left(1 - \frac{1}{im}\widehat{A}_1(x)\right)\widehat{U}_1(x,x) = -\widehat{A}_0(x)\widehat{f}_1^0$$

$$\left(1 + \frac{1}{im}\widehat{A}_{-1}(x)\right)\widehat{W}_1(x,x) = -\left(\widehat{A}_{-1}(x) + im\right)\widehat{f}_1^0$$

Alternatively, knowing $\widehat{U}_i(x,y)$ and $\widehat{W}_i(x,y)$, these equations allow us to reconstruct the connection $\widehat{A}_x(x,\lambda)$ and therefore the field $\varphi(x,t)$. In particular, the boundary conditions yield

$$e^{2i\beta\varphi(x)} = \frac{im + e^{i\pi Q}(\widehat{W}_1)_2(x,x)}{im + (\widehat{W}_1)_1(x,x)} \tag{13.21}$$

The Gelfand–Levitan–Marchenko equation which we will present below relates $\widehat{U}_i(x,y)$ and $\widehat{W}_i(x,y)$ directly to the scattering data.

13.3 Inverse scattering as a Riemann–Hilbert problem

One can of course define Jost solutions, f_3, f_4, analytic in the lower half-plane by choosing the appropriate boundary conditions:

$$f_3 \sim \begin{pmatrix} 1 \\ -e^{i\pi Q} \end{pmatrix} e^{-ik(\lambda)x}, \quad x \to +\infty; \quad f_4 \sim \begin{pmatrix} 1 \\ 1 \end{pmatrix} e^{ik(\lambda)x}, \quad x \to -\infty$$

$$\tag{13.22}$$

These solutions are linear combinations of f_1, f_2. By comparing at $x = \pm\infty$, we find the relations, valid for real λ:

$$f_3 = \frac{b^*(\lambda)}{a(\lambda)}e^{-iQ\pi}f_1 + \frac{e^{-iQ\pi}}{a(\lambda)}f_2, \quad f_4 = \frac{1}{a(\lambda)}f_1 + \frac{b(\lambda)}{a(\lambda)}f_2 \tag{13.23}$$

We can write $b(\lambda) = -\frac{1}{2}W(f_1, f_4)$. Since f_1 and f_4 are not analytic in the same half-plane, we recover the fact that $b(\lambda)$ cannot be extended outside the real axis in general. Let us define the matrices $\Theta_\pm(\lambda)$, analytic in the upper and lower half-plane respectively, by (recall that f_i are two-component vectors):

$$(f_2, f_1) = \Theta_+(\lambda)e^{-ik(\lambda)\sigma_3 x}, \quad \left(\frac{1}{a^*(\lambda)}e^{iQ\pi}f_3, \frac{1}{a^*(\lambda)}f_4\right) = \Theta_-(\lambda)e^{-ik(\lambda)\sigma_3 x}$$

The factors $e^{-ik(\lambda)x}$ are introduced so that $\Theta^\pm(\lambda)$ have finite limits when $\lambda \to 0$ and $\lambda \to \infty$ in their respective domains of analyticity. We can write eq. (13.23) as

$$\Theta_-^{-1}\Theta_+ = \begin{pmatrix} 1 & -b(\lambda)e^{-2ik(\lambda)x} \\ -b^*(\lambda)e^{2ik(\lambda)x} & 1 \end{pmatrix}$$

This is a Riemann–Hilbert problem, typical of a dressing transformation. Note, however, that the matrix Θ_+ is degenerate at the zeroes of $a(\lambda)$ in the upper half-plane and the matrix Θ_-^{-1} is degenerate at the zeroes of $a^*(\lambda)$ in the lower half-plane. We are thus led to a Riemann–Hilbert problem with zeroes, as discussed in Chapter 3. In the following, we propose another route to the solution of the inverse scattering problem, by transforming it to the Gelfand–Levitan–Marchenko linear integral equation.

13.4 Time evolution of the scattering data

In the previous sections, from a field $\varphi(x, 0)$ at time $t = 0$, we have defined the scattering data $a(\lambda)$, $b(\lambda)$, λ_n and c_n. The second step in the classical inverse scattering method is to compute the time evolution of the scattering data, which turns out to be beautifuly simple.

Proposition. *The time evolution of the sine-Gordon theory linearizes on the scattering data:*

$$\dot{a}(\lambda, t) = 0, \quad \dot{b}(\lambda, t) = 2im(\lambda + \lambda^{-1})b(\lambda, t) \tag{13.24}$$
$$\dot{\lambda}_n = 0, \quad \dot{c}_n = -2im(\lambda_n + \lambda_n^{-1})c_n$$

Proof. Recall that for the Jost solution $f_1(x, \lambda)$ we have

$$f_1(x, \lambda)|_{x \to +\infty} \sim \begin{pmatrix} 1 \\ e^{iQ\pi} \end{pmatrix} e^{ik(\lambda)x}$$

$$f_1(x, \lambda)|_{x \to -\infty} \sim a(\lambda)\begin{pmatrix} 1 \\ 1 \end{pmatrix} e^{ik(\lambda)x} - b(\lambda)\begin{pmatrix} 1 \\ -1 \end{pmatrix} e^{-ik(\lambda)x}$$

Consider now the time evolution of Ψ given by the second equation of the linear system: $\frac{\partial \Psi}{\partial t} - A_t \Psi = 0$. In the limit $x \to +\infty$, it reduces to

$$\frac{\partial \Psi}{\partial t} + ie^{iQ\pi} m(\lambda + \lambda^{-1}) \begin{pmatrix} 0 & 1 \\ 1 & 0 \end{pmatrix} \Psi = 0 \qquad (13.25)$$

Choose $\Psi = \alpha(t) f_1(x, t, \lambda)$. Then the asymptotic time evolution equation at $x \to +\infty$ gives $\dot{\alpha} = -im(\lambda + \lambda^{-1})\alpha$ and for $x \to -\infty$ it gives

$$\frac{d}{dt}(\alpha a) + im(\lambda + \lambda^{-1})(\alpha a) = 0$$

$$\frac{d}{dt}(\alpha b) - im(\lambda + \lambda^{-1})(\alpha b) = 0$$

Therefore $\dot{a}(\lambda, t) = 0$ and $\dot{b}(\lambda, t) = 2im(\lambda + \lambda^{-1})b(\lambda, t)$ as claimed in the proposition.

For bound states, we have by definition $a(\lambda_n) = 0$. So λ_n does not evolve. Consider now the wave function $f_n(x) \equiv f_2(x, \lambda_n) = c_n f_1(x, \lambda_n)$. We have

$$f_n(x)|_{x \to -\infty} \sim \begin{pmatrix} 1 \\ -1 \end{pmatrix} e^{-ik(\lambda_n)x}, \qquad f_n(x)|_{x \to +\infty} \sim c_n \begin{pmatrix} 1 \\ e^{iQ\pi} \end{pmatrix} e^{ik(\lambda_n)x}$$

Take $\Psi_n = \alpha(t) f_n$. For $x \to -\infty$ we have $\dot{\alpha} = im(\lambda_n + \lambda_n^{-1})\alpha$, while for $x \to +\infty$ we have $\dot{c}_n + 2im(\lambda_n + \lambda_n^{-1})c_n = 0$. ∎

Integrating eqs. (13.24), we get simple time evolutions of the scattering data:

$$a(\lambda, t) = a(\lambda, 0), \quad b(\lambda, t) = e^{+2im(\lambda + \lambda^{-1})t} b(\lambda, 0) \qquad (13.26)$$

$$\lambda_n(t) = \lambda_n(0), \quad c_n(t) = e^{-2im(\lambda_n + \lambda_n^{-1})t} c_n(0)$$

13.5 The Gelfand–Levitan–Marchenko equation

We now explain the inverse problem which amounts to reconstructing the potential from the scattering data. The Gelfand–Levitan–Marchenko equation is a linear integral equation which determines the kernels $\widehat{U}_i(x, y)$ and $\widehat{W}_i(x, y)$ from the scattering data. Once these kernels are known, the local fields are reconstructed from their boundary values $\widehat{U}_i(x, x)$, $\widehat{W}_i(x, x)$ by eq. (13.21).

Recall eq. (13.11), which we write in the form:

$$\frac{f_2(x, \lambda)}{a(\lambda)} = r(\lambda) f_1(x, \lambda) + i\overline{f}_1(x, \lambda) \quad \text{with} \quad r(\lambda) = -\frac{b^*(\lambda)}{a(\lambda)} \qquad (13.27)$$

To introduce the kernels $\widehat{U}_i(x, y), \widehat{W}_i(x, y)$, we need to rewrite this equation in terms of the functions $\widehat{f}_i$. Recall that $f_1 = g\widehat{f}_1$ with $g = \exp\left(\frac{i}{2}(\beta\varphi - Q\pi)\sigma_3\right)$. We multiply eq. (13.27) by g^{-1}. Notice that $g^{-1}f_2 = e^{i\frac{\pi Q}{2}\sigma_3}\overline{\widehat{f}}_2$ and $g^{-1}\overline{f}_1 = \overline{\widehat{f}}_1$, because $g^{-1}\overline{f}_1 = g^{-1}\sigma_2 f_1^* = g^{-1}\sigma_2 g^* \widehat{f}_1^*$ and $g^{-1}\sigma_2 g^* = \sigma_2$. We get:

$$e^{i\frac{\pi Q}{2}\sigma_3}\frac{1}{a(\lambda)}\widehat{f}_2(x, \lambda) = r(\lambda)\widehat{f}_1(x, \lambda) + i\overline{\widehat{f}}_1(x, \lambda) \qquad (13.28)$$

The Gelfand–Levitan–Marchenko equation is essentially the Fourier transform of this equation. To perform this transformation we need a lemma:

Lemma. *We have the relations*

$$\int_{-\infty}^{+\infty} e^{ik(\lambda)x}d\lambda = \frac{2\pi}{m}\delta(x)$$

$$\int_{-\infty}^{+\infty} e^{ik(\lambda)x}\frac{d\lambda}{\lambda} = 0 \qquad (13.29)$$

$$\int_{-\infty}^{+\infty} e^{ik(\lambda)x}\frac{d\lambda}{\lambda^2} = \frac{2\pi}{m}\delta(x)$$

Proof. Because of the singularities at $\lambda = 0, \infty$, one has to give a careful definition of these integrals on the real axis. We give a principal part definition, that is we set:

$$\int_{-\infty}^{\infty} d\lambda = \lim_{\epsilon \to 0}\left(\int_{-1/\epsilon}^{-\epsilon} d\lambda + \int_{\epsilon}^{1/\epsilon} d\lambda\right) \qquad (13.30)$$

Let us prove the first formula. Recall that $k(\lambda) = m(\lambda - \lambda^{-1})$, so that $k(-\lambda^{-1}) = k(\lambda)$. In the integral from $-1/\epsilon$ to $-\epsilon$, change $\lambda \to -\lambda^{-1}$ to get

$$\int_{-\infty}^{+\infty} e^{ik(\lambda)x}d\lambda = \lim_{\epsilon \to 0}\int_{\epsilon}^{1/\epsilon} e^{ik(\lambda)x}(1 + \lambda^{-2})d\lambda = \frac{1}{m}\int_{-\infty}^{+\infty} e^{ikx}dk = \frac{2\pi}{m}\delta(x)$$

The same technique applied to the second formula yields:

$$\int_{-\infty}^{+\infty} e^{ik(\lambda)x}\frac{d\lambda}{\lambda} = \lim_{\epsilon \to 0}\int_{\epsilon}^{1/\epsilon} e^{ik(\lambda)x}\left(-\frac{1}{\lambda} + \frac{1}{\lambda}\right)d\lambda = 0$$

Finally, the last equation is equivalent to the first one changing λ to $-1/\lambda$. ∎

Theorem. *The kernels* $\widehat{U}_1(x,y), \widehat{W}_1(x,y),$ $y \geq x$, *appearing in the Fourier transform of* $\widehat{f}_1$, *eq. (13.19), satisfy the linear integral equations:*

$$-\frac{2i\pi}{m}\overline{\widehat{U}}_1(x,y) = F_0(x+y)\widehat{f}_1^0$$

$$+ \int_x^\infty \Big(F_0(y+z)\widehat{U}_1(x,z) + F_{-1}(y+z)\widehat{W}_1(x,z) \Big) dz$$

$$-\frac{2i\pi}{m}\overline{\widehat{W}}_1(x,y) = F_{-1}(x+y)\widehat{f}_1^0 \qquad (13.31)$$

$$+ \int_x^\infty \Big(F_{-1}(y+z)\widehat{U}_1(x,z) + F_{-2}(y+z)\widehat{W}_1(x,z) \Big) dz$$

The functions $F_j(x)$ *are directly computed in terms of the scattering data by:*

$$F_j(x) = \int_{-\infty}^\infty d\lambda\, \lambda^j e^{ik(\lambda)x} r(\lambda) - 2i\pi \sum_n e^{ik(\lambda_n)x} \lambda_n^j m_n \qquad (13.32)$$

where we defined the parameters

$$m_n = \frac{c_n}{a'(\lambda_n)}$$

<u>Proof.</u> Note that, in eq. (13.31), $\widehat{U}_1$ and $\widehat{W}_1$ always appear with their second argument greater than the first one, in agreement with their definition. Hence the system of two equations (13.31) determines these two quantities in their domain of definition. We multiply eq. (13.28) by $\lambda^j e^{ik(\lambda)y}$ (for $j = 0, -1$) and integrate over λ from $-\infty$ to $+\infty$, with a principal part prescription, getting:

$$e^{i\frac{\pi Q}{2}\sigma_3} \int_{-\infty}^\infty d\lambda\, \lambda^j \frac{\widehat{f}_2(x,\lambda)}{a(\lambda)} e^{ik(\lambda)y}$$

$$= \int_{-\infty}^\infty d\lambda\, \lambda^j e^{ik(\lambda)y} \Big(r(\lambda)\widehat{f}_1(x,\lambda) + i\overline{\widehat{f}}_1(x,\lambda) \Big) \qquad (13.33)$$

Recall the Fourier representations:

$$\widehat{f}_1(x) = e^{ikx}\widehat{f}_1^0 + \int_x^\infty (\widehat{U}_1(x,z) + \lambda^{-1}\widehat{W}_1(x,z))e^{ikz}dz$$

$$\overline{\widehat{f}}_1(x) = e^{-ikx}\overline{\widehat{f}}_1^0 + \int_x^\infty (\overline{\widehat{U}}_1(x,z) + \lambda^{-1}\overline{\widehat{W}}_1(x,z))e^{-ikz}dz$$

where the second equation is derived from the first by complex conjugating and multiplying by σ_2. We evaluate the right-hand side of eq. (13.33) using the lemma. We find for $j = 0, -1$ respectively:

$$\int_{-\infty}^{\infty} d\lambda \, e^{iky} (r(\lambda) \widehat{f}_1(x) + i\overline{\widehat{f}}_1(x))$$

$$= R_0(x+y) \widehat{f}_1^0 + i\frac{2\pi}{m} \delta(x-y) \overline{\widehat{f}}_1^0 + i\frac{2\pi}{m} \theta(y-x) \overline{\widehat{U}}_1(x,y)$$

$$+ \int_x^{\infty} dz \left(R_0(y+z) \widehat{U}_1(x,z) + R_{-1}(y+z) \widehat{W}_1(x,z) \right) \quad (13.34)$$

and similarly:

$$\int_{-\infty}^{\infty} d\lambda \, \lambda^{-1} e^{iky} (r(\lambda) \widehat{f}_1(x) + i\overline{\widehat{f}}_1(x))$$

$$= R_{-1}(x+y) \widehat{f}_1^0 + i\frac{2\pi}{m} \theta(y-x) \overline{\widehat{W}}_1(x,y)$$

$$+ \int_x^{\infty} dz \left(R_{-1}(y+z) \widehat{U}_1(x,z) + R_{-2}(y+z) \widehat{W}_1(x,z) \right)$$

where we have introduced the notation:

$$R_j(x) = \int_{-\infty}^{\infty} d\lambda \, \lambda^j e^{ik(\lambda)x} r(\lambda), \quad j = 0, -1, -2$$

To evaluate the left-hand side of eq. (13.33) we use the analyticity properties of the Jost solutions, and the residue theorem. Recall that the λ integrals are defined with the prescription (13.30). We close the contour in the upper half-plane by introducing a small half-circle C_ϵ of center 0 and radius ϵ and a large half-circle $C_{1/\epsilon}$ of center 0 and radius $1/\epsilon$. In the upper half-plane , the integrand has poles at the zeroes λ_n of $a(\lambda)$. So the residue theorem gives:

$$e^{i\frac{\pi Q}{2}\sigma_3} \int_{-\infty}^{\infty} d\lambda \, \lambda^j \frac{\widehat{f}_2(x,\lambda)}{a(\lambda)} e^{ik(\lambda)y} = 2i\pi e^{i\frac{\pi Q}{2}\sigma_3} \sum_n e^{ik(\lambda_n)y} \lambda_n^j \frac{\widehat{f}_2(x,\lambda_n)}{a'(\lambda_n)}$$

$$+ e^{i\frac{\pi Q}{2}\sigma_3} \oint_{C_\epsilon + C_{1/\epsilon}} d\lambda \, \lambda^j \frac{\widehat{f}_2(x,\lambda)}{a(\lambda)} e^{ik(\lambda)y} \quad (13.35)$$

where $a'(\lambda) = \frac{da}{d\lambda}$. To proceed we need to evaluate the integrals on the half-circles. Using the asymptotic expansions eq. (13.18) we see that on these circles $\widehat{f}_2(x,\lambda) = e^{-ik(\lambda)x} \times$ regular, and similarly $a(\lambda)$ is regular. In particular, at ∞ we have

$$\frac{\widehat{f}_2(x,\lambda)}{a(\lambda)} = e^{\frac{i\pi Q}{2}} e^{-ik(\lambda)x} (\widehat{f}_2^0 + O(\lambda^{-1}))$$

We consider, first for $j = 0$, integrals of the form $\oint_C e^{ik(\lambda)(y-x)}d\lambda$, where k is large. The existence of such integrals *requires* $y \geq x$, otherwise the exponential explodes. Assuming in the following that this condition is satisfied, we have $|e^{ik(\lambda)(y-x)}| \leq 1$ for λ in the upper half-plane, hence the integral on C_ϵ is bounded by $\pi\epsilon$ and can be neglected when $\epsilon \to 0$. On the other hand, the integral over $C_{1/\epsilon}$ reduces by the residue theorem to the integral on the real axis:

$$\oint_{C_{1/\epsilon}} e^{ik(\lambda)(y-x)}d\lambda \sim \int_{-\infty}^{\infty} e^{im\lambda(y-x)}d\lambda = \frac{2\pi}{m}\delta(x-y)$$

so that the last term in eq. (13.35) is equal to $\frac{2\pi}{m}\delta(x-y)e^{\frac{i\pi Q(1+\sigma_3)}{2}}\widehat{f}_2^{\,0}$. Taking into account the value of $\overset{\rightarrow 0}{\widehat{f}_1}$, this term precisely cancels the $\delta(x-y)$ term in the right-hand side of eq. (13.34). For $j = -1$, we have to consider integrals of the form $\oint_{C_\epsilon} e^{ik(\lambda)x}d\lambda/\lambda$ with $x > 0$. Write $\lambda = \epsilon e^{i\theta}$ with $0 < \theta < \pi$, and fix $\eta > 0$ small enough. The integral takes the form:

$$\oint_{C_\epsilon} e^{ik(\lambda)x}\lambda^{-1}d\lambda \sim i\int_0^{\pi} d\theta\, e^{-\frac{mx}{\epsilon}\sin\theta - \frac{imx}{\epsilon}\cos\theta}$$

On the interval $[\eta, \pi - \eta]$ the integrand decays exponentially when $\epsilon \to 0$. On the intervals $[0, \eta]$ and $[\pi - \eta, \pi]$ the integrand is bounded by 1, and the integral by 2η. So the contribution on C_ϵ can be neglected. A similar analysis shows that the contribution on $C_{1/\epsilon}$ can also be neglected.

We now evaluate $\widehat{f}_2(x, \lambda_n)$ appearing in eq. (13.35). Precisely at the zeroes λ_n of $a(\lambda)$ we have $f_2(x, \lambda_n) = c_n f_1(x, \lambda_n)$, which translates into $\widehat{f}_2(x, \lambda_n) = c_n e^{-i\frac{1}{2}\pi Q\sigma_3}\widehat{f}_1(x, \lambda_n)$. Replacing $\widehat{f}_1(x, \lambda_n)$ by its expression, eq. (13.19), one gets:

$$e^{ik(\lambda_n)y}\lambda_n^j \frac{\widehat{f}_2(x, \lambda_n)}{a'(\lambda_n)} = e^{ik(\lambda_n)y}\frac{c_n\lambda_n^j}{a'(\lambda_n)}e^{-i\frac{1}{2}\pi Q\sigma_3}\left[e^{ik(\lambda_n)x}\widehat{f}_1^{\,0} \right.$$
$$\left. + \int_x^{\infty} dz\, e^{ik(\lambda_n)z}(\widehat{U}_1(x, z) + \lambda_n^{-1}\widehat{W}_1(x, z))\right]$$

Combining everything finally yields the Gelfand–Levitan–Marchenko equations. ∎

13.6 Soliton solutions

The solution of the Gelfand–Levitan–Marchenko equation (13.31) is particularly simple when we take $R(x) = 0$. This corresponds to $b(\lambda) = 0$, which means that there is no reflection in the auxiliary scattering problem. Corresponding potentials are called reflectionless potentials. Then the kernels F_j are degenerate and the Gelfand–Levitan–Marchenko equations reduce to a finite linear system. The sine-Gordon solutions $\varphi(x,t)$ we get in this way are just the multi-soliton solutions.

If there is no reflection, the scattering data are λ_n and m_n. The Gelfand–Levitan–Marchenko kernels read:

$$F_j(x,y) = -2i\pi \sum_n m_n \lambda_n^j e^{ik(\lambda_n)(x+y)}$$

Looking at the y dependence in the Gelfand–Levitan–Marchenko equation shows that $\widehat{U}_1(x,y)$ and $\widehat{W}_1(x,y)$ can be expanded as:

$$\widehat{U}_1(x,y) = \sum_n e^{-ik(\lambda_n^*)(x+y)} m_n^* u_n(x)$$

$$\widehat{W}_1(x,y) = \sum_n e^{-ik(\lambda_n^*)(x+y)} m_n^* w_n(x)$$

The y exponentials have been chosen to remain bounded when $y \to \infty$, so that the z integrals in the Gelfand–Levitan–Marchenko equations converge. The factor $m_n^* e^{-ik(\lambda_n^*)x}$ has been introduced to simplify later formulae.

Inserting these forms into the Gelfand–Levitan–Marchenko equations and identifying the coefficient of $\exp(ik(\lambda_n)y)$ on both sides yields:

$$\frac{1}{m}\bar{u}_n(x) = \widehat{f}_1^0 + \sum_p \frac{im_p^*}{k(\lambda_n) - k(\lambda_p^*)} e^{-2ik(\lambda_p^*)x}\left(u_p(x) + \lambda_n^{-1} w_p(x)\right)$$

$$\frac{\lambda_n}{m}\bar{w}_n(x) = \widehat{f}_1^0 + \sum_p \frac{im_p^*}{k(\lambda_n) - k(\lambda_p^*)} e^{-2ik(\lambda_p^*)x}\left(u_p(x) + \lambda_n^{-1} w_p(x)\right)$$

Since the right-hand sides of these equations are identical, we get $w_n(x) = \frac{1}{\lambda_n^*} u_n(x)$. Substituting back into the equation for u_n and noting that $k(\lambda_n) - k(\lambda_p^*) = m(\lambda_n - \lambda_p^*)(1 + 1/(\lambda_n \lambda_p^*))$, the equation simplifies to:

$$\bar{u}_n(x) = m\widehat{f}_1^0 + \sum_p \frac{im_p^*}{\lambda_n - \lambda_p^*} e^{-2ik(\lambda_p^*)x} u_p(x)$$

In the following, we restrict ourselves to pure soliton and antisoliton solutions (no breathers), so that the λ_p are pure imaginary. To connect with the notations of Chapter 12 we set $\lambda_p = i\mu_p$. The Gelfand–Levitan–Marchenko equation becomes, in matrix notation:

$$\bar{u} = mf + Vu \tag{13.36}$$

where u is the vector with components $u_n = \begin{pmatrix} u_{1n} \\ u_{2n} \end{pmatrix}$, f is the vector with components $f_n = \hat{f}_1^0 e_n$, where $e_n = 1$ for all n, and we have defined the matrix:

$$V_{np} = \frac{\mu_p}{\mu_n + \mu_p} X_p, \quad X_p = \frac{m_p}{\mu_p} e^{-2m(\mu_p + \mu_p^{-1})x} \qquad (13.37)$$

Note that, since $a(-\lambda^*) = e^{i\pi Q}(a(\lambda))^*$ and $c_n^* = -e^{i\pi Q}c_n$, we have $m_p^* = m_p$, and the matrix V is real in this pure solitonic case. Note that, if one includes the time dependence, i.e. $m_p \to m_p e^{2m(\mu_p - \mu_p^{-1})t}$, the exponential appearing in V becomes $\exp(-2m[\mu_p(x-t) + \mu_p^{-1}(x+t)])$, which is the familiar form encountered in Chapter 12.

Taking the bar of eq. (13.36) and using that $\bar{\bar{u}} = -u$, we get $-u = m\bar{f} + V\bar{u}$ so that, eliminating $\bar{u}$:

$$(1 + V^2)u = -m\bar{f} - mVf$$

Writing $(1 + V^2) = (1 + ie^{i\pi Q}V)(1 - ie^{i\pi Q}V)$, this is solved by:

$$u_1 = im e^{i\pi Q}(1 - ie^{i\pi Q}V)^{-1}e, \quad u_2 = -im(1 + ie^{i\pi Q}V)^{-1}e$$

We can then compute the field φ by eq. (13.21). This is done in the:

Proposition. *We have:*

$$im + (\widehat{W}_1)_1(x, x) = im\frac{\det(1 + ie^{i\pi Q}V)}{\det(1 - ie^{i\pi Q}V)} \qquad (13.38)$$

$$im + e^{i\pi Q}(\widehat{W}_1)_2(x, x) = im\frac{\det(1 - ie^{i\pi Q}V)}{\det(1 + ie^{i\pi Q}V)} \qquad (13.39)$$

so that $e^{-i\beta\varphi} = \frac{\tau_+}{\tau_-}$ *as in eqs. (12.67, 12.76) in Chapter 12.*

Proof. Let us prove the first equation. We have

$$(\widehat{W}_1)_1(x, x) = i\sum_n X_n(u_1)_n = -m e^{i\pi Q} \operatorname{Tr}(M), \quad M \equiv (1 - ie^{i\pi Q}V)^{-1}e \otimes X$$

where X is the vector of components X_n as in eq. (13.37). Hence

$$im + W_1(x, x) = im\left(1 + ie^{i\pi Q}\operatorname{Tr}(M)\right)$$

Note that M is of rank 1, so $1 + \operatorname{Tr}(ie^{i\pi Q}M) = \det(1 + ie^{i\pi Q}M)$. Moreover:

$$1 + ie^{i\pi Q}M = (1 - ie^{i\pi Q}V)^{-1}\left(1 - ie^{i\pi Q}(V - e \otimes X)\right)$$

Finally, we remark that $(V - e \otimes X) = -\mu V \mu^{-1}$ so that

$$1 + ie^{i\pi Q} M = (1 - ie^{i\pi Q} V)^{-1} \mu (1 + ie^{i\pi Q} V) \mu^{-1}$$

Taking the determinant proves eq. (13.38). Equation (13.39) is obtained similarly. Equation (13.21) gives:

$$e^{2i\beta\varphi} = \frac{\det^2 (1 - ie^{i\pi Q} V)}{\det^2 (1 + ie^{i\pi Q} V)}$$

which identifies with the tau-function formula. The parameters a_n, μ_n in Chapter 12 are related to m_n, λ_n in this chapter by:

$$\lambda_n = i\mu_n, \quad m_n = -2ie^{i\pi Q} \mu_n a_n \tag{13.40}$$

■

13.7 Poisson brackets of the scattering data

We now consider the sine-Gordon equation from the Hamiltonian point of view. The aim is to compute the Poisson brackets of the scattering data defined in the previous sections. We follow the method of the classical r-matrix introduced by Faddeev, Sklyanin and Takhtajan.

We start from the canonical Poisson brackets:

$$\{\pi(x), \varphi(y)\} = \delta(x - y)$$

and define the Hamiltonian on the interval $[-\ell, \ell]$, eventually ℓ will tend to ∞:

$$H = \int_{-\ell}^{\ell} dx \left(\frac{1}{2} \pi^2(x) + \frac{1}{2} (\partial_x \varphi)^2(x) + \frac{4m^2}{\beta^2} (1 - \cos(2\beta\varphi)) \right) dx$$

The equations of motion read $\dot{\varphi} = \pi$ and $\dot{\pi} = \varphi'' - \frac{8m^2}{\beta} \sin(2\beta\varphi)$, which reproduces eq. (13.1).

Consider the auxiliary linear problem eq. (13.7) on the interval $[-\ell, \ell]$. Let $\Psi(-\ell)$ be the value of a solution at $x = -\ell$ and $\Psi(\ell)$ its value at $x = \ell$. Then we can write:

$$\Psi(\ell) = T_\ell(\lambda) \Psi(-\ell)$$

where $T_\ell(\lambda)$ is the monodromy matrix.

Proposition. *The monodromy matrix and the scattering data are directly related by:*

$$
T(\lambda) \equiv \begin{pmatrix} a(\lambda) & b(\lambda) \\ -b^*(\lambda) & a^*(\lambda) \end{pmatrix}
$$
$$
= \frac{1}{2} \lim_{\ell \to \infty} \begin{pmatrix} e^{i\pi Q} e^{ik(\lambda)\ell} & -e^{ik(\lambda)\ell} \\ e^{-ik(\lambda)\ell} & e^{i\pi Q} e^{-ik(\lambda)\ell} \end{pmatrix} T_\ell(\lambda) \begin{pmatrix} e^{ik(\lambda)\ell} & e^{-ik(\lambda)\ell} \\ -e^{ik(\lambda)\ell} & e^{-ik(\lambda)\ell} \end{pmatrix}
$$

$$(13.41)$$

<u>Proof.</u> From the asymptotic form of the Jost solutions and the relation $|a|^2 + |b|^2 = 1$ we find

$$
\frac{1}{a} f_1 + \frac{b}{a} f_2 \sim \begin{cases} \begin{pmatrix} 1 \\ 1 \end{pmatrix} e^{ik(\lambda)x} & x \to -\infty \\ \\ a^* \begin{pmatrix} 1 \\ e^{i\pi Q} \end{pmatrix} e^{ik(\lambda)x} + b \begin{pmatrix} e^{i\pi Q} \\ -1 \end{pmatrix} e^{-ik(\lambda)x} & x \to +\infty \end{cases}
$$

Any solution $\Psi(x)$ can be written as $\psi_1(\frac{1}{a} f_1 + \frac{b}{a} f_2) + \psi_2 f_2$ and behaves at $x = -\ell$ as:

$$
\Psi(-\ell) = \begin{pmatrix} e^{-ik(\lambda)\ell} & e^{ik(\lambda)\ell} \\ e^{-ik(\lambda)\ell} & -e^{ik(\lambda)\ell} \end{pmatrix} \begin{pmatrix} \psi_1 \\ \psi_2 \end{pmatrix}
$$

while at $x = +\infty$ it behaves as:

$$
\Psi(\ell) = \begin{pmatrix} a^* e^{ik(\lambda)\ell} + b e^{i\pi Q} e^{-ik(\lambda)\ell} & -b^* e^{ik(\lambda)\ell} + a e^{i\pi Q} e^{-ik(\lambda)\ell} \\ a^* e^{i\pi Q} e^{ik(\lambda)\ell} - b e^{-ik(\lambda)\ell} & -b^* e^{i\pi Q} e^{ik(\lambda)\ell} - a e^{-ik(\lambda)\ell} \end{pmatrix} \begin{pmatrix} \psi_1 \\ \psi_2 \end{pmatrix}
$$

The result follows from writing $\Psi(\ell) = T_\ell(\lambda)\Psi(-\ell)$, and conjugating the relation obtained by σ_1. ∎

To compute the Poisson brackets of the scattering data we need the r-matrix relation:

Proposition.

$$
\{T_{\ell,1}(\lambda), T_{\ell,2}(\mu)\} = [r_{12}(\lambda, \mu), T_\ell(\lambda) \otimes T_\ell(\mu)] \tag{13.42}
$$

where the matrix $r_{12}(\lambda, \mu)$ is given by

$$
r_{12}(\lambda, \mu) = \frac{\beta^2}{4} \left(\frac{\lambda^2 + \mu^2}{\lambda^2 - \mu^2} H \otimes H + \frac{4\lambda\mu}{\lambda^2 - \mu^2}(E_+ \otimes E_- + E_- \otimes E_+) \right)
$$

$$(13.43)$$

<u>Proof.</u> As we know from Chapter 3, see eqs. (3.90, 3.91), it suffices to prove the much simpler local relation:

$$\{A_{x,1}(\lambda, x), A_{x,2}(\mu, y)\} = [r_{12}(\lambda, \mu), A_x(\lambda, x) \otimes I + I \otimes A_x(\mu, y)]\delta(x - y) \tag{13.44}$$

The r-matrix is obtained, up to a factor, by the general formula for Toda models,

$$r_{12} = \frac{1}{2}(r_{12}^+ + r_{12}^-)$$

The formulae for $r_{12}^\pm$ are given by eqs. (12.28, 12.29) in Chapter 12 in which we insert the root decomposition, eq. (12.59), in the same chapter. We get in the loop representation:

$$\begin{aligned}
r_{12}^+ = \frac{1}{4}H \otimes H + \sum_{n>0} \Big(\frac{1}{2}\lambda^{2n}H \otimes \mu^{-2n}H \\
+\lambda^{2n-1}E_+ \otimes \mu^{-2n+1}E_- + \lambda^{2n-1}E_- \otimes \mu^{-2n+1}E_+\Big)
\end{aligned}$$

(we take as invariant bilinear form on $sl(2)$ the trace in the 2×2 representation, so that $(H, H) = 2$ and $(E_+, E_-) = 1$. This accounts for the relative factors). Summing the geometric series yields eq. (13.43) in which $|\lambda| < |\mu|$. The formula for r_{12}^- is the same but with $|\lambda| > |\mu|$. The antisymmetric matrix r_{12} is just the half sum of these two identical rational functions for $\lambda \ne \mu$. It is easy to check that the relation eq. (13.44) holds true, and this allows us to adjust the factor β^2.   ∎

Proposition. *The complete list of Poisson brackets of the scattering data is:*
(1) continuum–continuum:

$$\{a(\lambda), b(\mu)\} = \beta^2 \frac{\lambda\mu}{(\lambda + \mu)(\lambda - \mu + i0)} a(\lambda)b(\mu)$$

$$\{a(\lambda), b^*(\mu)\} = -\beta^2 \frac{\lambda\mu}{(\lambda + \mu)(\lambda - \mu + i0)} a(\lambda)b^*(\mu)$$

$$\{b(\lambda), b^*(\mu)\} = -i\pi\beta^2 \, \lambda \, |a(\lambda)|^2 \, \delta(\lambda - \mu)$$

$$\{a(\lambda), a(\mu)\} = 0, \quad \{a(\lambda), a^*(\mu)\} = 0, \quad \{b(\lambda), b(\mu)\} = 0$$

(2) continuum–discrete:

$$\{a(\lambda), \lambda_n\} = 0, \quad \{a(\lambda), \lambda_n^*\} = 0$$

$$\{b(\lambda), \lambda_n\} = 0, \quad \{b(\lambda), \lambda_n^*\} = 0$$

$$\{b(\lambda), c_n\} = 0, \quad \{b(\lambda), c_n^*\} = 0$$

$$\{a(\lambda), c_n\} = -\beta^2 \frac{\lambda \lambda_n}{\lambda^2 - \lambda_n^2} a(\lambda) c_n, \quad \{a(\lambda), c_n^*\} = \beta^2 \frac{\lambda \lambda_n^*}{\lambda^2 - \lambda_n^{*2}} a(\lambda) c_n^*$$

(3) discrete–discrete:

$$\{\lambda_n, \lambda_m\} = 0, \quad \{\lambda_n, \lambda_m^*\} = 0$$

$$\{c_n, c_m\} = 0, \quad \{c_n, c_m^*\} = 0$$

$$\{\lambda_n, c_m\} = \frac{\beta^2}{2} \lambda_n c_m \delta_{nm}, \quad \{\lambda_n^*, c_m\} = 0 \text{ unless } \lambda_m = -\lambda_n^* \qquad (13.45)$$

<u>Proof.</u> Define $E_L(\lambda)$ and $E_R(\lambda)$ as the matrices:

$$E_L(\lambda) = \begin{pmatrix} e^{i\pi Q} e^{ik(\lambda)\ell} & -e^{ik(\lambda)\ell} \\ e^{-ik(\lambda)\ell} & e^{i\pi Q} e^{-ik(\lambda)\ell} \end{pmatrix}, \quad E_R(\lambda) = \begin{pmatrix} e^{ik(\lambda)\ell} & e^{-ik(\lambda)\ell} \\ -e^{ik(\lambda)\ell} & e^{-ik(\lambda)\ell} \end{pmatrix}$$

so that $T(\lambda) = \frac{1}{2} \lim_{\ell \to \infty} E_L(\lambda) T_\ell(\lambda) E_R(\lambda)$. Inserting this into eq. (13.42), we get the Poisson brackets of the elements of the matrix $T(\lambda)$:

$$\{T_1(\lambda), T_2(\mu)\} = r_+(\lambda, \mu) T_1(\lambda) T_2(\mu) - T_1(\lambda) T_2(\mu) r_-(\lambda, \mu)$$

where we have defined:

$$r_+(\lambda, \mu) = E_L(\lambda) \otimes E_L(\mu) r_{12}(\lambda, \mu) E_L^{-1}(\lambda) \otimes E_L^{-1}(\mu), \quad \ell \to \infty$$
$$r_-(\lambda, \mu) = E_R^{-1}(\lambda) \otimes E_R^{-1}(\mu) r_{12}(\lambda, \mu) E_R(\lambda) \otimes E_R(\mu), \quad \ell \to \infty$$

In order to evaluate these r-matrices, one first computes:

$$E_L H E_L^{-1} = e^{i\pi Q} \left(e^{2ik\ell} E_+ + e^{-2ik\ell} E_- \right)$$

$$E_L E_\pm E_L^{-1} = \frac{1}{2} \left(-e^{i\pi Q} H \pm e^{2ik\ell} E_+ \mp e^{-2ik\ell} E_- \right)$$

$$E_R^{-1} H E_R = e^{-2ik\ell} E_+ + e^{2ik\ell} E_-$$

$$E_R^{-1} E_\pm E_R = \frac{1}{2} \left(-H \pm e^{-2ik\ell} E_+ \mp e^{2ik\ell} E_- \right)$$

and then obtain $r_\pm$, before taking the $\ell \to \infty$ limit:

$$\frac{4}{\beta^2} r_\pm(\lambda, \mu) = \frac{2\lambda\mu}{\lambda^2 - \mu^2} H \otimes H$$

$$+ \frac{\lambda - \mu}{\lambda + \mu} e^{\pm 2i(k(\lambda) + k(\mu))\ell} E_+ \otimes E_+ + \frac{\lambda - \mu}{\lambda + \mu} e^{\mp 2i(k(\lambda) + k(\mu))\ell} E_- \otimes E_-$$

$$+ \frac{\lambda + \mu}{\lambda - \mu} e^{\pm 2i(k(\lambda) - k(\mu))\ell} E_+ \otimes E_- + \frac{\lambda + \mu}{\lambda - \mu} e^{\mp 2i(k(\lambda) - k(\mu))\ell} E_- \otimes E_+$$

We take the limit $\ell \to \infty$ in the sense of distribution theory. This is done using the formula:

$$\lim_{\ell \to \infty} P \frac{1}{x} e^{\pm i \ell x} = \pm i \pi \delta(x)$$

Indeed, if f is analytic with slow growth at ∞ and $x > 0$ one can compute

$$\left(\int_{-\infty}^{-\epsilon} + \int_{\epsilon}^{\infty} \right) \frac{1}{x} e^{\pm i \ell x} f(x) dx = i \pi f(0)$$

by considering the closed contour obtained by adding a half-circle C_ϵ and a half-circle at infinity. The integral on this last circle vanishes in the limit $\ell \to \infty$, and the integral on C_ϵ gives $-i\pi f(0)$. From this we deduce a formula more suited to our case,

$$\lim_{\ell \to \infty} P \frac{1}{\lambda - \mu} e^{\pm 2i(k(\lambda) - k(\mu))} = \pm i \pi \delta(\lambda - \mu)$$

which is obtained by the change of variables $\lambda - \mu \to 2(k(\lambda) - k(\mu))$ in the delta function. We have similar formulae for $\lambda + \mu$, but due to the symmetry properties under $\lambda \to -\lambda$, $a(-\lambda) = e^{i\pi Q} a^*(\lambda)$ and $b(-\lambda) = -e^{-i\pi Q} b^*(\lambda)$, we can restrict ourselves to $\lambda > 0$ and $\mu > 0$, and ignore the terms in $\delta(\lambda + \mu)$.

We can now take the limit $\ell \to \infty$ in $r_\pm$, getting:

$$r_\pm(\lambda, \mu) = \frac{\beta^2}{8} \left(P \frac{\lambda + \mu}{\lambda - \mu} - \frac{\lambda - \mu}{\lambda + \mu} \right) H \otimes H$$

$$\mp \frac{i\pi\beta^2}{4} (\lambda + \mu)(E_+ \otimes E_- - E_- \otimes E_+) \delta(\lambda - \mu)$$

From this we compute the Poisson brackets of the scattering data. For instance:

$$\{a(\lambda), b(\mu)\} = \beta^2 \left[\frac{\lambda\mu}{\lambda^2 - \mu^2} a(\lambda) b(\mu) - \frac{i\pi}{4} (\lambda + \mu) \delta(\lambda - \mu) b(\lambda) a(\mu) \right]$$

$$= \frac{\beta^2}{4} \left[(\lambda + \mu) \left(P \frac{1}{\lambda - \mu} - i\pi\delta(\mu - \lambda) \right) - \frac{\lambda - \mu}{\lambda + \mu} \right] a(\lambda) b(\mu)$$

$$= \beta^2 \frac{\lambda\mu}{(\lambda - \mu + i0)(\lambda + \mu)} a(\lambda) b(\mu)$$

where in the last step we have used the identity:

$$\frac{1}{x \pm i0} = P \frac{1}{x} \mp i\pi\delta(x)$$

Note that the left-hand side and the right-hand side are analytic in the upper λ half-plane , as it should be. The other Poisson brackets are computed similarly. There are sixteen Poisson brackets in $\{T_1, T_2\}$ but the independent ones are listed in the proposition.

The Poisson brackets for the discrete spectrum would require a detailed analysis, but they can be obtained quickly using the following trick. We insisted already on the fact that the function $b(\mu)$ cannot be analytically continued in the upper half-plane. The situation changes, however, if the field $\varphi(x)$ is *compactly supported*. Then $b(\mu)$ is analytic in the plane, and we have seen that:

$$c_n = -\frac{1}{b(\lambda_n)}$$

Assuming that we are in such a case, setting $\mu = \lambda_m$ into the equation for $\{a(\lambda), b(\mu)\}$, we get immediately $\{a(\lambda), c_m\} = -\beta^2 \lambda \lambda_m / (\lambda^2 - \lambda_m^2) a(\lambda) c_m$. Letting further $a(\lambda) = (\lambda - \lambda_n)\tilde{a}_n$ in this equation with $\lambda \to \lambda_n$, one gets $\{\lambda_n, c_m\} = \frac{\beta^2}{2} \lambda_n c_m \delta_{nm}$. The remaining Poisson brackets are computed similarly.   ∎

13.8 Action–angle variables

Due to the boundary conditions of the field φ, which differ at $x = -\infty$ and $x = +\infty$, the generating function of conserved quantities is a modified trace of the monodromy matrix, specifically $\text{Tr}\,(T_\ell(\lambda)\rho)$, where ρ is the diagonal matrix $\rho = \text{Diag}\,(e^{-\frac{i\pi Q}{2}}, e^{\frac{i\pi Q}{2}})$. This is a consequence of the zero curvature condition which implies, (see eq. (3.72) in Chapter 3):

$$\partial_t T_\ell(\lambda, t) = A_t(\lambda, \ell) T_\ell(\lambda, t) - T_\ell(\lambda, t) A_t(\lambda, -\ell)$$

and the explicit values $A_t(\lambda, \ell) = -im(\lambda + \lambda^{-1})e^{i\pi Q}\sigma_1$ and $A_t(\lambda, -\ell) = -im(\lambda + \lambda^{-1})\sigma_1$ for $\ell \to +\infty$. Computing $T_\ell(\lambda, t)$ from eq. (13.41), we can express $\text{Tr}\,(T_\ell(\lambda)\rho)$ in terms of the scattering data:

$$\text{Tr}\,(T_\ell(\lambda)\rho) \sim e^{\frac{i\pi Q}{2}} e^{-2ik(\lambda)\ell} a(\lambda) + e^{-\frac{i\pi Q}{2}} e^{2ik(\lambda)\ell} a^*(\lambda)$$

So the generating functional for conserved quantities can be taken as $a(\lambda)$. Since $\{a(\lambda), a(\mu)\} = 0$ these conserved quantities Poisson commute.

Remembering the asymptotics

$$a(\lambda) = e^{-\frac{i\pi Q}{2}} + O\left(\frac{1}{|\lambda|}\right), \quad |\lambda| \to \infty; \quad a(\lambda) = e^{\frac{i\pi Q}{2}} + O\left(|\lambda|\right), \quad |\lambda| \to 0$$

we see that we can expand $\log a(\lambda)$ around $\lambda = 0$ and $\lambda = \infty$:

$$\log a(\lambda) = -\frac{i\pi Q}{2} + \sum_{n=1}^{\infty} I_n^+ (i\lambda)^{-n}, \quad |\lambda| \to \infty \tag{13.46}$$

$$\log a(\lambda) = \frac{i\pi Q}{2} + \sum_{n=1}^{\infty} I_n^-(i\lambda)^n, \quad |\lambda| \to 0 \qquad (13.47)$$

We will now calculate the $I_n^\pm$ in two different ways. The first one will give $I_n^\pm$ in terms of the original sine-Gordon field φ, while the second one will express $I_n^\pm$ in terms of the scattering data.

In order to compute I_n^+ in terms of φ, we perform a gauge transformation $\Psi \to e^{i\frac{\beta}{2}\varphi\sigma_3}\tilde{\Psi}$, so that the gauge transformed connection $\tilde{A}_x$ reads:

$$\tilde{A}_x = i\frac{\beta}{2}(\dot{\varphi} - \varphi')H + im\lambda(1 - \lambda^{-2}e^{-2i\beta\varphi})E_+ + im\lambda(1 - \lambda^{-2}e^{2i\beta\varphi})E_-$$

Note that $\tilde{A}_x$ takes the same value at $x = \pm\infty$ and we have $\mathrm{Tr}\,(\tilde{T}_\ell(\lambda)) = \mathrm{Tr}\,(T_\ell(\lambda)\rho)$. To compute this trace we can directly apply eq. (11.8) in Chapter 11, where we have found that

$$\mathrm{Tr}\,(\tilde{T}_\ell(\lambda)) = e^{\mathcal{P}_\ell(\lambda)} + e^{-\mathcal{P}_\ell(\lambda)}$$

For smooth (C^∞) fields φ, the quantity $\mathcal{P}_\ell(\lambda)$ admits asymptotic expansions for $\lambda \to 0$ and $\lambda \to \infty$ which can be found using the Ricatti equation. The coefficients of this asymptotic expansion are integrals over local densities in the field φ containing higher and higher derivatives. Hence the smoothness condition is essential for their existence. On the other hand, for such smooth fields we have seen that $b(\lambda)$ vanishes at $\lambda = 0, \infty$ as well as all its derivatives. This means that $b(\lambda)$ has zero asymptotic expansion at these points. Since $|a|^2 = 1 - |b|^2$ we see that $|a| = 1$ in the asymptotic sense, or $a(\lambda)^* \sim 1/a(\lambda)$. Using this fact, we can compare the asymptotic expansions in both sides of the equation:

$$e^{\mathcal{P}_\ell(\lambda)} + e^{-\mathcal{P}_\ell(\lambda)} = e^{\frac{i\pi Q}{2}}e^{-2ik(\lambda)\ell}a(\lambda) + e^{-\frac{i\pi Q}{2}}e^{2ik(\lambda)\ell}a^*(\lambda)$$

and identify in the asymptotic sense:

$$\mathcal{P}_\ell(\lambda) = \frac{i\pi Q}{2} - 2ik(\lambda)\ell + \log a(\lambda), \quad \lambda \to \infty$$

To compute the left-hand side, we recall that $\mathcal{P}_\ell(\lambda) = \int_{-\ell}^{\ell} v(x, \lambda)dx$, where $v(x, \lambda)$ is a solution of the Ricatti equation $v' + v^2 = V$ and V is determined in eq. (11.8) in terms of $\tilde{A}_x$. We compute this expansion at the lowest nontrivial order, so that $O(\lambda^{-2})$ terms are neglected. We get:

$$V = -m^2\lambda^2 - \frac{\beta^2}{4}(\dot{\varphi} - \varphi')^2 + 2m^2\cos(2\beta\varphi) - i\frac{\beta}{2}(\dot{\varphi} - \varphi')' + O(\lambda^{-2})$$

One inserts $v = \pm im\lambda + \cdots$ and observes that there is no $O(1)$ term. Up to a choice of sign, and substituting $k(\lambda) = m\lambda - m\lambda^{-1}$, we have:

$$v = -ik(\lambda) - \frac{im}{\lambda}\left(\frac{\beta^2}{8m^2}(\dot\varphi - \varphi')^2 + 1 - \cos(2\beta\varphi) + \frac{i\beta}{4m^2}(\dot\varphi - \varphi')'\right) + O(\frac{1}{\lambda^2})$$

Since $(\dot\varphi - \varphi')$ vanishes at $x = \pm\infty$, we obtain:

$$\mathcal{P}_\ell(\lambda) = -2ik(\lambda)\ell - im\lambda^{-1}\frac{\beta^2}{4m^2}(H - P) + O(\lambda^{-2})$$

where:

$$H = \int_{-\ell}^{\ell}\left(\frac{1}{2}\dot\varphi^2 + \frac{1}{2}\varphi'^2 + \frac{4m^2}{\beta^2}(1 - \cos(2\beta\varphi))\right)dx$$

$$P = \int_{-\ell}^{\ell}\dot\varphi\varphi'\,dx$$

Similarly, to compute I_n^- ($\lambda \to 0$) we perform the gauge transformation $\Psi \to e^{-i\frac{\beta}{2}\varphi\sigma_3}\tilde\Psi$, so that the gauge transformed connection $\tilde{A}_x$ reads:

$$\tilde{A}_x = i\frac{\beta}{2}(\dot\varphi + \varphi')H - im\lambda^{-1}(1 - \lambda^2 e^{2i\beta\varphi})E_+ - im\lambda^{-1}(1 - \lambda^2 e^{-2i\beta\varphi})E_-$$

Note, however, that, due to the sign change in the gauge transformation, we now have $\operatorname{Tr}\tilde{T} = \operatorname{Tr}(T\rho^{-1}) = e^{i\pi Q}\operatorname{Tr}(T\rho)$. The same computation as above yields:

$$\mathcal{P}_\ell(\lambda) = -2ik(\lambda)\ell + im\lambda\frac{\beta^2}{4m^2}(H + P) + O(\lambda^2)$$

Comparing with the asymptotic expansions in:

$$\operatorname{Tr}\tilde{T} = e^{\mathcal{P}_\ell(\lambda)} + e^{-\mathcal{P}_\ell(\lambda)} = e^{-i\pi Q}(e^{\frac{i\pi Q}{2}}e^{-2ik(\lambda)\ell}a(\lambda) + e^{-\frac{i\pi Q}{2}}e^{2ik(\lambda)\ell}a^*(\lambda))$$

we find

$$\mathcal{P}_\ell(\lambda) = -\frac{i\pi Q}{2} - 2ik(\lambda)\ell + \log a(\lambda), \quad \lambda \to 0$$

from which we identify:

$$I_1^\pm = \frac{\beta^2}{4m}(H \mp P)$$

Now we reconstruct $a(\lambda)$ from its analyticity properties, and get alternative expressions for the quantities $I_n^\pm$. Recall that the function $a(\lambda)$ is analytic in the upper half-plane, behaves at $\lambda = 0, \infty$ as:

$$e^{\frac{i\pi Q}{2}}a(\lambda) = 1 + O\left(\frac{1}{|\lambda|}\right), \quad |\lambda| \to \infty; \qquad e^{\frac{i\pi Q}{2}}a(\lambda) = e^{i\pi Q} + O(|\lambda|), \quad |\lambda| \to 0$$

and obeys $(e^{\frac{i\pi Q}{2}} a(-\lambda^*))^* = e^{\frac{i\pi Q}{2}} a(\lambda)$. Therefore we can reconstruct $a(\lambda)$ if we know its modulus on the real axis and the position of its zeroes.

$$e^{\frac{i\pi Q}{2}} a(\lambda) = \prod_i \left(\frac{\lambda - \lambda_i}{\lambda - \lambda_i^*} \right) \exp \left[-\frac{1}{i\pi} \int_{-\infty}^{\infty} \frac{\log |a(\mu)|}{\lambda - \mu + i0} d\mu \right] \qquad (13.48)$$

The right-hand side is analytic in the *upper* half-plane. It is invariant under complex conjugation and changing $\lambda \to -\lambda^*$ because for each λ_i there is a $\lambda_j = -\lambda_i^*$ and one can change variables $\mu \to -\mu$ in the integral $(|a(-\mu)| = |a(\mu)|)$. For $\lambda \to \infty$ one gets the correct asymptotic value 1. Finally, for λ real, using:

$$\frac{1}{\lambda - \mu + i0} = P\frac{1}{\lambda - \mu} - i\pi\delta(\lambda - \mu)$$

the modulus of the right-hand side is $\exp \left(\int \log |a(\mu)| \delta(\lambda - \mu) d\mu \right) = |a(\lambda)|$. Using the asymptotic value at $\lambda = 0$, and noting that the integral vanishes for $\lambda = 0$ (by $\mu \to -\mu$) we must have $e^{i\pi Q} = \prod_i (\lambda_i / \lambda_i^*)$. Note that a pair of roots symmetric with respect to the imaginary axis contribute a $+1$ in this product, but pure imaginary roots contribute a -1, so, modulo 2, Q must be equal to the number of pure imaginary roots, i.e. the total number of solitons and antisolitons must have the same parity as the topological charge Q. This is as it should be since a soliton has $Q = 1$ and an antisoliton has $Q = -1$.

On the real axis we have $|a(\lambda)|^2 = 1 - |b(\lambda)|^2$ and we can replace $\frac{1}{\pi} \log |a(\lambda)|$ by

$$\rho(\lambda) = \frac{1}{2\pi} \log (1 - |b(\lambda)|^2)$$

in eq. (13.48). Thus $a(\lambda)$ can be reconstructed from the knowledge of $|b(\lambda)|$ on the real axis and the zeroes λ_n.

For a smooth field φ, $\rho(\mu)$ decreases fast at $\mu \to 0, \infty$, so that one can expand $1/(\lambda - \mu)$ in powers of λ/μ or μ/λ in the integral. We get asymptotic expansions at $\lambda \to \infty$:

$$\log a(\lambda) = -\frac{i\pi Q}{2} + \sum_{n=1}^{\infty} \frac{1}{\lambda^n} \left(-\sum_j \frac{\lambda_j^n - \lambda_j^{*n}}{n} + i \int_{-\infty}^{\infty} \mu^{n-1} \rho(\mu) d\mu \right)$$

and at $\lambda \to 0$:

$$\log a(\lambda) = \frac{i\pi Q}{2} + \sum_{n=1}^{\infty} \lambda^n \left(-\sum_j \frac{1}{n} \left(\frac{1}{\lambda_j^n} - \frac{1}{\lambda_j^{*n}} \right) - i \int_{-\infty}^{\infty} \mu^{-n-1} \rho(\mu) d\mu \right)$$

Comparing with eqs. (13.46, 13.47), we identify the $I_n^\pm$. We find $I_n^\pm = 0$ for n even, and for n odd $I_n^\pm = \pm I_{\pm n}$, where I_n is defined for $n \in \mathbb{Z}$ by:

$$I_n = (-1)^{(n-1)/2}\left(\sum_j \frac{\lambda_j^n - \lambda_j^{*n}}{in} - \int_0^\infty \mu^{n-1}\rho(\mu)d\mu\right) \in \mathbb{R}$$

In particular, for $n = \pm 1$ we obtain:

$$H \pm P = \frac{4m}{\beta^2}\left(\pm\sum_j \frac{\lambda_j^{\pm 1} - \lambda_j^{*\pm 1}}{i} - \int_{-\infty}^\infty \mu^{\pm 1}\rho(\mu)\frac{d\mu}{\mu}\right)$$

If, for simplicity, we don't consider breathers and set $\lambda_j = i\xi_j$, we get:

$$H = \frac{4m}{\beta^2}\left(\sum_j(\xi_j + \xi_j^{-1}) + \int_0^\infty (\mu + \mu^{-1})|\rho(\mu)|\frac{d\mu}{\mu}\right)$$

$$P = \frac{4m}{\beta^2}\left(\sum_j(\xi_j - \xi_j^{-1}) + \int_0^\infty (\mu - \mu^{-1})|\rho(\mu)|\frac{d\mu}{\mu}\right)$$

Setting $k_j = \frac{8}{\beta^2}\frac{m}{2}(\xi_j - \xi_j^{-1})$, $M = \frac{8}{\beta^2}m$ and $k = \frac{m}{2}(\mu - \mu^{-1})$, this can be written as:

$$H = \sum_j \sqrt{k_j^2 + M^2} + \frac{8}{\beta^2}\int_{-\infty}^\infty \sqrt{k^2 + m^2}|\rho(k)|\frac{dk}{\sqrt{k^2 + m^2}}$$

$$P = \sum_j k_j + \frac{8}{\beta^2}\int_{-\infty}^\infty k|\rho(k)|\frac{dk}{\sqrt{k^2 + m^2}}$$

which nicely exhibits the decomposition of the theory into a sum of relativistic modes. Note that the soliton j has mass M and momentum k_j, while the continuous spectrum is a superposition of modes of mass m.

Remark.   One can extract the Poisson brackets of the solitonic modes from eq. (13.45) and recover eq. (12.81) in Chapter 12. The parameters a_n, μ_n are related to m_n, λ_n by eq. (13.40). Moreover, from eqs. (13.32, 13.48), for purely solitonic solutions we have:

$$m_n = \frac{c_n}{a'(\lambda_n)}, \quad a(\lambda) = e^{-i\frac{\pi Q}{2}}\prod_j \frac{\lambda - \lambda_j}{\lambda + \lambda_j}$$

so that:

$$a_n = 2ie^{i\frac{\pi Q}{2}}c_n\prod_{j\neq n}\frac{\mu_n + \mu_j}{\mu_n - \mu_j}$$

Then a straightforward computation using eqs. (13.45), which mean that $\log \mu_n$ and $\log c_n$ are canonically conjugated variables, yields:

$$\{\mu_i, \mu_j\} = 0, \quad \{\mu_i, a_j\} = \frac{\beta^2}{2} \mu_i a_j \delta_{ij}, \quad \{a_i, a_j\} = \frac{\beta^2}{2} \left(\frac{4\mu_i\mu_j}{\mu_i^2 - \mu_j^2} \right) a_i a_j$$

which identifies with eq. (12.81) up to the factor β^2 (we should set $\beta = -i$ to compare the two formulae).

With this, we end this chapter on the classical inverse scattering method, thereby paying due tribute to Gardner, Greene, Kruskal and Miura, without whom this book would not exist.

References

[1] C.S. Gardner, J.M. Greene, M.D. Kruskal and R.M. Miura, Method for solving the Korteweg–de Vries equation. *Phys. Rev. Lett.* **19** (1967) 1095.

[2] E.K. Sklyanin, *On the complete integrability of the Landau–Lifchitz equation.* Preprint LOMI E-3-79. Leningrad (1979).

[3] S.Novikov, S.V. Manakov, L.P. Pitaevskii and V.E. Zakharov, *Theory of Solitons, the Inverse Scattering Method.* Consultants Bureau, NY (1984).

[4] L.D. Faddeev and L.A. Takhtajan, *Hamiltonian Methods in the Theory of Solitons.* Springer (1986).

14
Symplectic geometry

The aim of this chapter is to provide a concise presentation of classical mechanics in the framework of symplectic geometry. This geometrical approach of mechanics is essential to gain any understanding of integrable systems theory. We assume the reader has some basic knowledge of elementary differential geometry and differential forms, but we present all the symplectic theory we need. We then explain the notion of symplectic reduction under a Lie group action, a concept which frequently appears in our discussion of integrable systems. The chapter ends by a discussion of a more recent topic, Poisson–Lie groups, which is used in the analysis of dressing transformations, and whose importance has to be stressed in connection with quantum group theory.

14.1 Poisson manifolds and symplectic manifolds

In this chapter we investigate the formulation of classical mechanics using Poisson brackets. We work on a phase space M and consider the differentiable functions on M. We denote by $\mathcal{F}(M)$ the algebra of such functions. A Poisson bracket is a bilinear antisymmetric derivation of the algebra $\mathcal{F}(M)$: $\{\,,\,\} : \mathcal{F}(M) \times \mathcal{F}(M) \to \mathcal{F}(M)$ such that:

$$\{f_1, f_2\} = -\{f_2, f_1\}, \quad \text{antisymmetry}$$
$$\{f_1, \alpha f_2 + \beta f_3\} = \alpha\{f_1, f_2\} + \beta\{f_1, f_3\}, \quad \alpha, \beta \text{ constants}$$
$$\{f_1, f_2 f_3\} = \{f_1, f_2\}f_3 + f_2\{f_1, f_3\}, \quad \text{Leibnitz rule}$$
$$\{f_1, \{f_2, f_3\}\} + \{f_3, \{f_1, f_2\}\} + \{f_2, \{f_3, f_1\}\} = 0, \quad \text{Jacobi identity}$$

Since the Poisson bracket is linear in f and obeys the Leibnitz rule, with any function $H \in \mathcal{F}(M)$ we can associate a vector field X_H on M by

$X_H f = \{H, f\}$. When H is the Hamiltonian of the system, this vector field defines the time evolution by

$$\dot{f} = X_H f = \{H, f\} \quad \forall f \in \mathcal{F}(M)$$

Definition. *A Poisson manifold M is a manifold on which a Poisson bracket is defined.*

The important feature of Poisson brackets in classical mechanics is that if f_1 and f_2 are two conserved quantities, i.e. $\{H, f_1\} = \{H, f_2\} = 0$, then $\{f_1, f_2\}$ is also conserved due to the Jacobi identity.

In general a Poisson bracket is degenerate, which means that there are functions f on M such that $\{f, g\} = 0$ for all g. The set of such functions is called the centre of the Poisson algebra. If the centre is non-trivial, i.e. contains non-constant functions, one can reduce the dynamical system by setting all functions of the centre to constant values. This defines a foliation of M and the Poisson bracket is non-degenerate on the leaves.

In a local coordinate system x^j on M we can write:

$$\{f_1, f_2\}_M(x) = \sum_{ij} P^{ij}(x) \frac{\partial f_1}{\partial x^i} \frac{\partial f_2}{\partial x^j}$$

due to the bilinearity and Leibnitz rule. Antisymmetry requires that $P^{ij}(x) = -P^{ji}(x)$. The Jacobi identity reads:

$$P^{is} \partial_s P^{jk} + P^{ks} \partial_s P^{ij} + P^{js} \partial_s P^{ki} = 0$$

where $\partial_s = \partial_{x_s}$. Assume now that the matrix P^{ij} is invertible. In particular, this means that the centre of the Poisson algebra is trivial. This can occur only when the dimension of M is even. Denote by $(P^{-1})_{ij}$ the inverse matrix of P^{ij} so that $\partial_s P^{ij} = -P^{ia} \partial_s (P^{-1})_{ab} P^{bj}$. Inserting this into the Jacobi identity yields:

$$\sum_{s,a,b} P^{is} P^{ja} P^{kb} \left(\partial_s (P^{-1})_{ab} + \partial_b (P^{-1})_{sa} + \partial_a (P^{-1})_{bs} \right) = 0$$

Since P^{ij} is invertible, this is equivalent to the *linear* conditions:

$$\partial_a (P^{-1})_{bc} + \partial_b (P^{-1})_{ca} + \partial_c (P^{-1})_{ab} = 0$$

These conditions can be interpreted as the closedness of the 2-form:

$$\omega = -\sum_{i<j} (P^{-1})_{ij} dx^i \wedge dx^j$$

that is $d\omega = 0$. This 2-form is invariant under changes of coordinates, and so is globally defined on M. We emphasize that the matrices entering the definition of the Poisson bracket and the symplectic form are inverse to each other.

Definition. *A symplectic manifold* (M, ω) *is a manifold* M *equipped with a non-degenerate closed 2-form* ω, $d\omega = 0$.

Given a function H on a symplectic manifold, the Hamiltonian vector field X_H is defined using the interior product i_X by:

$$dH = -i_{X_H}\omega, \quad \text{i.e.} \quad dH = -\omega(X_H, \cdot)$$

Using local coordinates x^i on M we have

$$\omega = \sum_{i<j} \omega_{ij} dx^i \wedge dx^j$$

Setting $X_H = \sum_i X_H^i \partial_i$, we get $X_H^i = \omega^{ij}\partial_j H$, where ω^{ij} is the inverse matrix of ω_{ij}. Knowing the symplectic form ω one can reconstruct the Poisson bracket as follows:

$$\{f_1, \ f_2\} = X_{f_1}(f_2) = -X_{f_2}(f_1) = \omega(X_{f_1}, X_{f_2}) \qquad (14.1)$$

In components, we have

$$\{f_1, \ f_2\} = -\sum_{ij} \omega^{ij}\partial_i f_1 \partial_j f_2$$

On a symplectic space, one can define the notion of symplectic transformations. Consider a bijection $\gamma : M \to M$ and the transform of the symplectic form ω under γ. This is the form

$$(\gamma^*\omega)_m(V, W) = \omega_{\gamma(m)}(\gamma_* V, \gamma_* W)$$

where m is a point of M, and γ_* is the differential of γ, sending a tangent vector at m to a tangent vector at $\gamma(m)$. We say that the transformation γ is symplectic if $\gamma^*\omega = \omega$. For an infinitesimal transformation, γ is specified by a vector field X on M, and this condition is equivalent to $\mathcal{L}_X\omega = 0$, where $\mathcal{L}_X$ is the Lie derivative, $\mathcal{L}_X = di_X + i_X d$. We translate the symplecticity property of ω on Poisson brackets and show that it reads:

$$\gamma\{f, g\} = \{\gamma f, \gamma g\} \qquad (14.2)$$

We recall that the action of γ on functions is given by $(\gamma f)(m) = f(\gamma^{-1}(m))$. Note that we have introduced the inverse of γ so that the property $\gamma_1 \cdot (\gamma_2 \cdot f) = (\gamma_1 \gamma_2) \cdot f$ holds. If γ is symplectic, we have:

$$X_{\gamma f}(m) = \gamma_*\left(X_f(\gamma^{-1}(m))\right)$$

This is because d commutes with the pullback operation, i.e. $d(\gamma f) = \gamma^{-1*}df$, so applying this to $V \in T_m M$ we get

$$d(\gamma f)_m(V) = df_{\gamma^{-1}(m)}(\gamma_*^{-1}V)$$

that is:

$$\omega_m(X_{\gamma f}, V) = \omega_{\gamma^{-1}(m)}(X_f(\gamma^{-1}(m)), \gamma_*^{-1}V) = \omega_m(\gamma_* X_f \circ \gamma^{-1}(m), V)$$

where in the last step we have used that γ is symplectic. We use this to prove eq. (14.2). We have

$$\{\gamma f, \gamma g\}_m = \omega_m(X_{\gamma f}(m), X_{\gamma g}(m)) = \omega_{\gamma(\gamma^{-1}(m))}(\gamma_* X_f \circ \gamma^{-1}, \gamma_* X_g \circ \gamma^{-1})$$

Using that γ is symplectic, this is equal to

$$\omega_{\gamma^{-1}(m)}(X_f \circ \gamma^{-1}(m), X_g \circ \gamma^{-1}(m)) = \{f, g\}(\gamma^{-1}(m)) = (\gamma\{f, g\})(m)$$

Proposition. *Any Hamiltonian flow is a symplectic transformation.*

<u>Proof</u>. Let H be the Hamiltonian with associated vector field X_H such that $dH = -i_{X_H}\omega$. We have:

$$\mathcal{L}_{X_H}\omega = (i_{X_H}d + di_{X_H})\omega = d(i_{X_H}\omega) = -d^2 H = 0$$

where we have used $d\omega = 0$.   ∎

Example 1. A standard example of symplectic space is given by $M = \mathbb{R}^{2n}$ with coordinates (p_i, q_i), and symplectic form:

$$\omega = \sum_i dp_i \wedge dq_i$$

These coordinates are called canonical coordinates. The corresponding Poisson bracket reads:

$$\{q_i, q_j\} = 0, \quad \{p_i, p_j\} = 0, \quad \{p_i, q_j\} = \delta_{ij}$$

The Hamiltonian vector field corresponding to the function H is:

$$X_H = \sum_i -\frac{\partial H}{\partial q_i}\partial_{p_i} + \frac{\partial H}{\partial p_i}\partial_{q_i}$$

In fact this example is generic, at least locally. This is the Darboux theorem.

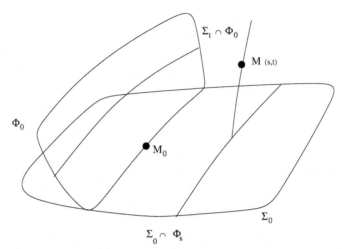

Fig. 14.1. Foliation of phase space by the surfaces Σ_t and Φ_s.

Theorem. *On any symplectic manifold (M, ω) one can introduce,* locally *around a point m_0, canonical coordinates (q_i, p_i) such that $\omega = \sum_i dp_i \wedge dq_i$. Moreover, one can choose p_1 to be any given function H on M such that $dH(m_0) \neq 0$.*

<u>Proof.</u> The proof will be done by induction on the dimension $(2n)$ of M. We choose a coordinate system $y \in \mathbb{R}^{2n}$ around m_0, and define $p_1(y) = H(y)$. We can assume that $p_1(m_0) = 0$. We then introduce the Hamiltonian vector field X_{p_1} associated with p_1, which is non-vanishing in the considered neighbourhood of m_0. Choose a hypersurface Σ_0 passing through m_0, and transverse to X_{p_1}, i.e. such that X_{p_1} is not tangent to Σ_0. We want to define q_1 such that $\{p_1, q_1\} = 1$ and $q_1 = 0$ on Σ_0. This means that on any trajectory of the flow of X_{p_1} we want to achieve $\dot{q}_1 = \{p_1, q_1\} = 1$. Hence $q_1(y) - q_1(z) = q_1(y) = t$, where z is the point at which the trajectory crosses Σ_0, that is $q_1(z) = 0$. By the assumption of transversality, for any y close to Σ_0 one defines $q_1(y)$ as the time needed to go from $z \in \Sigma_0$ to y along the trajectory of the flow X_{p_1}. The neighbourhood of m_0 is foliated by the hypersurfaces Σ_t where $q_1(y) = t$. On the other hand, it is also foliated by the hypersurfaces Φ_s where $p_1(y) = s$.

We now show that one can simultaneously solve:

$$\partial_s y = -X_{q_1}(y), \quad \partial_t y = X_{p_1}(y)$$

thereby allowing us to write $y = y(s, t, z)$ with $z \in \Sigma_0 \cap \Phi_0$. This is because for any function $f(y)$ one has by definition $\partial_s f = -X_{q_1} f = \{-q_1, f\}$ and

$\partial_t f = X_{p_1} f = \{p_1, f\}$ so that, using the Jacobi identity:

$$\partial_s(\partial_t f) - \partial_t(\partial_s f) = -\{q_1, \{p_1, f\}\} + \{p_1, \{q_1, f\}\} = \{\{p_1, q_1\}, f\} = 0$$

since $\{p_1, q_1\} = 1$ is constant.

The vector field X_{p_1} is tangent to Φ_0 (in fact to any Φ_s, because $X_{p_1} p_1 = \{p_1, p_1\} = 0$), and similarly X_{q_1} is tangent to Σ_t. They are both transverse to $\Sigma_0 \cap \Phi_0$ and independent, because $\{p_1, q_1\} = \omega(X_{p_1}, X_{q_1}) = 1 \neq 0$. For any vector V tangent to $\Sigma_0 \cap \Phi_0$ we have $\omega(V, X_{p_1}) = \omega(V, X_{q_1}) = 0$ because $\omega(V, X_{p_1}) = V \cdot p_1$, $\omega(V, X_{q_1}) = V \cdot q_1$, and $p_1 = q_1 = 0$ are constant on this intersection. It follows that the restriction of ω to $\Sigma_0 \cap \Phi_0$ is non-degenerate.

By the induction hypothesis we assume that we have found canonical coordinates p_i, q_i, $i \geq 2$ on the $(2n - 2)$ dimensional symplectic variety $\Sigma_0 \cap \Phi_0$. We extend these coordinates as functions on M by setting $p_j(y) = p_j(z)$ and $q_j(y) = q_j(z)$ for $j \geq 2$ and $y = y(p_1, q_1, z)$ with $z \in \Sigma_0 \cap \Phi_0$. This amounts to keeping them constant along the flows of X_{p_1} and X_{q_1} so that

$$\{p_1, p_j\} = \{p_1, q_j\} = \{q_1, p_j\} = \{q_1, q_j\} = 0$$

It remains to show that the symplectic form ω on M is equal to $\tilde{\omega} = \sum_{i=1}^n dp_i \wedge dq_i$. We first show that this is true at any point $p_1 = q_1 = 0$ of $\Sigma_0 \cap \Phi_0$. Any vector V tangent to M at this point can be decomposed as a sum of a vector V_1 on the space spanned by X_{p_1} and X_{q_1}, and a vector V_2 tangent to $\Sigma_0 \cap \Phi_0$. Computing $\omega(V_1 + V_2, W_1 + W_2)$ we have seen that $\omega(V_1, W_2) = \omega(V_2, W_1) = 0$, while by the induction hypothesis $\omega(V_2, W_2) = (\sum_{j \geq 2} dp_j \wedge dq_j)(V_2, W_2)$. Finally, $\omega(V_1, W_1) = (dp_1 \wedge dq_1)(V_1, W_1)$, because it is sufficient to compute this for $V_1 = X_{p_1}$ and $W_1 = X_{q_1}$, and both members are then equal to $\{p_1, q_1\} = 1$. To show that the equality holds on M, consider the Hamiltonian evolution $U_{(s,t)}$ under the flows of X_{p_1} and X_{q_1}. In the coordinates (p_i, q_i) it reads $(p_1, q_1, p_2, q_2, \ldots) \rightarrow (p_1 + s, q_1 + t, p_2, q_2, \ldots)$ so that the form $\tilde{\omega} = \sum_i dp_i \wedge dq_i$ is obviously invariant. On the other hand, the symplectic form ω is also invariant under Hamiltonian evolutions. Since we have shown that $\omega = \tilde{\omega}$ for $p_1 = q_1 = 0$, these two forms coincide everywhere. ∎

Example 2. Another very natural example of symplectic space is the cotangent bundle $M = T^*N$ of any differentiable manifold N. On this bundle is defined a canonical 1-form α given by:

$$\alpha_x(X) = p(\pi_* X)$$

where $X \in T_x(T^*N)$, and $x \in T^*N$. One can write $x = (q, p)$, with $q = \pi(x) \in N$ (π is the projection on N), and p is a 1-form belonging to the fibre of q. We define the symplectic form as the closed form $\omega = d\alpha$. To show that it is non-degenerate we express it in terms of local coordinates. If $q = (q_1, \ldots, q_n)$ is a system of local coordinates on N, a basis of the tangent space of N at q is $(\partial_{q_1}, \ldots, \partial_{q_n})$ and a basis of the cotangent space at q is $(dq_1, \ldots, dq_n)$. In particular, any point p in the fibre of T^*N above q is of the form $p = \sum_i p_i dq_i$. A tangent vector $X = (\delta q, \delta p)$ at the point (q, p) of T^*N has projection $\pi_* X = \delta q = \sum_i (\delta q)_i \partial_{q_i}$, so that $\alpha(X) = \sum_i p_i (\delta q)_i$, and the canonical form reads:

$$\alpha = p_1 dq_1 + \cdots + p_n dq_n$$

Then $\omega = d\alpha = \sum_i dp_i \wedge dq_i$ is non-degenerate.

14.2 Coadjoint orbits

Our aim is to present a non-trivial example of symplectic structure on the coadjoint orbits of a Lie group, which plays an important role in the study of integrable systems.

We first recall some notions about adjoint and coadjoint actions of Lie algebras and Lie groups. Let G be a connected Lie group with Lie algebra $\mathcal{G}$. The group G acts on $\mathcal{G}$ by the adjoint action denoted Ad:

$$X \longrightarrow (\operatorname{Ad} g)(X) = gXg^{-1}, \quad g \in G, \ X \in \mathcal{G}$$

Similarly, the coadjoint action of G on the dual $\mathcal{G}^*$ of the Lie algebra $\mathcal{G}$ (i.e. the vector space of linear forms on the Lie algebra) is defined by:

$$(\operatorname{Ad}^* g.\Xi)(X) = \Xi(\operatorname{Ad} g^{-1}(X)), \quad g \in G, \ \Xi \in \mathcal{G}^*, \ X \in \mathcal{G}$$

The infinitesimal version of these actions provides actions of the Lie algebra $\mathcal{G}$ on $\mathcal{G}$ and $\mathcal{G}^*$, denoted by ad and ad*, given by:

$$\operatorname{ad} X(Y) = [X, Y], \quad X, Y \in \mathcal{G},$$
$$\operatorname{ad}^* X.\Xi(Y) = -\Xi([X, Y]), \quad X, Y \in \mathcal{G}, \Xi \in \mathcal{G}^*$$

Coadjoint orbits in $\mathcal{G}^*$ are equipped with the canonical Kostant–Kirillov symplectic structure. Before defining it we need some simple facts concerning the functions on $\mathcal{G}^*$. Denote the space of such functions by $\mathcal{F}(\mathcal{G}^*)$. The coadjoint action of G on $\mathcal{G}^*$ induces a coadjoint action of G on functions on $\mathcal{G}^*$, also denoted by Ad*. If $g \in G$ and $h \in \mathcal{F}(\mathcal{G}^*)$,

$$\operatorname{Ad}^* g . h(\Xi) = h(\operatorname{Ad}^* g^{-1}(\Xi))$$

and a similar formula for the infinitesimal action ad*. For a function h on $\mathcal{G}^*$ the differential dh may be viewed as an element of $\mathcal{G}$. This is because,

$\mathcal{G}^*$ being a vector space, the differential dh is a linear form on $\mathcal{G}^*$, i.e. an element of $\mathcal{G}^{**} \sim \mathcal{G}$, and one can write for $\delta\Xi \in \mathcal{G}^*$:

$$h(\Xi + \delta\Xi) = h(\Xi) + \delta\Xi(dh) + \mathcal{O}\left((\delta\Xi)^2\right)$$

From the Lie algebra structure on $\mathcal{G}$, we can construct a Poisson bracket on the space $\mathcal{F}(\mathcal{G}^*)$ of functions on $\mathcal{G}^*$. This is the Kostant–Kirillov bracket.

Definition. *If h_1 and $h_2 \in \mathcal{F}(\mathcal{G}^*)$, the Kostant–Kirillov bracket is defined by:*

$$\{h_1, h_2\}(\Xi) = \Xi([dh_1, dh_2])$$

It is obvious that the bracket $\{\ ,\ \}$ is antisymmetric and verifies the Jacobi identity. Choosing two linear functions $h_1(\Xi) = \Xi(X)$ and $h_2(\Xi) = \Xi(Y)$ with $X, Y \in \mathcal{G}$, we have $dh_1 = X$ and $dh_2 = Y$, and the Kostant–Kirillov Poisson bracket reads:

$$\{\Xi(X), \Xi(Y)\} = \Xi([X, Y])$$

The right-hand side is the linear function $\Xi \to \Xi([X,Y])$. This Poisson bracket is very natural but one has to be aware that it is a degenerate Poisson bracket.

Proposition. *The kernel of the Kostant–Kirillov bracket $\{\ ,\ \}$ is the set $\mathcal{I}(\mathcal{G}^*)$ of Ad*-invariant functions.*

Proof. Let us first express the Ad*-invariance property of a function h on $\mathcal{G}^*$. Performing an infinitesimal transformation, we have:

$$h(\Xi + t\, \mathrm{ad}^* X \cdot \Xi) = h(\Xi) + t\,(\mathrm{ad}^* X \cdot \Xi)(dh) + \mathcal{O}(t^2)$$

so h is invariant if $(\mathrm{ad}^* X \cdot \Xi)(dh) = \Xi([dh, X]) = 0$ for all $\Xi \in \mathcal{G}^*$ and all $X \in \mathcal{G}$. Assuming now that $k \in \mathcal{F}(\mathcal{G}^*)$ is in the kernel of $\{\ ,\ \}$, we have $\{k, f\}(\Xi) = \Xi([dk, df]) = 0$, $\forall f \in \mathcal{F}(\mathcal{G}^*)$, in particular $f = \Xi(X)$ for any $X \in \mathcal{G}$. Thus k is ad*-invariant. The converse is obvious. ∎

This proposition means that the kernel of the Kostant–Kirillov bracket $\{\ ,\ \}$ is the set of functions which are constant on the orbits of the coadjoint action of G. Let $\mathcal{I}_1, \mathcal{I}_2, \ldots$ be the primitive invariant functions, i.e. any invariant function is a function of them, and denote by $I_1, I_2, \ldots$ the constant values they take on a specific orbit. We consider the ideal of the function algebra generated by the non-constant functions $\mathcal{I}_1 - I_1, \mathcal{I}_2 - I_2, \ldots$. It is also an ideal of the Poisson algebra. The quotient of the function algebra by this ideal can be identified with the functions on the orbit, and the quotient Poisson bracket yields a Poisson bracket on the orbit, which by construction is non-degenerate, and therefore defines a symplectic structure on the orbit. More explicitly:

Proposition. *For any two tangent vectors at the point Ξ of the orbit, $V_X = \mathrm{ad}^* X \cdot \Xi$ and $V_Y = \mathrm{ad}^* Y \cdot \Xi$, define*

$$\omega_K(V_X, V_Y) = \Xi([X, Y]) \tag{14.3}$$

This form is closed and non-degenerate on any G-orbit. It induces the Kostant–Kirillov bracket.

<u>Proof.</u> First note that since we are on an orbit of G, the vectors $\mathrm{ad}^* X \cdot \Xi$, $X \in \mathcal{G}$ describe all the tangent space at Ξ. To show that ω is closed, let us recall the definition of exterior differentiation:

$$d\eta(X_0, \ldots, X_p) = \sum_{j=0}^{p} (-1)^j X_j \cdot \eta(X_0, \ldots, \widehat{X}_j, \ldots, X_p)$$

$$+ \sum_{0 \leq i < j \leq p} (-1)^{i+j} \eta([X_i, X_j], X_0, \ldots, \widehat{X}_i, \ldots, \widehat{X}_j, \ldots, X_p)$$

In our case, the first term vanishes because $V_X \cdot \Xi([Y, Z]) = -\Xi([X, [Y, Z]])$, and we apply the Jacobi identity. The second term vanishes for the same reason once one notices that $[V_X, V_Y] = -V_{[X,Y]}$. This is because, by the general definition of the Lie derivative, we have:

$$\mathcal{L}_{X \cdot m} Y \cdot m \equiv [X \cdot m, Y \cdot m] = \frac{d}{dt}\left(e^{-Xt} Y.(e^{Xt} m)\right)|_{t=0} = -[X, Y] \cdot m \tag{14.4}$$

To show that the form ω is non-degenerate on the orbit, assume that the tangent vector V_X is such that $\omega(V_X, V_Y) = 0$ for all $Y \in \mathcal{G}$. This means $\Xi([X, Y]) = 0, \forall Y$, that is $\mathrm{ad}^* X \cdot \Xi(Y) = 0, \forall Y$. Hence $V_X = \mathrm{ad}^* X \cdot \Xi = 0$. To compute the Poisson bracket associated with ω_K, we need the Hamiltonian vector field of any function f:

$$X_f(\Xi) = -\mathrm{ad}^* df \cdot \Xi$$

To show it, note that $df(\mathrm{ad}^* Y \cdot \Xi) = \Xi([df, Y])$, which is also $\omega_K(\mathrm{ad}^* df \cdot \Xi, \mathrm{ad}^* Y \cdot \Xi)$. Hence $\omega_K(X_f, X_g)(\Xi) = \Xi([df, dg])$. ∎

The 2-form ω_K defining the Kostant–Kirillov symplectic structure on coadjoint orbits is closed, but not exact. However, the coadjoint action defines a map φ from the group G to the orbit $\mathcal{O}_{\Xi_0}$ by

$\varphi : g \to Ad^*g \cdot \Xi_0 \equiv \Xi$. The pullback $\omega = \varphi^*\omega_K$ of ω_K on G is exact and one can write $\omega = \delta\alpha$, with

$$\alpha = -\Xi_0(g^{-1}\delta g)$$

To check this, note that $\varphi_*(gX) = ad^*gXg^{-1} \cdot \Xi$. Hence

$$\varphi^*\omega_K(gX, gY) = \omega_K(\varphi_*(gX), \varphi_*(gY)) = \Xi(g[X, Y]g^{-1}) = \Xi_0([X, Y])$$

On the other hand,

$$\delta\alpha(gX, gY) = gX \cdot \alpha(gY) - gY \cdot \alpha(gX) - \alpha([gX, gY])$$

The first two terms vanish because they are derivatives of constant functions $(\alpha(gY) = -\Xi_0(Y))$. By definition of the Lie bracket $[gX, gY] = g[X, Y]$ so that the last term is equal to $\Xi_0([X, Y])$, hence $\varphi^*\omega = \delta\alpha$.

14.3 Symmetries and Hamiltonian reduction

Let G be a Lie group and $\mathcal{G}$ its Lie algebra. Consider a symplectic manifold M on which G operates. We say that the action of G is symplectic if for any $g \in G$ the transformation $m \to gm$ is symplectic. In view of eq. (14.2) this means:

$$\{f_1(gm), f_2(gm)\} = \{f_1, f_2\}(gm)$$

For any $X \in \mathcal{G}$, consider the one-parameter group $g^t = \exp(tX)$. In the limit $t \to 0$ we define the action of X on functions by:

$$X \cdot f(m) = \frac{d}{dt} f(e^{-tX} \cdot m)|_{t=0} \qquad (14.5)$$

so that we get a representation on functions of the Lie algebra $\mathcal{G}$:

$$X \cdot (Y \cdot f) - Y \cdot (X \cdot f) = [X, Y] \cdot f \qquad (14.6)$$

Notice that $X \cdot f = -\mathcal{L}_{X \cdot m}f$, where $\mathcal{L}$ is the Lie derivative.

Finally, the symplecticity condition reads:

$$\{X \cdot f_1, f_2\} + \{f_1, X \cdot f_2\} = X \cdot \{f_1, f_2\} \qquad (14.7)$$

Proposition. *Let G be a Lie group acting on M by symplectic diffeomorphism. The action of any one-parameter subgroup of G is locally Hamiltonian. This means that there exists a function H_X, locally defined on M, such that:*

$$X \cdot f = \{H_X, f\} \qquad (14.8)$$

<u>Proof.</u> The condition eq. (14.7) is obviously necessary to have $X \cdot f = \{H_X, f\}$. It is also sufficient. To show it, we use the canonical Darboux coordinates. Writing $X \cdot m = \sum_i (X^{p_i} \partial_{p_i} + X^{q^i} \partial_{q_i})$, we have:

$$\{X \cdot f, h\} + \{f, X \cdot h\} - X \cdot \{f, h\}$$
$$= \sum_{i,j} \left[(\partial_{p_j} X_{q_i} - \partial_{p_i} X_{q_j}) \partial_{q_i} f \partial_{q_j} h + (\partial_{p_j} X_{p_i} + \partial_{q_i} X_{q_j}) \partial_{p_i} f \partial_{q_j} h \right]$$
$$- \sum_{i,j} \left[(\partial_{q_j} X_{q_i} + \partial_{p_i} X_{p_j}) \partial_{q_i} f \partial_{p_j} h + (\partial_{q_j} X_{p_i} - \partial_{q_i} X_{p_j}) \partial_{p_i} f \partial_{p_j} h \right]$$

The condition that this vanishes identically for all f and h is equivalent to $d\Omega_X = 0$, where $\Omega_X = -i_{X \cdot m}\omega = \sum_i (X_{q_i} dp_i - X_{p_i} dq_i)$. So there exists, at least locally, a function H_X such that $\Omega_X = dH_X$ or:

$$X_{q_i} = \frac{\partial H_X}{\partial p_i} \qquad X_{p_i} = -\frac{\partial H_X}{\partial q_i}$$

Then

$$X \cdot f = \sum_i -\frac{\partial H_X}{\partial q_i}\frac{\partial f}{\partial p_i} + \frac{\partial H_X}{\partial p_i}\frac{\partial f}{\partial q_i} = \{H_X, f\}$$

This proves eq. (14.8). ∎

If one knows that there is an invariant 1-form α such that $\omega = d\alpha$, one can give an explicit formula for the function H_X, which is then globally defined:

$$H_X(m) = \alpha(X \cdot m) \tag{14.9}$$

Indeed, since α is invariant we have $\mathcal{L}_X \alpha = 0$. Then $0 = \mathcal{L}_X \alpha = (i_X d + di_X)\alpha = \omega(X, \cdot) + d\alpha(X)$, so that comparing with $dH_X = -i_X\omega$ we see that $H_X = \alpha(X \cdot m)$.

Using eq. (14.8) and the Jacobi identity, we find that $X \cdot (Y \cdot f) - Y \cdot (X \cdot f) = \{\{H_X, H_Y\}, f\}$, so that eq. (14.6) yields $\{H_{[X,Y]} - \{H_X, H_Y\}, f\} = 0$. Because constants commute with any function we cannot conclude that $H_{[X,Y]} = \{H_X, H_Y\}$. This motivates the:

Definition. *Consider a Lie group G acting on a symplectic manifold M by symplectic action. This action is said to be Poissonian if the Hamiltonians H_X of the one-parameter subgroups are globally defined, depend linearly on X, and are such that*

$$H_{[X,Y]} = \{H_X, H_Y\}$$

In the previous case, where there exists an invariant 1-form α, this property is always satisfied. In this case, we have $\{H_X, H_Y\} = \omega(X, Y) = d\alpha(X, Y) = X \cdot \alpha(Y) - Y \cdot \alpha(X) - \alpha([X, Y])$. By invariance of α we have $X \cdot \alpha(Y) = \mathcal{L}_X \alpha(Y) = \alpha([X, Y]) = -Y \cdot \alpha(X)$, so that $\{H_X, H_Y\} = \alpha([X, Y]) = H_{[X,Y]}$.

Example. This particular situation is important because it occurs in the case of a cotangent bundle. In this case we already know that the 1-form α exists and is globally defined. We consider particular diffeomorphisms of $M = T^*N$, namely those which are induced by a diffeomorphism of the base N. We are going to show that α is invariant under such diffeomorphisms, and in particular under those which are induced by group actions on N. Any diffeomorphism ϕ of N induces a transformation on T^*N as follows: a point (q, p) of T^*N is determined by a point $q \in N$ and a linear form p on the tangent space $T_q N$ to N at q. The differential of ϕ at q, which we denote by ϕ_*, maps $T_q N$ into $T_{\phi(q)} N$. Its transpose ϕ^* maps $T^*_{\phi(q)} N$ to $T^*_q N$, hence ϕ^{*-1} maps $T^*_q N$ to $T^*_{\phi(q)} N$. The induced transformation $\widehat{\phi}$ on $M = T^*N$ is given by

$$\widehat{\phi}(q, p) = (\phi(q), \phi^{*-1}(p))$$

Proposition. *The 1-form α on T^*N is invariant under transformations induced by transformations of the base manifold N.*

Proof. The transformation $\widehat{\phi}$ of M induces a transformation $\widehat{\phi}_*$ on TM given by $\widehat{\phi}_*(\delta q, \delta p) = (\phi_* \delta q, \phi^{*-1} \delta p)$. Recall the definition $\alpha(\delta q, \delta p) = p(\delta q)$. Quite generally, we have:

$$(\widehat{\phi}^* \alpha)_{(q,p)}(\delta q, \delta p) = \alpha_{\widehat{\phi}(q,p)}(\widehat{\phi}_*(\delta q, \delta p))$$
$$= \phi^{*-1}(p)(\phi_* \delta q) = p(\delta q) = \alpha(\delta q, \delta p)$$

■

In particular, assuming that a Lie group G acts on the base manifold N, this action lifts to a Poissonian action on $M = T^*N$, and for any $X \in \mathcal{G}$ we have $H_X(m) = \alpha(X \cdot m) = p(X \cdot q)$ for $m = (q, p)$.

When the action of a Lie group G on a symplectic manifold M is Poissonian, any $X \in \mathcal{G}$ is associated with a function H_X such that $X \cdot f = \{H_X, f\}$, and $X \to H_X$ is linear. Hence there exists a function $\mathcal{P} : M \longrightarrow \mathcal{G}^*$ such that one can write $H_X(m) = \langle \mathcal{P}(m), X \rangle$, where $\langle \rangle$ is the pairing between $\mathcal{G}$ and its dual.

Definition. *The application* $m \to \mathcal{P}(m) \in \mathcal{G}^*$ *is called the moment map.*

The moment map has the following covariance property with respect to the action of G:

Proposition. *The value of the moment at the point* $g \cdot m$ *is related to its value at the point* m *by:*

$$\mathcal{P}(g \cdot m) = \mathrm{Ad}_g^* \mathcal{P}(m)$$

<u>Proof.</u> This is equivalent to $H_X(g \cdot m) = H_{g^{-1}Xg}(m)$. Since we asssume that G is connected, it is sufficient to show this for an infinitesimal $g = 1 + \epsilon Y$. Then we have to show that $dH_X(Y \cdot m) = H_{[X,Y]}(m)$. Using eq. (14.5) we have $dH_X(Y \cdot m) = -Y \cdot H_X = -\{H_Y, H_X\}$, where we used eq. (14.8) in the last step. Since the group action is Poissonian, this is $H_{[X,Y]}$. ∎

The moment map associates conserved quantities with symmetries of the Hamiltonian. This is the Noether theorem.

Theorem. *Let G be a Lie group acting by Poissonian action on a symplectic manifold M, and let H be a Hamiltonian invariant under the action of G. Then the moment $\mathcal{P}$ is conserved under the flow of H.*

<u>Proof.</u> Let us fix $X \in \mathcal{G}$ and consider the function on M: $\langle \mathcal{P}(m), X \rangle = H_X(m)$. Its time derivative under the flow of H is $\partial_t H_X = \{H, H_X\} = -X \cdot H = 0$ because H is invariant under the group action. ∎

In the situation of the theorem, one can use the conserved quantities to reduce the number of degrees of freedom. This is called Hamiltonian reduction, because one is able to define a new symplectic variety of smaller dimension on which the reduced motion takes place.

Let M be a symplectic manifold and let G be a group acting on M by Poissonian action. Let $\mathcal{P}$ be the moment map. Let us fix a particular value μ of the moment and consider the set of points of phase space where $\mathcal{P}(m) = \mu$. By the Noether theorem the motion takes place on this set:

$$M_\mu \equiv \mathcal{P}^{-1}(\mu), \qquad \mu \in \mathcal{G}^*$$

We have to assume that μ is not a critical value of $\mathcal{P}$, that is, at all $m \in M_\mu$ we have $d\mathcal{P}(m) \neq 0$. Hence there exists a tangent space at m to M_μ. Since M_μ is defined by $\dim G$ equations, we have:

$$\dim M_\mu = \dim M - \dim G \tag{14.10}$$

Note that M_μ is not in general a symplectic variety, and is not in general of even dimension. However, there is a residual action of the group G

on M_μ. We have seen that the action of G is transformed by $\mathcal{P}$ into the coadjoint action on $\mathcal{G}^*$. So the stabilizer G_μ of the moment μ, that is the group of $g \in G$ such that $\mathrm{Ad}_g^* \mu = \mu$, preserves M_μ.

Definition. *The reduced phase space F_μ is the quotient:*

$$F_\mu \equiv M_\mu/G_\mu = \mathcal{P}^{-1}(\mu)/G_\mu$$

Here we assume that the quotient is well-defined as a differentiable manifold. In general there are particular values of μ where this quotient is ill-defined. However, for a generic situation we don't have to enter into such subtleties.

The nice feature of the reduced phase space is that it is naturally equipped with a symplectic structure, and in particular is of even dimension.

Proposition. *Let ξ and η be two vectors tangent to F_μ at the point f. Consider at a point $m \in M_\mu$ above f any two tangent vectors to M_μ, ξ' and η', projecting on ξ, η respectively. We then set:*

$$\omega_f(\xi, \eta) = \omega_m(\xi', \eta')$$

This is independent of the chosen representatives m, ξ', η' and defines a symplectic form on F_μ.

<u>Proof.</u> We first show that given m, $\omega_m(\xi', \eta')$ is independant of the choices of representatives ξ', η'. This amounts to showing that $\omega_m(V, W) = 0$ for V vertical, i.e. tangent to the orbit of G_μ, and W tangent to M_μ. Since V is vertical we can write $V = X \cdot m$ with $X \in \mathcal{G}_\mu$. We can consider the Hamiltonian H_X defined on M and note that $\omega_m(V, W) = -dH_X(W)$. But on M_μ we have $H_X = \mathcal{P}(X) = \mu(X)$ is constant. Since $dH_X(W)$ is the derivative of H_X in the direction of W, which is tangent to M_μ, this derivative vanishes.

The quantity $\omega_m(\xi', \eta')$ is independent of the choice of the point m above f since by invariance of ω we have for any $g \in G_\mu$, $\omega_{gm}(g\xi', g\eta') = \omega_m(\xi', \eta')$. This shows that ω_F, the form that we have defined on F_μ, is well-defined, bilinear and antisymmetric.

To show that ω_F is closed, note that the restriction $\omega|_{M_\mu}$ of ω to M_μ is obviously closed. On the other hand, if π is the projection $M_\mu \to M_\mu/G_\mu$ we have just shown that $\omega|_{M_\mu} = \pi^*\omega_F$. Since d commutes with π^* we have $\pi^* d\omega_F = 0$. This means that $d\omega_F(\pi_* X, \pi_* Y) = 0$ for all X, Y. Since π_* is surjective we have $d\omega_F = 0$.

Finally, we have to show that ω_F is non-degenerate. Vertical vectors are in the kernel of $\omega|_{M_\mu}$, in fact they are the whole kernel of $\omega|_{M_\mu}$ as we now explain. We have seen that vertical vectors are orthogonal under ω to $T_m(M_\mu)$. More precisely, we have

$$(\mathcal{G} \cdot m)^\perp = T_m(M_\mu)$$

because both sides have the same dimension, $\dim M - \dim G$, since ω is non-degenerate. But the kernel of the restriction of ω to M_μ is

$$\operatorname{Ker} \omega|_{M_\mu}(m) = T_m(M_\mu) \cap T_m^\perp(M_\mu) = \mathcal{G}_\mu \cdot m$$

These vectors project to 0 when we take the quotient under G_μ, hence ω_F is non-degenerate. ∎

We often need to compute the Poisson brackets of functions on the reduced phase space F_μ, knowing the Poisson brackets on M. Any function $\tilde{f}$ on F_μ uniquely extends to a G_μ invariant function on M_μ. However, to be able to compute Poisson brackets on M we have to extend this function further to a function f defined in a vicinity of M_μ in M. Even requiring complete G invariance is not sufficient to lift $\tilde{f}$ to M. This is because while $\dim M = \dim M_\mu + \dim G$, the fibre along G at $m \in M_\mu$ is not transverse to M_μ, since G_μ leaves M_μ invariant. The general procedure consists of choosing arbitrary extensions f of $\tilde{f}$ outside of M_μ. Two extensions differ by a function vanishing on M_μ. Then we will show how to compute the reduced Poisson bracket as some modification of $\{f, g\}$ (computed on M) independent of the arbitrary choices (see eq. (14.11) below). The difference of the Hamiltonian vector fields of two extensions of the same function on F_μ is controlled by the following:

Lemma. *Let f be a function defined in a vicinity of M_μ and vanishing on M_μ. Then the Hamiltonian vector field X_f associated with f is tangent to the orbit $G \cdot m$ at any point $m \in M_\mu$.*

<u>Proof.</u> The subvariety M_μ is defined by the equations $H_{X_i} = \mu_i$ for some basis X_i of $\mathcal{G}$. Since f vanishes on M_μ one can write $f = \sum (H_{X_i} - \mu_i) f_i$ for some functions f_i defined in the vicinity of M_μ. For any tangent vector V at a point $m \in M_\mu$ one has, using that $\sum (H_{X_i} - \mu_i) \partial f_i$ vanishes on M_μ:

$$df(m)(V) = \sum_i dH_{X_i}(m)(V) f_i(m) = -\omega \left(\sum_i f_i(m) X_i \cdot m, \, V \right)$$

where we used that the Hamiltonian vector field associated with H_{X_i} is $X_i.m$. Hence $X_f = \sum f_i(m) X_i \cdot m \in \mathcal{G} \cdot m$. ∎

As a consequence of this lemma we have a method for computing the reduced Poisson bracket. We take two functions defined on M_μ and invariant under G_μ and extend them arbitrarily. Then we compute their Hamiltonian vector fields on M. It turns out that they can be "projected" on the tangent space to M_μ by adding a vector tangent to the orbit $G \cdot m$. These projections are independent of the extensions and the reduced Poisson bracket is given by the value of the symplectic form on M acting on them.

Proposition. *Let f be a function defined in a vicinity of M_μ and G_μ-invariant on M_μ. At each point $m \in M_\mu$ one can choose a vector $V_f \cdot m \in G \cdot m$ such that $X_f + V_f \cdot m \in T_m(M_\mu)$ and $V_f \cdot m$ is determined up to a vector in $G_\mu \cdot m$.*

Proof. Recall that the symplectic orthogonal of $G \cdot m$ is exactly $T_m(M_\mu)$, so we want to solve $\omega \left(X_f + V_f \cdot m, X \cdot m \right) = 0$, $\forall X \in G$. Note that for $X, Y \in G$, $\omega(X \cdot m, Y \cdot m) = \{H_X, H_Y\} = H_{[X,Y]} = \mathcal{P}\left([X,Y]\right) = \mu\left([X,Y]\right)$ since the action is Poissonian and $m \in M_\mu$. So the equation to be solved reads $\mathcal{L}_{X \cdot m} f = \mu\left([V_f, X]\right)$. Both members are linear in $X \in G$ and vanish when $X \in G_\mu$ (the left-hand side because f is G_μ-invariant, the right-hand side because G_μ stabilizes μ), so the equation can be seen as an equation on G/G_μ. On this quotient, the mapping $(\bar{X}, \bar{Y}) \rightarrow \mu([X,Y])$ is a skew-symmetric non-degenerate (since we have quotiented by the kernel G_μ) bilinear form, hence the equation can be uniquely solved for $\bar{V}_f$. ∎

Proposition. *Let $\tilde{f}$, $\tilde{g}$ be two functions on the reduced phase space F_μ. We lift them to functions f, g defined on a vicinity of M_μ, and G_μ-invariant on M_μ. The reduced Poisson bracket is given by:*

$$\{\tilde{f}, \tilde{g}\}_{\text{red}} = \omega \left(X_f + V_f \cdot m, X_g + V_g \cdot m \right) = \{f, g\} - \mu([V_f, V_g]) \quad (14.11)$$

Proof. We want to show that the Hamiltonian vector field associated with $\tilde{f}$ for the reduced symplectic form is $\pi_*(X_f(m) + V_f \cdot m)$ for $m \in M_\mu$. Note that this is independent of the choice of V_f modulo G_μ, and that $X_f(m) + V_f \cdot m$ is by construction tangent to M_μ. If V is an arbitrary tangent vector to M_μ at m, by definition of the reduced symplectic form, one has to check that:

$$d\tilde{f}(\pi_* V) = -\omega_{\text{red}}(\pi_*(X_f(m) + V_f \cdot m), \pi_* V) = -\omega_m(X_f(m) + V_f \cdot m, V) \quad (14.12)$$

We have seen that the symplectic orthogonal of $G \cdot m$ is $T_m(M_\mu)$, so that $\omega_m(V_f \cdot m, V) = 0$. On the other hand, since f is G_μ-invariant we have $d\tilde{f}(\pi_* V) = df(V) = -\omega_m(X_f, V)$. This proves eq. (14.12) and the first equality in eq. (14.11). To get the second form of the reduced Poisson bracket, note that $\omega(X_f, V_g \cdot m) = -\mu([V_f, V_g])$. This is because, since $X_f + V_f \cdot m \in (\mathcal{G} \cdot m)^\perp$, one has, for $m \in M_\mu$:

$$\omega(X_f, V_g \cdot m) = -\omega(V_f \cdot m, V_g \cdot m) = -H_{[V_f, V_g]}(m) = -\mu([V_f, V_g])$$

∎

Remark. Note that if $f_{\mid M_\mu} = 0$ we have $X_f + V_f \cdot m \in \mathcal{G}_\mu \cdot m$, hence $\{\tilde{f}, \tilde{g}\}_{\text{red}} = \omega(X_f + V_f \cdot m, X_g + V_g \cdot m) = 0$, so that eq. (14.11) is independent of the arbitrary choices of f and g.

In the applications, we are often given functions f and g on M which are G-invariant. It is then obvious that $\{f, g\}$ is also G-invariant (invariance of ω), hence its restriction to M_μ is G_μ-invariant. Moreover, the associated Hamiltonian vector fields X_f, X_g are tangent to M_μ since the G-invariance of f implies:

$$0 = df(X \cdot m) = -\omega(X_f, X \cdot m) = -dH_X(X_f), \quad X \in \mathcal{G}$$

therefore the functions H_X are constant along X_f, i.e. X_f is tangent to M_μ. In that case, one can take $V_f = 0$. It follows that for such functions the reduced Poisson bracket on F_μ is simply given by the ordinary Poisson bracket on M.

Proposition. *If f and g are G-invariant functions on M they define functions $\tilde{f}$ and $\tilde{g}$ on the reduced phase space F_μ and we have:*

$$\{\tilde{f}, \tilde{g}\}_{\text{red}} = \{f, g\}$$

where the right-hand side is also G-invariant and defines a function on F_μ.

14.4 The case $M = T^*G$

Let us now apply these general considerations to the case where N is a Lie group G and $M = T^*G$. We have two actions of G on itself, namely

$$L_g : (g, n) \to gn, \quad R_g : (g, n) \to ng^{-1}$$

The first one is called left-action and the second one right-action. The differential L_{g*} of the map $n \to gn$ sends $T_n G \to T_{gn} G$. In particular if $n = e$, the unit element of G, this differential maps $T_e G = \mathcal{G}$ to $T_g G$. For any fixed $X \in \mathcal{G}$, we get a left-invariant vector field $g \cdot X$ on G. As above, the action on functions on G is defined by $(gf)(n) = f(g^{-1}n)$ so that $(g(hf)) = ((gh)f)$. For an infinitesimal transformation $g = \exp(tX)$ with t small we have:

$$X \cdot f(n) = \frac{d}{dt} f(e^{-tX} \cdot n)|_{t=0} = -\mathcal{L}_{X \cdot n} f$$

To build a phase space we need the cotangent bundle $M = T^* G$. We can use left translations to identify M with $G \times \mathcal{G}^*$, where $\mathcal{G}^*$ is the dual of the Lie algebra $\mathcal{G}$.

$$p \in T_g^* G \longrightarrow (g, \xi) \quad \text{where} \quad p = L_{g^{-1}}^* \xi$$

Indeed, $L_{g^{-1}*}$ maps $T_g G$ to $\mathcal{G}$, hence its transposed $L_{g^{-1}}^*$ maps $\mathcal{G}^*$ to $T_g^* G$. Explicitly, $p(V) = \xi(g^{-1}V)$, which is simply $\xi(X)$ when $V = g \cdot X$ is a left-invariant vector field. Note that if X_i is a basis of the Lie algebra $\mathcal{G}$, the left-invariant vector fields $g \cdot X_i$ provide a basis of each $T_g G$.

    Since $M = T^*G$ there is a canonical 1-form α defined as follows: if (v, κ) is a tangent vector to $T^* G$ at the point (g, ξ), so that $v \in T_g G$ and $\kappa \in \mathcal{G}^*$, we have:

$$\alpha(v, \kappa) = \xi(g^{-1}v)$$

This canonical 1-form is both left-invariant and right-invariant, because, according to the general construction, it is invariant under any diffeomorphism of the base $M = G$ of the cotangent bundle. Hence the symplectic form $\omega = d\alpha$ is invariant under both actions.

    We can compute the Hamiltonians which generate infinitesimal left and right translations on functions. We have seen that in the case of a cotangent bundle $H_X(m) = \alpha(-X \cdot m)$, where, $X \cdot m$ is the infinitesimal group action on M and the minus sign is introduced because we consider the action on functions. For right translations $g \to gh^{-1}$ and $h = e^{tX_R}$ we get

$$H_{X_R}(g, \xi) = \alpha(g \cdot X_R) = \xi(X_R)$$

For left translations $g \to hg$ and $h = e^{tX_L}$ we get

$$H_{X_L}(g, \xi) = \alpha(-X_L \cdot g) = -\xi(g^{-1}X_L g)$$

From this, one sees that the moment maps for left and right actions are given by

$$\mathcal{P}_L(X_L) = -(g\xi g^{-1}, X_L), \quad \mathcal{P}_R(X_R) = (\xi, X_R)$$

In applications, we consider the case when only subgroups H_L and H_R act by left and right translations on G. The moments live in the duals of the Lie algebra $\mathcal{H}_L$ and $\mathcal{H}_R$, hence require natural projections from $\mathcal{G}^*$, induced by restriction to the subalgebras. Specifically, if ξ is a linear form on G its restriction to $\mathcal{H}$ is an element of $\mathcal{H}^*$, and we denote it by $P_{\mathcal{H}^*}\xi$. We have shown the:

Proposition. *Let H_L and H_R be two subgroups of G acting by left and right translations respectively on T^*G. The moment maps associated with these actions are:*

$$\mathcal{P}_R(g,\xi) = P_{\mathcal{H}_R^*}\xi, \quad \mathcal{P}_L(g,\xi) = -P_{\mathcal{H}_L^*}\, g\xi g^{-1}$$

We often need to compute Poisson brackets of functions on T^*G. We have natural elementary functions on this phase space, namely the quantities $\xi(X)$ for any given $X \in \mathcal{G}$, and the matrix elements $\rho_{ij}(g)$ of g in any faithful representation of G. Any other function can be expressed as a function of these elementary ones. So it is enough to give the Poisson brackets of these elementary functions.

Proposition. *The Poisson brackets of the elementary functions on T^*G read:*

$$\{\rho_{ij}(g), \rho_{kl}(g)\} = 0, \quad \{\xi(X), \rho_{ij}(g)\} = \rho_{ij}(gX), \quad \{\xi(X), \xi(Y)\} = \xi([X,Y])$$

<u>Proof.</u> These relations are consequences of the fact that $H_X = \xi(X)$ is the Hamiltonian generating right translations. Since, by the general theory, the action is Poissonian, we have $\{H_X, H_Y\} = H_{[X,Y]}$ and this gives the last equation. Moreover, $\{H_X, \rho_{ij}(g)\}$ is the infinitesimal action of the right translation by X on the function $\rho_{ij}(g)$, namely the action $(h\rho_{ij})(g) = \rho_{ij}(gh)$ (note that we have here gh because this is an action on functions). So $\{H_X, \rho_{ij}(g)\} = \rho_{ij}(gX)$, proving the second equation. Finally, the first equation is obvious since the $\rho_{ij}(g)$ only depend on the position variables and not the momenta. ∎

The Poisson bracket on ξ is the Kirillov bracket. We often drop the explicit reference to the representation ρ in the above formulae, but it is important to keep in mind that the g occurring in these equations is a function on phase space, and not a point on the base.

14.5 Poisson–Lie groups

Consider two Poisson manifolds M_1 and M_2. The cartesian product $M_1 \times M_2$ is also equipped with a natural Poisson structure as follows. The space of functions on $M_1 \times M_2$ is the tensor product of the space of functions on M_1 and the space of functions on M_2. That is, one can write any such function in the form $f(x,y) = \sum_i f_i^{(1)}(x) f_i^{(2)}(y)$, where the sum is in general infinite and requires some topology for its precise definition. We then define for two functions $f(x, y)$ and $g(x, y)$ the Poisson bracket:

$$\{f, g\}_{M_1 \times M_2} = \sum_{ij} \{f_i^{(1)}, g_j^{(1)}\}_{M_1} f_i^{(2)} g_j^{(2)} + \{f_i^{(2)}, g_j^{(2)}\}_{M_2} f_i^{(1)} g_j^{(1)}$$

This obeys all the properties of a Poisson bracket, and implies that functions on M_1 Poisson commute with functions on M_2.

In particular, if G is a Lie group endowed with a Poisson structure, the product $G \times G$ has a Poisson structure and one may wonder whether the multiplication $(g, h) \to gh$ from $G \times G$ to G is compatible with the respective Poisson structures. More precisely, if we have two Poisson manifolds M and N and a map $\phi : M \to N$, this map is said to be Poisson if for any two functions $f_1, f_2,$ on N we have $\{f_1 \circ \phi, f_2 \circ \phi\}_M = \{f_1, f_2\}_N \circ \phi$. In our case the multiplication is Poisson if we have:

$$\{f_1(gh), f_2(gh)\}_{G \times G} = \{f_1, f_2\}_G(gh)$$

where in the left-hand side $f_{1,2}(gh)$ are to be viewed as functions on $G \times G$.

Definition. *A Poisson–Lie group G is a Lie group G equipped with a Poisson structure such that the multiplication in G, viewed as a map $G \times G \to G$, is a Poisson mapping.*

To describe the Poisson structure on G one can use the Lie algebra $\mathcal{G}$ to label the derivatives of any function at a point $g \in G$. We first choose a basis E_a of the Lie algebra $\mathcal{G}$:

$$[E_a, E_b]_{\mathcal{G}} = f_{ab}^c E_c \tag{14.13}$$

and consider the right-invariant vector fields ∇_a^R defined as:

$$\nabla_a^R f(g) = \frac{d}{dt} f(e^{tE_a} g) \Big|_{t=0}$$

which form a basis of the tangent space $T_g G$. The Poisson bracket of two functions f_1, f_2 on G can be written as a bilinear combination of derivatives with coefficients $\eta^{ab}(g)$ as:

$$\{f_1, f_2\}_G(g) = \sum_{a,b} \eta^{ab}(g)(\nabla_a^R f_1)(g)(\nabla_b^R f_2)(g) \tag{14.14}$$

The coefficients $\eta^{ab}(g)$ contain all the information on the Poisson structure, and we will express the Lie–Poisson condition on them. For this, it is convenient to introduce the element $\eta(g) \in \mathcal{G} \times \mathcal{G}$:

$$\eta(g) = \sum_{a,b} \eta^{ab}(g) E_a \otimes E_b$$

Any function on G can be expressed on elementary functions, i.e. the matrix elements of a faithful representation ρ of G. It is sufficient to express the Poisson brackets of such elementary functions. If we consider the function $g \to \rho(g)$ we have $\nabla_a^R \rho(g) = \rho(E_a g) = \rho(E_a)\rho(g)$. The Poisson bracket of two matrix elements then reads:

$$\{\rho_{ij}(g), \rho_{kl}(g)\}_G = \rho_{ij} \otimes \rho_{kl}\Big(\eta(g) g \otimes g\Big)$$

In the following we drop the explicit mention of the representation ρ and write this formula in the usual tensor notation:

$$\{g_1, g_2\}_G = \eta_{12}(g)\, g \otimes g \tag{14.15}$$

Note that the antisymmetry of the Poisson bracket (14.14) requires $\eta_{12} = -\eta_{21}$.

Proposition. *The Lie–Poisson property is equivalent to the following cocycle condition on η:*

$$\eta(gh) = \eta(g) + \mathrm{Ad}_g \cdot \eta(h)$$

where the adjoint action on $\mathcal{G} \otimes \mathcal{G}$ is defined as $\mathrm{Ad}_g \cdot \eta(h) = g \otimes g\, \eta(h)\, g^{-1} \otimes g^{-1}$.

Proof. We have $\{(gh)_1, (gh)_2\}_{G \times G} = \{g_1, g_2\}_G h_1 h_2 + g_1 g_2 \{h_1, h_2\}_G$ since $\rho(gh) = \rho(g)\rho(h)$. This has to be equal to $\eta(gh)(gh)_1(gh)_2$. Comparing the two expressions, the cocycle condition follows. ∎

This condition is called a cocycle condition because it means that η is a 1-cocycle for the Hochschild group cohomology of G with values in the representation $\mathcal{G} \otimes \mathcal{G}$. Looking at the vicinity of the identity e in G, that is writing $g = \exp(tX)$ and $h = \exp(t'Y)$, with t, t' small and $\eta(e^{tX}) = \eta(e) + t d_e \eta(X) + \cdots$, the cocycle condition implies, at order 0 in t, t', that $\eta(e) = 0$, and at second order that

$$d_e\eta([X,Y]) = [\Delta X, d_e\eta(Y)] - [\Delta Y, d_e\eta(X)] \tag{14.16}$$

with $\Delta X = X \otimes 1 + 1 \otimes X$. This means that the linear function $d_e\eta$ on $\mathcal{G}$ is a 1-cocycle for the Lie algebra cohomology with values in the same representation $\mathcal{G} \otimes \mathcal{G}$.

One can use $d_e\eta$ to introduce a Lie algebra structure on $\mathcal{G}^*$. For any function f on G the differential $d_e f$ at the identity is a linear function on $\mathcal{G}$, hence an element of $\mathcal{G}^*$. Considering two functions f_1 and f_2, eq. (14.14) shows that the differential $d_e\{f_1, f_2\}$ only depends on $d_e f_1$ and $d_e f_2$ and is proportional to $d_e\eta$, since $\eta(e) = 0$.

Definition. *The Poisson bracket $\{,\}_G$ defines a Lie algebra structure on $\mathcal{G}^*$ by the following formula:*

$$[d_e f_1, d_e f_2]_{\mathcal{G}^*} = d_e\{f_1, f_2\}_G \tag{14.17}$$

The Jacobi identity for the Lie algebra structure is direct consequence of the Jacobi identity for the Poisson bracket on G.

Introducing a basis (E^a) in $\mathcal{G}^*$, dual to the basis (E_a) in $\mathcal{G}$, the differential at the identity can written as $d_e f = \sum_a E^a(\nabla_a f) \in \mathcal{G}^*$, where $\nabla_a f = \frac{d}{dt} f(e^{tE_a})|_{t=0}$. Similarly,

$$d_e \eta = E^c \nabla_c \eta = C_c^{ab} E^c \otimes E_a \otimes E_b$$

With these notations eq. (14.17) reads:

$$[E^a, E^b]_{\mathcal{G}^*} = C_c^{ab} E^c \tag{14.18}$$

so the structure constants are $C_c^{ab} = (\nabla_c \eta^{ab})$. We shall denote by G^* the connected Lie group with Lie algebra $\mathcal{G}^*$.

Let G be a Poisson–Lie group and let $\mathcal{D} = \mathcal{G} \oplus \mathcal{G}^*$. This is called the classical double. One can introduce a Lie algebra structure on $\mathcal{D}$ which extends the Lie algebra structures on $\mathcal{G}$ and $\mathcal{G}^*$ and such that the elements of $\mathcal{G}$ and $\mathcal{G}^*$ do not commute. In terms of a basis $(E_a) \in \mathcal{G}$ and its dual $(E^a) \in \mathcal{G}^*$, this structure reads:

$$\begin{aligned}
[E_a, E_b] &= f_{ab}^c E_c \\
[E^a, E_b] &= f_{bc}^a E^c - C_b^{ac} E_c \\
[E^a, E^b] &= C_c^{ab} E^c
\end{aligned} \tag{14.19}$$

Proposition. *The above brackets define a Lie algebra structure on $\mathcal{D}$.*

<u>Proof.</u> One defines also $[E_b, E^a] = -[E^a, E_b]$ so the antisymmetry is obvious. We need to verify the Jacobi identity. Due to the Jacobi identity on $\mathcal{G}$ and $\mathcal{G}^*$ one has only to verify the two cases $[E^a, [E^b, E_c]] + [E_c, [E^a, E^b]] + [E^b, [E_c, E^a]] = 0$ and $[E^a, [E_b, E_c]] + [E_c, [E^a, E_b]] + [E_b, [E_c, E^a]] = 0$. These relations reduce to:

$$C_d^{ab} f_{cl}^d - C_l^{bd} f_{dc}^a + C_l^{ad} f_{dc}^b + C_c^{db} f_{ld}^a - C_c^{da} f_{ld}^b = 0$$

when using the Jacobi identity on the structure constants C_c^{ab} and f_{bc}^a. This is just the cocycle relation, eq. (14.16). ∎

Remark. With this bracket on the double, one can construct a solution of the Yang–Baxter equation:

$$r_{12} = \sum_a E_a \otimes E^a \quad \in \mathcal{G} \otimes \mathcal{G}^*$$

The Yang–Baxter equation:

$$[r_{12}, r_{13}] + [r_{12}, r_{23}] + [r_{13}, r_{23}]$$
$$= [E_a, E_b] \otimes E^a \otimes E^b + E_a \otimes [E^a, E_b] \otimes E^b + E_a \otimes E_b \otimes [E^a, E^b] = 0$$

is identically satisfied due to the definition of the commutators on $\mathcal{D}$.

14.6 Action of a Poisson–Lie group on a symplectic manifold

Let G be a Poisson–Lie group and M be a symplectic manifold. We assume that G acts on M. In this case the natural compatibility condition between the group action and the Poisson structures on G and M is:

Definition. *The action of a Poisson–Lie group on a symplectic manifold is a Lie–Poisson action if for any $g \in G$ and any functions f_1 and f_2 on M, we have:*

$$\{f_1(g \cdot m), f_2(g \cdot m)\}_{G \times M} = \{f_1, f_2\}_M(g \cdot m) \qquad (14.20)$$

Here the Poisson structure on $G \times M$ is the product Poisson structure.

At the infinitesimal level, let $X \in \mathcal{G}$ and denote by $X \cdot m$ the vector field on M corresponding to the infinitesimal transformation generated by X. For any function f on M we define the action of X on f by:

$$(X.f)(m) = \frac{d}{dt}f(e^{-tX} \cdot m)|_{t=0} = \langle \zeta_f(m), X \rangle$$

where $\langle \, , \, \rangle$ denotes the pairing between $\mathcal{G}$ and $\mathcal{G}^*$. This defines a function $\zeta_f : m \to \zeta_f(m) \in \mathcal{G}^*$.

Proposition. *The infinitesimal form of eq. (14.20) for a Lie–Poisson action is:*

$$\{X \cdot f_1, f_2\}_M + \{f_1, X \cdot f_2\}_M + \langle [\zeta_{f_1}, \zeta_{f_2}]_{\mathcal{G}^*}, X \rangle = X \cdot \{f_1, f_2\}_M \qquad (14.21)$$

or, equivalently, $\{\zeta_{f_1}, f_2\}_M + \{f_1, \zeta_{f_2}\}_M + [\zeta_{f_1}, \zeta_{f_2}]_{\mathcal{G}^*} = \zeta_{\{f_1, f_2\}_M}.$

<u>Proof.</u> The definition of the product Poisson bracket is equivalent to:

$$\{f_1(g \cdot m), f_2(g \cdot m)\}_{G \times M} = \{f_1(g \cdot m), f_2(g \cdot m)\}_M + \{f_1(g \cdot m), f_2(g \cdot m)\}_G$$

In the right-hand side of this formula, in the term $\{ \, , \, \}_M$ the functions $f_{1,2}(gm)$ are viewed as functions of m, and g is a parameter, while in the term $\{ \, , \, \}_G$ they are viewed as functions of g and m is a parameter. So the Lie–Poisson condition reads:

$$\{f_1(g \cdot m), f_2(g \cdot m)\}_M + \{f_1(g \cdot m), f_2(g \cdot m)\}_G = \{f_1, f_2\}_M(g \cdot m)$$

Setting $g = e^{-tX}$ and taking t infinitesimal, the first Poisson bracket becomes

$$-\{X \cdot f_1(m), f_2(m)\}_M - \{f_1(m), X \cdot f_2(m)\}_M$$

and the right-hand side becomes $-X \cdot \{f_1, f_2\}_M(m)$. The second Poisson bracket is $-\langle d_e\{f_1(g \cdot m), f_2(g \cdot m)\}_G, X \rangle$, where d_e is the differential of a function on G taken at $g = e$. By definition of the Lie bracket on $\mathcal{G}^*$ this is $-\langle [d_e f_1, d_e f_2]_{\mathcal{G}^*}, X \rangle$. One gets eq. (14.21) noting that $\zeta_f(m) = -d_e f(g \cdot m)$.

∎

Introducing two dual basis of the Lie algebras $\mathcal{G}$ and $\mathcal{G}^*$, $E_a \in \mathcal{G}$ and $E^a \in \mathcal{G}^*$ with $\langle E^a, E_b \rangle = \delta_b^a$, eq. (14.21) becomes, using $\zeta_f = \sum_a (E_a \cdot f) E^a$:

$$\{E_a \cdot f_1, f_2\}_M + \{f_1, E_a \cdot f_2\}_M - C_a^{bc}(E_b \cdot f_1)(E_c \cdot f_2) = E_a.\{f_1, f_2\}_M \tag{14.22}$$

It follows immediately from eq. (14.21) that a Lie–Poisson action cannot be symplectic if the algebra $\mathcal{G}^*$ is non-Abelian. Hence we cannot expect that infinitesimal group actions are locally generated by Hamiltonians as in the symplectic case. There is, however, a generalization of this notion, in the Lie–Poisson case, by what are called non-Abelian Hamiltonians.

Proposition. *Assume that a Poisson–Lie group G acts on M by a Lie–Poisson action. Then, there exists a function Γ, locally defined on M and taking values in the group G^*, with Lie algebra $\mathcal{G}^*$, such that for any function f on M,*

$$X.f = \langle \, \Gamma^{-1} \{f, \Gamma\}_M, X \rangle \quad , \quad \forall\, X \in \mathcal{G} \tag{14.23}$$

Equivalently, $\zeta_f(m) = \Gamma^{-1}\{f, \Gamma\}_M(m)$. We will refer to Γ as the non-Abelian Hamiltonian of the Lie–Poisson action.

Proof. Introduce the Darboux coordinates (q^i, p^i) on the symplectic manifold M. For any $X \in \mathcal{G}$ expand $X \cdot m = X^{p^i} \partial_{p^i} + X^{q^i} \partial_{q^i}$ and introduce the form $\Omega_X = \sum_i (X^{q^i} dp^i - X^{p^i} dq_i)$. Finally, let Ω be the $\mathcal{G}^*$-valued 1-form $\Omega = E^a \, \Omega_{E_a}$. As in the symplectic case, eq. (14.21) is then equivalent to the following zero-curvature condition for Ω:

$$d\Omega + [\Omega, \Omega]_{\mathcal{G}^*} = 0$$

Therefore, locally on M, $\Omega = \Gamma^{-1} d\Gamma$ with $\Gamma \in G^*$. This proves eq. (14.23).

∎

The converse is true: an action generated by a non-Abelian Hamiltonian as in eq. (14.23) is Lie–Poisson since we have:

$$X \cdot \{f_1, f_2\}_M - \{X \cdot f_1, f_2\}_M - \{f_1, X \cdot f_2\}_M$$
$$= \langle [\Gamma^{-1}\{f_1, \Gamma\}_M \, , \, \Gamma^{-1}\{f_2, \Gamma\}_M]_{\mathcal{G}^*}, X \rangle$$

In the Abelian case, we have $\Gamma(m) = \exp\left(-\mathcal{P}(m)\right)$, where $\mathcal{P}$ is the momentum taking values in the Abelian Lie algebra $\mathcal{G}^*$. This is because, in the symplectic case, eq. (14.23) becomes $X \cdot f = \langle \{\mathcal{P}, f\}, X \rangle = \{H_X, f\}$. Hence Γ is the non–Abelian generalization of the moment map.

14.7 The groups G and G^*

As a preparation for our study of dressing transformations, we apply the previous results to a more specific situation.

Let $\mathcal{G}$ be a Lie algebra with a bilinear invariant form denoted by Tr, with associated connected Lie group G. We denote by $\mathcal{C}$ the tensor Casimir in $\mathcal{G} \otimes \mathcal{G}$.

We equip G with a Lie–Poisson structure by choosing a cocycle $\eta(g)$. A simple way to fulfil the cocycle condition is to take for $\eta(g)$ a coboundary

$$\eta_{12}(g) = r_{12} - \mathrm{Ad}_g r_{12}$$

where r is a constant element in $\mathcal{G} \otimes \mathcal{G}$. Then eq. (14.15) becomes:

$$\{g_1, g_2\}_G = [r_{12}, g_1 g_2] \tag{14.24}$$

This is the Sklyanin bracket. In this case the Lie–Poisson property is easy to check. One has to show that:

$$\{(gh)_1, (gh)_2\}_{G \times G} = [r_{12}, (gh)_1 (gh)_2]$$

This is obvious when we notice that, in the product Poisson structure on $G \times G$, g and h Poisson commute. The computation then reduces to the computation we have done for the Sklyanin approach to the closed Toda chain, see Chapter 6. Since in $G \times G$, g and h Poisson commute, we have:

$$\{(gh)_1, (gh)_2\}_{G \times G} = \{g_1, g_2\}_G h_1 h_2 + g_1 g_2 \{h_1, h_2\}_G$$
$$= [r_{12}, g_1 g_2] h_1 h_2 + g_1 g_2 [r_{12}, h_1 h_2] = [r_{12}, g_1 h_1 g_2 h_2]$$

Note that r_{12} is defined only up to the addition of a multiple of the Casimir element which drops out of $\eta(g)$. We define $r_{12}^{\pm} = r_{12} \pm \frac{1}{2} \mathcal{C}_{12}$. Antisymmetry of the Poisson bracket is ensured by choosing r_{12} antisymmetric, so that $r_{12}^{\pm} = -r_{21}^{\mp}$. The Jacobi identity for the Poisson bracket is ensured by requiring that $r_{12}^{\pm}$ are solutions of the classical Yang–Baxter equation:

$$\left[r_{12}^{\pm}, r_{13}^{\pm}\right] + \left[r_{12}^{\pm}, r_{23}^{\pm}\right] + \left[r_{13}^{\pm}, r_{23}^{\pm}\right] = 0 \tag{14.25}$$

Equation (14.25) is an equation in $\mathcal{G} \otimes \mathcal{G} \otimes \mathcal{G}$, and the indices on $r^{\pm}$ refer to the copies of $\mathcal{G}$ on which $r^{\pm}$ is acting. Using the bilinear form Tr to identify the vector spaces $\mathcal{G}^*$ and $\mathcal{G}$, the elements $r_{12}^{\pm}$ of $\mathcal{G} \otimes \mathcal{G}$ can be mapped into elements $R^{\pm} \in \mathcal{G} \otimes \mathcal{G}^* \cong \text{End}\mathcal{G}$ defined by:

$$R^{\pm}(X) \;=\; \text{Tr}_2\left(r_{12}^{\pm}(1 \otimes X)\right), \quad \forall\, X \in \mathcal{G} \tag{14.26}$$

Note that we have $R^+ - R^- = \text{Id}$.

The Poisson bracket (14.24) on G induces a Lie algebra structure on $\mathcal{G}^*$ by eq. (14.17). Identifying the vector spaces $\mathcal{G}$ and $\mathcal{G}^*$ by Tr, the bracket on $\mathcal{G}^*$ is mapped to the R-bracket:

$$[\,X, Y\,]_R = \left[R^{\pm}(X), Y\right] + \left[X, R^{\mp}(Y)\right] = [R(X), Y] + [X, R(Y)] \tag{14.27}$$

with $R = \frac{1}{2}(R^+ + R^-)$. We gave a detailed analysis of this bracket in Chapter 4, and all the results apply here. We simply recall that because $R^+ - R^- = \text{Id}$, any $X \in \mathcal{G}$ admits a decomposition as:

$$X \;=\; X_+ - X_-, \quad X_{\pm} = R^{\pm}(X) = \text{Tr}_2\left(r_{12}^{\pm}(1 \otimes X)\right) \tag{14.28}$$

In terms of the components X_+ and X_-, the commutator in $\mathcal{G}^*$ becomes:

$$[\,X, Y\,]_R \;=\; [X_+, Y_+] - [X_-, Y_-]$$

In particular, the plus and minus components commute in $\mathcal{G}^*$. Moreover, $R^{\pm}$ are Lie algebra homomorphisms so that $X_{\pm}$ live in two subalgebras $\mathcal{G}_{\pm}$ of $\mathcal{G}$. Recall also that the image of $\mathcal{G}^*$ in $\mathcal{G}_- \oplus \mathcal{G}_+$ is the set of $X = X_+ - X_-$ such that $\theta(X_+) = X_-$, see Chapter 4.

By exponentiation, the subalgebras $\mathcal{G}_{\pm}$ correspond to connected Lie subgroups $G_{\pm}$ of G, and the group G^* can be viewed as the set of pairs (g_-, g_+), subjected to some condition $\theta(g_+) = g_-$, with product law:

$$(g_-, g_+) \cdot (h_-, h_+) = (g_- h_-, g_+ h_+) \tag{14.29}$$

Any element $g \in G$ (in a neighbourhood of the identity) admits a unique factorization as:

$$g \;=\; g_-^{-1} g_+ \tag{14.30}$$

with $\theta(g_+) = g_-$. This associates with $g \in G$ a unique element of $G^* = (g_-, g_+)$ through a factorization problem. So as sets, G and G^* are identified, but they have different group structures.

The group G^* itself becomes a Poisson–Lie group if we introduce on it the Semenov-Tian-Shansky Poisson bracket:

$$\{(g_+)_1, (g_+)_2\}_{G^*} \;=\; -\left[r_{12}^{\pm}, (g_+)_1\,(g_+)_2\right]$$
$$\{(g_-)_1, (g_-)_2\}_{G^*} \;=\; -\left[r_{12}^{\mp}, (g_-)_1\,(g_-)_2\right]$$
$$\{(g_-)_1, (g_+)_2\}_{G^*} \;=\; -\left[r_{12}^{-}, (g_-)_1\,(g_+)_2\right]$$
$$\{(g_+)_1, (g_-)_2\}_{G^*} \;=\; -\left[r_{12}^{+}, (g_+)_1\,(g_-)_2\right] \tag{14.31}$$

or, for the factorized element $g = g_-^{-1} g_+$:

$$\{g_1, g_2\}_{G^*} = -g_1 \, r_{12}^+ \, g_2 - g_2 \, r_{12}^- \, g_1 + g_1 \, g_2 \, r_{12}^\pm + r_{12}^\mp \, g_1 \, g_2 \qquad (14.32)$$

The multiplication in G^* is a Poisson map for the brackets (14.31). The group G^* is therefore a Poisson–Lie group.

14.8 The group of dressing transformations

We use the above results to understand the Poisson structure of the dressing transformations introduced in Chapter 3.

Definition. *Let G be a Poisson–Lie group associated with an r-matrix, and G^* its dual Poisson–Lie group. We define an action of G^* on G which we call a dressing transformation. We identify G^* to G as sets, via the factorization problem $g = g_-^{-1} g_+$. The dressing of $x \in G$ by $g \in G^*$ is defined by:*

$$(g = g_-^{-1} g_+ \in G^*, \ x \in G) : \longrightarrow {}^g x = (xgx^{-1})_\pm \, x \, g_\pm^{-1} \in G \quad (14.33)$$

In this equation $(xgx^{-1})_\pm$ refers to the factorization

$$(xgx^{-1})_-^{-1} (xgx^{-1})_+ = (xgx^{-1})$$

and this implies that the two signs give the same result for ${}^g x$.

Proposition. *The action $x \to {}^g x$ is a group action of G^* on G.*

<u>Proof.</u> We have to show that ${}^g({}^h x) = {}^{gh} x$, that is:

$$({}^h x \, g \, {}^h x^{-1})_\pm \, {}^h x \, g_\pm^{-1} = (x(gh)x^{-1})_\pm x(gh)_\pm^{-1} \qquad (14.34)$$

Introducing the notation, for any $h \in G^*$: $\Theta_\pm^h = (xhx^{-1})_\pm$, so that ${}^h x = \Theta_\pm^h x h_\pm^{-1}$, and using the sign freedom to write the first ${}^h x$ in $({}^h x \, g \, {}^h x^{-1})_\pm$ with the minus sign and the second one with the plus sign, the left-hand side of eq. (14.34) reads:

$$\left(\Theta_-^h \, x \, h_-^{-1} g h_+ \, x^{-1} \Theta_+^{h \, -1} \right)_\pm \Theta_\pm^h \, x \, h_\pm^{-1} g_\pm^{-1}$$

Since $g = g_-^{-1} g_+$, and due to the definition of the group law in G^*, $h_-^{-1} g h_+ = (gh)$. Moreover, since by definition $x(gh)x^{-1} = \Theta_-^{(gh) \, -1} \Theta_+^{(gh)}$ we get:

$$\Theta_-^h \, x \, h_-^{-1} g h_+ \, x^{-1} \Theta_+^{h \, -1} = \Theta_-^h \, \Theta_-^{(gh) \, -1} \Theta_+^{(gh)} \Theta_+^{h \, -1}$$

so that one reads the factorization:

$$\left(\Theta_-^h \, x \, h_-^{-1} g h_+ \, x^{-1} \, \Theta_+^{h\,-1}\right)_\pm = \Theta_\pm^{(gh)} \Theta_\pm^{h\,-1}$$

From this the result follows immediately. ∎

The infinitesimal form of eq. (14.33) is, for any $X \in \mathcal{G}$ with $X = X_+ - X_-$:

$$\delta_X \, x = Y_\pm \, x - x \, X_\pm \qquad \text{with} \qquad Y_\pm = (xXx^{-1})_\pm \qquad (14.35)$$

One of the main properties of this action is that it is a Lie–Poisson action of G^* on G if the groups G and G^* are equipped with the Poisson structures defined in eqs. (14.24) and (14.31), i.e. we have:

$$\{{}^g x_1, {}^g x_2\}_{G^* \times G} = \left[r_{12}^\pm \, , \, {}^g x_1 \, {}^g x_2\right]$$

We will prove this fact by exhibiting a non–Abelian Hamiltonian for dressing transformations, so that they are automatically Lie–Poisson.

Proposition. *The non–Abelian Hamiltonian of the dressing transformations (14.33), which is an element of $G^{**} \cong G$, is the identity function of G, i.e. it is x itself:*

$$X \cdot x = -\delta_X \, x = \text{Tr}_2 \left(x_2^{-1} \, \{x_2, x_1\}_G X_2\right) \qquad \forall X \in \mathcal{G}, x \in G \qquad (14.36)$$

<u>Proof.</u> First note that, as usual, the action on functions is defined with the inverse group element, so that the action of X on the function x is $-\delta_X x = -(xXx^{-1})_\pm x + xX_\pm$. From eq. (14.24), in which we take $\Gamma(x) = x$, we have:

$$
\begin{aligned}
\text{Tr}_2 \left(x_2^{-1} \, \{x_2, x_1\}_G X_2\right) &= -\text{Tr}_2 \left(x_2^{-1}[r_{12}^\pm, x_1 x_2]X_2\right) \\
&= -\text{Tr}_2 \left(r_{12}^\pm x_1 \, x_2 X_2 x_2^{-1}\right) + \text{Tr}_2 \left(x_1 r_{12}^\pm X_2\right) \\
&= -(xXx^{-1})_\pm \, x + x \, X_\pm = -\delta_X \, x
\end{aligned}
$$

∎

It is a remarkable fact that there exists a non-Abelian Hamiltonian, since the group G with the Sklyanin bracket is not a symplectic manifold, the bracket being degenerate.

References

[1] V. Arnold, *Méthodes mathématiques de la Mécanique classique*. MIR, Moscow (1976).

[2] R. Abraham and J. Marsden, *Foundations of Mechanics*. Benjamin, Reading, Massachusetts (1978).

[3] M. Semenov-Tian-Shansky, Dressing transformations and Poisson group actions. *Publ. RIMS* **21** (1985) 1237.

[4] V. Drinfeld, Hamiltonian structures on Lie groups, Lie bialgebras and the geometric meaning of classical Yang–Baxter equations. *Soviet Math. Dokl.* **27** (1983) no. 1, 68–71.

[5] J.H. Lu, *Multiplicative and Affine Poisson Structures on Lie Groups.* PhD Thesis, University of California at Berkeley (1990).

15

Riemann surfaces

Riemann surfaces play a ubiquitous role in the analytic study of integrable systems. Here it is fundamental to see Riemann surfaces both as smooth analytical one-dimensional varieties and as the desingularization of the locus of an algebraic equation $P(x, y) = 0$. We explain the notion of line bundle which arises naturally in the study of integrable systems. This allows us to provide a proof of the Riemann–Roch theorem. This theorem is the main enumerative tool in our applications. Riemann himself discovered the profound implications of theta functions and notably the geometry of the theta divisor in the subject. The starting point is Riemann's theorem which we use to exhibit explicit solutions of integrable systems in terms of theta-functions. We close the chapter by sketching Birkhoff's proof of the Riemann–Hilbert factorization theorem, which plays a central role throughout the book.

15.1 Smooth algebraic curves

Riemann surfaces are compact smooth analytic varieties of dimension 1. This means that around each point p there is a neighbourhood and a local parameter $z(p)$ mapping it homeomorphically to an open disc $|z| < 1$ of the complex numbers. Moreover, in the intersection of two such neighbourhoods the corresponding local parameters $z_1(p)$ and $z_2(p)$ must be related by an analytic bijection. Hence, locally a smooth curve looks like the complex line. Finally, a Riemann surface is compact, hence it is a closed surface without boundary.

For our purposes it is very important to look at Riemann surfaces from an algebraic viewpoint, that is as the locus in $\mathbb{C}^2$ of an algebraic equation $P(x, y) = 0$, where P is a polynomial in the complex variables x and y.

At a generic point of this locus one has $\partial_x P \neq 0$ and $\partial_y P \neq 0$, hence both x and y can be taken as analytic local parameters in the vicinity of this point. If P has degree N in y we can find N analytic solutions $y = f_j(x)$ for $j = 1, \ldots, N$ defined in some open set of the x variable. These are the N branches of the curve presented as an N-fold covering of the complex line of the x variable. This situation can be analytically continued until one gets to a point where some branches meet, that is the equation in y, $P(x, y) = 0$, has a multiple root, so that $\partial_y P = 0$. Generically one still has $\partial_x P \neq 0$ at such a point and one can choose y instead of x as local parameter in the vicinity of this point which is a perfectly smooth point on the curve. The covering projection $(x, y) \to x$, however, is branched at this point, that is several branches coalesce to one point. We say that $P(x, y) = 0$ expresses the curve as an N-fold branched covering of the complex line. Branch points occur when the discriminant of $P(x, y)$, viewed as a polynomial in y, vanishes. This is a polynomial in x, hence has a finite number of roots above which the covering projection branches.

A more complicated situation arises at points where both partial derivatives of $P(x, y)$ vanish. This means that locally the curve looks like the intersection of several lines, hence is not smooth. In order to make contact with our general definition of a Riemann surface one has to perform an operation called desingularization, which basically consists of replacing the singular point by several ordinary points while leaving the analytic structure of the neighbourhood untouched. An easy way to understand how this can be achieved is to consider birational transformations. They are mappings of the complex plane to itself $(x, y) \to (x' = x'(x, y), y' = y'(x, y))$, where x' and y' are rational functions of x and y such that one can also express x and y as rational functions of x' and y'. At a non-singular point of such a transformation it is clearly bijective and preserves the analytic structure. A simple example is the quadratic transformation $x' = 1/x$ and $y' = 1/y$, which is obviously bijective and analytic on its domain of definition. When transforming the equation $P(x, y) = 0$ under such a mapping chosen to have its singular set precisely at a singular point of the curve one may blow up the singular point into several ordinary points of the transformed curve.

Let us see how this works on a simple example. Consider the curve $y^2 = x^2 + x^3$, which is singular at $(0, 0)$, and the birational transformation $x = x'$, $y = x'y'$, which can be inverted by $y' = y/x$. It is obviously bijective except at the singular point $(0, 0)$ where y' is indeterminate. Substituting into the equation of the curve one gets $y'^2 = 1 + x'$ and we see that the two ordinary points $x' = 0$ and $y' = \pm 1$ project to the one singular point $x = y = 0$, while any other point of the transformed curve bijectively corresponds to just one point of the initial curve. This bijection preserves the analytic structure of both curves. A similar construction can be done around any singular point of an algebraic curve and then patched with the analytic structure around ordinary points to get a smooth Riemann surface associated with the equation $P(x,y) = 0$. We say that this Riemann surface is the desingularized curve of the equation $P(x,y) = 0$.

Finally, to get a compact surface one has to consider points at infinity. We transform the equation $P(x,y) = 0$ under the quadratic transformation $x = 1/x'$, $y = 1/y'$ so that a chart around ∞ becomes an analytic chart around 0. In general one gets a singular point at $(0,0)$ which has to be desingularized by the above method. We shall give a simple example of this procedure in the next section, in which we study a type of Riemann surface frequently occuring in integrable models: the hyperelliptic curves.

15.2 Hyperelliptic curves

A hyperelliptic curve is the locus of an equation of the form:

$$y^2 = P(x), \quad P(x) = \prod_i^N (x - a_i) \tag{15.1}$$

where $P(x)$ is a polynomial in x of degree N. In order not to have singular points we shall assume $a_i \neq a_j$ for $i \neq j$. There is an analytic involution of the curve into itself $(x,y) \to (x,-y)$ which is called the hyperelliptic involution, and the existence of such an automorphism in fact characterizes hyperelliptic curves. Of course the curve can be expressed as a two-sheeted covering of the complex line branched above the points $x = a_i$. Around such a point $(x = a_i, y = 0)$ one can take y as a perfectly smooth local parameter. For example, if $P(x) = xP_1(x)$ with $P_1(0) \neq 0$ one can express $x = y^2 + O(y^3)$ around $(0,0)$. Note that the situation would be entirely similar for an equation of the type $y^n = xP_1(x)$ around $(0,0)$ except that now n branches coalesce at the origin. The only tricky point is to consider the situation at infinity. Performing the transformation $x = 1/x'$, $y = 1/y'$ one gets

$$y'^2 \prod_{i=1}^N (1 - x'a_i) - x'^N = 0$$

and we see that the origin is *singular* for $N > 1$. It is now necessary to distinguish two cases, N odd and even.

When $N = 2g + 1$, for some integer g, one can perform the birational transformation $x' = x''$ and $y' = (x''^g / \prod(1 - x''a_i))y''$ and the transformed equation reads $y''^2 - x'' \prod(1 - x''a_i) = 0$. We see that the singular point $(x' = 0, y' = 0)$ gives rise to a single point $(x'' = 0, y'' = 0)$ on the desingularised curve, hence we say that there is just one point at infinity, but it is a branch point of the covering $(x, y) \to x$. The local parameter around ∞ is y'' and we have locally:

$$ x = \frac{1}{y''^2}(1 + O(y'')), \quad y = \frac{1}{y''^{2g+1}}(1 + O(y'')) $$

Note that x and y are meromorphic functions on the Riemann surface with poles only at ∞ where, for example, x has a pole of order 2. Finally, one frequently says that $\sqrt{x}$ is a local parameter around ∞ since this is equivalent to $1/y''$.

When $N = 2g + 2$ one performs the birational transformation $x' = x''$ and $y' = (x''^{(g+1)} / \prod(1 - x''a_i))y''$, which yields $y''^2 = \prod(1 - x''a_i)$. Hence we are in the generic situation, i.e. there are two points corresponding to the singular point $(x' = 0, y' = 0)$, specifically $\infty_+ = (x'' = 0, y'' = +1)$ and $\infty_- = (x'' = 0, y'' = -1)$. In other words, all the points of the desingularised curve bijectively and analytically correspond to points of the original curve except the two points $\infty_\pm$ which map to the same $\infty = (x' = 0, y' = 0)$. We say that the singular curve is obtained from the non-singular one by identifying two points. Around $\infty_\pm$ either x'' or y'' are good local parameters and one gets for example

$$ x = \frac{1}{x''}, \quad y = \frac{1}{x''^{(g+1)}}(1 + O(x'')) $$

We see that the meromorphic function x has two simple poles on the curve, one at ∞_+ and the other at its hyperelliptic conjugate ∞_-. The curve can be seen as a two-sheeted branched covering of the Riemann sphere branched over the $2g + 2$ points $(a_k, 0)$.

Let us remark that in both cases, $N = 2g + 1$ and $N = 2g + 2$, eq. (15.1) allows us to present the hyperelliptic curve as a two-sheeted branched covering of the Riemann sphere with $2g + 2$ branch points. In fact, for $N = 2g + 1$ we have the $2g + 1$ points $(a_k, 0)$ plus the single point at infinity ∞, while for $N = 2g + 2$ we have two points at infinity and they are not branch points of the covering. This will allow us to conclude that in both cases the underlying topological surface is of genus g as we now explain.

15.3 The Riemann–Hurwitz formula

Let us recall that for a triangulated surface where the triangulation has F faces, V vertices and A edges one defines the Euler–Poincaré characteristic

$$\chi = F - A + V = 2 - 2g$$

which is a topological invariant. Here g is called the genus of the Riemann surface. Any surface of genus g is homeomorphic to a sphere with g handles.

Let us now assume that a Riemann surface is presented as an N-sheeted branched covering of some base space of Euler characteristic χ_0. Choose a *triangulation* of the base space such that all branch points are included as vertices. Now, for each triangle consider its N pre-images under the covering projection, and similarly the N pre-images of each edge. While ordinary vertices have N pre-images, each base point corresponding to a branch point has fewer than N pre-images, and the reduction is given by the order of the branch point, i.e. the number of branches which coalesce at the branch point minus 1. This is also called the index of the branch point. Altogether one gets a triangulation of the Riemann surface with NF triangles, NA edges and $NV - B$ vertices, where B is the total number of branch points counted with their index on the Riemann surface. Hence the Euler characteristic χ of the Riemann surface is related to χ_0 by $\chi = N\chi_0 - B$. This is the Riemann–Hurwitz formula:

$$2g - 2 = N(2g_0 - 2) + B \tag{15.2}$$

Let us apply this formula to the computation of the genus of a hyperelliptic curve. The base is the Riemann sphere for which there is a triangulation with eight faces, twelve edges and six vertices, so that $\chi_0 = 2$ or $g_0 = 0$. We have seen that the covering has $B = 2g + 2$ branch points of index 1. Since $N = 2$, the genus computed by the Riemann–Hurwitz formula is precisely the number g parametrizing the degree of the polynomial $P(x)$ in the equation of the curve $y^2 = P(x)$.

15.4 The field of meromorphic functions of a Riemann surface

Consider a Riemann surface of genus g. A complex valued function on this surface is analytic around a point if its expression in terms of a local parameter is analytic. Similarly, it has a pole of order n if its expression in terms of a local parameter has a pole of order n, and so on. These definitions are invariant under analytic reparametrizations. It is impossible to get an everywhere non-constant analytic function on a compact

Riemann surface, this is basically the Liouville theorem. We are generally interested in meromorphic functions which have a finite number of poles. Obviously the meromorphic functions form a field which is called the function field of the surface. When the curve is given by an equation $P(x,y) = 0$, any rational function of x and y is a meromorphic function on the curve with poles located at arbitrary points and one can show that the most general meromorphic function can be so constructed. The field of meromorphic functions is just the field of rational functions of x and y modulo the equation of the curve. This allows us to give a completely algebraic description of Riemann surfaces.

Conversely, let f be a non-trivial meromorphic function on a Riemann surface. The Cauchy theorem still holds true on a Riemann surface. Integrating over a small circle whose interior contains no poles or zeroes of $f - a$, one gets:

$$0 = \frac{1}{2\pi i} \oint \frac{df}{f - a} = \text{ number of zeroes of } (f - a) - \text{number of poles of } f$$

which shows that f takes each value the same number of times, say N times. This allows us to present a general Riemann surface as an N-sheeted branched covering of the Riemann sphere by $p \to z = f(p)$. For any z on the Riemann sphere consider its N pre-images p_j under f on the Riemann surface. Now take any other meromorphic function h, and consider the elementary symmetric functions of $h(p_j)$. They do not depend on the order of the sheets, hence define meromorphic functions of z on the Riemann sphere, that is *rational* functions of z. It is then clear that h obeys a polynomial equation $P(f, h) = 0$. By similar arguments one can show that one can choose h so that the polynomial P is irreducible and then any other meromorphic function is a rational function of f and h. We see that any abstract Riemann surface may be viewed as a smooth compact algebraic curve.

Let us consider the example of a hyperelliptic curve $y^2 = P_{2g+1}(x)$, where P_{2g+1} is a polynomial of degree $2g + 1$. Any meromorphic function can be written as $f(x,y) = (A(x) + yB(x))/C(x)$ with A, B, C polynomials in x since one can eliminate y^2 using the equation of the curve and eliminate y in the denominator. In this case we get a simple construction of a meromorphic function with $g + 1$ poles at $(x_1, y_1), \ldots, (x_{g+1}, y_{g+1})$ by taking

$$f(x,y) = \frac{y + Q(x)}{\prod_i (x - x_i)}$$

where $Q(x)$ is a polynomial of degree g determined by requiring that $Q(x_i) = y_i$ so that the numerator of f vanishes at the point $(x_i, -y_i)$. Note that f vanishes at ∞ and has g zeroes at finite distance. Also note that the meromorphic function x has a double pole at ∞ hence takes each value twice. The existence of such a function characterizes hyperelliptic curves.

15.5 Line bundles on a Riemann surface

It is important to generalize the notion of function on a Riemann surface by considering line bundles on the surface, and sections of these bundles. Such line bundles occur naturally in the study of integrable systems. Let us consider a covering $\{U_\alpha\}$ of the Riemann surface by open sets U_α and assume that for each non-void intersection $U_\alpha \cap U_\beta$ some continuous functions (called transition functions) $t_{\alpha\beta}$ are given which are neither vanishing nor ∞ on the intersection. Moreover, we assume that on each non-void triple intersection $U_\alpha \cap U_\beta \cap U_\gamma$ one has $t_{\alpha\beta} t_{\beta\gamma} t_{\gamma\alpha} = 1$ and that $t_{\alpha\beta} t_{\beta\alpha} = 1$. This defines a line bundle ξ. When the transition functions are differentiable it is a differentiable bundle, while if the transition functions are analytic it is an analytic bundle. We shall be concerned with analytic bundles on a Riemann surface. A section of the bundle ξ is a collection of functions f_α on each U_α such that on each intersection we have $f_\alpha = t_{\alpha\beta} f_\beta$. If all functions f_α are holomorphic this is a holomorphic section of ξ and the space of such analytic sections will be denoted by $\Gamma(\xi)$. If all functions f_α are meromorphic it is called a meromorphic section. If f_α has a pole or a zero of order m at some point of $U_\alpha \cap U_\beta$, it is the same for f_β since $t_{\alpha\beta}$ is analytic without zero, and we say that the section has a zero or a pole of order m.

Geometrically, an analytic line bundle can be seen as a triple (E, B, π) such that for any point $b \in B$ there exists an open set U_b and an analytic isomorphism $\pi^{-1}(U_b) \simeq U_b \times \mathbb{C}$. Any point above U_α can be written (p, ϕ_α) with $p \in U_\alpha$ and ϕ_α a number. If, moreover, $p \in U_\beta$, the same point can be written (p, ϕ_β). The two descriptions patch if one can write $\phi_\alpha = t_{\alpha\beta}(p)\phi_\beta$ for some analytic non-vanishing function $t_{\alpha\beta}$ defined on $U_\alpha \cap U_\beta$. The triple intersection condition is obviously satisfied. Moreover, for any non-vanishing analytic functions f_α defined on U_α an equivalent description of the line bundle is obtained by sending the point (p, ϕ_α) of the line bundle to $(p, f_\alpha(p)\phi_\alpha)$. In the new description, the transition functions are now $t_{\alpha\beta}(p) f_\alpha(p)/f_\beta(p)$, so that transition functions differing by multiplication by a ratio f_α/f_β define the *same* line bundle ξ. A local or global section of the line bundle ξ can be viewed intrinsically as a map which associates with each point of the Riemann surface a point in the fibre above it. In a local trivialization $U_\alpha \times \mathbb{C}$ this point can be written $(p, f_\alpha(p))$. In the intersection $U_\alpha \cap U_\beta$ the two descriptions are related by $f_\alpha = t_{\alpha\beta} f_\beta$. We recover the above definition of sections.

Note that the quotient of two sections is a meromorphic *function* since the transition functions cancel, and the product of a section by a function is a section.

One can differentiate *differentiable* sections of ξ with respect to $\bar{z}$ since $\partial_{\bar{z}} f_\alpha = t_{\alpha\beta}\partial_{\bar{z}} f_\beta$, hence $\{\partial_{\bar{z}} f_\alpha\}$ is a section.

Example 1. A very important example of line bundle is provided by the *canonical bundle*. Consider a covering U_α such that each open set U_α is analytically isomorphic to a domain of the complex numbers by a coordinate $p \to z_\alpha(p)$. In each non-trivial intersection $U_\alpha \cap U_\beta$ the coordinate z_β can be expressed as an analytic function of the coordinate z_α, and conversely. One can define:

$$\kappa_{\alpha\beta}(p) = \frac{dz_\beta}{dz_\alpha}(p)$$

which is holomorphic non-vanishing in the intersection. The canonical bundle is defined by these transition functions. This definition is canonical since under change of local parameters $z_\alpha \to w_\alpha$ the transition functions transform according to $\kappa_{\alpha\beta} \to w'_\beta/w'_\alpha \cdot \kappa_{\alpha\beta}$ and the w'_α are analytic non-vanishing on U_α. They therefore define the same bundle.

A global section of this bundle is a collection of analytic functions f_α such that $f_\alpha = \kappa_{\alpha\beta} f_\beta$ on intersections, that is $f_\alpha dz_\alpha = f_\beta dz_\beta$. It can be viewed as a globally defined holomorphic differential form of the type $(1,0)$, i.e. a form involving dz only.

Example 2. Another important example is provided by the so-called point bundles. Let the covering be as above and assume that p is a point on the Riemann surface belonging to some U_α and to no other U_β. One can assume that $z_\alpha(p) = 0$. Define the transition functions $t_{\alpha\beta} = z_\alpha$ on non-trivial $U_\alpha \cap U_\beta$ and choose $t_{\beta\gamma} = 1$ on all other non-trivial intersections. This defines a line bundle ξ_p. Note that our point bundle admits at least one analytic section, σ_α, namely $\sigma_\alpha(z_\alpha) = z_\alpha$ and $\sigma_\beta = 1$. The patching conditions $\sigma_\beta = t_{\beta\gamma}\sigma_\gamma$ are obeyed for any β, γ including α. This section has just one zero at p. We see that an analytic line bundle may have non-trivial holomorphic sections while holomorphic functions on a compact Riemann surface are constant. Also, note that a bundle ξ is the trivial bundle if and only if it admits an analytic non-vanishing section f_α, since in this case one can write $t_{\alpha\beta} = f_\alpha/f_\beta$.

One can introduce a group structure on line bundles. Given two line bundles ξ and σ one can assume that they are defined on a common suitably refined covering U_α by transition functions $t_{\alpha\beta}$ and $s_{\alpha\beta}$. Then the product $\xi\sigma$ is defined by the transition functions $t_{\alpha\beta}s_{\alpha\beta}$ and the inverse ξ^{-1} is defined by $1/t_{\alpha\beta}$, while the neutral element is just the trivial bundle, i.e. the cartesian product of the Riemann surface by the complex numbers which can be defined by $t_{\alpha\beta} = 1$. Note that the above definitions are coherent with the redefinitions $t_{\alpha\beta} = f_\alpha/f_\beta \cdot t_{\alpha\beta}$. This group law is commutative, hence an *additive* notation is frequently used.

If $\{f_\alpha\}$ is a section of ξ and $\{g_\alpha\}$ a section of σ then $\{f_\alpha g_\alpha\}$ is a section of $\xi\sigma$ and $\{1/f_\alpha\}$ is a meromorphic section of ξ^{-1}.

15.6 Divisors

One can build more complicated bundles from point bundles at points $p_1,\ldots,p_k$ by taking their product, that is $\sum_j n_j\xi_{p_j}$. Here n_j are positive or negative integers. A formal sum of points p_j on the Riemann surface with multiplicities n_j is called a *divisor*, denoted by $\mathcal{D} = \sum_j n_j p_j$. The line bundle $\xi = \sum_j n_j\xi_{p_j}$ is the line bundle *associated* with the divisor $\mathcal{D}(\xi) = \sum_j n_j p_j$.

For any meromorphic function f on the Riemann surface with zeroes at p_j of order n_j and poles at q_k of order m_k one defines the divisor of the function f as the formal sum

$$\mathcal{D}(f) = \sum_j n_j p_j - \sum_k m_k q_k$$

The divisor of a section of a line bundle is similarly defined. A divisor is positive if all n_j are positive, hence a meromorphic section is analytic if and only if its divisor is positive.

Since ξ_p has a section with divisor p, the line bundle ξ associated with $\mathcal{D}$ has a section f_ξ with divisor $\mathcal{D}$. When the divisor $\mathcal{D}$ is the divisor of a meromorphic function f, the associated bundle ξ has a section f_ξ with divisor $\mathcal{D}$. The section $f^{-1}f_\xi$ is analytic non-vanishing so that ξ is the trivial bundle. Conversely, if ξ associated with $\mathcal{D}$ is trivial, the section f_ξ gives rise to a meromorphic function of divisor $\mathcal{D}$, since we can divide f_ξ by an analytic non-vanishing section. This allows us to introduce an equivalence relation between divisors: two divisors $\mathcal{D}_1$ and $\mathcal{D}_2$ are equivalent if their associated bundles ξ_1 and ξ_2 are such that $\xi_1 - \xi_2$ is the trivial bundle, i.e. if $\mathcal{D}_1 - \mathcal{D}_2$ is the divisor of a meromorphic function.

With a divisor $\mathcal{D} = \sum_j n_j p_j$ is associated a number

$$\deg(\mathcal{D}) = \sum_j n_j$$

called the degree of the divisor. Two equivalent divisors have the same degree since a meromorphic function has the same number of zeroes and poles. Note that if the line bundle ξ has a meromorphic section, σ, of divisor $\mathcal{D}$ and η is the line bundle associated with $\mathcal{D}$ with meromorphic section τ having the same divisor $\mathcal{D}$, $\xi - \eta$ has a holomorphic non-vanishing section σ/τ, hence $\xi = \eta$.

15.7 Chern class

It is well known from differential geometry that one can associate with differential vector bundles an integer called the Chern class which can be computed as the integral of some curvature form. For line bundles it can be shown that this integer classifies *differential* bundles. Moreover, the Chern class is compatible with the group structure, hence the Chern class of the "product" $\xi + \sigma$ is just the sum $c(\xi) + c(\sigma)$ of the Chern classes of ξ and σ. Simple proofs of these facts, adapted from the differential geometric case, can be found in the references. This leads to an index theorem stating that for any meromorphic section f of a line bundle ξ one has:

$$\deg\left(\mathcal{D}\left(f\right)\right) = c(\xi)$$

The trivial bundle has Chern class 0, hence this theorem reduces in this case to the above mentioned fact that a meromorphic function has the same number of zeroes and poles. This implies in particular that only bundles with positive Chern class may have holomorphic sections. Note that the point bundle ξ_p has Chern class 1 since it has a holomorphic section with just one zero at p.

15.8 Serre duality

Having introduced these natural definitions, we can now study the space of most interest to us, i.e. the space $\Gamma(\xi)$ of holomorphic sections of the line bundle ξ. In order to do that it turns out to be very useful to introduce new spaces called $H^1(\Sigma, \mathcal{O}(\xi))$ associated with the Riemann surface Σ and the line bundle ξ. Their definition is analogous to that of line bundles, except that the multiplicative structure is replaced by an additive one. Specifically, consider a covering U_α (and we assume that each U_α is connected and simply connected) and for each non-void intersection $U_\alpha \cap U_\beta$ a holomorphic local section $f_{\alpha\beta}$ of ξ. Such sections are assumed to obey $f_{\alpha\beta} + f_{\beta\gamma} + f_{\gamma\alpha} = 0$ on each non-void triple intersection $U_\alpha \cap U_\beta \cap U_\gamma$. The space $H^1(\Sigma, \mathcal{O}(\xi))$ is the space of $\{f_{\alpha\beta}\}$ modulo trivial ones, i.e. modulo those of the form $f_{\alpha\beta} = f_\alpha - f_\beta$ for holomorphic sections f_α of ξ over U_α. Note that there is no non-vanishing condition here. If trivialisations of the bundle are defined over the U_α one can represent these sections by complex valued functions related by transition functions. Let f_α and $f_{\alpha\beta}$ be represented over U_α by functions f^α_α and $f^\alpha_{\alpha\beta}$, while $f_{\alpha\beta}$ and f_β are represented on U_β by $f^\beta_{\alpha\beta}$ and f^β_β respectively. Then the equation expressing the triviality of $f_{\alpha\beta}$ reads on $U_\alpha \cap U_\beta$: $f^\alpha_{\alpha\beta} = f^\alpha_\alpha - t_{\alpha\beta}f^\beta_\beta$ or a similar expression for the β trivialization. Note that one can define a $\bar\partial$ operator on sections of ξ since $\bar\partial$ vanishes on transition functions.

Let us give an alternative description of $H^1(\Sigma, \mathcal{O}(\xi))$. One can always solve $f_{\alpha\beta} = f_\alpha - f_\beta$ with f_α *differentiable* sections of ξ on U_α by partition of unity arguments*. Consider the type $(0, 1)$ forms (i.e. forms in $d\bar z$)

$$\bar\partial f_\alpha = \frac{\partial f_\alpha}{\partial \bar z} d\bar z$$

On an intersection we have $\bar\partial f_\alpha - \bar\partial f_\beta = \bar\partial f_{\alpha\beta} = 0$, hence these forms patch to a globally defined section σ of ξ with values in type $(0, 1)$ forms.

Conversely, if such a differentiable section is given, one can write it as $\bar\partial f_\alpha$ on each U_α because, by an application of the Cauchy integral formula, one can always solve $\bar\partial f = \sigma$ on a connected set with f a differentiable function. Then one can define on an intersection $f_{\alpha\beta} = f_\alpha - f_\beta$ which is analytic since $\bar\partial$ vanishes on it. Of course $f_{\alpha\beta}$ vanishes if and only if the set of all f_α define a section of ξ, i.e. if $f_\alpha = f_\beta$. Hence $H^1(\Sigma, \mathcal{O}(\xi))$ can be identified with the set of differentiable sections of type $(0, 1)$ of ξ, denoted by $\Gamma^1_d(\xi)$, modulo the image by $\bar\partial$ of the set of differentiable sections of ξ.

$$H^1(\Sigma, \mathcal{O}(\xi)) \equiv \Gamma^1_d(\xi)/\bar\partial\Gamma_d(\xi)$$

It can be shown that $H^1(\Sigma, \mathcal{O}(\xi))$ is finite-dimensional.

This description is useful for understanding the Serre duality between the spaces $H^1(\Sigma, \mathcal{O}(\xi))$ and the space $\Gamma(\kappa - \xi)$, defined by a non-singular pairing:

$$H^1(\Sigma, \mathcal{O}(\xi)) \times \Gamma(\kappa - \xi) \to \mathbb{C}$$

As we have seen in the discussion of the canonical bundle, a section of $\kappa - \xi$ can be viewed as a holomorphic section of $-\xi$ with values in type $(1, 0)$ forms. Take a differentiable element of $\Gamma^1(\xi)$, locally of the form $f(z, \bar z)d\bar z$, and a holomorphic section of $\kappa - \xi$, locally of the form $g(z)dz$, and consider their wedge product, locally $f(z, \bar z)g(z)\, dz \wedge d\bar z$. In the product, the transition functions $t_{\alpha\beta}$ of ξ and $t^{-1}_{\alpha\beta}$ of $-\xi$ cancel so that one

---

* If $\sum_\alpha r_\alpha = 1$ with $\mathrm{supp}(r_\alpha) \in U_\alpha$, define $f_\alpha = \sum_\gamma r_\gamma f_{\alpha\gamma}$ where $r_\gamma f_{\alpha\gamma}$ is extended to 0 in $U_\alpha - U_\gamma$, and note that $f_\alpha - f_\beta = \sum_\gamma r_\gamma(f_{\alpha\gamma} - f_{\beta\gamma}) = \sum_\gamma r_\gamma f_{\alpha\beta} = f_{\alpha\beta}$.

ends up with a globally defined volume form on the Riemann surface
that we integrate. This defines the pairing. Note that if f is identified
with 0, that is if $f(z, \bar{z})d\bar{z} = \bar{\partial}h$, one can integrate by parts obtaining
the integral of $h\bar{\partial}g \wedge dz$ which vanishes since g is holomorphic. Hence the
pairing is well-defined between the considered spaces. Finally, the essential
point is that the pairing is non-singular. Indeed, any linear form acting
on $fd\bar{z}$ may be written $\int f(z, \bar{z})g(z, \bar{z})\, dz \wedge d\bar{z}$ for some distribution g,
since this form vanishes when $fd\bar{z} = \bar{\partial}h$, the distribution g is weakly analytic hence is an analytic function by the classical Weyl lemma. It follows
that $\Gamma(\kappa - \xi)$ and $H^1(\Sigma, \mathcal{O}(\xi))$ are finite-dimensional spaces of the same
dimension.

15.9 The Riemann–Roch theorem

Let $\Gamma(\xi)$ be the finite-dimensional space of holomorphic sections of the
line bundle ξ and $c(\xi)$ its Chern class.

Theorem. *On a Riemann surface of genus g with canonical bundle κ we
have, for any line bundle ξ:*

$$\dim\Gamma(\xi) - \dim\Gamma(\kappa - \xi) = c(\xi) + 1 - g \qquad (15.3)$$

<u>Proof.</u> We first show that $\chi(\xi) = \dim\Gamma(\xi) - \dim H^1(\Sigma, \mathcal{O}(\xi)) - c(\xi)$
is independent of the line bundle ξ. As a first step, we show that
$\chi(\xi + \xi_p) = \chi(\xi)$ for any point bundle ξ_p (see Example 2), with analytic
section σ_p vanishing at p. Note that any local or global analytic section
of ξ can be multiplied by σ_p to produce a section of $\xi + \xi_p$ vanishing at
p. This is clearly an injective homomorphism of the space of sections. It
fails to be surjective if some global sections of $\xi + \xi_p$ do not vanish at p.
Hence we have two cases:
(a) There exists a global section of $\xi + \xi_p$ non-vanishing at p. We have
    $\dim\Gamma(\xi + \xi_p) = \dim\Gamma(\xi) + 1$.
(b) All global sections of $\xi + \xi_p$ vanish at p. We have $\dim\Gamma(\xi + \xi_p) = \dim\Gamma(\xi)$.
Now let us consider local sections over intersections $U_\alpha \cap U_\beta$ (we assume
that p does not belong to such intersections and $p \in U_\alpha$). The homo-
morphism $f_{\alpha\beta} \to \sigma_p f_{\alpha\beta}$ is clearly bijective in this case since σ_p does not
vanish outside p and trivial elements are mapped to trivial elements. This
induces a surjective homomorphism $H^1(\Sigma, \mathcal{O}(\xi)) \to H^1(\Sigma, \mathcal{O}(\xi + \xi_p))$
which fails to be injective if there exists a non-trivial set of sections
$f_{\alpha\beta}$ of ξ mapping to a trivial set of sections $f'_\alpha - f'_\beta$ of $\xi + \xi_p$, that is
$\sigma_p f_{\alpha\beta} = f'_\alpha - f'_\beta$.

In case (a) above, we can substract the global non-vanishing section from all f'_γ so as to achieve $f'_\alpha(p) = 0$. Then one can divide all f'_γ by σ_p, getting f_α such that $f_{\alpha\beta} = f_\alpha - f_\beta$. Hence the mapping is bijective and we have $\dim H^1(\Sigma, \mathcal{O}(\xi)) = \dim H^1(\Sigma, \mathcal{O}(\xi + \xi_p))$.

In case (b) we will show that $\dim H^1(\Sigma, \mathcal{O}(\xi)) = \dim H^1(\Sigma, \mathcal{O}(\xi + \xi_p)) + 1$ by constructing a section $f_{\alpha\beta}$ which maps under σ_p on a trivial section $f'_{\alpha\beta} = f'_\alpha - f'_\beta$, where we have chosen $f'_\alpha(p) \neq 0$. If we had $f_{\alpha\beta} = f_\alpha - f_\beta$ we would get $f'_\alpha - \sigma_p f_\alpha = f'_\beta - \sigma_p f_\beta$, hence defining a global section of $\xi + \xi_p$ which by hypothesis vanishes at p, yielding $f'_\alpha(p) = 0$, a contradiction. If, however, $f'_\alpha(p) = 0$ one can divide it by σ_p and $f_{\alpha\beta}$ is trivial. This shows that non-trivial elements $\{f_{\alpha\beta}\}$ are parametrized by $f'_\alpha(p)$ and form a space of dimension 1 modulo trivial elements.

Since $c(\xi + \xi_p) = c(\xi) + 1$, in both cases we get $\chi(\xi + \xi_p) = \chi(\xi)$. Starting from $\xi - \xi_p$ we get $\chi(\xi) = \chi(\xi - \xi_p)$. Hence for any divisor $\mathcal{D}$ and line bundle η associated with $\mathcal{D}$ we have $\chi(\xi + \eta) = \chi(\xi)$.

The rest of the proof is easier. There exists $\mathcal{D}$ such that $\dim \Gamma(\xi + \eta) \neq 0$. Otherwise one gets $c(\eta) = C^{\text{ste}} - \dim H^1(\Sigma, \mathcal{O}(\xi + \eta)) \leq C^{\text{ste}}$, where $C^{\text{ste}} = \dim H^1(\Sigma, \mathcal{O}(\xi)) - \dim \Gamma(\xi)$ is independent of η. This is impossible since $c(\eta)$ can be arbitrarily large. Let σ be a non-trivial analytic section of $\xi + \eta$. Since η has a meromorphic section σ_η of divisor $\mathcal{D}$ we see that ξ has a non-trivial meromorphic section σ/σ_η. Hence ξ is the line bundle associated with this meromorphic section. We have obtained the:

Proposition. *Any line bundle on the Riemann surface Σ has a non-trivial meromorphic section of divisor $\mathcal{D}$ and is the line bundle associated with this divisor.*

We can now take for ξ the trivial bundle in $\chi(\xi + \eta) = \chi(\xi)$ and we see that $\chi(\eta)$ is independent of η, as previously claimed. Taking into account the Serre duality formula, we get:

$$\chi = \dim \Gamma(\xi) - \dim \Gamma(\kappa - \xi) - c(\xi)$$

When ξ is the trivial bundle $\dim \Gamma(\xi) = 1$ since global analytic functions on Σ are constants, hence $\chi = 1 - \dim \Gamma(\kappa)$. When $\xi = \kappa$ one gets $\chi = \dim \Gamma(\kappa) - 1 - c(\kappa)$, hence $\chi = -(1/2)c(\kappa)$. To compute $c(\kappa)$ we view the Riemann surface Σ as an n-sheeted covering of the Riemann sphere using a meromorphic function f on Σ. We start from a meromorphic differential ω on the Riemann sphere and take its pullback $f^*(\omega)$ on Σ. On the sphere we can take $\omega = dz$. Its divisor has degree -2, since it has a double pole at infinity. Hence its pullback has n poles of order 2 above ∞. Moreover, for each branch point of order ν the mapping f is locally $z = f(w) = w^\nu$, so that $f^*(dz) = \nu w^{\nu-1} dw$ has a zero of order $\nu - 1$ which is the multiplicity index of the branch point. So the total multiplicity of the zeroes is precisely the total multiplicity B of the branch points of the covering, which by the Riemann–Hurwitz formula, eq. (15.2), with $g_0 = 0$ is equal to $2g - 2 + 2n$. The Chern class of κ is therefore $2g - 2$. Hence $\chi = 1 - g$, yielding the Riemann–Roch theorem. Moreover, we see that $\dim \Gamma(\kappa) = g$ which means that the space of globally defined analytic one forms is of dimension g.   ∎

Consider now the meromorphic functions on Σ. Let $M(-\mathcal{D})$ be the set of meromorphic functions on Σ whose divisor is bigger than $-\mathcal{D}$; i.e. $f \in M(-\mathcal{D})$ iff the orders of its poles are less than or equal to those specified by $-\mathcal{D}$ and the orders of its zeroes are greater than or equal to those specified by $-\mathcal{D}$. Let ξ be the line bundle associated with the divisor $\mathcal{D}$. The space of holomorphic sections of ξ is isomorphic to $M(-\mathcal{D})$, because this line bundle has a meromorphic section of divisor $\mathcal{D}$ and any other section is obtained by multiplication by a meromorphic function. The section will be holomorphic iff the divisor of the function is greater than $-\mathcal{D}$. Hence $\dim M(-\mathcal{D}) = \dim \Gamma(\xi)$. We define $i(\mathcal{D}) = \dim \Gamma(\kappa - \xi)$. This is the dimension of the space of differentials with a divisor greater than $\mathcal{D}$. Recalling that $c(\xi) = \deg \mathcal{D}$, the Riemann–Roch theorem can be written as:

$$\dim M(-\mathcal{D}) = i(\mathcal{D}) + \deg \mathcal{D} - g + 1 \qquad (15.4)$$

In general the Riemann-Roch formula eq. (15.3) relates two unknown quantities. However, if $c(\xi) > 2g - 2$, we see that $c(\kappa - \xi) < 0$, hence $\dim \Gamma(\kappa - \xi) = 0$ (because $c(\kappa - \xi)$ is the degree of the divisor of any meromorphic section of $\kappa - \xi$ which has then necessarily poles), and we get:

$$\dim \Gamma(\xi) = c(\xi) + 1 - g, \quad \text{if} \quad c(\xi) > 2(g - 1) \qquad (15.5)$$

Corollary. *The dimension of the space of meromorphic functions having at most k prescribed poles and at least l prescribed zeroes is greater than or equal to $k - l + 1 - g$. Equality occurs when $k - l \geq 2g - 2$.*

If $k - l \geq g$ the equality is satisfied for generic positions of the prescribed zeroes and poles. Let us take for simplicity $l = 0$. Then $\dim \Gamma(\kappa - \xi)$ is the dimension of the space of holomorphic differentials having k prescribed zeroes. But the space of holomorphic differentials is of dimension g and we want to impose k linear conditions. This is generically impossible for $k \geq g$. Hence the useful statement:

$$\deg \mathcal{D} \geq g \Rightarrow i(\mathcal{D}) = 0 \quad \text{generically}$$

Note that we have previously illustrated this situation by constructing a meromorphic function with $g + 1$ prescribed poles on a hyperelliptic surface.

15.10 Abelian differentials

Consider a Riemann surface Σ of genus g. Let a_i, b_i be a basis of cycles on Σ with canonical intersection matrix $(a_i \cdot a_j) = (b_i \cdot b_j) = 0$, $(a_i \cdot b_j) = \delta_{ij}$. This means that one can take differentiable loops $t \to a_i(t)$ and $t \to b_i(t)$ such that there is no intersection between loops a_i and a_j or b_j with $j \neq i$, while a_i and b_i intersect at just one point p. Moreover, at p the tangent vectors a_i' and b_i' form a positively oriented basis of the tangent space at p (for the orientation given by the complex space structure). One can then continuously deform these loops without changing the intersection index which is the sum of signs ± 1 at each intersection according to the orientation of the tangent vectors. In particular, one can deform the loops a_i and b_i so that they have a common base point and then cut the Riemann surface along them, getting a polygon with some edges identified. The boundary of this polygon can be described as $a_1 \cdot b_1 \cdot a_1^{-1} \cdot b_1^{-1} \cdots a_g \cdot b_g \cdot a_g^{-1} \cdot b_g^{-1}$, where the identifications are obvious. The common base point becomes all the vertices of the polygon.

The globally defined analytic 1-forms on Σ are called *Abelian differentials of the first kind*. They form a space of dimension g over the complex numbers. Note that such a differential has $2g - 2$ zeroes on the Riemann surface since $c(\kappa) = 2g - 2$ and it has no pole. There is a natural pairing between these forms and loops obtained by integrating the form along the loop. It can be shown that the pairing between a-cycles and differentials is non-degenerate (note they have the same dimension g). This is a consequence of the Riemann bilinear identities that we shall describe below. We choose a basis of first kind Abelian differentials, which we denote by ω_j, $j = 1, \ldots, g$, normalized with respect to the a-cycles:

$$\oint_{a_j} \omega_i = \delta_{ij} \qquad (15.6)$$

The matrix of b-periods is then defined as the matrix $\mathcal{B}$ with matrix elements:

$$\mathcal{B}_{ij} = \oint_{b_i} \omega_j \qquad (15.7)$$

Taking the example of a hyperelliptic surface $y^2 = P_{2g+1}(x)$, where $P(x)$ is a polynomial of degree $2g + 1$, a basis of regular Abelian differentials is provided by the forms

$$\omega_j = x^j dx/y \quad \text{for } j = 0, \ldots, g-1$$

These forms are regular except perhaps at the branch points and at ∞. At a branch point the local parameter is y and we have $y^2 = a(x-b) + \cdots$, hence $x^j dx/y = (2b^j/a)(1 + \cdots)dy$ which is regular. At ∞ we take $x' = 1/x$ and $y' = y/x^{(g+1)}$ so that $y'^2 = ax' + \cdots$ and $x^j dx/y = by'^{2(g-j-1)}dy'$ which is regular for $j \le g-1$ since y' is the local parameter. Of course these forms are unnormalized. Note that their $(2g-2)$ zeroes are located at $x = 0$ and $x = \infty$.

Similarly *Abelian differentials of the second kind* are meromorphic differentials with poles of order greater than 2. Given a point p on Σ, there exists an Abelian differential of the second kind whose only singularity is a pole of second order at p. Indeed, by eq. (15.3) we see that $\dim \Gamma(\kappa+2\xi_p) \ge c(\kappa+2\xi_p)+1-g = g+1$. An element in $\Gamma(\kappa+2\xi_p)$ comes from an Abelian differential multiplied by σ_p^2, where σ_p is the section of the point bundle ξ_p vanishing at p. Since the space of regular differentials is of dimension g we see that there exists at least one section in $\Gamma(\kappa+2\xi_p)$ whose division by σ_p^2 has a pole at p, which is necessarily of second order. Adding a proper combination of differentials of the first kind, one can always ensure that all a-periods of the second kind differential vanish. Such a second kind differential is called normalized.

We can apply the Cauchy theorem to a meromorphic globally defined 1-form of type (1,0), yielding the vanishing of the sum of residues of first order poles. Note that the residue of a meromorphic 1-form is intrinsically defined since $\text{Res} = (1/2\pi i) \oint \omega$, where the contour is a small loop around the given singularity. So we define *Abelian differentials of the third kind* as general meromorphic differentials with first order poles whose sum of residues vanish. Given two points p and q there exists a unique normalized (all a-periods vanish) third kind differential whose only singularities are a pole of order 1 at p with residue 1, and a pole of order 1 at q with residue -1.

15.11 Riemann bilinear identities

On a Riemann surface on which we have chosen canonical cycles there is a pairing between meromorphic differentials. Specifically, let Ω_1 and Ω_2 be two meromorphic differentials on Σ. The pairing $(\Omega_1 \bullet \Omega_2)$ is defined by integrating them along the canonical cycles as follows:

$$(\Omega_1 \bullet \Omega_2) = \sum_{i=1}^{g} \left(\oint_{a_j} \Omega_1 \oint_{b_j} \Omega_2 - \oint_{a_j} \Omega_2 \oint_{b_j} \Omega_1 \right)$$

The Riemann bilinear identity expresses this quantity in terms of residues:

Proposition. *Let g_1 be a function defined on the Riemann surface, cut along the canonical cycles, and such that $dg_1 = \Omega_1$. We have:*

$$(\Omega_1 \bullet \Omega_2) = 2i\pi \sum_{\text{poles}} \text{res}(g_1\Omega_2) \tag{15.8}$$

<u>Proof.</u> One computes $\oint g_1\Omega_2$ on the boundary of the polygon representing the Riemann surface. On one hand, this produces the sum of residues in eq. (15.8). On the other hand, we compute this integral explicitly:

$$\oint_{a_j a_j^{-1}} g_1\Omega_2 = \oint_{a_j} g_1\Omega_2 - \oint_{a_j} \left(g_1 + \int_{b_j} \Omega_1\right)\Omega_2$$

since g_1 is shifted by $\sum_{b_j} \Omega_1$ when crossing the cut a_j. Similarly, one gets:

$$\oint_{b_j b_j^{-1}} g_1\Omega_2 = \oint_{b_j} g_1\Omega_2 - \oint_{b_j} \left(g_1 - \oint_{a_j} \Omega_1\right)\Omega_2$$

Adding the contributions for all j one gets $(\Omega_1 \bullet \Omega_2)$. ∎

Corollary. *The matrix of b-periods $\mathcal{B}$ is symmetric.*

<u>Proof.</u> The pairing between the normalized holomorphic differentials is trivial: $(\omega_i \bullet \omega_j) = 0$ for $i, j = 1, \ldots, g$. This reads $\mathcal{B}_{ij} = \mathcal{B}_{ji}$. ∎

Corollary. *Let Ω_2 be a normalized differential of the second kind with a pole of order n, with principal part $z^{-n}dz$ at $z = 0$ for some local parameter z. Let $\Omega_1 = \omega_k$ be a normalized holomorphic differential expanded as*

$$\omega_k = \left(\sum_{i=0}^{\infty} c_i z^i\right)dz$$

around $z = 0$. One has:

$$\oint_{b_k} \Omega_2 = 2\pi i \frac{c_{n-2}}{n-1}$$

By linearity, if $\Omega^{(P)}$ is a normalized second kind differential with principal part $dP(z)$, where $P(z) = \sum_{n=1}^{N} p_n z^{-n}$, then we have

$$\frac{1}{2i\pi} \oint_{b_k} \Omega^{(P)} = -\text{Res}\,(\omega_k P) \tag{15.9}$$

15.12 Jacobi variety

We have seen that differential line bundles on a Riemann surface are classified by their Chern class. This is not so for analytic line bundles. To describe their classification, it is sufficient to consider the different analytic structures on line bundles of Chern class 0. The space of such structures is called the Jacobi variety. Since analytic line bundles are the same as equivalence classes of divisors on the Riemann surface, the Jacobi variety identifies with equivalence classes of divisors of degree 0. It is thus necessary to characterize the divisors of meromorphic functions.

In order to do that, consider a divisor of degree 0 which can always be written $\mathcal{D} = \sum_i (p_i - q_i)$, with non-necessarily distinct points. Choose paths γ_i from q_i to p_i and associate with $\mathcal{D}$ the point in $\mathbb{C}^g$ of coordinates:

$$\rho_k(\mathcal{D}) = \sum_i \int_{\gamma_i} \omega_k, \quad k = 1, \dots, g$$

Such sums are called Abel sums. If the paths are homotopically deformed these integrals remain constant by the Cauchy theorem. If one makes a loop around a_k, then $\rho_l \to \rho_l + \delta_{kl}$. If one makes a loop around b_k, then $\rho_l \to \rho_l + \mathcal{B}_{kl}$. Hence the maps ρ_k give a well-defined point on the torus:

$$J(\Sigma) = \mathbb{C}^g / (\mathbb{Z}^g + \mathcal{B}\mathbb{Z}^g) \tag{15.10}$$

where $\mathcal{B}$ is the matrix of the b-periods. If one permutes the points p_i and q_i independently the point in the torus does not change. To see it, let the paths γ_1' connect q_1 to p_2, γ_2' connect q_2 to p_1 and σ connect q_1 to q_2. One has $\int_{\gamma_1} \omega = \int_{\gamma_1'} \omega - \int_\sigma \omega$ up to periods and $\int_{\gamma_2} \omega = \int_{\gamma_2'} \omega + \int_\sigma \omega$ up to periods, so $\int_\sigma \omega$ cancels in the sum. Note that $J(\Sigma)$ is an additive group and that the above map from line bundles to points of $J(\Sigma)$ is a homomorphism.

The theorems of Abel and Jacobi state that this point on the torus $J(\Sigma)$ characterizes the divisor $\mathcal{D}$ up to equivalence, so that the Jacobian variety can be identified with the g-dimensional complex torus $J(\Sigma)$.

Theorem (Abel). *A divisor $\mathcal{D} = \sum_i (p_i - q_i)$ is the divisor of a meromorphic function if and only if, for any first kind Abelian differential ω, the Abel sum $\sum_i \int_{\gamma_i} \omega$ vanishes modulo $\mathbb{Z} + \mathcal{B}\mathbb{Z}$ for any choice of paths γ_i from q_i to p_i.*

Theorem (Jacobi). *For any point $\lambda \in J(\Sigma)$ and a fixed reference divisor $\mathcal{D}_0 = \sum_{i=1}^g q_i$, one can find a divisor of g points $\mathcal{D} = p_1 + \cdots + p_g$ on Σ such that $\rho_k(\mathcal{D} - \mathcal{D}_0)$ maps to λ. Moreover, for generic λ the divisor $\mathcal{D}$ is unique.*

<u>Proof.</u> Let f be a meromorphic function with divisor $\mathcal{D} = \sum_i (p_i - q_i)$ and consider for $\lambda \in \mathbb{C}$ the pencil of divisors $\mathcal{D}^\lambda = \sum_i (p_i^\lambda - q_i)$ of the meromorphic functions $f + \lambda$ (the poles are those of f, the zeroes vary analytically with λ). Finally, let $\phi(\lambda)$ be the point in $\mathbb{C}/(\mathbb{Z}+\mathcal{B}\mathbb{Z})$ obtained by integrating ω along paths from q_i to p_i^λ. The map ϕ is obviously analytic and can be extended to an analytic map on the Riemann sphere because when $\lambda \to \infty$, we have $p^\lambda \to q$. But such a map from the Riemann sphere to a torus is necessarily constant, since dz is a regular differential on the torus, hence $\phi^*(dz)$ has to vanish on the Riemann sphere. Since $\phi(\infty) = 0$ one gets $\phi = 0$.

In order to prove the converse and the Jacobi theorem one has to exhibit particular functions or divisors. This will also be a consequence of the powerful Riemann theorem that we will show later on. To show the generic uniqueness of the g points mapping to a given λ note that we have g equations for g unknowns, hence the space of solutions is generically of dimension 0. If we have two solutions, there exists a meromorphic function f whose poles and zeroes are respectively these solutions. Then $f + \lambda$ for $\lambda \in \mathbb{C}$ relates the divisor of poles to a one-parameter family of equivalent divisors, hence the solution space would be of dimension 1, a contradiction. ∎

One can embed the Riemann surface Σ into its Jacobian $J(\Sigma)$ by the Abel map. Specifically, choose a point $q_0 \in \Sigma$ and define the vector $\mathcal{A}(p)$ with coordinates $\mathcal{A}_k(p)$ modulo the lattice of periods:

$$\mathcal{A} : \Sigma \longmapsto J(\Sigma) \tag{15.11}$$

$$\mathcal{A}_k(p) \;=\; \int_{q_0}^{p} \omega_k \tag{15.12}$$

Clearly, the Abel map depends on the point q_0. But changing this point just amounts to a translation in $J(\Sigma)$. The Abel map is an analytic embedding of the Riemann surface into the g-dimensional torus $J(\Sigma)$, i.e. is injective.

15.13 Theta-functions

One can show using Riemann bilinear type identities that the imaginary part of the period matrix $\mathcal{B}$ is a positive definite quadratic form. This allows us to define the Riemann theta-function:

$$\theta(z_1, \ldots, z_g) = \sum_{m \in \mathbb{Z}^g} e^{2\pi i (m,z) + \pi i (\mathcal{B}m, m)} \tag{15.13}$$

Since the series is convergent, it defines an analytic function on $\mathbb{C}^g$.

The theta-function has simple automorphy properties with respect to the period lattice of the Riemann surface: for any $l \in \mathbb{Z}^g$ and $z \in \mathbb{C}^g$

$$\theta(z + l) = \theta(z)$$
$$\theta(z + \mathcal{B}l) = \exp[-i\pi(\mathcal{B}l, l) - 2i\pi(l, z)]\theta(z) \tag{15.14}$$

The divisor of the theta-function is the set of points in the Jacobian torus where $\theta(z) = 0$. Note that this is an analytic subvariety of dimension $g-1$ of the torus, well-defined due to the automorphy property.

The fundamental theorem of Riemann expresses the intersection of the image of the embedding of Σ into $J(\Sigma)$ with the divisor of the theta-function.

Theorem. *Let $w = (w_1, \ldots, w_g) \in \mathbb{C}^g$, arbitrary. Either the function $\theta(\mathcal{A}(p) - w)$ vanishes identically for $p \in \Sigma$ or it has exactly g zeroes $p_1, \ldots, p_g$ such that:*

$$\mathcal{A}(p_1) + \cdots + \mathcal{A}(p_g) = w - \mathcal{K} \tag{15.15}$$

where $\mathcal{K}$ is the so-called vector of Riemann's constants, depending on the curve Σ and the point q_0 but independent of w.

Proof. We first dissect the Riemann surface Σ as explained above, obtaining a polygon with boundary $a_1 \cdot b_1 \cdot a_1^{-1} \cdot b_1^{-1} \cdots a_g \cdot b_g \cdot a_g^{-1} \cdot b_g^{-1}$ in $\mathbb{C}$. Consider the analytic function on the polygon (or more precisely on the Riemann surface cut along the previous loops):

$$f(p) = \theta(\mathcal{A}(p) - w)$$

Assuming that f does not vanish identically, it has discrete zeroes p_i. Since it has no pole the number of these zeroes is given by:

$$\text{number of zeroes} = \frac{1}{2\pi i} \oint \frac{df}{f}$$

where the integral is taken on the boundary of the polygon. This integral is a sum of terms on the arcs a_k and b_k. The integrals on the arcs b_k and b_k^{-1} are related by a translation by the a_k period, hence cancel. Similarly the difference of the integrals on a_k and a_k^{-1} reduces by the automorphy property to $2i\pi \int_{a_k} \omega_k = 2i\pi$. Thus number of zeroes $= g$.

To prove the second identity we proceed similarly by considering the integral:

$$\frac{1}{2\pi i} \oint g_k \frac{df}{f}, \quad \text{with } dg_k = \omega_k, \text{ and } g_k(q_0) = 0$$

computed on the edge of the polygon. On one hand, this integral is equal to the sum of residues which occur at the zeroes of f, and produces:

$$\sum_{j=1}^{g} g_k(p_j) = \sum_j \mathcal{A}_k(p_j)$$

On the other hand, it can be computed as a sum over arcs using the automorphy properties of the theta-function:

$$\oint_{a_j a_j^{-1}} g_k d(\log f) = - \oint_{a_j} (g_k + \mathcal{B}_{jk})(d(\log f) - 2i\pi d\mathcal{A}_j(p))$$

$$+ \oint_{a_j} g_k d(\log f) \equiv 2i\pi \oint_{a_j} g_k \omega_j$$

modulo $\mathcal{B}$ periods. But we have:

$$\oint_{b_j b_j^{-1}} g_k d(\log f) = \oint_{b_j} (g_k + \delta_{jk}) d(\log f) - \oint_{b_j} g_k d(\log f)$$

$$\equiv \delta_{jk}(2i\pi w_j - 2i\pi \mathcal{A}_j(q_1))$$

modulo periods, where q_1 is the base point of all the loops at the boundary. Putting everything together one gets the Riemann formula with $\mathcal{K}$ given by a complicated expression, independent of w. ∎

Corollary (Jacobi's theorem). *Any point in the Jacobian $J(\Sigma)$ is the image of some degree g divisor $p_1 + \cdots + p_g$ on Σ.*

Proof. One has to find g points such that $\mathcal{A}(p_1) + \cdots + \mathcal{A}(p_g) = z$ modulo periods for given z. We find these points by solving the equation $\theta(\mathcal{A}(p) - \mathcal{K} - z) = 0$. ∎

The divisor of the zeroes of a theta-function has also a nice characterization in terms of points on the Riemann surface.

Theorem. *The zero divisor of a theta-function can be written as*

$$X = \mathcal{K} - \sum_{i=1}^{g-1} \mathcal{A}(\eta_i)$$

Proof. Let X be a point of the Θ divisor, i.e. $\theta(X) = 0$. Consider the function

$$\theta(\mathcal{A}(p) + X) \tag{15.16}$$

By Riemann's theorem, the zeroes of this function are such that (if it does not vanish identically)

$$\mathcal{A}(p_1) + \cdots + \mathcal{A}(p_g) + X = -\mathcal{K}$$

Among these zeroes, one has necessarily q_0, the base point of Abel's map (since $\theta(X) = 0$), say $p_g = q_0$. Hence we have $X = -\mathcal{K} - \sum_{i=1}^{g-1} \mathcal{A}(p_i)$.

Conversely, if X is of this form, one solves eq. (15.16) producing a divisor of g points $\sum_{i=1}^{g} q_i$ such that $\sum_{i=1}^{g-1} \mathcal{A}(p_i) = \sum_{i=1}^{g} \mathcal{A}(q_i)$. A solution is obviously $p_i = q_i$ for $i = 1, \ldots, g-1$ and $q_g = q_0$. This solution is generically unique up to permutation due to the Jacobi theorem. Equation (15.16) for $p = q_0$ reads $\theta(X) = 0$. Note that the space of points of the form $X = -\mathcal{K} - \sum_{i=1}^{g-1} \mathcal{A}(p_i)$ and the solution of $\theta(X) = 0$ are both closed in the Jacobian, hence they are equal.   ∎

The Riemann theorem can be used to express meromorphic functions in terms of theta-functions. Let f be a meromorphic function with g poles at points $\delta_1, \ldots, \delta_g$ and an additional pole at the point q^+ and one of its zeroes at a specified point q^-; i.e. with divisor greater than $D = -\delta_1 - \cdots - \delta_g - q^+ + q^-$. By the Riemann–Roch theorem there is a unique such function generically. Let w, w^+, w^-, w^0 be vectors defined by the formulae:

$$w = \sum_{s=1}^{g} \mathcal{A}(\delta_s) + \mathcal{K}$$

$$w^+ = \mathcal{A}(q^+) + \sum_{s=2}^{g} \mathcal{A}(\delta_s) + \mathcal{K}$$

$$w^- = \mathcal{A}(q^-) + \sum_{s=2}^{g} \mathcal{A}(\delta_s) + \mathcal{K}$$

$$w^0 = w + w^+ - w^-.$$

Let us define the function

$$f(p) = \frac{\theta(\mathcal{A}(p) - w^-)\theta(\mathcal{A}(p) - w^0)}{\theta(\mathcal{A}(p) - w)\theta(\mathcal{A}(p) - w^+)} \tag{15.17}$$

From the Riemann theorem it follows that the two factors in the denominator vanish generically at the points $\delta_1, \ldots, \delta_g$ and $q^+, \delta_2, \ldots, \delta_g$, respectively. Similarly, the two factors in the numerator vanish at $q^-, \delta_2, \ldots, \delta_g$ and g other points. The zeroes at $\delta_2, \ldots, \delta_g$ cancel between the numerator and the denominator, thereby leaving us with the correct divisor of zeroes and poles. It remains to show that the function f is well-defined on Σ. This is because, due to the definition of w^0, the automorphy factors of the theta functions in eq. (15.14) cancel between the numerator and the denominator when p describes b-cycles on Σ. The converse of Abel's theorem results from an analogous construction.

15.14 The genus 1 case

The application of these ideas to the genus 1 case is the classical theory of elliptic functions. A genus 1 analytic surface can be viewed as the quotient of $\mathbb{C}$ by a lattice, whose periods are denoted by a classical convention $2\omega_1, 2\omega_2$. In this case, the theorems of Abel and Jacobi identify the curve and its Jacobi variety. Note that dz is a well-defined regular analytic differential on this torus, and spans the one-dimensional space of Abelian differentials. For any meromorphic function f on the torus, i.e. periodic with respect to the lattice, the differential $f dz$ is a meromorphic differential, so the sum of its residues vanishes. Thus f has at least two-poles or a pole of order 2, and a meromorphic function with just one pole is impossible. The main example is the Weierstrass $\wp$-function:

$$\wp(z) = \frac{1}{z^2} + \sum_{m,n\neq 0} \left\{ \frac{1}{(z - 2m\omega_1 - 2n\omega_2)^2} - \frac{1}{(2m\omega_1 + 2n\omega_2)^2} \right\}$$

An analogue of the theta-function is provided by the Weierstrass sigma-function:

$$\sigma(z) = z \prod_{m,n\neq 0} \left(1 - \frac{z}{\omega_{mn}}\right) \exp\left[\frac{z}{\omega_{mn}} + \frac{1}{2}\left(\frac{z}{\omega_{mn}}\right)^2\right] \qquad (15.18)$$

with $\omega_{mn} = 2m\omega_1 + 2n\omega_2$. This function is related to the $\wp$-function by the equations:

$$\zeta(z) = \frac{\sigma'(z)}{\sigma(z)}, \qquad \wp(z) = -\zeta'(z), \qquad (15.19)$$

The $\wp$-function is doubly periodic, and the sigma-function and zeta-functions transform according to:

$$\zeta(z + 2\omega_l) = \zeta(z) + 2\eta_l, \qquad \sigma(z + 2\omega_l) = -\sigma(z)e^{2\eta_l(z+\omega_l)}$$

The Riemann bilinear identity applied to the forms $\Omega_1 = -\wp(z)dz$ and $\Omega_2 = dz$ yields $2(\eta_1\omega_2 - \eta_2\omega_1) = i\pi$. These functions have the symmetries

$$\sigma(-z) = -\sigma(z), \qquad \zeta(-z) = -\zeta(z), \qquad \wp(-z) = \wp(z).$$

Their behaviour at the neighbourhood of zero is

$$\sigma(z) = z + O(z^5), \quad \zeta(z) = z^{-1} + O(z^3), \quad \wp(z) = z^{-2} + O(z^2)$$

It is useful for the study of the elliptic Calogero model to introduce the Lamé function:

$$\Phi(x, z) = \frac{\sigma(z - x)}{\sigma(x)\,\sigma(z)} e^{\zeta(z)x} \qquad (15.20)$$

It has the symmetry property $\Phi(-x, z) = -\Phi(x, -z)$. The function $\Phi(x, z)$ is a doubly-periodic function of the variable z, $\Phi(x, z + 2\omega_l) = \Phi(x, z)$, and has an expansion of the form:

$$\Phi(x, z) = (-z^{-1} + \zeta(x) + O(z))e^{\zeta(z)x} \qquad (15.21)$$

at the point $z = 0$. As a function of x, it has the following monodromy properties:

$$\Phi(x + 2\omega_l, z) = \Phi(x, z)\exp 2(\zeta(z)\omega_l - \eta_l z). \qquad (15.22)$$

and has a pole at the point $x = 0$: $\Phi(x, z) = x^{-1} + O(x)$. The function Φ is a solution of the Lamé equation:

$$\left(\frac{d^2}{dx^2} - 2\wp(x)\right)\Phi(x, z) = \wp(z)\Phi(x, z) \qquad (15.23)$$

Choosing the periods $\omega_1 = \infty$ and $\omega_2 = i\frac{\pi}{2}$, we obtain the hyperbolic functions:

$$\sigma(z) \to \sinh(z)\exp\left(-\frac{z^2}{6}\right), \quad \zeta(z) \to \coth(z) - \frac{z}{3}, \quad \wp(z) \to \frac{1}{\sinh^2(z)} + \frac{1}{3}$$

and

$$\Phi(x, z) \to \frac{\sinh(z - x)}{\sinh(z)\sinh(x)} e^{x\coth z} \qquad (15.24)$$

In the rational limit, we have

$$\sigma(z) \to z, \quad \zeta(z) \to \frac{1}{z}, \quad \wp(z) \to \frac{1}{z^2}, \quad \Phi(x, z) \to \left(\frac{1}{x} - \frac{1}{z}\right)e^{\frac{x}{z}}$$

15.15 The Riemann–Hilbert factorization problem

In this section, we give the proof of the Riemann–Hilbert theorem on the Riemann sphere, see eq. (3.49) in Chapter 3. Let U_+ be the disc $|x| < 1+\eta$, and U_- be the disc $|x| > 1 - \eta$. Let C be the circle $|x| = 1$.

Let us give ourselves a matrix A analytic in the ring $U_+ \cap U_-$, such that $\det A \neq 0$. We consider the kernel:

$$K(x,t) = A^{-1}(x)A(t) - 1$$

which is analytic in the above ring and vanishes for $x = t$. On continuous functions on C we define an operator $\mathcal{F}$

$$(\mathcal{F}F)(x) = F(x) + \frac{1}{2i\pi} \oint_C \frac{K(x,t)}{x-t} F(t) dt$$

This is a Fredholm operator because it is of the form 1 plus a compact operator. Hence, Im $\mathcal{F}$ is closed and of finite codimension. So one can choose a matrix of polynomials $P(x)$ such that $\det P(x) \neq 0$ for $|x|$ sufficiently large and such that there exists a function F satisfying

$$(\mathcal{F}F)(x) = A^{-1}(x)P(x), \quad x \in C$$

The function $F(x)$ has an analytical extension on the ring $U_+ \cap U_-$ given by

$$F(x) = A^{-1}(x)P(x) + \frac{1}{2i\pi} \oint_C \frac{K(x,t)}{t-x} F(t) dt \qquad (15.25)$$

It follows that one can expand $F(x)$ in a Laurent series and write $F(x) = F_+(x) + F_-(x)$, where $F_\pm(x)$ are analytic in $U_\pm$ respectively and $F_-(x)$ vanishes at $x = \infty$. Note that for $|x| > 1$, one has

$$F_-(x) = -\frac{1}{2i\pi} \oint_C \frac{F(t)}{t-x} dt$$

Subtracting from eq. (15.25), we get for $1 < |x| < 1 + \eta$:

$$F_+(x) = A^{-1}(x)P(x) + \frac{1}{2i\pi} \oint_C \frac{F(t) + K(x,t)F(t)}{t-x} dt$$

Multiplying by $A(x)$, we get in the same domain

$$A(x)F_+(x) = P(x) + \frac{1}{2i\pi} \oint_C \frac{A(t)F(t)}{t-x} dt \equiv \tilde{F}_-(x)$$

The function $\tilde{F}_-(x)$ is in fact analytic for $|x| > 1$ and behaves as $P(x)$ for $x \to \infty$, so that its determinant does not vanish for $|x|$ sufficiently large. It follows that $\det F_+(x)$ and $\det \tilde{F}_-(x)$ do not vanish identically and therefore have a finite number of zeroes at finite distance.

We remove each zero successively by the following procedure. Consider an equation of the form

$$AG_+ = G_- \qquad (15.26)$$

Suppose that $\det G_+(x)$ has a simple zero at x_0 in U_+. This means that there is a linear combination of its columns vanishing at x_0. This combination is realized by multiplying on the right by a constant matrix M such that $\det M \neq 0$. We have of course $AG_+M = G_-M$. Let us assume that it is the first column of G_+M which vanishes at x_0. Multiplying on the right by $\mathrm{diag}(1/(x-x_0), 1, \ldots, 1)$, we remove the zero x_0 without modifying the analytic properties of the right-hand side. Iterating the procedure, we get a pair G_+, G_- satisfying eq. (15.26) with $\det G_+(x) \neq 0$ for $x \in U_+$. It follows that the zeroes of $\det G_-(x)$ are not in U_+ and can be removed by the same procedure without modifying the analytic properties of $G_+(x)$. At the end we get a matrix $G_-(x)$ behaving at ∞ as

$$G_-(x) = A_-(x)\mathrm{diag}\,(x^{\kappa_1}, \ldots, x^{\kappa_N}), \quad \det A_-(\infty) \neq 0$$

Setting $A_+ = G_+$, we have finally factorized

$$A = A_-\mathrm{diag}\,(x^{\kappa_1}, \ldots, x^{\kappa_N})A_+^{-1}, \quad \det A_\pm \neq 0$$

Remark. When A is close to the identity matrix one can write $A = 1 + \epsilon A'$ and the map $\mathcal{F}$ is surjective for sufficiently small ϵ. In that case one can take $P(x) = 1$, and it is then clear that $F_+(x) = 1 + O(\epsilon)$, $\tilde{F}_-(x) = 1 + O(\epsilon)$, so that their determinants do not vanish. Hence all the indices κ_i vanish.

References

[1] G. Springer, *Introduction to Riemann surfaces*. Addison–Wesley (1957).

[2] R.C. Gunning, *Lectures on Riemann surfaces*. Princeton University Press (1966).

[3] P. Griffiths and J. Harris, *Principles of Algebraic Geometry*. Wiley (1978).

[4] D. Mumford, *Tata Lectures on Theta*. Vols. I and II. Birkhauser (1983).

[5] J.P. Serre, *Algebraic Groups and Class Fields*. Springer (1997).

[6] E.T. Whittaker and G.N. Watson, *A Course of Modern Analysis*. Cambridge University Press (1902).

[7] J. Fay, *Theta Functions on Riemann Surfaces*. Springer lectures notes (1973).

16

Lie algebras

We present basic facts about Lie groups and Lie algebras. We describe semi-simple Lie algebras and their representations which can be characterized in terms of roots and weights. We discuss infinite-dimensional Lie algebras, called affine Kac–Moody algebras, which are at the heart of the study of field theoretical integrable systems. In particular we construct the so-called level one representations using the techniques of Fock spaces and vertex operators introduced in Chapter 9.

16.1 Lie groups and Lie algebras

A Lie group is a group G which is at the same time a differentiable manifold, and such that the group operation $(g, h) \to gh^{-1}$ is differentiable. Due to a theorem of Montgomery and Zippin, the differentiable structure is automatically real analytic.

The maps $h \to gh$ and $h \to hg$ are called respectively left and right translations by g. Their differentials at the point h map the tangent space $T_h(G)$ respectively to $T_{gh}(G)$ and $T_{hg}(G)$. We will denote by $g \cdot X$ and $X \cdot g$ the images of $X \in T_h(G)$ by these maps. This notation is coherent because, differentiating the associativity condition in G, one gets $(g \cdot X) \cdot h = g \cdot (X \cdot h)$, and $g \cdot (h \cdot X) = (gh) \cdot X$.

In particular, this last relation shows that, for any X in the tangent space of G at the unit element e, the vector field with value $g \cdot X$ at g is invariant under any left translation. Conversely, any such left-invariant vector field is of the form $g \cdot X$. So in the following we identify left-invariant vector fields on G and elements of $T_e(G)$. This finite-dimensional vector space is called the Lie algebra of G and will be denoted by $\mathcal{G}$. Alternatively, one can define the vector field $X(g)$ by its action on any function f, that is $(X \cdot f)(g)$ is the derivative of f along the tangent vector X at g. This defines a new function $X \cdot f$. The left invariance of the vector field $X(g)$ means that $(X \cdot f)(hg) = X \cdot {}^h f(g)$, where ${}^h f : g \to f(hg)$. In other words, the differential operator X commutes with left translations. More generally, for $X_1, X_2, \ldots \in \mathcal{G}$, one can consider linear combinations of differential operators (of any order) of the form $X_1 \cdot (X_2 \cdots (X_k \cdot f) \cdots)$ which obviously form an associative algebra of left-invariant differential operators. The Lie algebra is embedded into this associative algebra as the set of first order differential operators. One defines the Lie bracket $[X, Y] = XY - YX$, in terms of the associative algebra product. The main point is that, while XY and YX are second order differential operators, their commutator is a first order differential operator, so that $[X, Y]$ belongs to the Lie algebra. It is clear from this definition that $(X, Y) \to [X, Y]$ is bilinear antisymmetric and obeys the Jacobi identity:

$$[[X, Y], Z] + [[Z, X], Y] + [[Y, Z], X] = 0 \qquad (16.1)$$

The associative algebra of left-invariant differential operators on G is called the universal enveloping algebra of the Lie algebra $\mathcal{G}$ and will be denoted by $\mathcal{U}(\mathcal{G})$.

Finally, there is a natural action of the Lie group G on its Lie algebra $\mathcal{G}$ called the adjoint action. Note that for any X in the tangent space at e and any g, $g \cdot X \cdot g^{-1}$ is also in the tangent space at e to G. We define the adjoint action:

$$\text{Ad}\,(g)(X) = g \cdot X \cdot g^{-1}, \quad X \in \mathcal{G} \qquad (16.2)$$

and note that $\text{Ad}_{gh} = \text{Ad}_g \text{Ad}_h$, so this is a group action of G on $\mathcal{G}$.

Let $X \in \mathcal{G}$ and consider the left-invariant vector field $g \cdot X$. For small t we can solve the differential equation $\dot{g} = g \cdot X$ with initial condition $g(0) = e$. The solution $g(t) \in G$ is such that $g(s + t) = g(s)g(t)$ for s, t small. This is because both members solve the above differential equation with initial value $g(s)$ for $t = 0$. One can then use this property to show that the solution of the differential equation extends to a domain larger than initially defined (if we have a solution for $|s| \leq \epsilon$ then $g(s + t)$ is a solution for $|s + t| \leq 2\epsilon$ and still solves the equation there), and successively extends to all of $\mathbb{R}$. The solution, defined for all t, is denoted by $\exp{(tX)}$. In particular, for $t = 1$ this defines the exponential map from $\mathcal{G}$ to G (obviously $\exp{(tX)}$ belongs to the connected component of e in G, so in the following we assume that G is connected).

The exponential map allows us to relate subgroups of G to subalgebras of $\mathcal{G}$. Let $\mathcal{H}$ be a subalgebra of $\mathcal{G}$, and H be the smallest subgroup of G containing all the $\exp{(X)}$ for $X \in \mathcal{H}$. One can show that H is a Lie subgroup of G with Lie algebra $\mathcal{H}$. The precise definition of a Lie subgroup is quite tricky, in particular H may be embedded in a very complicated way in G, as shown by the simple example below. Conversely, given H its Lie algebra $\mathcal{H}$ is the set of $X \in \mathcal{G}$ such that $t \to e^{tX}$ is a continuous curve in H.

Example. Consider the torus $\mathbb{R}^2/\mathbb{Z}^2$ with the Abelian group law induced by the addition in $\mathbb{R}^2$. This is a Lie group G, with Abelian Lie algebra $\mathbb{R}^2$. Consider the one-parameter subgroup $H = \{\exp(tX)|t \in \mathbb{R}\}$, which is a Lie subgroup. When X has irrational slope, this subgroup is dense in G. In particular, any neighbourhood U of e contains infinititely many components of H.

A much nicer situation is obtained when H is closed in G, and fortunately this is the situation of interest for our purposes. We are mostly interested in the case when the Lie group G acts differentiably on a manifold M, and H is the stabilizer of a point $m \in M$. In this case, since the operation is continuous, H is automatically closed in G. Remarkably, this is sufficient to ensure that H is a closed Lie subgroup of G, thanks to a theorem of E. Cartan.

Theorem. *If H is a closed subgroup of a Lie group G, there exists a unique analytic structure on H such that H is a Lie subgroup of G.*

Proof. We sketch the proof. The idea is to define

$$\mathcal{H} = \{X \in \mathcal{G}|\exp(tX) \in H, \ \forall t\}$$

and show that this is a subalgebra of $\mathcal{G}$. Easy computations show that:

$$\left(\exp\frac{t}{n}X \exp\frac{t}{n}Y\right)^n = \exp\left(t(X+Y) + O(1/n)\right)$$

$$\left(\exp\frac{-t}{n}X \exp\frac{-t}{n}Y \exp\frac{t}{n}X \exp\frac{t}{n}Y\right)^{n^2} = \exp\left(t^2[X,Y] + O(1/n)\right)$$

so that, if $X, Y \in \mathcal{H}$, both $(X+Y)$ and $[X,Y]$ are in $\mathcal{H}$ since H is closed. One then uses the closedness of H to show that the Lie subgroup of G of Lie subalgebra $\mathcal{H}$ is in fact equal to H (more precisely to the connected component of the identity in H). ∎

In the situation described above where G acts on M, one can show that the application $g \to g \cdot m$ of G onto the orbit $\mathcal{O}_m$ of m is open so that $\mathcal{O}_m$ is isomorphic to the homogeneous space G/H.

Examples. The most natural examples of Lie groups are provided by so-called algebraic subgroups of the general linear group $GL(n)$. These are subvarieties of $GL(n)$ defined by polynomial equations compatible with the multiplication law. For example, the special linear group is defined by the equation $\det g = 1$, and the product of two such matrices has determinant 1. Hence these are naturally closed Lie subgroups of $GL(n)$. The other standard examples are the subgroups of orthogonal and symplectic matrices.

For any Lie group G, a representation ρ on a vector space V is a differentiable group homomorphism $G \to GL(V)$. For $g \in G$ and $v \in V$ one denotes $g \cdot v = \rho(g)(v)$ so that $(gh) \cdot v = g \cdot (h \cdot v)$. The differential of ρ at e maps the Lie algebra $\mathcal{G}$ to the Lie algebra $gl(V)$ of $GL(V)$. Similarly, left-invariant vector fields on G are mapped on left-invariant vector fields on $GL(V)$, and so are their Lie brackets. Hence we get a representation of $\mathcal{G}$ on $gl(V)$, which we shall also denote by ρ. In other words, we have $[\rho(X), \rho(Y)] = \rho([X, Y])$. Such a representation is faithful if $\rho : \mathcal{G} \to gl(V)$ is injective (i.e. for $X \neq 0$ there exists $v \in V$ such that $\rho(X)(v) \neq 0$). In this case $\mathcal{G}$ may be seen as a subalgebra of $gl(V)$. There is a natural representation of any Lie group on its Lie algebra, i.e. $V = \mathcal{G}$, given by the adjoint representation. This induces a representation of $\mathcal{G}$ on $\mathcal{G}$, also called the adjoint representation:

$$\mathrm{ad}_X(Y) = [X, Y] \tag{16.3}$$

It is easy to check that this is a representation of $\mathcal{G}$, using the Jacobi identity. Almost all results on Lie algebras are obtained by studying this representation.

16.2 Semi-simple Lie algebras

Because there is such an interplay between Lie groups and Lie algebras, we study here Lie algebras from an algebraic viewpoint. In this section we consider Lie algebras over the complex numbers, e.g. complexifications $\mathcal{G}_{\mathbb{R}} \otimes \mathbb{C}$ of real Lie algebras.

We will often use a basis (T_a), $a = 1, \dots, \dim \mathcal{G}$, on the complex Lie algebra $\mathcal{G}$. The Lie bracket is then expressed as:

$$[T_a, T_b] = f_{ab}^c \, T_c$$

The coefficients f_{ab}^c are called structure constants. Note that in this basis the matrix elements of the adjoint representation are $(\mathrm{ad}_{T_a})_b^c = f_{ab}^c$.

The adjoint representation eq. (16.3) allows us to define a natural bilinear form on $\mathcal{G}$, also called the Killing form, by:

$$(X, Y) = \text{Tr}\,(\text{ad}_X\,\text{ad}_Y)$$

This bilinear form is *invariant* in the sense that:

$$([X, Y], Z) = (X, [Y, Z])$$

This results immediately from the cyclic invariance of the trace and the fact that $X \to \text{ad}_X$ is a representation. The invariance property also means that $(\text{ad}_Y X, Z) + (X, \text{ad}_Y Z) = 0$.

A Lie algebra is said to be semi-simple if it does not contain any non-trivial Abelian ideal. The Cartan criterion says that this is the case if and only if the Killing form is non-degenerate. In one direction this is easy. If X belongs to an Abelian ideal $\mathcal{I}$, and Y is arbitrary, choose a basis T_a of $\mathcal{G}$ such that T_a is a basis of $\mathcal{I}$ for $i = 1, \ldots, r$. Then $\text{ad}_Y\text{ad}_X(T_a)$ vanishes for $a \leq r$ and belongs to $\mathcal{I}$ for $a > r$, hence $\text{ad}_Y\,\text{ad}_X$ has no diagonal element in this basis, and its trace vanishes. We see that any Abelian ideal is in the kernel of the Killing form.

A Lie algebra $\mathcal{G}$ is called simple if it is semi-simple and its only ideals are either $\{0\}$ or the algebra $\mathcal{G}$ itself. For any ideal $\mathcal{I}$ in a semi-simple algebra, its orthogonal under the Killing form is also an ideal. Moreover, by invariance and non-degeneracy of the Killing form one sees that $\mathcal{I} \cap \mathcal{I}^\perp$ is an Abelian ideal (for $X \in \mathcal{I}$, $Y \in \mathcal{I}^\perp$, and Z arbitrary $([X, Y], Z) = (X, [Y, Z]) = 0$), hence vanishes. It follows that $\mathcal{G}$ is the direct sum of its simple ideals, this being an orthogonal decomposition.

We now introduce the concept of a Cartan subalgebra in a semi-simple Lie algebra. First, an element X of $\mathcal{G}$ is called *semi-simple* if ad_X is a diagonalizable matrix in the adjoint representation. A Cartan subalgebra $\mathcal{H}$ of a semi-simple Lie algebra $\mathcal{G}$ is a maximal Abelian subalgebra of $\mathcal{G}$ whose elements are all semi-simple. The existence of such an algebra is a very non-trivial result. To construct it one starts with a *regular* element, that is an element of $\mathcal{G}$ such that ad_X has a maximal number of distinct eigenvalues (as a matrix in the adjoint representation). Then the subalgebra of $\mathcal{G}$ on which ad_X is nilpotent is a Cartan subalgebra. One can show that any two Cartan subalgebras are related by a Lie algebra automorphism, and their common dimension is called the rank of the Lie algebra and will be denoted by $\text{rank}\,\mathcal{G}$. In the adjoint representation, the endomorphisms $\text{ad}\,(H)$ for $H \in \mathcal{H}$ form a system of commuting diagonalizable endomorphisms. We can thus diagonalize them *simultaneously*. Let $E_\alpha \in \mathcal{G}$ be the common eigenvectors:

$$\text{ad}\,(H) \cdot E_\alpha = \alpha(H)\,E_\alpha$$

The application

$$\alpha : H \in \mathcal{H} \to \alpha(H) \in R$$

is a linear form defined over $\mathcal{H}$. That is, α belongs to the dual of the Cartan algebra: $\alpha \in \mathcal{H}^*$. These forms are called the *roots* of the Lie algebra. We shall denote their set by Δ. They satisfy a few simple properties: (i) if α is a root then so is $-\alpha$, (ii) a non-zero root is non-degenerate (i.e. the eigenspace is of dimension 1), (iii) if α is a root and $t \in \mathbb{C}$, $t\alpha$ is not a root, except for $t = \pm 1$.

Let $\{H_i\}$ be a basis of the Cartan subalgebra. Then $\{H_i, E_\alpha\}$ form a basis of the Lie algebra $\mathcal{G}$, on which the Killing form has a very simple structure, namely:

$$(H_i, E_\alpha) = 0, \quad (E_\alpha, E_\beta) = 0, \quad \alpha + \beta \neq 0 \qquad (16.4)$$

This is because $(H, [H', E_\alpha]) = \alpha(H')(H, E_\alpha) = ([H, H'], E_\alpha) = 0$, and $([H, E_\alpha], E_\beta) = \alpha(H)(E_\alpha, E_\beta) = (-([H, E_\beta], E_\alpha) = -\beta(H)(E_\alpha, E_\beta)$. As a consequence, the restriction of the Killing form to the Cartan subalgebra is non-degenerate, otherwise the Killing form would be degenerate on the Lie algebra.

It is convenient to introduce the isomorphism between $\mathcal{H}$ and its dual $\mathcal{H}^*$ induced by the Killing form:

$$\alpha \in \mathcal{H}^* \to H_\alpha \in \mathcal{H} \quad \text{with} \quad \alpha(H) = (H_\alpha, H), \quad \forall H \in \mathcal{H}$$

This defines, for any $\alpha \in \mathcal{H}^*$, an element $H_\alpha \in \mathcal{H}$ depending linearly on α. We may then define a bilinear form on $\mathcal{H}^*$ by:

$$(\alpha, \beta) = (H_\alpha, H_\beta) = \alpha(H_\beta), \quad \alpha, \beta \in \mathcal{H}^*$$

This form is non-degenerate because the Killing form is non-degenerate on the Cartan subalgebra. Moreover, H_α for $\alpha \in \Delta$ span the Cartan subalgebra.

In the basis $\{H_i, E_\alpha\}$ the Lie bracket reads:

$$
\begin{aligned}
&[H_i, H_j] = 0 \\
&[H_i, E_\alpha] = \alpha(H_i)E_\alpha \\
&[E_\alpha, E_\beta] = \begin{cases} (E_\alpha, E_{-\alpha})H_\alpha & \text{if } \alpha + \beta = 0 \\ C_{\alpha,\beta}E_{\alpha+\beta} & \text{if } \alpha + \beta \in \Delta \\ 0 & \text{if } \alpha + \beta \notin \Delta \end{cases}
\end{aligned}
\qquad (16.5)
$$

with $C_{\alpha,\beta}$ some structure constants. Here we remark that $[E_\alpha, E_\beta]$ either vanishes or is proportional to $E_{\alpha+\beta}$ if $\alpha + \beta \neq 0$ is a root, because $[H, [E_\alpha, E_\beta]] = [[H, E_\alpha], E_\beta] + [E_\alpha, [H, E_\beta]] = (\alpha(H) + \beta(H))[E_\alpha, E_\beta]$. If, however, $\alpha + \beta = 0$, this shows that $[E_\alpha, E_{-\alpha}]$ is in the Cartan subalgebra, and we have $(H, [E_\alpha, E_{-\alpha}]) = ([H, E_\alpha], E_{-\alpha}) = \alpha(H)(E_\alpha, E_{-\alpha}) = (H, (E_\alpha, E_{-\alpha})H_\alpha)$.

For each root α the three generators $H_\alpha, E_\alpha, E_{-\alpha}$ form an $sl(2)$ subalgebra of $\mathcal{G}$. This allows us to study the α-chain through β, that is the set of roots of the form $\beta + n\alpha$, using the commutation relations:

$$[H_\alpha, E_{\pm\alpha}] = \pm\alpha(H_\alpha)E_{\pm\alpha}, \quad [E_\alpha, E_{-\alpha}] = (E_\alpha, E_{-\alpha})H_\alpha$$

The vectors $\left(\mathrm{ad}E_{\pm\alpha}\right)^j E_\beta$ for $j \in \mathbb{N}$ are obviously linearly independent root vectors in $\mathcal{G}$, for the roots $\beta \pm j\alpha$, if they don't vanish. They span a representation space for the considered $sl(2)$ and since this representation is of finite dimension, the chain must be of finite length. Let $p \le 0$ be the minimal index such that $\beta + p\alpha$ is a root, and $q \ge 0$ be the maximal index such that $\beta + q\alpha$ is a root. Let $\beta' = \beta + p\alpha$ and consider the vectors $v_j = \left(\mathrm{ad}E_\alpha\right)^j E_{\beta'}$ for $j \in \mathbb{N}$. By the minimality of p, we have $\mathrm{ad}\,E_{-\alpha}v_0 = 0$. Using this property, we can compute:

$$\mathrm{ad}H_\alpha v_j = \left(\beta'(H_\alpha) + j\alpha(H_\alpha)\right)v_j$$

$$\mathrm{ad}E_{-\alpha}v_j = -j(E_\alpha, E_{-\alpha})\left(\beta'(H_\alpha) + (j-1)\frac{\alpha(H_\alpha)}{2}\right)v_{j-1} \quad (16.6)$$

Since v_{q-p+1} vanishes, but v_{q-p} does not, we have $\beta'(H_\alpha) + (q - p)\alpha(H_\alpha)/2 = 0$ or:

$$2\frac{(\beta, \alpha)}{(\alpha, \alpha)} = -(p+q) \in \mathbb{Z} \quad (16.7)$$

This result allows us to show that the Killing form induces a *positive definite* scalar product on the *real* vector space $\sum_{\alpha \in \Delta} \mathbb{R}H_\alpha$. By duality this defines a positive definite scalar product on $\sum_{\alpha \in \Delta} \mathbb{R}\alpha$. Indeed, computing the Killing form on the basis of $\mathcal{G}$ provided by the H_i and the $E_{\pm\alpha}$, we have

$$(\alpha, \beta) = (H_\alpha, H_\beta) = \mathrm{Tr}\,(\mathrm{ad}\,H_\alpha \mathrm{ad}\,H_\beta) = \sum_\gamma (\alpha, \gamma)(\beta, \gamma)$$

Taking $\alpha = \beta$ and dividing by $(\alpha, \alpha)^2$, we get:

$$\frac{1}{(\alpha, \alpha)} = \sum_\gamma \left(\frac{(\alpha, \gamma)}{(\alpha, \alpha)}\right)^2$$

so that (α, α) is a rational number. It follows that (α, β) is a rational number, hence is real. Then for any $x = \sum_\alpha x_\alpha \alpha$ with $x_\alpha \in \mathbb{R}$ we have $(x, x) = \sum_\gamma (x, \gamma)^2 \ge 0$ and this vanishes only if $x = 0$.

For later use let us write another consequence of eq. (16.6):

$$[E_{-\alpha}[E_\alpha, E_\beta]] = (E_\alpha, E_{-\alpha})q(1-p)\frac{(\alpha, \alpha)}{2}E_\beta \qquad (16.8)$$

Both members are homogeneous in the normalizations of $E_{\pm\alpha}$ and E_β so one can replace $E_\beta = v_{-p}$ with the notations of eq. (16.6), and then use eq. (16.7).

With any root α one can associate a reflection w_α acting on $\mathcal{H}^*$ by:

$$w_\alpha(x) = x - 2\frac{(\alpha, x)}{(\alpha, \alpha)}\alpha$$

These orthogonal reflections are called Weyl reflections. They preserve the root system, i.e. if β is a root so is $w_\alpha(\beta)$. This is because, using eq. (16.7), $w_\alpha(\beta) = \beta + (p + q)\alpha$ is in the chain $\beta + p\alpha, \dots, \beta + q\alpha$. The Weyl group is the discrete group generated by these reflections.

While roots span $\mathcal{H}^*$, they are not linearly independent in general, and one can choose a subset of them which forms a basis. There exists a subset Π of roots α_i, $i = 1, \dots, r$, such that any other root α can be written $\alpha = \sum_i n_i \alpha_i$, where the n_i are integers all of the same sign. When all n_i are ≥ 0 we say that α is a positive root and otherwise α is called a negative root. The α_i are called *simple roots*. So they are positive roots which cannot be written as the sum of two positive roots. The choice of Π is not unique, but any two such choices are related by a unique transformation of the Weyl group.

To show the existence of a basis of simple roots, choose a hyperplane in $\mathcal{H}^*$ which does not contain any root. Half of the roots are then on one side of this hyperplane, and we call them positive roots. If a positive root can be written as a sum of two positive roots we call it decomposable, otherwise we call it simple. Obviously, any positive root can be written as a linear combination of simple roots with positive integer coefficients. In particular the simple roots span $\mathcal{H}^*$. We show that they are linearly independent. Note that for two simple roots α and β their difference $(\beta - \alpha)$ is not a root, because if $(\beta - \alpha)$ is a positive root we can decompose $\beta = (\beta - \alpha) + \alpha$ as sum of positive roots, in contradiction with the simplicity of β, while if $(\alpha - \beta)$ is positive, we get similarly a contradiction with the simplicity of α. Hence the simple root condition means that $p = 0$ in eq. (16.7), so that:

$$-2\frac{(\alpha, \beta)}{(\alpha, \alpha)} = n \in \mathbb{N}$$

In particular, the scalar product (α, β) is negative for α, β simple roots. The α-chain through β consists of the roots of the form $\beta + n\alpha$

for $n = 0, 1, \ldots, -2(\alpha, \beta)/(\alpha, \alpha)$. Assume now that there is a linear relation between the simple roots that we can write in the form $\sum_s r_s \alpha_s = \sum_{s'} r_{s'} \alpha_{s'}$ with r_s and $r_{s'}$ real and positive, and the set $\{s\}$ disjoint from the set $\{s'\}$. From this equality one gets $(\sum_s r_s \alpha_s)^2 = \sum_{ss'} r_s r_{s'} (\alpha_s, \alpha_{s'})$. The left-hand side is obviously positive, while the right-hand side is obviously negative because $s \neq s'$ so $(\alpha_s, \alpha_{s'}) \leq 0$. It follows that $r_s = r_{s'} = 0$, so the simple roots are linearly independent.

At the extreme opposite of the simple roots are *highest* roots. They are of the form $\theta = \sum_i n_i \alpha_i$, where the n_i are maximal ≥ 0 integers. For a *simple* Lie algebra one can show that the highest root is unique, and that all $n_i > 0$.

Let α_i be a set of simple roots. One defines the *Cartan matrix*, which is independent of the choice of basis (since two bases are related by the Weyl group) by:

$$a_{ij} = \frac{2(\alpha_j, \alpha_i)}{(\alpha_i, \alpha_i)} \quad , \quad i, j = 1, \ldots, \operatorname{rank} \mathcal{G} \tag{16.9}$$

It is such that $a_{ii} = 2$ and $a_{ij} \leq 0$, $a_{ij} = 0 \Rightarrow a_{ji} = 0$ for $i \neq j$. Moreover, the a_{ij} for $i \neq j$ are negative integers such that $0 \leq a_{ij} a_{ji} \leq 4$. This last condition comes from the fact that the scalar product is positive definite on $\mathcal{H}^*$.

The Cartan matrix is non-degenerate: $\det(a) \neq 0$ because $\det(a)$ is proportional to the determinant of the matrix of scalar products of simple roots which are linearly independent.

With the Cartan matrix, we can give a presentation of the Lie algebra $\mathcal{G}$, by generators and relations. For each simple root α_i the elements:

$$h_i = \frac{2}{(\alpha_i, \alpha_i)} H_{\alpha_i}, \quad e_i^+ = E_{\alpha_i}, \quad e_i^- = \frac{1}{(E_{\alpha_i}, E_{-\alpha_i})} E_{-\alpha_i}$$

generate an $sl(2)$ subalgebra with standard commutation relations. The Cartan matrix allows us to reconstruct the Lie algebra from this set of $sl(2)$ subalgebras. Given a Cartan matrix a_{ij} satisfying the properties mentioned above, we may define $\mathcal{G}$ as the Lie algebra generated by the sets (h_i, e_i^+, e_i^-) with the relations (called the Serre relations):

$$[h_i, h_j] = 0$$
$$\left[h_i, e_j^\pm\right] = \pm a_{ij} e_j^\pm \tag{16.10}$$
$$\left[e_i^+, e_j^-\right] = \delta_{ij} h_i$$
$$(\operatorname{ad} e_i^\pm)^{1 - a_{ij}} \cdot e_j^\pm = 0 \quad \text{for} \quad i \neq j$$

The last condition is just the condition that the α_i-chain starting at α_j is of length $-a_{ij}$. The elements h_i generate the Cartan subalgebra $\mathcal{H}$. The fact that these relations yield a finite-dimensional Lie algebra is a theorem by J.P. Serre.

Let $\mathcal{N}_\pm$ be the subalgebra generated by the $e_i^\pm$. We have:

$$\mathcal{G} = \mathcal{N}_- \oplus \mathcal{H} \oplus \mathcal{N}_+$$

The subalgebras $\mathcal{B}_\pm = \mathcal{H} \oplus \mathcal{N}_\pm$ are called Borel subalgebras.

The classification of finite-dimensional simple Lie algebras is then reduced to the classification of finite-dimensional Cartan matrices satisfying the mentioned properties. This leads to four infinite series $A_n = sl(n+1)$, $B_n = so(2n + 1)$, $C_n = sp(2n)$, $D_n = so(2n)$ and a few exceptional algebras called E_6, E_7, E_8 and F_4, G_2 (see the References).

A consequence of Serre's theorem is the existence of an involutive automorphism ω of the Lie algebra $\mathcal{G}$, called the Chevalley automorphism. To define it we give its action on the generators:

$$\omega(h_i) = -h_i, \quad \omega(e_i^+) = -e_i^-, \quad \omega(e_i^-) = -e_i^+$$

and check that it preserves the relations. Hence it extends to the whole Lie algebra. For any root α of the Lie algebra one can *choose* $E_{-\alpha} = -\omega(E_\alpha)$. Changing $E_\alpha \to \lambda E_\alpha$, we have $(E_\alpha, E_{-\alpha}) \to \lambda^2(E_\alpha, E_{-\alpha})$ so that we can always achieve the condition $(E_\alpha, E_{-\alpha}) = 1$. Notice that in general λ will be a complex number.

16.3 Linear representations

Recall that a linear representation on a vector space V, of a finite-dimensional Lie algebra $\mathcal{G}$, is a homomorphism ρ from $\mathcal{G}$ to $\mathrm{End}\,V$. We can define the sum and the product of two representations (ρ_1, V_1) and (ρ_2, V_2). The sum is the representation on the direct sum $V_1 \oplus V_2$ such that $\rho_{V_1 \oplus V_2}$ maps elements of $\mathcal{G}$ into block diagonal endomorphisms whose restrictions to $V_{1,2}$ coincide with their images under $\rho_{1,2}$, in other words $\rho_{V_1 \oplus V_2}$ is block diagonal. The product is the representation on the tensor product $V_1 \otimes V_2$ with $\rho_{V_1 \otimes V_2}(X) = \rho_1(X) \otimes 1 + 1 \otimes \rho_2(X)$ for any element $X \in \mathcal{G}$. A representation on a vector space V is said to be indecomposable if it cannot be decomposed into the sum of subrepresentations. Note that for a general algebra $\mathcal{A}$ the sum of two representations is a representation, but the tensor product is not. The Lie algebra case appears as very special, and this is due to the existence of the algebra homomorphism, called the coproduct:

$$\Delta : \mathcal{U}(\mathcal{G}) \to \mathcal{U}(\mathcal{G}) \otimes \mathcal{U}(\mathcal{G}), \quad \Delta(X) = X \otimes 1 + 1 \otimes X \qquad (16.11)$$

The elements of the Cartan subalgebra are represented by a family of commuting diagonalizable endomorphisms. They can thus be simultaneously diagonalized. Let $|\lambda\rangle$ be an eigenvector, and $\lambda(H)$ be the corresponding eigenvalues, which depend linearly on H. The various λ are linear forms acting on the Cartan subalgebra $\mathcal{H}$, i.e. $\lambda \in \mathcal{H}^*$, and are called weights. The weights $\lambda(H)$ may have multiplicities, so we denote by $|\lambda_a\rangle$ the weight vectors with the same weight λ.

$$H|\lambda_a\rangle = \lambda(H)|\lambda_a\rangle$$

The weight vectors $|\lambda_a\rangle$ form a basis of the representation space V. Their number, degeneracy included, is the dimension of the representation. The representation space V contains representations of the $sl(2)$ subalgebras generated by $(H_\alpha, E_\alpha, E_{-\alpha})$ for any root α. From the knowledge of finite-dimensional representations of $sl(2)$ we get the basic integrality condition:

$$2\frac{(\lambda, \alpha)}{(\alpha, \alpha)} \in \mathbb{Z} \quad \text{for all} \quad \alpha \in \Delta$$

Note that the weight system of a representation is invariant under the action of the Weyl group. In other words, if λ is a weight, so is $w_\alpha(\lambda)$ for any root α.

The difference of two weights of a given irreducible representation always belongs to the root system. We may thus introduce an order between weights of a representation by $\lambda_1 > \lambda_2$ iff $\lambda_1 - \lambda_2 > 0$. Any finite-dimensional representation possesses a highest weight since its number of weights is finite. This vector is unique for irreducible representations. Let Λ be this *highest weight*, which is thus non-degenerate. The corresponding eigenvector $|\Lambda\rangle$ is called the highest weight vector of the representation. It is such that:

$$H|\Lambda\rangle = \Lambda(H)|\Lambda\rangle, \quad E_{\alpha_i}|\Lambda\rangle = 0$$

for $H \in \mathcal{H}$ and α_i any simple positive root. This follows from the fact that since Λ is a highest weight $\Lambda + \alpha_i$ is not a weight.

Given a representation with highest weight Λ, one defines its Dynkin indices δ_i by:

$$\delta_i = 2\frac{(\Lambda, \alpha_i)}{(\alpha_i, \alpha_i)} \in \mathbb{N}$$

with α_i the simple roots. The proof that this is a positive integer is the same as in the case of the adjoint representation. By definition, the *fundamental weights* Λ_j are the highest weights with Dynkin indices δ_{ij}. They are specified by:

$$2\frac{(\Lambda_j, \alpha_i)}{(\alpha_i, \alpha_i)} = \delta_{ij}$$

The number of fundamental weights is equal to the rank of $\mathcal{G}$, and there exist representations with highest weights the fundamental weights (called fundamental repesentations). Any highest weight Λ of a finite-dimensional representation may be decomposed on the fundamental weights: $\Lambda = \sum_j \delta_j \Lambda_j$ with δ_j the Dynkin indices. More generally, the weight lattice is the set of $\lambda \in \mathcal{H}^*$ such that $(\lambda, \alpha) \in \mathbb{Z}$ for any root α. Any weight of any representation is on the weight lattice. As a $\mathbb{Z}$-module, the weight lattice has a basis provided by the fundamental weights, such that $(\lambda_i, \alpha_j) = \delta_{ij}$, for any simple root α_j.

The highest weight representations may also be viewed as quotient of Verma modules. The Verma module associated with a highest weight vector $|\Lambda\rangle$ is the space $(U(\mathcal{N}_-)|\Lambda\rangle)$ with $U(\mathcal{N}_-)$ the enveloping algebra of $\mathcal{N}_-$. Then the irreducible representation with highest weight vector Λ is isomorphic to the quotient:

$$(U(\mathcal{N}_-)|\Lambda\rangle) / \mathcal{M}_\Lambda$$

where $\mathcal{M}_\Lambda$ the maximal submodule of $U(\mathcal{N}_-)|\Lambda\rangle$, which is shown to exist, and is unique (by maximality). The above quotient is finite-dimensional when Λ belongs to the weight lattice. The Verma module construction shows that any weight of the weight lattice is conjugated by the Weyl group to the highest weight of some representation. The roots themselves generate a lattice called the root lattice. It is a sublattice of the weight lattice.

On the universal enveloping algebra, one can define a $\star$ operation such that $(XY)^\star = Y^\star X^\star$ and $(\lambda X)^\star = \bar{\lambda} X^\star$. In a Cartan–Weyl basis it reads $H^\star = H$, $E_\alpha^\star = E_{-\alpha}$, which is compatible with the commutation relations since $C_{-\alpha,-\beta} = -C_{\alpha,\beta}$ and $C_{\alpha,\beta}$ and the $\alpha(H)$ are real. In the highest weight representation, the operation $\star$ is just Hermitian conjugation and allows us to introduce complex conjugated representations. In particular, the state $\langle\Lambda|$, dual to the highest weight $|\Lambda\rangle$, satisfies:

$$\langle\Lambda|H = \Lambda(H)\langle\Lambda|, \quad \langle\Lambda|E_{-\alpha} = 0 \text{ for } \alpha > 0$$

since $\langle\Lambda|E_{-\alpha} = (E_\alpha|\Lambda\rangle)^\star = 0$.

The Casimir operator C is the following operator, quadratic in the Lie algebra generators (T_a) forming a basis of the Lie algebra, hence living in the universal enveloping algebra of $\mathcal{G}$:

$$C = \sum_{a,b} T_a K^{ab} T_b$$

with K^{ab} the matrix inverse of the Killing form K_{ab}. Its main property is that it is in the centre of the enveloping algebra, so that, in any given representation, the Casimir operator commutes with the endomorphisms representing the elements of the Lie algebra. It thus acts proportionally to the identity on irreducible representations. If Λ is the highest weight of the representation, its value is:

$$C(\Lambda) = (\Lambda, \Lambda + \rho)$$

with ρ the Weyl vector equal to the sum of the fundamental weights:

$$\rho = \sum_j \Lambda_j$$

We frequently meet the tensor Casimir operator living in $\mathcal{G} \otimes \mathcal{G}$ given by $C_{12} = \sum_{a,b} K^{ab} T_a \otimes T_b$. Note that we have, usinq eq. (16.11):

$$C_{12} = \frac{1}{2}(\Delta C - C \otimes 1 - 1 \otimes C)$$

The main property of the tensor Casimir is that $[C_{12}, \Delta(X)] = 0$ for any $X \in \mathcal{G}$.

16.4 Real Lie algebras

Up to now, we considered complex Lie algebras. Examples are provided by complexification of real Lie algebras. More precisely, let $\mathcal{G}$ be a real Lie algebra. This means that we have a basis X_a of the Lie algebra such that the structure constants are real, and we consider linear combinations of the X_a with real coefficients. Its complexification $\mathcal{G}_{\mathbb{C}}$ is the set of elements $Z = X + iY$ with $X, Y \in \mathcal{G}$. On $\mathcal{G}_{\mathbb{C}}$ we define a conjugation $c : X + iY \rightarrow X - iY$. We have

$$c^2 = 1, \quad [c(Z_1), c(Z_2)] = c([Z_1, Z_2]), \quad c(\lambda Z) = \bar{\lambda} c(Z)$$

Conversely, given any such conjugation, c, we can write $\mathcal{G}_{\mathbb{C}} = \mathcal{G}_+ \oplus \mathcal{G}_-$, where $c|_{\mathcal{G}_\pm} = \pm 1$ and $\mathcal{G}_{\mathbb{C}}$ can be viewed as the complexification of $\mathcal{G}_+$ which is a real Lie algebra.

Different real Lie algebras may have the same complexification. For example, $sl(2, \mathbb{C})$ is the common complexification of the two real Lie algebras $sl(2, \mathbb{R})$ and $su(2)$. The algebra $sl(2, \mathbb{R})$ is the Lie algebra of 2×2 traceless real matrices, with basis:

$$H = \begin{pmatrix} 1 & 0 \\ 0 & -1 \end{pmatrix} \quad E_+ = \begin{pmatrix} 0 & 1 \\ 0 & 0 \end{pmatrix} \quad E_- = \begin{pmatrix} 0 & 0 \\ 1 & 0 \end{pmatrix}$$

and commutation relations $[H, E_\pm] = \pm 2E_\pm$ and $[E_+, E_-] = H$. On the other hand, $su(2)$ is the Lie algebra of antihermitean traceless 2×2 matrices, i.e. linear combinations with *real* coefficients of the matrices $t_k = i\sigma_k$, where σ_k are the Pauli matrices:

$$\sigma_3 = \begin{pmatrix} 1 & 0 \\ 0 & -1 \end{pmatrix} \quad \sigma_2 = \begin{pmatrix} 0 & -i \\ i & 0 \end{pmatrix} \quad \sigma_1 = \begin{pmatrix} 0 & 1 \\ 1 & 0 \end{pmatrix}$$

The commutation relations are $[t_i, t_j] = -2\epsilon_{ijk} t_k$. We see that the structure constants of $su(2)$ are *real*. The Lie algebras $sl(2, \mathbb{R})$ and $su(2)$ are referred to as the non-compact and compact real forms of $sl(2, \mathbb{C})$ respectively. Notice that although the algebra $su(2)$ is real, the matrices representing it have complex entries.

It is an important problem to classify the real forms of a given complex Lie algebra. This amounts to classifying the conjugations c. For this purpose note that one can build a basis of the Lie algebra $\mathcal{G}_\mathbb{C}$ such that all the structure constants are real, as follows. Choose the basis $E_{\pm\alpha}$, H_α such that $\omega(E_\alpha) = -E_{-\alpha}$ (where ω is the Chevalley automorphism) and $(E_\alpha, E_{-\alpha}) = 1$. Then $[E_\alpha, E_{-\alpha}] = H_\alpha$, and $[H_\alpha, E_{\pm\beta}] = \pm\beta(H_\alpha)E_{\pm\beta}$, where $\beta(H_\alpha)$ are real. Setting $[E_\alpha, E_\beta] = C_{\alpha,\beta} E_{\alpha+\beta}$ and applying ω, we get the relation:

$$C_{-\alpha,-\beta} = -C_{\alpha,\beta} \tag{16.12}$$

To show that these structure constants are real, we compute:

$$[[E_\alpha, E_\beta], [E_{-\alpha}, E_{-\beta}]] = -C_{\alpha,\beta}^2 H_{\alpha+\beta} = -C_{\alpha,\beta}^2 (H_\alpha + H_\beta)$$

On the other hand, using the Jacobi identity and eq. (16.8), we get:

$$[[E_\alpha, E_\beta], [E_{-\alpha}, E_{-\beta}]] = [E_{-\beta}, [E_{-\alpha}, [E_\alpha, E_\beta]]] + [E_{-\alpha}, [E_{-\beta}, [E_\beta, E_\alpha]]]$$
$$= -q(1-p)\frac{(\alpha, \alpha)}{2} H_\alpha - q'(1-p')\frac{(\beta, \beta)}{2} H_\beta$$

Here p' and q' refer to the β-chain through α. If $C_{\alpha,\beta} \neq 0$, i.e. $\alpha + \beta$ is a root, H_α and H_β are linearly independent, so identifying the coefficients we get:

$$C_{\alpha,\beta}^2 = q(1-p)\frac{(\alpha, \alpha)}{2}$$

and a similar formula with p' and q'. Recalling that $p < 0$ and $(\alpha, \alpha) > 0$, we see that $C_{\alpha,\beta}$ is real. The basis of the Lie algebra that we have constructed is called a *Weyl basis*.

The real Lie algebra $\mathcal{G}'$ spanned over $\mathbb{R}$ by $E_{\pm\alpha}$ and H_α is the analogue of the non-compact $sl(2,\mathbb{R})$ in the general case. We obviously have $\mathcal{G}_\mathbb{C} = \mathcal{G}' + i\mathcal{G}'$ and the real form is selected by the conjugation $c'(X+iY) = X - iY$.

It is a theorem by H. Weyl that for any semi-simple complex Lie algebra, there exists a real form which is the Lie algebra of a compact Lie group. The Lie algebra of a semi-simple compact Lie group is characterized by the fact that its Killing form is negative definite. Indeed, if G is a compact Lie group, choose any positive definite scalar product on $\mathcal{G}$, its Lie algebra. Since G is compact, one can use the Haar integral on G to take the average of this bilinear form and obtain a positive definite *invariant* scalar product on $\mathcal{G}$. This means that the Lie group G acts by orthogonal matrices in the adjoint representation (for this scalar product). Hence, in an orthonormal basis the matrices ad_X are antisymmetric and the Killing form $(X,X) = -\sum_{ij}(\mathrm{ad}_X)_{ij}^2 \le 0$ vanishes only when $\mathrm{ad}_X = 0$, i.e. when X is in the centre of $\mathcal{G}$. We see that $\mathcal{G}$ decomposes as the orthogonal sum of its centre and a semi-simple algebra $[\mathcal{G},\mathcal{G}]$ on which the Killing form is negative definite.

Conversely, starting from the Weyl basis, one can construct the compact form as follows: consider the generators

$$X_\alpha = (E_\alpha - E_{-\alpha}), \quad Y_\alpha = i(E_\alpha + E_{-\alpha}), \quad Z_\alpha = iH_\alpha \qquad (16.13)$$

The real vector space spanned by these elements is a real Lie algebra thanks to eq. (16.12). Moreover, it has a negative definite Killing form. To show this recall the orthogonality relations, eq. (16.4), so it is sufficient to look at each subspace $(X_\alpha, Y_\alpha, Z_\alpha)$ independently, where the check is simple, thereby proving the Weyl theorem. For instance, $(X_\alpha, X_\alpha) = (E_\alpha - E_{-\alpha}, E_\alpha - E_{-\alpha}) = -2(E_\alpha, E_{-\alpha}) = -2$ and $(X_\alpha, Y_\alpha) = (E_\alpha - E_{-\alpha}, i(E_\alpha + E_{-\alpha})) = 0$, and so on. This compact real form corresponds to the conjugation:

$$c(E_{\pm\alpha}) = -E_{\mp\alpha}, \quad c(H_\alpha) = -H_\alpha \qquad (16.14)$$

This conjugation selects the analogue of $su(2)$ in the $sl(2)$ case.

In general, the representations of a (real) compact Lie group G may be complex. Choosing on the representation space V an arbitrary sesquilinear form and averaging it by the group G, one gets an invariant sesquilinear form, i.e. $\langle gv, gw \rangle = \langle v, w \rangle$. Hence all elements of G are represented by unitary matrices and elements of $\mathcal{G}$ by antihermitean matrices. The generators eq. (16.13) are such that $X_\alpha^+ = -X_\alpha$, $Y_\alpha^+ = -Y_\alpha$ and $Z_\alpha^+ = -Z_\alpha$. This also reads $E_\alpha^+ = E_{-\alpha}$, $H_\alpha^+ = H_\alpha$, or more abstractly for any X in the complexified Lie algebra, $X^+ = -c(X)$.

In particular, any maximal Abelian subalgebra of $\mathcal{G}$ is a Cartan sub-algebra, because antihermitean matrices are always diagonalizable with eigenvalues purely imaginary. Its image by the exponential map is the Weyl torus. One can choose a basis H_j of the Cartan algebra such that any element of the torus is of the form $h = \exp\left(\sum_j \theta_j H_j\right)$ with $\exp(2\pi H_j) = 1$ for all j. If $|\lambda\rangle$ is a weight vector in V, we have $h|\lambda\rangle = \chi_\lambda(h)|\lambda\rangle$, where $\chi_\lambda(h)$ is a character, i.e. $\chi_\lambda(hh') = \chi_\lambda(h)\chi_\lambda(h')$. So we have $\chi_\lambda(h) = \exp\left(\sum_j \theta_j \lambda(H_j)\right)$. The condition $\exp(2\pi H_j) = 1$ gives $\lambda(H_j) \in i\mathbb{Z}$ for all j. This defines a lattice in $\mathcal{H}^*$ called the weight lattice of the *group* G, which is a sublattice of the weight lattice of the Lie algebra. Moreover, since the adjoint representation of G is well-defined, the root lattice is a sublattice of the weight lattice of G.

In general the weight lattice of G is a sublattice of the weight lattice of $\mathcal{G}$, and is equal to it only when G is simply connected. This is because any positive weight of the weight lattice of $\mathcal{G}$ is the highest weight of some representation of $\mathcal{G}$ which can then be lifted to G. Hence, for a compact semi-simple simply connected Lie group, the weight lattices of the group and its Lie algebra are the same. Note that $2\lambda(H_\alpha)/(\alpha,\alpha) \in \mathbb{Z}$ for any weight λ of $\mathcal{G}$, so in this case the elements H_j such that $\exp(2\pi H_j) = 1$ are of the form $H_j = i2H_{\alpha_j}/(\alpha_j, \alpha_j)$, where α_j are the simple roots.

When G is a semi-simple compact connected Lie group, its centre is a finite group contained in all maximal tori of G. Assuming that the centre of G is trivial, its weight lattice is equal to the root lattice, because in this case the adjoint representation is a faithful representation of G and so generates all representations of G, taking tensor products. It follows that the root lattice generates the weight lattice.

This allows us to describe the various compact Lie groups G with Lie algebra $\mathcal{G}$. Starting from the universal cover of any one of them, which we call G (and can be shown to be compact), having centre Z, the other compact Lie groups are of the form G/D, where D is any subgroup of the discrete Abelian group Z. They have centre Z/D isomorphic to the quotient of their weight lattice by the root lattice. Moreover, their first homotopy group is isomorphic to the quotient of the weight lattice of $\mathcal{G}$ by their own weight lattice. We see that global topological properties of compact Lie groups are remarkably encoded in the structure of their tangent space at the unit element.

The classification of real forms of complex Lie algebras is also the basis of the study of symmetric spaces. We have obtained two conjugations c' and c which select non-compact and compact real forms of $\mathcal{G}_\mathbb{C}$. Note that c' commutes with the conjugation c defined in eq. (16.14). Hence we can diagonalize c in the eigenspaces of c' and conversely. This yields the decompositions, called Cartan decompositions of the real Lie algebras $\mathcal{G}$ and $\mathcal{G}'$:

$$\mathcal{G} = t \oplus p, \quad \mathcal{G}' = t \oplus ip$$

The Lie algebra $t = \mathcal{G} \cap \mathcal{G}'$ is generated by the $(E_\alpha - E_{-\alpha})$, and p is spanned by the $i(E_\alpha + E_{-\alpha})$ and the iH_α. Moreover, we have the relations:

$$[t,t] \subset t, \quad [t,p] \subset p, \quad [p,p] \subset t$$

and similarly with $p \to ip$. For example in the case of $sl(n)$, this corresponds to the decomposition into symmetric and antisymmetric matrices. At the Lie group level, with the algebra t corresponds a compact group K, and with the Lie algebras $\mathcal{G}$ and $\mathcal{G}'$ correspond appropriate Lie groups G and G', respectively compact and non-compact. One gets symmetric spaces G/K and G'/K of the compact and non-compact type respectively. This is the situation we have encountered in Chapter 7. Many more conjugations exist, but we will not enter into this subject.

16.5 Affine Kac–Moody algebras

We start from a finite-dimensional simple Lie algebra, $\mathcal{G}$, and construct the loop algebra which consists of formal Laurent polynomials $\tilde{\mathcal{G}} = \mathcal{G} \otimes \mathbb{C}[\lambda, \lambda^{-1}]$ with Lie bracket:

$$[X \otimes \lambda^n, Y \otimes \lambda^m] = [X,Y] \otimes \lambda^{n+m}$$

The affine Kac–Moody algebra, $\widehat{\mathcal{G}}$, is the central extension of the loop algebra $\tilde{\mathcal{G}}$ by a central element denoted by K (this means that the formal element K commutes with all other elements). It is convenient to further extend this algebra by including the derivation $d = \lambda \partial_\lambda$. Thus, the affine Kac–Moody algebra is:

$$\widehat{\mathcal{G}} = \tilde{\mathcal{G}} \oplus \mathbb{C}K \oplus \mathbb{C}d$$

and the Lie brackets are defined (with (X,Y) the Killing form on $\mathcal{G}$) by:

$$\begin{aligned}
[X \otimes \lambda^n, Y \otimes \lambda^m] &= [X,Y] \otimes \lambda^{n+m} + \tfrac{1}{2}n\delta_{m+n,0}(X,Y)K \\
[d, X \otimes \lambda^n] &= n\,X \otimes \lambda^n \\
[K, X \otimes \lambda^n] &= [K,d] = 0
\end{aligned} \qquad (16.15)$$

Note that the coefficient

$$\omega(X \otimes \lambda^n, Y \otimes \lambda^m) \equiv \tfrac{1}{2}n\delta_{m+n,0}(X,Y) \qquad (16.16)$$

of the central element K satisfies the cocycle condition $\omega([X,Y],Z) + \omega([Z,X],Y) + \omega([Y,Z],X) = 0$, ensuring that the Jacobi identity is satisfied.

An invariant bilinear form on the Kac–Moody algebra is given by:

$$(X \otimes \lambda^n, Y \otimes \lambda^m) = (X,Y)\delta_{n+m,0}, \quad (K,K) = (d,d) = 0, \quad (K,d) = 1$$

and $(K, X \otimes \lambda^n) = (d, X \otimes \lambda^n) = 0$. The fact that this form is invariant is easy to check by direct computation.

Alternatively, denoting a general element of $\widehat{\mathcal{G}}$ by

$$\widehat{X} = \tilde{X}(\lambda) + X_K\, K + X_d\, d$$

we have in the affine Kac–Moody algebra:

$$\left[\widehat{X}, \widehat{Y}\right] = \left[\tilde{X}, \tilde{Y}\right] + \tfrac{1}{2} \oint \frac{d\lambda}{2i\pi}(\partial_\lambda \tilde{X}(\lambda), \tilde{Y}(\lambda))\, K \qquad (16.17)$$

$$\left[K, \widehat{X}\right] = [K,d] = 0$$

$$\left[d, \widehat{X}\right] = \lambda \frac{d}{d\lambda} \tilde{X}(\lambda)$$

It is worth noticing that affine Kac–Moody algebras are subalgebras of the Lie algebra $gl(\infty)$ introduced in Chapter 9. To see it, one has to associate an infinite-dimensional matrix with $\lambda^n X$, where X is a $k \times k$ matrix. We represent λ by the shift operator S with matrix elements $S_{IJ} = \delta_{I+1,J}$ for $I, J \in \mathbb{Z}$ and $\lambda^n X$ by $X \otimes S^n$. In other words, one has

$$(\lambda^n X)_{i+kI, j+kJ} = X_{ij}\delta_{I+n,J}$$

The loop algebra structure is obviously preserved by this identification, moreover, one can check that the cocycles eq. (9.12) and eq. (16.16) also match. Hence $\widehat{gl}(k)$ is embedded into $gl(\infty)$ as the subalgebra of infinite matrices with period k along the diagonal

Let α_i be the simple roots of the finite-dimensional simple Lie algebra $\mathcal{G}$ and θ its highest root. The affine Kac–Moody algebra $\mathcal{G} \otimes \mathbb{C}(\lambda, \lambda^{-1}) \oplus \mathbb{C}K$ is generated by the following elements:

$$(E_{\alpha_i}, H_{\alpha_i}, E_{-\alpha_i}), \quad i = 1, \ldots, \text{rank}\,\mathcal{G}, \quad (E_{-\theta} \otimes \lambda, K - H_\theta, E_\theta \otimes \lambda^{-1}) \tag{16.18}$$

Each triplet form an sl_2 subalgebra. These triplets are associated with the simple roots of the affine Kac–Moody algebra. The derivation $d = \lambda \partial_\lambda$ is not in the algebra generated by these elements and has to be added by hand. The λ dependence in this presentation corresponds to what is called the homogeneous gradation. The gradation is defined by the degree in λ, which is counted by d.

A slight modification of this construction allows us to define the twisted affine Kac–Moody algebras. Assume that $\mathcal{G}$ has an automorphism τ of order N, i.e. $\tau^N = 1$. Let $\zeta = e^{2i\pi/N}$. One extends τ to an automorphism $\hat{\tau}$ of the Kac–Moody algebra by setting:

$$\hat{\tau}(X \otimes \lambda^n) = \tau(X) \otimes (\zeta\lambda)^n, \quad \hat{\tau}(K) = K, \quad \hat{\tau}(d) = d \qquad (16.19)$$

Since $\hat{\tau}$ is an automorphism, the set of its fixed points is a Lie algebra, which is called the twisted affine Kac–Moody algebra associated with τ, and denoted by $\hat{\mathcal{G}}_\tau$.

If the automorphism τ is an inner automorphism, $\hat{\mathcal{G}}_\tau$ is isomorphic to the untwisted algebra $\hat{\mathcal{G}}$. If, however, τ is not an inner automorphism, one gets an essentially different algebra. It is known that this situation occurs only when $N = 2$ or $N = 3$ and only for particular simple Lie algebras.

Let us illustrate the use of an inner automorphism to obtain the presentation of the affine Kac–Moody algebra in the principal gradation. Consider the Weyl vector $\rho = \sum_i \Lambda_i$, where the Λ_i are the fundamental weights of $\mathcal{G}$. So we have $(\rho, \alpha_i) = 1$ for any simple root α_i of $\mathcal{G}$. Moreover, if θ is the highest root of $\mathcal{G}$ we define the dual Coxeter number h^* by $(\rho, \theta) = h^* - 1$. Let τ be the inner automorphism of $\mathcal{G}$:

$$\tau(X) = e^{-\frac{2i\pi}{h^*}H_\rho}Xe^{\frac{2i\pi}{h^*}H_\rho}, \quad \tau(H) = H, \quad \tau(E_\alpha) = e^{-\frac{2i\pi}{h^*}(\rho,\alpha)}E_\alpha$$

and extend it to an automorphism $\hat{\tau}$ of the affine algebra as in eq. (16.19) with $\zeta = e^{\frac{2i\pi}{h^*}}$. The algebra of its fixed points is isomorphic to our affine algebra, and is linearly generated by the $H \otimes \lambda^{mh^*}$ and the $E_\alpha \otimes \lambda^{mh^*+(\rho,\alpha)}$. It follows that the elements of degree $\pm 1, 0$ are:

$$(E_{\alpha_i} \otimes \lambda, H_{\alpha_i}, E_{-\alpha_i} \otimes \lambda^{-1}), \ i = 1, \ldots, \operatorname{rank}\mathcal{G}, \quad (E_{-\theta} \otimes \lambda, K - H_\theta, E_\theta \otimes \lambda^{-1}) \qquad (16.20)$$

These elements generate the whole fixed point algebra. This presentation differs from the presentation in eq. (16.18) by the way in which the degrees in λ are distributed.

Affine Kac–Moody algebra may also be presented by generators and relations using Cartan matrices and their associated set of generators. Specifically, an affine Cartan matrix is a finite-dimensional matrix a_{ij} such that: $a_{ii} = 2$; $a_{ij} \leq 0$; $a_{ij} = 0 \Rightarrow a_{ji} = 0$, the a_{ij} are negative integers for $i \neq j$, and the dimension of its kernel is 1. Note that the only difference with the Cartan matrix of a semi-simple algebra is that its determinant vanishes. A classification of such Cartan matrices may be found in the References, where it is shown that irreducible ones yield exactly the standard and twisted algebras constructed above, for any simple Lie algebra $\mathcal{G}$. In analogy to the finite-dimensional case, the affine Kac–Moody algebra with Cartan matrix a_{ij} is defined as the Lie algebra generated by the elements (e_i^+, e_i^-, h_i) with Serre relations:

$$[h_i, h_j] = 0$$
$$\left[h_i, e_j^\pm\right] = \pm a_{ij} e_j^\pm \qquad (16.21)$$
$$\left[e_i^+, e_j^-\right] = \delta_{ij} h_i$$
$$(\text{ad } e_i^\pm)^{1-a_{ij}} \cdot e_j^\pm = 0 \quad \text{for} \quad i \neq j$$

One gets an infinite algebra because $\det(a) = 0$. The elements h_i generate the Cartan subalgebra $\mathcal{H}$. Let n_j be such that $\sum_i n_i a_{ij} = 0$. Since, by hypothesis, the kernel of a_{ij} is one-dimensional, such coefficients are unique up to a multiplicative constant. It is usually convenient to normalize them such that $\sum_j n_j = h^*$ with h^* the dual Coxeter number; the coefficients n_j are then all non-negative integers. By construction, the element K,

$$K = \sum_i n_i h_i$$

is central. The derivation d is not an element of the algebra generated by the (e_i^+, e_i^-, h_i). It has to be added by hand. Its commutation relations depends on the gradation one chooses. For example, the principal gradation obtained in eq. (16.20) corresponds to the choice:

$$[d, e_i^+] = 1, \quad [d, h_i] = 0, \quad [d, e_i^-] = -1$$

In particular, the rank of the (untwisted) affine Kac–Moody algebra $\hat{\mathcal{G}}$ is $(1 + \text{rank } \mathcal{G})$ if one does not include the derivation in its definition and is $(2 + \text{rank } \mathcal{G})$ if one does include it.

As for finite-dimensional Lie algebras, one has the decomposition:

$$\hat{\mathcal{G}} = \mathcal{N}_- \oplus \mathcal{H} \oplus \mathcal{N}_+$$

with $\mathcal{N}_\pm$ the subalgebra generated by the $e_i^\pm$ and $\mathcal{H}$ the Cartan subalgebra.

One may also introduce roots, which are points in the dual $\mathcal{H}^*$ of the Cartan subalgebra $\mathcal{H}$, and systems of simple roots. However, in contrast to the finite-dimensional case, the number of roots is infinite and roots may have multiplicities.

Weights are elements of $\mathcal{H}^*$. The fundamental weight vectors Λ_j are such that

$$\Lambda_j(h_i) = \delta_{ij} \qquad (16.22)$$

By definition, *integrable highest vectors* Λ are integer linear combinations of the fundamental weights: $\Lambda = \sum_j \delta_j \Lambda_j$ with δ_j integers. The coefficients δ_j are called Dynkin indices.

Unitary highest weight representations may be defined as in the finite-dimensional case. Let Λ be an integrable highest weight vector and $|\Lambda\rangle$ be the corresponding highest weight vector. By definition, one assumes that

$$h_i|\Lambda\rangle = \Lambda(h_i)|\Lambda\rangle, \quad e_j^+|\Lambda\rangle = 0$$

The highest weight representation $V(\Lambda)$, with highest weight vector Λ, is then defined as:

$$V(\Lambda) = (U(\mathcal{N}_-)|\Lambda\rangle)/\mathcal{M}_\Lambda$$

with the $U(\mathcal{N}_-)$ the universal enveloping algebra of $\mathcal{N}_-$ and $\mathcal{M}_\Lambda$ the maximal submodule of $U(\mathcal{N}_-)|\Lambda\rangle$. More concretely, vectors in $V(\Lambda)$ are obtained by multiple action of the generators e_i^- on the highest weight vector $|\Lambda\rangle$. Note that if $\Lambda = \sum_i \delta_i \Lambda_i$, then the central element $K = \sum_i n_i h_i$ acts on $V(\lambda)$ as the $\mathbb{C}$-number $K = \sum_i n_i \delta_i$. This number is called the level of the representation. Note that the adjoint representation is not a highest weight representation.

Let us present the affine Kac–Moody algebra $\widehat{sl(2)}$ in more detail. Let E_+, E_-, H be the three generators of the Lie algebra $sl(2)$:

$$[H, E_\pm] = \pm 2E_\pm, \quad [E_+, E_-] = H$$

We normalize the Killing form on $sl(2)$ by $(H, H) = 2$, $(E_+, E_-) = 1$. The loop algebra $\widetilde{sl}(2)$ is the Lie algebra of traceless 2×2 matrices with entries Laurent polynomials in λ: $\widetilde{sl}(2) = sl(2) \otimes \mathbb{C}(\lambda, \lambda^{-1})$. The affine Lie algebra $\widehat{sl(2)}$ is the central extension of $\widetilde{sl}(2)$: $\widehat{sl(2)} = \widetilde{sl}(2) \oplus \mathbb{C}K \oplus \mathbb{C}d$, with K the central element and d the derivation $d = \lambda\frac{\partial}{\partial\lambda}$. Let us write the decomposition: $\widehat{sl(2)} = \widehat{\mathcal{N}}_- \oplus \widehat{\mathcal{H}} \oplus \widehat{\mathcal{N}}_+$. First one can choose, as in eq. (16.18), the simple root vectors $E_{\alpha_1} = E_+$, $E_{\alpha_2} = \lambda E_-$. Together with the Cartan algebra generators H, K, d and $E_{-\alpha_1} = E_-$, $E_{-\alpha_2} = \lambda^{-1}E_+$ they generate the whole algebra. The simple root vectors are of degree 0 and 1. This is called the homogeneous gradation. It is more convenient to define a gradation such that simple root vectors have degree 1, the so-called principal gradation. To do that we choose simple root vectors $E_{\alpha_1} = \lambda E_+$, $E_{\alpha_2} = \lambda E_-$, and $E_{-\alpha_1} = \lambda^{-1}E_-$, $E_{-\alpha_2} = \lambda^{-1}E_+$. Together with the Cartan algebra generators H, K, d, they generate the algebra. The degree 0 elements are:

$$\widehat{\mathcal{H}} = \{H, d, K\}$$

The positive degree ones $(n > 0)$ are:

$$\widehat{\mathcal{N}}_+ = \{E_+^{(2n-1)} = E_+ \otimes \lambda^{2n-1}, E_-^{(2n-1)} = E_- \otimes \lambda^{2n-1}, H^{(2n)} = H \otimes \lambda^{2n}\}$$

and the negative degree ones $(n < 0)$ are:

$$\widehat{\mathcal{N}}_- = \{E_+^{(2n+1)} = E_+ \otimes \lambda^{2n+1}, E_-^{(2n+1)} = E_- \otimes \lambda^{2n+1}, H^{(2n)} = H \otimes \lambda^{2n}\}$$

We can exhibit the isomorphism between the affine algebra $\widehat{sl(2)}$ in the homogeneous gradation and this presentation which corresponds to the principal gradation. First, replace the parameter λ by λ^2, in the homogeneous presentation. Then the simple root vectors of the homogeneous gradation are E_+ and $\lambda^2 E_-$. Then perform a conjugation by $\exp(\log(\lambda)H/2)$. This conjugation sends E_+ to λE_+ and E_- to $\lambda^{-1}E_-$ and extends to an isomorphism of the two algebras. Note that $d \to d - \frac{1}{2}H$. In the principal gradation, the commutation relations read:

$$\left[H^{(r)}, H^{(s)}\right] = r\,\delta_{r+s,0}K$$

$$\left[H^{(r)}, E_\pm^{(s)}\right] = \pm 2E_\pm^{(r+s)} \tag{16.23}$$

$$\left[E_+^{(r)}, E_-^{(s)}\right] = H^{(r+s)} + \frac{1}{2}r\,\delta_{r+s,0}K$$

In the notation of eq. (16.21), we have $h_1 = H + \frac{1}{2}K$ and $h_2 = -H + \frac{1}{2}K$. Let H^*, K^*, d^* be the dual basis of the basis H, K, d of the Cartan algebra. With the root vectors $E_\pm^{(r)}$ are associated the roots $\pm 2H^* + rd^*$, and with the root vectors $H^{(r)}$ are associated the roots rd^*. Let us draw the root diagram of $\widehat{sl(2)}$, see Fig 16.1.

The affine $\widehat{sl(2)}$ algebra possesses two fundamental highest weights, denoted by Λ^+ and Λ^-. They are characterized by eq. (16.22). Expanding on the dual basis H^*, K^*, one gets $\Lambda^\pm = \pm\frac{1}{2}H^* + K^*$, or equivalently:

$$\Lambda^\pm(H) = \pm\frac{1}{2}; \quad \Lambda^\pm(K) = 1; \quad \Lambda^\pm(d) = 0$$

Note that the levels of these fundamental representations are equal to one, i.e. K takes the value 1 on them.

16.6 Vertex operator representations

We now recall the vertex operator construction of the level one representations of $\widehat{sl(2)}$, in the principal gradation. We introduce oscillators p_n for n odd, such that

$$[p_m, p_n] = m\delta_{n+m,0}$$

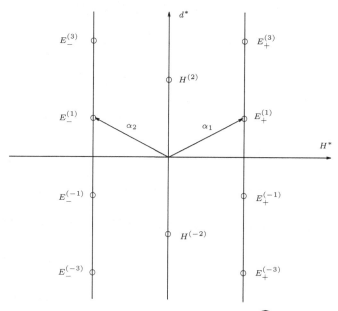

Fig. 16.1. The root diagram of $\widehat{sl(2)}$.

Note that the choice n odd ensures that there is no centre in this algebra. Assume $p_n^+ = p_{-n}$. The vacuum $|0\rangle$ is defined by $p_n|0\rangle = 0$ for $n > 0$. Its dual $\langle 0|$ is defined by $\langle 0|p_n = 0$ for $n < 0$, and the normalization condition $\langle 0|0\rangle = 1$. This allows us to compute the vacuum expectation value, $\langle 0|\mathcal{O}|0\rangle$, of any operator $\mathcal{O}$. The representation space is the Fock space generated by the p_{-n} acting on the vacuum. We define the normal ordering on monomials of the p_n by putting p_n, $n > 0$ to the right. We denote it by $:\ :$. Define the operators acting on the Fock space:

$$Q(z) = -i\sqrt{2} \sum_{n \text{ odd}} p_{-n} \frac{z^n}{n}$$

We have

$$\langle 0|Q(z_1)Q(z_2)|0\rangle = \log\left(\frac{z_1 + z_2}{z_1 - z_2}\right), \qquad |z_1| > |z_2| \qquad (16.24)$$

This is because

$$\langle 0|Q(z_1)Q(z_2)|0\rangle = -2 \sum_{n_1 n_2} \langle 0|p_{-n_1}p_{-n_2}|0\rangle \frac{z_1^{n_1} z_2^{n_2}}{n_1 n_2}$$

Here n_1, n_2 are odd integers, but the properties of the vacuum select $n_1 < 0$ and $n_2 > 0$ and we have $\langle 0|p_{-n_1}p_{-n_2}|0\rangle = -n_1\delta_{n_1+n_2,0}$ using

$p_{-n_1}p_{-n_2} = p_{-n_2}p_{-n_1} - n_1\delta_{n_1+n_2,0}$. The sum reduces to:

$$\langle 0|Q(z_1)Q(z_2)|0\rangle = 2\sum_{n_2>0}\frac{1}{n_2}\left(\frac{z_2}{z_1}\right)^{n_2} = \log\left(\frac{z_1+z_2}{z_1-z_2}\right), \quad |z_1|>|z_2|$$

The vertex operator $V(r,z)$ is defined by:

$$V(r,z) = \frac{1}{2} : \exp\left(irQ(z)\right):$$

Proposition. *The normal ordered form of a product of two vertex operators is given by:*

$$V(r,z_1)V(s,z_2) = \left(\frac{z_1-z_2}{z_1+z_2}\right)^{rs} : V(r,z_1)V(s,z_2): \quad |z_1|>|z_2| \quad (16.25)$$

<u>Proof.</u> Let

$$Q_\pm = -i\sqrt{2}\sum_{\mp n>0} p_{-n}z^n/n$$

so that $Q = Q_+ + Q_-$ and $Q_+|0\rangle = 0$. Then, by definition of the normal order, $V(r,z) = \frac{1}{2}e^{irQ_-(z)}e^{irQ_+(z)}$. To compute $: V(r,z_1)V(s,z_2):$ we need to commute $\exp\left(irQ_+(z_1)\right)$ to the right of $\exp\left(isQ_-(z_2)\right)$. Now it is clear that the commutator of $Q_+(z_1)$ and $Q_-(z_2)$ is a $\mathbb{C}$-number. To evaluate this number one can take its vacuum expectation value. One has, using $Q_+|0\rangle = 0$ and $\langle 0|Q_- = 0$,

$$\langle 0|[Q_+(z_1),Q_-(z_2)]|0\rangle = \langle 0|Q_+(z_1)Q_-(z_2)|0\rangle = \langle 0|Q(z_1)Q(z_2)|0\rangle$$

which is given by eq. (16.24). Moreover, if A and B are two operators such that $[A,B]$ is a $\mathbb{C}$-number, one has $e^Ae^B = e^Be^Ae^{[A,B]}$. So we arrive at eq. (16.25) since $e^{[A,B]} = ((z_1-z_2)/(z_1+z_2))^{rs}$. ∎

The level one vertex operator representations of the Lie algebra $\widehat{sl(2)}$ are obtained as follows:

$$\sum_{n\ odd} z^{-n}(E_+^{(n)}+E_-^{(n)}) = P(z) \equiv \sum_n p_nz^{-n} \quad (16.26)$$

$$\sum_{n\ even} z^{-n}H^{(n)} + \sum_{n\ odd} z^{-n}(E_+^{(n)}-E_-^{(n)}) = \pm V(z) \quad (16.27)$$

where $V(z)$ denotes the vertex operator:

$$V(z) = V(-\sqrt{2},z) = \frac{1}{2} : e^{-i\sqrt{2}Q(z)} := \frac{1}{2} : e^{i\sqrt{2}Q(-z)}: \quad (16.28)$$

Proposition. *The operators $E_{\pm}^{(n)}$ and $H^{(n)}$ defined in eq. (16.27) provide representations of the affine algebra eq. (16.23) with $K = 1$. These representations correspond to the fundamental highest weights $\Lambda^{\pm}$ according to the sign in eq. (16.27). They are the fundamental level one representations of $\widehat{sl(2)}$.*

<u>Proof.</u> We derive eq. (16.25) with respect to z_1 and get for $|z_1| > |z_2|$:

$$\frac{1}{r}\frac{\partial}{\partial z_1}V(r, z_1)V(s, z_2) = \frac{i}{2} : \frac{dQ(z_1)}{dz_1}e^{irQ(z_1)} : V(s, z_2)$$

$$= \frac{2sz_2}{z_1^2 - z_2^2}\left(\frac{z_1 - z_2}{z_1 + z_2}\right)^{rs} : V(r, z_1)V(s, z_2) :$$

$$+ \frac{i}{2}\left(\frac{z_1 - z_2}{z_1 + z_2}\right)^{rs} : \frac{dQ(z_1)}{dz_1}e^{irQ(z_1)}V(s, z_2) :$$

Here we have used the fact that in the normal ordered product everything commutes so that one can derive the exponential straightforwardly. We then set $r = 0$. Defining:

$$\Gamma(z_1, z_2) = \frac{\sqrt{2}sz_1z_2}{z_1^2 - z_2^2}V(s, z_2) + : P(z_1)V(s, z_2) :$$

we get:

$$P(z_1)V(s, z_2) = \Gamma(z_1, z_2), \quad |z_1| > |z_2|$$

Similarly, we derive eq. (16.25) with respect to z_2 and then perform the exchange $(z_1, z_2, r) \to (z_2, z_1, s)$. We get:

$$V(s, z_2)P(z_1) = \Gamma(z_1, z_2), \quad |z_1| < |z_2|$$

Expanding $P(z) = \sum_n p_n z^{-n}$, where n is odd, so that

$$p_n = E_{+}^{(n)} + E_{-}^{(n)} = \oint_C \frac{dz}{2i\pi} z^{n-1}P(z)$$

we can write the commutator $[p_n, V(s, z_2)]$ as:

$$[p_n, V(s, z_2)] = \oint_{C_1 - C_2} \frac{dz_1}{2i\pi} z_1^{n-1}\Gamma(z_1, z_2)$$

where C_1 is a circle around the origin with $|z_1| > |z_2|$ while C_2 is a circle around the origin with $|z_1| < |z_2|$. This contour integral is given by the residues at the two-poles $z_1 = \pm z_2$ and we finally obtain, setting $s = -\sqrt{2}$:

$$[E_{+}^{(n)} + E_{-}^{(n)}, V(z)] = -2z^n V(z) \tag{16.29}$$

Similarly, starting from eq. (16.25) and setting $V(z) = \sum_n V^{(n)} z^{-n}$, one gets:

$$[V_n, V(z_2)] = \oint_{C_1 - C_2} \frac{dz_1}{2i\pi} \, z_1^{n-1} \left(\frac{z_1 - z_2}{z_1 + z_2} \right)^2 : V(z_1) V(z_2) :$$

The residue is at $z_1 = -z_2$ and is easily computed, noting that: $V(z_2) V(-z_2) := 1/4$. One finds $[V_n, V(z)] = 2(-1)^n z^n P(z) + (-1)^n n z^n$. Separating n even and odd this reads (with ϵ the sign of eq. (16.27)):

$$\epsilon[H^{(n)}, V(z)] = 2z^n P(z) + n z^n, \quad \epsilon[E_+^{(n)} + E_-^{(n)}, V(z)] = -2z^n P(z) - n z^n \tag{16.30}$$

From eqs. (16.29, 16.30), one gets by expanding $V(z)$ into its components (note that ϵ cancels):

$$[H^{(n)}, E_\pm^{(m)}] = \pm 2 E_\pm^{(n+m)} \tag{16.31}$$

$$[E_\pm^{(n)}, E_+^{(m)} - E_-^{(m)}] = -H^{(n+m)} \mp \frac{1}{2} n \delta_{n+m,0} \tag{16.32}$$

Finally, we have $P(z) = \sum_n p_n z^{-n}$ so that $E_+^{(n)} + E_-^{(n)} = p_n$, and from the commutation relations of p_n we get $[E_+^{(n)} + E_-^{(n)}, E_+^{(m)} + E_-^{(m)}] = n \delta_{n+m,0}$. Combining with eq. (16.30), this gives:

$$[E_\pm^{(n)}, E_+^{(m)} + E_-^{(m)}] = \pm H^{(n+m)} + \frac{1}{2} n \delta_{n+m,0} \tag{16.33}$$

Equations (16.31, 16.32, 16.33) are equivalent to the commutation relations eq. (16.23) for $K = 1$. We have obtained a level one repesentation of $\widehat{sl(2)}$. It remains to identify the highest weight. It is provided by the vacuum vector $|0\rangle$ because

$$V(z)|0\rangle = \frac{1}{2} e^{-i\sqrt{2} Q_-(z)} |0\rangle \tag{16.34}$$

contains only positive odd powers of z. It follows that $(E_+^{(n)} - E_-^{(n)})|0\rangle = 0$ and $H^{(n)}|0\rangle = 0$ for $n > 0$. Moreover, since $E_+^{(n)} + E_-^{(n)} = p_n$ we have $(E_+^{(n)} + E_-^{(n)})|0\rangle = 0$ for $n > 0$. This implies that the vacuum is annihilated by all positive root vectors, hence is a highest weight vector. Equation (16.34) also gives $H^{(0)}|0\rangle = \epsilon \frac{1}{2}|0\rangle$ so that the corresponding weight is $\epsilon \frac{1}{2}$. It is known that this representation on Fock space is irreducible. ∎

References

[1] S. Helgason, *Differential Geometry and Symmetric spaces.* Academic Press (1962).

[2] J.E. Humphreys, *Introduction to Lie Algebras and Representation Theory.* Springer (1972).

[3] V. G. Kac, *Infinite Dimensional Lie Algebras.* Cambridge University Press (1985).

Index